T0221719

Analysis of Monge–Ampère Equations

GRADUATE STUDIES
IN MATHEMATICS **240**

Analysis of Monge–Ampère Equations

Nam Q. Le

AMERICAN
MATHEMATICAL
SOCIETY
Providence, Rhode Island

EDITORIAL COMMITTEE

Matthew Baker
Marco Gualtieri
Gigliola Staffilani (Chair)
Jeff A. Viaclovsky
Rachel Ward

2020 *Mathematics Subject Classification.* Primary 35-XX; Secondary 52-XX, 49-XX.

For additional information and updates on this book, visit
www.ams.org/bookpages/gsm-240

Library of Congress Cataloging-in-Publication Data

Names: Lê, Nam Q., author.
Title: Analysis of Monge-Ampère equations / Nam Q. Le.
Description: Providence, Rhode Island : American Mathematical Society, [2024] | Series: Graduate studies in mathematics, 1065-7339 ; volume 240 | Includes bibliographical references and index.
Identifiers: LCCN 2023046004 | ISBN 9781470474201 (hardback) | ISBN 9781470476250 (paperback) | ISBN 9781470476243 (ebook)
Subjects: LCSH: Monge-Ampère equations. | Differential equations, Elliptic. | Generalized spaces. | AMS: Partial differential equations. | Convex and discrete geometry. | Calculus of variations and optimal control; optimization.
Classification: LCC QA377 .L3876 2024 | DDC 515/.3533–dc23/eng/20231023
LC record available at https://lccn.loc.gov/2023046004

Copying and reprinting. Individual readers of this publication, and nonprofit libraries acting for them, are permitted to make fair use of the material, such as to copy select pages for use in teaching or research. Permission is granted to quote brief passages from this publication in reviews, provided the customary acknowledgment of the source is given.

Republication, systematic copying, or multiple reproduction of any material in this publication is permitted only under license from the American Mathematical Society. Requests for permission to reuse portions of AMS publication content are handled by the Copyright Clearance Center. For more information, please visit www.ams.org/publications/pubpermissions.
Send requests for translation rights and licensed reprints to reprint-permission@ams.org.

© 2024 by the American Mathematical Society. All rights reserved.
The American Mathematical Society retains all rights
except those granted to the United States Government.
Printed in the United States of America.

♾ The paper used in this book is acid-free and falls within the guidelines
established to ensure permanence and durability.
Visit the AMS home page at https://www.ams.org/

10 9 8 7 6 5 4 3 2 1   29 28 27 26 25 24

To Chi and Minh

Contents

Part 1. The Monge–Ampère Equation

Preface

This book aims to cover basic results in the Monge–Ampère equation, the linearized Monge–Ampère equation, and their applications. It expands upon materials of special topic courses (M742) that I taught at Indiana University, Bloomington, in the spring semesters of 2016 and 2020.

The Monge–Ampère equation has been studied for a long time and is still a center of intensive interest nowadays because of its significant role in many important problems in analysis, geometry, physics, partial differential equations, and applications. Its linearization, that is, the linearized Monge–Ampère equation, has attracted considerable attention in recent years. It arises from several fundamental problems in different subjects such as affine geometry, complex geometry, meteorology, and the calculus of variations.

For the Monge–Ampère equation, I present a wide range of fundamental results and phenomena in the solvability and regularity theory for Aleksandrov solutions with more or less optimal assumptions on the data. The regularity theory is considered both in the interior and at the boundary. Thus, this book complements the interior regularity theory treated in standard textbooks on the modern theory of the Monge–Ampère equation including the classical book *The Monge–Ampère Equation* by Gutiérrez and the recent, elegant book *The Monge–Ampère Equation and Its Application* by Figalli. In addition, it also introduces the spectral theory of the Monge–Ampère operator. Moreover, due to their role in diverse applications, considerable attention is paid to the Monge–Ampère equations in their very natural setting, that is, general convex domains that are not strictly convex.

For the linearized Monge–Ampère equation, my goal here is to give a more or less complete account of the local and global regularity theories in Hölder spaces for solutions. It should be noted that the above-mentioned book by Gutiérrez was also the first to cover the interior Harnack inequality for the linearized Monge–Ampère equation—the fundamental work of Caffarelli and Gutiérrez—which initiated many developments and interesting applications. Around the same time, the book *Dynamical and Geometric Aspects of Hamilton–Jacobi and Linearized Monge–Ampère Equations* by Le, Mitake, and Tran gave an expository account of the Caffarelli–Gutiérrez interior Harnack inequality and a brief introduction to boundary estimates. This book significantly expands the scopes and methods of these two books in its treatment of the linearized Monge–Ampère equation. Furthermore, it also discusses in depth their applications in various areas of mathematics.

As with most literature on the subject, my expositions here on Monge–Ampère equations are primarily based on measure-theoretic, convex geometric, or maximum-principle-based methods. However, alternative methods are also given. For instance, I include discussion of variational methods, and, in addition to the nondivergence form nature of Monge–Ampère equations, I also discuss them from divergence form perspectives.

Except for a few classical results that will be recalled in the second chapter, the rest of this book is self-contained. I have tried to make it a textbook rather than a research monograph. An effort has been made to give motivations and examples of many concepts introduced in the book. Students having a basic background in linear algebra, real analysis, and partial differential equations can follow the book without much difficulties. To facilitate reading for beginners to the subject, complete details of the proofs of all theoretical results are presented while many subtle constructions are explained. These contribute to the length of the book. I hope that researchers in analysis, geometry, partial differential equations, mathematical physics, and the calculus of variations will find this book a useful reference. There are about a hundred and fifty problems for the interested reader. More details on the contents of the book will be presented in the first chapter.

I am extremely grateful to several anonymous reviewers for their critical comments and suggestions. They tremendously helped improve the presentation of the initial manuscript. I would like to thank the National Science Foundation for support via grants DMS-1764248 and DMS-2054686 during the writing of this book. Finally, I would like to thank the AMS, especially Sergei Gelfand, Arlene O'Sean, and Christine Thivierge, for bringing this book into publication.

Nam Q. LÊ

Fall 2023, Bloomington

Notation

We collect here some standard notation and convention used in the book. Unless otherwise indicated, we will work with real-valued functions on real Euclidean spaces.

Geometric notation.

- $\mathbb{N}$ = the set of positive integers; $\mathbb{N}_0 = \mathbb{N} \cup \{0\}$.

- $\mathbb{R} = \mathbb{R}^1$ = the set of real numbers.

- $\mathbb{R}^n = n$-dimensional real Euclidean space $(n \geq 1)$. A typical point in $\mathbb{R}^n$ is denoted by $x = (x_1, \ldots, x_n)$. We will also, depending on the context, regard x as a row or column vector. Unless otherwise indicated, we write $x = (x', x_n)$ where $x' = (x_1, \ldots, x_{n-1})$. Sometimes, when there is no possibility of confusion, we also use x_1, x_2, etc., to denote points in $\mathbb{R}^n$.

- The dot product and Euclidean norm on $\mathbb{R}^n$: If $x = (x_1, \ldots, x_n)$ and $y = (y_1, \ldots, y_n)$ belong to $\mathbb{R}^n$, then

$$
x \cdot y = \sum_{i=1}^{n} x_i y_i, \quad |x| = (x \cdot x)^{1/2} = \left(\sum_{i=1}^{n} x_i^2 \right)^{1/2}.
$$

- $\mathbb{R}_+^n$ = open upper half-space = $\{x = (x_1, \ldots, x_n) \in \mathbb{R}^n : x_n > 0\}$.

- The ith coordinate vector in $\mathbb{R}^n$: $e_i = (0, \ldots, 0, 1, 0, \ldots, 0)$ with 1 in the ith slot.

- $B_r(x) = \{y \in \mathbb{R}^n : |y - x| < r\}$ = the open ball in $\mathbb{R}^n$ with center x and radius $r > 0$. When $x = 0$, we simply write B_r for $B_r(0)$.

- $B_r^+(0) = B_r(0) \cap \mathbb{R}_+^n$ denotes the upper half of $B_r(0)$ where $r > 0$.

- δ_x (Dirac measure): For $x \in \mathbb{R}^n$, we usually use δ_x to denote the Dirac measure on $\mathbb{R}^n$ giving the unit mass to the point x.

- Ω usually denotes an open subset of $\mathbb{R}^n$.

- Domain: a nonempty open, connected subset of $\mathbb{R}^n$ (not necessarily bounded).

- δ_{ij} = the Kronecker symbol: $\delta_{ij} = 1$ if $i = j$, and $\delta_{ij} = 0$ if $i \neq j$.

- ω_n = volume of the unit ball in $\mathbb{R}^n = \frac{\pi^{n/2}}{\Gamma(1+\frac{n}{2})}$.

- $|E|$ denotes the Lebesgue measure of a measurable set $E \subset \mathbb{R}^n$.

- ∂E and $\overline{E}$: the boundary and the closure of a set $E \subset \mathbb{R}^n$.

- Compact inclusion: If $A \subset B \subset \mathbb{R}^n$, $\overline{A} \subset B$, and $\overline{A}$ is compact, then we write $A \Subset B$ and say A is *compactly contained* in B.

- $\mathrm{diam}(E):$ the diameter of a bounded set E.

- $\mathrm{dist}(\cdot, E):$ the distance function from a closed set E.

- The Minkowski sum of two subsets $A, B \subset \mathbb{R}^n$ is denoted by

$$A + B = \{x + y : x \in A, y \in B\}.$$

- d_H: the Hausdorff distance; see Definition 2.13.

- Given a set $E \subset \mathbb{R}^n$, its dilation by a factor $\gamma > 0$ is denoted by

$$\gamma E := \{\gamma x : x \in E\}.$$

  Its image under an affine map $T : \mathbb{R}^n \to \mathbb{R}^n$ is often denoted by

$$TE := \{Tx : x \in E\}.$$

Notation for functions.

- $u|_E$ is the restriction of a function u to the set E.

- Characteristic function χ_E of a set $E \subset \mathbb{R}^n$: $\chi_E(x) = 1$ if $x \in E$, and $\chi_E(x) = 0$ if $x \notin E$.

- The support of a function $u : \Omega \to \mathbb{R}$ is defined by

$$\mathrm{spt}(u) = \overline{\{x \in \Omega : u(x) \neq 0\}}.$$

- The convolution of two integrable functions $f, g : \mathbb{R}^n \to \mathbb{R}$ is denoted by $f * g$ where

$$(f * g)(x) = \int_{\mathbb{R}^n} f(x - y)g(y)\,dy \quad \text{for } x \in \mathbb{R}^n.$$

- $\mathrm{osc}_\Omega\, u = \sup_\Omega u - \inf_\Omega u$ = oscillation of a function $u : \Omega \to \mathbb{R}$.

- The modulus of continuity of a continuous function $f : \Omega \to \mathbb{R}$ is the function $\omega(f, \Omega; \cdot) : (0, \infty) \to [0, \infty)$ defined by

$$\omega(f, \Omega; r) = \sup\{|f(x) - f(y)| : x, y \in \Omega, |x - y| < r\}.$$

Notation for Derivatives.

- Partial differentiations on $\mathbb{R}^n$:

$$D_i = \frac{\partial}{\partial x_i} \quad \text{and} \quad D_{ij} = \frac{\partial^2}{\partial x_i \partial x_j} \quad (i, j = 1, \ldots, n).$$

- Partial differentiations using multi-indices: For a multi-index $\beta = (\beta_1, \ldots, \beta_n)$ of order $|\beta| = \beta_1 + \cdots + \beta_n$, where $\beta_i \in \mathbb{N}_0$, we define

$$D^\beta u(x) := \frac{\partial^{|\beta|} u(x)}{\partial x_1^{\beta_1} \cdots \partial x_n^{\beta_n}} = \partial_{x_1}^{\beta_1} \cdots \partial_{x_n}^{\beta_n} u(x).$$

- If $k \in \mathbb{N}_0$ and $u : \mathbb{R}^n \to \mathbb{R}$, then the set of all partial derivatives of order k of u at $x \in \mathbb{R}^n$ is denoted by $D^k u(x) := \{D^\beta u(x) : |\beta| = k\}$.

- For a function u of n variables $x_1, \ldots, x_n$, we define its

$$\text{gradient vector} \ = Du = \left(\frac{\partial u}{\partial x_1}, \ldots, \frac{\partial u}{\partial x_n} \right) = (D_1 u, \ldots, D_n u),$$

$$\text{Hessian matrix} \ = D^2 u = \left(\frac{\partial^2 u}{\partial x_i \partial x_j} \right)_{1 \le i, j \le n} = (D_{ij} u)_{1 \le i, j \le n},$$

$$\text{Laplacian} \ = \Delta u = \operatorname{div} Du = \operatorname{trace}(D^2 u) = \sum_{i=1}^{n} D_{ii} u.$$

- If u is an affine function, then Du is also called its slope.

- Sometimes we use a subscript attached to the symbols D, D^2 to indicate the variables being differentiated. For example, if $u = u(x, y)$ $(x \in \mathbb{R}^n, y \in \mathbb{R}^m)$, then $D_x u = (\frac{\partial u}{\partial x_1}, \ldots, \frac{\partial u}{\partial x_n})$.

- Divergence of a vector field: If $\mathbf{F} = (F^1, \ldots, F^n) : \mathbb{R}^n \to \mathbb{R}^n$ is a vector field, then $D \cdot \mathbf{F} = \operatorname{div} \mathbf{F} = \sum_{i=1}^{n} \frac{\partial F^i}{\partial x_i}$.

- Repeated indices are summed. For example,

$$a^{ij} D_{ik} u = \sum_i a^{ij} D_{ik} u.$$

Notation for function spaces.

- $C(\Omega) = \{u : \Omega \to \mathbb{R} : u \text{ is continuous}\}$ = the set of continuous functions on Ω. We also denote $C^0(\Omega) = C(\Omega)$.

- $C(\overline{\Omega}) = \{u : \overline{\Omega} \to \mathbb{R} : u \text{ is continuous}\}$ = the set of continuous functions on $\overline{\Omega}$. We also denote $C^0(\overline{\Omega}) = C(\overline{\Omega})$.

- $C^k(\Omega)$ = the set of functions $u : \Omega \to \mathbb{R}$ all of whose partial derivatives of order $\le k$ are continuous in Ω. Here $k \in \mathbb{N}_0$ or $k = \infty$.

- $C^k(\overline{\Omega})$ = the set of functions $u \in C^k(\Omega)$ whose $D^\beta u$ continuously extends to $\overline{\Omega}$ for each $|\beta| \le k$. The norm on $C^k(\overline{\Omega})$ is

$$\|u\|_{C^k(\overline{\Omega})} = \sum_{0 \le |\beta| \le k} \sup_{\Omega} |D^\beta u|.$$

- $C_c(\Omega), C_c^k(\Omega)$, etc., denote the spaces of functions in $C(\Omega), C^k(\Omega)$, etc., that have *compact supports* in Ω. We also write $C^{k,0}$ for C^k.

- Lebesgue space: $L^p(\Omega)$ ($1 \le p \le \infty$) is the Banach space consisting of Lebesgue measurable functions $f : \Omega \to \mathbb{R}$ with $\|f\|_{L^p(\Omega)} < \infty$, where the L^p norm of f is defined by

$$\|f\|_{L^p(\Omega)} = \begin{cases} \left(\displaystyle\int_\Omega |f|^p \, dx \right)^{1/p} & \text{if } 1 \le p < \infty, \\ \text{ess sup}_\Omega |u| & \text{if } p = \infty. \end{cases}$$

- Sobolev spaces:

$$W^{k,p}(\Omega) = \left\{ u \in L^p(\Omega) : D^\beta u \in L^p(\Omega) \text{ for all } |\beta| \le k \right\};$$
$$W_0^{1,2}(\Omega) = \{ u \in W^{1,2}(\Omega) : u = 0 \text{ on } \partial\Omega \text{ in the trace sense} \}.$$

- Local Lebesgue and Sobolev spaces (with convention $L^p = W^{0,p}$):

$$W_{\text{loc}}^{k,p}(\Omega) = \{ f \in W^{k,p}(\Omega') \text{ for each open set } \Omega' \Subset \Omega \}.$$

- We identify functions in $W^{k,p}(\Omega)$ which agree almost everywhere.
- $W^{k,p}(\Omega; \mathbb{R}^n) = \left\{ \mathbf{F} = (F^1, \dots, F^n) : \Omega \to \mathbb{R}^n \text{ with } F^i \in W^{k,p}(\Omega) \right\}$.
- Lipschitz continuous function: A function $u \in C(\Omega)$ is called Lipschitz continuous if there exists a constant $C > 0$ such that

$$|u(x) - u(y)| \le C|x - y| \quad \text{for all } x, y \in \Omega.$$

- Uniformly Hölder continuous function: A function $u : E \to \mathbb{R}$ on a bounded set E in $\mathbb{R}^n$ is called uniformly Hölder continuous with exponent $\alpha \in (0, 1]$ in E if there is a constant $C > 0$ such that

$$|u(x) - u(y)| \le C|x - y|^\alpha \quad \text{for all } x, y \in E;$$

equivalently, the seminorm $[u]_{C^{0,\alpha}(E)}$ of u defined below is finite:

$$[u]_{C^{0,\alpha}(E)} := \sup_{x \ne y \in E} \frac{|u(x) - u(y)|}{|x - y|^\alpha} < \infty.$$

We call u locally Hölder continuous with exponent $\alpha \in (0, 1]$ in a subset E of $\mathbb{R}^n$ if it is uniformly Hölder continuous with exponent α on compact subsets of E.

- Hölder spaces: $C^{k,\alpha}(\overline{\Omega})$ $(C^{k,\alpha}(\Omega))$ consists of $C^k(\overline{\Omega})$ $(C^k(\Omega))$ functions whose kth-order partial derivatives, where $k \in \mathbb{N}_0$, are uniformly Hölder continuous (locally Hölder continuous) with exponent $\alpha \in (0,1]$ in Ω. The $C^{k,\alpha}(\overline{\Omega})$ norm of $u \in C^{k,\alpha}(\overline{\Omega})$ is

$$\|u\|_{C^{k,\alpha}(\overline{\Omega})} := \sum_{0 \leq |\beta| \leq k} \sup_{\Omega} |D^\beta u| + \sup_{|\beta|=k} [D^\beta u]_{C^{0,\alpha}(\Omega)}.$$

  $C^{0,1}(\overline{\Omega})$ is the space of Lipschitz continuous functions on $\overline{\Omega}$. We also write $C^\alpha(\overline{\Omega}) = C^{0,\alpha}(\overline{\Omega})$ and $C^\alpha(\Omega) = C^{0,\alpha}(\Omega)$, for $0 < \alpha < 1$.

- Locality of Hölder spaces: Hölder spaces $C^{k,\alpha}(\Omega)$ defined in this book are local spaces, including $\alpha = 1$, and Ω possibly unbounded. For example, $C^{1,1}(\Omega)$ here corresponds to $C^{1,1}_{\text{loc}}(\Omega)$ in some texts.

Notation for matrices.

- $\mathbb{M}^{n\times n}$: the space of real $n \times n$ matrices. If $A = (a_{ij})_{1\leq i,j\leq n} \in \mathbb{M}^{n\times n}$, then for brevity, we also write $A = (a_{ij})$.

- $\mathbb{S}^n$: the space of real, symmetric $n \times n$ matrices.

- I_n is the identity $n \times n$ matrix.

- $\operatorname{diag}(d_1,\ldots,d_n)=$diagonal matrix with diagonal entries $d_1,\ldots,d_n$.

- If $A = (a_{ij})_{1\leq i,j\leq n} \in \mathbb{S}^n$ and $x = (x_1,\ldots,x_n) \in \mathbb{R}^n$, then the corresponding quadratic form is $(Ax)\cdot x = x^t A x \equiv \sum_{i,j=1}^n a_{ij} x_i x_j$.

- We write $A > (\geq) 0$ if $A \in \mathbb{S}^n$ is positive (nonnegative) definite.

- We write $A \geq B$ for $A, B \in \mathbb{S}^n$ if $A - B$ is nonnegative definite. In particular, we write $A \geq \lambda I_n$ if $(Ax)\cdot x \geq \lambda|x|^2$ for all $x \in \mathbb{R}^n$.

- A^t: the transpose of the matrix A.

- $\operatorname{trace}(A)$ or $\operatorname{trace} A$: the trace of a square matrix A.

- $\det A$: the determinant of a square matrix A.

- If $T : \mathbb{R}^n \to \mathbb{R}^n$ is an affine transformation on $\mathbb{R}^n$ given by $Tx = Ax+b$ where $A \in \mathbb{M}^{n\times n}$ and $b \in \mathbb{R}^n$, then we denote $\det T = \det A$.

- $\operatorname{Cof} A$: the cofactor matrix of a matrix $A \in \mathbb{M}^{n\times n}$. The (i,j)th entry of $\operatorname{Cof} A$ is $(\operatorname{Cof} A)^{ij} = (-1)^{i+j}\det A(i \mid j)$ where $A(i \mid j) \in \mathbb{M}^{(n-1)\times(n-1)}$ is obtained by deleting the ith row and jth column of A. If $A \in \mathbb{S}^n$ is positive definite, then $\operatorname{Cof} A = (\det A)A^{-1}$.

- $\|A\| = \sqrt{\operatorname{trace}(A^t A)}$: the Hilbert-Schmidt norm of a matrix A.

Notation with convexity.

- Normal mapping: For a function $u : \Omega \subset \mathbb{R}^n \to \mathbb{R}$, we use $\partial u(x)$ to denote the normal mapping (or subdifferential) of u at $x \in \Omega$:

$$\partial u(x) = \{p \in \mathbb{R}^n : u(y) \geq u(x) + p\cdot(y-x) \quad \text{for all } y \in \Omega\}.$$

- μ_u denotes the Monge–Ampère measure of a convex function u; see Definition 3.1.

- Section: For an open set Ω in $\mathbb{R}^n$ and a function $u : \overline{\Omega} \to \mathbb{R} \cup \{+\infty\}$ which is convex and finite in Ω, the section of u with center at x_0, height $h > 0$, and slope $p \in \partial u(x_0)$ is denoted by

$$S_u(x_0, p, h) = \{x \in \overline{\Omega} : u(x) < u(x_0) + p \cdot (x - x_0) + h\}.$$

When a convex function u is differentiable at x_0, its section with center at x_0 and height $h > 0$ is denoted by

$$S_u(x_0, h) = \{x \in \overline{\Omega} : u(x) < u(x_0) + Du(x_0) \cdot (x - x_0) + h\}.$$

- Special subdifferential $\partial_s u(x)$ at a boundary point; see Definition 4.1.

Notation for estimates.

- Constants in estimates: We use $C = C(*, \ldots, \star)$ to denote a *positive constant* C depending on the quantities $*$, $\ldots$, $\star$ appearing in the parentheses. C can be either computed explicitly or its existence comes from compactness arguments. In general, its value may change from line to line in a given context.

- (Big-oh notation) For two functions f and g defined in a neighborhood of $x_0 \in \mathbb{R}^n$, we write

$$f = O(g) \quad \text{as } x \to x_0$$

if there is a constant $C > 0$ such that

$$|f(x)| \le C|g(x)| \quad \text{for all } x \text{ sufficiently close to } x_0.$$

- (Little-oh notation) For two functions f and g defined in a neighborhood of $x_0 \in \mathbb{R}^n$, we write

$$f = o(g) \quad \text{as } x \to x_0$$

if

$$\lim_{x \to x_0} |f(x)|/|g(x)| = 0.$$

Chapter 1

Introduction

In this book, we study the Monge–Ampère equation

$$(1.1) \qquad \det D^2 u = f \quad \text{in } \Omega \subset \mathbb{R}^n,$$

for an unknown function $u : \Omega \to \mathbb{R}$, and its linearization, that is, the linearized Monge–Ampère equation

$$(1.2) \qquad \sum_{i,j=1}^{n} U^{ij} D_{ij} v = g \quad \text{in } \Omega \subset \mathbb{R}^n,$$

where $(U^{ij})_{1 \le i,j \le n}$ is the cofactor matrix of the Hessian matrix $D^2 u$ of a given potential function $u : \Omega \to \mathbb{R}$ where now $v : \Omega \to \mathbb{R}$ is the unknown.

The first equation is fully nonlinear in u while the second equation is linear in v under a "nonlinear background" due to the structure of the coefficient matrix. These equations can be put into the abstract form

$$(1.3) \qquad F(D^2 w(x), x) = 0 \quad \text{in } \Omega \subset \mathbb{R}^n,$$

with $F(r, x) : \mathbb{S}^n \times \Omega \to \mathbb{R}$, where $\mathbb{S}^n$ is the space of real, symmetric $n \times n$ matrices. We assume $n \ge 2$ in this chapter.

Besides the existence theory for solutions under suitable data, our main focus is on regularity theory in the framework of elliptic partial differential equations (PDEs). For (1.3) to be elliptic, the usual requirement that

$$F(r + A, x) > F(r) \quad \text{for all positive definite matrices } A \in \mathbb{S}^n$$

turns into requiring $\left(\frac{\partial F}{\partial r_{ij}} \right)_{1 \le i,j \le n}$—the coefficient matrix of the linearization of F—to be positive definite. Specializing this to (1.1) where

$$F(r, x) = \det r - f(x)$$

and (1.2) where

$$F(r, x) = \sum_{i,j=1}^{n} U^{ij} r_{ij} - g(x), \quad \text{with } r = (r_{ij})_{1 \le i,j \le n},$$

we are led to requiring that $U = (U^{ij})_{1 \le i,j \le n}$ be positive definite. In this case, due to $U = (\det D^2 u)(D^2 u)^{-1}$, it is natural to consider $D^2 u$ to be positive definite. This is the reason why we will study (1.1) and (1.2) among the class of convex functions u. In particular, the function f in (1.1) must be nonnegative. This book will study mostly the case of f being bounded between two positive constants.

1.1. The Monge–Ampère Equation

The Monge–Ampère equation in the general form

$$\det D^2 u(x) = G(x, u(x), Du(x))$$

arises in many important problems in analysis, partial differential equations, geometry, mathematical physics, and applications. The study of the Monge–Ampère equation began in two dimensions with the works of Monge [**Mon**] in 1784 and Ampère [**Amp**] in 1820. It seems that Lie was the first to use the terminology "Monge–Ampère equation" in his 1877 paper [**Li**]. In this section, we provide some examples where the Monge–Ampère equation appears, and we give some impressionistic features and difficulties of this equation. This is not meant to be a survey.

1.1.1. Examples. We list below some examples concerning the Monge–Ampère equation.

Example 1.1 (Prescribed Gauss curvature equation). Let $u \in C^2(\Omega)$ where $\Omega \subset \mathbb{R}^n$. Suppose that the graph of u has Gauss curvature $K(x)$ at the point $(x, u(x))$ where $x \in \Omega$. Then u satisfies the equation

$$\det D^2 u = K(1 + |Du|^2)^{(n+2)/2}.$$

Example 1.2 (Affine sphere of parabolic type). The classification of complete affine spheres of parabolic type is based on the classification of global convex solutions to

$$\det D^2 u = 1 \quad \text{in } \mathbb{R}^n.$$

Example 1.3 (Special Lagrangian equation). Let $\lambda_i(D^2 u)$ $(i = 1, \ldots, n)$ be the eigenvalues of the Hessian matrix $D^2 u$ of a function $u : \mathbb{R}^n \to \mathbb{R}$. The special Lagrangian equation

$$\sum_{i=1}^{n} \arctan \lambda_i(D^2 u) = \text{constant } c$$

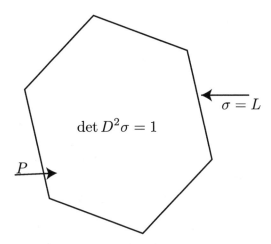

Figure 1.1. Surface tension in the dimer model solves a Monge–Ampère equation on a planar polygon with piecewise affine boundary data.

has close connections with the Monge–Ampère equation. When $n = 2$ and $c = \pi/2$, it is exactly

$$\det D^2 u = 1.$$

When $n = 3$ and $c = \pi$, it becomes $\det D^2 u = \Delta u$.

Example 1.4 (Surface tension in the dimer model). In the dimer model in combinatorics and statistical physics, the surface tension σ is a convex solution to the Monge–Ampère equation

$$\begin{cases} \det D^2 \sigma = 1 & \text{in } P, \\ \sigma = L & \text{on } \partial P. \end{cases}$$

Here $P \subset \mathbb{R}^2$ is a convex polygon in the plane, and L is a piecewise affine function on the boundary ∂P. See Figure 1.1.

Example 1.5 (Affine hyperbolic sphere). Let Ω be a bounded, convex domain in $\mathbb{R}^n$. If $u \in C(\overline{\Omega})$ is a convex solution to

$$\det D^2 u = |u|^{-n-2} \quad \text{in } \Omega, \qquad u = 0 \quad \text{on } \partial \Omega,$$

then the Legendre transform of u is a complete affine hyperbolic sphere.

Example 1.6 (The Monge–Ampère eigenvalue problem). Let Ω be a convex, bounded domain in $\mathbb{R}^n$. The Monge–Ampère eigenvalue problem consists of finding constants $\lambda > 0$ and nonzero convex functions $u \in C(\overline{\Omega})$ such that

$$\begin{cases} \det D^2 u = \lambda |u|^n & \text{in } \Omega, \\ u = 0 & \text{on } \partial \Omega. \end{cases}$$

Example 1.7 (The Minkowski problem). Assume we are given a bounded, positive function K on the unit sphere $S^n = \{x \in \mathbb{R}^{n+1} : |x| = 1\}$, satisfying the integral conditions

$$\int_{S^n} \frac{x_i}{K} \, d\mathcal{H}^n = 0, \quad \text{for all } i = 1, \dots, n+1,$$

where each x_i is the coordinate function. The Minkowski problem asks to find a closed convex hypersurface $\mathcal{M} \subset \mathbb{R}^{n+1}$ such that its Gauss curvature at $p \in \mathcal{M}$ is equal to $K(\nu_p)$ where ν_p is the outer unit normal of $\mathcal{M}$ at p. This problem reduces to solving the Monge–Ampère equation

$$\det(\nabla^2 u + uI) = K^{-1} \quad \text{on } S^n,$$

where $u(x) = \sup\{x \cdot p : p \in \mathcal{M}\}$ is the support function of $\mathcal{M}$, I is the identity matrix, and ∇ is the covariant derivative in a local orthogonal frame.

Example 1.8 (The Monge–Kantorovich optimal transportation problem). Suppose we are given two bounded, convex domains Ω and Ω' in $\mathbb{R}^n$ and bounded, positive functions f_1 and f_2 in Ω and Ω', respectively, with the normalized compatibility condition

$$\int_\Omega f_1(x) \, dx = \int_{\Omega'} f_2(y) \, dy = 1.$$

Thus, $\mu(x) = f_1(x) \, dx$ and $\nu(y) = f_2(y) \, dy$ are probability measures on Ω and Ω', respectively, with densities f_1 and f_2. The *optimal transportation problem* with quadratic cost between (Ω, μ) and (Ω', ν) consists of finding a measurable map $T : \Omega \to \Omega'$ such that $T_\sharp \mu = \nu$, that is, $\nu(A) = \mu(T^{-1}(A))$ for all measurable subsets $A \subset \Omega'$ (equivalently, $\int_{\Omega'} f \, d\nu = \int_\Omega f \circ T \, d\mu$ for all $f \in L^1(\Omega')$) such that T minimizes the quadratic transportation cost:

$$\int_\Omega |x - T(x)|^2 \, d\mu(x) = \min_{S_\sharp \mu = \nu} \int_\Omega |x - S(x)|^2 \, d\mu(x),$$

where the minimum is taken over all measurable maps $S : \Omega \to \Omega'$ such that $S_\sharp \mu = \nu$. A classical theorem of Brenier [**Br**] shows that the optimal map T exists and is given by the gradient of a convex function u; that is, $T = Du$. Moreover, u is a weak solution to the second boundary value problem for the Monge–Ampère equation:

$$\det D^2 u = f_1/(f_2 \circ Du) \quad \text{in } \Omega \quad \text{and} \quad Du(\Omega) = \Omega'.$$

Example 1.9 (Two-dimensional incompressible Navier–Stokes equations). Consider the two-dimensional incompressible Navier–Stokes equations

$$\begin{cases} \partial_t \mathbf{u} + (\mathbf{u} \cdot D)\mathbf{u} - \nu \Delta \mathbf{u} + Dp - \mathbf{f} = 0 & \text{in } \mathbb{R}^2 \times (0, \infty), \\ \operatorname{div} \mathbf{u} = 0 & \text{in } \mathbb{R}^2 \times (0, \infty), \\ \mathbf{u}(x, 0) = \mathbf{u}_0(x) & \text{in } \mathbb{R}^2. \end{cases}$$

Here, $\mathbf{u}(x,t) = (u^1(x,t), u^2(x,t))$ is the velocity vector field, $\nu \geq 0$ is the viscosity, p is the pressure, $\mathbf{f}$ is the force, and $\mathbf{u}_0$ is a divergence-free initial velocity. The incompressibility condition div $\mathbf{u} = 0$ implies the existence of a stream function $\Psi : \mathbb{R}^2 \times [0, \infty) \to \mathbb{R}$ such that $\mathbf{u} = (D\Psi)^\perp \equiv (-D_2\Psi, D_1\Psi)$. Taking the divergence of the Navier–Stokes equations yields

$$\text{div}[(\mathbf{u} \cdot D)\mathbf{u}] + \Delta p - \text{div}\,\mathbf{f} = 0.$$

Since $\text{div}[(\mathbf{u}\cdot D)\mathbf{u}] = -2\det D^2\Psi$, we obtain from the incompressible Navier–Stokes equations the following Monge–Ampère equation:

$$\det D^2\Psi = (\Delta p - \text{div}\,\mathbf{f})/2.$$

Example 1.10 (Dual semigeostrophic equations in meteorology). Let ρ^0 be a probability measure on the two-dimensional torus $\mathbb{T}^2 = \mathbb{R}^2/\mathbb{Z}^2$. The *dual semigeostrophic equations* of the semigeostrophic equations on $\mathbb{T}^2$ are the following system of nonlinear transport equations for (ρ_t, P_t^*):

$$\begin{cases} \partial_t \rho_t(x) + \text{div}[\rho_t(x)(x - DP_t^*(x))^\perp] = 0, & (x,t) \in \mathbb{T}^2 \times [0, \infty), \\ \det D^2 P_t^*(x) = \rho_t(x), & (x,t) \in \mathbb{T}^2 \times [0, \infty), \\ P_t^*(x) \text{ convex}, & (x,t) \in \mathbb{T}^2 \times [0, \infty), \\ \rho_0(x) = \rho^0(x), & x \in \mathbb{T}^2, \end{cases}$$

with $P_t^*(x) - |x|^2/2$ being $\mathbb{Z}^2$-periodic. Here, we use the notation $\rho_t(x) = \rho(x,t)$, $P_t^*(x) = P^*(x,t)$, and $w^\perp = (-w_2, w_1)$ for $w = (w_1, w_2) \in \mathbb{R}^2$.

1.1.2. Existence of solutions. We will introduce various concepts of weak solutions to (1.1) but will work mostly with Aleksandrov solutions. The other types of solutions are viscosity and variational solutions. Using the concept of normal mapping

$$\partial u(x) = \{p \in \mathbb{R}^n : u(y) \geq u(x) + p \cdot (y - x) \quad \text{for all } y \in \Omega\},$$

we can define for each convex function u the Monge–Ampère measure μ_u via

$$\mu_u(E) = \left| \bigcup_{x \in E} \partial u(x) \right| \quad \text{for each Borel subset } E \subset \Omega.$$

Here $|K|$ denotes the Lebesgue measure of $K \subset \mathbb{R}^n$. We say that u is an Aleksandrov solution to (1.1) if, as Borel measures, we have

$$\mu_u = f.$$

Here we identify an integrable function f with its Borel measure $f\,dx$. Therefore, it is natural to consider (1.1) for f being a nonnegative Borel measure.

Using the Perron method and maximum principles including those of Aleksandrov, we can solve the Monge–Ampère equation (1.1) with suitable boundary data φ on $\partial\Omega$ for any nonnegative Borel measure f with finite mass. This general construction does not give much information about the smoothness of solutions when the data is smooth. In other words, can we solve (1.1) in the classical sense? The interior smoothness issue was settled in the works of Pogorelov [**P2, P5**] and Cheng and Yau [**CY**]. Ivočkina [**Iv**], Krylov [**K3**], and Caffarelli, Nirenberg, and Spruck [**CNS1**] provided a positive answer with regards to global smoothness to this question. They proved the existence of smooth, uniformly convex solutions to the Monge–Ampère equation based on higher-order a priori global estimates.

1.1.3. Interior regularity theory of Aleksandrov solutions. We will study the regularity of Aleksandrov solutions to (1.1) both in the interior and at the boundary, especially when

$$(1.4) \qquad\qquad 0 < \lambda \le f \le \Lambda,$$

where λ and Λ are constants. Particular attentions will be paid to the right-hand side f not having much regularity. Our study somehow mirrors the weak solutions and regularity theory of Poisson's equation

$$(1.5) \qquad\qquad \Delta w = g \quad \text{in } \Omega \subset \mathbb{R}^n,$$

but as the book progresses, we will see dramatic differences between the two equations in the regime of limited regularity of the data. Note that, however, if u in (1.1) is known to be $C^{2,\alpha}$, then any partial derivative $v := D_k u$ of u is a solution of $U^{ij} D_{ij} v = D_k f$ which is locally uniformly elliptic with C^α coefficients, for which the classical Schauder theory is applicable. Therefore, it suffices to focus our attention on estimates up to second-order derivatives and their continuity.

Convex Aleksandrov solutions to (1.1) are locally Lipschitz but going beyond Lipschitz regularity turns out to be quite a challenging task. Moreover, the heuristic picture that "solutions of Poisson's equation have two derivatives more than the right-hand side" fails spectacularly for the Monge–Ampère equation. Pogorelov [**P5**] constructed an example in all dimensions at least three with f being a positive analytic function and a non-C^2 solution to (1.1). One reason for this failure of regularity is the existence of a line segment in the graph of solutions to (1.1). Ruling out this behavior, which means showing that the solutions are strictly convex, is a key to the interior regularity. A deeper reason for this failure comes from a crucial geometric fact that the Monge–Ampère equation (1.1) is invariant under affine transformations: We can "stretch" u in one direction and at the same time "contract" it in other directions to get another solution. Thus, it is possible that some singular eigenvalues of $D^2 u$ are compensated by zero eigenvalues.

When the right-hand side of (1.1) is C^2, Pogorelov [**P2**] discovered very important second derivative estimates for strictly convex solutions. Then,

Calabi's third derivative estimates [**Cal1**] yield interior $C^{2,\alpha}$ estimates when f is C^3, and higher-order regularity estimates follow for more regular f. If f is only C^2, then interior $C^{2,\alpha}$ estimates can also be obtained via the use of the Evans–Krylov Theorem for concave, uniformly elliptic equations [**E1, K2**]. All these estimates rely on differentiating the equation, so it is not clear if the above estimates are valid when f is less regular.

In 1990, Caffarelli proved a striking localization theorem for the Monge–Ampère equation [**C2**] which implies strict convexity of solutions under quite general conditions on f and the boundary data, and he established a large part of the interior regularity theory for Aleksandrov solutions.

Consider (1.1) with (1.4). Then Caffarelli [**C2, C3**] showed that strictly convex solutions to (1.1) are $C^{1,\alpha}$ while they are $C^{2,\alpha}$ when f is further assumed to be C^α. Moreover, when f is close to a positive constant or continuous, strictly convex solutions to (1.1) belong to $W^{2,p}_{\text{loc}}$ for all finite p. In Caffarelli's deep regularity theory, sections (or cross sections) of solutions to the Monge–Ampère equation plays an important role. They are defined as sublevel sets of convex solutions after subtracting their supporting hyperplanes. As it turns out in further developments of the Monge–Ampère and linearized Monge–Ampère equations, sections have the same fundamental role that Euclidean balls have in the classical theory of second-order, uniformly elliptic equations. In general, sections have degenerate geometry. However, the affine invariance property of the Monge–Ampère equation in combination with John's Lemma allows us to focus on settings where sections are roughly Euclidean balls.

When f is allowed to have arbitrarily large oscillations, Wang's counterexamples [**W3**] show that strictly convex solutions to (1.1) are not expected to be in $W^{2,p}_{\text{loc}}$ for any fixed exponent $p > 1$. It was an open question whether strictly convex solutions to (1.1) with (1.4) belong to $W^{2,1+\varepsilon}_{\text{loc}}$ for some positive constant ε depending only on n, λ, and Λ. This was not resolved until 2013 when De Philippis, Figalli, and Savin [**DPFS**] and Schmidt [**Schm**] independently established this result. These papers built on the groundbreaking work of De Philippis and Figalli [**DPF1**] on $W^{2,1}_{\text{loc}}$ regularity.

1.1.4. Boundary regularity theory.

Another example of the dramatic difference between (1.1) and (1.5) is concerned with boundary regularity theory. When Ω is a triangle in the plane and $g \equiv 1$, the solution to (1.5) with zero boundary data is C^∞ around any boundary point P away from the vertices; however, the corresponding convex solution to (1.1) is not even Lipschitz (it is only log-Lipschitz) around P, though it is analytic in the interior! To establish global $W^{2,p}$ estimates for (1.5) where $1 < p < \infty$, it suffices to require $\partial\Omega$ to be C^2 and the boundary value of w on $\partial\Omega$ to be $W^{2,p}$. However, for the Monge–Ampère equation, for $p \geq 3$, there is a counterexample of Wang [**W4**] in the case of either $\partial\Omega$ or the boundary value of u failing to be C^3. In the linear, uniformly elliptic equations, to prove a boundary regularity result, we can typically guess what the optimal regularity assumptions on the boundary data and domain should be. This is not the case for the Monge–Ampère equation. In this regard, we have important results of Trudinger and Wang [**TW5**] and Savin [**S4, S5**] on global second derivative estimates. In particular, Savin [**S4**] established a boundary localization theorem for the Monge–Ampère equation under sharp conditions. This theorem paved the way for extending previous interior regularity results to the boundary.

1.2. The Linearized Monge–Ampère Equation

The linearized Monge–Ampère equation arises from several fundamental problems in different subjects including: the affine maximal surface equation in affine geometry, Abreu's equation in the problem of finding Kähler metrics of constant scalar curvature in complex geometry, the semigeostrophic equations in fluid mechanics, and the approximation of minimizers of convex functionals with a convexity constraint in the calculus of variations, to name a few.

The linearized Monge–Ampère equation associated with a C^2 strictly *convex potential* u defined on some convex domain of $\mathbb{R}^n$ is of the form

$$(1.6) \qquad L_u v := \sum_{i,j=1}^{n} U^{ij} D_{ij} v \equiv \text{trace}(U D^2 v) = g.$$

Throughout, the cofactor matrix $\text{Cof } D^2 u$ of the Hessian matrix $D^2 u = (D_{ij} u)_{1 \leq i,j \leq n}$ is denoted by

$$U = (U^{ij})_{1 \leq i,j \leq n} = (\det D^2 u)(D^2 u)^{-1}.$$

The coefficient matrix U of L_u arises from linearizing the Monge–Ampère operator $F(r) = \det r$ where $r = (r_{ij})_{1 \leq i,j \leq n} \in \mathbb{S}^n$ because $U^{ij} = \frac{\partial F}{\partial r_{ij}}(D^2 u)$.

Put differently, $L_u v$ is the coefficient of t in the expansion

$$\det D^2(u + tv) = \det D^2 u + t \,\text{trace}(U D^2 v) + \cdots + t^n \det D^2 v.$$

In this book, we mainly focus on the case that the convex potential u solves the Monge–Ampère equation

$$(1.7) \quad \det D^2 u = f \text{ for some function } f \text{ satisfying } 0 < \lambda \leq f \leq \Lambda < \infty.$$

Given the bounds in (1.7), U is a positive definite matrix, but we cannot expect to obtain structural bounds on its eigenvalues, as indicated by the discussions in Section 1.1.3. Hence, L_u is a linear elliptic partial differential operator which can be both degenerate and singular. *Contrary to what the name might suggest, the linearized Monge–Ampère equation is actually more general than the Monge–Ampère equation. In fact, it includes the Monge–Ampère equation as a special case* because of the identity

$$L_u u = n \det D^2 u,$$

but the difficulty here lies in the fact that one typically does not require any convexity of solutions to the linearized Monge–Ampère equation. On the other hand, in the special case where u is the quadratic polynomial $u(x) = |x|^2/2$, L_u becomes the Laplace operator:

$$L_{|x|^2/2} = \Delta = \sum_{i=1}^{n} \frac{\partial^2}{\partial x_i^2}.$$

Thus, the linearized Monge–Ampère operator L_u captures two of the most important second-order equations in PDEs from the simplest linear equation to one of the most significant fully nonlinear equations.

As U is divergence-free, that is, $\sum_{i=1}^{n} D_i U^{ij} = 0$ for all $j = 1, \dots, n$, the linearized Monge–Ampère equation can be written in both divergence and double divergence form:

$$L_u v = \sum_{i,j=1}^{n} D_i(U^{ij} D_j v) = \sum_{i,j=1}^{n} D_{ij}(U^{ij} v).$$

1.2.1. Examples. We list here some examples where the linearized Monge–Ampère operator appears.

Example 1.11 (Affine maximal surface equation).

$$\sum_{i,j=1}^{n} U^{ij} D_{ij}\left[(\det D^2 u)^{-\frac{n+1}{n+2}}\right] = 0.$$

This equation arises in affine geometry.

Example 1.12 (Abreu's equation).

$$\sum_{i,j=1}^{n} U^{ij} D_{ij}\left[(\det D^2 u)^{-1}\right] = -1.$$

This equation arises in the problem of finding Kähler metrics of constant scalar curvature in complex geometry. A more familiar form of Abreu's equation is

$$\sum_{i,j=1}^{n} \frac{\partial^2 u^{ij}}{\partial x_i \partial x_j} = -1,$$

where $(u^{ij})_{1 \le i,j \le n} = (D^2 u)^{-1}$ is the inverse matrix of $D^2 u$.

Example 1.13 (Dual semigeostrophic equations). The time derivative $\partial_t P_t^*$ of the potential function P_t^* in the two-dimensional semigeostrophic equations in Example 1.10 satisfies

$$\sum_{i,j=1}^{2} (\operatorname{Cof} D^2 P_t^*)^{ij} D_{ij}(\partial_t P_t^*) = \operatorname{div}\left[\det D^2 P_t^* (DP_t^*(x) - x)^\perp\right],$$

where $w^\perp$ denotes the vector $(-w_2, w_1)$ for $w = (w_1, w_2) \in \mathbb{R}^2$.

Example 1.14 (Singular Abreu equation).

$$\sum_{i,j=1}^{n} U^{ij} D_{ij}[(\det D^2 u)^{-1}] = (-\Delta u)\chi_{\Omega_0} + \frac{1}{\delta}(u - \varphi)\chi_{\Omega\setminus\Omega_0} \quad \text{in } \Omega.$$

Here $\delta > 0$ and Ω_0, Ω are smooth, bounded, convex domains in $\mathbb{R}^n$ such that $\Omega_0 \Subset \Omega$. This equation arises from approximating minimizers of convex functionals subjected to a convexity constraint.

1.2.2. Mixed Monge–Ampère measure. We define the mixed Monge–Ampère measure $\tilde{\mu}_{u_1,\ldots,u_n}$ of n convex functions $u_1, \ldots, u_n$ on $\mathbb{R}^n$ by the polarization formula

$$\tilde{\mu}_{u_1,\ldots,u_n} = \frac{1}{n!}\sum_{k=1}^{n}\sum_{1\le i_1<\cdots<i_k\le n}(-1)^{n-k}\mu_{u_{i_1}+\cdots+u_{i_k}}.$$

Clearly,

$$(1.8) \qquad\qquad\qquad\qquad \mu_u = \tilde{\mu}_{u,\ldots,u}.$$

Example 1.15 (The linearized Monge–Ampère operator and mixed Monge–Ampère measures). The linearized Monge–Ampère operator is a special case of mixed Monge–Ampère measures since

$$(1.9) \qquad\qquad\qquad\qquad U^{ij} D_{ij} v = n\tilde{\mu}_{u,\ldots,u,v}.$$

An interesting application of the mixed Monge–Ampère measures to *amoeba* can be found in Passare–Rullgård [**PR**]. More general estimates concerning the mixed Hessian can be found in Trudinger–Wang [**TW3**].

1.2.3. The Caffarelli–Gutiérrez theory and beyond. The regularity theory for the linearized Monge–Ampère equation was initiated in the fundamental paper [**CG2**] by Caffarelli and Gutiérrez. It is worth mentioning that one of their motivations was Lagrangian models of atmospheric and oceanic flows in the work of Cullen, Norbury, and Purser [**CNP**], including the dual semigeostrophic equations in Examples 1.10 and 1.13 that we will discuss in detail in Section 15.4. Caffarelli and Gutiérrez developed an interior Harnack inequality theory for nonnegative solutions of the homogeneous equation $L_u v = 0$ in terms of the structure, called the A_∞-condition, of the

Monge–Ampère measure μ_u of the convex potential function u. This A_∞-condition is clearly satisfied when we have the pinching $\lambda \le \det D^2 u \le \Lambda$ of the Hessian determinant. Their approach is based on that of Krylov and Safonov [**KS1, KS2**] on Hölder estimates for linear, uniformly elliptic equations in nondivergence form, with sections of u replacing Euclidean balls.

When the right-hand side f of the Monge–Ampère equation is only bounded between two positive constants, sections of Aleksandrov solutions to $\det D^2 u = f$ can have degenerate geometry. The key issue in the regularity of the linearized Monge–Ampère equation is to prove that these sections have properties similar to Euclidean balls as in uniformly elliptic equations. These are covered in Chapter 5 for interior sections and in Chapter 9 for boundary sections.

After [**CG2**], many works on the interior regularity of the linearized Monge–Ampère equation have appeared, including: a Liouville theorem [**S2**], interior $W^{2,\delta}, C^{1,\alpha}$, and $W^{2,p}$ estimates [**GTo, GN1, GN2**], interior Hölder estimates under minimal geometric conditions or to equations with lower-order terms [**Md2, Md3, L5**], Green's function and Monge–Ampère Sobolev inequality [**TiW, Md1, Md4, L2, L6**], a Harnack inequality for the parabolic linearized Monge–Ampère equation [**Hu1, Md5**]. Corresponding global Hölder, $C^{1,\alpha}$, $W^{1,p}$, and $W^{2,p}$ estimates for the linearized Monge–Ampère equation were established in [**L1, LS1, LS3, LN2, LN3**]. Many results obtained in the linearized Monge–Ampère equation parallel their analogues in the uniformly elliptic equations. *The overall picture is that information on individual eigenvalues of the coefficient matrix of the uniformly elliptic equations is replaced by information on the Hessian determinant of the convex potential in the linearized Monge–Ampère equation.*

The Caffarelli–Gutiérrez interior Harnack inequality plays a crucial role in Trudinger and Wang's resolution of Chern's conjecture in affine geometry concerning affine maximal surfaces in three dimensions. Trudinger and Wang proved in 2000 that locally uniformly convex and smooth solutions u to the affine maximal surface equation in Example 1.11 must be quadratic functions when $n = 2$. In addition to applications to Chern's conjecture and the dual semigeostrophic equations, this book will discuss other applications arising in affine geometry and the monopolist's problem in economics.

1.2.4. Geometric tools and convexity of the potential functions.

In general, for estimations of solutions to the linearized Monge–Ampère equation that depend only on structural quantities but not on the smoothness of the potential u, we work with its sections instead of Euclidean balls. Moreover, their deep geometric properties will be of vital importance. In particular, for interior estimates, we rely on the results in Chapter 5, while for boundary estimates, we rely on the results of Chapters 8 and 9.

In this book, we focus on the linearized Monge–Ampère equations associated with a convex potential function. Of course, there are equations involving the linearized Monge–Ampère operator where no convexity on the potential function is assumed. One example, as described in Berger [**Bg**], is concerned with von Kármán equations in elasticity. These are two fourth-order quasi-linear elliptic equations that describe the equilibrium states of thin shallow elastic shells under the action of applied forces.

Example 1.16 (Thin elastic shells)**.** Consider a thin shallow shell S of arbitrary shape in three-dimensional $\mathbb{R}^3$ space whose plane projection is a bounded domain Ω in the x_1x_2-plane with boundary $\partial\Omega$. Suppose the shell is acted on by an external force $F(x)$ $(x \in \mathbb{R}^2)$ and by forces along $\partial\Omega$. Then, subject to appropriate boundary conditions, the equilibrium states of S will be determined by solving the following system of fourth-order equations:

$$\begin{cases} \Delta^2 f = -\det D^2 w - D_1(k_1 D_1 w) - D_2(k_2 D_2 w) & \text{in } \Omega, \\ \Delta^2 w = L_f w + D_1(k_1 D_1 f) + D_2(k_2 D_2 f) + F & \text{in } \Omega, \end{cases}$$

where Δ^2 denotes the biharmonic operator on $\mathbb{R}^2$. Here $w(x)$ represents the deflection of the shell from its initial state, $f(x)$ is the Airy stress function, and k_1 and k_2 denote the initial curvatures of the shell in cross sections parallel to the x_1x_3- and x_2x_3-planes, respectively. The right-hand sides of the above system contain the linearized Monge–Ampère operator

$$L_f w = D_{11} f D_{22} w + D_{22} f D_{11} w - 2 D_{12} f D_{12} w.$$

1.3. Plan of the Book

This book consists of a preliminary Chapter 2 and two main parts. In **Part 1** which consists of Chapters 3–11, we cover basic interior and boundary regularity theories of the Monge–Ampère equation. The spectral theory of the Monge–Ampère operator is also introduced. A few applications are presented in Chapters 4 and 7. **Part 2**, consisting of Chapters 12–15, is devoted to the regularity theory of the linearized Monge–Ampère equation and some select applications. The list of applications where the linearized Monge–Ampère equation appears naturally is far from being comprehensive. The applications presented here reflect the author's interests and they cover several problems in geometry, physics, and economics.

To make the book self-contained, we start from scratch the modern theory of the Monge–Ampère equation. Thus, Chapters 3–7 overlap with the books by Figalli [**F2**], Gutiérrez [**G2**], Han [**H**], Le, Mitake, and Tran [**LMT**] and survey papers by De Philippis and Figalli [**DPF2**], Liu and Wang [**LW**], and Trudinger and Wang [**TW6**], though some proofs here are different. The reader may consult the books by Bakelman [**Ba5**] and Pogorelov [**P1**, **P4**,

P5] for earlier treatments of the Monge–Ampère equation, and the book by Schulz [**Schz**] for the Monge–Ampère equation in two dimensions. Many books such as Aubin [**Au2**], Figalli [**F2**], and Villani [**Vi**] cover diverse applications of the Monge–Ampère equation so we will not repeat them here. Our applications are mostly concerned with the linearized Monge–Ampère equation.

The book will address the following **major themes**:

- For the Monge–Ampère equation:
  - (i) *basic concepts and existence results for solutions* (Chapters 3 and 4);
  - (ii) *localization theorems* (Chapters 5 and 8);
  - (iii) *geometry of sublevel sets (or sections)* (Chapters 5 and 9);
  - (iv) *regularity results in the interior and at the boundary* (Chapters 5, 6, and 10);
  - (v) *classification of global solutions* (Chapter 7);
  - (vi) *spectral theory* (Chapter 11).

- For the linearized Monge–Ampère equation:
  - (i) *regularity results in the interior and at the boundary* (Chapters 12 and 13);
  - (ii) *analysis from nondivergence form and divergence form perspectives* (Chapters 14 and 15).

Below is a more detailed description of each chapter.

Chapter 2 collects basic facts and techniques on convex sets, convex functions, analysis, and classical PDE theory that we will use in studying Monge–Ampère equations. Topics in convex sets include: existence of supporting hyperplanes at boundary points, extremal points, John's Lemma, the Hausdorff distance, the Blaschke Selection Theorem, defining function, and approximation of convex bodies by smooth, uniformly convex sets. Topics in convex functions include: supporting hyperplane, the normal mapping, Lipschitz continuity, and Legendre transform. Topics in analysis include: the Hausdorff measure, the area formula, the change of variables formula, the Sobolev Embedding Theorem, and the Implicit Function Theorem. We review the classical PDE theory including: the maximum principle, Perron's method, Schauder estimates, Calderon–Zygmund estimates, Evans–Krylov estimates, pointwise estimates, and perturbation arguments.

Part 1: The Monge–Ampère Equation

Chapter 3 lays the foundations for our investigation of Aleksandrov solutions to the Monge–Ampère equation. We will introduce the important notion of the Monge–Ampère measure of a convex function. With this notion, we can speak of Aleksandrov solutions to a Monge–Ampère equation without the requirement of having two classical derivatives everywhere for the solutions. Then we will prove the Aleksandrov maximum principle, the Aleksandrov–Bakelman–Pucci maximum principle, and compactness results. We will prove the comparison principle and establish various optimal global Hölder estimates. Then we discuss the solvability of the inhomogeneous Dirichlet problem with continuous boundary data via Perron's method.

Chapter 4 discusses the solvability of classical solutions to the Monge–Ampère equation. We will prove the theorem of Ivočkina, Krylov, and Caffarelli, Nirenberg, and Spruck on the existence of smooth, uniformly convex solutions to the Monge–Ampère equation. To prepare for later study of boundary regularity theory, we introduce the notion of the *special subdifferential* of a convex function at a boundary point of a convex domain. We establish the quadratic separation of the function from its tangent hyperplanes at boundary points under suitable hypotheses. Using the existence of classical solutions and the Aleksandrov maximum principle, we give a proof of the classical isoperimetric inequality. We also prove a nonlinear integration by parts for the Monge–Ampère operator. We will discuss Pogorelov's counterexamples to interior regularity which demonstrate that there are no purely interior estimates for the Monge–Ampère equation.

Chapter 5 introduces the notion of sections of convex functions. This notion plays a crucial role in the study of geometry of solutions to the Monge–Ampère equation. We discuss properties of sections including: volume estimates, the size of sections, the engulfing property of sections, the inclusion and exclusion properties of sections, and a localization property of intersecting sections. Two real analysis results for sections will be proved: Vitali's Covering Lemma, and the Crawling of Ink-spots Lemma. The existence of centered sections will be also established. We start the interior regularity theory of the Monge–Ampère equation by proving the important Localization Theorem of Caffarelli. From this together with properties of sections, we obtain Caffarelli's $C^{1,\alpha}$ regularity of strictly convex solutions. We will present Caffarelli's counterexamples to show the failure of $C^{1,\alpha}$ regularity when solutions are not strictly convex.

Chapter 6 is the culmination of the interior regularity theory of the Monge–Ampère equation where we establish critical estimates for second-order derivatives. We will prove Pogorelov's second derivative estimates, Caffarelli's $W^{2,p}$, $C^{2,\alpha}$ estimates, and De Philippis–Figalli–Savin–Schmidt $W^{2,1+\varepsilon}$ estimates. An application is given to interior regularity and uniqueness of degenerate Monge–Ampère equations. We will discuss Wang's counterexample to interior $W^{2,p}$ estimates and Mooney's counterexample to $W^{2,1}$ estimates.

In Chapter 7, we introduce the concept of viscosity solutions to the Monge–Ampère equation and show that Aleksandrov solutions are also viscosity solutions, and vice versa (for a strictly positive, continuous right-hand side). We will prove a classical theorem of Jörgens, Calabi, and Pogorelov on the classification of global solutions to the Monge–Ampère equation with constant right-hand side. We will discuss the relation between Jörgens's Theorem and other Liouville-type theorems in two dimensions such as Bernstein's Theorem on linearity of solutions to the minimal surface equation. We also give a brief introduction to the Legendre–Lewy rotation.

Chapter 8 is devoted to proving Savin's Boundary Localization Theorem. This theorem, which is concerned with the shape of boundary sections, will play a very important role in the boundary regularity theory for the Monge–Ampère equation. The proof uses compactness arguments and viscosity solutions for the Monge–Ampère equation with boundary discontinuities whose groundwork has been built in Chapter 7.

Chapter 9 proves the global $C^{1,\alpha}$ estimates for the Monge–Ampère equation and investigates several important geometric properties of boundary sections and maximal interior sections of convex solutions to the Monge–Ampère equation. We will establish the following properties of boundary sections: dichotomy of sections, volume estimates, engulfing and separating properties, inclusion and exclusion properties, and a chain property. As applications, we prove Besicovitch's Covering Lemma and employ it to prove a covering theorem and a strong-type (p,p) estimate for the maximal function with respect to boundary sections. Moreover, we will introduce a quasi-distance induced by boundary sections and show that the structure of our Monge–Ampère equation gives rise to a space of homogeneous type.

Chapter 10 proves several boundary second derivative estimates for the Monge–Ampère equation using Savin's Boundary Localization Theorem. We will prove the global $W^{2,1+\varepsilon}$ and $W^{2,p}$ estimates, the pointwise $C^{2,\alpha}$ estimates at the boundary, and global $C^{2,\alpha}$ estimates. The latter estimates were previously established by Trudinger and Wang with a different method. We will present Wang's examples to show that the conditions imposed on the boundary in our analysis are in fact optimal.

Chapter 11 is concerned with the Monge–Ampère eigenvalue and eigenfunctions. Similar to the variational characterization of the first Laplace eigenvalue using Rayleigh quotient, we will show that the Monge–Ampère eigenvalue also has a variational characterization using a Monge–Ampère functional. We will study basic properties of this Monge–Ampère functional. The existence of the Monge–Ampère eigenfunctions is established using compactness and solutions to certain degenerate Monge–Ampère equations. In turn, these solutions are found using a parabolic Monge–Ampère flow.

Part 2: The Linearized Monge–Ampère Equation

Chapter 12 proves the Caffarelli–Gutiérrez interior Harnack inequality for the linearized Monge–Ampère equation with the aid of Savin's method of sliding paraboloids. Then we obtain interior Hölder estimates for the inhomogeneous linearized Monge–Ampère equations with L^n right-hand side. Using the Caffarelli–Gutiérrez interior Hölder estimates and Caffarelli's interior $C^{2,\alpha}$ estimates, we sketch the Trudinger–Wang resolution of Chern's conjecture that a locally uniformly convex solution to the affine maximal surface equation in two dimensions must be a quadratic polynomial.

Chapter 13 establishes boundary Hölder, Harnack, and gradient estimates and regularity for solutions to the linearized Monge–Ampère equations under natural assumptions on the domain, Monge–Ampère measures, and boundary data. We briefly describe an application of the boundary Hölder estimates to the Sobolev solvability of global solutions to the second boundary value problems of the prescribed affine mean curvature equation and Abreu's equation. We also briefly describe applications of the boundary Hölder gradient estimates to problems in the calculus of variations concerning minimizers of linear functionals with prescribed determinant.

Chapter 14 is concerned with the Green's function of the linearized Monge–Ampère operator. We prove sharp pointwise estimates and study the integrability of the Green's function and its gradient. In particular, we show that the Green's function of the linearized Monge–Ampère operator has the same integrability range as that of the Green's function of the Laplace operator. The Monge–Ampère Sobolev inequality will be proved. We apply properties of the Green's function to establish local and global Hölder estimates for solutions to the linearized Monge–Ampère equation in terms of the low integrability norm of the inhomogeneity.

Chapter 15 focuses on Hölder estimates for solutions to the inhomogeneous linearized Monge–Ampère equation with right-hand side being the divergence of a bounded vector field. They will be applied to prove the Hölder estimates for the dual semigeostrophic equations in two dimensions when the initial density is bounded away from zero and infinity. We also apply these estimates and those in Chapter 14 to study the solvability of second boundary value problems of singular fourth-order equations of Abreu type arising from the approximation of several convex functionals whose Lagrangians depend on the gradient variable, subject to a convexity constraint.

1.4. Notes

As can be seen in Examples 1.9 and 1.16, solutions of Monge–Ampère equations arising in fluid mechanics and elasticity are not a priori required to be convex. The Monge–Ampère equation with nonconvex solutions is an important direction but so far, it is not well understood. However, we refer the reader to interesting papers [**LMP**, **LP**, **Lwk**] and their references for recent developments in this direction. In this book, the only result where we treat the Monge–Ampère equation with nonconvex solutions is Theorem 3.43 at the cost of requiring the high integrability of the Hessian of the solutions.

For boundary value problems concerning the Monge–Ampère and linearized Monge–Ampère equations, we only consider the Dirichlet boundary condition. There is a vast literature on other boundary conditions. As representatives, we refer the reader to [**LTU**] for the Neumann boundary condition, [**U4**] for oblique boundary condition, [**C8**, **CLW**, **SY2**, **U3**] for the second boundary condition, and [**Hua2**, **Ru**] for the Guillemin boundary condition.

It should be emphasized that we cover in **Part 1** the regularity theory for the Monge–Ampère equations mostly for Monge–Ampère measures with density bounded between two positive constants. The general case of degenerate right-hand side is not covered though this is an important topic in itself. However, only a special case of degeneracy will be treated as it will be used for the treatment of the Monge–Ampère eigenvalue in Chapter 11. Likewise, for the linearized Monge–Ampère equations in **Part 2**, we focus on the equations whose convex potentials have Hessian determinants being bounded between two positive constants. These potentials appear in many applications and they form a general class of settings for the linearized Monge–Ampère equation. To keep the book at a reasonable length, we omit some interesting results dealing with potentials having more regularities such as those having continuous Monge–Ampère measures or potentials having less regularities such as those having Monge–Ampère measures satisfying only a doubling condition. Moreover, we only focus on regularity estimates for smooth solutions with smooth potentials, and we manage to obtain estimates that are independent of the assumed smoothness. These are sufficient for applications where solvability relies on regularity estimates and fixed point arguments.

1.5. Problems

Problem 1.1. Let Ω be a bounded, measurable set in $\mathbb{R}^n$. Show that for any real $n \times n$ matrix A we have
$$|A(\Omega)| = |\det A||\Omega|.$$

Problem 1.2. Let ω_n be the volume of the unit ball in $\mathbb{R}^n$. Show that
$$\int_{\mathbb{R}^n} \frac{1}{\left(1 + |x|^2\right)^{\frac{n+2}{2}}} \, dx = \omega_n.$$

Problem 1.3. Let Ω be a bounded domain in $\mathbb{R}^n$, and let $K, \varphi \in C(\overline{\Omega})$. Suppose that there is a convex solution $u \in C^{0,1}(\overline{\Omega}) \cap C^2(\Omega)$ to

$$\det D^2 u = K(1 + |Du|^2)^{(n+2)/2} \quad \text{in } \Omega, \quad u = \varphi \quad \text{on } \partial\Omega.$$

Show that K must satisfy $\int_\Omega K \, dx < \omega_n$.

Hint: Observe that

$$\int_\Omega K \, dx = \int_{Du(\Omega)} \frac{1}{(1 + |y|^2)^{\frac{n+2}{2}}} \, dy.$$

(This problem is based upon Trudinger–Urbas [**TU**].)

Problem 1.4. In Example 1.12, deduce the second form of Abreu's equation from the original one.

Problem 1.5. Verify formulae (1.8) and (1.9).

Problem 1.6. Let u be a C^2 function on $\mathbb{R}^n$. Recall $D_j u = \frac{\partial u}{\partial x_j}$. Prove that, as differential forms,

$$dD_1 u \wedge \cdots \wedge dD_n u = (\det D^2 u) dx_1 \wedge \cdots \wedge dx_n.$$

Problem 1.7. Consider the function $u(x) = |x| - 1$ on $B_1(0) \subset \mathbb{R}^n$ where $n \geq 2$. Show that

(a) $\det D^2 u(x) = 0$ for all $x \neq 0$, $u = 0$ on $\partial B_1(0)$, and

(b) $u \in W^{2,p}(B_1(0))$ for $1 \leq p < \infty$, if and only if $p < n$.

Chapter 2

Geometric and Analytic Preliminaries

In this chapter, we collect basic definitions and results concerning convex sets, convex functions, real analysis, and classical PDE theory that will be used throughout the book. For convex sets, our discussion includes convex hull, supporting hyperplane at boundary points, Carathéodory's Theorem, John's Lemma, the Hausdorff distance and compactness, the Blaschke Selection Theorem, boundary curvatures, defining function, and approximation by smooth, uniformly convex sets. For convex functions, we study supporting hyperplanes, normal mapping, local Lipschitz property, almost everywhere differentiability, locally uniform convergence, and the Legendre transform. We recall several topics in analysis including: the Hausdorff measure, the area formula, the change of variables formula, the Sobolev Embedding Theorem, and the Implicit Function Theorem. Our review of the classical PDE theory includes the maximum principle, Perron's method, barriers, Schauder estimates, Calderon–Zygmund estimates, Evans–Krylov estimates, pointwise estimates, and perturbation arguments.

2.1. Convex Sets

In this book, convex sets are among the most recurring notions.

Definition 2.1 (Convex set and convex body). A subset E of $\mathbb{R}^n$ is called *convex* if it contains the line segment joining any two points in it; that is,

$$tx + (1-t)y \in E, \quad \text{for all } x, y \in E, \text{ and } t \in [0,1].$$

A nonempty, compact, and convex subset of $\mathbb{R}^n$ is called a *convex body*.

We implicitly assume that, *unless otherwise indicated, all sets in the statements of definitions, theorems, lemmas in this book are nonempty, and all convex sets in statements in Chapters 3–15 have nonempty interiors.*

It is convenient to denote the line segment between x and y by

$$[x, y] := \{tx + (1 - t)y : 0 \le t \le 1\}.$$

Any point of the form $\sum_{i=1}^{N} \lambda_i x^i$ where $x^1, \ldots, x^N \in E$, $\lambda_i \ge 0$ $(1 \le i \le N)$, with $\sum_{i=1}^{N} \lambda_i = 1$, is called a *convex combination* of points in E.

It can be verified that if a set E in $\mathbb{R}^n$ is convex, then so is its closure $\overline{E}$; furthermore, E contains all convex combinations of its points; that is,

$$(2.1) \qquad \sum_{i=1}^{N} \lambda_i x^i \in E \quad \text{for all } x^1, \ldots, x^N \in E, \lambda_i \ge 0 \text{ with } \sum_{i=1}^{N} \lambda_i = 1.$$

From Definition 2.1, it is clear that the intersection of a family of convex sets in $\mathbb{R}^n$ is also a convex set. This leads to:

Definition 2.2 (Convex hull). Let E be a subset of $\mathbb{R}^n$. The intersection of all convex subsets of $\mathbb{R}^n$ containing E is a convex set, called the *convex hull* of E, and it is denoted by $\text{conv}(E)$.

Example 2.3. We list here some examples of convex hulls.

    (a) The convex hull of $(n + 1)$ equally spaced, distinct points in $\mathbb{R}^n$ is called an n-dimensional regular simplex.

    (b) The convex hull of 2^n points $\{(\pm a, \ldots, \pm a)\}$ in $\mathbb{R}^n$ where $a > 0$ is a cube of size length $2a$.

Lemma 2.4 (Characterization of convex hull). *If E is a subset of $\mathbb{R}^n$, then*

$$(2.2) \quad \text{conv}(E) = \left\{ \sum_{i=1}^{N} \lambda_i x^i : x^1, \ldots, x^N \in E, \lambda_i \ge 0 \text{ with } \sum_{i=1}^{N} \lambda_i = 1 \right\}.$$

Proof. Let K be the set on the right-hand side of (2.2). Then $E \subset K$, and K consists of all convex combinations of points in E. Thus, K lies in any convex subset of $\mathbb{R}^n$ containing E. Therefore, $K \subset \text{conv}(E)$. For the other inclusion $\text{conv}(E) \subset K$, it suffices to show that K is convex, but this is a consequence of the simple fact that a convex combination of two convex combinations of points in E is also a convex combination of points in E. $\qquad \square$

We have the following theorem on the generation of the convex hull.

Theorem 2.5 (Carathéodory's Theorem). *If E is a subset of $\mathbb{R}^n$, then each point in $\text{conv}(E)$ is a convex combination of $(n + 1)$ or fewer points in E.*

Proof. Let $x \in \text{conv}(E)$. Then, by Lemma 2.4, x can be represented as

$$x = \sum_{i=1}^{N} \lambda_i x^i, \text{ where } x^1, \ldots, x^N \in E, \lambda_i \geq 0 \text{ with } \sum_{i=1}^{N} \lambda_i = 1,$$

and we can assume that $N \geq 1$ is minimal. The case $N = 1$ is obvious. Assume now that $N \geq 2$. We show that $(N-1)$ vectors $x^2 - x^1, \ldots, x^N - x^1$ are linearly independent, and therefore $N - 1 \leq n$ which concludes the theorem. Suppose otherwise that the above vectors are linearly dependent. We can assume that $x^N - x^1 = \sum_{i=2}^{N-1} \beta_i (x^i - x^1)$ where $\beta_i \in \mathbb{R}$. Thus, the numbers

$$\alpha_1 = -1 + \beta_2 + \cdots + \beta_{N-1}, \ \alpha_2 = -\beta_2, \ldots, \ \alpha_{N-1} = -\beta_{N-1}, \ \alpha_N = 1$$

satisfy $\sum_{i=1}^{N} \alpha_i x^i = 0$, and $\sum_{i=1}^{N} \alpha_i = 0$. Choose k $(1 \leq k \leq N)$ such that λ_k / α_k is smallest among all positive ratios λ_i / α_i. Now, we represent x as

$$x = \sum_{i=1}^{N} \lambda_i' x^i, \quad \text{where } \lambda_i' = \lambda_i - \frac{\lambda_k}{\alpha_k} \alpha_i.$$

From the choice of λ_k / α_k, we find that $\lambda_i' \geq 0$ for all $i = 1, \ldots, N$, while $\lambda_k' = 0$. This contradicts the minimality of N, and the theorem is proved. $\square$

From Theorem 2.5, we easily deduce the following result.

Lemma 2.6. *If E is a compact set in $\mathbb{R}^n$, then $\text{conv}(E)$ is also compact.*

Definition 2.7 (Supporting hyperplane at a boundary point). Let E be a closed, convex set in $\mathbb{R}^n$, and let $x_0 \in \partial E$. A hyperplane H is called a *supporting hyperplane* to E at x_0 if $x_0 \in H$ and E lies entirely in one of the closed half-spaces bounded by H. See Figure 2.1.

Definition 2.8 (Strictly convex domain). A convex domain Ω in $\mathbb{R}^n$ is said to be strictly convex if for each $x_0 \in \partial\Omega$, any supporting hyperplane to $\overline{\Omega}$ at x_0 intersects $\overline{\Omega}$ only at x_0. In other words, strictly convex domains do not contain line segments on their boundary. See Figure 2.1.

We will use the *nearest-point map* to prove the existence of supporting hyperplanes at each boundary point x_0 of a closed, convex set E in $\mathbb{R}^n$. The idea is to show that there exists $x \in \mathbb{R}^n \setminus E$ such that x_0 is the nearest point from E to x. Then the hyperplane orthogonal to the line segment $[x_0, x]$ at x_0 is a supporting hyperplane to E.

Lemma 2.9 (Nearest-point map and its properties). *Let A be a closed, convex set in $\mathbb{R}^n$. Let $x \in \mathbb{R}^n \setminus A$. Then, the following assertions hold.*

(i) *There is a unique point $p(A, x) \in A$ such that*

$$|x - p(A, x)| = \min\{|x - y| : y \in A\}.$$

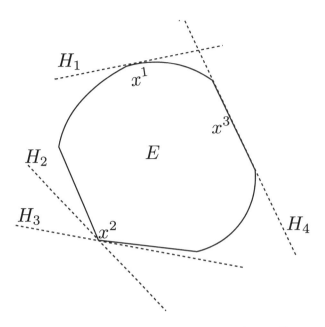

Figure 2.1. The supporting hyperplanes to E at x^1 and x^3 are, respectively, H_1 and H_4. There are infinitely many supporting hyperplanes to E at x^2. The points x^1 and x^2 are extremal points while x^3 is not. The set E is strictly convex at x^1 but not at x^2 and x^3.

(ii) *The hyperplane that is orthogonal to the line segment $[x, p(A, x)]$ at $p(A, x)$ is a supporting hyperplane to A. Moreover, the affine function*

$$l(y) := 2(x - p(A, x)) \cdot (y - p(A, x)) - |x - p(A, x)|^2$$

verifies

$$l(x) = |x - p(A, x)|^2, \text{ and } l(y) \le -|x - p(A, x)|^2 \quad \text{for all } y \in A,$$

so l strongly separates A and x.

(iii) *If $y \in \mathbb{R}^n \setminus A$, then $|p(A, x) - p(A, y)| \le |x - y|$.*

(iv) *Consider the ray $\overrightarrow{p(A, x)x} := \{p(A, x) + t(x - p(A, x)) : t \ge 0\}$. Then $p(A, y) = p(A, x)$ for all $y \in \overrightarrow{p(A, x)x}$.*

(v) *If, moreover, A is compactly contained in the interior of a bounded convex domain $B \subset \mathbb{R}^n$, then $p(A, \partial B) = \partial A$.*

Proof. We prove assertion (i). There is $r > 0$ such that $E := \overline{B_r(x)} \cap A$ is nonempty and compact. The continuous function $d(y) := |x - y|$ $(y \in E)$ then attains its minimum value at some $y_0 \in E$. Clearly $|x - y_0| \le |x - y|$ for all $y \in A$. To show the uniqueness of y_0, suppose that there is $y^0 \in A \setminus \{y_0\}$

such that $|x-y_0| = |x-y^0|$. Then, by the convexity of A, $z = (y_0+y^0)/2 \in A$ but $|x - z| < |x - y_0|$, a contradiction. Now, set $p(A,x) = y_0$.

To prove assertion (ii), let $y \in A \setminus \{p(A,x)\}$ and $z = y - p(A,x)$. Then, for any $t \in [0,1]$, we have $p(A,x)+tz = ty+(1-t)p(A,x) \in A$. The function $|x - (p(A,x)+tz)|^2/2$ has a minimum value at 0 so its right-hand derivative at 0 is nonnegative. This translates to $(x - p(A,x)) \cdot z \le 0$, or

$$(2.3) \qquad (x - p(A,x)) \cdot (y - p(A,x)) \le 0 \quad \text{for all } y \in A.$$

Hence, the hyperplane orthogonal to the line segment $[x, p(A,x)]$ at $p(A,x)$ is a supporting hyperplane to A, and the properties of l easily follow.

Now, we prove assertion (iii). For $y \in \mathbb{R}^n \setminus A$, let $z = p(A,y) - p(A,x)$. Then, by (2.3), $(x - p(A,x)) \cdot z \le 0$, and $(y - p(A,y)) \cdot z \ge 0$. We only consider the case $z \ne 0$. Then, the line segment $[x,y]$ meets the two hyperplanes orthogonal to z that pass through $p(A,x)$ and $p(A,y)$, respectively, so $|p(A,x) - p(A,y)| \le |x - y|$.

We prove assertion (iv). Let $y \in \overrightarrow{p(A,x)x}$. Then, there is $t \ge 0$ such that $y = p(A,x) + t(x - p(A,x))$. Let $z := p(A,y) - p(A,x)$. By (2.3), we have $(x - p(A,x)) \cdot z \le 0$ and $(y - p(A,y)) \cdot z \ge 0$. These give

$$(y - p(A,x)) \cdot z \le 0 \quad \text{and} \quad (p(A,y) - y) \cdot z \le 0.$$

Adding these inequalities yields $z \cdot z \le 0$ which shows that $z = 0$, as desired.

Finally, we prove assertion (v). Clearly, $p(A, \partial B) \subset \partial A$ so we only need to prove that $\partial A \subset p(A, \partial B)$. Let $x \in \partial A$ be arbitrary. Then, for each $i \in \mathbb{N}$, there exists $x^i \in B_{1/i}(x) \setminus A$. The point $p(A, x^i) \in A$ satisfies $|x^i - p(A, x^i)| \le |x^i - x| < 1/i$. Hence

$$|x - p(A, x^i)| \le |x - x^i| + |x^i - p(A, x^i)| < 2/i.$$

Therefore, $p(A, x^i) \to x$ when $i \to \infty$. Now, the ray $\overrightarrow{p(A, x^i)x^i}$ intersects ∂B at some y^i. Up to extracting a subsequence, y^i converges to $y \in \partial B$. By the continuity of $p(A, \cdot)$ from assertion (iii), we infer that $p(A, y^i) \to p(A, y)$ when $i \to \infty$. By (iv), $p(A, y^i) = p(A, x^i)$ so $p(A, y^i) \to x$ when $i \to \infty$. Therefore, $x = p(A, y) \in p(A, \partial B)$, which shows that $\partial A \subset p(A, \partial B)$. $\quad\square$

The strong separation result in Lemma 2.9(ii) implies that any closed convex set in $\mathbb{R}^n$ is the intersection of the closed half-spaces that contain it. This is a sort of external representation of a closed convex set. In Theorem 2.12, we will see a different representation for compact convex sets.

Theorem 2.10 (Existence of supporting hyperplanes at boundary points). *Let E be a closed, convex set in $\mathbb{R}^n$, and let $x_0 \in \partial E$. Then, there is a supporting hyperplane to E at x_0.*

Proof. Consider first the case E is bounded. Then by Lemma 2.9(v), there is $x \in \mathbb{R}^n \setminus E$ such that $p(E, x) = x_0$. By Lemma 2.9(ii), the hyperplane orthogonal to the line segment $[x, x_0]$ at x_0 is a supporting hyperplane to E.

For the general case, we know from the above argument that for the bounded, closed, and convex set $A := E \cap \overline{B_1(x_0)}$, there is a supporting hyperplane $H = \{x \in \mathbb{R}^n : l(x) = 0\}$ to A at x_0, where l is an affine function. This is also a supporting hyperplane to E at x_0 by the convexity of E. Indeed, suppose A lies in the half-space $\{l \geq 0\}$ bounded by H. If $x \in E \setminus \overline{B_1(x_0)}$, then the line segment $[x_0, x]$ lies in E from its convexity while $[x_0, x]$ intersects $\partial B_1(x_0)$ at a point $y = tx_0 + (1 - t)x \in A$ where $0 < t < 1$. Since l is affine, $0 \leq l(y) = tl(x_0) + (1 - t)l(x) = (1 - t)l(x)$ which shows that $l(x) \geq 0$ and hence x lies in the half-space $\{l \geq 0\}$. □

Definition 2.11. (Extremal points and exposed points) Let Ω be a convex set in $\mathbb{R}^n$. A point $x_0 \in \partial\Omega$ is called

(a) an *extremal point* of Ω if x_0 is not a convex combination of other points in $\overline{\Omega}$ (see Figure 2.1);

(b) an *exposed point* of Ω if there is a supporting hyperplane H in $\mathbb{R}^n$ to $\overline{\Omega}$ such that $H \cap \overline{\Omega} = \{x_0\}$.

For a given convex set Ω in $\mathbb{R}^n$, an exposed point of Ω is also an extremal point. The converse is not true in general. The convex set

$$\Omega := \{(x_1, x_2) \in \mathbb{R}^2 : |x_1| \leq 1, -(1 - x_1^2)^{1/2} \leq x_2 \leq 1\}$$

has $(\pm 1, 0)$ as extremal points, but they are not exposed points.

The following theorem gives a sort of internal representation of a compact convex set in $\mathbb{R}^n$ in terms of its extremal points or exposed points.

Theorem 2.12 (Extremal representation)**.** *Let K be a nonempty closed convex subset of $\mathbb{R}^n$. Then the following assertions hold.*

(i) *(Minkowski's Theorem) Assume that K is also compact. Then, the set E of extremal points of K is nonempty. Moreover, each point in K is a convex combination of at most $(n + 1)$ points in E.*

(ii) *(Straszewicz's Theorem) The set F of exposed points of K is dense in the set E of its extremal points.*

Proof. We prove Minkowski's Theorem by induction on n. If $n = 1$, then K is a closed interval $[a, b]$ and $E = \{a, b\}$, so the conclusions of the theorem are obvious. Suppose the theorem has been proved for dimensions up to $n - 1 \geq 1$. We now prove it for dimension n. Let $x_0 \in \partial K$. By Theorem 2.10, there is a supporting hyperplane H to K at x_0. Let $K_1 := H \cap K$. Then K_1 is convex, compact, $x_0 \in K_1$, and K_1 lies in an $(n - 1)$-dimensional subspace of $\mathbb{R}^n$. Clearly, any extremal point of K_1 is also an extremal point of K. By the induction hypothesis, the set E_1 of extremal points of K_1 is nonempty, and x_0 is a convex combination of at most n points in E_1. Therefore, the set E of extremal points of K is nonempty and any boundary point of K is a convex combination of at most n points in E. Now, consider an interior point x of K. Let $z \in E$. Then the ray $\overrightarrow{z\hat{x}}$ intersects ∂K at some x_0. Since x is a convex combination of z and x_0 while x_0 is a convex combination of at most n points in E, we find that x is a convex combination of at most $(n + 1)$ points in E. This completes the proof of Minkowski's Theorem.

We prove Straszewicz's Theorem by showing that given $x_0 \in E$ and $\varepsilon > 0$, the set $\partial K \cap B_\varepsilon(x_0)$ contains an exposed point. Indeed, replacing K by $K \cap \overline{B_1(x_0)}$ if necessary, we can assume that K is bounded. Since x_0 is an extremal point of K, $x_0 \notin A := \operatorname{conv}(K \setminus B_\varepsilon(x_0))$. Because A is convex and closed (Lemma 2.6), by Lemma 2.9, there is a hyperplane H that strongly separates A and x_0. By a transformation of coordinates, we can assume that

$$H = \{x_n = 0\}, \quad A \subset \{x \in \mathbb{R}^n : x_n \geq 0\}, \quad \text{and} \quad x_0 = -se_n$$

for some $s \in (0, \varepsilon)$. Thus, $B_\varepsilon(x_0)$ contains the cap $\{x \in \mathbb{R}^n : x_n < 0\} \cap K$.

Now, for any $t > 0$, let $x^t = te_n$. Then, there is $z^t \in \partial K$ such that $r(t) := |x^t - z^t| = \max_{z \in K} |x^t - z|$. Clearly, the supporting hyperplane to $\overline{B_{r(t)}(x^t)}$ at z^t intersects ∂K only at z^t so z^t is an exposed point of K; that is, $z^t \in F$. Write $z^t = ((z^t)', z_n^t)$. From the boundedness of K and

$$(t + s)^2 = |x^t - x_0|^2 \leq |x^t - z^t|^2 = |(z^t)'|^2 + (t - z_n^t)^2,$$

we find that $z_n^t < 0$ when t is large. In this case, $z^t \in \{x_n < 0\} \cap K \subset B_\varepsilon(x_0)$, and this completes the proof of Straszewicz's Theorem. $\qquad\square$

2.2. The Hausdorff Distance

Several central compactness arguments in this book involve convergence of bounded convex domains so it is important to carefully discuss the distance between these sets. An appropriate notion of distance is that of Hausdorff.

Definition 2.13 (Hausdorff distance). The *Hausdorff distance* $d_H(A, B)$ between two nonempty subsets A and B of $\mathbb{R}^n$ is defined by

$$d_H(A, B) = \max \left\{ \sup_{x \in A} \inf_{y \in B} |x - y|, \sup_{x \in B} \inf_{y \in A} |x - y| \right\}.$$

Recall that $A + B := \{x + y : x \in A, y \in B\}$ is the Minkowski sum of two subsets $A, B \subset \mathbb{R}^n$. We have the following useful neighborhood characterization of d_H.

Lemma 2.14. *Let A and B be subsets of $\mathbb{R}^n$ with $d_H(A, B) < \infty$. Then*

$$d_H(A, B) = \inf\{\lambda \geq 0 : A \subset B + \lambda B_1(0), \ B \subset A + \lambda B_1(0)\}$$
$$= \inf\{\lambda \geq 0 : A \subset B + \lambda \overline{B_1(0)}, \ B \subset A + \lambda \overline{B_1(0)}\}.$$

In particular, if A and B are nonempty, compact subsets of $\mathbb{R}^n$, then

$$(2.4) \qquad d_H(A, B) = \min\{\lambda \geq 0 : A \subset B + \lambda \overline{B_1(0)}, \ B \subset A + \lambda \overline{B_1(0)}\}.$$

Proof. Let

$$\gamma := \inf\{\lambda \geq 0 : A \subset B + \lambda B_1(0), \ B \subset A + \lambda B_1(0)\}.$$

If $d_H(A, B) < \infty$, then it follows from the definition of d_H that $\gamma \leq d_H(A, B) + 1 < \infty$. Let $\lambda \geq 0$ be such that $A \subset B + \lambda B_1(0)$ and $B \subset A + \lambda B_1(0)$. If $x \in A$, then $x \in B + \lambda B_1(0)$ so there is $y_0 \in B$ and $z \in B_1(0)$ such that $x = y_0 + \lambda z$. It follows that $\inf_{y \in B} |x - y| \leq |x - y_0| = \lambda |z| \leq \lambda$. Since $x \in A$ is arbitrary, we have $\sup_{x \in A} \inf_{y \in B} |x - y| \leq \lambda$. Similarly, from $B \subset A + \lambda B_1(0)$, we also have $\sup_{x \in B} \inf_{y \in A} |x - y| \leq \lambda$. Thus, $d_H(A, B) \leq \lambda$. Taking the infimum over λ yields $d_H(A, B) \leq \gamma$.

Now, let $0 \leq \mu < \gamma$. By the definition of γ, we have either $A \not\subset B + \mu B_1(0)$ or $B \not\subset A + \mu B_1(0)$. Assume $A \not\subset B + \mu B_1(0)$. Then, there is $x_0 \in A$ such that $x_0 \notin B + \mu B_1(0)$. Hence, $|x_0 - y| \geq \mu$ for all $y \in B$. This shows that $\sup_{x \in A} \inf_{y \in B} |x - y| \geq \inf_{y \in B} |x_0 - y| \geq \mu$ and $d_H(A, B) \geq \mu$. Since $0 \leq \mu < \gamma$ is arbitrary, we must have $d_H(A, B) \geq \gamma$. Combining this with $d_H(A, B) \leq \gamma$, we obtain $d_H(A, B) = \gamma$.

If in the definition of γ we replace $B_1(0)$ by its closure, then the same arguments also give the asserted equality. $\qquad\square$

Remark 2.15. Using (2.4), we can verify that d_H is a metric on the collection of nonempty compact subsets in $\mathbb{R}^n$. Sometimes, it is convenient to adopt the following convention: We say that a sequence of open convex sets $\{S_i\}_{i=1}^{\infty} \subset \mathbb{R}^n$ converges in the Hausdorff distance to an open convex set S_∞ if $\{\overline{S_i}\}_{i=1}^{\infty}$ converges in the Hausdorff distance to $\overline{S_\infty}$.

In this book, a basic principle that we repeatedly use is the Blaschke Selection Theorem, Theorem 2.20, on subsequent convergence of bounded sequences of convex bodies. We first prove some preparatory results on the Hausdorff metric d_H on nonempty compact subsets in $\mathbb{R}^n$.

A sequence $\{A_i\}_{i=1}^{\infty} \subset \mathbb{R}^n$ is called *decreasing* if $A_{i+1} \subset A_i$ for all $i \geq 1$.

Lemma 2.16. *Let $\{K_i\}_{i=1}^{\infty}$ be a decreasing sequence of nonempty compact subsets of $\mathbb{R}^n$. Then*

$$\lim_{i \to \infty} d_H\left(K_i, \bigcap_{j=1}^{\infty} K_j\right) = 0.$$

Proof. By Cantor's Intersection Theorem, $K := \bigcap_{j=1}^{\infty} K_j$ is compact and nonempty. The sequence $\{d_H(K_i, K)\}_{i=1}^{\infty}$ is decreasing and has a limit $2\lambda \geq 0$. Assume $\lambda > 0$. Then, $d_H(K_m, K) > \lambda$ for all $m \geq 1$. By (2.4), this implies that $K_m \not\subset K + \lambda \overline{B_1(0)}$. Let $A_m := K_m \setminus \mathrm{int}(K + \lambda \overline{B_1(0)})$, where $\mathrm{int}(E)$ denotes the interior of a set $E \subset \mathbb{R}^n$. Then, $\{A_m\}_{m=1}^{\infty}$ is a decreasing sequence of nonempty compact subsets of $\mathbb{R}^n$. Therefore, by Cantor's Intersection Theorem, $A := \bigcap_{m=1}^{\infty} A_m$ is compact and nonempty. Since $A_m \cap K = \emptyset$, we have $A \cap K = \emptyset$. On the other hand, we also have $A_m \subset K_m$ so their intersections satisfy $A \subset K$ which gives us a contradiction. $\qquad\square$

The following theorem says that the space of nonempty compact subsets of $\mathbb{R}^n$ equipped with the Hausdorff metric is a complete metric space.

Theorem 2.17 (Completeness). *Let $\{K_i\}_{i=1}^{\infty}$ be a Cauchy sequence of nonempty compact subsets of $\mathbb{R}^n$ in the metric d_H. Then, it converges in d_H to a nonempty compact subset of $\mathbb{R}^n$.*

Proof. Since $d_H(K_i, K_j) \to 0$ when $i, j \to \infty$, $\{d_H(K_1, K_i)\}_{i=1}^{\infty}$ is bounded by a constant $M > 0$. Hence $K_i \subset K_1 + M\overline{B_1(0)}$ and $\bigcup_{i=1}^{\infty} K_i \subset K_1 + M\overline{B_1(0)}$. Let $A_m := \overline{\bigcup_{i=m}^{\infty} K_i}$. Then $\{A_m\}_{m=1}^{\infty}$ is a decreasing sequence of nonempty compact subsets of $\mathbb{R}^n$. By Lemma 2.16, A_m converges in the metric d_H to the nonempty compact set $A := \bigcap_{m=1}^{\infty} A_m$.

We now show that K_i converges to A when $i \to \infty$. Indeed, for any $\varepsilon > 0$, there is $m(\varepsilon) \in \mathbb{N}$ such that $K_m \subset A_m \subset A + \varepsilon B_1(0)$ for all $m \geq m(\varepsilon)$. Since $d_H(K_i, K_j) \to 0$ when $i, j \to \infty$, there is $n(\varepsilon) \geq m(\varepsilon)$ such that $K_i \subset K_m + \varepsilon \overline{B_1(0)}$ for all $i, m \geq n(\varepsilon)$. Thus, $A_m = \overline{\bigcup_{i=m}^{\infty} K_i} \subset K_m + \varepsilon \overline{B_1(0)}$ for all $m \geq n(\varepsilon)$. It follows that $A \subset K_m + \varepsilon \overline{B_1(0)}$, and hence $d_H(K_m, A) \leq \varepsilon$ for all $m \geq n(\varepsilon)$. The theorem is proved. $\qquad\square$

Theorem 2.18 (Compactness). *Let $\{K_i\}_{i=1}^{\infty}$ be a sequence of nonempty compact subsets of $\mathbb{R}^n$ that is contained in a bounded domain. Then, it has a convergent subsequence in the Hausdorff metric to a nonempty compact subset of $\mathbb{R}^n$.*

Proof. By Theorem 2.17, it suffices to show that $\{K_i\}_{i=1}^{\infty}$ has a Cauchy subsequence in the metric d_H. Suppose $\{K_i\}_{i=1}^{\infty}$ is contained in a cube $\mathcal{C}$ of size length L.

We introduce the following notation. For each positive integer m, we can divide $\mathcal{C}$ into 2^{mn} subcubes of size length $2^{-m}L$, and for each nonempty compact subset K of $\mathbb{R}^n$, we let $A_m(K)$ be the union of those subcubes of $\mathcal{C}$ that intersect with K. Obviously, the number of subcubes in $A_m(K)$ is always bounded from above by 2^{nm}. Moreover, if $A_m(K) = A_m(Q)$, then

$d_H(K, Q)$ is not greater than the diameter of each subcube. Therefore,

$$(2.5) \qquad A_m(K) = A_m(Q) \quad \text{implies } d_H(K, Q) \leq \sqrt{n}2^{-m}L.$$

Back to our proof. We can find a subsequence $\{K_i^1\}_{i=1}^\infty$ of $\{K_i\}_{i=1}^\infty$ such that $A_1(K_i^1) := K^1$ for all i. Now, for $m = 2$, there is a subsequence $\{K_i^2\}_{i=1}^\infty$ of $\{K_i^1\}_{i=1}^\infty$ such that $A_2(K_i^2) := K^2$ for all i. Inductively, we construct subsequences $\{K_i^m\}_{i=1}^\infty$ for each positive integer $m \geq 1$ such that

$$(2.6) \qquad \{K_i^m\}_{i=1}^\infty \text{ is a subsequence of } \{K_i^k\}_{i=1}^\infty \quad \text{for all } k < m$$

and $A_m(K_i^m) := K^m$ for all i. The last equality and (2.5) give

$$d_H(K_i^m, K_j^m) \leq \sqrt{n}2^{-m}L \quad \text{for all } m, i, j.$$

This estimate and (2.6) give $d_H(K_k^k, K_m^m) \leq \sqrt{n}2^{-k}L$ for all $m > k$, and $\{K_m^m\}_{m=1}^\infty$ is a Cauchy subsequence of the original sequence. $\qquad \square$

A consequence of Theorem 2.18 is that when restricting d_H to the nonempty closed subsets of a compact set, we obtain a compact metric space. Observe that the Hausdorff distance preserves convexity in the limit.

Lemma 2.19 (Convexity passes to the limit). *Suppose that $\{K_i\}_{i=1}^\infty$ is a sequence of nonempty closed convex subsets of $\mathbb{R}^n$ that converges in the Hausdorff distance to a closed set K. Then K is convex.*

Proof. Since $d_H(K_i, K) \to 0$ when $i \to \infty$, we use Lemma 2.14 to find for each $\varepsilon > 0$ a positive integer $i(\varepsilon)$ such that $K_i \subset K + \varepsilon B_1(0)$ and $K \subset K_i + \varepsilon B_1(0)$ for all $i \geq i(\varepsilon)$. Now, let $x, y \in K$. Then $x, y \in K_i + \varepsilon B_1(0)$ for each $i \geq i(\varepsilon)$, and hence, by the convexity of K_i and $B_1(0)$, the line segment $[x, y] \subset K_i + \varepsilon B_1(0)$. Therefore, $[x, y] \subset K + 2\varepsilon B_1(0)$. This holds for all $\varepsilon > 0$, so $[x, y] \subset \overline{K} = K$ since K is closed. Therefore, K is convex. $\quad \square$

Combining Theorem 2.18 with Lemma 2.19, we obtain:

Theorem 2.20 (Blaschke Selection Theorem). *Every sequence of convex bodies in $\mathbb{R}^n$ that is contained in a bounded domain has a subsequence that converges in the Hausdorff distance to a convex body.*

2.3. Convex Functions and the Normal Mapping

2.3.1. Convex functions. We will mostly work on convex functions defined on convex open sets whose usual definition is similar to the one below for $\mathbb{R}^n$. However, in many situations, it is also convenient to discuss convex functions on general open sets.

Definition 2.21 (Convex function). Let Ω be an open subset in $\mathbb{R}^n$.

(a) A function $v : \mathbb{R}^n \to \mathbb{R} \cup \{+\infty\}$ is said to be *convex* if
$$v(tx + (1-t)y) \le tv(x) + (1-t)v(y) \quad \text{for all } x, y \in \mathbb{R}^n \text{ and } t \in [0,1].$$

(b) A function $u : \Omega \to \mathbb{R}$ is called *convex* if it can be extended to a convex function, possibly taking value $+\infty$, on all of $\mathbb{R}^n$. In particular, the convexity of u on Ω implies that, for all $x^1, \ldots, x^N \in \Omega$ and $\lambda_i \ge 0$ with $\sum_{i=1}^{N} \lambda_i = 1$ such that $\sum_{i=1}^{N} \lambda_i x^i \in \Omega$, we have
$$u\Big(\sum_{i=1}^{N} \lambda_i x^i\Big) \le \sum_{i=1}^{N} \lambda_i u(x^i).$$

Clearly, in (b), we can also allow u to take $+\infty$ values. A function $v : \Omega \to \mathbb{R} \cup \{-\infty\}$ on an open set $\Omega \subset \mathbb{R}^n$ is called *concave* if $-v$ is convex on Ω. *Affine functions* are convex functions that will appear frequently. They are of the form $l(x) = p \cdot x + a$ where $p \in \mathbb{R}^n$ is called the *slope* of l.

If Ω is convex and $u : \Omega \to \mathbb{R}$ is convex in the usual definition of convexity, then by setting $u = +\infty$ on $\mathbb{R}^n \setminus \Omega$, we obtain a convex function on $\mathbb{R}^n$. In general, convex extensions to $\mathbb{R}^n$ of a convex function on an open set are not unique. For example, if $u(x) = 1/x$ on $(0, 1) \subset \mathbb{R}$, then we can extend u to a convex function on $\mathbb{R}$ by setting $u(1) = a \ge 1$ and $u(x) = +\infty$ if $x \le 0$ or $x > 1$. The extension with $a = 1$ is minimal among all possible convex extensions. We will return to this issue in Chapter 7.

Remark 2.22. Let Ω be a bounded, open set in $\mathbb{R}^n$.

- If $u \in C(\overline{\Omega})$ and u is convex on Ω, then we usually write "$u \in C(\overline{\Omega})$, convex". Moreover, for all $x, y \in \overline{\Omega}$ and $t \in [0, 1]$, we have
$$u(tx + (1-t)y) \le tu(x) + (1-t)u(y) \text{ provided } tx + (1-t)y \in \overline{\Omega}.$$

- If $v : \Omega \to \mathbb{R}$ is convex, then we usually extend v to $\partial\Omega$ by restricting to $\partial\Omega$ its minimal convex extension to $\mathbb{R}^n$. This gives the usual boundary values of v when Ω is convex and $v \in C(\overline{\Omega})$ a priori.

From Definition 2.21 and Lemma 2.4, we obtain:

Lemma 2.23 (Maximum principle for convex functions). *Assume that* $u : \Omega \to \mathbb{R}$ *is convex where* Ω *is an open subset of* $\mathbb{R}^n$.

(i) *If* $E \subset \mathbb{R}^n$ *satisfies* $\text{conv}(E) \subset \Omega$, *then* $\sup_{\text{conv}(E)} u = \sup_E u$.

(ii) *If* Ω *is bounded and* $u \in C(\overline{\Omega})$, *then* $\max_{\overline{\Omega}} u = \max_{\partial\Omega} u$.

Assertion (ii) uses the fact that any interior point of a bounded domain lies in an interval with one endpoint on the boundary. The following simple lemma is a useful device to produce convex functions; see Problem 2.1.

Lemma 2.24 (Supremum of convex functions). *Let $\{u_\alpha\}_{\alpha\in\mathcal{A}}$ be a family of convex functions on an open set $\Omega \subset \mathbb{R}^n$. Define $u : \Omega \to \mathbb{R} \cup \{+\infty\}$ by $u(x) = \sup_{\alpha\in\mathcal{A}} u_\alpha(x)$ for $x \in \Omega$. Then u is convex on Ω.*

Now, we introduce the concepts of *pointwise Lipschitz* and *locally Lipschitz* continuous functions; see Definition 2.88 for pointwise C^k and $C^{k,\alpha}$.

Definition 2.25 (Pointwise Lipschitz and locally Lipschitz continuous functions). Let Ω be a subset of $\mathbb{R}^n$ and let $u : \Omega \to \mathbb{R}$. We say that u is

(a) pointwise Lipschitz continuous at $x_0 \in \Omega$ if there is a ball $B_r(x_0) \subset \Omega$ and a constant $C > 0$ such that
$$|u(x) - u(x_0)| \le C|x - x_0| \quad \text{for all } x \in B_r(x_0);$$

(b) locally Lipschitz continuous in Ω if for all compact sets $K \subset \Omega$, there is a constant $C_K > 0$ such that
$$|u(x) - u(y)| \le C_K|x - y| \quad \text{for all } x, y \in K.$$

We have the following basic theorem on the pointwise Lipschitz continuity of a convex function in the interior of the set where it is finite.

Theorem 2.26 (Pointwise Lipschitz continuity of a convex function). *Let $f : \mathbb{R}^n \to \mathbb{R} \cup \{+\infty\}$ be a convex function. Then f is pointwise Lipschitz continuous in the interior of the set where it is finite.*

Proof. Let $\Omega = \{x \in \mathbb{R}^n : f(x) < \infty\}$. Suppose $B_{2r}(z) \subset \Omega$ where $z = (z_1, \ldots, z_n)$. Let K be the set of 2^n points $\{(z_1 \pm r/\sqrt{n}, \ldots, z_n \pm r/\sqrt{n})\}$ on $\partial B_r(z)$. By Lemma 2.23, $M := \sup_{\text{conv}(K)} f = \sup_K f < \infty$ due to K being finite. It suffices to show that f is pointwise Lipschitz continuous in the interior E of $\text{conv}(K)$. Indeed, let $x \in E$. Then there is $s > 0$ such that $B_s(x) \subset E$. Consider $y \in B_s(0)$. Then, $x \pm y \in E$. Let $h \in (0, 1]$. By the convexity of f, we have $f(x + hy) \le hf(x + y) + (1 - h)f(x)$. Hence
$$[f(x + hy) - f(x)]/h \le f(x + y) - f(x) \le \sup_{\text{conv}(K)} f - f(x) = M - f(x).$$
Similarly, we have
$$[f(x + hy) - f(x)]/h \ge f(x) - f(x - y) \ge f(x) - M.$$
Now, the pointwise Lipschitz continuity of f at x follows from
$$[f(x) - M]h \le f(x + hy) - f(x) \le h[M - f(x)] \quad \text{for all } y \in B_s(0).$$
The theorem is proved. $\qquad\square$

Associated with each convex function is a convex set, called the *epigraph*, which is also known as the *upper graph*. See Figure 2.2.

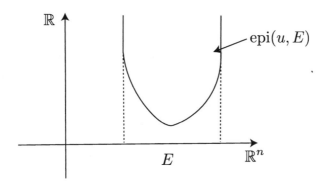

Figure 2.2. The epigraph of a convex function u over E.

Definition 2.27 (Epigraph). Let E be a subset of $\mathbb{R}^n$ and let $u : E \to \mathbb{R} \cup \{+\infty\}$. The epigraph of u over E is denoted by

$$\mathrm{epi}(u, E) := \{(x, x_{n+1}) \in E \times \mathbb{R} : x_{n+1} \geq u(x)\}.$$

Clearly, if Ω is a convex domain in $\mathbb{R}^n$ and $u : \Omega \to \mathbb{R}$ is a convex function, then $\mathrm{epi}(u, \Omega)$ is a convex set in $\mathbb{R}^{n+1}$. In Section 7.2, we will use epigraphs to focus our discussion on convex sets and obtain results on convex functions as corollaries.

We just saw that to each convex function there is an associated convex set which is its epigraph. Vice versa, it is also possible to associate a convex function to each compact convex set. This is done via the important concept of the *gauge function* of a convex body, due to Minkowski.

Theorem 2.28 (Convex body and its gauge function). *Let K be a convex body in $\mathbb{R}^n$ containing the origin in its interior. Let p_K be its gauge function, which is defined by*

$$p_K(x) = \inf\{t > 0 : x/t \in K\}.$$

Then p_K is a convex positively homogeneous function, which is positive (and could be infinite), except at the origin; that is,

(2.7)
$$\begin{cases} \text{(a)} & p_K(\lambda x) = \lambda p_K(x), \text{ for all } \lambda > 0, \text{ and } x \in \mathbb{R}^n, \\ \text{(b)} & p_K(x) > 0, \text{ for all } x \neq 0, \quad p_K(0) = 0, \\ \text{(c)} & p_K(x + y) \leq p_K(x) + p_K(y) \text{ for all } x, y \in \mathbb{R}^n. \end{cases}$$

Moreover

(2.8)
$$K = \{x \in \mathbb{R}^n : p_K(x) \leq 1\}.$$

Conversely, if p_K is a function satisfying (2.7) and K is defined by (2.8), then p_K is the gauge function of the convex body K.

Proof. Suppose K is a convex body in $\mathbb{R}^n$ containing the origin in its interior. It is obvious from the definition of p_K that (2.8) and (2.7)(a) and (b) hold. To verify (2.7)(c), it suffices to consider the case when $p_K(x)$ and $p_K(y)$ are positive and finite. For any $\varepsilon > 0$, there are $s, t > 0$ and $X, Y \in K$ such that $s \leq p_K(x) + \varepsilon$, $t \leq p_K(y) + \varepsilon$, and $x = sX, y = tY$. Since K is convex,

$$\frac{x+y}{s+t} = \frac{s}{s+t}X + \frac{t}{s+t}Y \in K.$$

It follows that $p_K(x + y) \leq s + t \leq p_K(x) + p_K(y) + 2\varepsilon$. Since $\varepsilon > 0$ is arbitrary, $p_K(x + y) \leq p_K(x) + p_K(y)$.

Conversely, assume p_K satisfies (2.7). Then, p_K is convex. Thus, the set defined by (2.8) is a convex body in $\mathbb{R}^n$ containing the origin in its interior, and $p_K(x) = \inf\{t > 0 : x/t \in K\}$ so p_K is its gauge function. $\square$

2.3.2. Normal mapping. Let Ω be an open subset in $\mathbb{R}^n$ and let $u : \Omega \to \mathbb{R}$.

Definition 2.29 (Supporting hyperplane). Given $x_0 \in \Omega$, a *supporting hyperplane* to the graph of u at $(x_0, u(x_0))$ is an affine function of the form $l(x) = u(x_0) + p \cdot (x - x_0)$ where $p \in \mathbb{R}^n$ such that $u(x) \geq l(x)$ for all $x \in \Omega$. See Figure 2.3.

It is possible to have infinitely many supporting hyperplanes to the graph of u at $(x_0, u(x_0))$. See Figure 2.4.

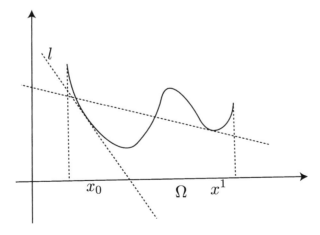

Figure 2.3. Supporting hyperplane l to the graph of u at $(x_0, u(x_0))$. There are no supporting hyperplanes at $(x^1, u(x^1))$.

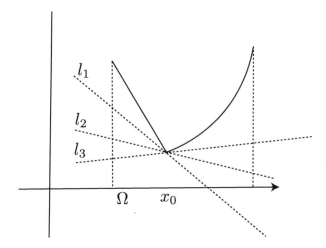

Figure 2.4. There are infinitely many supporting hyperplanes to the graph of u at $(x_0, u(x_0))$.

Definition 2.30 (The normal mapping). The *normal mapping* (also called *subdifferential*) $\partial u(x_0)$ of u at $x_0 \in \Omega$ is the set of slopes of supporting hyperplanes to the graph of u at $(x_0, u(x_0))$:

$$\partial u(x_0) = \{p \in \mathbb{R}^n : u(x) \geq u(x_0) + p \cdot (x - x_0) \text{ for all } x \in \Omega\}.$$

As will be seen below, the normal mapping provides a multi-valued generalization of the gradient map for convex functions. The normal mapping a priori depends on the domain under consideration. However, it is a *local* concept for convex functions. This is because, geometrically, if a hyperplane is locally below the graph of a convex function, then it is also globally below. We record below this locality property.

Remark 2.31 (Locality property of the normal mapping). Let $u : \Omega \to \mathbb{R}$ be a convex function on an open set Ω in $\mathbb{R}^n$. If $x_0 \in \Omega$, $p \in \mathbb{R}^n$, and $l(x) := u(x_0) + p \cdot (x - x_0)$ is such that $u \geq l$ in an open set $\Omega' \subset \Omega$ containing x_0, then $u \geq l$ in Ω.

The existence of supporting hyperplanes for convex functions is guaranteed by the following theorem.

Theorem 2.32 (Existence of supporting hyperplanes). *Let Ω be an open set in $\mathbb{R}^n$, and let $u : \Omega \to \mathbb{R}$ be a convex function. If $x_0 \in \Omega$, then $\partial u(x_0) \neq \emptyset$.*

Proof. Since Ω is open and $x_0 \in \Omega$, there is a closed ball $E := \overline{B_r(x_0)} \subset \Omega$. The epigraph $\text{epi}(u, E) := \{(x, x_{n+1}) \in E \times \mathbb{R} : x_{n+1} \geq u(x)\}$ of u on E is closed and convex due to the convexity of u. By Theorem 2.10 and the fact that x_0 is an interior point of E, there is a nonvertical supporting hyperplane of the form $l(x) = u(x_0) + p \cdot (x - x_0)$ to epi(u, E) at the boundary point $(x_0, u(x_0))$. Then $p \in \partial u(x_0)$ by Remark 2.31.   □

The assumption that Ω is *open* in Theorem 2.32 is important and cannot be removed in general; see Problem 2.3.

The following lemma relates the normal mapping $\partial u(x_0)$ to local differentiability, up to second order, of u at x_0.

Lemma 2.33. *Let Ω be an open set in $\mathbb{R}^n$. Let $u : \Omega \to \mathbb{R}$, and let $x_0 \in \Omega$.*

(i) *Suppose $Du(x_0)$ exists. If $\partial u(x_0) \neq \emptyset$, then $\partial u(x_0) = \{Du(x_0)\}$.*

(ii) *If $u \in C^2(\Omega)$ and $\partial u(x_0) \neq \emptyset$, then $D^2 u(x_0) \geq 0$; that is, the Hessian matrix $D^2 u(x_0)$ is nonnegative definite.*

Proof. We prove assertion (i). Let $p = (p_1, \ldots, p_n) \in \partial u(x_0)$. Note that $B_r(x_0) \subset \Omega$ for some $r > 0$ and that $u(x) \geq u(x_0) + p \cdot (x - x_0)$ in $B_r(x_0)$. Thus, $u(x_0 + te_i) \geq u(x_0) + tp_i$, for each $i = 1, \ldots, n$ and $|t| < r$. It follows that

$$\lim_{t \to 0^+} \frac{u(x_0 + te_i) - u(x_0)}{t} \geq p_i \quad \text{and} \quad \lim_{t \to 0^-} \frac{u(x_0 + te_i) - u(x_0)}{t} \leq p_i.$$

Since $D_i u(x_0)$ exists, we have $D_i u(x_0) = p_i$. Hence $p = Du(x_0)$ which gives $\partial u(x_0) = \{Du(x_0)\}$.

To prove assertion (ii), we use Taylor's expansion. For all $h \in \mathbb{R}^n \setminus \{0\}$ with $|h|$ small, there is $\xi \in [x_0, x_0 + h]$ such that

$$u(x_0 + h) - u(x_0) - Du(x_0) \cdot h = (D^2 u(\xi)h) \cdot h/2.$$

Since $\partial u(x_0) = \{Du(x_0)\}$ by assertion (i), we infer from the left-hand side being nonnegative that $D^2 u(\xi) \geq 0$. Letting $h \to 0$ yields $D^2 u(x_0) \geq 0$.   □

Remark 2.34. Let Ω be an open set in $\mathbb{R}^n$. Let $u : \Omega \to \mathbb{R}$.

- If u has a minimum value at $x_0 \in \Omega$, then clearly $0 \in \partial u(x_0)$. On the other hand, if $\partial u(x_0) \neq \emptyset$, then $\partial u(x_0)$ is a closed convex set.

- Note that $\partial u(x_0)$ can be empty. Consider the double-well potential $u(x) = (x^2 - 1)^2$ on $\mathbb{R}$. Then $\partial u(0) = \emptyset$. Indeed, if otherwise, then, by Lemma 2.33(ii), $u''(0) \geq 0$, contradicting $u''(0) = -4$.

- (Monotonicity of normal mapping) The normal mapping ∂u is *monotone* in the following sense: If $p \in \partial u(x)$ and $q \in \partial u(y)$ where $x, y \in \Omega$, then $(p - q) \cdot (x - y) \geq 0$. Indeed, this follows from adding the inequalities $u(y) \geq u(x) + p \cdot (y - x)$ and $u(x) \geq u(y) + q \cdot (x - y)$.

Example 2.35 (Normal mapping of a cone).

(a) Let $\Omega = B_R(z)$ and $u(x) = K|x - z|$ for $x \in \Omega$ where $K > 0$. Then

$$\partial u(x) = \begin{cases} \{K(x - z)/|x - z|\} & \text{if } 0 < |x - z| < R, \\ \overline{B_K(0)} & \text{if } x = z. \end{cases}$$

(b) More generally, let Ω be a bounded convex domain in $\mathbb{R}^n$, $z \in \Omega$, and $M > 0$. Let v be the convex function whose graph is the cone with vertex $(z, -M)$ and with the base Ω, with $v = 0$ on $\partial \Omega$. Then

$$\partial v(z) = \{Mx : x \cdot (y - z) \leq 1 \quad \text{for all } y \in \overline{\Omega}\}.$$

Proof. We verify part (a). If $0 < |x - z| < R$, then u is differentiable at x with $Du(x) = K(x - z)/|x - z|$. Hence $\partial u(x) = \{K(x - z)/|x - z|\}$.

On the other hand, $p \in \partial u(z)$ if and only if

$$K|x - z| = u(x) \geq p \cdot (x - z) + u(z) = p \cdot (x - z) \quad \text{for all } x \in \Omega,$$

or, equivalently, $|p| \leq K$. Thus, $\partial u(z) = \overline{B_K(0)}$.

The verification of part (b) is similar so we omit it. Note that $\partial v(z)$ is related to the polar body of $\overline{\Omega}$ with respect to z; see Problem 2.10. □

The following lemma shows that if a convex function on $\mathbb{R}^n$ is affine on a line, then its normal mappings are contained in a hyperplane.

Lemma 2.36. *Let $u : \mathbb{R}^n \to \mathbb{R}$ be a convex function. Suppose that*

$$u(z) = q \cdot z + a \quad \text{on the line} \quad L := \{z = t\omega + y : t \in \mathbb{R}\},$$

where $q, \omega, y \in \mathbb{R}^n$ and $a \in \mathbb{R}$. Then

$$\partial u(\mathbb{R}^n) \subset q + \omega^\perp := \{p \in \mathbb{R}^n : (p - q) \cdot \omega = 0\};$$

that is, $\partial u(\mathbb{R}^n)$ is contained in a hyperplane orthogonal to L.

Proof. Fix $x \in \mathbb{R}^n$. Let $p \in \partial u(x)$. Then $u(z) \geq u(x) + p \cdot (z - x)$ for all $z \in L$. Substituting $u(z) = q \cdot z + a$ and $z = t\omega + y$ then yields

$$q \cdot y + a - u(x) - p \cdot (y - x) \geq t(p - q) \cdot \omega \quad \text{for all } t \in \mathbb{R}.$$

Since the left-hand side is bounded and the above inequality holds for all $t \in \mathbb{R}$, we must have $(p - q) \cdot \omega = 0$. This proves the lemma. □

We will frequently use the following estimates for convex functions and the size of normal mappings in the interior of the domain.

Lemma 2.37 (Estimates for convex functions and size of normal mappings)**.** *Let $u \in C(\overline{\Omega})$ be a convex function on a bounded convex domain Ω in $\mathbb{R}^n$.*

(i) *If $u = 0$ on $\partial\Omega$, then $|u(x)| \geq \frac{\text{dist}(x, \partial\Omega)}{\text{diam}(\Omega)} \|u\|_{L^\infty(\Omega)}$ for all $x \in \Omega$.*

(ii) *If $u \leq 0$ on $\partial\Omega$, then $\|u\|_{L^\infty(\Omega)} \leq \frac{n+1}{|\Omega|} \int_\Omega |u| \, dx$.*

(iii) *If $p \in \partial u(x)$ where $x \in \Omega$ (not necessarily a convex set), then*

$$|p| \leq \max_{y \in \partial\Omega} \frac{u(y) - u(x)}{|y - x|} \leq \frac{\max_{\partial\Omega} u - u(x)}{\text{dist}(x, \partial\Omega)}.$$

(iv) *If there exist $M > 0$ and $\alpha \in (0, 1]$ such that*

$$|u(x) - u(y)| \leq M|x - y|^\alpha \quad \text{for all } x \in \Omega, y \in \partial\Omega,$$

then $u \in C^{0,\alpha}(\overline{\Omega})$ with

$$\|u\|_{C^{0,\alpha}(\overline{\Omega})} \leq \|u\|_{L^\infty(\partial\Omega)} + (1 + [\text{diam}(\Omega)]^\alpha)M.$$

(v) *If $u \in C^1(\overline{\Omega})$, then $|Du|$ attains its maximum on the boundary.*

Proof. We prove assertion (i). It suffices to consider $\|u\|_{L^\infty(\Omega)} > 0$. By considering $\|u\|_{L^\infty(\Omega)}^{-1} u$, we can assume that $\|u\|_{L^\infty(\Omega)} = 1$. Then $u(x_0) = -1$ for some $x_0 \in \Omega$. For any $x \in \Omega \setminus \{x_0\}$, the ray $\overrightarrow{x_0 x}$ intersects $\partial\Omega$ at z and $x = \alpha x_0 + (1 - \alpha)z$ for some $\alpha \in [0, 1)$. Since u is convex and $u = 0$ on $\partial\Omega$, we get $u(x) \leq \alpha u(x_0) + (1 - \alpha)u(z) = -\alpha$. It follows that

$$|u(x)| \geq \alpha = \frac{|x - z|}{|x_0 - z|} \geq \frac{\text{dist}(x, \partial\Omega)}{\text{diam}(\Omega)} = \frac{\text{dist}(x, \partial\Omega)}{\text{diam}(\Omega)} \|u\|_{L^\infty(\Omega)}.$$

Now, we prove (ii). By convexity, $u \leq \max_{\partial\Omega} u \leq 0$. Thus, we can assume $|u|$ attains its maximum at $x_0 \in \Omega$. Let $\mathcal{C}$ be the cone with base Ω and vertex at $(x_0, u(x_0))$. Then, assertion (ii) follows from these estimates:

$$\frac{1}{n+1} \|u\|_{L^\infty(\Omega)} |\Omega| = \frac{1}{n+1} |u(x_0)| |\Omega| = \text{Volume of } \mathcal{C} \leq \int_\Omega |u| \, dx.$$

To prove (iii), note that there exist $y \in \partial\Omega$ and $t > 0$ such that $y = x + tp$ if $p \neq 0$. From $u(y) \geq u(x) + p \cdot (y - x) = u(x) + |p||y - x|$, we obtain (iii).

For assertion (iv), it remains to show that $|u(x) - u(y)| \leq M|x - y|^\alpha$ for all $x \neq y \in \overline{\Omega}$. Indeed, if $y \in \partial\Omega$ and $x \in \overline{\Omega}$, then this follows from the assumption. Suppose now $x, y \in \Omega$. Then the ray $\overrightarrow{yx}$ intersects $\partial\Omega$ at z and $x = ty + (1 - t)z$ for some $t \in (0, 1)$. By convexity, $u(x) \leq tu(y) + (1 - t)u(z)$. The assertion now follows from

$$u(x) - u(y) \leq (1 - t)(u(z) - u(y)) \leq M(1 - t)|z - y|^\alpha = M(1 - t)^{1-\alpha}|x - y|^\alpha.$$

Finally, we prove assertion (v). Consider any $x_0 \in \Omega$ with $Du(x_0) \neq 0$. Let $\tau = Du(x_0)/|Du(x_0)|$. Let $f(t) := u(x_0 + t\tau)$ where $t \in [t_1, t_2]$ with $t_1 < 0 < t_2$, $z_1 := x_0 + t_1 \tau \in \partial\Omega$, $z_2 := x_0 + t_2 \tau \in \partial\Omega$, and $[z_1, z_2] \subset \overline{\Omega}$. Then

f is convex in t so $f'(t)$ is nondecreasing. In particular $f'(0) \leq f'(t_2)$, or $|Du(x_0)| \leq Du(z_2) \cdot \tau$, showing that $|Du(x_0)| \leq |Du(z_2)| \leq \max_{\partial\Omega} |Du|$. $\qquad \square$

We record a relationship between the normal mapping and differentiability of a convex function.

Theorem 2.38. *Let $u : \Omega \to \mathbb{R}$ be a convex function on an open set $\Omega \subset \mathbb{R}^n$.*

    (i) *If $\partial u(x_0)$ has only one element at $x_0 \in \Omega$, then u is differentiable at x_0.*

    (ii) *If u is differentiable at every point of Ω, then $u \in C^1(\Omega)$.*

Proof. We prove part (i). Assume that $\partial u(x_0) = \{p_0\}$. By subtracting $u(x_0) + p_0 \cdot (x - x_0)$ from u and translating coordinates, we can assume that $x_0 = 0$, $u(0) = 0$, $u \geq 0$, $\partial u(0) = \{0\}$. For the differentiability of u at 0, we need to show that

$$(2.9) \qquad u(x) = o(|x|) \quad \text{when } x \to 0.$$

Note that $A := \overline{B_{2r}(0)} \subset \Omega$ for some $r > 0$. Let $x^i \in \overline{B_r(0)}$, $x^i \to 0$, and let $p^i \in \partial u(x^i)$. We first show that $p^i \to 0$ when $i \to \infty$.

Indeed, we have

$$(2.10) \qquad u(x) \geq u(x^i) + p^i \cdot (x - x^i) \quad \text{for all } x \in A.$$

Note that u is continuous in A and $|p^i| \leq C(\|u\|_{L^\infty(A)}, r)$ by Lemma 2.37. Thus, for any convergent subsequence $\{p^{i_j}\}_{j=1}^\infty$ of $\{p^i\}_{i=1}^\infty$ that converges to some $p \in \mathbb{R}^n$, we obtain from (2.10) that $u(x) \geq p \cdot x$ for all $x \in A$. By the locality property of the normal mapping, we have $p \in \partial u(0)$ and hence $p = 0$. The uniqueness of p shows that $p^i \to 0$.

Note that (2.10) holds for any $x^i \in \overline{B_r(0)}$. Letting $x = 0$ gives $0 \geq u(x^i) - p^i \cdot x^i$. This combined with $u \geq 0$ implies that

$$0 \leq u(x^i) \leq p^i \cdot x^i \leq |p^i||x^i|.$$

Since $p^i \to 0$ whenever $x^i \to 0$, we obtain (2.9) as desired.

We now prove part (ii). Since u is convex and differentiable in Ω, by Theorem 2.32 and Lemma 2.33, $\partial u(x) = \{Du(x)\}$ for each $x \in \Omega$. We establish the continuity of Du by showing that for each $x_0 \in \Omega$ and all sequences $\{x^i\}_{i=1}^\infty \subset \Omega$ such that $x^i \to x_0$, we have $Du(x^i) \to Du(x_0)$. Indeed, applying (2.10) to $p^i = Du(x^i) \in \partial u(x^i)$ and then letting $i \to \infty$, we obtain as before that $Du(x^i)$ converges to $p \in \partial u(x_0) = \{Du(x_0)\}$, so $p = Du(x_0)$. This confirms the C^1 character of u. $\qquad \square$

We now define various classes of convex functions.

Definition 2.39. Let Ω be an open set in $\mathbb{R}^n$.

(a) (Strictly convex function) A convex function $u : \Omega \to \mathbb{R}$ is said to be *strictly convex* at $x_0 \in \Omega$ if any supporting hyperplane to u at x_0 touches its graph at only $(x_0, u(x_0))$. That is, if $p \in \partial u(x_0)$, then

$$u(x) > u(x_0) + p \cdot (x - x_0) \quad \text{for all } x \in \Omega \setminus \{x_0\}.$$

A convex function u on Ω is said to be strictly convex in Ω if it is strictly convex at every $x_0 \in \Omega$.

(b) (Uniformly convex function) A function $u \in C^2(\Omega)$ is said to be *uniformly convex* in Ω if there is a positive constant $\theta > 0$ such that $D^2 u \geq \theta I_n$ in Ω.

To illustrate the difference between these classes of convex functions, consider $u(x) = \sqrt{1 + |x|^2}$. It is strictly convex on $\mathbb{R}^n$ but it is not uniformly convex on $\mathbb{R}^n$. However, it is *locally uniformly convex* on $\mathbb{R}^n$ in the sense that it is uniformly convex on each compact subset of $\mathbb{R}^n$. Clearly, a locally uniformly convex function u on a convex domain Ω is strictly convex.

2.3.3. Lipschitz estimates and compactness. We refine Theorem 2.26:

Lemma 2.40 (Lipschitz estimates and differentiability). *Let $u : \Omega \to \mathbb{R}$ be a convex function in an open set $\Omega \subset \mathbb{R}^n$. Let $E \Subset \Omega' \Subset \Omega$. Then:*

(i) *$\partial u(\overline{E})$ is compact, and u is uniformly Lipschitz in E with estimate*

$$|u(x) - u(y)| \leq \left([\mathrm{dist}(\overline{E}, \partial \Omega')]^{-1} \operatorname*{osc}_{\Omega'} u\right)|x - y| \quad \text{for all } x, y \in E.$$

(ii) *u is differentiable almost everywhere in Ω.*

(iii) *$\|Du\|_{L^\infty(E)} \leq [\mathrm{dist}(\overline{E}, \partial \Omega')]^{-1} \operatorname{osc}_{\Omega'} u.$*

Proof. We prove part (i). From the compactness of $\overline{E}$ and the continuity of u, we can show that $\partial u(\overline{E})$ is a closed set in $\mathbb{R}^n$. By Lemma 2.37, $\partial u(\overline{E})$ is bounded. Therefore, $\partial u(\overline{E})$ is a compact set.

Now, we prove that u is uniformly Lipschitz in E. Since u is convex, by Theorem 2.32, $\partial u(x) \neq \emptyset$ for each $x \in E$. If $p \in \partial u(x)$, then by Lemma 2.37, $|p| \leq C$ where $C = [\mathrm{dist}(\overline{E}, \partial \Omega')]^{-1} \operatorname{osc}_{\Omega'} u$. Therefore,

$$u(y) \geq u(x) + p \cdot (y - x) \geq u(x) - C|y - x| \quad \text{for all } y \in E.$$

Reversing the roles of x and y, we obtain the Lipschitz estimate in part (i).

Now, we prove part (ii). By part (i), u is locally Lipschitz in Ω.

Step 1. $Du(x)$ exists for almost every $x \in \Omega$. Since differentiability is a local concept, we can assume that $\Omega = B_r(0)$ and u is Lipschitz on $B_r(0)$. It suffices to show that for each unit vector $\omega \in \mathbb{R}^n$, $|\omega| = 1$, the limit

$$D_\omega u(x) \equiv \lim_{t \to 0} \frac{u(x + t\omega) - u(x)}{t}$$

exists for almost every $x \in \Omega$. Indeed, since u is continuous,

$$\overline{D}_\omega u(x) \equiv \limsup_{t \to 0} \frac{u(x + t\omega) - u(x)}{t}$$

$$= \lim_{k \to \infty} \sup_{0 < |t| < 1/k, \ t \text{ rational}} \frac{u(x + t\omega) - u(x)}{t}$$

is Borel measurable, and likewise

$$\underline{D}_\omega u(x) \equiv \liminf_{t \to 0} \frac{u(x + t\omega) - u(x)}{t}.$$

Therefore,

$$J_\omega \equiv \{x \in \Omega : D_\omega u(x) \text{ does not exist}\} = \{x \in \Omega : \underline{D}_\omega u(x) < \overline{D}_\omega u(x)\}$$

is Borel measurable. To show that $|J_\omega| = 0$, by Fubini's Theorem, it suffices to show that the one-dimensional Hausdorff measure $\mathcal{H}^1$ of $J_\omega \cap L$ is 0 for any line L parallel to ω. (In this case, it is also the one-dimensional Lebesgue measure; see Section 2.7 for more on Hausdorff measures.) This follows from the property of convex functions of one variable. To see this, for $x \in \Omega$ and $|\omega| = 1$, we define $f(t) = u(x + t\omega)$ for t in an interval $I = I(x, \omega)$ so that $x + t\omega \in \Omega$. Then f is Lipschitz on I and hence absolutely continuous; that is, for any $\varepsilon > 0$, there is $\delta > 0$ such that for any finite number of nonoverlapping subintervals $\{[a_i, b_i]\}_{i=1}^N \subset I$ with $\sum_{i=1}^N (b_i - a_i) < \delta$, we have $\sum_{i=1}^N |f(b_i) - f(a_i)| < \varepsilon$. The absolute continuity of f implies that it is differentiable for almost every $t \in I$, and hence $\mathcal{H}^1(J_\omega \cap L) = 0$.

Step 2. Conclusion. By Step 1, $Du(x)$ exists for almost every $x \in \Omega$. At each such x, by Theorem 2.32 and Lemma 2.33, $\partial u(x) = \{Du(x)\}$ which implies the differentiability of u at x, by Theorem 2.38. This proves (ii).

Finally, we prove part (iii). It follows from part (ii) and Lemmas 2.33 and 2.37. By part (ii), for almost every $x \in E$, $Du(x)$ exists, and $Du(x) \in \partial u(x)$ by Lemma 2.33(i). By Lemma 2.37, $|Du(x)| \leq [\text{dist}(\overline{E}, \partial\Omega')]^{-1} \text{osc}_{\Omega'} u$. $\quad\square$

Alternatively, we can infer the almost differentiability of u in Lemma 2.40(ii) from Rademacher's Theorem which states that any locally Lipschitz function $f : \mathbb{R}^n \to \mathbb{R}$ is differentiable almost everywhere. A proof of this theorem can be found in Evans [**E2**, Theorem 6, p. 298].

Next, we recall the following compactness theorem.

Theorem 2.41 (Arzelà–Ascoli Theorem). *Let $\{f_k\}_{k=1}^{\infty}$ be a sequence of real-valued functions on an open set $\Omega \subset \mathbb{R}^n$, such that $\sup_k \|f_k\|_{L^\infty(\Omega)} < \infty$ and such that the functions f_k are uniformly equicontinuous; that is, for each $\varepsilon > 0$, there exists $\delta > 0$ such that*

$$|x - y| < \delta \ (x, y \in \Omega) \quad implies \ |f_k(x) - f_k(y)| < \varepsilon, \quad for \ all \ k = 1, \ldots.$$

Then there exist a subsequence $\{f_{k_j}\}_{j=1}^{\infty} \subset \{f_k\}_{k=1}^{\infty}$ and a continuous function f, such that $f_{k_j} \to f$ uniformly on compact subsets of Ω.

Combining the Lipschitz estimate in Lemma 2.40(i) with the Arzelà–Ascoli Theorem, we obtain the following result which can be viewed as an analogue of the Blaschke Selection Theorem (Theorem 2.20) for convex functions.

Theorem 2.42 (Compactness of uniformly bounded convex functions). *Let $\{u_k\}_{k=1}^{n}$ be a sequence of convex functions in a convex domain Ω in $\mathbb{R}^n$ such that $\sup_k \|u_k\|_{L^\infty(\Omega)} < \infty$. Then there is a subsequence $\{u_{k_i}\}_{i=1}^{\infty}$ that converges uniformly on compact subsets of Ω to a convex function u on Ω.*

2.3.4. Legendre transform. A useful concept of duality for convex functions is that of Legendre transform. Let Ω be an open set in $\mathbb{R}^n$.

Definition 2.43 (Legendre transform). The Legendre transform of a function $u : \Omega \to \mathbb{R}$ is the function $u^* : \mathbb{R}^n \to \mathbb{R} \cup \{+\infty\}$ defined by

$$u^*(p) = \sup_{x \in \Omega}[x \cdot p - u(x)].$$

Since u^* is a supremum of affine functions, it is a convex function on $\mathbb{R}^n$ by Lemma 2.24. Moreover, if $u_1 \leq u_2$, then $u_1^* \geq u_2^*$, so the Legendre transform is *order-reversing*. Clearly, if Ω is bounded and u is bounded from below on Ω, then u^* is finite on $\mathbb{R}^n$. The latter condition holds for convex functions $u : \Omega \to \mathbb{R}$; see Problem 2.2.

We record here a relationship between the normal mapping and the Legendre transform.

Lemma 2.44 (Normal mapping and the Legendre transform). *Let u^* be the Legendre transform of $u : \Omega \to \mathbb{R}$. If $p \in \partial u(x)$ where $x \in \Omega$, then $u(x) + u^*(p) = x \cdot p$ and $x \in \partial u^*(p)$.*

Proof. Since $p \in \partial u(x)$, $u(y) \geq u(x) + p \cdot (y - x)$ for all $y \in \Omega$, and hence $p \cdot x - u(x) \geq p \cdot y - u(y)$. Thus, $u^*(p) = \sup_{y \in \Omega}[p \cdot y - u(y)] = p \cdot x - u(x)$, which gives $u(x) + u^*(p) = x \cdot p$. Moreover, for any $z \in \mathbb{R}^n$, we also have

$$u^*(z) \geq x \cdot z - u(x) = p \cdot x - u(x) + x \cdot (z - p) = u^*(p) + x \cdot (z - p),$$

which shows that $x \in \partial u^*(p)$. $\square$

A consequence of Lemma 2.44 is the following result of Aleksandrov:

Lemma 2.45 (Aleksandrov's Lemma). *Let $\Omega \subset \mathbb{R}^n$ be open, and let $u \in C(\Omega)$. Then, the set $\mathfrak{S}$ of slopes of* singular supporting hyperplanes *(those that touch the graph of u at more than one point) has Lebesgue measure zero.*

Proof. We first prove the lemma for the case when Ω is bounded and u is bounded on Ω. In this case, $u^* : \mathbb{R}^n \to \mathbb{R}$ is a convex function. Consider $p \in \mathfrak{S}$. Then $p \in \partial u(x) \cap \partial u(y)$ where $x \neq y \in \Omega$. In view of Lemma 2.44, $x, y \in \partial u^*(p)$. Lemma 2.33 then asserts that u^* is not differentiable at any $p \in \mathfrak{S}$. Thus, by Lemma 2.40, $\mathfrak{S}$ has Lebesgue measure zero.

We next prove the lemma for the general case. Let $\{\Omega_k\}_{k=1}^\infty$ be a sequence of open sets such that $\bigcup_{k=1}^\infty \Omega_k = \Omega$ and $\Omega_k \Subset \Omega_{k+1}$ for each $k \geq 1$. Then, for each k, Ω_k is bounded, and $u_k := u|_{\Omega_k}$ is bounded on Ω_k. Owing to the previous case, the corresponding set $\mathfrak{S}_k$ for u_k has Lebesgue measure zero. We now show that $\mathfrak{S} \subset \bigcup_{k=1}^\infty \mathfrak{S}_k$, which will imply that $\mathfrak{S}$ also has Lebesgue measure zero. Indeed, let $p \in \mathfrak{S}$. Then $p \in \partial u(x) \cap \partial u(y)$ where $x \neq y \in \Omega$. Therefore, there is an integer k such that $x, y \in \Omega_k$. With this k, we have $p \in \partial u_k(x) \cap \partial u_k(y)$, so $p \in \mathfrak{S}_k$. Hence $\mathfrak{S} \subset \bigcup_{k=1}^\infty \mathfrak{S}_k$ as desired. $\square$

Lemma 2.44 suggests using $\partial u(\Omega)$ as another possibility for the domain of definition Ω^* of u^* when u is convex. In this case, components of the normal mapping become new independent variables. Furthermore, it can be verified that $(u^*)^* = u$, so the Legendre transform is an *involution*. Remarkably, involution and order-reversion characterize the Legendre transform. Artstein-Avidan and Milman [**AAM**] proved that any involution on the class of lower semicontinuous convex functions on the whole space which is order-reversing must be, up to linear terms, the Legendre transform.

When $u \in C^2(\overline{\Omega})$ is a uniformly convex function on a bounded convex domain Ω in $\mathbb{R}^n$, we can obtain relations between the Hessians of u and its Legendre transform. For example, we take up the above discussion and restrict the domain of the Legendre transform u^* of u to $\Omega^* = Du(\Omega)$, as this is useful in several applications (see, for example, Section 12.4). Then

$$u^*(y) = \sup_{z \in \Omega}(z \cdot y - u(z)) = x \cdot y - u(x), \quad \text{for } y \in \Omega^* = Du(\Omega),$$

where $x \in \Omega$ is uniquely determined by $y = Du(x)$.

The Legendre transform u^* is a uniformly convex, C^2 smooth function in Ω^*. It satisfies $(u^*)^* = u$. From $y = Du(x)$, we have $x = Du^*(y)$ and $D^2u(x) = (D^2u^*(y))^{-1}$. The last relation follows from $y = Du(Du^*(y))$ and

$$\delta_{ik} = \frac{\partial y_i}{\partial y_k} = \frac{\partial}{\partial y_k}D_iu(x) = D_{ij}u(x)\frac{\partial x_j}{\partial y_k} = D_{ij}u(x)D_{jk}u^*(y).$$

Now, if $F : (0, \infty) \to \mathbb{R}$, then the change of variables $x = Du^*(y)$ gives

$$(2.11) \qquad \int_\Omega F(\det D^2 u(x)) \, dx = \int_{\Omega^*} F([\det D^2 u^*(y)]^{-1}) \det D^2 u^*(y) \, dy.$$

2.4. Boundary Principal Curvatures and Uniform Convexity

In the boundary analysis of partial differential equations, it is natural (and sometimes necessary) to start with the case when the domain boundaries and functions on them are sufficiently nice. This section assembles most relevant concepts for Monge–Ampère equations.

Definition 2.46 ($C^{k,\alpha}$ boundary and $C^{k,\alpha}$ functions on the boundary). Let Ω be a bounded open set in $\mathbb{R}^n$, $k \in \{1, 2, \ldots\}$, and $\alpha \in [0, 1]$. We say:

(a) The boundary $\partial\Omega$ is $C^{k,\alpha}$ at $z \in \partial\Omega$ if there are $r_z > 0$ and a $C^{k,\alpha}$ function $\gamma_z : \mathbb{R}^{n-1} \to \mathbb{R}$ such that, upon relabeling and reorienting the coordinates axes if necessary, we have

$$\Omega \cap B_{r_z}(z) = \{x \in B_{r_z}(z) : x_n > \gamma_z(x_1, \ldots, x_{n-1})\}.$$

We will use the notation $[\partial\Omega]_{C^{k,\alpha}(z)}$ to encode the pointwise $C^{k,\alpha}$ information of $\partial\Omega$ at z which consists of k, α, γ_z, r_z defined above.

(b) $\partial\Omega$ is $C^{k,\alpha}$, and we write $\partial\Omega \in C^{k,\alpha}$ if it is $C^{k,\alpha}$ at each $z \in \partial\Omega$. By the $C^{k,\alpha}$ regularity of $\partial\Omega$, we mean the collection of k, α, γ_z, r_z for some finite covering $\bigcup B_{r_z}(z)$ of $\partial\Omega$ where r_z and γ_z are as in part (a). We say that $\partial\Omega$ is C^∞ if $\partial\Omega$ is C^k for all $k \geq 1$.

(c) Ω satisfies the *interior ball condition* with radius ρ at $z \in \partial\Omega$ if there is a ball $B := B_\rho(\hat{z}) \subset \Omega$ such that $z \in \partial B$.

(d) A function $\varphi : \partial\Omega \to \mathbb{R}$ is $C^{k,\alpha}$ if it is the restriction to $\partial\Omega$ of a $C^{k,\alpha}(\overline{\Omega})$ function Φ. We define $\|\varphi\|_{C^{k,\alpha}(\partial\Omega)}$ as the infimum of $\|\Phi\|_{C^{k,\alpha}(\overline{\Omega})}$ among all such Φ.

Due to convention (d), we will mostly use globally defined $\varphi \in C^{k,\alpha}(\overline{\Omega})$ functions when discussing $C^{k,\alpha}$ boundary values and estimates for $k \geq 1$.

Clearly, if Ω is convex and satisfies the interior ball condition at $z \in \partial\Omega$, then $\partial\Omega$ is $C^{1,1}$ at z. Vice versa, Ω satisfies the interior ball condition at $z \in \partial\Omega$ if $\partial\Omega$ is $C^{1,1}$ at z. To see this, assume that $z = 0$, $\Omega \cap B_r(0) = \{x \in B_r(0) : x_n > f(x')\}$ where $f(0) = 0$, $Df(0) = 0$, and

$$|Df(x') - Df(y')| \leq K|x' - y'| \text{ for all } x', y' \in \mathbb{R}^{n-1} \text{ with } \max\{|x'|, |y'|\} \leq r.$$

Then, Taylor's formula gives $|f(x')| \leq K|x'|^2$ for $|x'| \leq r$. Let $B := B_\rho(\rho e_n)$. If $x = (x', x_n) \in B$, then $x_n \geq |x'|^2/(2\rho)$, so $x_n \geq f(x')$ and $x \in \Omega$ if $\rho \leq \min\{1/K, r\}/2$. With this choice of ρ, we have $B \subset \Omega$ and $0 \in \partial B$.

2.4.1. Principal coordinate system. Let Ω be a bounded domain in $\mathbb{R}^n$ with C^2 boundary $\partial\Omega$. Let $x_0 \in \partial\Omega$ with outer unit normal vector ν_{x_0}. We describe the principal curvatures of $\partial\Omega$ at x_0 as follows.

By rotating coordinates, we can assume that $x_0 = 0$ and $\nu_{x_0} = -e_n$. Thus $x_n = 0$ is the tangent hyperplane to $\partial\Omega$ at 0. In some neighborhood $\mathcal{N}$ of 0, $\partial\Omega$ is given by $x_n = \psi(x')$ where ψ is a C^2 function in the variables $x' = (x_1, \ldots, x_{n-1})$ in some neighborhood $\mathcal{N}'$ of 0 in $\mathbb{R}^{n-1}$ with $\psi(0) = 0$ and $D\psi(0) = 0$. The eigenvalues $\kappa_1, \ldots, \kappa_{n-1}$ of $D^2\psi(0)$ are called the *principal curvatures* of $\partial\Omega$ at 0. The eigenvectors corresponding to κ_i are called the *principal directions* of $\partial\Omega$ at 0. By a further rotation of coordinates, we can assume that the $x_1, \ldots, x_{n-1}$ axes lie along the principal directions corresponding to $\kappa_1, \ldots, \kappa_{n-1}$. We call such coordinate system a *principal coordinate system* at x_0. Thus, in a principal coordinate system at $x_0 = 0$,

$$\partial\Omega \cap \mathcal{N} = \left\{ (x', x_n) : x_n = \psi(x') = \sum_{1 \leq i \leq n-1} \frac{\kappa_i}{2} x_i^2 + o(|x'|^2) \quad \text{as } x' \to 0 \right\}.$$

If Ω is locally convex at 0, then $x_n = 0$ is a supporting hyperplane to the graph of ψ at 0. Therefore, $\psi(x') \geq 0$ for all $x' \in \mathcal{N}'$ and hence $\kappa_i \geq 0$ for all $i = 1, \ldots, n-1$. On the other hand, if Ω is not convex at 0, then ψ takes negative values in any small neighborhood $\mathcal{N}'$ which shows that at least one of the κ_i is negative. Thus, we have proved the following relationship between convexity and principal curvatures of the boundary of a domain:

Proposition 2.47. *A domain Ω in $\mathbb{R}^n$ with C^2 boundary is convex if and only if the principal curvatures are nonnegative at every boundary point.*

Definition 2.48 (Uniformly convex domain). Let Ω be a bounded, convex domain in $\mathbb{R}^n$. We say that Ω (or interchangeably $\partial\Omega$) is uniformly convex

  (a) at $z \in \partial\Omega$ if there is an enclosing ball $B_{R_z}(x) \supset \Omega$ where R_z is the *pointwise uniform convexity radius* and $z \in \partial B_{R_z}(x)$;

  (b) if there is a *uniform convexity radius* R such that at each $z \in \partial\Omega$, there is an enclosing ball $B_R(x) \supset \Omega$ and $z \in \partial B_R(x)$.

If $\partial\Omega \in C^2$, then Ω is uniformly convex if and only if the principal curvatures of $\partial\Omega$ are bounded below by a positive constant.

If $\Omega = \{y \in \mathbb{R}^n : u(y) < a\}$ is a sublevel set of a C^2, uniformly convex function u (so $D^2 u \geq \theta I_n$ on $\mathbb{R}^n$ for some $\theta > 0$), then Ω is uniformly convex. Indeed, at each $z \in \partial\Omega$, we obtain from Taylor's formula that

$$\Omega \subset B_{R_z}(x) := \{y \in \mathbb{R}^n : u(z) + Du(z) \cdot (y - z) + \theta |y - z|^2 / 2 < a\},$$

so $R_z = |Du(z)|/\theta$, and we can take $R := \sup_{z \in \partial\Omega} R_z$ in Definition 2.48.

Convex bodies can be approximated by smooth, uniformly convex sets. For this purpose, we use gauge functions and their *mollifications*.

Definition 2.49 (Standard mollifiers). Define the standard mollifier $\varphi \in C_c^\infty(\mathbb{R}^n)$ with $0 \leq \varphi$ and $\mathrm{spt}(\varphi) \subset \overline{B_1(0)}$ by

$$\varphi(x) := C(n)e^{\frac{1}{|x|^2-1}} \quad \text{if } |x| < 1 \quad \text{and} \quad \varphi(x) = 0 \quad \text{if } |x| \geq 1,$$

where the constant $C(n)$ is chosen so that $\int_{\mathbb{R}^n} \varphi \, dx = 1$.

For each $\varepsilon > 0$, define $\varphi_\varepsilon(x) := \varepsilon^{-n}\varphi(x/\varepsilon)$. Then $\varphi_\varepsilon \in C_c^\infty(\mathbb{R}^n)$, $\mathrm{spt}(\varphi_\varepsilon) \subset \overline{B_\varepsilon(0)}$, and $\int_{\mathbb{R}^n} \varphi_\varepsilon \, dx = 1$.

Theorem 2.50 (Properties of mollifications). *Let Ω be an open set in $\mathbb{R}^n$. For each $f : \Omega \to \mathbb{R}$ which is locally integrable, we define its mollifications by $f^\varepsilon := \varphi_\varepsilon * f$ in $\Omega_\varepsilon := \{x \in \Omega : \mathrm{dist}(x, \partial\Omega) > \varepsilon\}$ (where $\varepsilon > 0$); that is,*

$$f^\varepsilon(x) = \int_\Omega \varphi_\varepsilon(x-y)f(y) \, dy = \int_{B_\varepsilon(0)} \varphi_\varepsilon(y)f(x-y) \, dy \quad \text{for } x \in \Omega_\varepsilon.$$

Then, the following assertions hold.

(i) *$f^\varepsilon \to f$ almost everywhere as $\varepsilon \to 0$, and $f^\varepsilon \in C^\infty(\Omega_\varepsilon)$.*

(ii) *If $f \in C(\Omega)$, then $f^\varepsilon \to f$ uniformly on compact subsets of Ω.*

(iii) *If $f \in W_{\mathrm{loc}}^{k,p}(\Omega)$ ($k \in \mathbb{N}$ and $p \in [1, \infty)$), then $f^\varepsilon \to f$ in $W_{\mathrm{loc}}^{k,p}(\Omega)$.*

(iv) *If $f \in C(\Omega)$ is convex and Ω is convex, then f^ε is convex on Ω_ε.*

(v) *If $f \in C^2(\Omega)$ with $D^2 f \geq \theta I_n$ in Ω where $\theta \in \mathbb{R}$, then $D^2 f^\varepsilon \geq \theta I_n$ in Ω_ε.*

Proof. Note that the first part of assertion (i) follows from the Lebesgue Differentiation Theorem, while the second part follows from the differentiation under the integral sign formula

$$D^\alpha f^\varepsilon(x) = \int_\Omega D^\alpha \varphi_\varepsilon(x-y)f(y) \, dy \quad \text{for all multi-indices } \alpha.$$

For the proofs of assertions (ii) and (iii), see Evans–Gariepy [**EG**, Section 4.2] or Gilbarg–Trudinger [**GT**, Sections 7.2 and 7.6].

For assertion (iv), let $x, z \in \Omega_\varepsilon$ and $t \in [0, 1]$. Using the convexity of f,

$$f(tx + (1-t)z - y) \leq tf(x-y) + (1-t)f(z-y) \quad \text{for all } y \in B_\varepsilon(0),$$

and then integrating appropriate functions over $B_\varepsilon := B_\varepsilon(0)$, we find

$$f^\varepsilon(tx + (1-t)z) = \int_{B_\varepsilon} \varphi_\varepsilon(y)f(tx + (1-t)z - y) \, dy \leq tf^\varepsilon(x) + (1-t)f^\varepsilon(z).$$

Thus, f^ε is convex on Ω_ε, and assertion (iv) is proved.

Finally, for all $\xi \in \mathbb{R}^n$ and $x \in \Omega_\varepsilon$, we have from $D^2 f \geq \theta I_n$ that

$$(D^2 f^\varepsilon(x)\xi) \cdot \xi = \int_{B_\varepsilon} \left(\rho_\varepsilon(y)D^2 f(x-y)\xi\right) \cdot \xi \, dy \geq \int_{B_\varepsilon} \rho_\varepsilon(y)(\theta|\xi|^2) \, dy = \theta|\xi|^2,$$

proving assertion (v). $\qquad\square$

Theorem 2.51 (Approximation of convex bodies by smooth, uniformly convex sets). *Let K be a convex body in $\mathbb{R}^n$ with nonempty interior. Then there are sequences $\{K_m^1\}_{m=1}^\infty \subset K$ and $\{K_m^2\}_{m=1}^\infty \supset K$ of uniformly convex domains with C^∞ boundaries that converge to K in the Hausdorff distance.*

Proof. We show the existence of the sequence $\{K_m^2\}_{m=1}^\infty$ because that of $\{K_m^1\}_{m=1}^\infty$ is similar. For this, it suffices to prove the following: For any open, convex neighborhood Ω of K, there is a uniformly convex domain K' with C^∞ boundary $\partial K'$ such that $K \subset K' \subset \Omega$.

We might assume that K contains the origin in its interior. Let p_K be its gauge function; see Theorem 2.28. Let $\varphi \in C_c^\infty(\mathbb{R}^n)$ be a standard mollifier, so $0 \leq \varphi$, $\mathrm{spt}(\varphi) \subset \overline{B_1(0)}$, and $\int_{\mathbb{R}^n} \varphi \, dx = 1$.

For $0 < \varepsilon < \mathrm{dist}(K, \partial\Omega)/2$, consider the C^∞ function

$$p_{K,\varepsilon}(x) = \varepsilon |x|^2 + \int_{B_1(0)} p_K(x - \varepsilon y) \varphi(y) \, dy \quad (x \in \mathbb{R}^n).$$

Since p_K is convex, $p_{K,\varepsilon}$ is uniformly convex, with $D^2 p_{K,\varepsilon} \geq 2\varepsilon I_n$ on $\mathbb{R}^n$.

If $x \in K$, then for $y \in B_1(0)$, we have

$$p_K(x - \varepsilon y) \leq p_K(x) + \varepsilon p_K(-y) \leq 1 + \varepsilon \sup_{B_1(0)} p_K$$

from which, denoting $C_1 := \max_K |x|^2 + \sup_{B_1(0)} p_K$, we find

$$p_{K,\varepsilon}(x) \leq 1 + \varepsilon C_1(K) \quad \text{and} \quad p_{K,\varepsilon}(0) \leq \varepsilon C_1.$$

If $x \notin \Omega$, then, for $y \in B_1(0)$, we have $\mathrm{dist}(x - \varepsilon y, K) \geq C(K, \Omega) > 0$ and $p_K(x - \varepsilon y) \geq C_2(K, \Omega) > 1$, from which we find

$$p_{K,\varepsilon}(x) \geq \int_{B_1(0)} p_K(x - \varepsilon y)\varphi(y) \, dy \geq C_2 \int_{B_1(0)} \varphi(y) \, dy = C_2.$$

Therefore, for ε small and $\varepsilon < (C_2 - 1)/(2C_1)$, the convex body

$$K' := \{x \in \mathbb{R}^n : p_{K,\varepsilon}(x) \leq 1 + 2C_1\varepsilon\}$$

satisfies $K \subset K' \subset \Omega$. Moreover, K' has C^∞ boundary, and it is uniformly convex since $p_{K,\varepsilon}$ is uniformly convex. $\qquad\square$

2.4.2. Defining functions. In the following calculations, we use the principal coordinate system at 0 as described in Section 2.4.1. Suppose that $u = \varphi$ on $\partial\Omega \cap \mathcal{N}$ where $\varphi \in C^2(\mathcal{N})$. Then

$$(2.12) \qquad u(x', \psi(x')) = \varphi(x', \psi(x')) \quad \text{in } \mathcal{N}'.$$

Differentiating (2.12) with respect to x', we get at $(x', \psi(x'))$

$$D_{x'}u + D_n u D_{x'}\psi = D_{x'}\varphi + D_n\varphi D_{x'}\psi.$$

Again, differentiating the above equation with respect to x', then evaluating the resulting equation at 0 and recalling $D_{x'}\psi(0) = 0$, we obtain

$$(2.13) \qquad D^2_{x'}u(0) + D_n u(0)D^2_{x'}\psi(0) = D^2_{x'}\varphi(0) + D_n\varphi(0)D^2_{x'}\psi(0).$$

If $\varphi \equiv 0$, then from (2.13), we find that

$$D^2_{x'}u(0) = -D_n u(0)D^2_{x'}\psi(0) = -D_n u(0)\mathrm{diag}(\kappa_1, \ldots, \kappa_{n-1})$$

and hence

$$\det D^2 u(0) = [-D_n u(0)]^{n-2} \prod_{i=1}^{n-1} \kappa_i \left\{ [-D_n u(0)]D_{nn}u(0) - \sum_{j=1}^{n-1} \frac{[D_{jn}u(0)]^2}{\kappa_j} \right\}.$$

Suppose now that u is convex and Ω is also convex. Assume further that $\varphi(0) = 0$ and $D_{x'}\varphi(0) = 0$. Let $l(x) := D_n u(0)x_n$ be the supporting hyperplane to the graph of u at 0. Let

$$\hat{\varphi}(x') = \varphi(x', \psi(x')) - l(x', \psi(x')) = \varphi(x', \psi(x')) - D_n u(0)\psi(x')$$

be the restriction of $\varphi - l$ on $\partial\Omega \cap \mathcal{N}$. Then, from (2.13), we find that

$$(2.14) \qquad\qquad\qquad D^2_{x'}u(0) = D^2_{x'}\hat{\varphi}(0).$$

Lemma 2.52 (Smoothness and Hessian of distance function). *Let Ω be a bounded open set in $\mathbb{R}^n$ with $\partial\Omega \in C^k$ for $k \geq 2$. Let $d(x) := \mathrm{dist}(x, \partial\Omega)$ be the distance function, and let $\Gamma_\mu := \{x \in \overline{\Omega} : d(x) < \mu\}$. Then:*

(i) *There exists a positive constant $\mu(\Omega) \leq 1/4$ such that $d \in C^k(\Gamma_\mu)$.*

(ii) *With μ in part (i), for any $x_0 \in \Gamma_\mu$, let $y_0 \in \partial\Omega$ be such that $|x_0 - y_0| = d(x_0)$. Then, in terms of a principal coordinate system at y_0 with principal curvatures $\kappa_1, \ldots, \kappa_{n-1}$, we have*

$$D^2 d(x_0) = \mathrm{diag}\left(\frac{-\kappa_1}{1 - \kappa_1 d(x_0)}, \ldots, \frac{-\kappa_{n-1}}{1 - \kappa_{n-1}d(x_0)}, 0\right).$$

As a consequence, letting $r(x) := -d(x) + d^2(x)$, then

$$D^2 r(x_0) = \mathrm{diag}\left(\frac{\kappa_1(1 - 2d(x_0))}{1 - \kappa_1 d(x_0)}, \ldots, \frac{\kappa_{n-1}(1 - 2d(x_0))}{1 - \kappa_{n-1}d(x_0)}, 2\right).$$

(iii) *If Ω is uniformly convex, then there is an extension of the function r in part (ii) from Γ_μ to a uniformly convex function $w \in C^k(\overline{\Omega})$.*

Proof. The proofs of parts (i) and (ii) can be found in [**GT**, Section 14.6].

We now prove part (iii) where Ω is uniformly convex. By part (ii), there is $\kappa(\partial\Omega, k) > 0$ such that $D^2 r \geq \kappa I_n$ in Γ_μ if $\mu > 0$ is small. Fix such μ. Then, $r \in C^k(\Gamma_\mu)$ is uniformly convex in Γ_μ.

Set $\delta := \mu/4$. Let $f, \zeta \in C^\infty((-\infty, 0])$ be such that

- $f', f'' \geq 0$, and $f(r) = r$ for $r \geq -\delta$, $f(r) = -2\delta$ for $r \leq -4\delta$, $f'(-3\delta) \geq c_0 > 0$ for some small c_0 depending only on δ,

- $\zeta \geq 0$, $\zeta(r) = 0$ for $r \geq -2\delta$, $\zeta(r) = 1$ for $r < -3\delta$, $|\zeta'(r)| + |\zeta''(r)| \leq C_1$ for $-3\delta \leq r \leq -2\delta$ where C_1 depends only on δ.

Consider the extension of r from Γ_μ to $\overline{\Omega}$ given by

$$w(x) = f(r(x)) + \varepsilon\zeta(r(x))|x|^2$$

for $\varepsilon > 0$ small to be chosen so that $D^2 w \geq cI_n$ for a suitable $c > 0$. Then

$$D_{ij}w = [f'(r) + \varepsilon\zeta'(r)|x|^2]D_{ij}r + [f''(r) + \varepsilon\zeta''(r)|x|^2]D_i r D_j r$$
$$+ 2\varepsilon\zeta'(r)(x_i D_j r + x_j D_i r) + 2\varepsilon\zeta(r)\delta_{ij}.$$

Since $\mu \leq 1/4$, we have $|D_i r| = |(2d-1)D_i d| \leq |2d-1| \leq 1$ in Γ_μ.

Case 1. $r > -2\delta$ or $r < -3\delta$. In this case $\zeta'(r) = \zeta''(r) = 0$ and

$$D_{ij}w = f'(r)D_{ij}r + f''(r)D_i r D_j r + 2\varepsilon\zeta(r)\delta_{ij}.$$

Clearly $D^2 w \geq 2\varepsilon I_n$ if $r < -3\delta$. If $r > -2\delta$, then $f'(r) \geq f'(-3\delta) \geq c_0$ and $D^2 w \geq f'(r)D^2 r \geq c_0 \kappa I_n$.

Case 2. $-3\delta \leq r \leq -2\delta$. Let $M = \sup_{x\in\Omega}(|x|+|x|^2)$. Then, for any $\xi \in \mathbb{R}^n$,

$$(D^2 w\xi) \cdot \xi \geq (c_0 - \varepsilon C_1 M)\kappa|\xi|^2 - \varepsilon C_1 M|\xi|^2 - 4\varepsilon C_1 M|\xi|^2 \geq (c_0/2)\kappa|\xi|^2$$

if ε is small. With this ε, w satisfies the required conditions. $\quad\square$

Lemma 2.52(iii) asserted the existence of uniformly convex *defining functions* for any C^2 uniformly convex domain. The smoothness of defining functions is that of the domain.

Definition 2.53 (Defining function). Let Ω be a uniformly convex domain in $\mathbb{R}^n$ with boundary $\partial\Omega \in C^k$ for $k \geq 2$. A uniformly convex function $\rho \in C^k(\overline{\Omega})$ is called a defining function of Ω if

$$\rho < 0 \quad \text{in } \Omega, \quad \rho = 0 \quad \text{on } \partial\Omega, \quad \text{and} \quad D\rho \neq 0 \quad \text{on } \partial\Omega.$$

The defining function will be useful in extending smooth functions on the boundary of a uniformly convex domain to uniformly convex functions.

Lemma 2.54 (Uniformly convex extension). *Let $k \geq 2$, $\alpha \in [0,1)$, and $\lambda > 0$. Let Ω be a uniformly convex domain in $\mathbb{R}^n$ with boundary $\partial\Omega \in C^{k,\alpha}$. Let $\varphi \in C^{k,\alpha}(\overline{\Omega})$. For small $\delta > 0$, let $\Omega_\delta := \{x \in \Omega : \text{dist}(x,\partial\Omega) < \delta\}$. Then, there is a uniformly convex function $v \in C^{k,\alpha}(\overline{\Omega})$ such that*

$$\det D^2 v \geq \lambda \quad \text{in } \Omega, \quad v = \varphi \quad \text{on } \partial\Omega, \quad \text{and} \quad v \leq -\lambda \quad \text{on } \partial\Omega_\delta \cap \Omega.$$

Proof. Let ρ be a $C^{k,\alpha}(\overline{\Omega})$ uniformly convex defining function of Ω as in Definition 2.53. Then, there are $\eta(\Omega) > 0$ and $c(\delta,\Omega) > 0$ such that

$$D^2\rho \geq \eta I_n \text{ and } \rho \geq -\eta^{-1} \quad \text{in } \Omega, \quad \text{and } \rho \leq -c \quad \text{on } \partial\Omega_\delta \cap \Omega.$$

Consider the function $v := \varphi + m(e^\rho - 1)$, for $m > 0$ to be determined.

Then $v = \varphi$ on $\partial\Omega$. From $v \leq \|\varphi\|_{L^\infty(\Omega)} + m(e^{-c} - 1)$ on $\partial\Omega_\delta \cap \Omega$ and

$$D^2 v = D^2\varphi + me^\rho(D^2\rho + D\rho \otimes D\rho) \geq D^2\varphi + m\eta e^{-\eta^{-1}}I_n \quad \text{in } \Omega,$$

we find that for $m = m(n, \lambda, \Omega, \delta, \|\varphi\|_{C^{1,1}(\overline{\Omega})}) > 0$ large, $v \in C^{k,\alpha}(\overline{\Omega})$ is uniformly convex with $\det D^2 v \geq \lambda$ in Ω, and $v \leq -\lambda$ on $\partial\Omega_\delta \cap \Omega$. $\qquad\square$

2.5. Calculus with Determinant

This section will be useful in the analysis of Monge–Ampère equations where we need to do calculus with determinants.

Lemma 2.55 (Partial derivatives of the determinant). *Let $F(r) = \log \det r$ for $r = (r_{ij})_{1 \leq i,j \leq n} \in \mathbb{S}^n$ with $r > 0$. Denote $r^{-1} = (r^{ij})_{1 \leq i,j \leq n}$. Then*

(i) $\frac{\partial F}{\partial r_{ij}}(r) = r^{ij} = r^{ji}$ *for all $1 \leq i, j \leq n$;*

(ii) $\frac{\partial^2 F}{\partial r_{ij}\partial r_{kl}}(r) = -r^{ik}r^{lj}$ *for all $1 \leq i, j, k, l \leq n$.*

Proof. Let $R = (R^{ij})_{1 \leq i,j \leq n}$ be the cofactor matrix of r. Then $Rr = e^{F(r)}I_n$, and $e^{F(r)} = \sum_{k=1}^n R^{jk}r_{kj}$ for each $j = 1, \ldots, n$. Hence, using the fact that R^{jk} does not contain r_{ij}, we easily get assertion (i) from

$$\frac{\partial F}{\partial r_{ij}}(r)e^{F(r)} = \frac{\partial}{\partial r_{ij}}[e^{F(r)}] = R^{ij} = e^{F(r)}r^{ij}.$$

Next, differentiating $\delta_{im} = r^{ij}r_{jm} = \frac{\partial F}{\partial r_{ij}}(r)r_{jm}$ with respect to r_{kl}, we get

$$0 = \frac{\partial^2 F}{\partial r_{ij}\partial r_{kl}}(r)r_{jm} + r^{ij}\frac{\partial r_{jm}}{\partial r_{kl}}.$$

Multiplying both sides by r^{mp} and summing with respect to m, we obtain

$$0 = \frac{\partial^2 F}{\partial r_{ij}\partial r_{kl}}(r)\delta_{jp} + r^{ij}r^{mp}\frac{\partial r_{jm}}{\partial r_{kl}} = \frac{\partial^2 F}{\partial r_{ip}\partial r_{kl}}(r) + r^{ik}r^{lp},$$

proving the claimed identity in assertion (ii). $\qquad\square$

Specializing to Hessian matrices of convex functions, we obtain:

Lemma 2.56. *Let u be a C^2 convex function on a domain Ω in $\mathbb{R}^n$. Let $U = (U^{ij})_{1 \leq i,j \leq n}$ be the cofactor matrix of $D^2 u$. We view $\det D^2 u$ as the function $f(D^2 u)$ where $f(r) = \det r$ for $r = (r_{ij})_{1 \leq i,j \leq n} \in \mathbb{S}^n$. Then:*

(i) $\frac{\partial}{\partial r_{ij}}\det D^2 u = U^{ji} = U^{ij}$. *Thus, the cofactor matrix U of $D^2 u$ arises from linearizing the Monge–Ampère operator $\det D^2 u$.*

(ii) *If $D^2 u > 0$, then, denoting $(D^2 u)^{-1} = (u^{ij})_{1 \leq i,j \leq n}$, we have*

$$\frac{\partial u^{ij}}{\partial r_{kl}} = -u^{ik}u^{lj} \quad \text{for all } 1 \leq i, j, k, l \leq n.$$

(iii) *If $u \in C^3$, then U is* divergence-free; *that is,* $\sum_{k=1}^{n} \frac{\partial}{\partial x_k} U^{ik} = 0$ *for each $i = 1, \ldots, n$.*

Proof. Note that assertions (i) and (ii) easily follow from Lemma 2.55.

Now, we prove assertion (iii). By considering $u_\varepsilon = u + \varepsilon |x|^2$ for $\varepsilon > 0$ and proving (iii) for u_ε and then letting $\varepsilon \to 0$, it suffices to consider the case $D^2 u$ is positive definite. In this case, $U^{ik} = (\det D^2 u) u^{ik}$. Thus, from

$$\frac{\partial}{\partial x_k} \det D^2 u = U^{ls} D_{lsk} u, \quad U^{ik} = (\det D^2 u) u^{ik}, \quad \text{and} \quad D_{sk} u u^{ik} = \delta_{si},$$

we have for each $i = 1, \ldots, n$,

$$\begin{aligned}
\frac{\partial}{\partial x_k} U^{ik} &= U^{ls} D_{lsk} u u^{ik} + (\det D^2 u) D_k(u^{ik}) \\
&= U^{ls} D_l(D_{sk} u u^{ik}) - U^{ls} D_{sk} u D_l(u^{ik}) + (\det D^2 u) D_k(u^{ik}) \\
&= -(\det D^2 u) \delta_{lk} D_l(u^{ik}) + (\det D^2 u) D_k(u^{ik}) = 0.
\end{aligned}$$

The lemma is proved. $\qquad\square$

The next three lemmas will be concerned with matrix inequalities.

Lemma 2.57. *Let $A, B \in \mathbb{S}^n$ be two nonnegative definite matrices. Then*

$$\mathrm{trace}(AB) \geq n (\det A)^{1/n} (\det B)^{1/n}.$$

Proof. Observe that if $M \in \mathbb{S}^n$ is nonnegative definite, then

$$(2.15) \qquad\qquad \mathrm{trace}(M) \geq n (\det M)^{1/n}.$$

Indeed, let $\alpha_1, \ldots, \alpha_n$ be eigenvalues of M. They are nonnegative. Then, the arithmetic-geometric inequality gives

$$\mathrm{trace}(M) = \sum_{i=1}^{n} \alpha_i \geq n \prod_{i=1}^{n} \alpha_i^{1/n} = n (\det M)^{1/n}.$$

Now, applying (2.15) to $M := B^{1/2} A B^{1/2}$, we obtain

$$\begin{aligned}
\mathrm{trace}(AB) = \mathrm{trace}(B^{1/2} A B^{1/2}) &\geq n [\det(B^{1/2} A B^{1/2})]^{1/n} \\
&= n (\det A)^{1/n} (\det B)^{1/n},
\end{aligned}$$

completing the proof of the lemma. $\qquad\square$

Lemma 2.58. *Let $A \in \mathbb{S}^n$ be nonnegative definite. Then, the following hold.*

(i) $\|A\| = \max_{\xi \in \mathbb{R}^n, |\xi|=1} (A\xi) \cdot \xi$.

(ii) $\mathrm{trace}(ABAB) \geq 0$ *for any real, symmetric $n \times n$ matrix B.*

(iii) $(A\xi) \cdot \xi \geq |\xi|^2 / \operatorname{trace}(A^{-1})$ *for any* $\xi \in \mathbb{R}^n$, *if* $A > 0$.
  As a consequence, for any C^2 *convex function* u *with* $\det D^2 u > 0$
  and any C^1 *function* v, *denoting* $(U^{ij})_{1 \leq i,j \leq n} = \operatorname{Cof} D^2 u$, *we have*

$$U^{ij} D_i v D_j v \geq |Dv|^2 \det D^2 u / \Delta u.$$

Proof. Let $\lambda_1, \ldots, \lambda_n$ be the nonnegative eigenvalues of A. There is an orthogonal $n \times n$ matrix P such that $A = P \Lambda P^t$ where $\Lambda = \operatorname{diag}(\lambda_1, \ldots, \lambda_n)$.

We prove assertion (i). For any $\xi \in \mathbb{R}^n$, we have

$$(A\xi) \cdot \xi = (P \Lambda P^t \xi) \cdot \xi = (\Lambda P^t \xi) \cdot (P^t \xi) = \sum_{i=1}^n \lambda_i (P^t \xi)_i.$$

It follows that

$$\max_{\xi \in \mathbb{R}^n, |\xi|=1} (A\xi) \cdot \xi = \max_{\xi \in \mathbb{R}^n, |\xi|=1} \sum_{i=1}^n \lambda_i (P^t \xi)_i = \left(\sum_{i=1}^n \lambda_i^2 \right)^{1/2} = \|A\|.$$

For assertion (ii), if $B \in \mathbb{S}^n$, then $D = P^t B P = (d_{ij})_{1 \leq i,j \leq n} \in \mathbb{S}^n$, and

$$\operatorname{trace}(ABAB) = \operatorname{trace}(P \Lambda P^t B P \Lambda P^t B)$$
$$= \operatorname{trace}(\Lambda D \Lambda D) = \lambda_i d_{il} \lambda_l d_{li} = \lambda_i \lambda_l d_{il}^2 \geq 0.$$

Finally, we prove assertion (iii). As before, $(A\xi) \cdot \xi = (\Lambda P^t \xi) \cdot (P^t \xi)$. Because $|P^t \xi| = |\xi|$ and $\operatorname{trace}(A^{-1}) = \operatorname{trace}(\Lambda^{-1})$, it suffices to prove (iii) for the case $A = \Lambda$, for which it is equivalent to the obvious inequality

$$\left(\sum_{i=1}^n \lambda_i^{-1} \right) \left(\sum_{i=1}^n \lambda_i \xi_i^2 \right) \geq \sum_{i=1}^n \xi_i^2.$$

Applying $(A\xi) \cdot \xi \geq |\xi|^2 / \operatorname{trace}(A^{-1})$ to $A = (U^{ij})_{1 \leq i,j \leq n} = (\det D^2 u)(D^2 u)^{-1}$ where $\operatorname{trace}(A^{-1}) = \Delta u / \det D^2 u$ and $\xi = Dv$ yields the consequence. $\qquad \square$

Lemma 2.59 (Concavity of $\det^{1/n}$). *Let* $\theta \in [0, 1/n]$. *Let* A *and* B *be two symmetric nonnegative definite* $n \times n$ *matrices, and let* $\lambda \in [0, 1]$. *Then*

$$(2.16) \qquad [\det(\lambda A + (1 - \lambda) B)]^\theta \geq \lambda (\det A)^\theta + (1 - \lambda)(\det B)^\theta.$$

When $\theta = 1/n$, (2.16) *is equivalent to* Minkowski's determinant inequality:

$$(2.17) \qquad [\det(A + B)]^{\frac{1}{n}} \geq (\det A)^{\frac{1}{n}} + (\det B)^{\frac{1}{n}} \quad \textit{for all } A, B \geq 0.$$

Moreover, if $\theta = 1/n$, $\lambda \in (0, 1)$, *and both* A *and* B *are positive definite, then the equality holds in* (2.16) *if and only* $B = \kappa A$ *for some* $\kappa > 0$.

Proof. We first prove the lemma for $\theta = 1/n$. Since $\det(\lambda M) = \lambda^n \det M$ for all $M \in \mathbb{M}^{n \times n}$, (2.16) is equivalent to (2.17). It suffices to prove (2.17) for A being invertible, for, in the general case, we apply the result in the invertible case to $A + \varepsilon I_n$ and B for each $\varepsilon > 0$, and then we let $\varepsilon \to 0$.

Let us assume now that A is invertible. In view of the identities

$$A + B = A^{1/2}(I_n + D)A^{1/2}, \quad D = A^{-1/2}BA^{-1/2},$$

where D is nonnegative definite and $\det(MN) = (\det M)(\det N)$ for all $M, N \in \mathbb{M}^{n \times n}$, (2.17) reduces to $[\det(I_n + D)]^{\frac{1}{n}} \geq 1 + (\det D)^{\frac{1}{n}}$ which becomes

$$\prod_{i=1}^{n}(1 + \lambda_i)^{\frac{1}{n}} \geq 1 + \prod_{i=1}^{n}\lambda_i^{\frac{1}{n}},$$

where $\lambda_1, \ldots, \lambda_n$ are nonnegative eigenvalues of D. But this follows from the arithmetic-geometric inequality, since

$$\prod_{i=1}^{n}\left(\frac{1}{1+\lambda_i}\right)^{\frac{1}{n}} + \prod_{i=1}^{n}\left(\frac{\lambda_i}{1+\lambda_i}\right)^{\frac{1}{n}} \leq \frac{1}{n}\sum_{i=1}^{n}\frac{1}{1+\lambda_i} + \frac{1}{n}\sum_{i=1}^{n}\frac{\lambda_i}{1+\lambda_i} = 1,$$

with equality if and only if $\lambda_1 = \cdots = \lambda_n = \kappa \geq 0$. If both A and B are positive definite, this occurs if and only if $\kappa > 0$ and $B = \kappa A$.

Now, we prove the lemma for $0 \leq \theta \leq 1/n$. Using the result for the case $\theta = 1/n$, together with the concavity of the map $t \mapsto t^{n\theta}$ on $(0, \infty)$, we get

$$[\det(\lambda A + (1 - \lambda)B)]^{\theta} \geq \left[\lambda(\det A)^{\frac{1}{n}} + (1 - \lambda)(\det B)^{\frac{1}{n}}\right]^{n\theta}$$
$$\geq \lambda(\det A)^{\theta} + (1 - \lambda)(\det B)^{\theta}.$$

The lemma is proved. $\qquad\square$

In several quantitative analyses of the Monge–Ampère equation, we need to calculate the Hessian determinant of functions having certain symmetries.

Lemma 2.60 (Hessian determinant of radial functions). *Let $u : \mathbb{R}^n \to \mathbb{R}$.*

(i) *Assume that u is radial; that is, $u(x) = f(r)$ where $r = |x|$ and $f : [0, \infty) \to \mathbb{R}$. Then, in a suitable coordinate system, we have*

$$D^2 u = \text{diag}\left(f'', \frac{f'}{r}, \ldots, \frac{f'}{r}\right).$$

(ii) *Let k, s be positive integers such that $k + s = n$. Assume that*

$$u(x_1, \ldots, x_k, y_1, \ldots, y_s) = |x|^{\alpha} f(|y|) \quad (x \in \mathbb{R}^k, \ y \in \mathbb{R}^s),$$

for some $\alpha \in \mathbb{R}$. Then

$$\det D^2 u = \alpha^k |x|^{n\alpha - 2k} f^{k-1} (f')^{s-1} |y|^{-s+1} \left((\alpha - 1)f f'' - \alpha(f')^2\right).$$

Proof. We begin with the proof of part (i). We can compute

$$\frac{\partial^2 u}{\partial x_j \partial x_i} = \frac{\partial}{\partial x_j}\left(f'(r)\frac{x_i}{r}\right) = \left(f''(r) - \frac{f'(r)}{r}\right)\frac{x_i x_j}{r^2} + \delta_{ij}\frac{f'(r)}{r}.$$

Note that the matrix $X := (x_i x_j)_{1\le i,j\le n}$ is similar to $\mathrm{diag}(|x|^2, 0, \ldots, 0)$ for any $x \in \mathbb{R}^n$. Indeed, it suffices to consider $x \ne 0$. Then, $|x|^2$ is the only nonzero eigenvalue of X (with eigenvector x). Since $\mathrm{trace}(X) = |x|^2$, the claim follows. Therefore, there is an orthogonal $n \times n$ matrix P such that $P^t X P = \mathrm{diag}(|x|^2, 0, \ldots, 0)$. Geometrically, transforming the coordinate system using P, x becomes $(|x|, 0, \ldots, 0)$. Now, part (i) follows from

$$P^t D^2 u P = P^t \left(\left(f''(r) - \frac{f'(r)}{r}\right)\frac{X}{r^2} + \frac{f'(r)}{r}I_n\right)P$$

$$= \left(f'' - \frac{f'}{r}\right)\mathrm{diag}(1, 0, \ldots, 0) + \frac{f'}{r}I_n = \mathrm{diag}\left(f'', \frac{f'}{r}, \ldots, \frac{f'}{r}\right).$$

Next, we prove part (ii). Since u depends only on $|x|$ and $|y|$, by part (i), it suffices to compute $\det D^2 u(x, y)$ for $x = (|x|, 0, \ldots, 0) \in \mathbb{R}^k$ and $y = (0, \ldots, 0, |y|) \in \mathbb{R}^s$. Then, denoting $M = \alpha|x|^{\alpha-2}f$, we have

$$D^2 u = \left(\begin{array}{cccc|cccc}
(\alpha-1)M & 0 & \cdots & 0 & 0 & \cdots & 0 & \alpha|x|^{\alpha-1}f' \\
0 & M & 0 & \cdots & 0 & 0 & \cdots & 0 \\
0 & 0 & \cdots & M & 0 & 0 & \cdots & 0 \\
\hline
0 & 0 & \cdots & 0 & \frac{|x|^\alpha f'}{|y|} & 0 & 0 & 0 \\
0 & 0 & \cdots & 0 & 0 & \frac{|x|^\alpha f'}{|y|} & \cdots & 0 \\
0 & 0 & \cdots & 0 & \cdots & \cdots & \cdots & \cdots \\
\alpha|x|^{\alpha-1}f' & 0 & \cdots & 0 & 0 & \cdots & 0 & |x|^\alpha f''
\end{array}\right).$$

Now, a short calculation gives the asserted formula for $\det D^2 u$. $\square$

2.6. John's Lemma

In this section, we will prove a crucial result, due to John [**Jn**], in the investigation of Monge–Ampère equations as it allows us to normalize their solutions to make them close to quadratic functions in some sense. John's Lemma says that all convex bodies (by our convention, they have nonempty interiors) are equivalent to balls modulo affine transformations. We will present two versions of John's Lemma for a convex body K: The first version is concerned with a closed ellipsoid of maximal volume contained in K while the second version is concerned with a closed ellipsoid of minimal volume enclosing K and having the same center of mass as K.

Definition 2.61 (Ellipsoid). An open ellipsoid E in $\mathbb{R}^n$ is the image of the unit ball $B_1(0)$ in $\mathbb{R}^n$ under an invertible affine map; that is,

$$E = AB_1(0) + b$$

where $A \in \mathbb{M}^{n \times n}$ with $\det A \neq 0$ and $b \in \mathbb{R}^n$. We call b the center of E. A closed ellipsoid is defined similarly when $B_1(0)$ is replaced by $\overline{B_1(0)}$.

In practice, it is convenient to know that for each ellipsoid as defined above, we can assume the matrix A to be positive definite.

Lemma 2.62 (Representation of ellipsoids by positive definite matrices). *Let E be an ellipsoid in $\mathbb{R}^n$ so that $E = AB_1(0) + b$ where $A \in \mathbb{M}^{n \times n}$ with $\det A \neq 0$ and $b \in \mathbb{R}^n$. Then in this representation we can take A to be positive definite.*

Proof. By the Polar Decomposition Theorem in linear algebra, each invertible matrix $A \in \mathbb{M}^{n \times n}$ can be factored into $A = SP$ where $P \in \mathbb{M}^{n \times n}$ is orthogonal and $S = (AA^t)^{1/2} \in \mathbb{S}^n$ is positive definite. We have

$$E = AB_1(0) + b = SPB_1(0) + b = SB_1(0) + b,$$

since $PB_1(0) = B_1(0)$. Thus, we can replace the matrix A by the positive definite matrix S in the representation $E = AB_1(0) + b$. $\qquad \square$

Lemma 2.63 (John's Lemma). *Let K be a convex body in $\mathbb{R}^n$. Then there is a unique closed ellipsoid E of maximal volume contained in K. Moreover, letting b be the center of E, then the following hold.*

(i) $E \subset K \subset b + n(E - b)$.

(ii) *If, in addition, K is centrally symmetric, that is, $-x \in K$ whenever $x \in K$, then $E \subset K \subset b + \sqrt{n}(E - b)$.*

Note that the prefactors n and $\sqrt{n}$ in Lemma 2.63 are sharp, as can be seen from K being a regular simplex (Example 2.3), and a cube, respectively.

Proof. We proceed in several steps.

Step 1. K contains an ellipsoid of maximal volume. We identify the set of ordered pairs (A, b) where $A \in \mathbb{M}^{n \times n}$ and $b \in \mathbb{R}^n$ as $\mathbb{R}^{n^2 + n}$. Let

$$\mathcal{E} := \{(A, b) \in \mathbb{R}^{n^2 + n} : A\overline{B_1(0)} + b \subset K\}.$$

Then $\mathcal{E}$ is a nonempty compact set of $\mathbb{R}^{n^2 + n}$. The volume map

$$V(A, b) := |A\overline{B_1(0)} + b| \equiv \omega_n |\det A|$$

is a continuous function on $\mathcal{E}$. Thus there is an $(A_0, b_0) \in \mathcal{E}$ that maximizes V on $\mathcal{E}$. This proves Step 1.

Since K has nonempty interior, we have $\omega_n |\det A_0| > 0$ so A_0 is invertible. We claim that $E := A_0 \overline{B_1(0)} + b_0$ is the desired ellipsoid. Indeed, replacing K by $A_0^{-1}(K - b_0)$ if necessary, we can assume that $\overline{B_1(0)} = A_0^{-1}(E - b_0)$ is an ellipsoid of maximal volume in K.

Step 2. Proof of assertion (i). It suffices to show that if $p \in K$, then $|p| \leq n$. Assume by contradiction that there is a point $p \in K$ with $a := |p| > n$. We can choose an orthogonal coordinate system $(x_1, x_2, \ldots, x_n)$ on $\mathbb{R}^n$ such that $p = (a, 0, \ldots, 0)$. The idea is to shrink the unit ball $B_1(0)$ in the $x_2, \ldots, x_n$ directions while stretching it in the x_1 direction to get an enclosed ellipsoid with larger volume. For $t, \lambda > 0$, consider the affine map

$$\mathbf{\Psi}_t^\lambda(x_1, x_2, \ldots, x_n) = (-1 + e^t(x_1 + 1), e^{-\lambda t} x_2, \ldots, e^{-\lambda t} x_n).$$

We claim that for small $t > 0$,

$$\mathbf{\Psi}_t^\lambda(\overline{B_1(0)}) \subset C_a^n := \operatorname{conv}\big(\overline{B_1(0)} \cup \{p\}\big) \subset K,$$

provided that $\lambda > 1/(a-1)$. Recall that $\operatorname{conv}(E)$ is the convex hull of E.

Granted the claim, then the volume of the ellipsoid $\mathbf{\Psi}_t^\lambda(\overline{B_1(0)})$ is

$$|\mathbf{\Psi}_t^\lambda(B_1(0))| = e^{(1-(n-1)\lambda)t}|B_1(0)| > |B_1(0)|$$

provided that $\lambda < 1/(n-1)$ and $t > 0$ is small. Now, for $a > n$, we can choose λ satisfying $1/(a-1) < \lambda < 1/(n-1)$, and this choice of λ contradicts the maximality of the volume of $\overline{B_1(0)}$.

It remains to prove the claim. By symmetry, it suffices to consider $n = 2$. Observe that one tangent line l from $(a, 0)$ to $\overline{B_1(0)} \subset \mathbb{R}^2$ intersects $\partial B_1(0)$ at the point $z = (z_1, z_2) = (1/a, \sqrt{a^2-1}/a)$. The equation for l is $(x_1 - a)/\sqrt{a^2-1} + x_2 = 0$. To prove the inclusion in the claim, we only need to show that $\mathbf{\Psi}_t^\lambda(z) \equiv (-1 + e^t(z_1 + 1), e^{-\lambda t} z_2)$ is below l (see Figure 2.5) for small $t > 0$; that is,

$$\varphi(t) := \frac{e^t(z_1 + 1) - 1 - a}{\sqrt{a^2-1}} + e^{-\lambda t} z_2 < 0.$$

We have

$$\varphi'(0) = \frac{z_1 + 1}{\sqrt{a^2-1}} - \lambda z_2 = \frac{1+a}{a\sqrt{a^2-1}} - \lambda \frac{\sqrt{a^2-1}}{a} < 0,$$

provided $\lambda > 1/(a-1)$. Thus $\varphi(t) < \varphi(0) = 0$ for small $t > 0$ as claimed.

Step 3. Proof of assertion (ii). We argue by contradiction as in the proof of assertion (i). Assume there is $p \in K$ with $a := |p| > \sqrt{n}$. We repeat the above arguments, replacing C_a^n by $\tilde{C}_a^n := \operatorname{conv}\big(\overline{B_1(0)} \cup \{\pm p\}\big) \subset K$, $\mathbf{\Psi}_t^\lambda$ by

$$\tilde{\mathbf{\Psi}}_t^\lambda(x_1, x_2, \ldots, x_n) = (e^t x_1, e^{-\lambda t} x_2, \ldots, e^{-\lambda t} x_n),$$

and $\varphi(t)$ by $\tilde{\varphi}(t) = (e^t z_1 - a)/\sqrt{a^2-1} + e^{-\lambda t} z_2$. If $\lambda > 1/(a^2-1)$, then

$$\tilde{\varphi}'(0) = \frac{z_1}{\sqrt{a^2-1}} - \lambda z_2 = \frac{1}{a\sqrt{a^2-1}} - \lambda \frac{\sqrt{a^2-1}}{a} < 0,$$

and we obtain a contradiction as in assertion (i) if $a > \sqrt{n}$.

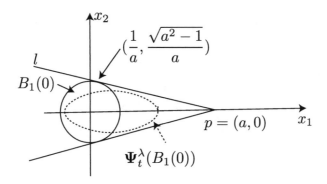

Figure 2.5. When $\lambda > \frac{1}{a-1}$, $\Psi_t^\lambda(\overline{B_1(0)}) \subset C_a^n$ when $t > 0$ is small.

Step 4. Uniqueness of E. Suppose $E_1 = A_1 B_1(0) + b_1$ and $E_2 = A_2 B_1(0) + b_2$ are two ellipsoids of maximal volume V contained in K, where $A_1, A_2 \in \mathbb{M}^{n \times n}$ are symmetric and positive definite (see Lemma 2.62). Note that $V = \omega_n \det A_1 = \omega_n \det A_2$. Since K is convex, the ellipsoid

$$E = \frac{1}{2}(A_1 + A_2)B_1(0) + \frac{1}{2}(b_1 + b_2)$$

is contained in K. By the definition of V and by Lemma 2.59, we have

$$V \geq |E| = \omega_n \det\left(\frac{A_1 + A_2}{2}\right) \geq \omega_n \left[\frac{1}{2}(\det A_1)^{1/n} + \frac{1}{2}(\det A_2)^{1/n}\right]^n = V.$$

Thus, the above inequality becomes an equality. By the equality case of Lemma 2.59, $A_1 = \kappa A_2$ for some positive constant κ. Since $\det A_1 = \det A_2$, we must have $\kappa = 1$ and $A_1 = A_2$. It follows that $E_2 = E_1 + (b_2 - b_1)$.

It remains to show that $b_2 = b_1$. By an affine transformation, we can assume that both E_1 and E_2 are balls. Since K contains the convex hull of E_1 and E_2, K contains a translate of E_1 centered at $b_3 = (b_1 + b_2)/2$. If $b_1 \neq b_2$, then b_1, b_2, and b_3 are all different points, and we can easily find an ellipsoid E contained in K with volume larger than that of E_1, a contradiction. Therefore, $b_1 = b_2$ and this proves the asserted uniqueness. □

Lemma 2.63 motivates the following definition.

Definition 2.64 (Normalized convex set). If Ω is a bounded convex domain with nonempty interior in $\mathbb{R}^n$, then there is an affine transformation $Tx = Ax + b$ for $x \in \mathbb{R}^n$, where $A \in \mathbb{S}^n$ is positive definite and $b \in \mathbb{R}^n$, such that

$$B_1(0) \subset T(\Omega) \subset B_n(0).$$

We say that T *normalizes* Ω. A convex open set Ω in $\mathbb{R}^n$ is called *normalized* if

$$B_1(z) \subset \Omega \subset B_n(z) \quad \text{for some } z \in \mathbb{R}^n.$$

Definition 2.65 (Center of mass). Let S be a bounded and measurable set in $\mathbb{R}^n$ with $|S| > 0$. We define the center of mass of S to be the point

$$\mathbf{c}(S) = \frac{1}{|S|} \int_S x \, dx.$$

Lemma 2.66 (John's Lemma, center of mass version). *Let K be a convex body in $\mathbb{R}^n$. Then there is a unique closed ellipsoid E of minimal volume enclosing K with center $\mathbf{c}(E)$ at the center of mass $\mathbf{c}(K)$ of K. Moreover, for some $\alpha(n) > 0$,*

$$\mathbf{c}(E) + \alpha(n)(E - \mathbf{c}(E)) := \Big\{ \mathbf{c}(E) + \alpha(n)(x - \mathbf{c}(E)) : x \in E \Big\} \subset K \subset E.$$

Proof. The proof is based on two facts:

- For a cone in $\mathbb{R}^n$, the distance from its vertex to the base is $(n+1)$ times the distance from its center of mass to the base.
- The cross section of a ball through its center can be enclosed by an ellipsoid of half the volume of the ball.

We now implement these ideas without optimizing the value of $\alpha(n)$. By a continuity argument as in Step 1 in the proof of Lemma 2.63, we can show the existence of a closed ellipsoid E of minimal volume enclosing K with center at the center of mass $\mathbf{c}(K)$ of K. Using affine transformations, we can assume that $\mathbf{c}(K)$ is at the origin and that $E = \overline{B_1(0)}$.

Step 1. Existence of $\alpha(n)$. Rotating coordinates, we can assume that the point on ∂K closest to the origin $0 \in \mathbb{R}^n$ is $P = me_1$ on the x_1-axis where $m > 0$. Since K is convex, the hyperplane (H) with equation $x_1 = m$ is a supporting hyperplane to K at P. Let $P' = -Me_1$ $(M > 0)$ be another intersection of K with the x_1-axis. We claim that

(2.18) $M/m \leq n.$

To prove this, consider the cross section $S = \{x \in K : x_1 = 0\}$ of K with the hyperplane $x_1 = 0$. Let (C) be a cone with vertex P' that passes through S and is contained in the slab $\{-M \leq x_1 \leq m\}$. See Figure 2.6.

Since $\mathbf{c}(K) = 0$, we have $\int_K x_1 \, dx = 0$. Because $x_1 < 0$ in $K \setminus (C)$, we deduce that $\int_{(C) \cap K} x_1 \, dx > 0$. Thus, the center of mass of $(C) \cap K$ has positive x_1-coordinate. From $(C) \cap K \subset (C)$, we infer that the x_1-coordinate of the center of mass $\mathbf{c}(C)$ of (C) is also positive. Therefore

$$\frac{m}{m + M} = \frac{\text{dist}(0, (H))}{\text{dist}(P', (H))} \geq \frac{\text{dist}(\mathbf{c}(C), (H))}{\text{dist}(P', (H))} = \frac{1}{n + 1},$$

from which (2.18) follows.

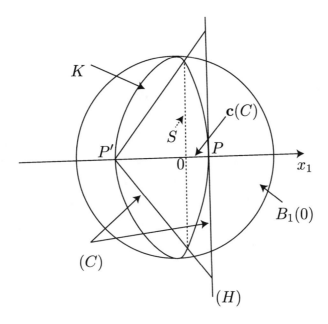

Figure 2.6. Configuration in the proof of Lemma 2.66.

To show the existence of $\alpha(n)$ satisfying the conclusion of the lemma, it suffices to show that $M \geq \beta(n) > 0$ for this gives $m \geq \alpha(n) := n^{-1}\beta(n)$ and hence $\alpha(n)E := \{\alpha(x)x : x \in E\} = \overline{B_{\alpha(n)}(0)} \subset K$.

We note that

$$K \subset S_M := \{x \in \mathbb{R}^n : -M \leq x_1 \leq M\} \cap B_1(0).$$

Consider the ellipsoid

$$E_n = \left\{x \in \mathbb{R}^n : \frac{x_1^2}{a^2} + \frac{x_2^2 + \cdots + x_n^2}{b^2} \leq 1\right\},$$

where $a = 2^{-n}, b = 2$ with volume $|E_n| = ab^{n-1}|B_1(0)| = |B_1(0)|/2$. Clearly, the intersection $E_n \cap B_1(0)$ contains $S_{\beta(n)}$ for some $\beta(n) > 0$.

If $M < \beta(n)$, then we can enclose K by the ellipsoid E_n with smaller volume than that of E, and this contradicts the minimality of E. Therefore $M \geq \beta(n)$ as desired.

Step 2. Uniqueness of E. Again, we assume that $\mathbf{c}(K) = 0$. Note that $E = A^{-1}\overline{B_1(0)}$ for some positive definite matrix $A \in \mathbb{S}^n$. Suppose there is another ellipsoid $E' = A'^{-1}\overline{B_1(0)}$ of minimal volume containing K where

$$|E| = |E'| = \frac{\omega_n}{\det A} = \frac{\omega_n}{\det A'}.$$

It follows that $\det A = \det A' := D > 0$. Since $AK, A'K \subset \overline{B_1(0)}$, we have $\frac{1}{2}(A + A')K \subset \overline{B_1(0)}$. Therefore, the ellipsoid $\tilde{E} = 2(A + A')^{-1}\overline{B_1(0)}$

contains K with volume $|\tilde{E}| = 2^n \omega_n / \det(A + A')$. By Lemma 2.59, we have

$$(2.19) \qquad \det(A + A') \geq 2^{n-1}(\det A + \det A') = 2^n D,$$

which shows that $|\tilde{E}| \leq \omega_n / D = |E|$. By the minimality of E, we must have $|\tilde{E}| = |E|$ and (2.19) is an equality. Therefore, by Lemma 2.59, there must be $\kappa > 0$ such that $A' = \kappa A$. Since $\det A = \det A'$, we have $\kappa = 1$ and hence $A = A'$ which shows that $E \equiv E'$. $\qquad \square$

For convenience, we can assume that $\alpha(n)$ is nonincreasing in n. From Lemma 2.66, we see that a convex body K is balanced around its center of mass $x^* = \mathbf{c}(K)$. This is also true for any of its cross section through x^*.

Lemma 2.67 (Balance around center of mass). *Let* $K \subset \mathbb{R}^n$ *be a convex body with center of mass* x^*. *Let* $\alpha(n)$ *be as in Lemma 2.66.*

   (i) *(Balance of convex body) Let* $y \in K \setminus \{x^*\}$, *and let* z *be the intersection of the ray* $\overrightarrow{yx^*}$ *with* ∂K. *Then,* $|z - x^*|/|z - y| \geq \alpha(n)/2$.

   (ii) *(Balance of cross section) Let* H *be a hyperplane going through* x^*. *Then, there is an ellipsoid* $\hat{E} \subset (H - x^*)$ *with center at the origin such that*

$$[\alpha^2(n)/4]\hat{E} \subset (K \cap H) - x^* \subset \hat{E}.$$

Proof. We prove part (i). By Lemma 2.66, there is an ellipsoid E with center at the origin such that $\alpha(n)E \subset K - x^* \subset E$. By an affine transformation, we can assume that $E = \overline{B}_r(0)$ for some $r > 0$. Then $|z - x^*| \geq r\alpha(n)$ while $|z - y| \leq |z - x^*| + |x^* - y| \leq 2r$. The asserted estimate follows.

Next, we prove part (ii). Let y^* be the center of mass of $K' := K \cap H$. Applying Lemma 2.66 to K', we find an ellipsoid $E' \subset (H - x^*)$ with center at the origin such that $\alpha(n)E' \subset K' - y^* \subset E'$.

If $y^* = x^*$, then we are done. Assume now that $y^* \neq x^*$. Let z^* be the intersection of the ray $\overrightarrow{y^*x^*}$ with $\partial K'$. Let $\tau := |z^* - x^*|/|z^* - y^*|$. By convexity, we have

$$\alpha(n)\tau E' + x^* \subset K'.$$

Indeed, for any $w_1 = x^* + \alpha(n)\tau w$ where $w \in E'$, we have

$$w_2 := y^* + \tau^{-1}(w_1 - x^*) \in y^* + \alpha(n)E' \subset K'.$$

Moreover, w_1 lies in the interval $[z^*, w_2] \subset K'$. Therefore, $w_1 \in K'$.

Since $x^* \in K' \subset y^* + E'$, there is $z \in E'$ such that $x^* = y^* + z$. Hence $y^* = x^* - z \in x^* + E'$ since E' is symmetric around 0. Therefore $y^* + E' \subset x^* + 2E'$. Hence $\alpha(n)\tau E' + x^* \subset K' \subset x^* + 2E'$. By part (i), we have $\tau \geq \alpha(n)/2$. Thus, the conclusion follows with $\hat{E} = 2E'$. $\qquad \square$

2.7. Review of Measure Theory and Functional Analysis

2.7.1. Measures. Recall that the Borel σ-algebra of an open set $\Omega \subset \mathbb{R}^n$ is the smallest σ-algebra of Ω containing the open subsets of Ω. A regular Borel measure on $\mathbb{R}^n$ which is finite on compact subsets of $\mathbb{R}^n$ is called a *Radon measure*.

Theorem 2.68 (Weak convergence of Radon measures). *Let Ω be an open set in $\mathbb{R}^n$. Let μ, μ_k $(k = 1, 2, \ldots)$ be Radon measures on Ω. Then the following statements are equivalent:*

(i) $\lim_{k\to\infty} \int_\Omega f \, d\mu_k = \int_\Omega f \, d\mu$ *for all $f \in C_c(\Omega)$.*

(ii) $\limsup_{k\to\infty} \mu_k(K) \leq \mu(K)$ *for each compact set $K \subset \Omega$, and $\mu(\mathcal{O}) \leq \liminf_{k\to\infty} \mu_k(\mathcal{O})$ for each open set $\mathcal{O} \subset \Omega$.*

When one of the statements in Theorem 2.68 holds, we have the weak* convergence of μ_k to μ and we write $\mu_k \overset{*}{\rightharpoonup} \mu$. For later references, we record the following important weak compactness result for Radon measures.

Theorem 2.69 (Weak compactness). *Let Ω be an open set in $\mathbb{R}^n$. Let $\{\mu_k\}_{k=1}^\infty$ be a sequence of Radon measures on Ω satisfying $\sup_k \mu_k(\Omega) < \infty$. Then there exist a subsequence $\{\mu_{k_j}\}_{j=1}^\infty$ and a Radon measure μ on Ω such that $\mu_{k_j} \overset{*}{\rightharpoonup} \mu$.*

The n-dimensional Lebesgue measure on $\mathbb{R}^n$ is denoted by $\mathcal{L}^n$. It is an n-fold product of the one-dimensional Lebesgue measure $\mathcal{L}^1$. We will write "dx", "dy", etc., instead of $d\mathcal{L}^n$ in integrals taken with respect to $\mathcal{L}^n$.

We use $|\Omega|$ to denote the n-dimensional Lebesgue measure $\mathcal{L}^n(\Omega)$ of a Lebesgue measurable set $\Omega \subset \mathbb{R}^n$. From the definition of $\mathcal{L}^n$, we then have the following characterization: For any $\delta > 0$,

$$(2.20) \qquad |\Omega| = \inf \left\{ \sum_{i=1}^\infty |Q_i| : Q_i \text{ cubes}, \ \Omega \subset \bigcup_{i=1}^\infty Q_i, \ \mathrm{diam}(Q_i) \leq \delta \right\}.$$

Sometimes, we work with surface-like measures of lower-dimensional subsets of $\mathbb{R}^n$. The relevant concept here is that of the Hausdorff measure.

Definition 2.70 (Hausdorff measure). Fix $0 \leq s < \infty$.

(a) Let $E \subset \mathbb{R}^n$ and $0 < \delta \leq \infty$. Let $\omega_s = \pi^{s/2}/\Gamma(\frac{s}{2} + 1)$. Define

$$\mathcal{H}_\delta^s(E) := \inf \left\{ \omega_s \sum_{j=1}^\infty \left(\frac{\mathrm{diam}(E_j)}{2} \right)^2 : E \subset \bigcup_{j=1}^\infty E_j, \mathrm{diam}(E_j) \leq \delta \right\}.$$

(b) Let $E \subset \mathbb{R}^n$. We define

$$\mathcal{H}^s(E) := \lim_{\delta \to 0} \mathcal{H}^s_\delta(E) = \sup_{\delta > 0} \mathcal{H}^s_\delta(E).$$

We call $\mathcal{H}^s$ s-dimensional Hausdorff measure on $\mathbb{R}^n$.

Clearly, $\mathcal{H}^s(\lambda E) = \lambda^s \mathcal{H}^s(E)$ for all $\lambda > 0$. Note that $\mathcal{H}^0(E)$ is equal to the number of elements of E and is also denoted by $\#E$.

We have the following monotonicity property of $\mathcal{H}^{n-1}$ when restricted to the boundaries of convex sets.

Lemma 2.71. *Let A and B be two bounded convex domains in $\mathbb{R}^n$ with $A \subset B$. Then*

$$\mathcal{H}^{n-1}(\partial A) \leq \mathcal{H}^{n-1}(\partial B).$$

Proof. It suffices to consider $A \Subset B$. Fix $\delta > 0$, and consider an arbitrary covering $\{E_j\}_{j=1}^\infty$ of ∂B, that is, $\partial B \subset \bigcup_{j=1}^\infty E_j$, with $\mathrm{diam}(E_j) \leq \delta$ for all j. Let $p(\bar{A}, \cdot)$ be the projection map on $\bar{A}$ as in Lemma 2.9. By this lemma, $p(\bar{A}, \partial B) = \partial A$. It follows that $\partial A \subset \bigcup_{j=1}^\infty \tilde{E}_j$ where $\tilde{E}_j = p(\bar{A}, E_j)$. By Lemma 2.9, we have

$$\mathrm{diam}(\tilde{E}_j) = \mathrm{diam}(p(\bar{A}, E_j)) \leq \mathrm{diam}(E_j) \leq \delta.$$

Hence

$$\mathcal{H}^{n-1}_\delta(\partial A) \leq \omega_{n-1} \sum_{j=1}^\infty \left(\frac{\mathrm{diam}(\tilde{E}_j)}{2} \right)^2 \leq \omega_{n-1} \sum_{j=1}^\infty \left(\frac{\mathrm{diam}(E_j)}{2} \right)^2.$$

Since this holds for an arbitrary covering $\{E_j\}_{j=1}^\infty$ of ∂B, we infer that $\mathcal{H}^{n-1}_\delta(\partial A) \leq \mathcal{H}^{n-1}_\delta(\partial B)$. Letting $\delta \to 0$ yields $\mathcal{H}^{n-1}(\partial A) \leq \mathcal{H}^{n-1}(\partial B)$. $\square$

2.7.2. Area formula. We recall the following area formula for Lipschitz maps; see Evans–Gariepy [**EG**, Theorem 1, p. 96].

Theorem 2.72 (Area formula). *Let Ω be an open set in $\mathbb{R}^n$. Let $\mathbf{f} = (f^1, \ldots, f^m) : \Omega \to \mathbb{R}^m$ be Lipschitz where $m \geq n$. Then for each measurable subset $A \subset \Omega$, we have*

$$\int_A J\mathbf{f}(x) \, dx = \int_{\mathbf{f}(A)} \#(A \cap \mathbf{f}^{-1}(y)) \, d\mathcal{H}^n(y),$$

where $J\mathbf{f} \geq 0$ is the Jacobian of $\mathbf{f}$. It is defined as follows: $(J\mathbf{f})^2$ is the sum of the squares of the determinant of each $(n \times n)$-submatrix of the $(m \times n)$-gradient matrix $D\mathbf{f} = (D_i f^j)_{1 \leq i \leq n, 1 \leq j \leq m}$ which exists by Rademacher's Theorem. In particular, when $m = n$, we have

$$\int_A |\det(D_i f^j(x))_{1 \leq i,j \leq n}| \, dx = \int_{\mathbf{f}(A)} \#(x \in A : \mathbf{f}(x) = y) \, dy.$$

In the last formula of Theorem 2.72, we used a nontrivial fact that on $\mathbb{R}^n$, $\mathcal{H}^n = \mathcal{L}^n$. For a proof of this fact, see [**EG**, Theorem 2, p. 70].

The next two results are important consequences of the area formula.

Lemma 2.73. *If $n \in \{1, \dots, m-1\}$ is an integer and E is an n-dimensional Lipschitz surface in $\mathbb{R}^m$, then $\mathcal{H}^n(E)$ coincides with the classical n-dimensional surface area of E.*

Proof. Suppose that $E = \mathbf{f}(A)$ where $\mathbf{f} = (f^1, \dots, f^n) : A \to E$ is a one-to-one and Lipschitz map. Let $J\mathbf{f}$ be the Jacobian of $\mathbf{f}$. By the classical formula for surface area, the n-dimensional surface area of E is $\int_A J\mathbf{f}(x)\, dx$. Since $\mathbf{f}$ is one-to-one on A, the area formula in Theorem 2.72 gives

$$\int_A J\mathbf{f}(x)\, dx = \int_{\mathbf{f}(A)} d\mathcal{H}^n(y) = \int_E d\mathcal{H}^n(y) = \mathcal{H}^n(E),$$

which proves the lemma. $\qquad\square$

Theorem 2.74 (Change of variables formula for $C^{1,1}$ convex functions). *Let Ω be an open bounded set in $\mathbb{R}^n$. Let $u : \Omega \to \mathbb{R}$ be a function of class $C^{1,1}$. Then for all Borel sets $E \subset \Omega$, we have*

$$\int_E |\det D^2 u|\, dx \quad \begin{cases} \geq |Du(E)|, \\ = |Du(E)| & \text{if } u \text{ is convex.} \end{cases}$$

Proof. It suffices to prove the theorem for all Borel sets $E \Subset \Omega$. Let $w \in C^{1,1}(\Omega)$. We apply Theorem 2.72 to the Lipschitz map Dw with $|\det D_i(D_j w)_{1 \leq i,j \leq n}| = |\det D^2 w|$, and we obtain for all Borel sets $E \Subset \Omega$,

$$\int_E |\det D^2 w|\, dx = \int_{Dw(E)} \#(x \in E : Dw(x) = y)\, dy \geq |Dw(E)|,$$

which proves the stated inequality for any $C^{1,1}(\Omega)$ function u.

If $w = u$ is convex, then we have $|\det D^2 u| = \det D^2 u$, and by Lemma 2.45, $\#(x \in E : Du(x) = y) = 1$ for almost every $y \in Du(E)$. Therefore

$$\int_E \det D^2 u\, dx = \int_E |\det D^2 u|\, dx = \int_{Du(E)} dy = |Du(E)|. \qquad\square$$

Lemma 2.75 (Layer-cake formula). *Let $v : E \to \mathbb{R}$ be a measurable function on a Borel set E in $\mathbb{R}^n$. Let $\mu : E \to \mathbb{R}$ be an integrable, almost everywhere positive function. We also use μ for the measure with density $\mu(x)$. Let $F : (0, \infty) \to (0, \infty)$ be an increasing, differentiable bijection. Then*

$$\int_E F(|v|)\mu(x)\, dx = \int_0^\infty F'(t)\mu(E \cap \{|v| > t\})\, dt.$$

In particular, when $\mu \equiv 1$ and $F(t) = t^p$ where $p \geq 1$, we have

$$\int_E |v|^p \, dx = p \int_0^\infty t^{p-1} |E \cap \{|v| > t\}| \, dt.$$

Proof. Let $f(x) := F(|v(x)|)$. For all $x \in E$, we have the identity

$$f(x) = \int_0^{f(x)} d\tau = \int_0^\infty \chi_{\{f > \tau\}}(x) \, d\tau,$$

where χ denotes the characteristic function. Therefore

$$\int_E F(|v|)\mu(x) \, dx = \int_E f(x)\mu(x) \, dx = \int_E \int_0^\infty \chi_{\{f > \tau\}}(x)\mu(x) \, d\tau \, dx$$

$$= \int_0^\infty \mu(\{E \cap \{f(x) > \tau\}\}) \, d\tau,$$

where we used Fubini's Theorem in the last equality. Using the change of variables $\tau = F(t)$ and observing that $\{f(x) > \tau\} = \{|v(x)| > t\}$, we obtain the layer-cake formula as asserted. $\qquad\square$

2.7.3. Inequalities. Many inequalities in this text are of an elementary nature though their derivations come from sophisticated arguments. We recall here two popular ones: Hölder's and Sobolev's inequalities.

Let $\Omega \subset \mathbb{R}^n$, $1 \leq p, q \leq \infty$ with $p^{-1} + q^{-1} = 1$, and $u \in L^p(\Omega), v \in L^q(\Omega)$. Then, we have the Hölder inequality:

$$\|uv\|_{L^1(\Omega)} \leq \|u\|_{L^p(\Omega)} \|v\|_{L^q(\Omega)}.$$

Theorem 2.76 (Sobolev Embedding Theorem). *Let Ω be a bounded open subset of $\mathbb{R}^n$ with C^1 boundary. Let $v \in W^{k,p}(\Omega)$ where $p \geq 1$ and $k \in \mathbb{N}$.*

(i) *If $kp < n$, then $v \in L^{np/(n-pk)}(\Omega)$, and we have the estimate*

$$\|v\|_{L^{np/(n-pk)}(\Omega)} \leq C(n, p, k, \Omega)\|v\|_{W^{k,p}(\Omega)}.$$

When $k = 1$ and $v \in W_0^{1,p}(\Omega)$ where $p < n$, we can replace $\|v\|_{W^{1,p}(\Omega)}$ by $\|Dv\|_{L^p(\Omega)}$ in the above estimate.

(ii) *If $kp > n$, then $v \in C^{k-1-[n/p],\gamma}(\overline{\Omega})$, where*

$$\gamma = \begin{cases} [n/p] + 1 - n/p & \text{if } n/p \text{ is not an integer,} \\ \text{any positive number} < 1 & \text{if } n/p \text{ is an integer} \end{cases}$$

with $[t]$ denoting the largest integer not exceeding t. Moreover, we have the estimate

$$\|v\|_{C^{k-1-[n/p],\gamma}(\overline{\Omega})} \leq C(n, p, k, \gamma, \Omega)\|v\|_{W^{k,p}(\Omega)}.$$

2.7.4. Implicit Function Theorem. Let $\mathfrak{B}_1$ and $\mathfrak{B}_2$ be real Banach spaces. Let $E(\mathfrak{B}_1, \mathfrak{B}_2)$ denote the Banach space of bounded linear mappings L from $\mathfrak{B}_1$ into $\mathfrak{B}_2$ with norm given by $\|L\| = \sup_{u \in \mathfrak{B}_1, u \neq 0} \|Lu\|_{\mathfrak{B}_2} / \|u\|_{\mathfrak{B}_1}$. Let F be a mapping from an open set $\mathfrak{U}$ in $\mathfrak{B}_1$ into $\mathfrak{B}_2$. The mapping F is called *Fréchet differentiable* at $u \in \mathfrak{B}_1$ if there is a bounded linear mapping $L \in E(\mathfrak{B}_1, \mathfrak{B}_2)$ such that

$$\frac{\|F(u + h) - F(u) - Lh\|_{\mathfrak{B}_2}}{\|h\|_{\mathfrak{B}_1}} \to 0 \quad \text{when } h \to 0 \quad \text{in } \mathfrak{B}_1.$$

The linear mapping L is then called the *Fréchet derivative* of F at u and we denote $L = F_u$. We call F continuously differentiable at u if F is Fréchet differentiable in a neighborhood of u and the mapping $v \mapsto F_v \in E(\mathfrak{B}_1, \mathfrak{B}_2)$ is continuous at u.

Let X be a real Banach space. Let G be a mapping from an open subset of $\mathfrak{B}_1 \times X$ onto $\mathfrak{B}_2$. If $\hat{L} \in E(\mathfrak{B}_1, \mathfrak{B}_2)$ and satisfies

$$\frac{\|G(u + h, \sigma) - G(u, \sigma) - \hat{L}h\|_{\mathfrak{B}_2}}{\|h\|_{\mathfrak{B}_1}} \to 0 \quad \text{when } h \to 0 \quad \text{in } \mathfrak{B}_1,$$

then we call $\hat{L}$ the *partial Fréchet derivative* of G at $(u, \sigma) \in \mathfrak{B}_1 \times X$ with respect to u, and we denote $\hat{L} = G_u(u, \sigma)$.

We have the following theorem; see Deimling [**De**, Theorem 15.1 and Corollary 15.1] for a proof.

Theorem 2.77 (Implicit Function Theorem). *Let $\mathfrak{B}_1, \mathfrak{B}_2$, and X be real Banach spaces. Let G be a mapping from an open subset of $\mathfrak{B}_1 \times X$ onto $\mathfrak{B}_2$. Let $(\hat{u}, \hat{\sigma}) \in \mathfrak{B}_1 \times X$ be such that*

(i) *$G(\hat{u}, \hat{\sigma}) = 0$, G is continuously differentiable at $(\hat{u}, \hat{\sigma})$, and*

(ii) *the partial Fréchet derivative $G_u(\hat{u}, \hat{\sigma})$ is invertible.*

Then, there exists a neighborhood $\mathcal{N}$ of $\hat{\sigma}$ in X such that the following holds: For all $\sigma \in \mathcal{N}$, there is $u_\sigma \in \mathfrak{B}_1$ such that

$$\mathcal{N} \ni \sigma \mapsto u_\sigma \in \mathfrak{B}_1 \text{ is continuously differentiable and } G(u_\sigma, \sigma) = 0.$$

2.8. Review of Classical PDE Theory

We review in this section basic concepts regarding elliptic equations of second order and classical PDE estimates that will be used later.

2.8.1. Elliptic equations of second order and maximum principle.
We consider equations in the simple form

$$F(D^2 u) = f \quad \text{in } \Omega$$

in some domain $\Omega \subset \mathbb{R}^n$ where $F : \mathbb{S}^n \to \mathbb{R}$. This equation is called linear if F is linear in $D^2 u$, and it is called fully nonlinear if F is nonlinear in $D^2 u$. One example of a fully nonlinear operator F is $F(r) = \det r$ for $r \in \mathbb{S}^n$.

For $r = (r_{ij})_{1 \le i,j \le n} \in \mathbb{S}^n$, we denote

$$F_{ij}(r) = \frac{\partial F}{\partial r_{ij}}(r), \quad F_{ij,kl}(r) = \frac{\partial^2 F}{\partial r_{ij} \partial r_{kl}}(r) \quad (1 \le i, j, k, l \le n).$$

The linearized operator of F at u is defined by

$$L_u v := \left. \frac{d}{dt} \right|_{t=0} F(D^2 u + t D^2 v) = F_{ij}(D^2 u) D_{ij} v.$$

The operator F is called *elliptic* in a subset $\mathcal{A} \subset \mathbb{S}^n$ if the symmetric matrix $(F_{ij}(A))_{1 \le i,j \le n}$ is positive definite for each $A \in \mathcal{A}$. If $u \in C^2(\Omega)$ and F is elliptic on the range of $D^2 u$, then we say that F is elliptic at u. The same can be said for the more general operator $F + k(Du, u, x)$.

We say that the operator F is *concave* at $A \in \mathbb{S}^n$ if for any $B = (b_{ij})_{1 \le i,j \le n} \in \mathbb{S}^n$,

$$\left. \frac{d^2}{dt^2} \right|_{t=0} F(A + tB) = F_{ij,kl}(A) b_{ij} b_{kl} \le 0.$$

Example 2.78. Let us verify the ellipticity and concavity of $F(D^2 u) = \log \det D^2 u$ over the set of strictly convex C^2 functions u. By Lemma 2.56, $F_{ij}(D^2 u) = u^{ij}$, where $(u^{ij})_{1 \le i,j \le n} = (D^2 u)^{-1}$, so $(F_{ij}(D^2 u))_{1 \le i,j \le n}$ is positive definite. Thus F is elliptic. On the other hand, $F_{ij,kl}(D^2 u) = -u^{ik} u^{lj}$. Thus, F is concave, because if $B = (b_{ij})_{1 \le i,j \le n} \in \mathbb{S}^n$, then Lemma 2.58 gives

$$F_{ij,kl}(D^2 u) b_{ij} b_{kl} = -u^{ik} b_{kl} u^{lj} b_{ji} = -\text{trace}\left((D^2 u)^{-1} B (D^2 u)^{-1} B\right) \le 0.$$

We recall the following maximum principle for elliptic operators.

Theorem 2.79 (Maximum principle for possibly degenerate second-order elliptic operator). *Let Ω be a bounded domain in $\mathbb{R}^n$. Consider the elliptic operator $a^{ij} D_{ij} + b^i D_i$ where $a^{ij}, b^i \in C(\Omega)$ for $i, j = 1, \dots, n$ and $(a^{ij})_{1 \le i,j \le n}$ is nonnegative definite in Ω. If $u \in C^2(\Omega) \cap C(\overline{\Omega})$ satisfies*

$$a^{ij} D_{ij} u + b^i D_i u > 0 \quad \text{in } \Omega,$$

then

$$\max_{\overline{\Omega}} u = \max_{\partial \Omega} u.$$

The proof of Theorem 2.79 is quite simple. Assume otherwise that u attains its maximum value on $\overline{\Omega}$ at some $z \in \Omega$. Then $Du(z) = 0$ and $D^2 u(z) \le 0$. From $A(z) := (a^{ij}(z))_{1 \le i,j \le n} \ge 0$ and Lemma 2.57, we find $a^{ij}(z) D_{ij} u(z) = \text{trace}(A(z) D^2 u(z)) \le 0$, which contradicts the assumption

$$a^{ij}(z) D_{ij} u(z) = a^{ij}(z) D_{ij} u(z) + b^i(z) D_i u(z) > 0.$$

2.8.2. Perron's method for Laplace's equation. We review here Perron's method [**Per**] which was designed to solve the Dirichlet problem for Laplace's equation with continuous boundary data. Let Ω be a bounded domain in $\mathbb{R}^n$, and let $\varphi \in C(\partial\Omega)$. To find a function $u \in C(\overline{\Omega})$ solving

$$\Delta u = 0 \quad \text{in } \Omega, \qquad u = \varphi \quad \text{on } \partial\Omega,$$

we look at the pointwise maximum of all subsolutions, that is, $u(x) = \sup_{v \in S_\varphi} v(x)$, where

$$S_\varphi = \{v \in C(\overline{\Omega}) : v \le \varphi \text{ on } \partial\Omega \text{ and } v \text{ is subharmonic; that is, } \Delta v \ge 0 \text{ in } \Omega\}.$$

Then, we can show that u is harmonic in Ω. For this, the following properties are key ingredients:

(a) The maximum principle for harmonic functions.

(b) The solvability of the Dirichlet problem for Ω being any ball B and φ being any continuous function on ∂B. This uses essentially the Poisson integral of φ.

(c) The stability of the lifting of subharmonic and harmonic functions. More precisely, suppose $\bar{u}$ is harmonic in a ball $B \Subset \Omega$ and u is subharmonic in Ω with $u = \bar{u}$ on ∂B. Define the lifting of $\bar{u}$ and u by

$$\check{u}(x) = \bar{u}(x) \quad \text{for } x \in B \quad \text{and} \quad \check{u}(x) = u(x) \quad \text{for } x \in \Omega \setminus B.$$

Then $\check{u}$ is also a subharmonic function in Ω.

A useful feature of Perron's method is that it separates the interior existence problem from that of the boundary behavior of solutions. An important question regarding Perron's method for Laplace's equation is whether u just defined satisfies $u = \varphi$ on $\partial\Omega$. The answer to this question depends on the local behavior of $\partial\Omega$ near each boundary point $x_0 \in \partial\Omega$. But the answer is always in the affirmative if Ω is convex. This is based on the concept of *barriers*; see Gilbarg–Trudinger [**GT**, Chapters 2 and 6] for more details.

Theorem 2.80 (Solvability of Laplace's equation on convex domain). *Let Ω be a bounded convex domain in $\mathbb{R}^n$. Then for any $\varphi \in C(\partial\Omega)$, there is a unique function $u \in C^\infty(\Omega) \cap C(\overline{\Omega})$ satisfying*

$$\Delta u = 0 \quad \text{in } \Omega, \qquad u = \varphi \quad \text{on } \partial\Omega.$$

We give here some of the flavor of this concept of barriers for Laplace's equation. Let $x_0 \in \partial\Omega$. A barrier at x_0 relative to Ω is a superharmonic function $w \in C(\overline{\Omega})$ (that is, $\Delta w \le 0$ in Ω) such that $w(x_0) = 0$ and $w > 0$ in $\overline{\Omega} \setminus \{x_0\}$. Suppose such w exists. Then, for each $\varepsilon > 0$, there is $\delta = \delta(\varepsilon, \varphi)$

and $k = k(\varepsilon, \varphi)$ such that $|\varphi(x) - \varphi(x_0)| \leq \varepsilon$ for $x \in \partial\Omega$ with $|x - x_0| \leq \delta$, while $kw(x) \geq 2\|\varphi\|_{L^\infty(\Omega)}$ for $x \in \partial\Omega$ with $|x - x_0| \geq \delta$. From these, we find

$$\varphi(x_0) - \varepsilon - kw \leq \varphi \leq \varphi(x_0) + \varepsilon + kw \quad \text{on } \partial\Omega.$$

Using the maximum principle, we deduce the following estimates for the harmonic function u constructed by the Perron method:

$$\varphi(x_0) - \varepsilon - kw(x) \leq u(x) \leq \varphi(x_0) + \varepsilon + kw(x) \quad \text{in } \Omega.$$

When $x \to x_0$, these inequalities show not only $u(x_0) = \varphi(x_0)$ but also the continuity of u at x_0. Our treatment of the Monge–Ampère and linearized Monge–Ampère equations will use some variants of this barrier concept.

Finally, we note that Perron's method can be extended easily to more general classes of second-order elliptic equations and the solutions can be understood in the viscosity sense.

2.8.3. Classical Schauder and Calderon–Zygmund theories.

We recall the classical Schauder [**Sch**] and Calderon–Zygmund [**CZ1**] theories of second-order, linear, uniformly elliptic equations.

The basic results in the Schauder theory are the following $C^{k+2,\alpha}$ estimates; see [**GT**, Theorems 6.2 and 6.17] for interior $C^{k+2,\alpha}$ estimates and [**GT**, Theorems 3.7, 6.6, 6.14, and 6.19] for global $C^{k+2,\alpha}$ estimates.

Theorem 2.81 (Schauder interior $C^{k+2,\alpha}$ estimates). *Let Ω be a bounded domain in $\mathbb{R}^n$. Let $k \in \mathbb{N}_0$, and let $0 < \alpha < 1$. Let $f \in C^{k,\alpha}(\overline{\Omega})$. Suppose that $u \in C^2(\Omega)$ satisfies*

$$a^{ij}(x)D_{ij}u + b^i(x)D_i u + c(x)u = f \quad \text{in } \Omega,$$

where the coefficient matrix $(a^{ij})_{1 \leq i,j \leq n}$ and the functions c, $(b^i)_{1 \leq i \leq n}$ satisfy

$$\lambda I_n \leq (a^{ij})_{1 \leq i,j \leq n} \quad \text{and} \quad \max\left\{\|a^{ij}\|_{C^{k,\alpha}(\overline{\Omega})}, \|b^i\|_{C^{k,\alpha}(\overline{\Omega})}, \|c\|_{C^{k,\alpha}(\overline{\Omega})}\right\} \leq \Lambda$$

for positive constants λ, Λ. Then, $u \in C^{k+2,\alpha}(\Omega)$, and for any subdomain $\Omega' \Subset \Omega$, the following estimate holds:

$$\|u\|_{C^{k+2,\alpha}(\overline{\Omega'})} \leq C(\|f\|_{C^{k,\alpha}(\overline{\Omega})} + \|u\|_{L^\infty(\Omega)}),$$

where C depends only on $n, k, \alpha, \lambda, \Lambda, \Omega'$, and $\mathrm{dist}(\Omega', \partial\Omega)$.

Theorem 2.82 (Existence of and global $C^{k+2,\alpha}$ estimates). *Let Ω be a bounded domain in $\mathbb{R}^n$ with $C^{k+2,\alpha}$ boundary, where $k \in \mathbb{N}_0$ and $0 < \alpha < 1$. Consider the operator $Lu = a^{ij}(x)D_{ij}u + b^i(x)D_i u + c(x)u$ where $c \leq 0$ and where the coefficient matrix $(a^{ij})_{1 \leq i,j \leq n}$ and the vector field $(b^i)_{1 \leq i \leq n}$ satisfy*

$$\lambda I_n \leq (a^{ij})_{1 \leq i,j \leq n} \quad \text{and} \quad \max\left\{\|a^{ij}\|_{C^{k,\alpha}(\overline{\Omega})}, \|b^i\|_{C^{k,\alpha}(\overline{\Omega})}, \|c\|_{C^{k,\alpha}(\overline{\Omega})}\right\} \leq \Lambda$$

for positive constants λ, Λ. Then, if $f \in C^{k,\alpha}(\overline{\Omega})$ and $\varphi \in C^{k+2,\alpha}(\overline{\Omega})$, the Dirichlet problem

$$Lu = f \quad in \ \Omega, \quad u = \varphi \quad on \ \partial\Omega,$$

has a unique solution $u \in C^{k+2,\alpha}(\overline{\Omega})$ with the estimate

$$\|u\|_{C^{k+2,\alpha}(\overline{\Omega})} \leq C(n,k,\lambda,\Lambda,\Omega,\alpha)(\|f\|_{C^{k,\alpha}(\overline{\Omega})} + \|\varphi\|_{C^{k+2,\alpha}(\overline{\Omega})}).$$

We next state a basic result in the Calderon–Zygmund theory concerning global $W^{2,p}$ estimates; see [**GT**, Theorems 9.13 and 9.15].

Theorem 2.83 (Global $W^{2,p}$ estimates). *Let Ω be a bounded domain in $\mathbb{R}^n$ with $C^{1,1}$ boundary. Consider the operator $Lu = a^{ij}(x)D_{ij}u$ where the coefficient matrix satisfies*

$$a^{ij} \in C(\overline{\Omega}), \lambda I_n \leq (a^{ij})_{1 \leq i,j \leq n} \leq \Lambda I_n \quad for \ positive \ constants \ \lambda, \Lambda.$$

Then, if $f \in L^p(\Omega)$ and $\varphi \in W^{2,p}(\Omega)$, with $1 < p < \infty$, the Dirichlet problem

$$Lu = f \quad in \ \Omega, \quad u = \varphi \quad on \ \partial\Omega,$$

has a unique solution $u \in W^{2,p}(\Omega)$ with the estimate

$$\|u\|_{W^{2,p}(\Omega)} \leq C(\|u\|_{L^p(\Omega)} + \|f\|_{L^p(\Omega)} + \|\varphi\|_{W^{2,p}(\Omega)})$$

where C depends on $n, p, \lambda, \Lambda, \Omega$, and the moduli of continuity of a^{ij} in $\overline{\Omega}$.

A domain Ω in $\mathbb{R}^n$ is said to satisfy an *exterior cone condition* at $x_0 \in \partial\Omega$ if there is a finite right circular cone V with vertex x_0 such that $\overline{\Omega} \cap V = \{x_0\}$. We have the following solvability result; see [**GT**, Theorem 9.30].

Theorem 2.84 (Solvability in $W^{2,p}_{\text{loc}}(\Omega) \cap C(\overline{\Omega})$). *Let Ω be a bounded domain in $\mathbb{R}^n$ satisfying an exterior cone condition at every boundary point. Consider the operator $Lu = a^{ij}(x)D_{ij}u + b^i(x)D_iu + cu$ where the coefficients satisfy $0 < (a^{ij})_{1 \leq i,j \leq n}$ in Ω, $a^{ij} \in C(\Omega) \cap L^\infty(\Omega)$, $b^i, c \in L^\infty(\Omega)$, and $c \leq 0$. Let $f \in L^p(\Omega)$ for some $p \geq n$, and let $\varphi \in C(\partial\Omega)$. Then the Dirichlet problem*

$$Lu = f \quad in \ \Omega, \quad u = \varphi \quad on \ \partial\Omega,$$

has a unique solution $u \in W^{2,p}_{\text{loc}}(\Omega) \cap C(\overline{\Omega})$.

2.8.4. Evans–Krylov $C^{2,\alpha}$ estimates. Here we recall the classical Evans–Krylov Theorems which give $C^{2,\alpha}$ estimates for $C^{1,1}$ solutions to concave, fully nonlinear uniformly elliptic equations. The first theorem, due to Evans [**E1**] and Krylov [**K2**], is concerned with interior estimates.

Theorem 2.85 (Evans–Krylov Theorem: interior $C^{2,\alpha}$ estimates). *Let Ω be an open bounded set in $\mathbb{R}^n$. Let $F \in C^2(\mathbb{S}^n)$, $f \in C^2(\Omega)$, and let $u \in C^4(\Omega)$ be a solution of*

$$F(D^2u) = f \quad in \ \Omega.$$

Assume that the following conditions hold:

(a) *F is uniformly elliptic with respect to u inside Ω; that is, there exist positive constants λ, Λ such that*

$$\lambda|\xi|^2 \leq F_{ij}(D^2 u(x))\xi_i\xi_j \leq \Lambda|\xi|^2 \quad \text{for all } \xi = (\xi_1, \ldots, \xi_n) \in \mathbb{R}^n \text{ and } x \in \Omega.$$

(b) *F is concave in the range of $D^2 u$ inside Ω; that is,*

$$F_{ij,kl}(D^2 u(x))b_{ij}b_{kl} \leq 0 \text{ for all } x \in \Omega \text{ and for all } B = (b_{ij})_{1 \leq i,j \leq n} \in \mathbb{S}^n.$$

Assume that $B_r(x_0) \subset \Omega \subset B_R(x_0)$, for some $x_0 \in \mathbb{R}^n$ and $0 < r \leq R$. Then, for any open set $\Omega' \Subset \Omega$, we have the estimates

$$[D^2 u]_{C^\alpha(\Omega')} \leq C\big(\|D^2 u\|_{L^\infty(\Omega)} + \|Df\|_{L^\infty(\Omega)} + \|D^2 f\|_{L^\infty(\Omega)}\big)$$

where $\alpha \in (0,1)$ depends only on n, λ, Λ and C depends only on $n, \lambda, \Lambda, r, R$, and $\mathrm{dist}(\Omega', \partial\Omega)$.

The second theorem, due to Krylov [**K3**] (see also Caffarelli–Nirenberg–Spruck [**CNS1**]) is concerned with global estimates.

Theorem 2.86 (Evans–Krylov Theorem: global $C^{2,\alpha}$ estimates). *Let Ω be a bounded domain in $\mathbb{R}^n$ with C^3 boundary $\partial\Omega$. Let $\phi \in C^3(\overline{\Omega})$, $F \in C^2(\mathbb{S}^n)$, $f \in C^2(\overline{\Omega})$. Suppose that $u \in C^4(\Omega) \cap C^3(\overline{\Omega})$ is a solution of*

$$F(D^2 u) = f \quad \text{in } \Omega, \quad u = \phi \quad \text{on } \partial\Omega.$$

Assume that the uniform ellipticity condition (a) *and concavity condition* (b) *in Theorem 2.85 hold. Then, for some $\alpha(n, \lambda, \Lambda) \in (0,1)$, we have*

$$[D^2 u]_{C^\alpha(\Omega)} \leq C(n, \lambda, \Lambda, \Omega)\big(\|u\|_{C^2(\overline{\Omega})} + \|\phi\|_{C^3(\overline{\Omega})} + \|f\|_{C^2(\overline{\Omega})}\big).$$

Remark 2.87. Let $u \in C^2(\Omega)$ be a convex function on $\Omega \subset \mathbb{R}^n$ such that

$$\det D^2 u \geq c_0 > 0 \quad \text{in } \Omega \quad \text{and} \quad \|D^2 u\|_{L^\infty(\Omega)} \leq C_1.$$

- Then $C_2^{-1} I_n \leq D^2 u \leq C_2 I_n$ in Ω for some $C_2(n, c_0, C_1) > 0$. Indeed, fix $x \in \Omega$. Let $\lambda_1 \leq \cdots \leq \lambda_n$ be the positive eigenvalues of $D^2 u(x)$. From $\sum_{i=1}^n \lambda_i^2 = \mathrm{trace}(D^2 u(D^2 u)^t) \leq n^2 C_1^2$, we find that $\lambda_n \leq nC_1$. Hence $\lambda_1 \geq \det D^2 u(x)/\lambda_n^{n-1} \geq c_0/(nC_1)^{n-1}$. The claimed inequalities hold for $C_2 := \max\{nC_1, (nC_1)^{n-1}c_0^{-1}\}$.

- Thus, in view of Example 2.78, the operator $F(D^2 v) := \log \det D^2 v$ is concave and uniformly elliptic at u.

- Once we know that $u \in C^{2,\alpha}(\overline{\Omega})$ (by an application of Theorem 2.86, for example), we can obtain global higher-order regularity for u in Hölder or Sobolev spaces from those of $\det D^2 u$ and the boundary data. For this, we straighten out the boundary, differentiate, and then apply Theorem 2.82 or 2.83.

2.9. Pointwise Estimates and Perturbation Argument

In this section, we present two general concepts that prove to be quite powerful for local quantitative analysis in many PDEs.

2.9.1. Pointwise estimates. We begin with the concept of pointwise $C^{k,\alpha}$.

Definition 2.88 (Pointwise C^k and $C^{k,\alpha}$). Let $k \geq 0$ be an integer, and let $\alpha \in (0,1]$. Let $\mathfrak{P}_k$ be the set of all polynomials on $\mathbb{R}^n$ with real coefficients and degree not greater than k. We say the following:

(a) u is pointwise $C^{k,\alpha}$ at x_0 and we write $u \in C^{k,\alpha}(x_0)$ if there exist $P \in \mathfrak{P}_k$ and $M, r > 0$ such that, in the domain of definition of u,

$$|u(x) - P(x)| \leq M|x - x_0|^{k+\alpha} \quad \text{for all } |x - x_0| < r.$$

(b) u is pointwise C^k at x_0 and we write $u \in C^k(x_0)$ if there exists $P \in \mathfrak{P}_k$ such that, in the domain of definition of u,

$$u(x) = P(x) + o(|x - x_0|^k) \quad \text{as } x \to x_0.$$

We use the notation $[u]_{C^{k,\alpha}(x_0)}$ ($[u]_{C^k(x_0)}$) to encode the pointwise $C^{k,\alpha}$ (pointwise C^k) information of u at x_0 which consists of P, M, r, k, α in part (a) (respectively, P, k in part (b)).

Convex functions provide examples for the concept of pointwise C^2. This is due to a classical result of Aleksandrov which asserts that any convex function is twice differentiable almost everywhere.

Theorem 2.89 (Aleksandrov's Theorem). *Let Ω be a convex domain in $\mathbb{R}^n$ and let $u : \Omega \to \mathbb{R}$ be a convex function. Then u is twice differentiable almost everywhere in Ω. More precisely, for almost every $x_0 \in \Omega$, we have*

$$u(x) = u(x_0) + Du(x_0) \cdot (x - x_0) + \frac{1}{2}\big(D^2 u(x_0)(x - x_0)\big) \cdot (x - x_0)$$
$$+ o(|x - x_0|^2) \quad \text{as } x \to x_0.$$

To relate pointwise estimates to uniform estimates, we can use the following lemma, due to Calderon and Zygmund [**CZ2**].

Lemma 2.90 (Calderon–Zygmund Lemma). *Given an integer $m \geq 0$, there exists a function $\phi \in C_c^\infty(B_1(0))$, such that for every polynomial P on $\mathbb{R}^n$ of degree not greater than m and every $\varepsilon > 0$, we have*

$$(\phi_\varepsilon * P)(x) := \int_{\mathbb{R}^n} \varepsilon^{-n}\phi((x - y)/\varepsilon)P(y)\, dy = P(x) \quad \text{for all } x \in \mathbb{R}^n.$$

Proof. By changing variables, we have

$$(\phi_\varepsilon * P)(x) = \int_{\mathbb{R}^n} \phi(z)P(x - \varepsilon z)\, dz.$$

Given $x \in \mathbb{R}^n$ and $\varepsilon > 0$, let $Q(z) = P(x - \varepsilon z)$. Then, the desired equation becomes $\int_{\mathbb{R}^n} \phi Q \, dz = Q(0)$. Note that Q is a polynomial of degree not greater than m. Therefore, to prove the lemma, it suffices to show the existence of a function $\phi \in C_c^\infty(B_1(0))$ such that

$$(2.21) \quad \int_{\mathbb{R}^n} \phi(x) \, dx = 1 \quad \text{and} \quad \int_{\mathbb{R}^n} \phi(x) x^\alpha \, dx = 0 \quad \text{for all } 0 < |\alpha| \le m.$$

Here, we denote $x^\alpha = x_1^{\alpha_1} \ldots x_n^{\alpha_n}$ for $x \in \mathbb{R}^n$ and $\alpha = (\alpha_1, \ldots, \alpha_n) \in \mathbb{N}_0^n$.

Let $N = N(n, m)$ be the number of multi-indices $\alpha \in \mathbb{N}_0^n$ with length $|\alpha| \le m$. Consider the mapping $\xi = (\xi_\alpha)_{0 \le |\alpha| \le m}$ from the class of all functions $\phi \in C_c^\infty(B_1(0))$ into $\mathbb{R}^N$ given by $\xi_\alpha = \int_{\mathbb{R}^n} \phi(x) x^\alpha \, dx$.

To establish (2.21), we show that the range of ξ is $\mathbb{R}^N$. Were this not the case, there would exist $\eta = (\eta_\alpha)_{0 \le |\alpha| \le m} \in \mathbb{R}^N \setminus \{0\}$ such that for all $\phi \in C_c^\infty(B_1(0))$, we have

$$\sum_{0 \le |\alpha| \le m} \eta_\alpha \xi_\alpha = \int_{\mathbb{R}^n} \phi(x) \sum_{0 \le |\alpha| \le m} \eta_\alpha x^\alpha \, dx = 0.$$

Inserting into the above equation $\phi(x) = \psi(x) \sum_{0 \le |\alpha| \le m} \eta_\alpha x^\alpha$, where ψ satisfies $\mathrm{spt}(\psi) \subset \overline{B_{1/2}(0)}$ and $\psi > 0$ in $B_{1/2}(0)$, we obtain $\sum_{0 \le |\alpha| \le m} \eta_\alpha x^\alpha = 0$ in $B_{1/2}(0)$. Consequently, $\eta \equiv 0$, contradicting our assumption on η. $\qquad\square$

Theorem 2.91 (From pointwise $C^{m,\alpha}$ to uniform $C^{m,\alpha}$ estimates). *Let Ω be a bounded domain in $\mathbb{R}^n$, and let $\Omega' \Subset \Omega$. Let $m \in \mathbb{N}_0$, $\alpha \in (0, 1]$, $M \ge 0$, and $0 < r < \mathrm{dist}(\overline{\Omega'}, \partial\Omega)$. Let $u \in C(\Omega)$. Assume that for each $z \in \Omega'$, there is a polynomial P_z of degree not greater than m such that*

$$|u(x) - P_z(x)| \le M|x - z|^{m+\alpha} \quad \text{for all } x \in B_r(z).$$

Then $u \in C^{m,\alpha}(\overline{\Omega'})$ with

(i) *$D^\beta u(z) = D^\beta P_z(z)$ for all $z \in \Omega'$ and multi-indices β with $|\beta| \le m$ and*

(ii) *$[D^m u]_{C^{0,\alpha}(\Omega')} \le C(\mathrm{diam}(\Omega'), n, m, \alpha, r)M$.*

Proof. Clearly, u is m times differentiable in Ω' and assertion (i) holds.

It remains to prove assertion (ii). Let ϕ be as in Lemma 2.90, and let $\phi_\varepsilon(x) = \varepsilon^{-n} \phi(x/\varepsilon)$ for $\varepsilon > 0$.

Let $y \in \Omega'$. Consider $z \in B_r(y) \cap \Omega'$ and $z \ne y$. The case $|z - y| > r$ can be reduced to the previous case by a covering of balls of radius r connecting to y. Note that P_y and P_z are the Taylor expansions of order m of u at y and z. Then, for all $x \in \Omega$ and $\varepsilon > 0$, we have

$$(u * \phi_\varepsilon)(x) = ((u - P_y) * \phi_\varepsilon)(x) + (P_y * \phi_\varepsilon)(x)$$
$$= ((u - P_y) * \phi_\varepsilon)(x) + P_y(x),$$

from which we find that, noting also $D^m P_y(x) = D^m P_y(y) = D^m u(y)$,

$$D^m(u * \phi_\varepsilon)(x) = ((u - P_y) * D^m \phi_\varepsilon)(x) + D^m u(y).$$

Replacing y by z and subtracting the resulting equations, we obtain

$$D^m u(y) - D^m u(z) = ((u - P_z) * D^m \phi_\varepsilon)(x) - ((u - P_y) * D^m \phi_\varepsilon)(x).$$

Now, we estimate $|D^m u(y) - D^m u(z)|$ in terms of $|y - z|^\alpha$. For this, we choose $x = (y + z)/2$ and $\varepsilon = |y - z|/2$. Then, $B_\varepsilon(x) \subset B_{2\varepsilon}(y) \subset B_r(y)$ and $B_\varepsilon(x) \subset B_{2\varepsilon}(z) \subset B_r(z)$. It suffices to estimate $I_y := |((u - P_y) * D^m \phi_\varepsilon)(x)|$ in terms of $|y - z|^\alpha$. Using $|u(w) - P_y(w)| \leq M|w - y|^{m+\alpha}$ in $B_r(y)$, we have

$$I_y \leq \varepsilon^{-n-m} \int_{B_\varepsilon(x)} |(u - P_y)(w)||(D^m \phi)((x - w)/\varepsilon)| \, dw$$

$$\leq M\varepsilon^{-n-m} \|D^m \phi\|_{L^\infty(B_1(0))} \int_{B_{2\varepsilon}(y)} |w - y|^{m+\alpha} \, dw$$

$$= C(n, m, \alpha) M \varepsilon^\alpha = CM|y - z|^\alpha.$$

This completes the proof of assertion (ii). The theorem is proved. $\square$

2.9.2. Perturbation argument.

Several $C^{2,\alpha}$ estimates in this book, both in the interior and at the boundary, will be proved using the perturbation argument of Caffarelli [**C1, CC**] (see also related ideas in Safonov [**Saf1, Saf2**]). Since this argument is quite powerful with wide applicability, we present here its simple version for Poisson's equation for the benefit of readers who are not familiar with it. The starting point is higher-order derivative estimates for the model equation. In our situation, it is Laplace's equation and C^3 estimates suffice. In particular, we will use the following local estimates on derivatives for harmonic functions which is a consequence of the Mean Value Theorem: If $u \in C^2(\Omega)$ is harmonic in $\Omega \subset \mathbb{R}^n$, that is, $\Delta u = 0$, then for each ball $B_r(x_0) \subset \Omega$ and $k \in \mathbb{N}_0$, we have

$$(2.22) \qquad \|D^k u\|_{L^\infty(B_{r/2}(x_0))} \leq \frac{C(n, k)}{r^{n+k}} \|u\|_{L^1(B_r(x_0))}.$$

Poisson's equation $\Delta u = f$ can be appropriately viewed as a perturbation of Laplace's equation when f has certain continuity.

The following version of the perturbation argument, due to Caffarelli [**C1**], is concerned with pointwise $C^{2,\alpha}$ estimates for Poisson's equation.

Theorem 2.92 (Pointwise $C^{2,\alpha}$ estimates for Poisson's equation). *Suppose that $u \in C^2(B_1(0))$ solves Poisson's equation $\Delta u = f$ in the unit ball $B_1(0)$ in $\mathbb{R}^n$. If f is pointwise C^α at 0 where $0 < \alpha < 1$, then u is pointwise $C^{2,\alpha}$ at 0. More precisely, if, for some $M > 0$ and for $0 < r \leq 1$, we have*

$$|f(x) - f(0)| \leq M|x|^\alpha \qquad in \ B_r(0),$$

then there is a quadratic polynomial P such that

$$|P(0)| + |DP(0)| + \|D^2 P\| \le C \quad and \quad |u(x) - P(x)| \le C|x|^{2+\alpha} \ in \ B_{r_0}(0),$$

where $r_0(n, r, \alpha) > 0$ and $C > 0$ depends on $n, \alpha, M, f(0),$ and $\|u\|_{L^\infty(B_1(0))}$.

Note carefully that no other regularity of f is used.

Proof. We write B_t for $B_t(0)$. First, we make several simplifications so as to assume that

$$(2.23) \qquad r = 1, \quad f(0) = 0, \quad \|u\|_{L^\infty(B_1)} \le 1, \quad M \le \varepsilon,$$

where $\varepsilon > 0$ is to be determined. Indeed, by using the rescalings

$$u(x) \mapsto u(rx)/r^2, \quad f(x) \mapsto f(rx), \quad M \mapsto Mr^\alpha,$$

we can assume $r = 1$. Next, by using the translations

$$u \mapsto u - f(0)|x|^2/2n, \quad f \mapsto f - f(0),$$

we can assume that $f(0) = 0$. Finally, by dividing both u and f by $\|u\|_{L^\infty(B_1)} + \varepsilon^{-1}M$, we can assume the last two inequalities in (2.23) hold.

Step 1. We show that for a suitable $\varepsilon > 0$ depending only on n and α, there exist $r_0(n, \alpha) \in (0, 1/4)$ and a sequence of harmonic quadratic polynomials

$$P_m(x) = P_m(0) + DP_m(0) \cdot x + (D^2 P_m x) \cdot x/2,$$

with $P_0 \equiv 0$ such that for all integers $m \ge 1$, we have

$$(2.24) \qquad \qquad \|u - P_m\|_{L^\infty(B_{r_0^m})} \le r_0^{(2+\alpha)m}$$

and

$$(2.25) \quad \begin{aligned} |P_m(0) - P_{m-1}(0)| + r_0^{m-1}|DP_m(0) - DP_{m-1}(0)| \\ + r_0^{2m-2}\|D^2 P_m - D^2 P_{m-1}\| \le C(n) r_0^{(2+\alpha)(m-1)}. \end{aligned}$$

The choice of ε comes from constructing P_1. Let $Q(x) := (1/4 - |x|^2)/(2n)$, and let w be the harmonic function in $B_{1/2}$ satisfying $w = u$ on $\partial B_{1/2}$. Then,

$$\Delta(\varepsilon Q) = -\varepsilon \le \Delta(u - w) = f \le \varepsilon = \Delta(-\varepsilon Q) \quad in \ B_{1/2}.$$

By the maximum principle, we have in $B_{1/2}$

$$|u - w| \le \varepsilon Q = \varepsilon(1/4 - |x|^2)/(2n) \le \varepsilon.$$

On the other hand,

$$\|w\|_{L^\infty(B_{1/2})} \le \|w\|_{L^\infty(\partial B_{1/2})} = \|u\|_{L^\infty(\partial B_{1/2})} \le 1.$$

Thus, by (2.22), we have

$$\|D^k w\|_{L^\infty(B_{1/4})} \le C_0(n), \quad for \ all \ k = 0, 1, 2, 3.$$

Let us consider the quadratic Taylor polynomial of w at 0 given by

$$P_1(x) = w(0) + Dw(0) \cdot x + (D^2 w(0)x) \cdot x/2.$$

Then P_1 is harmonic since $\Delta P_1 = \Delta w(0) = 0$. Moreover, in $B_{1/4}$, we have

$$|w(x) - P_1(x)| \leq \|D^3 w\|_{L^\infty(B_{1/4})} |x|^3 \leq C_0(n)|x|^3.$$

Thus, for $r_0 \leq 1/4$, we have in B_{r_0}

$$|u - P_1| \leq |u - w| + |w - P_1| \leq \varepsilon + C_0(n) r_0^3 \leq r_0^{2+\alpha}$$

if we first choose r_0 such that $C_0(n) r_0^3 \leq r_0^{2+\alpha}/2$, and then choose ε such that $\varepsilon \leq r_0^{2+\alpha}/2$. Here r_0 and ε depend only on n and α.

With the above choices of ε and r_0, we prove (2.24) and (2.25) by induction. The case $m = 1$ has been confirmed where $C(n) = 3C_0(n)$.

Suppose that these estimates hold for $m \geq 1$. We prove them for $m + 1$. Consider

$$v(x) = r_0^{-(2+\alpha)m}(u - P_m)(r_0^m x), \quad g(x) = r_0^{-\alpha m} f(r_0^m x) \quad \text{in } B_1.$$

Then, in B_1,

$$\Delta v = g, \quad \|v\|_{L^\infty(B_1)} \leq 1, \quad |g(x)| \leq r_0^{-\alpha m}(\varepsilon |r_0^m x|^\alpha) = \varepsilon |x|^\alpha.$$

We are exactly in the situation of constructing P_1. Thus, we can find a harmonic quadratic polynomial $\tilde{P}_m$ such that

$$|v - \tilde{P}_m| \leq r_0^{2+\alpha} \quad \text{in } B_{r_0} \quad \text{and} \quad \max\{|\tilde{P}_m(0)|, |D\tilde{P}_m(0)|, \|D^2\tilde{P}_m\|\} \leq C_0(n).$$

Let

$$P_{m+1}(x) = P_m(x) + r_0^{(2+\alpha)m} \tilde{P}_m(r_0^{-m} x).$$

Then, clearly, (2.24) and (2.25) hold for $m + 1$.

Step 2. Conclusion. From (2.25), we find that $\{P_m\}_{m=1}^\infty$ converges uniformly in B_1 to a quadratic polynomial P satisfying

$$|P(0)| + |DP(0)| + \|D^2 P\| \leq C(n, \alpha).$$

Now, for any integer $m \geq 1$, using (2.24) and (2.25), we have

(2.26)
$$\|u - P\|_{L^\infty(B_{r_0^m})} \leq \|u - P_m\|_{L^\infty(B_{r_0^m})} + \sum_{i=m}^\infty \|P_{i+1} - P_i\|_{L^\infty(B_{r_0^m})}$$

$$\leq r_0^{m(2+\alpha)} + \sum_{i=m}^\infty 3C(n) r_0^{\alpha i + 2m} \leq C(n, \alpha) r_0^{m(2+\alpha)}.$$

For any $x \in B_{r_0}(0)$, let $m \in \mathbb{N}$ satisfy $r_0^{m+1} < |x| \leq r_0^m$. Then, (2.26) gives

$$|u(x) - P(x)| \leq C(n, \alpha) r_0^{-(2+\alpha)} |x|^{2+\alpha},$$

which completes the proof of the theorem. $\qquad\square$

Note carefully that in Theorem 2.92, estimates for solutions to Poisson's equation are obtained at a point from the information of the data only at that point. This justifies the terminology *pointwise estimates* and also reflects the local nature of the equation under consideration. When applying the perturbation argument to prove $C^{2,\alpha}$ estimates for the Monge–Ampère equation, we will replace the local derivative estimates (2.22) by appropriate Pogorelov estimates in Chapters 6 and 10.

If in Poisson's equation $\Delta u = f$ in $B_1(0)$, the right-hand side $f \in C^\alpha(B_{1/2}(0))$, then we obtain from Theorem 2.92 that u is pointwise $C^{2,\alpha}$ at each point $x_0 \in B_{1/4}(0)$, and as a consequence of Theorem 2.91, we can conclude that $u \in C^{2,\alpha}(B_{1/4}(0))$. Moreover, as pointed out in [**C1**, **CC**], the perturbation can be measured in L^n instead of L^∞; see Problem 2.18.

Using the perturbation argument, Wang [**W5**] gave a simple proof of classical $C^{2,\alpha}$ estimates for Poisson's equation with C^α right-hand side f. His proof allows estimating the continuity of the Hessian in terms of the modulus of continuity of f. The following $C^{2,\alpha}$ estimates from [**W5**] also have their Monge–Ampère counterparts in Jin–Wang [**JW1**].

Theorem 2.93 ($C^{2,\alpha}$ estimates for Poisson's equation). *Let $u \in C^2(B_1(0))$ be a solution to Poisson's equation $\Delta u = f$ in the unit ball $B_1(0)$ in $\mathbb{R}^n$ where $f \in C(B_1(0))$ with modulus of continuity*

$$\omega_f(r) := \omega(f, B_1(0); r) = \sup\left\{|f(x) - f(y)| : x, y \in B_1(0), |x - y| < r\right\}.$$

Then, for all $x, y \in B_{1/2}(0)$, we have the following estimate for the modulus of continuity of $D^2 u$ in terms of ω_f, $d := |x - y|$, and $M := \sup_{B_1(0)} |u|$:

$$(2.27) \quad |D^2 u(x) - D^2 u(y)| \leq C(n)\left[Md + \int_0^d \frac{\omega_f(r)}{r}\, dr + d \int_d^1 \frac{\omega_f(r)}{r^2}\, dr\right].$$

It follows that if $f \in C^\alpha(B_1(0))$ where $\alpha \in (0,1)$, then

$$\|u\|_{C^{2,\alpha}(B_{1/2}(0))} \leq C(n)\left[\sup_{B_1(0)} |u| + \frac{\|f\|_{C^\alpha(B_1(0))}}{\alpha(1 - \alpha)}\right].$$

2.10. Problems

Problem 2.1. Prove Lemma 2.24.

Problem 2.2. Let Ω be a bounded open set in $\mathbb{R}^n$. Let $u : \Omega \to \mathbb{R}$ be a convex function. Show that $\inf_\Omega u > -\infty$.

Problem 2.3. Let $u(x) = -(1 - |x|^2)^{1/2}$ where $x \in \overline{B_1(0)} \subset \mathbb{R}^n$.

    (a) Show that u is convex in $\overline{B_1(0)}$ but $\partial u(x_0) = \emptyset$ for all $x_0 \in \partial B_1(0)$.

    (b) Compute u^* and $(u^*)^*$.

Problem 2.4 (Cyclic monotonicity of normal mapping). Let Ω be a subset of $\mathbb{R}^n$. Let $u : \Omega \to \mathbb{R}$. Recall the monotonicity inequality in Remark 2.34 which can be rewritten as $p \cdot (y - x) + q \cdot (x - y) \leq 0$ if $p \in \partial u(x)$ and $q \in \partial u(y)$ where $x, y \in \Omega$. More generally, show that ∂u is *cyclically monotone*: For $m \geq 1$, if $y^i \in \partial u(x^i)$ where $x^i \in \Omega$ for $i = 1, \ldots, m$, then $\sum_{i=1}^{m} y^i \cdot (x^{i+1} - x^i) \leq 0$, with the convention $x^{m+1} = x^1$.

Problem 2.5 (Rockafellar's Theorem). Let $\Gamma \subset \mathbb{R}^n \times \mathbb{R}^n$ be a cyclically monotone subset in the following sense: For $m \geq 1$, if $(x^i, y^i) \in \Gamma$ ($i = 1, \ldots, m$), then $\sum_{i=1}^{m} y^i \cdot (x^{i+1} - x^i) \leq 0$, with the convention $x^{m+1} = x^1$. Show that there exists a convex function $\varphi : \mathbb{R}^n \to \mathbb{R} \cup \{+\infty\}$ with $\varphi \not\equiv \infty$ such that Γ is included in the graph of the normal mapping of φ; that is, if $(x, y) \in \Gamma$, then $y \in \partial \varphi(x)$.
Hint: Pick $(x^0, y^0) \in \Gamma$. Define for any $x \in \mathbb{R}^n$
$$\varphi(x) = \sup \left\{ y^m \cdot (x - x^m) + \cdots + y^0 \cdot (x^1 - x^0) : (x^1, y^1), \ldots, (x^m, y^m) \in \Gamma \right\}.$$

Problem 2.6. Let $u(x) = -(1 - |x|^p)^s$ where $p, s > 0$, $x \in B_1(0) \subset \mathbb{R}^n$. Find conditions on p and s so that u is convex in $B_1(0)$.

Problem 2.7. Let $k \geq 2$ and $\alpha \in [0, 1)$. Let Ω be a uniformly convex domain in $\mathbb{R}^n$ with $C^{k,\alpha}$ boundary. Let $\varphi \in C^{k,\alpha}(\overline{\Omega})$. Suppose that $f(x, z, p) : \overline{\Omega} \times \mathbb{R} \times \mathbb{R}^n$ satisfies
$$0 \leq f(x, z, p) \leq M(1 + |p|^2)^{n/2} \quad \text{for } x \in \overline{\Omega} \text{ and } z \leq \max_{\Omega} \varphi.$$
Show that there is a uniformly convex function $v \in C^{k,\alpha}(\overline{\Omega})$ such that
$$\det D^2 v \geq f(x, v, Dv) \quad \text{in } \Omega \quad \text{and} \quad v = \varphi \quad \text{on } \partial \Omega.$$

Problem 2.8. Let Ω be a bounded domain in $\mathbb{R}^n$. Consider the operator
$$F(D^2 u) = \det \left((\Delta u) I_n - D^2 u \right)$$
for $u \in C^2(\Omega)$. Show that F is elliptic at u if $(\Delta u) I_n - D^2 u > 0$ in Ω.

Problem 2.9. Let A be a symmetric nonnegative definite $n \times n$ matrix. Use Lemma 2.57 to prove that
$$(\det A)^{1/n} = \inf \left\{ \frac{1}{n} \text{trace}(AB) : B \in \mathbb{S}^n, \text{ positive definite with } \det B = 1 \right\}.$$
Use this result to give another proof of the concavity of $(\det D^2 u)^{1/n}$ over the set of C^2 convex functions u.

Problem 2.10 (Polar body of a convex body). Let K be a convex body in $\mathbb{R}^n$ containing z in its interior. We define the *polar body* K^z of K with respect to z by
$$K^z = \{x \in \mathbb{R}^n : x \cdot (y - z) \leq 1 \quad \text{for all } y \in K\}.$$

The normal mapping in Example 2.35(b) can be expressed as $\partial v(z) = M\overline{\Omega}^z$.

(a) Let K be a convex body in $\mathbb{R}^n$ containing the origin in its interior. Show that $(K^0)^0 = K$.

(b) (Bambah's inequality) Assume that the convex body K in $\mathbb{R}^n$ is centrally symmetric. Show that $|K||K^0| \geq n^{-n/2}\omega_n^2$.

Problem 2.11. Let $\alpha(n)$ be the constant in Lemma 2.66.

(a) Let $u : \mathbb{R}^n \to \mathbb{R}$ be a convex function such that $S = \{x \in \mathbb{R}^n : u(x) < 0\}$ is bounded with positive volume. Let y and z be, respectively, the minimum point of u and the center of mass of S. Show that $1 \leq u(y)/u(z) \leq [\alpha(n)]^{-1} + 1$.

(b) Determine the value of $\alpha(n)$ in the proof of Lemma 2.66.

Problem 2.12. Let $A(x) = (a^{ij}(x))_{1 \leq i,j \leq n}$ be a smooth, symmetric, positive definite matrix in $B_1(0) \subset \mathbb{R}^n$ with $\det A(x) > 1$ for all $x \in B_1(0)$. Suppose that $u \in C^2(B_1(0)) \cap C(\overline{B_1(0)})$ solves the boundary value problem

$$-a^{ij}D_{ij}u = 1 \quad \text{in } B_1(0), \quad u = 0 \quad \text{on } \partial B_1(0).$$

Prove that

$$0 \leq u(x) \leq (1 - |x|^2)/(2n) \text{ for all } x \in B_1(0).$$

Problem 2.13. Let Ω be a bounded convex domain in $\mathbb{R}^n$. Let $x_0 \in \partial\Omega$ and $B_R(y)$ be such that $\overline{B_R(y)} \cap \overline{\Omega} = \{x_0\}$. Show that, for all $\sigma > n - 2$, $w(x) = R^{-\sigma} - |x - y|^{-\sigma}$ is a barrier at x_0 relative to Ω for the Laplace operator; that is, $\Delta w \leq 0$ in Ω, $w(x_0) = 0$, and $w > 0$ in $\overline{\Omega} \setminus \{x_0\}$.

Problem 2.14. Let Ω be a bounded domain in $\mathbb{R}^n$ with C^3 boundary $\partial\Omega$. Let $u \in C^3(\overline{\Omega})$ with $u = 0$ on $\partial\Omega$.

(a) Assume u is convex. Show that

$$\int_\Omega \|D^2u\|^p\,dx \leq \int_\Omega (\Delta u)^p\,dx \quad \text{for all } p \geq 1.$$

(b) (Kadlec's inequality) Assume Ω is convex. Show that

$$\int_\Omega \|D^2u\|^2\,dx \leq \int_\Omega (\Delta u)^2\,dx.$$

Problem 2.15. Let u be a C^3 function on a domain Ω in $\mathbb{R}^n$. From Lemma 2.56, we know that the Monge–Ampère operator $\sigma_n(D^2u) = \det D^2u$ can be written in both nondivergence and divergence forms:

$$\sigma_n(D^2u) = (1/n)\sigma_{n,ij}(D^2u)D_{ij}u = (1/n)D_i(\sigma_{n,ij}(D^2u)D_ju).$$

Here we adopt the notation in Section 2.8.1. Show that the operator

$$\sigma_2(D^2u) = \left[\text{trace}(D^2u)\right]^2 - \|D^2u\|^2$$

has analogous properties; more precisely,

$$\sigma_2(D^2 u) = (1/2)\sigma_{2,ij}(D^2 u)D_{ij}u = (1/2)D_i(\sigma_{2,ij}(D^2 u)D_j u).$$

Problem 2.16. Give a direct proof of Theorem 2.91 without using the Calderon–Zygmund Lemma (Lemma 2.90). (See also Step 2 in the proof of Theorem 10.17 and Lemma 10.18.)

Problem 2.17. Prove (2.27) in Theorem 2.93.

Problem 2.18. Prove that the conclusion of Theorem 2.92 still holds if the condition

$$|f(x) - f(0)| \le M|x|^\alpha \quad \text{in } B_r(0)$$

is replaced by

$$\|f - f(0)\|_{L^n(B_s(0))} \le Ms^{1+\alpha} \quad \text{for all } 0 \le s \le r.$$

2.11. Notes

We rely on various sources for this chapter. For more on convex sets and convex functions, see Bakelman [**Ba5**], Hörmander [**Hor**], Rockafellar [**Roc**], and Schneider [**Schn**]; see also Sections 1.1 and 1.9 in Gutiérrez [**G2**] and the Appendix in Figalli [**F2**]. For more on classical PDE theory, see Caffarelli–Cabré [**CC**], Evans [**E2**], Gilbarg–Trudinger [**GT**], Han–Lin [**HL**], Han [**H**], and Krylov [**K4**]. For more on measure theory and Hausdroff measure, see Evans–Gariepy [**EG**].

In Section 2.1, Theorem 2.5 is from [**Schn**, Section 1.1], Lemma 2.9 is from [**Schn**, Section 1.2], Theorem 2.10 is from [**Schn**, Section 1.3], and Theorem 2.12 is from [**Hor**, Section 2.1], [**Roc**, Section 18], and [**Schn**, Section 1.4].

Section 2.2 is based on [**Schn**, Section 1.8].

In Section 2.3, Theorems 2.26 and 2.28 are from [**Hor**, Section 2.1], Lemma 2.36 is from [**F2**, Section A.4], Theorem 2.38 is from [**Roc**, Chapter 25]. Most of the materials on convex functions in Sections 2.3.2–2.3.4 are well known; many are contained in [**G2**, Sections 1.1 and 1.9] and [**F2**, Section A.4]. The proof of Theorem 2.38(i) is from [**H**, Section 8.1]. Step 1 in the proof of Lemma 2.40(ii) is from [**EG**, pp. 81–82].

In Section 2.4, Proposition 2.47 is from [**Hor**, Section 2.1], Theorem 2.51 is from [**Hor**, Section 2.3]. The arguments in the proof of part (iii) in Lemma 2.52 are from Caffarelli–Kohn–Nirenberg–Spruck [**CKNS**]. See [**GT**, Section 14.6] for more on boundary curvatures and the distance function.

In Section 2.5, identities regarding determinants are standard. The last inequality in Lemma 2.58(iii) is in Caffarelli–Gutiérrez [**CG2**]. Lemma 2.60 is in Caffarelli [**C7**].

In Section 2.6, the proof of Lemma 2.63 is based on the notes of Howard [**Hw**]; see also [**F2**, Section A.3]. The proof of Lemma 2.66 is based on [**G2**, Section 1.8] and Pogorelov [**P5**, §6]. Regarding the proof of Lemma 2.62, see Horn–Johnson [**HJ**, Section 7.3] for the Polar Decomposition Theorem in linear algebra. Further references on matrices can be found in [**HJ**].

In Section 2.7, proofs of Theorems 2.68 and 2.69 can be found in [**EG**, Section 1.9]. Lemma 2.71 and Theorem 2.74 are contained in [**F2**, Sections A.3–A.4]. For a proof of Theorem 2.76, see [**E2**, Theorem 6, pp. 287–288].

In Section 2.8, see [**GT**, Chapter 3] for more on the classical maximum principles. For a proof of Theorem 2.85, see [**CC**, Chapters 6 and 8], [**GT**, Section 17.4], and [**H**, Section 5.4]. For a proof of Theorem 2.86, see [**CC**, Chapter 9], [**GT**, Section 17.8], and [**H**, Section 5.5].

In Section 2.9, see [**EG**, pp. 242–245] for a proof of Aleksandrov's Theorem (Theorem 2.89). The notion of *pointwise estimates* in PDEs first appeared in Calderon–Zygmund [**CZ2**]. The idea for the proof of Theorem 2.91 using Lemma 2.90 is from [**G2**, Theorem 5.4.8]. The method of using the Calderon–Zygmund Lemma (Lemma 2.90) to show Hölder continuity first appeared in Gutiérrez [**G1**]. See [**E2**, Theorem 7, p. 29] for a proof of (2.22).

In Section 2.10, for Rockafellar's Theorem in Problem 2.5, see [**Roc**, Section 24]. Problem 2.7 is from Lions [**Ln1**] and Caffarelli–Nirenberg–Spruck [**CNS1**]. Problem 2.9 originates from Gaveau [**Gav**]. Problem 2.10(b) is taken from Bambah [**Bam**]. Problem 2.14(b) is taken from Grisvard [**Grv**, Section 3.1].

Part 1

The Monge–Ampère Equation

Chapter 3

Aleksandrov Solutions and Maximum Principles

This chapter aims to answer very basic questions in the theory of the Monge–Ampère equation: *What are suitable concepts of solutions to a Monge–Ampère equation? Given a Monge–Ampère equation, does there exist a solution to it?*

We motivate various notions of convex solutions to the Monge–Ampère equation. We will introduce the important notion of Monge–Ampère measure of a convex function. With this notion, we can define Aleksandrov solutions to a Monge–Ampère equation without requiring them to be twice differentiable everywhere. Then we will prove the Aleksandrov maximum principle which is a basic estimate in the Monge–Ampère equation. When coupled with the weak continuity property of Monge–Ampère measure, it gives a compactness result for the Monge–Ampère equation which will turn out to be vital in later chapters. Various maximum principles will also be established including the Aleksandrov–Bakelman–Pucci maximum principle. We will prove the comparison principle and establish various optimal global Hölder estimates. Then we discuss the solvability of the inhomogeneous Dirichlet problem with continuous boundary data via Perron's method. A comparison principle for $W^{2,n}$ nonconvex functions is also provided.

3.1. Motivations and Heuristics

In linear, second-order PDEs in divergence form, we can define a notion of weak solutions having less than two derivatives using integration by parts. As an example, consider Poisson's equation on a domain Ω in $\mathbb{R}^n$:

$$(3.1) \qquad \Delta u = f \quad \text{in } \Omega, \qquad u = 0 \quad \text{on } \partial\Omega.$$

If u and v are smooth up to the boundary of Ω and $v = 0$ on $\partial\Omega$, then multiplying both sides of (3.1) by v and integrating by parts, we get

$$(3.2) \qquad -\int_\Omega Du \cdot Dv \, dx = \int_\Omega fv \, dx.$$

Identity (3.2) allows us to define a notion of weak solutions to (3.1) with u having only one derivative in a weak sense. Precisely, we require that $u \in W_0^{1,2}(\Omega)$, f belongs to the dual of $W_0^{1,2}(\Omega)$, and (3.2) holds for all test functions $v \in W_0^{1,2}(\Omega)$. The theory of linear, second-order PDEs in divergence form is intimately connected to Sobolev spaces.

Now, we would like to define a suitable concept of weak solutions to the Monge–Ampère equation

$$(3.3) \qquad \det D^2 u = f \geq 0 \quad \text{in } \Omega,$$

which is similar to that of (3.1), and we will try to use less than two derivatives for possible convex solutions $u : \Omega \to \mathbb{R}$. (We note that u is twice differentiable almost everywhere by Aleksandrov's Theorem (Theorem 2.89). However, using this almost everywhere pointwise information will not give us a good notion of solutions as it lacks uniqueness. For example, in this notion, the equation $\det D^2 u = 0$ on $B_1(0) \subset \mathbb{R}^n$ with zero boundary data has infinitely many solutions, including 0 and $a(|x| - 1)$ for $a > 0$.) Of course, when f is continuous, we can also speak of a *classical solution* to (3.3) where the convex function u is required to be C^2. Although the Monge–Ampère equation can be written in divergence form $D_i(U^{ij} D_j u) = nf$, we cannot mimic the integration by parts as in (3.2) because this would give

$$(3.4) \qquad -\int_\Omega U^{ij} D_j u D_i v \, dx = \int_\Omega nfv \, dx,$$

which still requires two derivatives for u to define the cofactor matrix $U = (U^{ij})_{1 \leq i,j \leq n}$ of the Hessian matrix $D^2 u$ with certain integrability, unless we are in one dimension. Thus, we need to proceed differently.

We first explain Aleksandrov's idea. Assume that in (3.3) we have $u \in C^2(\Omega)$ with positive definite Hessian $D^2 u > 0$ in Ω. Then for any nice set $E \subset \Omega$, we obtain from integrating both sides of (3.3) over E

$$\int_E f \, dx = \int_E \det D^2 u(x) \, dx = \int_{Du(E)} dy = |Du(E)|,$$

where the second equality follows from the change of variables $y = Du(x)$. Thus

$$(3.5) \qquad \int_E f \, dx = |Du(E)|.$$

Upon inspecting (3.5), we observe that:

(1) $\int_E f \, dx$ can be defined for any Borel subset $E \subset \Omega$ if $\mu = f \, dx$ is a nonnegative Borel measure.

(2) $|Du(E)|$ can be defined if $Du(E)$ is Lebesgue measurable for any Borel subset $E \subset \Omega$, which is the case if $u \in C^1(\Omega)$.

The problem with observation (2) is that if u is a convex function, then u is generally only Lipschitz but not C^1. Thus, we barely fail. One example of a convex but not C^1 function in $\mathbb{R}^n$ is $u(x) = |x|$. In this case, u is not C^1 at 0. Geometrically, there are infinitely many supporting hyperplanes to the graph of u at 0. The set of slopes of these hyperplanes is the closed unit ball $\overline{B_1(0)}$ in $\mathbb{R}^n$.

Aleksandrov's key insight is to replace $Du(E)$ by $\partial u(E) = \bigcup_{x_0 \in E} \partial u(x_0)$ where $\partial u(x_0)$ is the set of all slopes of supporting hyperplanes to the graph of u at $(x_0, u(x_0))$. Then define the Monge–Ampère measure μ_u of u to be

$$\mu_u(E) = |\partial u(E)| \quad \text{for each Borel set } E.$$

Now, u is called an *Aleksandrov solution* to (3.3) if $\mu_u = f \, dx$ in the sense of measures. Section 3.2 will carry out this insight in detail. This is based on the *measure-theoretic* method, with deep *geometric* insights.

We can also define a concept of solutions to the Monge–Ampère equation which is based on the *maximum principle*. Let $u \in C^2(\Omega)$ be a convex solution to (3.3) on a domain Ω in $\mathbb{R}^n$. Suppose $\phi \in C^2(\Omega)$ is convex, and let $x_0 \in \Omega$ be such that $u - \phi$ has a local minimum at x_0. Then, $D^2u(x_0) \geq D^2\phi(x_0)$ and hence from $f(x_0) = \det D^2u(x_0)$, we obtain

$$(3.6) \qquad \det D^2\phi(x_0) \leq f(x_0).$$

Similarly, if $\varphi \in C^2(\Omega)$ and $x_0 \in \Omega$ are such that $u - \varphi$ has a local maximum at x_0, then

$$(3.7) \qquad \det D^2\varphi(x_0) \geq f(x_0).$$

Inspecting the arguments leading to (3.6) and (3.7), we see that all we need concerning u and f is their continuity. Thus, we can define a concept of *viscosity solutions* to (3.3) based on (3.6) and (3.7). This will be carried out in Section 7.1.

Having defined notions of weak solutions to (3.3), we can now proceed to solve it, say on a bounded convex domain Ω in $\mathbb{R}^n$ with Dirichlet boundary condition $u = 0$ on $\partial\Omega$. As in the classical theory for solving (3.1), we can use Perron's method which is based on *the maximum principle and subsolutions*. For (3.3), this amounts to taking

$$u(x) := \sup\{v(x) : v \in C(\overline{\Omega}) \text{ is convex in } \Omega, v = 0 \text{ on } \partial\Omega, \text{ and } \mu_v \geq f\}.$$

We will discuss this in detail in Section 3.7.

For certain Monge–Ampère-type equations where the right-hand sides also depend on the unknown solution u, the maximum principle might not apply. In some cases, the *variational method* can be useful. To illustrate this idea, let us come back to Poisson's equation (3.1) with the identity (3.2) for its solution. The last equation serves as a starting point for solving (3.1) by looking for a minimizer, in $W_0^{1,2}(\Omega)$, of the functional

$$J_1[u] := \int_\Omega \left(\frac{1}{2}|Du|^2 + fu\right) dx.$$

If we try to mimic this procedure for (3.3), we end up with the analogue of (3.2), that is, (3.4). Thus, we might consider functionals involving the quantity $I_n[u] := \int_\Omega U^{ij} D_i u D_j u \, dx$ for a convex function u vanishing on $\partial\Omega$. As mentioned above, this functional requires two derivatives for u to define the cofactor matrix $(U^{ij})_{1 \leq i,j \leq n}$ of $D^2 u$. As a curious fact, in 1937, without mentioning functional spaces and required regularities for u, Courant and Hilbert [**CH1**, p. 278] (see also [**CH2**, p. 326]) suggested looking for minimizers of the following functional in two dimensions:

$$\int_\Omega [(D_1 u)^2 D_{22} u - 2D_1 u D_2 u D_{12} u + (D_2 u)^2 D_{22} u + 6fu] \, dx = I_n[u] + \int_\Omega 6fu \, dx.$$

In the original German edition [**CH1**], the term $6fu$ appeared as $-4fu$. A similar functional was considered by Gillis [**Gil**] in 1950 where $6fu$ was replaced by $6F(x, Du)$. It seems that we could not make much use of the variational method for the Monge–Ampère equation due to the serious difficulty in defining the quantity $I_n[u]$ for a general convex function u. However, there is a twist to this. Note that if u is smooth and vanishes on $\partial\Omega$, then, by integrating by parts and using that $(U^{ij})_{1 \leq i,j \leq n}$ is divergence-free, we find

$$-I_n[u] = \int_\Omega D_j(U^{ij} D_i u)u \, dx = \int_\Omega U^{ij} D_{ij} uu \, dx = \int_\Omega n(\det D^2 u)u \, dx.$$

The last term can be defined for a general convex function u where it is replaced by $\int_\Omega nu \, d\mu_u$. With this, we can consider the following analogue of J_1 in solving (3.3):

$$J_n[u] := \int_\Omega \frac{1}{n+1}(-u) \, d\mu_u + \int_\Omega fu \, dx.$$

Minimizing this functional over all convex functions u vanishing on $\partial\Omega$ gives the unique solution to (3.3) with zero Dirichlet boundary condition; see Bakelman [**Ba3**] in two dimensions, Aubin [**Au1**] (in radially symmetric settings), and Bakelman [**Ba4**] in higher dimensions. Since J_n is lower semicontinuous (as can be seen from Proposition 11.2), the existence of a minimizer $u_{\min}$ of J_n is always guaranteed. Showing that $u_{\min}$ is an Aleksandrov solution of (3.3) is, however, a very nontrivial task. The computation of the first variation of J_n at a critical point $u_{\min}$, which is only convex and has no other a priori smoothness properties, relies on deep geometric ideas. These include dual convex hypersurfaces and the theory of Minkowski mixed volumes (see Schneider [**Schn**]). These are beyond the scope of this book.

On the other hand, the variational analysis motivates Tso [**Ts**] to solve the Monge–Ampère-type equation

$$\det D^2 u = |u|^q \quad \text{in } \Omega, \quad u = 0 \quad \text{on } \partial\Omega,$$

where $q > 0$, using the functional

$$\int_\Omega \frac{1}{n+1}(-u)\,d\mu_u - \int_\Omega \frac{1}{q+1}|u|^{q+1}\,dx.$$

We will study this functional and its minimizers in a more classical setting in Chapter 11. This is possible due to the smoothness property of solutions, if any, to the above equation. In Chapter 14, we will again encounter the quadratic form $\int_\Omega U^{ij} D_i v D_j v\,dx$ when discussing the Monge–Ampère Sobolev inequality.

Since Aleksandrov's geometric method does not seem to have a complex Monge–Ampère equation counterpart where the natural class of solutions consists of plurisubharmonic functions, Bedford and Taylor [**BT**] introduced a powerful *analytic method*, based on the theory of currents (they are differential forms with distribution coefficients) to define the Monge–Ampère measure for a large class of plurisubharmonic functions in $\mathbb{C}^n$, including continuous plurisubharmonic functions. Their construction gives the following formula to define a Monge–Ampère measure $\mathcal{M}_u$ of a convex function u on $\mathbb{R}^n$:

$$\mathcal{M}_u = dD_1 u \wedge \cdots \wedge dD_n u.$$

We will not discuss this analytic method in this book. We just mention that, due to Rauch and Taylor [**RT**], $\mathcal{M}_u$ coincides with μ_u; see also Problem 1.6.

3.2. The Monge–Ampère Measure and Aleksandrov Solutions

A central notion in the theory of the Monge–Ampère equation is the Monge–Ampère measure. The following definition and its content are due to Aleksandrov [**Al1**] (see also Bakelman [**Ba1**] in two dimensions).

Definition 3.1 (The Monge–Ampère measure). Let Ω be an open set in $\mathbb{R}^n$. Let $u : \Omega \to \mathbb{R}$ be a convex function. Given $E \subset \Omega$, we define

$$\mu_u(E) = |\partial u(E)| \quad \text{where } \partial u(E) = \bigcup_{x \in E} \partial u(x).$$

Then $\mu_u : \mathcal{S} \to [0, \infty]$ is a measure, finite on compact subsets of Ω where $\mathcal{S}$ is a Borel σ-algebra defined by

$$\mathcal{S} = \{E \subset \Omega : \partial u(E) \text{ is Lebesgue measurable}\}.$$

We call μ_u the Monge–Ampère measure associated with u.

Justification of μ_u being a measure. For this, our main observation is the following fact, which is an easy consequence of Aleksandrov's Lemma (Lemma 2.45): If A and B are disjoint subsets of Ω, then $|\partial u(A) \cap \partial u(B)| = 0$, so $\partial u(A)$ and $\partial u(B)$ are also *disjoint in the measure-theoretic sense*.

Let $E \subset \Omega$ be compact. Then, by Lemma 2.40(i), $\partial u(E)$ is compact; hence, it is Lebesgue measurable. Thus $E \in \mathcal{S}$ and $\mu_u(E) = |\partial u(E)| < \infty$. Now, by writing Ω as a union of compact sets, we deduce that $\Omega \in \mathcal{S}$.

Next, we show that $\Omega \backslash E \in \mathcal{S}$ if $E \in \mathcal{S}$. Indeed, since $E, \Omega \in \mathcal{S}$, the set $\partial u(\Omega) \backslash \partial u(E)$ is Lebesgue measurable. From $|\partial u(\Omega \backslash E) \cap \partial u(E)| = 0$ and

$$\partial u(\Omega \backslash E) = (\partial u(\Omega) \backslash \partial u(E)) \cup (\partial u(\Omega \backslash E) \cap \partial u(E)),$$

we find that $\partial u(\Omega \backslash E)$ is Lebesgue measurable and hence $\Omega \backslash E \in \mathcal{S}$.

Finally, we show that μ_u is σ-additive. Let $\{E_i\}_{i=1}^{\infty}$ be a sequence of disjoint sets in $\mathcal{S}$. We need to show that $\left|\partial u\left(\bigcup_{i=1}^{\infty} E_i\right)\right| = \sum_{i=1}^{\infty} |\partial u(E_i)|$. Indeed, this easily follows from the identities

$$\partial u\left(\bigcup_{i=1}^{\infty} E_i\right) = \bigcup_{i=1}^{\infty} \partial u(E_i) = \partial u(E_1) \cup \bigcup_{i=2}^{\infty}\left(\partial u(E_i) \backslash \bigcup_{k=1}^{i-1} \partial u(E_k)\right)$$

and the fact that $\{\partial u(E_i)\}_{i=1}^{\infty}$ are disjoint in the measure-theoretic sense. $\square$

Example 3.2 (The Monge–Ampère measure of a cone). Let $\Omega = B_R(x_0)$ and $u(x) = r|x - x_0|$ for $x \in \Omega$ where $r > 0$. By Example 2.35, we have

$$\partial u(x) = \begin{cases} \{r(x - x_0)/|x - x_0|\} & \text{if } 0 < |x - x_0| < R, \\ \overline{B_r(0)} & \text{if } x = x_0. \end{cases}$$

Therefore, $\mu_u = |B_r(0)|\delta_{x_0}$ where δ_{x_0} is the Dirac measure at $x_0 \in \mathbb{R}^n$.

Remark 3.3. Functions with cone-like graphs in Example 3.2 play the role of fundamental solutions to the Monge–Ampère equation; see Section 3.7 on the solvability of the Dirichlet problem. They are consistent with the notion of the fundamental solution of Laplace's equation when $n = 1$ as $\det D^2 u = u''$ in one dimension. Recall that the function

$$\Phi_{(n)}(x) := \begin{cases} -\dfrac{1}{2\pi} \log |x| & \text{if } n = 2, \\[2ex] \dfrac{1}{n(n-2)\omega_n} \dfrac{1}{|x|^{n-2}} & \text{if } n \geq 3 \end{cases}$$

defined for $x \in \mathbb{R}^n \setminus \{0\}$ is the fundamental solution of Laplace's equation as we can write $-\Delta \Phi_{(n)} = \delta_0$ in $\mathbb{R}^n$, where δ_0 is the Dirac measure at 0. When $n = 1$, $\Phi_{(1)}(x) = -|x|/2$ with $[\Phi_{(1)}(x)]'' = -\delta_0$ becomes the fundamental solution of the one-dimensional Laplace equation. If $n = 1$ and $u(x) = |x|/2$, then we obtain from Example 3.2 that $\mu_u = \delta_0$. For an interesting application of the concavity of the fundamental solution of the one-dimensional Laplace equation, see Problem 3.7.

Example 3.4. Let Ω be an open set in $\mathbb{R}^n$. If $u \in C^{1,1}(\Omega)$ is convex, then $\mu_u = (\det D^2 u)\, dx$ in Ω.

Proof. The result follows from the classical change of variables formula in calculus if u is C^2 with positive definite Hessian $D^2 u$. In our situation with less smoothness, we will use the change of variables formula in Theorem 2.74. Since $u \in C^{1,1}(\Omega)$ is convex, we have $\partial u(x) = \{Du(x)\}$ for all $x \in \Omega$. Let $E \subset \Omega$ be a Borel set. Then, by Theorem 2.74, we have

$$\mu_u(E) = |\partial u(E)| = |Du(E)| = \int_E \det D^2 u\, dx.$$

This shows that $\mu_u = (\det D^2 u)\, dx$. $\qquad\square$

Example 3.5. Let $u \in C(\overline{\Omega})$ be a convex function on a bounded domain Ω in $\mathbb{R}^n$. Then for any subdomain $\Omega' \Subset \Omega$, Lemma 2.37 gives $\partial u(\Omega') \subset \overline{B_r(0)}$ where $r = [\operatorname{dist}(\overline{\Omega'}, \partial\Omega)]^{-1} \operatorname{osc}_\Omega u$, and hence

$$\mu_u(\Omega') \leq \omega_n [\operatorname{dist}(\overline{\Omega'}, \partial\Omega)]^{-n} (\operatorname*{osc}_\Omega u)^n.$$

Remark 3.6 (Approximation by compact sets). Let $u : \Omega \to \mathbb{R}$ be a convex function on an open set Ω in $\mathbb{R}^n$. From the definition of μ_u, we find that it has the approximation by compact sets property. That is, for each open subset $\mathcal{O} \Subset \Omega$, we have

$$\mu_u(\mathcal{O}) = \sup_{K \subset \mathcal{O},\ K \text{ is compact}} \mu_u(K).$$

As such, μ_u is a Radon measure, that is, a measure which is Borel regular and finite on compact sets.

A very basic fact of the Monge–Ampère measure is its weak continuity property:

Theorem 3.7 (Weak continuity of Monge–Ampère measure). *Let $\{u_k\}_{k=1}^{\infty}$ be a sequence of convex functions on an open set Ω in $\mathbb{R}^n$ which converges to a convex function $u : \Omega \to \mathbb{R}$ uniformly on compact subsets of Ω. Then μ_{u_k} converges weakly* to μ_u and we write $\mu_{u_k} \overset{*}{\rightharpoonup} \mu_u$; that is,*

$$\lim_{k \to \infty} \int_{\Omega} f \, d\mu_{u_k} = \int_{\Omega} f \, d\mu_u \quad \text{for all } f \in C_c(\Omega).$$

Proof. Due to Remark 3.6 and Theorem 2.68, we only need to verify the following assertions:

(1) If $K \subset \Omega$ is a compact set, then $\limsup_{k \to \infty} \mu_{u_k}(K) \le \mu_u(K)$.

(2) If $\mathcal{O} \Subset \Omega$ is an open set, then $\mu_u(\mathcal{O}) \le \liminf_{k \to \infty} \mu_{u_k}(\mathcal{O})$.

From Remark 3.6, instead of proving assertion (2), we only need to prove:

(2$'$) If K is a compact set and $\mathcal{O}$ is open such that $K \subset \mathcal{O} \Subset \Omega$, then

$$\mu_u(K) \le \liminf_{k \to \infty} \mu_{u_k}(\mathcal{O}).$$

The proof of assertions (1) and (2$'$) uses Aleksandrov's Lemma (Lemma 2.45) together with the Dominated Convergence Theorem and the following inclusions:

(3) $\limsup_{k \to \infty} \partial u_k(K) := \bigcap_{i=1}^{\infty} \bigcup_{k=i}^{\infty} \partial u_k(K) \subset \partial u(K)$.

(4) $\partial u(K) \backslash \mathfrak{S} \subset \liminf_{k \to \infty} \partial u_k(\mathcal{O}) := \bigcup_{i=1}^{\infty} \bigcap_{k=i}^{\infty} \partial u_k(\mathcal{O})$, where

$\mathfrak{S} = \{p \in \mathbb{R}^n : \text{there are } x \ne y \in \Omega \text{ such that } p \in \partial u(x) \cap \partial u(y)\}$.

For completeness, we indicate, for instance, how assertion (3) and the Dominated Convergence Theorem give assertion (1). By definition, we have

$$\limsup_{k \to \infty} \mu_{u_k}(K) = \limsup_{k \to \infty} |\partial u_k(K)| := \lim_{i \to \infty} \sup_{k \ge i} |\partial u_k(K)|$$

$$\le \lim_{i \to \infty} \Big| \bigcup_{k=i}^{\infty} \partial u_k(K) \Big|.$$

By assertion (3), we have

$$\Big| \bigcap_{i=1}^{\infty} \bigcup_{k=i}^{\infty} \partial u_k(K) \Big| \le |\partial u(K)| = \mu_u(K).$$

Now, we will use the Dominated Convergence Theorem to show that

$$\lim_{i \to \infty} \Big| \bigcup_{k=i}^{\infty} \partial u_k(K) \Big| = \Big| \bigcap_{i=1}^{\infty} \bigcup_{k=i}^{\infty} \partial u_k(K) \Big|.$$

Indeed, let $f_i(p) = \chi_{\bigcup_{k=i}^{\infty} \partial u_k(K)}(p)$ for $p \in \mathbb{R}^n$. Then

$$\lim_{i \to \infty} f_i(p) = \chi_{\bigcap_{i=1}^{\infty} \bigcup_{k=i}^{\infty} \partial u_k(K)}(p).$$

So, it suffices to show that each f_i is bounded from above by an integrable function g on $\mathbb{R}^n$. To do this, let $K' \Subset \Omega$ be a compact set containing K such that $r := \mathrm{dist}(K, \partial K') > 0$. By the uniform convergence of u_k to u on K', there is a constant $M > 0$ such that $\|u_k\|_{L^\infty(K')} \le M$ for all k. Now, if $p \in \partial u_k(x)$ for some $x \in K$, then by Lemma 2.37, we have $|p| \le [\mathrm{dist}(x, \partial K')]^{-1} \mathrm{osc}_{K'} u_k \le 2Mr^{-1} := r'$. It follows that $f_i \le g := \chi_{B_{r'}(0)}$ and this g is the desired integrable function on $\mathbb{R}^n$.

To prove our theorem, it remains to prove assertions (3) and (4).

For assertion (3), let $p \in \limsup_{k \to \infty} \partial u_k(K)$. Then for each i, there are k_i and $z^{k_i} \in K$ such that $p \in \partial u_{k_i}(z^{k_i})$. Since K is compact, extracting a subsequence of $\{z^{k_i}\}_{i=1}^{\infty}$, still labeled $\{z^{k_i}\}_{i=1}^{\infty}$, we have $z^{k_i} \to z \in K$. Thus, using $u_{k_i}(x) \ge u_{k_i}(z^{k_i}) + p \cdot (x - z^{k_i})$ for all $x \in \Omega$ and the uniform convergence of u_{k_i} to u on compact subsets of Ω, we obtain

$$u(x) \ge u(z) + p \cdot (x - z) \quad \text{for all } x \in \Omega$$

and therefore, $p \in \partial u(z) \subset \partial u(K)$.

For assertion (4), let $p \in \partial u(x_0) \subset \partial u(K) \backslash \mathfrak{S}$. Then, $u(x) - l(x)(\ge 0)$ is strictly convex at x_0, where $l(x) = u(x_0) + p \cdot (x - x_0)$; see Definition 2.39. By subtracting $l(x)$ from u_k and u, we can assume $p = 0$, $u(x_0) = 0$, and we need to show that $0 \in \partial u_k(z^k)$ for all k large and some $z^k \in \mathcal{O}$. Recalling Remark 2.34, we prove this by choosing a minimum point z^k of the continuous function u_k in the compact set $\bar{\mathcal{O}}$. It remains to show that $z^k \notin \partial \mathcal{O}$ when k is large. Indeed, from the strict convexity of u at $x_0 \in K \subset \mathcal{O}$, we can find some $\gamma > 0$ such that $u(x) \ge \gamma$ on $\partial \mathcal{O}$. Hence, from the uniform convergence of u_k to u on compact sets, we find that $u_k \ge \gamma/2$ on $\partial \mathcal{O}$ if k is large. On the other hand, since $u(x_0) = 0$, we also find that $u_k(x_0) \le \gamma/4$ when k is large. Therefore, $z^k \notin \partial \mathcal{O}$ when k is large. $\square$

Definition 3.8 (Aleksandrov solutions). Given an open set $\Omega \subset \mathbb{R}^n$ and a Borel measure ν on Ω, a convex function $u : \Omega \to \mathbb{R}$ is called an *Aleksandrov solution* to the Monge–Ampère equation

$$\det D^2 u = \nu$$

if $\mu_u = \nu$ as Borel measures. When $\nu = f\, dx$, we will simply say that u solves

$$\det D^2 u = f,$$

and this is the notation we use in the book.

Similarly, when writing $\det D^2 u \ge \lambda$ ($\le \Lambda$) in the sense of Aleksandrov, we mean that $\mu_u \ge \lambda\, dx$ ($\le \Lambda\, dx$).

Notes. For possibly nonconvex $v \in W^{2,n}_{\mathrm{loc}}(\Omega)$ $(C^{1,1}(\Omega))$, $\det D^2 v$ is interpreted in the usual pointwise sense, and it belongs to $L^1_{\mathrm{loc}}(\Omega)$ $(L^\infty_{\mathrm{loc}}(\Omega))$.

The following lemma is useful in studying the Monge–Ampère equation:

Lemma 3.9 (Monge–Ampère measure under affine transformations). *Let $u : \Omega \to \mathbb{R}$ be a convex function on an open set Ω in $\mathbb{R}^n$. Let $A \in \mathbb{M}^{n \times n}$ be invertible, $\mathbf{b}, \mathbf{c} \in \mathbb{R}^n$, $d \in \mathbb{R}$, and $\gamma > 0$. Denote $Tx = Ax + \mathbf{b}$. Let*

$$v(x) = \gamma u(Tx) + \mathbf{c} \cdot x + d \quad \text{for } x \in T^{-1}\Omega.$$

(i) *If u is $C^2(\Omega)$, then*

$$\det D^2 v(x) = \gamma^n (\det A)^2 \det D^2 u(Tx).$$

(ii) *If, in the sense of Aleksandrov,*

$$\lambda \leq \det D^2 u \leq \Lambda \quad \text{in } \Omega,$$

then we also have, in the sense of Aleksandrov,

$$\lambda \gamma^n |\det A|^2 \leq \det D^2 v \leq \Lambda \gamma^n |\det A|^2 \quad \text{in } T^{-1}\Omega.$$

Proof. If $u \in C^2(\Omega)$, then

$$Dv(x) = \gamma A^t Du(Tx) + \mathbf{c} \quad \text{and} \quad D^2 v(x) = \gamma A^t (D^2 u(Tx)) A.$$

Taking determinants proves assertion (i).

We now prove (ii) using the normal mapping. From Definition 2.30, we infer that

(3.8) $\partial v(x) = \gamma A^t \partial u(Tx) + \mathbf{c} \quad \text{for any } x \in T^{-1}(\Omega).$

Now, let $E \subset T^{-1}\Omega$ be a Borel set. Then, by (3.8) and $\det A^t = \det A$,

$$\mu_v(E) = |\partial v(E)| = |\gamma A^t \partial u(T(E)) + \mathbf{c}| = \gamma^n |\det A| |\mu_u(T(E))|.$$

Because $\lambda \leq \det D^2 u \leq \Lambda$ in the sense of Aleksandrov, we have

$$\lambda |\det A| |E| = \lambda |T(E)| \leq |\mu_u(T(E))| \leq \Lambda |T(E)| = \Lambda |\det A| |E|.$$

The assertion (ii) then follows from

$$\lambda \gamma^n |\det A|^2 |E| \leq \mu_v(E) \leq \Lambda \gamma^n |\det A|^2 |E|.$$

The lemma is proved. $\qquad \square$

We now show that the Monge–Ampère measure is superadditive. It is a consequence of the concavity of $(\det D^2 u)^{1/n}$ on the set of C^2 convex functions.

Lemma 3.10 (Multiplicative and superadditive properties of the Monge–Ampère measure). *Let $u, v : \Omega \to \mathbb{R}$ be convex functions on an open set Ω in $\mathbb{R}^n$. Then*

$$\mu_{\lambda u} = \lambda^n \mu_u \quad \text{for } \lambda > 0 \quad \text{and} \quad \mu_{u+v} \geq \mu_u + \mu_v.$$

Proof. The multiplicative property is a consequence of Definition 3.1 and the fact that $\partial(\lambda u)(E) = \lambda \partial u(E)$ for all Borel sets $E \subset \Omega$.

Now we prove the superadditive property. We will reduce the proof to the case where u and v are C^2 because in this case, by Lemma 2.59,

$$\det D^2(u + v) \geq ((\det D^2 u)^{1/n} + (\det D^2 v)^{1/n})^n \geq \det D^2 u + \det D^2 v.$$

In general, u and v may not be C^2. We will use an approximation argument together with Theorem 3.7. Extend u and v to convex functions on $\mathbb{R}^n$. Let $\varphi_{1/k}$ be a standard mollifier with support in $\overline{B_{1/k}(0)}$ and $\int_{\mathbb{R}^n} \varphi_{1/k} \, dx = 1$; see Definition 2.49. Consider $u_k := u * \varphi_{1/k}$ and $v_k = v * \varphi_{1/k}$. Then, by Theorem 2.50, $u_k, v_k \in C^\infty(\Omega)$ are convex functions and hence,

$$\mu_{u_k + v_k} \geq \mu_{u_k} + \mu_{v_k}.$$

Since $u_k \to u$ and $v_k \to v$ uniformly on compact subsets of Ω, by Theorem 3.7, we have

$$\mu_{u_k + v_k} \overset{*}{\rightharpoonup} \mu_{u+v}, \quad \mu_{u_k} \overset{*}{\rightharpoonup} \mu_u, \quad \mu_{v_k} \overset{*}{\rightharpoonup} \mu_v.$$

Therefore, we must have $\mu_{u+v} \geq \mu_u + \mu_v$, as asserted. $\qquad\qquad\square$

3.3. Maximum Principles

In the analysis of Monge–Ampère equations, maximum principles are indispensable tools. In this section, we discuss the most basic ones involving convex functions and also nonconvex functions.

The following basic maximum principle implies that if two convex functions defined on the same domain and having the same boundary values, the one below the other will have larger image of the normal mapping.

Lemma 3.11. *Let Ω be a bounded open set in $\mathbb{R}^n$, and let $u, v \in C(\overline{\Omega})$.*

    (i) *(Pointwise maximum principle) If $u \geq v$ on $\partial\Omega$ and $v(x_0) \geq u(x_0)$ where $x_0 \in \Omega$, then $\partial v(x_0) \subset \partial u(\Omega)$.*

    (ii) *(Maximum principle) If $u = v$ on $\partial\Omega$ and $u \leq v$ in Ω, then $\partial v(\Omega) \subset \partial u(\Omega)$.*

Proof. Since assertion (ii) is a consequence of assertion (i), we only need to prove (i). Let $p \in \partial v(x_0)$. Then, p is the slope of a supporting hyperplane $l_0(x)$ to the graph of v at $(x_0, v(x_0))$; that is,

$$v(x) \geq l_0(x) := v(x_0) + p \cdot (x - x_0) \quad \text{for all } x \in \overline{\Omega}.$$

We will slide down l_0 by the amount

$$a := \max_{\overline{\Omega}}(l_0 - u) \geq l_0(x_0) - u(x_0) = v(x_0) - u(x_0) \geq 0$$

to obtain a supporting hyperplane $l(x) := l_0(x) - a$ to the graph of u at some point $(z, u(z))$ where $z \in \Omega$, and hence $p \in \partial u(\Omega)$. See Figure 3.1. To

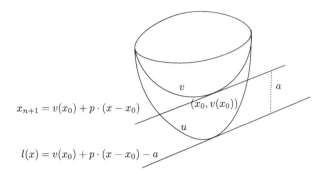

Figure 3.1. Sliding down the hyperplane $v(x_0) + p \cdot (x - x_0)$ to obtain a supporting hyperplane for the graph of u.

see this, we first note that $a \geq l_0(x) - u(x)$ for any $x \in \Omega$, so

$$u(x) \geq l_0(x) - a = l(x) \quad \text{in } \Omega.$$

It suffices to find $z \in \Omega$ such that $u(z) = l(z)$. By our assumptions,

$$\sup_{\partial\Omega}(l_0 - u) \leq \sup_{\partial\Omega}(v - u) \leq 0.$$

If $a = 0$, then $v(x_0) = u(x_0)$, and we can take $z = x_0$. Now, consider $a > 0$. Let $\bar{x} \in \overline{\Omega}$ be such that $l_0(\bar{x}) - u(\bar{x}) = a > 0$. Then, $\bar{x} \notin \partial\Omega$, and $u(\bar{x}) = l_0(\bar{x}) - a = l(\bar{x})$, so we can take $z = \bar{x} \in \Omega$. $\qquad\square$

Observe that, for the maximum principles in Lemma 3.11, no convexity on the functions nor the domain is assumed.

The following maximum principle, due to Aleksandrov [**Al4**], is of fundamental importance in the theory of Monge–Ampère equations. It quantifies how much a convex function, with finite total Monge–Ampère measure, will drop its values when stepping inside the domain.

Theorem 3.12 (Aleksandrov's maximum principle). *Let Ω be a bounded and convex domain in $\mathbb{R}^n$. Let $u \in C(\overline{\Omega})$ be a convex function with $u = 0$ on $\partial\Omega$. Consider the diameter-like function $D(x) := \max_{y \in \partial\Omega}|x - y|$. Then for all $x_0 \in \Omega$, we have*

$$|u(x_0)|^n \leq \min\left\{n\omega_{n-1}^{-1}D^{n-1}(x_0)\operatorname{dist}(x_0, \partial\Omega)|\partial u(\Omega)|, \ \omega_n^{-1}D^n(x_0)|\partial u(\Omega)|\right\}.$$

Proof. Let $v \in C(\overline{\Omega})$ be the convex function whose graph is the cone with vertex $(x_0, u(x_0))$ and the base Ω, with $v = 0$ on $\partial\Omega$. See Figure 3.2. Explicitly, we have $v(tx_0 + (1 - t)x) = tu(x_0)$ for all $x \in \partial\Omega$ and $t \in [0, 1]$.

Since u is convex, $0 \geq v \geq u$ in Ω. By the maximum principle in Lemma 3.11, $\partial v(\Omega) \subset \partial u(\Omega)$. Our proof is now based on the following observations:

(1) $\partial v(\Omega) = \partial v(x_0)$, and thus $\partial v(\Omega)$ is convex.

(2) $\partial v(\Omega)$ contains $B_{\frac{|u(x_0)|}{D(x_0)}}(0)$.

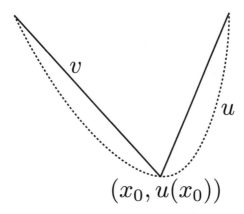

Figure 3.2. Graphs of u and v. The graph of v is the cone with vertex $(x_0, u(x_0))$ and $v = 0$ on $\partial\Omega$.

(3) Take $z \in \partial\Omega$ such that $|z - x_0| = \operatorname{dist}(x_0, \partial\Omega)$. Then

$$p_0 = -u(x_0)\frac{z - x_0}{|z - x_0|^2} \in \partial v(\Omega).$$

Assuming assertions (1)–(3), we see that $\partial v(\Omega)$ contains the convex hull of p_0 and $B_{\frac{|u(x_0)|}{D(x_0)}}(0)$. This convex hull has measure at least

$$\max\left\{\omega_n \left(\frac{|u(x_0)|}{D(x_0)}\right)^n, \frac{\omega_{n-1}}{n}\left(\frac{|u(x_0)|}{D(x_0)}\right)^{n-1}|p_0|\right\}.$$

Since $|\partial v(\Omega)| \le |\partial u(\Omega)|$ and $|p_0| = |u(x_0)|/\operatorname{dist}(x_0, \partial\Omega)$, the theorem follows.

Let us now verify assertions (1)–(3). To see assertion (1), we note that if $p \in \partial v(\Omega)$, then $p = \partial v(z)$ for some $z \in \Omega$. It suffices to consider the case $z \ne x_0$. Since the graph of v is a cone, $v(x) + p \cdot (x - z)$ is a supporting hyperplane to the graph of v at $(x_0, v(x_0))$; that is, $p \in \partial v(x_0)$.

For assertions (2) and (3), we use the fact from v being a cone that $p \in \partial v(x_0)$ if and only if $v(x) \ge v(x_0) + p \cdot (x - x_0)$ for all $x \in \partial\Omega$. Thus, assertion (2) is clearly valid.

To prove assertion (3), we note that for any $x \in \partial\Omega$, $(x - x_0) \cdot \frac{z - x_0}{|z - x_0|}$ is the vector projection of $x - x_0$ onto the ray $\overrightarrow{x_0 z}$. Since Ω is convex,

$$(x - x_0) \cdot \frac{z - x_0}{|z - x_0|} \le |z - x_0|,$$

and hence, from the formula for p_0, we find that for all $x \in \partial\Omega$,

$$0 = v(x) = u(x_0) + |p_0||z - x_0| \ge v(x_0) + p_0 \cdot (x - x_0).$$

Therefore $p_0 \in \partial v(x_0)$ as claimed. The theorem is proved. $\qquad\square$

Remark 3.13. Theorem 3.12 is sharp. If $\Omega = B_R(x_0)$ and $u(x) = |x - x_0| - R$ (see Example 3.2), then the inequality in Theorem 3.12 becomes equality.

A simple consequence of Aleksandrov's maximum principle is:

Corollary 3.14. Let Ω be a bounded and convex domain in $\mathbb{R}^n$, and let $u \in C(\overline{\Omega})$ be a convex function. Then, for all $x \in \Omega$, we have

$$\inf_{\partial\Omega} u - C(n)[\mathrm{dist}(x, \partial\Omega)]^{1/n}[\mathrm{diam}(\Omega)]^{\frac{n-1}{n}}|\partial u(\Omega)|^{1/n} \le u(x) \le \sup_{\partial\Omega} u.$$

Proof. Note that the inequality $u(x) \le \sup_{\partial\Omega} u$ just follows from the convexity of u. For the lower bound on u, let $v(x) = u(x) - \inf_{\partial\Omega} u$. Then $v \ge 0$ on $\partial\Omega$ and $\partial v(\Omega) = \partial u(\Omega)$. If $v(x) \ge 0$ for all $x \in \Omega$, then we are done. If this is not the case, then $E = \{x \in \Omega : v(x) < 0\}$ is a convex domain, with $v = 0$ on ∂E. We apply Theorem 3.12 to conclude that for each $x \in E$,

$$(-v(x))^n = |v(x)|^n \le C(n)[\mathrm{diam}(E)]^{n-1}\mathrm{dist}(x, \partial E)|\partial v(E)|$$
$$\le C(n)[\mathrm{diam}(\Omega)]^{n-1}\mathrm{dist}(x, \partial\Omega)|\partial u(\Omega)|, \quad C(n) = n\omega_{n-1}^{-1}.$$

The corollary follows. □

The conclusion of Theorem 3.12 raises the following question: Will a convex function drop its value when stepping inside the domain? Clearly, without a lower bound on the Monge–Ampère measure, the answer is in the negative as can be seen from the constant 0. However, we will provide a positive answer in Lemma 3.24 when the density of the Monge–Ampère measure μ_u has a positive lower bound. By convexity, it suffices to obtain a positive lower bound for $\|u\|_{L^\infty(\Omega)}$ (see Lemma 2.37(i)).

For functions that are not necessarily convex, we will establish in Theorem 3.16 the Aleksandrov–Bakelman–Pucci (ABP) maximum principle which is of fundamental importance in fully nonlinear elliptic equations. Its proof is based on following Aleksandrov-type estimates whose formulation involves the notion of the *upper contact set* Γ^+ of a continuous function u defined on a domain Ω in $\mathbb{R}^n$. It is defined as follows (see Figure 3.3):

$$\Gamma^+ = \{y \in \Omega : u(x) \le u(y) + p \cdot (x - y) \text{ for all } x \in \Omega, \text{ for some } p = p(y) \in \mathbb{R}^n\}.$$

Lemma 3.15 (Aleksandrov-type estimates). *Let Ω be a bounded domain in $\mathbb{R}^n$. Then, for $u \in C^2(\Omega) \cap C(\overline{\Omega})$, we have*

$$(3.9) \qquad \sup_\Omega u \le \sup_{\partial\Omega} u + \omega_n^{-1/n} \, \mathrm{diam}(\Omega)\left(\int_{\Gamma^+} |\det D^2 u| \, dx\right)^{1/n},$$

where Γ^+ is the upper contact set of u in Ω.

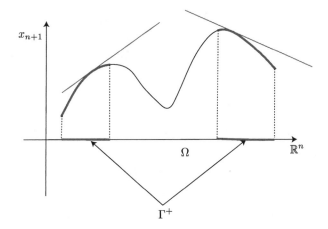

Figure 3.3. The upper contact set Γ^+.

Proof. By considering $u - \sup_{\partial\Omega} u$ instead of u, we can assume $\sup_{\partial\Omega} u = 0$. It suffices to prove (3.9) when the maximum of u on $\overline{\Omega}$ is attained at $x_0 \in \Omega$ with $u(x_0) > 0$. Our main observation is that

$$(3.10) \qquad\qquad B_{u(x_0)/d} \subset Du(\Gamma^+), \quad d = \operatorname{diam}(\Omega).$$

To see this, let $p \in B_{u(x_0)/d}$. Then, the function $v(x) := u(x) - p \cdot (x - x_0)$ satisfies $v(x_0) = u(x_0)$, while on $\partial\Omega$, $v(x) \le |p||x - x_0| < u(x_0)$. Thus, v attains its maximum value on $\overline{\Omega}$ at some point $y \in \Omega$. Therefore, $Dv(y) = 0$ which gives $p = Du(y)$. From $v(x) \le v(y)$ for all $x \in \Omega$, we deduce that $y \in \Gamma^+$. Hence, $p \in Du(T^+)$. The arbitrariness of p establishes (3.10).

Now, taking volumes in (3.10) and then invoking Theorem 2.74, we get

$$\omega_n \left(\frac{u(x_0)}{d}\right)^n \le |Du(\Gamma^+)| \le \int_{\Gamma^+} |\det D^2 u| \, dx.$$

This implies (3.9) and the proof of the lemma is complete. $\qquad\square$

Inclusions of the type (3.10) are the basis for the ABP method in proving geometric inequalities. See Brendle [**Bre**], Cabré [**Cab2**], and Trudinger [**Tr2**]. We refer to Problem 3.6 and Theorem 4.16 for illustrations of this method.

Now, we can establish the Aleksandrov–Bakelman–Pucci maximum principle for elliptic second-order operators with no lower-order terms. *Only the determinant of the coefficient matrix, but not the bounds on its eigenvalues, enters into the ABP estimate.* This is very crucial in many applications.

Theorem 3.16 (Aleksandrov–Bakelman–Pucci maximum principle). *Let Ω be a bounded domain in $\mathbb{R}^n$. Let the coefficient matrix $A(x) = (a^{ij}(x))_{1 \le i,j \le n}$ be measurable and positive definite in Ω. Let $Lu = a^{ij}D_{ij}u$ and $(Lu)^- = \max\{0, -Lu\}$. Then, the following statements hold:*

(i) *For $u \in C^2(\Omega) \cap C(\overline{\Omega})$, we have*

$$\sup_\Omega u \le \sup_{\partial\Omega} u + \frac{\operatorname{diam}(\Omega)}{n\omega_n^{1/n}} \left\| \frac{(Lu)^-}{(\det A)^{1/n}} \right\|_{L^n(\Gamma^+)},$$

where Γ^+ is the upper contact set of u in Ω.

(ii) *For $u \in W^{2,n}_{\text{loc}}(\Omega) \cap C(\overline{\Omega})$, we have*

$$(3.11) \qquad \sup_\Omega u \le \sup_{\partial\Omega} u + \frac{\operatorname{diam}(\Omega)}{n\omega_n^{1/n}} \left\| \frac{(Lu)^-}{(\det A)^{1/n}} \right\|_{L^n(\Omega)}.$$

Proof. We prove part (i). Since $u \in C^2(\Omega)$, $D^2 u \le 0$ on the upper contact set Γ^+. Using Lemma 2.57 for $A = (a^{ij})$ and $B = -D^2 u$, we have on Γ^+,

$$|\det D^2 u| = \det(-D^2 u) \le \frac{1}{\det A}\left(\frac{-a^{ij}D_{ij}u}{n}\right)^n \le \frac{1}{\det A}\left(\frac{(Lu)^-}{n}\right)^n.$$

Integrating over Γ^+ and then applying Lemma 3.15, we obtain part (i).

Next, we prove part (ii). If $u \in C^2(\Omega) \cap C(\overline{\Omega})$, then (3.11) follows from part (i). Now, we prove (3.11) for $u \in W^{2,n}_{\text{loc}}(\Omega) \cap C(\overline{\Omega})$ by approximation arguments. Let $\Lambda(x)$ be the largest eigenvalue of $A(x)$. For $\delta > 0$, let $A_\delta = \delta \Lambda I_n + A$ and $L_\delta u = \operatorname{trace}(A_\delta D^2 u)$.

Let $\{u_m\}_{m=1}^\infty \subset C^2(\Omega)$ be a sequence of functions converging to u in $W^{2,n}_{\text{loc}}(\Omega)$; see Theorem 2.50. For any $\varepsilon > 0$, we can assume that u_m converges to u in $W^{2,n}(\Omega_\varepsilon)$ and $u_m \le \varepsilon + \max_{\partial\Omega} u$ on $\partial\Omega_\varepsilon$ for some domain $\Omega_\varepsilon \Subset \Omega$ where Ω_ε is chosen so that it converges to Ω in the Hausdorff distance as $\varepsilon \to 0$. Let $C_\varepsilon = n^{-1}\omega_n^{-1/n}\operatorname{diam}(\Omega_\varepsilon)$. We can now apply (3.11) to u_m with

$$L_\delta u_m = L_\delta(u_m - u) + L_\delta u$$

in Ω_ε. Observing that $\|A_\delta/[\det A_\delta]^{1/n}\| \le n(\delta+1)/\delta$ in Ω, we then get

$$(3.12) \qquad \begin{aligned} \max_{\overline{\Omega_\varepsilon}} u_m &\le \varepsilon + \max_{\partial\Omega} u + C_\varepsilon n\frac{\delta+1}{\delta}\|D^2(u_m - u)\|_{L^n(\Omega_\varepsilon)} \\ &\quad + C_\varepsilon \left\| \frac{(L_\delta u)^-}{(\det A_\delta)^{1/n}} \right\|_{L^n(\Omega_\varepsilon)}. \end{aligned}$$

Letting $m \to \infty$ in (3.12) and recalling that $\|D^2(u_m - u)\|_{L^n(\Omega_\varepsilon)}$ converges to 0 and $\{u_m\}$ converges uniformly to u on Ω_ε, we have

$$\max_{\overline{\Omega_\varepsilon}} u \le \varepsilon + \max_{\partial\Omega} u + C_\varepsilon \left\| \frac{\delta\Lambda\Delta u}{(\det A_\delta)^{1/n}} \right\|_{L^n(\Omega_\varepsilon)} + C_\varepsilon \left\| \frac{(Lu)^-}{(\det A_\delta)^{1/n}} \right\|_{L^n(\Omega_\varepsilon)}.$$

Letting $\delta \to 0$ and using the Dominated Convergence Theorem, we obtain

$$\max_{\overline{\Omega}_\varepsilon} u \le \varepsilon + \max_{\partial\Omega} u + \frac{\text{diam}(\Omega_\varepsilon)}{n\omega_n^{1/n}} \left\| \frac{(Lu)^-}{(\det A)^{1/n}} \right\|_{L^n(\Omega_\varepsilon)}.$$

Letting $\varepsilon \to 0$, we obtain (3.11). The theorem is proved. $\qquad\qquad\square$

As an application of John's Lemma, we can refine the ABP maximum principle for convex domains where the diameter in Theorem 3.16 is now replaced by the nth root of the volume of the domain.

Lemma 3.17 (ABP estimate for convex domains). *Let Ω be a bounded, convex domain in $\mathbb{R}^n$. Let the $n \times n$ coefficient matrix $A(x)$ be measurable and positive definite in Ω. Then, for all $u \in W^{2,n}_{\text{loc}}(\Omega) \cap C(\overline{\Omega})$, we have*

$$\max_{\overline{\Omega}} u \le \max_{\partial\Omega} u + C(n)|\Omega|^{1/n} \left\| \frac{\text{trace}(AD^2 u)}{(\det A)^{1/n}} \right\|_{L^n(\Omega)}.$$

Proof. By Lemmas 2.62 and 2.63, there is an affine transformation $T(x) = Mx + b$, where $M \in \mathbb{S}^n$ is symmetric, positive definite and $b \in \mathbb{R}^n$ such that

$$(3.13) \qquad\qquad B_1(0) \subset T(\Omega) \subset B_n(0).$$

Let $Lu = \text{trace}(AD^2 u)$. For $x \in T(\Omega)$, we define

$$v(x) = u(T^{-1}x), \quad \tilde{A}(x) = MA(T^{-1}x)M^t, \quad \text{and} \quad \tilde{L}v = \text{trace}(\tilde{A}D^2 v).$$

Then, $D^2 v(x) = (M^{-1})^t D^2 u(T^{-1}x)M^{-1}$, and $\tilde{L}v(x) = Lu(T^{-1}x)$. Applying the ABP estimate to v and $\tilde{L}v(x)$ on $T(\Omega)$, we find

$$(3.14) \qquad \begin{aligned} \max_{\overline{T(\Omega)}} v &\le \max_{\partial T(\Omega)} v + C_1(n)\text{diam}(T(\Omega)) \left\| \frac{\tilde{L}v}{(\det \tilde{A})^{1/n}} \right\|_{L^n(T(\Omega))} \\ &= \max_{\partial\Omega} u + \frac{C_1(n)\text{diam}(T(\Omega))}{(\det M)^{1/n}} \left\| \frac{Lu}{(\det A)^{1/n}} \right\|_{L^n(\Omega)}. \end{aligned}$$

From (3.13), we have $\det M \ge \omega_n |\Omega|^{-1}$ and $\text{diam}(T(\Omega)) \le 2n$. Using these estimates in (3.14), we obtain the conclusion of the lemma. $\qquad\square$

It is important to emphasize that Theorem 3.16 extends the maximum principle in Theorem 2.79 from C^2 functions to $W^{2,n}_{\text{loc}}$ functions.

3.4. Global Hölder Estimates and Compactness

From the Aleksandrov maximum principle, we deduce a global $C^{0,1/n}$ regularity result for convex functions with finite total Monge–Ampère measure.

Lemma 3.18. *Let Ω be a bounded convex domain in $\mathbb{R}^n$. Let $u \in C(\overline{\Omega})$ be a convex function with $u = 0$ on $\partial\Omega$ and $\mu_u(\Omega) < \infty$. Then $u \in C^{0,1/n}(\overline{\Omega})$ with*

$$\|u\|_{C^{0,1/n}(\overline{\Omega})} \le C(n, \text{diam}(\Omega), \mu_u(\Omega)).$$

Proof. By Theorem 3.12, we have the boundary Hölder estimate

$$|u(x)| \leq C(n, \operatorname{diam}(\Omega), \mu_u(\Omega)) \operatorname{dist}^{\frac{1}{n}}(x, \partial\Omega) \quad \text{for all } x \in \Omega.$$

The lemma now follows from Lemma 2.37(iv).   □

Note that the exponent $1/n$ in Lemma 3.18 is sharp; see Problem 3.16.

Another consequence of Aleksandrov's maximum principle is the compactness of a family of convex functions on convex domains lying between two fixed balls with zero boundary values and having total Monge–Ampère measures bounded from above.

Theorem 3.19 (Compactness of solutions to the Monge–Ampère equation). *Let λ, Λ, M be positive numbers. Let*

$$\mathcal{C}_{\Lambda,M} = \big\{(\Omega, u, \nu) : \Omega \subset \mathbb{R}^n \text{ is a convex domain}, B_{1/M}(0) \subset \Omega \subset B_M(0),$$
$$u \in C(\overline{\Omega}) \text{ is convex}, u|_{\partial\Omega} = 0, \mu_u = \nu, \nu(\Omega) \leq \Lambda \big\}.$$

Then the set $\mathcal{C}_{\Lambda,M}$ is compact in the following sense: For any sequence $\{(\Omega_i, u_i, \nu_i)\}_{i=1}^\infty$ in $\mathcal{C}_{\Lambda,M}$, there exist a subsequence, which is still labeled as $\{(\Omega_i, u_i, \nu_i)\}_{i=1}^\infty$, and $(\Omega, u, \nu) \in \mathcal{C}_{\Lambda,M}$ such that

(1) *$\overline{\Omega}_i$ converges to $\overline{\Omega}$ in the Hausdorff distance,*

(2) *u_i converges to u locally uniformly in Ω, and*

(3) *$\nu_i = \mu_{u_i}$ converges weakly* to $\nu = \mu_u$.*

As a consequence, we obtain the compactness of each the following sets:

$$C_{\Lambda,M} = \big\{(\Omega, u) : \Omega \subset \mathbb{R}^n \text{ is a convex domain}, B_{1/M}(0) \subset \Omega \subset B_M(0),$$
$$u \in C(\overline{\Omega}) \text{ is convex}, u|_{\partial\Omega} = 0, \det D^2 u \leq \Lambda \big\}$$

and

$$C_{\lambda,\Lambda,M} = \big\{(\Omega, u) : \Omega \subset \mathbb{R}^n \text{ is a convex domain}, B_{1/M}(0) \subset \Omega \subset B_M(0),$$
$$u \in C(\overline{\Omega}) \text{ is convex}, u|_{\partial\Omega} = 0, \lambda \leq \det D^2 u \leq \Lambda \big\}.$$

Proof. Suppose we are given a sequence $\{(\Omega_i, u_i, \nu_i)\}_{i=1}^\infty \subset \mathcal{C}_{\Lambda,M}$. By the Blaschke Selection Theorem (Theorem 2.20), we can find a subsequence of $\{\Omega_i\}_{i=1}^\infty$, still labeled $\{\Omega_i\}_{i=1}^\infty$, such that $\overline{\Omega}_i$ converges to a convex body $\overline{\Omega}$ in the Hausdorff distance. Clearly, $\overline{B_{1/M}(0)} \subset \overline{\Omega} \subset \overline{B_M(0)}$ so $B_{1/M}(0) \subset \Omega \subset B_M(0)$.

By Lemma 3.18, we can find a constant $C(n, \Lambda, M) > 0$ such that

$$\|u_i\|_{C^{0,1/n}(\overline{\Omega}_i)} \leq C \quad \text{and} \quad |u_i(x)| \leq C[\operatorname{dist}(x, \partial\Omega_i)]^{1/n} \quad \text{for all } x \in \Omega_i.$$

It follows from the Arzelà–Ascoli Theorem (Theorem 2.41) that, up to extracting a further subsequence, u_i converges locally uniformly in Ω to a convex function $u \in C^{0,1/n}(\Omega)$ satisfying

$$|u(x)| \leq C[\mathrm{dist}(x, \partial\Omega)]^{1/n} \quad \text{for all } x \in \Omega.$$

Therefore $u \in C(\overline{\Omega})$ and $u = 0$ on $\partial\Omega$.

Since $u_i \to u$ locally uniformly in Ω, Theorem 3.7 gives that the restrictions of $\nu_i = \mu_{u_i}$ on Ω converge weakly* to $\nu = \mu_u$. Since $\overline{\Omega_i}$ converges to $\overline{\Omega}$ in the Hausdorff distance, we deduce that ν_i on Ω converges weakly* to ν. Thus, we have $(\Omega, u, \nu) \in \mathcal{C}_{\Lambda,M}$ with properties (1)–(3). $\qquad \square$

A consequence of the above theorem and the uniqueness of solutions to the Dirichlet problem (Corollary 3.23 below) is the following:

Theorem 3.20 (Compactness of solutions to the Monge–Ampère equation: zero boundary data). *Let $\{\Omega_k\}_{k=1}^{\infty} \subset \mathbb{R}^n$ be a sequence of bounded convex domains that converges to a bounded convex domain Ω in the Hausdorff distance. Let $\{\mu_k\}_{k=1}^{\infty}$ be a sequence of nonnegative Borel measures with $\sup_k \mu_k(\Omega_k) < \infty$ and which converges weakly* to a Borel measure μ on Ω. For each k, let $u_k \in C(\overline{\Omega_k})$ be the convex Aleksandrov solution of*

$$\det D^2 u_k = \mu_k \quad \text{in } \Omega_k, \qquad u_k = 0 \quad \text{on } \partial\Omega_k.$$

Then u_k converges locally uniformly in Ω to the convex Aleksandrov solution of

$$\det D^2 u = \mu \quad \text{in } \Omega, \qquad u = 0 \quad \text{on } \partial\Omega.$$

3.5. Comparison Principle and Global Lipschitz Estimates

We prove in this section an important converse of Lemma 3.11. It implies that if two convex functions have the same boundary values, the function with larger Monge–Ampère measure is in fact smaller than the other one in the interior. This property roughly says that $(\det D^2)$ is a negative operator on the set of convex functions, which is a nonlinear analogue of the Laplace operator $\Delta = (\mathrm{trace}\, D^2)$; see also Theorem 2.79.

Theorem 3.21 (Comparison principle for the Monge–Ampère equation). *Let $u, v \in C(\overline{\Omega})$ be convex functions on a bounded open set Ω in $\mathbb{R}^n$ such that*

$$\det D^2 v \geq \det D^2 u \quad \text{in } \Omega,$$

in the sense of Aleksandrov; that is, $\mu_v \geq \mu_u$. Then

$$\min_{\overline{\Omega}}(u - v) = \min_{\partial\Omega}(u - v).$$

In particular, if $u \geq v$ on $\partial\Omega$, then $u \geq v$ in Ω.

Proof. By adding a constant to v, we can assume that $\min_{\partial\Omega}(u-v) = 0$ and hence $u \geq v$ on $\partial\Omega$. We need to show that $u \geq v$ in Ω. Suppose otherwise that $u - v$ attains its minimum at $x_0 \in \Omega$ with $u(x_0) - v(x_0) = -m < 0$. Choose $\delta > 0$ such that $\delta(\mathrm{diam}(\Omega))^2 < m/2$. Consider

$$w(x) := u(x) - v(x) - \delta|x - x_0|^2.$$

If $x \in \partial\Omega$, then $w(x) \geq -\delta(\mathrm{diam}(\Omega))^2 \geq -m/2$, while $w(x_0) = -m < -m/2$. Thus, w attains its minimum value at $y \in \Omega$. Let

$$E = \{x \in \overline{\Omega} : w(x) < -3m/4\}.$$

Then E is open with $y \in E$, and $\partial E = \{x \in \overline{\Omega} : w(x) = -3m/4\} \subset \Omega$. The function u is below $v + \delta|x - x_0|^2 - 3m/4$ in E, but they coincide on ∂E. Consequently, the maximum principle in Lemma 3.11 gives

$$\partial u(E) \supset \partial(v + \delta|x - x_0|^2 - 3m/4)(E) = \partial(v + \delta|x - x_0|^2)(E).$$

It follows from Lemma 3.10 that

$$\begin{aligned}
\mu_u(E) = |\partial u(E)| &\geq |\partial(v + \delta|x - x_0|^2)(E)| \\
&\geq |\partial v(E)| + |\partial(\delta|x - x_0|^2)(E)| \\
&= |\partial v(E)| + (2\delta)^n|E| > |\partial v(E)| = \mu_v(E),
\end{aligned}$$

which is a contradiction. The theorem is thus proved. $\qquad\square$

The proof of Theorem 3.21 also implies the following result.

Lemma 3.22. *Let $u, v \in C(\overline{\Omega})$ be convex functions on an open set Ω in $\mathbb{R}^n$ such that $\mu_v \geq \mu_u$. Then for any $x_0 \in \Omega$ and $\delta > 0$, the function $w(x) := u(x) - v(x) - \delta|x - x_0|^2$ cannot attain its minimum value on $\overline{\Omega}$ at an interior point in Ω.*

Proof. Suppose that w attains its minimum value on $\overline{\Omega}$ at $y \in \Omega$. Let $\bar{w}(x) := w(x) + (\delta/2)|x - y|^2$. Then $\bar{w}(y) = w(y) < \min_{\partial\Omega} \bar{w}$. Note that

$$\bar{w}(x) = u(x) - v(x) - (\delta/2)|x|^2 + \text{affine function in } x.$$

The proof of Theorem 3.21 applied to $\bar{w}$ gives a contradiction. $\qquad\square$

An immediate consequence of the comparison principle is the uniqueness of solutions to the Dirichlet problem.

Corollary 3.23 (Uniqueness of solutions to the Dirichlet problem). *Let Ω be an open bounded set in $\mathbb{R}^n$. Let $g \in C(\partial\Omega)$ and let ν be a Borel measure in Ω. Then, there is at most one convex Aleksandrov solution $u \in C(\overline{\Omega})$ to the Dirichlet problem*

$$\det D^2 u = \nu \quad \text{in } \Omega, \qquad u = g \quad \text{on } \partial\Omega.$$

Next, we give a positive answer to the question raised after the proof of Corollary 3.14. The following lemma quantifies how much a convex function with a positive lower bound on the density of its Monge–Ampère measure will drop its values when stepping inside the domain.

Lemma 3.24. *Let Ω be an open set in $\mathbb{R}^n$ that satisfies $B_r(b) \subset \Omega \subset B_R(b)$ for some $b \in \mathbb{R}^n$ and $0 < r \leq R$. Assume that the convex function $u \in C(\overline{\Omega})$ satisfies, in the sense of Aleksandrov,*

$$\lambda \leq \det D^2 u \leq \Lambda \quad in \ \Omega, \quad u = \varphi \quad on \ \partial\Omega,$$

where $0 < \lambda \leq \Lambda$ and $\varphi \in C(\partial\Omega)$. Then

$$(3.15) \quad (\Lambda^{1/n}/2)(|y - b|^2 - R^2) + \min_{\partial\Omega} \varphi \leq u(y)$$

$$\leq (\lambda^{1/n}/2)(|y - b|^2 - r^2) + \max_{\partial\Omega} \varphi \quad in \ \Omega.$$

In particular, if $B_1(b) \subset \Omega \subset B_n(b)$ and $\varphi = 0$ on $\partial\Omega$, then

$$c(\lambda, n) \equiv \lambda^{1/n}/2 \leq |\min_{\Omega} u| \leq C(\Lambda, n) \equiv \Lambda^{1/n} n^2/2.$$

Proof. We prove

$$u(y) \leq v(y) := (\lambda^{1/n}/2)(|y - b|^2 - r^2) + \max_{\partial\Omega} \varphi,$$

the other inequality in (3.15) being similar. Indeed, since $B_r(b) \subset \Omega$,

$$v \geq \max_{\partial\Omega} \varphi \geq u \quad on \ \partial\Omega \quad and \quad \det D^2 v = \lambda \leq \det D^2 u \quad in \ \Omega.$$

Thus, by Theorem 3.21, we have $u \leq v$ in Ω as asserted.

Finally, in the special case of $r = 1$, $R = n$, and $\varphi = 0$, (3.15) gives

$$-\Lambda^{1/n} n^2/2 \leq \min_{\Omega} u \leq u(b) \leq -\lambda^{1/n}/2,$$

and the last assertion of the lemma follows. $\qquad\square$

We next observe that if a domain is the sublevel set of a convex function u with Monge–Ampère measure bounded from below and above, then the domain is balanced around the minimum point of u; see also Lemma 5.6.

Lemma 3.25 (Balancing). *Let Ω be a bounded, convex domain in $\mathbb{R}^n$. Suppose that the convex function $u \in C(\overline{\Omega})$ satisfies $u = 0$ on $\partial\Omega$ and $0 < \lambda \leq \det D^2 u \leq \Lambda$ in Ω, in the sense of Aleksandrov.*

(i) *Let l be a line segment in Ω with two endpoints $z', z'' \in \partial\Omega$. Let $z \in l$, and let $\alpha \in (0,1)$ be such that $u(z) \leq \alpha \inf_\Omega u$. Then, there is $c(n, \alpha, \lambda, \Lambda) > 0$ such that $|z' - z| \geq c(n, \alpha, \lambda, \Lambda)|z'' - z'|$.*

(ii) *Let $x_0 \in \Omega$ be the minimum point of u in $\overline{\Omega}$. Let H_1 and H_2 be two supporting hyperplanes to $\partial\Omega$ that are parallel. Then, there is $c_0(n, \lambda, \Lambda) > 0$ such that $\mathrm{dist}(x_0, H_1)/\mathrm{dist}(x_0, H_2) \geq c_0$.*

Proof. First, we discuss rescalings using John's Lemma. By Lemma 2.63, there is an affine transformation T such that $B_1(0) \subset T^{-1}\Omega \subset B_n(0)$. Let $v(x) = (\det T)^{-2/n} u(Tx)$ for $x \in T^{-1}\Omega$. Then, $\lambda \, dx \leq \mu_v \leq \Lambda \, dx$ in $T^{-1}\Omega$, by Lemma 3.9.

We now prove part (i). Since the ratio $|z' - z|/|z'' - z'|$ is invariant under affine transformations, with the above rescaling, we may assume that $B_1(0) \subset \Omega \subset B_n(0)$. By Lemma 3.24, $\inf_\Omega u \leq -c(n, \lambda)$. Hence, when

$$u(z) \leq \alpha \inf_\Omega u \leq -\alpha c(n, \lambda),$$

we have $\text{dist}(z, \partial\Omega) \geq c_1$ for some $c_1 = c_1(n, \alpha, \lambda, \Lambda) > 0$ by Aleksandrov's maximum principle, Theorem 3.12. It follows that

$$\frac{|z' - z|}{|z'' - z'|} \geq \frac{\text{dist}(z, \partial\Omega)}{\text{diam}(\Omega)} \geq \frac{c_1}{2n}.$$

The proof of part (ii) is similar to the above proof so we skip it. $\qquad \square$

Let us now state a strengthening of Aleksandrov's maximum principle for convex functions with Monge–Ampère measure having density bounded from above. A consequence of this result is that the global $C^{0,1/n}$ estimates in Lemma 3.18 are now improved to global $C^{0,2/n}$ estimates, with a slight exception in two dimensions.

Lemma 3.26. *Let Ω be a bounded, convex domain in $\mathbb{R}^n$, and let $u \in C(\overline{\Omega})$ be a convex function satisfying $u = 0$ on $\partial\Omega$ and, in the sense of Aleksandrov, $\det D^2 u \leq 1$ in Ω. Let $\alpha = 2/n$ if $n \geq 3$ and $\alpha \in (0, 1)$ if $n = 2$. Then $u \in C^{0,\alpha}(\overline{\Omega})$ with the estimate*

$$(3.16) \qquad |u(z)| \leq C(n, \alpha, \text{diam}(\Omega))[\text{dist}(z, \partial\Omega)]^\alpha \quad \text{for all } z \in \Omega.$$

Proof. We only need to prove the boundary Hölder estimate (3.16) since the global regularity $u \in C^{0,\alpha}(\overline{\Omega})$ then follows from the convexity of u and Lemma 2.37(iv).

Let $z \in \Omega$ be arbitrary. By translation and rotation of coordinates, we can assume the following: The origin 0 of $\mathbb{R}^n$ lies on $\partial\Omega$, the x_n-axis points inward Ω, z lies on the x_n-axis, and $\text{dist}(z, \partial\Omega) = z_n$.

To prove (3.16), it suffices to prove that for all $x = (x', x_n) \in \Omega$, we have

$$(3.17) \qquad |u(x)| \leq C(n, \alpha, \text{diam}(\Omega)) x_n^\alpha.$$

Let us consider, for $\alpha \in (0, 1)$ and $x \in \Omega$,

$$(3.18) \qquad \phi_\alpha(x) = x_n^\alpha (|x'|^2 - C_\alpha) \quad \text{where } C_\alpha = \frac{1 + 2[\text{diam}(\Omega)]^2}{\alpha(1 - \alpha)}.$$

Then, as a special case of Lemma 2.60 with $k = 1$ and $s = n - 1$, we have

$$\det D^2 \phi_\alpha(x) = 2^{n-1} x_n^{n\alpha-2} [\alpha(1-\alpha) C_\alpha - (\alpha^2 + \alpha)|x'|^2] \geq 2^{n-1} x_n^{n\alpha-2} \quad \text{in } \Omega.$$

Moreover, $D^2_{x'}\phi_\alpha = 2x^\alpha_n I_{n-1}$ is positive definite in Ω. Therefore, ϕ_α is convex in Ω with

(3.19) $\qquad \det D^2\phi_\alpha(x) \geq 2x^{n\alpha-2}_n$ in Ω and $\phi_\alpha \leq 0$ on $\partial\Omega$.

Consider now the case $n \geq 3$; the case $n = 2$ is similar.

Then $\det D^2\phi_{2/n} \geq 2 > \det D^2 u$ in Ω, while on $\partial\Omega$, $u = 0 \geq \phi_{2/n}$. By the comparison principle (Theorem 3.21), we have $u \geq \phi_{2/n}$ in Ω. Therefore,

$$|u(x)| = |u(x', x_n)| \leq -\phi_{2/n}(x', x_n) \leq C_{2/n} x^{2/n}_n \quad \text{for all } x \in \Omega,$$

from which (3.17) follows. The lemma is proved. $\qquad\square$

Remark 3.27. When $n \geq 3$, the exponent $2/n$ in Lemma 3.26 is optimal; see Proposition 3.42.

Remark 3.28. In the proof of Lemma 3.26, for $u \geq C\phi_\alpha$, we only need the following: u is a viscosity supersolution (see Section 7.1) of $\det D^2 u \leq f$ where $f \in C(\Omega), f \geq 0$ is bounded, and $u \geq 0$ on $\partial\Omega$. This is due to the fact that viscosity solutions satisfy the comparison principle with smooth functions; see Lemma 7.9. Under these relaxed conditions, we have $u(x) \geq -C\text{dist}^\alpha(x, \partial\Omega)$, which improves Corollary 3.14 in the case $\det D^2 u \leq f$.

In two dimensions, we can strengthen the estimate in Lemma 3.26 to a global log-Lipschitz estimate which is sharp (see Example 3.32). This follows from the following general result.

Lemma 3.29 (Global log-Lipschitz estimate). *Let Ω be a bounded convex domain in $\mathbb{R}^n$ ($n \geq 2$). Let $u \in C(\overline{\Omega})$ be a convex function satisfying $u = 0$ on $\partial\Omega$ and $\det D^2 u \leq M\text{dist}^{n-2}(\cdot, \partial\Omega)$ in Ω, in the sense of Aleksandrov, where $M > 0$. Then*

$$|u(z)| \leq C(M, \text{diam}(\Omega))\,\text{dist}(z, \partial\Omega)(1 + |\log \text{dist}(z, \partial\Omega)|) \quad \text{for all } z \in \Omega.$$

Proof. As in the proof of Lemma 3.26, we will construct an appropriate log-Lipschitz convex subsolution. Let $z = (z', z_n)$ be an arbitrary point in Ω. By translation and rotation of coordinates, we can assume the following: $0 \in \partial\Omega$, $\Omega \subset \mathbb{R}^n_+ = \{x = (x', x_n) \in \mathbb{R}^n : x_n > 0\}$, the x_n-axis points inward Ω, z lies on the x_n-axis, and $z_n = \text{dist}(z, \partial\Omega)$. Let $d = \text{diam}(\Omega)$ and

$$v(x) = (M + 2d^2)x_n \log(x_n/d) + x_n(|x'|^2 - d^2).$$

Then, $v \leq 0$ on $\partial\Omega$, and

$$D^2 v = \begin{pmatrix} 2x_n & 0 & \cdots & 0 & 2x_1 \\ 0 & 2x_n & \cdots & 0 & 2x_2 \\ \vdots & \vdots & \ddots & \vdots & \vdots \\ 0 & 0 & \cdots & 2x_n & 2x_{n-1} \\ 2x_1 & 2x_2 & \cdots & 2x_{n-1} & \frac{M+2d^2}{x_n} \end{pmatrix}.$$

By induction in n, we find that v is convex in Ω and

$$\det D^2 v(x) = 2^{n-1} x_n^{n-2} (M + 2d^2 - 2|x'|^2) \geq 2M \text{dist}^{n-2}(x, \partial\Omega).$$

By the comparison principle in Theorem 3.21, we have $u \geq v$ in Ω. Thus

$$|u(z)| \leq |v(z)| = (M + 2d^2) z_n \log(d/z_n) + d^2 z_n \leq C(M, d) z_n (1 + |\log z_n|).$$

Since $z_n = \text{dist}(z, \partial\Omega)$, the lemma is proved. □

We conclude this section by establishing global Lipschitz estimates under suitable conditions on the domain and boundary data.

Theorem 3.30 (Global Lipschitz estimate). *Let Ω be a uniformly convex domain in $\mathbb{R}^n$ with C^2 boundary $\partial\Omega$. Let $\varphi \in C^{1,1}(\overline{\Omega})$. Let $u \in C(\overline{\Omega})$ be the convex solution, in the sense of Aleksandrov, to*

$$\det D^2 u \leq \Lambda \quad in\ \Omega, \quad u = \varphi \quad on\ \partial\Omega.$$

Then, there is a constant $C = C(n, \Lambda, \Omega, \|\varphi\|_{C^{1,1}(\overline{\Omega})})$ such that

$$\|u\|_{C^{0,1}(\overline{\Omega})} := \sup_{\overline{\Omega}} |u| + \sup_{x \neq y \in \Omega} \frac{|u(x) - u(y)|}{|x - y|} \leq C.$$

Proof. By Lemma 3.24 and the convexity of u, we have

$$C(n, \Lambda, \min_{\partial\Omega} \varphi, \text{diam}(\Omega)) \leq u \leq \max_{\partial\Omega} \varphi \quad in\ \overline{\Omega}.$$

For the Lipschitz estimate, thanks to Lemma 2.37(iv), we only need to prove

$$|u(x) - u(x_0)| \leq C|x - x_0| \quad \text{for all } x \in \Omega, x_0 \in \partial\Omega.$$

First, we note that the ray $\overrightarrow{x_0 x}$ intersects $\partial\Omega$ at z and $x = tx_0 + (1-t)z$ for some $t \in (0, 1)$. It follows that $x - x_0 = (1-t)(z - x_0)$. By convexity, $u(x) \leq tu(x_0) + (1-t)u(z)$. Thus, using $u = \varphi$ on $\partial\Omega$, we deduce

$$u(x) - u(x_0) \leq (1-t)(\varphi(z) - \varphi(x_0)) \leq (1-t)\|D\varphi\|_{L^\infty(\Omega)}|z - x_0|.$$

Therefore, $u(x) - u(x_0) \leq \|D\varphi\|_{L^\infty(\Omega)}|x - x_0|$.

Thus, it remains to show $u(x) - u(x_0) \geq -C|x - x_0|$.

Let $K := \|\varphi\|_{C^{1,1}(\overline{\Omega})}$. Let ρ be a $C^2(\overline{\Omega})$ uniformly convex defining function of Ω (see Definition 2.53); that is, $\rho < 0$ in Ω, $\rho = 0$ and $D\rho \neq 0$ on $\partial\Omega$. As in the proof of Lemma 2.54, for $\mu(n, \Lambda, \Omega, K) > 0$ large, the function

$$v = \varphi + \mu(e^\rho - 1)$$

is convex and $\det D^2 v \geq \Lambda \geq \det D^2 u$ in Ω. With this μ and $u = v$ on $\partial\Omega$, we have $u \geq v$ in Ω by the comparison principle in Theorem 3.21.

Using $u = \varphi$ on $\partial\Omega$ and $e^\rho - 1 \geq \rho$, we have, for all $x_0 \in \partial\Omega$ and $x \in \Omega$,

$$u(x) - u(x_0) \geq \varphi(x) - \varphi(x_0) + \mu(\rho(x) - \rho(x_0)) \geq -C|x - x_0|$$

where $C = \|D\varphi\|_{L^\infty(\Omega)} + \mu\|D\rho\|_{L^\infty(\Omega)}$. This concludes the proof. □

3.6. Explicit Solutions

We will see in Section 3.7 general results regarding solvability of the Dirichlet problem for the Monge–Ampère equation. However, due to its highly nonlinear nature, there are very few examples of explicit solutions of the Monge–Ampère equation $\det D^2 u = 1$ with zero boundary condition on bounded convex domains Ω in $\mathbb{R}^n$. We list here two examples.

Example 3.31 (Solution on a ball). Consider the unit ball $B_1(0)$ in $\mathbb{R}^n$. Then, the function $u(x) = (|x|^2 - 1)/2$ is the unique convex solution to
$$\det D^2 u = 1 \quad \text{in } B_1(0), \quad u = 0 \quad \text{on } \partial B_1(0).$$

Example 3.32 (Solution on a triangle). Let T be the open triangle in the plane $\mathbb{R}^2$ with vertices at $(0,0), (1,0),$ and $(0,1)$. Consider the Monge–Ampère equation

$$(3.20) \qquad \det D^2 \sigma_T = 1 \quad \text{in } T, \quad \sigma_T = 0 \quad \text{on } \partial T.$$

We can verify that the solution $\sigma_T \in C(\overline{T})$ is given by

$$\sigma_T(x) = -\frac{1}{\pi^2}\big[\mathscr{L}(\pi x_1) + \mathscr{L}(\pi x_2) + \mathscr{L}(\pi(1 - x_1 - x_2))\big],$$

for $x = (x_1, x_2) \in T$, where

$$\mathscr{L}(\theta) = -\int_0^\theta \log|2\sin u|\, du$$

is the Lobachevsky function. This remarkable formula for σ_T comes from the study of the *dimer model* (also known as the lozenge tiling model) in the works of Kenyon, Okounkov, and Sheffield [**Kn, KO, KOS**]. The function σ_T is called a *surface tension*.

Verification of Example 3.32. By direct calculation, we have

$$(3.21) \qquad D\sigma_T(x) = \frac{1}{\pi}\left(\log\left(\frac{\sin(\pi x_1)}{\sin(\pi x_1 + \pi x_2)}\right), \log\left(\frac{\sin(\pi x_2)}{\sin(\pi x_1 + \pi x_2)}\right)\right)$$

and

$$D^2\sigma_T = \begin{pmatrix} \cot(\pi x_1) - \cot(\pi x_1 + \pi x_2) & -\cot(\pi x_1 + \pi x_2) \\ -\cot(\pi x_1 + \pi x_2) & \cot(\pi x_2) - \cot(\pi x_1 + \pi x_2) \end{pmatrix}.$$

Therefore, using the cotangent formula, we find $\det D^2 \sigma_T = 1$.

To verify the boundary condition, we use the fact that

$$\mathscr{L}(\pi) = -\int_0^\pi \log|2\sin u|\, du = 0;$$

see [**Ah**, p. 161]. From this, we see that $\mathscr{L}$ is odd and π-periodic. Thus, for example, on the x_1-axis part of ∂T, we have

$$\sigma_T(x_1, 0) = -\pi^{-2}[\mathscr{L}(\pi x_1) + \mathscr{L}(\pi - \pi x_1)] = 0.$$

Note that $|\sigma_T|$ grows at a log-Lipschitz rate $\text{dist}(\cdot, \partial T)|\log \text{dist}(\cdot, \partial T)|$ from the boundary of T. In view of Lemma 3.29, this is the sharp rate for convex functions with bounded Monge–Ampère measure and which vanish on the boundary of a convex domain in the plane.

3.7. The Dirichlet Problem and Perron's Method

In this section, using the Perron method, we discuss the solvability of the inhomogeneous Dirichlet problem for the Monge–Ampère equation

$$(3.22) \qquad \det D^2 u = \nu \quad \text{in } \Omega, \qquad u = g \quad \text{on } \partial\Omega$$

on a bounded convex domain Ω in $\mathbb{R}^n$, with continuous boundary data and right-hand side being a Borel measure with finite mass. This problem was first solved by Aleksandrov and Bakelman.

3.7.1. Homogeneous Dirichlet problem. Before solving the Dirichlet problem with inhomogeneous right-hand side, we consider a simpler problem regarding the solvability of the homogeneous Dirichlet problem for the Monge–Ampère equation with continuous boundary data.

Theorem 3.33 (Homogeneous Dirichlet problem). *Let Ω be a bounded and strictly convex domain in $\mathbb{R}^n$. Then, for any $g \in C(\partial\Omega)$, the problem*

$$\det D^2 u = 0 \quad \text{in } \Omega, \quad u = g \quad \text{on } \partial\Omega$$

has a unique convex Aleksandrov solution $u \in C(\overline{\Omega})$.

Remark 3.34. It should be emphasized that Theorem 3.33 asserts the solvability of the Dirichlet problem for all continuous boundary data. For this, the strict convexity of the domain cannot be dropped. For example, let $\Omega \subset \mathbb{R}^2$ be the interior of a triangle with vertices z_1, z_2, z_3. Let $g \in C(\partial\Omega)$ be such that $g(z_1) = g(z_2) = g(z_3) = 0$ and $g((z_1 + z_2)/2) = 1$. Then, there is no convex function $u \in C(\overline{\Omega})$ such that $u = g$ on $\partial\Omega$. Hence, the Dirichlet problem with this boundary data is not solvable.

Proof of Theorem 3.33. The uniqueness of Aleksandrov solutions follows from the comparison principle; see also Corollary 3.23. Now, we show the existence. Heuristically, the sought-after solution has the smallest possible Monge–Ampère measure among all convex functions with fixed boundary value g. The pointwise maximum principle in Lemma 3.11 suggests looking for the maximum of all convex functions with boundary values not exceeding g. This is exactly the Perron method since these convex functions are subsolutions of our equation.

Let us consider

$$\mathcal{S} = \{v \in C(\overline{\Omega}) : v \text{ is convex and } v \leq g \text{ on } \partial\Omega\}.$$

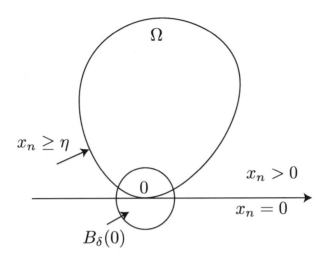

Figure 3.4. $x_n \geq \eta$ for all $x \in \partial\Omega \setminus B_\delta(0)$.

Since g is continuous, $\mathcal{S} \neq \emptyset$ because $v \equiv \min_{\partial\Omega} g \in \mathcal{S}$. Let

$$u(x) = \sup_{v \in \mathcal{S}} v(x), \ x \in \overline{\Omega}.$$

Then, by Lemma 2.24, u is convex in Ω, and $u \leq g$ on $\partial\Omega$ so $u : \overline{\Omega} \to \mathbb{R}$ is finite. We show that u is the desired solution in the following steps.

Step 1. For each $x_0 \in \partial\Omega$, we trap the boundary data g between two affine functions that are close to g near x_0. By a translation of coordinates, we can assume that $x_0 = 0 \in \partial\Omega$ and that $\Omega \subset \{x \in \mathbb{R}^n : x_n > 0\}$. From the continuity of g, given $\varepsilon > 0$, there exists $\delta > 0$ such that

(3.23) $|g(x) - g(0)| < \varepsilon$ for all $x \in \partial\Omega \cap B_\delta(0)$.

Recall that Ω is strictly convex. Thus, there exists $\eta > 0$ such that $x_n \geq \eta$ for all $x \in \partial\Omega \setminus B_\delta(0)$. See Figure 3.4.

It can be readily verified that, for $C_1 = 2\|g\|_{L^\infty(\partial\Omega)}/\eta$, we have

(3.24) $a_1(x) := g(0) - \varepsilon - C_1 x_n \leq g(x)$
$$\leq g(0) + \varepsilon + C_1 x_n := A(x) \quad \text{for all } x \in \partial\Omega.$$

For $a_1 \leq g$ to hold, it suffices to assume that g is lower semicontinuous at 0.

Step 2. $u = g$ on $\partial\Omega$. Let $x_0 \in \partial\Omega$. We only need to show $u(x_0) \geq g(x_0)$. Given $\varepsilon > 0$, let us assume that $x_0 = 0 \in \partial\Omega$ and also Ω is as in Step 1. From (3.24), we find $a_1(x) \in \mathcal{S}$. By the definition of u, we have $u(0) \geq a_1(0) = g(0) - \varepsilon$. This holds for all $\varepsilon > 0$ so $u(0) \geq g(0)$. Step 2 is proved.

Step 3. $u \in C(\overline{\Omega})$. Since u is convex in Ω, it is continuous there, so we only need to prove that u is continuous on $\partial\Omega$. Let us assume that $x_0 = 0 \in \partial\Omega$ and also Ω is as in Step 1. Given $\varepsilon > 0$, let a_1 and A be as in Step 1. Then $u \geq a_1$. On the other hand, if $v \in \mathcal{S}$, then $v \leq g \leq A$ on $\partial\Omega$. Since $v - A$ is convex, this implies $v - A \leq 0$ in Ω, so $v \leq A$ on $\overline{\Omega}$. By taking the supremum over $v \in \mathcal{S}$, we find $u \leq A$ on $\overline{\Omega}$. Therefore

$$(3.25) \qquad a_1 \leq u \leq A \quad \text{in } \Omega.$$

Since

$$g(0) - \varepsilon \leq \liminf_{x \to 0} a_1(x) \leq \limsup_{x \to 0} A(x) = g(0) + \varepsilon$$

and $\varepsilon > 0$ is arbitrary, we must have $\lim_{x \to 0} u(x) = g(0) = u(0)$.

Step 4. We show

$$\partial u(\Omega) \subset \{p \in \mathbb{R}^n : \text{there are } x \neq y \in \Omega \text{ such that } p \in \partial u(x) \cap \partial u(y)\}.$$

Let $p \in \partial u(\Omega)$. Then $p \in \partial u(x_0)$ for some $x_0 \in \Omega$, and hence

$$(3.26) \qquad u(x) \geq u(x_0) + p \cdot (x - x_0) := a(x) \quad \text{in } \Omega.$$

We first show that there is $y \in \partial\Omega$ such that $g(y) = a(y)$. Indeed, from (3.26), together with $u \equiv g$ on $\partial\Omega$ by Step 2, and the continuity of both u and g, we find $g \geq a$ on $\partial\Omega$. If such $y \in \partial\Omega$ does not exist, then by the continuity of g and a, there is $\varepsilon > 0$ such that $g \geq a + \varepsilon$ on $\partial\Omega$. Therefore, $a + \varepsilon \in \mathcal{S}$, which implies $u \geq a + \varepsilon$ on Ω, but this contradicts $u(x_0) = a(x_0)$.

Next, we complete the proof of Step 4 by showing that $a(x)$ is a supporting hyperplane to the graph of u at $(z, u(z))$ for z on a whole open segment I connecting $x_0 \in \Omega$ to $y \in \partial\Omega$ and so $p \in \partial u(z)$ for all $z \in I$. To prove this, we show that $u(z) \leq a(z)$ for all $z \in I$ because we already have $u \geq a$ in Ω. Let $z = t x_0 + (1 - t)y$ where $0 \leq t \leq 1$. By the convexity of u and the fact that a is affine with $a(y) = g(y) = u(y)$, we finally have

$$u(z) \leq t u(x_0) + (1 - t)u(y) = t a(x_0) + (1 - t)a(y) = a(z).$$

From Step 4 and Aleksandrov's Lemma (Lemma 2.45), we have $|\partial u(\Omega)| = 0$, so $\det D^2 u = 0$ in Ω in the sense of Aleksandrov. This completes the proof. $\qquad \square$

3.7.2. Inhomogeneous Dirichlet problem. We again use the Perron method to solve (3.22) as in the case $\nu \equiv 0$. Let

$$\mathcal{S}(\nu, g) = \{v \in C(\overline{\Omega}) : v \text{ convex}, \det D^2 v \geq \nu \text{ in } \Omega, v = g \text{ on } \partial\Omega\},$$

and let $u(x) = \sup_{v \in \mathcal{S}(\nu, g)} v(x)$. We wish to show that u is the desired solution.

For this, we first need to show that $\mathcal{S}(\nu, g) \neq \emptyset$. So far, the Monge–Ampère measures of convex functions that we have seen are either Dirac measures, Hessian determinants of $C^{1,1}$ convex functions, or zero. In general, it is not practical to construct explicitly an element in $\mathcal{S}(\nu, g) \neq \emptyset$. Here we will use the superadditivity property of the Monge–Ampère measure to show that $\mathcal{S}(\nu, g) \neq \emptyset$ when ν is a finite combination of Dirac measures with positive coefficients. In fact, we will solve (3.22) in this case first. This is the content of Theorem 3.36. The general case of Borel measure ν follows from Theorem 3.36 and the following compactness result.

Theorem 3.35 (Compactness of solutions to the Monge–Ampère equation: continuous boundary data). *Let Ω be a bounded and strictly convex domain in $\mathbb{R}^n$. Let $\{\mu_j\}_{j=1}^{\infty}$, μ be Borel measures in Ω such that $\mu_j(\Omega) \leq A < \infty$ and μ_j converges weakly* to μ in Ω. Let $\{g_j\}_{j=1}^{\infty} \subset C(\partial\Omega)$ be such that g_j converges uniformly to g in $C(\partial\Omega)$. Assume that $u_j \in C(\overline{\Omega})$ is the unique convex Aleksandrov solution to*

$$\det D^2 u_j = \mu_j \ \ in \ \Omega, \quad u_j = g_j \ \ on \ \partial\Omega.$$

Then $\{u_j\}_{j=1}^{\infty}$ contains a subsequence that converges uniformly on compact subsets of Ω to the unique convex Aleksandrov solution $u \in C(\overline{\Omega})$ to

$$\det D^2 u = \mu \ \ in \ \Omega, \quad u = g \ \ on \ \partial\Omega.$$

Proof. From Corollary 3.14, we have for all j

$$-\|g_j\|_{L^{\infty}(\partial\Omega)} - C(n)\mathrm{diam}(\Omega)|\partial u_j(\Omega)|^{1/n} \leq u_j \leq \|g_j\|_{L^{\infty}(\partial\Omega)} \ \ \text{in} \ \Omega.$$

Due to $\mu_j(\Omega) \leq A < \infty$ and g_j converging uniformly to g in $C(\partial\Omega)$, we deduce that the sequence $\{u_j\}_{j=1}^{\infty}$ is uniformly bounded. Now, Theorem 2.42 implies that $\{u_j\}_{j=1}^{\infty}$ contains a subsequence, also denoted by $\{u_j\}_{j=1}^{\infty}$, such that u_j converges uniformly on compact subsets of Ω to a convex function u in Ω. By the weak continuity property of the Monge–Ampère measure in Theorem 3.7, we have $\det D^2 u = \mu$ in Ω.

It remains to show that $u \in C(\overline{\Omega})$ and $u = g$ on $\partial\Omega$. Consider an arbitrary point $x_0 \in \partial\Omega$. We can assume $x_0 = 0$ and $\Omega \subset \{x \in \mathbb{R}^n : x_n > 0\}$. We will show that u is continuous at 0 and $u(0) = g(0)$.

By Theorem 3.33, there is a unique Aleksandrov solution $U_j \in C(\overline{\Omega})$ to

$$\det D^2 U_j = 0 \ \ in \ \Omega, \quad U_j = g_j \ \ on \ \partial\Omega.$$

Since $\det D^2 U_j \leq \det D^2 u_j$ in Ω and $U_j = u_j$ on $\partial\Omega$, the comparison principle in Theorem 3.21 gives

$$(3.27) \qquad\qquad u_j \leq U_j \ \ \text{in} \ \Omega.$$

Now, we try to obtain a good lower bound for u_j from below that matches U_j locally. Due to the continuity of g, given $\varepsilon > 0$, there exists $\delta > 0$ such that

$|g(x) - g(0)| < \varepsilon/3$ for all $x \in \partial\Omega \cap B_\delta(0)$. From the uniform convergence of g_j to g, there is $j(\varepsilon) > 0$ such that

$$|g_j(x) - g_j(0)| < \varepsilon \quad \text{for all } x \in \partial\Omega \cap B_\delta(0) \text{ and all } j \geq j(\varepsilon).$$

From now on, we only consider $j \geq j(\varepsilon)$. Since Ω is strictly convex, there exists $\eta > 0$ such that $x_n \geq \eta$ for all $x \in \partial\Omega \setminus B_\delta(0)$. As in (3.25), we have

$$(3.28) \qquad g_j(0) - \varepsilon - C_1 x_n \leq U_j(x) \leq g_j(0) + \varepsilon + C_1 x_n,$$

where $C_1 = 2\max_j\{\|g_j\|_{L^\infty(\Omega)}\}/\eta < \infty$. Now, consider

$$v_j(x) = u_j(x) - [g_j(0) - \varepsilon - C_1 x_n].$$

Then $v_j \geq 0$ on $\partial\Omega$ and $\det D^2 v_j = \det D^2 u_j = \mu_j$. By Corollary 3.14,

$$v_j(x) \geq -C(n)[\text{dist}(x, \partial\Omega)]^{1/n}(\text{diam}(\Omega))^{\frac{n-1}{n}}|\partial v_j(\Omega)|^{1/n} \quad \text{in } \Omega.$$

Since $|\partial v_j(\Omega)| = |\partial u_j(\Omega)| = \mu_j(\Omega) \leq A$ and $\text{dist}(x, \partial\Omega) \leq x_n$, we obtain

$$(3.29) \quad u_j(x) \geq g_j(0) - \varepsilon - C_1 x_n - C(n)x_n^{1/n}(\text{diam}(\Omega))^{\frac{n-1}{n}}A^{1/n} \quad \text{in } \Omega.$$

From (3.27)–(3.29), upon letting $j \to \infty$, we find that u is continuous at 0 and $u(0) = g(0)$. This completes the proof. $\qquad\square$

As mentioned earlier, we first solve the inhomogeneous problem (3.22) when ν is a finite combination of Dirac measures with positive coefficients.

Theorem 3.36 (The Dirichlet problem with Dirac measures). *Let Ω be a bounded and strictly convex domain in $\mathbb{R}^n$, and let $g \in C(\partial\Omega)$. Let $\nu = \sum_{i=1}^N a_i \delta_{x_i}$ where $a_i > 0$ and δ_{x_i} is the Dirac measure at $x_i \in \Omega$. Then, there is a unique convex Aleksandrov solution $u \in C(\overline{\Omega})$ to the problem*

$$\det D^2 u = \nu \quad \text{in } \Omega, \quad u = g \quad \text{on } \partial\Omega.$$

Proof. Let

$$\mathcal{S}(\nu, g) = \{v \in C(\overline{\Omega}) : v \text{ convex}, \ \det D^2 v \geq \nu \text{ in } \Omega, \ v = g \text{ on } \partial\Omega\},$$

and let

$$(3.30) \qquad\qquad u(x) = \sup_{v \in \mathcal{S}(\nu, g)} v(x).$$

We show that u is the desired solution in the following steps.

Step 1. There is $v_0 \in \mathcal{S}(\nu, g)$, and u defined by (3.30) is bounded. Note that the convex function $|z - x_i|$ on $\mathbb{R}^n$ has Monge–Ampère measure $\mu_{|z-x_i|} = \omega_n \delta_{x_i}$. Let $u_0(z) = \omega_n^{-1/n}\sum_{i=1}^N a_i^{1/n}|z - x_i|$. Then, the superadditivity property in Lemma 3.10 gives $\mu_{u_0} \geq \nu$.

By Theorem 3.33, there is a unique Aleksandrov solution $U_1 \in C(\overline{\Omega})$ to

$$\det D^2 U_1 = 0 \quad \text{in } \Omega, \quad U_1 = g - u_0 \quad \text{on } \partial\Omega.$$

Let $v_0 = u_0 + U_1$. Then $v_0 \in C(\overline{\Omega})$, $v_0 = g$ on $\partial\Omega$, and by Lemma 3.10

$$\mu_{v_0} = \mu_{u_0+U_1} \geq \mu_{u_0} + \mu_{U_1} \geq \nu.$$

Therefore, $v_0 \in \mathcal{S}(\nu, g)$.

For u defined by (3.30), we have $u \geq v_0$ so it is bounded from below. Now, we use Theorem 3.33 to obtain a unique convex solution $W \in C(\overline{\Omega})$ to

$$\mu_W = 0 \quad \text{in } \Omega, \qquad W = g \quad \text{on } \partial\Omega.$$

For any $v \in \mathcal{S}(\nu, g)$, we have $v \leq W$ by the comparison principle in Theorem 3.21. In particular, v is uniformly bounded from above and so is u.

Step 2. If $v_1, v_2 \in \mathcal{S}(\nu, g)$, then $v := \max\{v_1, v_2\} \in \mathcal{S}(\nu, g)$. Indeed, given a Borel set $E \subset \Omega$, we write $E = E_0 \cup E_1 \cup E_2$, $E_i = E \cap \Omega_i$, where Ω_i ($i = 0, 1, 2$) are the following subsets of Ω:

$$\Omega_0 = \{v_1 = v_2\}, \quad \Omega_1 = \{v_1 > v_2\}, \quad \Omega_2 = \{v_1 < v_2\}.$$

We show that $\mu_v(E_i) \geq \nu(E_i)$ for each $i = 0, 1, 2$. The cases $i = 1, 2$ are similar, so we only consider $i = 1$. It suffices to show that $\partial v_1(E_1) \subset \partial v(E_1)$. Indeed, if $p \in \partial v_1(x)$ where $x \in E_1$, then $p \in \partial v(x)$. This is because $v(x) = v_1(x)$ and for all $z \in \Omega$, we have

$$v(z) \geq v_1(z) \geq v_1(x) + p \cdot (z - x) = v(x) + p \cdot (z - x).$$

For the case $i = 0$, the same argument as above shows that $\partial v_1(E_0) \subset \partial v(E_0)$ and $\partial v_2(E_0) \subset \partial v(E_0)$, and we are done.

Step 3. Approximation property of u defined by (3.30). We show that

- for each $y \in \Omega$, there exists a sequence $\{v_m\}_{m=1}^\infty \subset \mathcal{S}(\nu, g)$ that converges uniformly on compact subsets of Ω to a function $w \in \mathcal{S}(\nu, g)$ so that $w(y) = u(y)$;
- $u \in C(\overline{\Omega})$.

Let $y \in \Omega$. Then, by the definition of u, there is a sequence $\{\bar{v}_m\}_{m=1}^\infty \subset \mathcal{S}(\nu, g)$ such that $\bar{v}_m(y) \to u(y)$ as $m \to \infty$. Let

$$v_m = \max\{v_0, \max_{1 \leq k \leq m} \bar{v}_k\}.$$

By Step 2 (applied finitely many times), we have $v_m \in \mathcal{S}(\nu, g)$. Moreover, $v_m \leq u$ in Ω while $\bar{v}_m(y) \leq v_m(y) \leq u(y)$. Hence, $v_m(y) \to u(y)$ as $m \to \infty$.

Since $v_0 \leq v_m \leq v_{m+1} \leq W$, Theorem 2.42 implies that $\{v_m\}_{m=1}^\infty$ contains a subsequence that converges uniformly on compact subsets of Ω to a convex function $w \in C(\Omega)$. Clearly, $w(y) = u(y)$. By Theorem 3.7, $\mu_w \geq \nu$.

Let w be g on $\partial\Omega$. We show that $w \in C(\overline{\Omega})$. Indeed, assume that $x_0 = 0 \in \partial\Omega$ and $\Omega \subset \{x \in \mathbb{R}^n : x_n > 0\}$ as in the proof of Theorem 3.33.

Given $\varepsilon > 0$, then, as in (3.24), there is $C_1(\varepsilon, g, \Omega) > 0$ such that

$$g(0) - \varepsilon - C_1 x_n \le g(x) \le g(0) + \varepsilon + C_1 x_n \quad \text{for } x \in \partial\Omega.$$

Using the comparison principle, we find that $v \le g(0) + \varepsilon + C_1 x_n$ in Ω for any $v \in \mathcal{S}(\nu, g)$. In particular, this is true for v_m and hence its limit w. Thus, $w \le g(0) + \varepsilon + C_1 x_n$ in Ω. Since $w \ge v_0$, we have

$$(3.31) \qquad v_0(x) - \varepsilon - C_1 x_n \le w(x) \le g(0) + \varepsilon + C_1 x_n \quad \text{in } \Omega.$$

It follows from $v_0(0) = g(0)$ that

$$-\varepsilon \le \liminf_{x \to 0}[w(x) - g(0)] \le \limsup_{x \to 0}[w(x) - g(0)] \le \varepsilon.$$

Since ε is arbitrary, w is continuous at 0. Thus, $w \in \mathcal{S}(\nu, g)$.

To show that $u \in C(\overline{\Omega})$, it suffices to show that u is continuous on $\partial\Omega$. With the above notation, we apply (3.31) at y where $w(y) = u(y)$, and then we vary y over Ω to conclude that

$$v_0(x) - \varepsilon - C_1 x_n \le u(x) \le g(0) + \varepsilon + C_1 x_n \quad \text{in } \Omega,$$

so u is continuous at 0.

Step 4. $u \in \mathcal{S}(\nu, g)$. From Step 3, it remains to prove that $\mu_u(\{x_i\}) \ge a_i$ for each $i = 1, \ldots, N$. We prove this for $i = 1$. By Step 3, there is a sequence $\{v_m\}_{m=1}^{\infty} \subset \mathcal{S}(\nu, g)$ that converges uniformly on compact subsets of Ω to a convex function w with $\mu_w \ge \nu$ so that $w(x_1) = u(x_1)$ and $w \le u$ in Ω. Thus $\mu_w(\{x_1\}) \ge a_1$. If $p \in \partial w(x_1)$, then $p \in \partial u(x_1)$ because for all $x \in \Omega$,

$$u(x) \ge w(x) \ge w(x_1) + p \cdot (x - x_1) = u(x_1) + p \cdot (x - x_1).$$

Therefore $\partial u(x_1) \supset \partial w(x_1)$ and hence

$$\mu_u(\{x_1\}) = |\partial u(x_1)| \ge |\partial w(x_1)| = \mu_w(\{x_1\}) \ge a_1.$$

Step 5. μ_u is concentrated on the set $X = \{x_1, \ldots, x_N\}$. For this, we use a lifting argument. Let $z \in \Omega \setminus X$. We can choose $r > 0$ such that $B_{2r}(z) \subset \Omega \setminus X$. Let $B = B_r(z)$ and let $v \in C(\overline{B})$ be the convex solution to

$$\mu_v = 0 \text{ in } B, \quad v = u \text{ on } \partial B.$$

Define the lifting w of u and v by

$$w(x) = v(x) \text{ if } x \in B \quad \text{and} \quad w(x) = u(x) \text{ if } x \in \Omega \setminus B.$$

Then $w \in C(\overline{\Omega})$ with $w = g$ on $\partial\Omega$. We claim that $w \in \mathcal{S}(\nu, g)$. Since $\mu_u \ge 0 = \mu_v$ in B and $u = v$ on ∂B, we have $v \ge u$ in B. Thus w is convex.

We now verify that $\mu_w(E) \ge \nu(E)$ for each Borel set $E \subset \Omega$. Let $E_1 = E \cap B$ and $E_2 = E \cap (\Omega \setminus B)$. As in Step 2, we have $\mu_w(E_1) \ge \mu_v(E_1)$ and $\mu_w(E_2) \ge \mu_u(E_2)$. Hence,

$$\mu_w(E) \ge \mu_v(E_1) + \mu_u(E_2) = \mu_u(E_2) \ge \nu(E_2) \ge \nu(E \cap X) = \nu(E).$$

This shows that $w \in \mathcal{S}(\nu, g)$. From the definition of u, we have $w \leq u$ in Ω. By the above argument, we have $w = v \geq u$ in B which implies that $u = v$ in B. It follows that $\mu_u(B) = 0$ for any ball $B = B_r(z)$ with $B_{2r}(z) \subset \Omega \setminus X$. Hence, if E is a Borel set with $E \cap X = \emptyset$, then $\mu_u(E) = 0$ by the regularity of μ_u. Therefore, μ_u is concentrated on the set X; that is, $\mu_u = \sum_{i=1}^{n} \lambda_i a_i \delta_{x_i}$, with $\lambda_i \geq 1$ for all $i = 1, \ldots, N$.

Step 6. $\mu_u = \nu$ in Ω. For this, we show that $\lambda_i = 1$ for all i. Suppose otherwise that $\lambda_i > 1$ for some i. For simplicity, we can assume that $a_i = 1$ and in some ball, say $B_r(0)$, we have $\mu_u = \lambda \delta_0$ with $\lambda > 1$ while $\nu = \delta_0$. We will locally insert a cone with Monge–Ampère measure δ_0 that is above u, and this will contradict the maximality of u.

Since $\partial u(0)$ is convex with measure $\lambda > 1$, there is a ball $B_{2\varepsilon}(p_0) \subset \partial u(0)$. Then $u(x) \geq u(0) + p \cdot x$ for all $p \in B_{2\varepsilon}(p_0)$ and all $x \in \Omega$. By subtracting $p_0 \cdot x$ from u and g, we can assume that for all $x \in \Omega$,

$$(3.32) \qquad u(x) \geq u(0) + \varepsilon |x|.$$

Indeed, let $w(x) := u(x) - p_0 \cdot x$. Then $w(x) \geq w(0) + (p - p_0) \cdot x$ for all $p \in B_{2\varepsilon}(p_0)$ and all $x \in \Omega$. For any $x \in \Omega \setminus \{0\}$, we take $p - p_0 = \varepsilon x/|x|$ and obtain $w(x) \geq w(0) + \varepsilon |x|$.

Now, assume (3.32) holds. By subtracting a constant, say $u(0) + \varepsilon r/2$ for small $r > 0$, from u and g we can assume that $u(0) < 0$ but $|u(0)|$ is small while $u(x) \geq 0$ for $|x| \geq r$. Then, the convex set $D = \{x \in \Omega : u(x) < 0\}$ contains a neighborhood of 0. On D, we have $\mu_{\lambda^{-1/n} u} = \delta_0$. We now define the lifting v of u and $\lambda^{-1/n} u$ by

$$v(x) = u(x) \quad \text{if } x \in \Omega \setminus D \qquad \text{and} \qquad v(x) = \lambda^{-1/n} u(x) \quad \text{if } x \in D.$$

As in Step 5, we have $v \in \mathcal{S}(\nu, g)$ but $v(0) = \lambda^{-1/n} u(0) > u(0)$, contradicting the definition of u. Therefore $\lambda = 1$, and this completes the proof. $\square$

We are now ready to solve the inhomogeneous Dirichlet problem for the Monge–Ampère equation with continuous boundary data and right-hand side being a Borel measure with finite mass.

Theorem 3.37 (The Dirichlet problem on strictly convex domains). *Let Ω be a bounded and strictly convex domain in $\mathbb{R}^n$. Let μ be a Borel measure in Ω with $\mu(\Omega) < \infty$. Then for any $g \in C(\partial \Omega)$, there is a unique convex Aleksandrov solution $u \in C(\overline{\Omega})$ to the Dirichlet problem*

$$\begin{cases} \det D^2 u = \mu & \text{in } \Omega, \\ u = g & \text{on } \partial \Omega. \end{cases}$$

Proof. Note that there exists a sequence of Borel measures $\{\mu_j\}_{j=1}^{\infty}$ converging weakly* to μ such that each μ_j is a finite combination of Dirac measures with positive coefficients and $\mu_j(\Omega) \leq A < \infty$ for some constant A. Indeed, using dyadic cubes, we can find for each $j \in \mathbb{N}$ a partition of Ω into Borel sets $\Omega = \bigcup_{i=1}^{N_j} B_{j,i}$ where $\mathrm{diam}(B_{j,i}) \leq 1/j$. Then set $\mu_j := \sum_{i=1}^{N_j} \mu(B_{j,i})\delta_{x_{j,i}}$ where $x_{j,i} \in B_{j,i}$.

For each j, by Theorem 3.36, there exists a unique Aleksandrov solution $u_j \in C(\overline{\Omega})$ to

$$\det D^2 u_j = \mu_j \ \text{ in } \Omega, \qquad u_j = g \ \text{ on } \partial\Omega.$$

The conclusion of the theorem now follows from Theorem 3.35, by letting $j \to \infty$ along a suitable subsequence. $\qquad\square$

The strict convexity of Ω was used in Theorems 3.35 and 3.36 to assert the continuity up to the boundary of u and to assert that $u = g$ on $\partial\Omega$ when g is an arbitrary continuous function. In the special case of $g \equiv 0$, these properties of u follow from the Aleksandrov maximum principle in Theorem 3.12. Thus, we have the following basic existence and uniqueness result for the Dirichlet problem with zero boundary data on a general bounded convex domain.

Theorem 3.38 (The Dirichlet problem with zero boundary data). *Let Ω be a bounded convex domain in $\mathbb{R}^n$, and let μ be a nonnegative Borel measure in Ω with $\mu(\Omega) < \infty$. Then there exists a unique convex function $u \in C(\overline{\Omega})$ that is an Aleksandrov solution of*

$$\det D^2 u = \mu \ \text{ in } \Omega, \qquad u = 0 \ \text{ on } \partial\Omega.$$

More generally, the zero boundary data can be replaced by the trace of a convex function on $\overline{\Omega}$. This is also a necessary condition for the solvability of the Dirichlet problem since the boundary data is the trace of the Aleksandrov solution, if any. We have the following theorem.

Theorem 3.39 (The Dirichlet problem with convex boundary data). *Let Ω be a bounded open convex domain in $\mathbb{R}^n$, and let μ be a nonnegative Borel measure in Ω with $\mu(\Omega) < \infty$. Let $\varphi \in C(\overline{\Omega})$ be a convex function. Then there exists a unique convex Aleksandrov solution $u \in C(\overline{\Omega})$ to*

$$\begin{cases} \det D^2 u = \mu & \text{in } \Omega, \\ \quad\ \ u = \varphi & \text{on } \partial\Omega. \end{cases}$$

In the proof of Theorem 3.39, we will use the following consequence of the maximum principle for harmonic functions.

Lemma 3.40. *Let Ω be a bounded convex domain in $\mathbb{R}^n$. Let $u \in C(\overline{\Omega})$ be convex, and let $h \in C(\overline{\Omega})$ be harmonic. If $u \leq h$ on $\partial\Omega$, then $u \leq h$ in Ω.*

Proof. For any $x_0 \in \Omega$, take $p \in \partial u(x_0)$. Then $l(x) := u(x_0) + p \cdot (x - x_0)$ satisfies $l \leq u$ in Ω, and also in $\overline{\Omega}$ by continuity. From $u \leq h$ on $\partial\Omega$, it follows that $h - l \geq 0$ on $\partial\Omega$. Note that $h - l$ is harmonic in Ω. Thus, by the maximum principle for harmonic functions, $h - l \geq 0$ in Ω. In particular $h(x_0) \geq l(x_0) = u(x_0)$. Since x_0 is arbitrary, we infer that $h \geq u$ in Ω. $\square$

Proof of Theorem 3.39. Let $h \in C(\overline{\Omega})$ be the harmonic function with $h = \varphi$ on $\partial\Omega$; see Theorem 2.80.

Step 1. We first prove the theorem for $\mu \equiv 0$. Define u as in the proof of Theorem 3.33; that is, $u(x) = \sup_{v \in \mathcal{S}} v(x)$ for $x \in \overline{\Omega}$, where

$$\mathcal{S} = \{v \in C(\overline{\Omega}) : v \text{ is a convex function and } v \leq \varphi \text{ on } \partial\Omega\}.$$

Instead of Steps 1–3 in the proof of Theorem 3.33, we only need to show that $\varphi \leq u \leq h$ in Ω and the rest is similar to Step 4 there. If $v \in \mathcal{S}$, then v is convex in Ω, with $v \leq h$ on $\partial\Omega$. Thus, by Lemma 3.40, $v \leq h$ in Ω. Hence $u \leq h$. To show that $u \geq \varphi$, let us fix $x_0 \in \Omega$ and take $p \in \partial\varphi(x_0)$. Then $\bar{v}(x) := \varphi(x_0) + p \cdot (x - x_0) \leq \varphi(x)$ for all $x \in \Omega$. This implies that $\bar{v} \in \mathcal{S}$ and $u(x_0) \geq \bar{v}(x_0) = \varphi(x_0)$. Therefore $u \geq \varphi$ as desired.

Step 2. Now, we prove the theorem for general μ. There exists a sequence of measures $\{\mu_j\}_{j=1}^\infty$ converging weakly* to μ such that each μ_j is a finite combination of Dirac measures with positive coefficients and $\mu_j(\Omega) \leq A < \infty$ for some constant A. By Theorem 2.51, there exists an increasing sequence of uniformly convex domains $\Omega_j \subset \Omega$ such that all points of mass concentration of μ_j lie in Ω_j and Ω_j converges to Ω in the Hausdorff distance.

For each j, we will apply Theorem 3.37 on Ω_j. There exists a unique convex function $u_j \in C(\overline{\Omega_j})$ that is an Aleksandrov solution of

$$\det D^2 u_j = \mu_j \quad \text{in } \Omega_j, \qquad u_j = \varphi \quad \text{on } \partial\Omega_j.$$

Let $w_j \in C(\overline{\Omega_j})$ be the Aleksandrov solution of

$$\det D^2 w_j = 0 \quad \text{in } \Omega_j, \qquad w_j = \varphi \quad \text{on } \partial\Omega_j.$$

Let $v_j \in C(\overline{\Omega_j})$ be the Aleksandrov solution of

$$\det D^2 v_j = \mu_j \quad \text{in } \Omega_j, \qquad v_j = 0 \quad \text{on } \partial\Omega_j.$$

Then, in view of Lemma 3.10, $\mu_{v_j + w_j} \geq \mu_{v_j} + \mu_{w_j} = \mu_{u_j}$. By the comparison principle in Theorem 3.21, $v_j + w_j \leq u_j$ in Ω_j.

Since $h + \max_{\partial\Omega_j}(\varphi - h)$ is harmonic in Ω_j and not less than the convex function u_j on $\partial\Omega_j$ (which is φ), Lemma 3.40 yields

$$u_j \leq h + \max_{\partial\Omega_j}(\varphi - h) \quad \text{in } \Omega_j.$$

By the comparison principle, we have $w_j \geq \varphi$ in Ω_j. Therefore, all together,

$$\varphi + v_j \leq u_j \leq h + \max_{\partial \Omega_j}(\varphi - h) \quad \text{in } \Omega_j.$$

Now, using the Aleksandrov maximum principle for v_j, we find

$$\varphi - C(n, \Omega)[\text{dist}(\cdot, \partial \Omega_j)]^{1/n} A^{1/n} \leq u_j \leq h + \max_{\partial \Omega_j}(\varphi - h) \text{ in } \Omega_j.$$

As in the first paragraph of the proof of Theorem 3.35, letting $j \to \infty$ along a subsequence, we obtain $u \in C(\overline{\Omega})$, solving our Dirichlet problem.   □

3.8. Comparison Principle with Nonconvex Functions

We first record a simple comparison principle for C^2 functions that are possibly nonconvex.

Lemma 3.41. *Let Ω be a bounded and open set in $\mathbb{R}^n$. Let $u, v \in C^2(\Omega) \cap C(\overline{\Omega})$ be functions such that v is convex and $\det D^2 v > \det D^2 u$ in Ω. Then*

$$\min_{\overline{\Omega}}(u - v) = \min_{\partial \Omega}(u - v).$$

Proof. Assume that $u - v$ attains its minimum value on $\overline{\Omega}$ at $x_0 \in \Omega$. Then $D^2 u \geq D^2 v(x_0) \geq 0$, so $\det D^2 u(x_0) \geq \det D^2 v(x_0)$, a contradiction.   □

Now, we use Lemma 3.41 to show that the global Hölder estimates in Lemma 3.26 are optimal in dimensions at least 3.

Proposition 3.42. *Let $n \geq 3$, $\alpha = 2/n$, and*

$$\Omega := \{(x', x_n) : |x'| < 2^{-n}, 0 < x_n < (2^{-2n} - |x'|^2)^{\frac{1}{1-\alpha}}\} \subset \mathbb{R}^n.$$

Let $u \in C(\overline{\Omega})$ be the Aleksandrov solution to

$$\det D^2 u = 1 \quad \text{in } \Omega, \qquad u = 0 \quad \text{on } \partial \Omega.$$

Then for $x = (0, x_n) \in \Omega$ sufficiently close to $\partial \Omega$, we have

$$|u(x)| \geq c(n) \, \text{dist}^{2/n}(x, \partial \Omega).$$

Proof. Let $v(x) = x_n + x_n^\alpha(|x'|^2 - 2^{-2n})$. Then, $v \in C^\infty(\Omega)$, and $v = 0$ on $\partial \Omega$. Computing as in the proof of Lemma 3.26, we have

$$\det D^2 v = 2^{n-1} x_n^{n\alpha - 2}\left[\alpha(1 - \alpha)2^{-2n} - (\alpha^2 + \alpha)|x'|^2\right] \leq 2^{n-1}2^{-2n} \leq 1/4.$$

By using Theorems 5.13 and 6.5 in later chapters, we have $u \in C^\infty(\Omega)$. Thus, Lemma 3.41 tells us that $v \geq u$ in Ω. Hence, for $x = (0, x_n)$, we have

$$|u(x)| \geq |v(x)| = x_n^\alpha(2^{-2n} - x_n^{1-\alpha}) \geq x_n^\alpha 2^{-2n-1} = 2^{-2n-1}\text{dist}^{2/n}(x, \partial \Omega)$$

if $x_n > 0$ is sufficiently small.   □

The solvability of the Dirichlet problem for the Monge–Ampère equation on strictly convex domains with continuous boundary data allows us to extend the comparison principle in Theorem 3.21 to $W^{2,n}$ nonconvex functions on possibly nonconvex domains.

Theorem 3.43 (Comparison principle for $W^{2,n}$ nonconvex functions). *Let Ω be a bounded and open set in $\mathbb{R}^n$. Let $u \in W^{2,n}(\Omega)$. Let $v \in C(\overline{\Omega})$ be a convex function satisfying $\mu_v(\Omega) < \infty$ and*

$$\mu_v \geq \max\{\det D^2 u, 0\}\, dx \quad in\ \Omega.$$

Then

$$\min_{\overline{\Omega}}(u - v) = \min_{\partial\Omega}(u - v).$$

Proof. To simplify, let us denote $f^+ = \max\{0, f\}$.

Step 1. We first prove the theorem for $u \in C^2(\Omega)$. Assume by contradiction that $u - v$ attains its minimum value on $\overline{\Omega}$ at $x_0 \in \Omega$ with

$$u(x_0) - v(x_0) < \min_{\partial\Omega}(u - v).$$

Since Ω is bounded, we can find $\delta > 0$ small such that the function

$$w(x) := v(x) + \delta|x - x_0|^2$$

satisfies the following: $u(x) - w(x)$ attains its minimum value on $\overline{\Omega}$ at $y \in \Omega$ with

$$u(y) - w(y) < \min_{\partial\Omega}(u - w).$$

Note that

$$\mu_w \geq \mu_v \geq (\det D^2 u)^+\, dx \geq \det D^2 u\, dx.$$

Observe that $D^2 u(y)$ cannot be positive definite; otherwise we obtain a contradiction when applying Lemma 3.22 to u and v in $B_r(y)$ where $D^2 u$ is positive definite and $\mu_u = \det D^2 u\, dx$. Hence $D^2 u(y)$ has an eigenvalue $\lambda \leq 0$ with a corresponding eigenvector $z \in \partial B_1(0)$. By the Taylor expansion,

$$(3.33) \qquad u(y + tz) - u(y) = At + (\lambda/2)t^2 + o(t^2) \quad \text{as } t \to 0,$$

where $A := Du(y) \cdot z$. We also have from the convexity of v

$$
\begin{aligned}
w(y + tz) - w(y) &= v(y + tz) - v(y) + \delta[|y + tz - x_0|^2 - |y - x_0|^2]\\
&\geq p \cdot (tz) + \delta[|y + tz - x_0|^2 - |y - x_0|^2] \quad (p \in \partial v(y))\\
&\geq \tilde{A}t + \delta t^2 \quad (\tilde{A} = p \cdot z + 2\delta(y - x_0) \cdot z).
\end{aligned}
$$

From the minimality of $u - w$ at y and (3.33), we find

$$
\begin{aligned}
u(y) - w(y) &\leq u(y + tz) - w(y + tz)\\
&\leq u(y) - w(y) + (A - \tilde{A})t + (\lambda/2 - \delta)t^2 + o(t^2), \text{ as } t \to 0.
\end{aligned}
$$

Therefore $(A-\tilde{A})t+(\lambda/2-\delta)t^2 \geq 0$ for all $|t|$ small, so $A = \tilde{A}$ and $\lambda/2-\delta \geq 0$. This contradicts $\lambda \leq 0 < \delta$.

Step 2. Finally, we prove the theorem for $u \in W^{2,n}(\Omega)$. We use an approximation argument and the solvability of the Dirichlet problem for the Monge–Ampère equation on a ball with continuous boundary data.

Let y and w be as above. Let $r > 0$ be such that $B := \overline{B_r(y)} \subset \Omega$. Then

$$\int_B (\det D^2 u)^+ \, dx \leq \mu_v(B) \leq \mu_v(\Omega) < \infty.$$

Let $\{u_k\}_{k=1}^{\infty} \subset C_c^{\infty}(\mathbb{R}^n)$ be a sequence of smooth functions that satisfies $\lim_{k\to\infty} \|u_k - u\|_{W^{2,n}(B)} = 0$; see Theorem 2.50. From the Sobolev Embedding Theorem (Theorem 2.76), we also have $\lim_{k\to\infty} \|u_k - u\|_{C(\overline{B})} = 0$. For each k, by Theorem 3.37, there exists a unique convex solution $\tilde{u}_k \in C(\overline{B})$ to

$$\mu_{\tilde{u}_k} = (\det D^2 u_k)^+ \, dx \quad \text{in } B, \quad \tilde{u}_k = u_k \quad \text{on } \partial B.$$

Since

$$\lim_{k\to\infty} \|(\det D^2 u_k)^+ - (\det D^2 u)^+\|_{L^1(B)} = 0 \quad \text{and} \quad \lim_{k\to\infty} \|u_k - u\|_{C(\partial B)} = 0,$$

we can let $k \to \infty$ and then use the compactness result in Theorem 3.35. We find that, up to extracting a subsequence, $\tilde{u}_k \to \tilde{u}$, uniformly on compact subsets of B, where $\tilde{u} \in C(\overline{B})$ is the convex solution to

$$\mu_{\tilde{u}} = (\det D^2 u)^+ \, dx \quad \text{in } B, \quad \tilde{u} = u \quad \text{on } \partial B.$$

Because $u_k \in C^2(\Omega)$, Step 1 gives $\tilde{u}_k \leq u_k$ in B which implies that $\tilde{u} \leq u$ on $\overline{B}$. Hence $\tilde{u}(y) \leq u(y)$ and

$$\tilde{u}(y) - w(y) \leq u(y) - w(y) \leq \min_{\overline{\Omega}}(u - w)$$

(3.34)
$$\leq \min_{\partial B}(u - w) = \min_{\partial B}(\tilde{u} - w).$$

On B, $\mu_{\tilde{u}} = (\det D^2 u)^+ \, dx \leq \mu_v \leq \mu_w$, and both $\tilde{u}$ and w are convex. Thus, the comparison principle in Theorem 3.21 implies that

$$\min_{\overline{B}}(\tilde{u} - w) = \min_{\partial B}(\tilde{u} - w).$$

This combined with (3.34) shows that $\tilde{u} - w$ has its minimum value over $\overline{B}$ at an interior point y in B. However, this contradicts Lemma 3.22. $\square$

We can use the comparison principle in Theorem 3.43 to obtain the uniqueness of $W^{2,n}$ solutions to the homogenous Monge–Ampère equation with zero boundary values; see Problem 3.17. In view of Problem 1.7, the exponent n in $W^{2,n}$ is optimal.

3.9. Problems

In Problems 3.3–3.5, we use the concept of polar body of a convex body defined in Problem 2.10; moreover, δ_z is the Dirac measure at $z \in \mathbb{R}^n$.

Problem 3.1. Let $p^1, \dots, p^{n+1}$ be the vertices of a convex polygon in $\mathbb{R}^2$, and let $u(x) = \sup_{1 \le i \le n+1}\{x \cdot p^i\}$ where $x \in \mathbb{R}^2$. Find $\partial u(\mathbb{R}^2)$ and μ_u.

Problem 3.2. Let $f : [0,1) \to (0, \infty)$ be an integrable function. Find the Aleksandrov solution to the Monge–Ampère equation

$$\det D^2 u(x) = f(|x|) \quad \text{in } B_1(0) \subset \mathbb{R}^n, \quad u = 0 \quad \text{on } \partial B_1(0).$$

Problem 3.3 (Polar body and fundamental solution of the Monge–Ampère equation). Let Ω be a bounded, convex domain in $\mathbb{R}^n$. For $z \in \Omega$, let $u^z \in C(\overline{\Omega})$ be the convex Aleksandrov solution to the Monge–Ampère equation

$$\det D^2 u^z = \delta_z \quad \text{in } \Omega, \qquad u^z = 0 \quad \text{on } \partial\Omega.$$

Let $\overline{\Omega}^z = \{x \in \mathbb{R}^n : x \cdot (y - z) \le 1 \text{ for all } y \in \overline{\Omega}\}$ be the polar body of $\overline{\Omega}$ with respect to $z \in \Omega$. Show that

$$\partial u^z(z) = |u^z(z)|\overline{\Omega}^z \quad \text{and} \quad |u^z(z)|^n = |\overline{\Omega}^z|^{-1}.$$

Problem 3.4 (Santaló point). Let K be a convex body in $\mathbb{R}^n$. Prove that there is a unique interior point $s(K)$ of K such that $|K^{s(K)}| \le |K^z|$ for all interior points z of K. We call $s(K)$ the Santaló point of K.
Hint: Use Problem 3.3, Aleksandrov's maximum principle, and compactness of the Monge–Ampère equation.

Problem 3.5. Let Ω be a bounded, convex domain in $\mathbb{R}^n$ that is centrally symmetric, so that $-x \in \Omega$ whenever $x \in \Omega$. Let $u \in C(\overline{\Omega})$ be the Aleksandrov solution to the Monge–Ampère equation

$$\det D^2 u = \delta_0 \quad \text{in } \Omega, \qquad u = 0 \quad \text{on } \partial\Omega.$$

(a) Show that $\|u\|_{L^\infty(\Omega)}^n = |u(0)|^n \le n^{n/2}|\Omega|$.

(b) The *symmetric Mahler conjecture* states that if K is a centrally symmetric convex body in $\mathbb{R}^n$, then

$$|K||K^0| \ge \frac{4^n}{n!}.$$

This conjecture was proved by Mahler [**Mah**] for $n = 2$ and by Iriyeh and Shibata [**ISh**] for $n = 3$. Prove that the symmetric Mahler conjecture is equivalent to the following estimate for the fundamental solution of the Monge–Ampère equation:

$$|u(0)|^n \le \frac{n!}{4^n}|\Omega|.$$

Problem 3.6. Let Ω be a bounded domain in $\mathbb{R}^n$ with smooth boundary. This problem aims to prove the classical isoperimetric inequality

$$\frac{\mathcal{H}^{n-1}(\partial\Omega)}{|\Omega|^{(n-1)/n}} \geq \frac{\mathcal{H}^{n-1}(\partial B_1(0))}{|B_1(0)|^{(n-1)/n}} = n\omega_n^{1/n}$$

via the ABP method, following Cabré [**Cab2**]. Proceed as follows.

(a) Let ν be the outer unit normal to $\partial\Omega$. Show that there is a solution $u \in C^2(\overline{\Omega})$ to the Neumann boundary value problem

$$\Delta u = \mathcal{H}^{n-1}(\partial\Omega)/|\Omega| \quad \text{in } \Omega, \quad Du \cdot \nu = 1 \quad \text{on } \partial\Omega.$$

(b) Let Γ^- be the lower contact set of u:

$$\Gamma^- = \{y \in \Omega : u(x) \geq u(y) + p \cdot (x - y) \text{ for all } x \in \Omega \text{ for some } p(y) \in \mathbb{R}^n\}.$$

Show that

$$B_1(0) \subset Du(\Gamma^-) \cap \{x \in \Omega : |Du(x)| < 1\}.$$

(c) Take volumes in the inclusion $B_1(0) \subset Du(\Gamma^-)$, invoke Theorem 2.74 and use $0 \leq \det D^2 u \leq (\Delta u/n)^n$ on Γ^- to conclude.

Problem 3.7. Let $A(x)$ be a smooth function on $(-1,1)$. Consider the following fourth-order boundary value problem for a convex function u:

$$\left(\frac{1}{u''}\right)'' = -A \quad \text{on } (-1,1), \quad u''(x) \to \frac{2}{1-x^2} + w(x) \quad \text{as } x \to \pm 1,$$

for some (also unknown) smooth function w on $[-1,1]$. This is a *one-dimensional version of the Abreu equation with Guillemin boundary conditions*. Geometrically, this equation corresponds to a rotationally invariant metric on the two-dimensional sphere whose scalar curvature is a given function $A(h)$ of the Hamiltonian h for the circle action.

(a) Suppose that a smooth convex solution u exists on $(-1,1)$. Show that A must satisfy the moment conditions

$$\int_{-1}^{1} A(x)\,dx = 2, \qquad \int_{-1}^{1} xA(x)\,dx = 0.$$

(b) Assume now that $A(x)$ satisfies the above moment conditions. Then, there exists a convex solution u if and only if

$$\mathcal{L}_A(f) := f(1) + f(-1) - \int_{-1}^{1} fA\,dx > 0$$

for all nonaffine convex functions f on $[-1,1]$. In this case, show that the solution is an absolute minimum of the functional

$$\mathcal{F}_A(f) := f(1) + f(-1) - \int_{-1}^{1} \left(fA + \log(f'')\right)dx.$$

Hint for the existence of solutions in part (b): Let $v = 1/u''$. Then $v'' = -A$ with $v(x) \to 0$ and $v'(x) \to \mp 1$ as $x \to \pm 1$. Thus v can be expressed in terms of A and the Green's function $G(x, y)$ of the one-dimensional Laplace equation via $v(x) = \int_{-1}^{1} G(x, y) A(y) \, dy$. Here, for a fixed $x \in (-1, 1)$, the function $G(x, \cdot)$ is concave; it is a linear function of y on the intervals $(-1, x)$ and $(x, 1)$, vanishing at ± 1. Suppose that $\mathcal{L}_A(f) > 0$ for all nonaffine convex functions f on $[-1, 1]$. Then $v(x) = \mathcal{L}_A(-G(x, \cdot))$ is positive in $(-1, 1)$. Integrate twice to get a solution u from $u'' = v^{-1}$.

Problem 3.8. Let Ω be a bounded domain in $\mathbb{R}^n$.

(a) Let $L(u; \Omega) = \{(x, Du(x)) : x \in \Omega\}$ be the gradient graph of a function $u : \Omega \to \mathbb{R}$. Prove that if $u \in C^2$, then

$$\int_{\Omega} |\det D^2 u| \, dx = \lim_{\lambda \to \infty} \frac{\mathcal{H}^n(L(\lambda u; \Omega))}{\lambda^n}.$$

Hint: Show that $\mathcal{H}^n(L(u; \Omega)) = \int_{\Omega} \sqrt{\det(I_n + (D^2 u)^2)} \, dx$.

(b) Let $u \in C^2(\mathbb{R}^n)$ be a function with support in Ω. Prove the following ABP-type estimate:

$$\sup_{\Omega} |u| \leq C(n) |\Omega|^{1/n} \left(\int_{\Omega} |\det D^2 u| \, dx \right)^{1/n}.$$

Compare with Lemmas 3.15 and 3.17.

Problem 3.9. Let $u, v \in C(\overline{\Omega})$ be convex functions in a bounded, convex domain Ω in $\mathbb{R}^n$ with $u = v = 0$ on $\partial \Omega$. Assume that for some $0 < \varepsilon < 1$,

$$1 - \varepsilon \leq \det D^2 u \leq 1 + \varepsilon \quad \text{in } \Omega \quad \text{and} \quad \det D^2 v = 1 \quad \text{in } \Omega.$$

Prove that there is a positive constant $C(n, |\Omega|)$ such that

$$|u(x) - v(x)| \leq C\varepsilon \quad \text{for all } x \in \Omega.$$

Problem 3.10. Verify that $u(x) = -(1 - |x|^2)^{\frac{1}{2}}$ is the unique convex solution to the following singular Monge–Ampère equation on the unit ball:

$$\det D^2 u = |u|^{-(n+2)} \quad \text{in } B_1(0), \qquad u = 0 \quad \text{on } \partial B_1(0).$$

Problem 3.11 (Nonconvex supersolution). Assume $p > 0$. Let

$$\Omega := \{(x', x_n) : |x'| < 1, 0 < x_n < (1 - |x'|^2)^{\frac{n}{n+p-2}}\} \subset \mathbb{R}^n.$$

Show that the function

$$u(x) = \left[x_n - x_n^{\frac{2}{n+p}} (1 - |x'|^2)^{\frac{n}{n+p}} \right] / 2$$

is smooth in Ω and satisfies

$$\det D^2 u \leq |u|^{-p} \quad \text{in } \Omega \quad \text{and} \quad u = 0 \quad \text{on } \partial \Omega.$$

Problem 3.12 (Supersolutions for a singular Monge–Ampère equation). Assume $p \geq 1$. Let $\Omega = \{(x', x_n) : |x'| < 1, 0 < x_n < 1 - |x'|^2\} \subset \mathbb{R}^n$, $n \geq 2$. Show that there is a constant $C = C(n, p)$ such that the function

$$u = Cx_n - Cx_n^{\frac{2}{n+p}}(1 - |x'|^2)^{\frac{n+p-2}{n+p}}$$

is smooth, convex in Ω, and satisfies

$$\det D^2 u \leq |u|^{-p} \quad \text{in } \Omega \qquad \text{and} \qquad u = 0 \quad \text{on } \partial\Omega.$$

Problem 3.13. Let Ω be a bounded, convex domain in $\mathbb{R}^n$. Let $p > 0$. Assume that $u \in C(\overline{\Omega}) \cap C^\infty(\Omega)$ solves the singular Monge–Ampère equation

$$\det D^2 u = |u|^{-p} \quad \text{in } \Omega, \qquad u = 0 \quad \text{on } \partial\Omega.$$

(a) Show that $u \in C^{\frac{2}{n+p}}(\overline{\Omega})$ with the estimate

$$|u(x)| \leq C(n, p, \Omega)\operatorname{dist}^{\frac{2}{n+p}}(x, \partial\Omega) \quad \text{for all } x \in \Omega.$$

(b) Show that the global Hölder regularity in part (a) is optimal in the following sense: There exist bounded convex domains Ω in $\mathbb{R}^n$ such that $u \notin C^\beta(\overline{\Omega})$ for any $\beta > \frac{2}{n+p}$.
    Hint: Use Lemma 3.41 and Problem 3.11.

Problem 3.14. Let Ω be a bounded, convex domain in $\mathbb{R}^n$. Let $q > 0$ and $\lambda(\Omega) > 0$. Assume that $u \in C(\overline{\Omega}) \cap C^\infty(\Omega)$ is a solution to the degenerate Monge–Ampère equation

$$\det D^2 u = \lambda(\Omega)|u|^q \quad \text{in } \Omega, \qquad u = 0 \quad \text{on } \partial\Omega.$$

(a) Assume $q \geq n - 2$. Show that $u \in C^\beta(\overline{\Omega})$ for all $\beta \in (0, 1)$.

(b) Assume $0 < q < n - 2$. Show that $u \in C^\beta(\overline{\Omega})$ for all $\beta \in (0, \frac{2}{n-q})$.

Problem 3.15. Let Δ be the open triangle in $\mathbb{R}^2$ with vertices at $(0, 0)$, $(1, 0)$, and $(1/2, \sqrt{3}/2)$. Write down an explicit formula for the solution to

$$\det D^2 u = 1 \quad \text{in } \Delta, \qquad u = 0 \quad \text{on } \partial\Delta.$$

Problem 3.16. Let $n \geq 2$ and $\Omega = \{(x', x_n) : |x'| < 1, 0 < x_n < 1 - |x'|^2\}$. For $0 < p < n$, consider $u(x) = x_n - x_n^{\frac{2}{n+p}}(1 - |x'|^2)^{\frac{n+p-2}{n+p}}$. Show that

(a) u is smooth, convex in Ω with $u = 0$ on $\partial\Omega$, $\det D^2 u \in L^1(\Omega)$, and

(b) $u \notin C^\alpha(\overline{\Omega})$ for any $\alpha > \frac{2}{n+p}$.

It follows that for $1/n < \alpha < 1$, we can choose $\max\{2/\alpha - n, 1\} < p < n$, and the above function shows that the exponent $1/n$ in Lemma 3.18 is sharp.

Problem 3.17. Let Ω be a bounded domain in $\mathbb{R}^n$. Suppose that $u \in W^{2,n}(\Omega)$ satisfies

$$\det D^2 u = 0 \quad \text{in } \Omega, \quad u = 0 \quad \text{on } \partial\Omega.$$

Show that $u \equiv 0$.

Problem 3.18. Let Ω be a bounded convex domain in $\mathbb{R}^n$. Let $u \in W^{2,n}(\Omega)$ be a convex function. Show that $\mu_u = (\det D^2 u)\, dx$.

3.10. Notes

Most materials in this chapter are classical for which we largely rely on Figalli [**F2**] and Gutiérrez [**G2**]. In particular, most results and arguments in Sections 3.2, 3.3, 3.5, and 3.7 can be found in [**G2**, Chapter 1] and [**F2**, Chapter 2]; see also Le–Mitake–Tran [**LMT**, Chapter 3]. Quite different from the real Monge–Ampère case where the Monge–Ampère measure can be defined for all convex functions, the complex Monge–Ampère measure cannot be defined for all plurisubharmonic functions on $\mathbb{C}^n$; see Bedford–Taylor [**BT**] for examples.

In Section 3.3, the sharp form of the Aleksandrov maximum principle in Theorem 3.12 is implicit in Trudinger [**Tr3**]. It can be used to give a proof of the classical isoperimetric inequality; see Theorem 4.16. The proof of Theorem 3.16 is based on Gilbarg–Trudinger [**GT**, Section 9.1]. For the Aleksandrov–Bakelman–Pucci maximum principle in Theorem 3.16, we refer to the original papers [**Al2**, **Al3**, **Ba2**, **Pu**]. Kuo and Trudinger [**KT**] extended the ABP estimates to $a^{ij} D_{ij} u$ being in L^q where $q < n$, and the new estimates depend on restrictions on the eigenvalues of the coefficient matrix $(a^{ij})_{1 \le i,j \le n}$. Brendle [**Bre**] recently proved the isoperimetric inequality for a minimal submanifold in Euclidean space, using a method inspired by the ABP maximum principle.

In Section 3.5, Lemma 3.25 is from Caffarelli [**C2**]; see also Trudinger–Wang [**TW6**, Section 3.2]. Lemma 3.26 is from [**C2**]. Lemma 3.29 is from Le [**L13**].

Example 3.32 in Section 3.6 and Proposition 3.42 in Section 3.8 show that many barehanded estimates based on the Aleksandrov maximum principle are essentially sharp.

For the Dirichlet problem in Section 3.7, see Bakelman [**Ba5**, Chapter 11] for a historical account; see also Rauch–Taylor [**RT**] for original motivations. Theorems 3.33, 3.36, and 3.37 were first established by Aleksandrov and Bakelman. Our proofs here follow [**RT**]. Theorem 3.39 is independently due to Błocki [**Bl**] and Hartenstine [**Ha1**]; it is also contained in Trudinger–Wang [**TW6**, Section 2.3] and Han [**H**, Section 8.2] with different proofs.

Theorem 3.43 in Section 3.8 is due to Ozanski [**Oz**]; it was first established for strictly convex domains by Rauch and Taylor [**RT**]. Ozanski [**Oz**] also discusses an interesting application of Theorem 3.43 in the theory of the two-dimensional Navier–Stokes equations.

Problem 3.7 is taken from Donaldson [**D2**]. Problem 3.8 is based on Viterbo [**Vb**] where he discovered that the ABP estimate for the Monge–Ampère equation is essentially equivalent to an isoperimetric inequality in symplectic geometry. Problems 3.12–3.14(b) are taken from Le [**L12**] while Problem 3.14(a) is taken from [**L7**]. Problems 3.11 and 3.16 are taken from [**L13**]. For a different construction of examples showing the sharpness of the exponent $1/n$ in Lemma 3.18, see Jerrard [**Jer**] in the context of Monge–Ampère functions.

Chapter 4

Classical Solutions

Having defined the concept of convex Aleksandrov solutions to a Monge–Ampère equation and having proved some basic existence results in Chapter 3, we now turn to investigate the regularity issue. One of the most basic regularity questions is: *How smooth are convex Aleksandrov solutions to a Monge–Ampère equation in terms of the data?* One particular question is whether all convex Aleksandrov solutions u to the Monge–Ampère equation $\det D^2 u = 1$ on a convex domain Ω in $\mathbb{R}^n (n \geq 2)$ are smooth. Surprisingly, the answer is negative in all dimensions $n \geq 3$. On the other hand, when solutions are constrained by suitably smooth boundary data, the answer is positive and we have classical solutions (those that are at least continuously twice differentiable). We will discuss these answers in this chapter.

For classical solutions, we establish higher-order a priori boundary estimates which are essentially due to Ivočkina, Krylov, Caffarelli, Nirenberg, and Spruck. They will be used in conjunction with the method of continuity to study the classical solvability of the Dirichlet problem. As a consequence of this solvability result and the maximum principle, we obtain the classical isoperimetric inequality and a nonlinear integration by parts inequality for the Monge–Ampère operator. We will discuss Pogorelov's counterexamples to interior regularity which demonstrate that there are no purely interior estimates for Aleksandrov solutions to the Monge–Ampère equation.

Several boundary estimates up to second-order derivatives can be established in less regular settings, and they will be used in later chapters. In so doing, we introduce the notion of the *special subdifferential* of a convex function at a boundary point of a convex domain. This notion will be relevant for the boundary regularity theory of the Monge–Ampère equation.

We then establish the quadratic separation of a convex function from its tangent hyperplanes at boundary points under suitable hypotheses.

Notation. Throughout this chapter, $n \geq 2$. We denote a point $x = (x_1, \ldots, x_{n-1}, x_n) \in \mathbb{R}^n$ by $x = (x', x_n)$ where $x' = (x_1, \ldots, x_{n-1})$. We refer to Definitions 2.46 and 2.88 for pointwise concepts used in this chapter.

4.1. Special Subdifferential at the Boundary

In this section, we motivate the definition of the *special subdifferential* of a convex function $u \in C(\overline{\Omega})$ at a boundary point $x_0 \in \partial\Omega$ of a bounded convex domain Ω. This notion is very convenient for studying the boundary regularity theory of the Monge–Ampère equation with optimal boundary conditions; see Chapter 8. Recall from Lemma 2.33 that the multi-valued subdifferential (or normal mapping) $\partial u(x_0)$ defined in Definition 2.30 reduces to the classical gradient $Du(x_0)$ if u is differentiable at an interior point $x_0 \in \Omega$. Due to the one-sided nature of $\partial\Omega$ with respect to $\overline{\Omega}$, this uniqueness does not hold when x_0 is a boundary point. Assume that $\partial\Omega$ is C^1 at $x_0 \in \partial\Omega$. Suppose that $p \in \partial u(x_0)$, so

$$u(x) \geq u(x_0) + p \cdot (x - x_0) \quad \text{for all } x \in \overline{\Omega}.$$

Then, for all $\gamma > 0$, due to the convexity of Ω, we also have $p + \gamma\nu_{x_0} \in \partial u(x_0)$, where ν_{x_0} is the outer unit normal to $\partial\Omega$ at x_0. In other words, *if $\partial u(x_0)$ is not empty, then it always contains a ray.* See Figure 4.1. Thus, it is not convenient to work directly with the whole large set $\partial u(x_0)$.

Now, suppose furthermore that

$$u(x) \leq u(x_0) + q \cdot (x - x_0) + o(|x - x_0|) \quad \text{when } x \in \overline{\Omega} \quad \text{and} \quad x \to x_0.$$

The last assumption was motivated by our goal of eventually establishing the global C^1 regularity for u with q being the classical gradient $Du(x_0)$. Then, $(q - p) \cdot (x - x_0) \geq 0$ for all $x \in \overline{\Omega}$ close to x_0, from which we find $q = p - \lambda\nu_{x_0}$, for some $\lambda \geq 0$. This implies a minimality property of $q \cdot \nu_{x_0}$.

The following definition will pick up the boundary gradient of u when u is globally C^1.

Definition 4.1 (Special subdifferential at the boundary). Let $u \in C(\overline{\Omega})$ be a convex function on a bounded convex domain Ω in $\mathbb{R}^n$. Suppose that $\partial\Omega$ is C^1 at $x_0 \in \partial\Omega$. Let ν_{x_0} be the outer unit normal to $\partial\Omega$ at x_0. We use $\partial_s u(x_0)$ to denote all vectors $p \in \mathbb{R}^n$ with the following property:

$$x_{n+1} = u(x_0) + p \cdot (x - x_0)$$

is a supporting hyperplane for the graph of u in $\overline{\Omega}$ at $(x_0, u(x_0))$, but for any $\varepsilon > 0$, $x_{n+1} = u(x_0) + (p - \varepsilon\nu_{x_0}) \cdot (x - x_0)$ is not a supporting hyperplane.

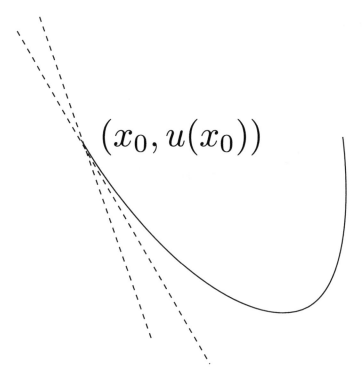

$$(x_0, u(x_0))$$

Figure 4.1. If there is a supporting hyperplane at a boundary point x_0, then there are infinitely many.

When $p \in \partial_s u(x_0)$, we call $x_{n+1} = u(x_0) + p \cdot (x - x_0)$ the *tangent hyperplane* to u at x_0.

In general, the special subdifferential is multi-valued.

Example 4.2. Let Ω be the upper half-ball $\{(x_1, x_2) : x_1^2 + x_2^2 < 1, x_2 > 0\}$ in the plane, and let $u(x_1, x_2) = |x_1| + \sqrt{x_1^2 + x_2^2}$ on $\overline{\Omega}$. Then

$$\partial u(0,0) \supset \{(a + p_1, p_2 - t) : |a| \leq 1, p_1^2 + p_2^2 \leq 1, t \geq 0\},$$

and

$$\partial_s u(0,0) \supset \{(a, 1) : |a| \leq 1\} \cup \{(2,0)\}.$$

Next, following Savin [**S3**, **S4**], we introduce the concept of quadratic separation of a convex function from its tangent hyperplanes. Roughly speaking, the quadratic separation gives a sort of uniform convexity in the directions tangent to the boundary.

Definition 4.3 (Quadratic separation from tangent hyperplanes). Let u, Ω, $x_0 \in \partial \Omega$ be as in Definition 4.1. Let L be a supporting hyperplane to $\overline{\Omega}$ at x_0. We say that u separates quadratically from a tangent hyperplane at

x_0 if there exist $p \in \partial_s u(x_0)$ and $\rho > 0$ such that

$$\rho |x - x_0|^2 \leq u(x) - u(x_0) - p \cdot (x - x_0) \leq \rho^{-1} |x - x_0|^2 \quad \text{on } \partial \Omega \cap \{\text{dist}(\cdot, L) \leq \rho\}.$$

We now prove the existence of the special subdifferential at the boundary under suitable pointwise assumptions; see also Problem 4.2.

Proposition 4.4 (Existence of special subdifferential at the boundary).
Let Ω be a bounded convex domain in $\mathbb{R}^n$ such that $\partial \Omega$ is pointwise C^1 at $x_0 \in \partial \Omega$ and uniformly convex at x_0. Let $u \in C(\overline{\Omega})$ be a convex function satisfying $\det D^2 u \leq \Lambda$ in the sense of Aleksandrov in Ω. Assume that $u = \varphi$ on $\partial \Omega$, where $\varphi \in C(\partial \Omega)$ and φ is $C^{1,1}$ at x_0. Then

(i) *$\partial_s u(x_0)$ is not empty, and*

(ii) *there is a positive constant C depending only on $n, \Lambda, [\varphi]_{C^{1,1}(x_0)}$, $[\partial \Omega]_{C^1(x_0)}, \|\varphi\|_{L^\infty(\partial \Omega)}$, and the pointwise uniform convexity radius R of $\partial \Omega$ at x_0 such that*

$$|p| \leq C \quad \text{for all } p \in \partial_s u(x_0).$$

Proof. Since $\partial \Omega$ is C^1 at x_0, by changing coordinates, we assume that Ω is in the upper half-space, $\partial \Omega$ is tangent to $\{x_n = 0\}$ at $x_0 = 0 \in \partial \Omega$, and there is $\rho_1 > 0$ depending on $[\partial \Omega]_{C^1(0)}$ such that $x_n \leq \rho_1^{-1} |x'|$ on $\partial \Omega \cap B_{\rho_1}$, where we denote $B_r = B_r(0)$. Since φ is $C^{1,1}$ at 0, there is an affine function l such that $|\varphi(x) - l(x)| \leq \rho_2^{-1} |x|^2$ on $\partial \Omega \cap B_{\rho_2}$ where $\rho_2 > 0$ depends on $[\varphi]_{C^{1,1}(0)}$. Subtracting l from φ and u and choosing ρ small, depending on ρ_1, ρ_2, we can assume that

$$|\varphi(x)| \leq \rho^{-1} |x'|^2 \quad \text{on } B_\rho \cap \partial \Omega.$$

Reducing ρ further, depending on R, we also have

$$B_\rho \cap \Omega \subset \{|x'|^2 + (x_n - \rho^{-1})^2 \leq \rho^{-2}\}.$$

It suffices to find $A \in \mathbb{R}$ with $|A| \leq C$, where C depends on the quantities stated in the proposition, such that $x_{n+1} = A x_n$ is supporting hyperplane for the graph of u, but for any $\varepsilon > 0$, $x_{n+1} = (A + \varepsilon) x_n$ is not a supporting hyperplane. For this, we show that

(4.1) $-C x_n \leq u(x) \leq C x_n \quad \text{in } \overline{\Omega}$

and then define

$$A := \max\{a \in \mathbb{R} : u(x) \geq a x_n \quad \text{for all } x \in \overline{\Omega}\}.$$

Note that, by Corollary 3.14 and $\text{diam}(\Omega) \leq 2R$,

$$\|u\|_{L^\infty(\Omega)} \leq C(n, \Lambda, R, \sup_{\partial \Omega} \varphi, \inf_{\partial \Omega} \varphi).$$

Thus, it suffices to establish (4.1) in a small neighborhood of the origin.

It is easy to see that $x_n \geq \rho|x'|^2/2$ in $B_\rho \cap \Omega$ and $x_n \geq \rho^3/2$ on $\partial B_\rho \cap \Omega$. On $\partial\Omega \cap B_\rho$, we have $u(x) = \varphi(x) \leq 2\rho^{-2}x_n$. Therefore, for

$$M := 2\rho^{-2} + 2\rho^{-3}\|u\|_{L^\infty(\Omega)},$$

we have $u(x) \leq Mx_n$ on $\partial(\Omega \cap B_\rho)$ which, by convexity, also implies that $u(x) \leq Mx_n$ in $\Omega \cap B_\rho$.

For the other inequality in (4.1), consider

$$v(x) := \Lambda^{1/n}[|x'|^2 + (x_n - \rho^{-1})^2 - \rho^{-2}] - Mx_n.$$

Then

$$\det D^2 v = 2^n \Lambda \geq \det D^2 u \quad \text{in } B_\rho \cap \Omega.$$

On $B_\rho \cap \partial\Omega$, we have

$$v(x) \leq -Mx_n \leq -\rho^{-1}|x'|^2 \leq \varphi(x) = u(x).$$

On $\partial B_\rho \cap \Omega$, we have $x_n \geq \rho^3/2$ and hence

$$v(x) \leq -Mx_n \leq -M\rho^3/2 \leq u(x).$$

By the comparison principle in Theorem 3.21, $u \geq v$ in $B_\rho \cap \Omega$. It follows that $u \geq -(M + 2\rho^{-1}\Lambda^{1/n})x_n$ in $B_\rho \cap \Omega$. Hence, (4.1) is proved. $\square$

4.2. Quadratic Separation at the Boundary

This section is concerned with the quadratic separation of convex solutions to the Monge–Ampère equation from their tangent hyperplanes at boundary points. We begin with a pointwise result.

Proposition 4.5 (Quadratic separation at a boundary point). *Let Ω be a bounded convex domain in $\mathbb{R}_+^n := \mathbb{R}^n \cap \{x_n > 0\}$ with $0 \in \partial\Omega$. Let $u \in C(\overline{\Omega})$ be a convex function satisfying, in the sense of Aleksandrov,*

$$0 < \lambda \leq \det D^2 u \leq \Lambda \quad \text{in } \Omega.$$

Assume that $u|_{\partial\Omega}$ and $\partial\Omega$ are pointwise C^3 at 0, and assume that $\partial\Omega$ is uniformly convex at 0. Then, on $\partial\Omega$, u separates quadratically from its tangent hyperplanes to $\partial\Omega$ at 0. This means that for any $p \in \partial_s u(0)$, we have

$$(4.2) \qquad \rho|x|^2 \leq u(x) - u(0) - p \cdot x \leq \rho^{-1}|x|^2 \quad \text{on } \partial\Omega \cap \{x_n \leq \rho\},$$

for some small constant ρ depending only on $n, \lambda, \Lambda, \Omega$, and $u|_{\partial\Omega}$.

Proof. By Corollary 3.14, $\|u\|_{L^\infty(\Omega)} \leq C_1(n, \Lambda, d, \min_{\partial\Omega} u, \max_{\partial\Omega} u)$, where $d := \text{diam}(\Omega)$. From Proposition 4.4, we find that the tangent hyperplane l_0 to the graph of u at the origin (in the sense of Definition 4.1) has slope bounded by $C_2(n, \Lambda, [\partial\Omega]_{C^3(0)}, [u|_{\partial\Omega}]_{C^3(0)}, K_0, \min_{\partial\Omega} u, \max_{\partial\Omega} u)$, where K_0 is the pointwise uniform convexity radius of Ω at 0. Note that $d \leq 2K_0$.

After subtracting l_0 from u and $\phi := u|_{\partial\Omega}$, we may assume $l_0 = 0$. Thus, $u \geq 0$ and it suffices to show that

$$(4.3) \qquad \rho|x|^2 \leq u(x) \leq \rho^{-1}|x|^2 \quad \text{on } \partial\Omega \cap \{x_n \leq \rho\},$$

for some $\rho > 0$ small. By the boundedness of u, we only need to prove (4.3) for $|x| \leq c$ where c is small. In this proof, *all small quantities* depend only on $n, \lambda, \Lambda, K_0, [\partial\Omega]_{C^3(0)}$, and $[u|_{\partial\Omega}]_{C^3(0)}$. Below, when writing $o(t^m)$ where $m > 0$, we implicitly understand that $t \to 0$.

Step 1. Since $\phi, \partial\Omega$ are C^3 at 0, there is a cubic polynomial Q_0 such that

$$(4.4) \qquad \phi(x) = Q_0(x') + o(|x'|^3) \quad \text{for } x = (x', x_n) \in \partial\Omega.$$

Indeed, for some $c > 0$ small, $\partial\Omega \cap B_c(0)$ is given by the graph

$$\partial\Omega \cap B_c(0) = \{(x', x_n) : x_n = \psi(x')\}$$

of a C^3 function $\psi(x')$. Thus, for $(x', x_n) \in \partial\Omega \cap B_c(0)$ we can write

$$(4.5) \qquad x_n = Q_1(x') + o(|x'|^3),$$

where Q_1 is a cubic polynomial with no affine part, due to $\partial\Omega \cap \{x_n = 0\} = \{0\}$. Since ϕ is C^3 at 0, we can again find a cubic polynomial Q_2 such that

$$\phi(x) = Q_2(x) + o(|x|^3) \quad \text{for } x \in \partial\Omega \quad \text{near } 0.$$

Substituting (4.5) into this equation, we obtain (4.4) as claimed.

Because $u = \phi \geq 0$ on $\partial\Omega$, Q_0 in (4.4) has no linear part and its quadratic part $Q_0^{(2)}$ is given by, say in a suitable coordinate system (*not necessary the principal one*), at 0,

$$Q_0^{(2)}(x') = \sum_{i<n} \frac{\mu_i}{2} x_i^2, \quad \text{with} \quad \mu_i \geq 0.$$

We need to show that $\mu_i \geq \mu(n, \lambda, \Lambda, [\partial\Omega]_{C^3(0)}, [u|_{\partial\Omega}]_{C^3(0)}, K_0) > 0$. For example, let us establish a quantitative positive lower bound for μ_1. By a dilation of coordinates in the x' variables, we can assume that

$$x_n = Q_1(x') + o(|x'|^3) = \frac{1}{2}|x'|^2 + \text{cubic in } x' + o(|x'|^3) \quad \text{on } \partial\Omega \cap B_c(0).$$

Let $\tilde{u}(x) = u(x) - \mu_1 x_n$. We will show that

$$(4.6) \qquad \tilde{u}(0, \delta) \leq -\delta^{3/2}\gamma^2/2,$$

for suitably small $\delta, \gamma > 0$. The last estimate together with $\tilde{u}(0, \delta) \geq -\mu_1\delta$ gives the desired lower bound $\mu_1 \geq \delta^{1/2}\gamma^2/2$, which proves the proposition.

Step 2. We now prove (4.6). Note that, on $\partial\Omega \cap B_c(0)$, we have

$$\tilde{u}(x', x_n) = \sum_{i=2}^{n-1} \frac{\mu_i - \mu_1}{2} x_i^2 + A x_1^3 + \sum_{i,j,k \leq n-1; i+j+k \geq 4} A_{ijk} x_i x_j x_k + o(|x'|^3).$$

Let us denote $\tilde{x} = (x_2, \ldots, x_{n-1})$. If $i = j = 1$ and $2 \leq k \leq n-1$, then $|x_i x_j x_k| = x_1^2 |x_k| \leq x_1^4 + x_k^2$. On the other hand, if $j, k \geq 2$, then on $\partial\Omega \cap B_c(0)$, we have $|x_i x_j x_k| \leq c|\tilde{x}|^2$. From these inequalities, we find $M(n, \Lambda, [\partial\Omega]_{C^3(0)}, [u|_{\partial\Omega}]_{C^3(0)}, K_0) \geq 2$ such that, on $\partial\Omega \cap B_c(0)$,

$$\tilde{u} \leq \frac{M}{2}|\tilde{x}|^2 + A x_1^3 + o(|x'|^3).$$

Since the coefficient of x_1^2 in $\tilde{u}$ is 0, the above analysis for u suggests that we can improve the above estimate in some region $D \subset \Omega$ where A can be taken to be 0. Consider

$$D = \left\{ x \in \mathbb{R}^n : |x_1| < \sqrt{\delta}/8, \quad |\tilde{x}| < \gamma\delta^{3/4}, \quad \delta/2 < x_n < 3\delta/2 \right\}$$

where δ and γ are small positive constants to be chosen later. When γ and δ are small, we have $D \Subset \Omega$.

Step 2(a). First, we show that

(4.7) $$\tilde{u} \leq \frac{M}{2}|\tilde{x}|^2 + o(\delta^{3/2}) \quad \text{on } \partial D.$$

This is the improvement mentioned above. By convexity, we only need to prove the above inequality for $x \in \partial D \cap \{x_1 = \pm\sqrt{\delta}/8\}$. We can assume without loss of generality that $A \geq 0$. For each $x \in \partial D \cap \{x_1 = \pm\sqrt{\delta}/8\}$ and $K > 1$ to be chosen (independent of δ and γ), the line

$$l_x = \left\{ x + t(1, 0, -K\delta^{1/2}) = (x_1 + t, \tilde{x}, x_n - K\delta^{1/2}t) : t \in \mathbb{R} \right\}$$

intersects $\partial\Omega$ at x^- and x^+ where $x_1^- < x_1^+$. See Figure 4.2.

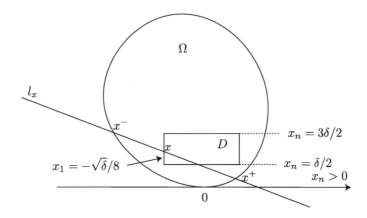

Figure 4.2. The set D and the line l_x for $x \in \partial D \cap \{x_1 = \pm\sqrt{\delta}/8\}$.

Using the equation preceding (4.6) for $\partial\Omega$ near 0, we can deduce that

$$\begin{cases} -3K\delta^{1/2} - o(\delta^{1/2}) \le x_1^- \le -K\delta^{1/2} + o(\delta^{1/2}), \\ \qquad -\dfrac{1}{8}\delta^{1/2} \le x_1^+ \le 3\delta^{1/2} + o(\delta^{1/2}), \end{cases}$$

and

$$\frac{1}{3K}\delta^{1/2} + o(\delta^{1/2}) < x_1^+ - x_1$$

if K is suitably large, depending on $[\partial\Omega]_{C^3(0)}$. Hence

$$\frac{x_1^+ - x_1}{x_1^+ - x_1^-} \ge \frac{1}{9K(K+1)} + o(\delta^{1/2}).$$

Recall from the boundary behavior of $\tilde{u}$ that

$$\tilde{u}(x^-) \le -AK^3\delta^{3/2} + \frac{M}{2}|\tilde{x}|^2 + o(\delta^{3/2}),$$

$$\tilde{u}(x^+) \le 27A\delta^{3/2} + \frac{M}{2}|\tilde{x}|^2 + o(\delta^{3/2}).$$

By the convexity of $\tilde{u}$, we have

$$\tilde{u}(x) \le \frac{x_1^+ - x_1}{x_1^+ - x_1^-}\tilde{u}(x^-) + \frac{x_1 - x_1^-}{x_1^+ - x_1^-}\tilde{u}(x^+)$$

$$\le \left(\frac{-K^2}{9(K+1)} + 27\right)A\delta^{3/2} + \frac{M}{2}|\tilde{x}|^2 + o(\delta^{3/2}) \le \frac{M}{2}|\tilde{x}|^2 + o(\delta^{3/2})$$

if $K \ge 250$. Therefore, (4.7) holds.

Step 2(b). Next, we construct an upper barrier for $\tilde{u}$ on D. Let

$$v(x) = \lambda(2M)^{-n}M\delta^{1/2}x_1^2 + M|\tilde{x}|^2 + M\delta^{-1/2}|x_n - \delta|^2 - \frac{1}{2}\gamma^2\delta^{3/2}.$$

Then $\det D^2 v = \lambda \le \det D^2\tilde{u}$ in D. On ∂D, we have

$$v \ge \min\left\{\frac{\lambda(2M)^{-n}M\delta^{3/2}}{64}, \frac{M\delta^{3/2}}{4}, M\gamma^2\delta^{3/2}\right\} - \frac{1}{2}\gamma^2\delta^{3/2}.$$

Thus, recalling (4.7),

$$v \ge \frac{3}{4}M\gamma^2\delta^{3/2} \ge \frac{1}{2}M\gamma^2\delta^{3/2} + o(\delta^{3/2}) \ge \tilde{u} \quad \text{on } \partial D$$

if we first choose γ suitably small and then δ small. With these choices of δ, γ, we have from the comparison principle in Theorem 3.21 that $\tilde{u} \le v$ in D. In particular, $\tilde{u}(0, \delta) \le v(0, \delta) = -\gamma^2\delta^{3/2}/2$, as claimed in (4.6). $\quad\square$

We can also use the volume estimates for sections in Chapter 5 to give a quick proof of $\mu_1 > 0$ without a quantitative lower bound as follows. Suppose that $\mu_1 = 0$. Then the coefficient of x_1^3 is 0 in Q_0. Denote $\tilde{x} = (x_2, \ldots, x_{n-1})$. Now, if we restrict to $\partial\Omega$ in a small neighborhood near the origin, then for all small h the set $\{\phi < h\}$ contains $\{|x_1| \le b(h)h^{1/3}\} \cap \{|\tilde{x}| \le c_1 h^{1/2}\}$ for some $c_1([\partial\Omega]_{C^3(0)}, [u|_{\partial\Omega}]_{C^3(0)}, K_0) > 0$, and $b(h) \to \infty$ as $h \to 0$.

Since $S_u(0, h) := \{x \in \overline{\Omega} : u(x) < h\}$ contains the convex set generated by $\{\phi < h\}$ and since $x_n \ge |x'|^2/(2K_0)$ in $S_h(0, h)$ because Ω is uniformly convex, we have

$$|S_u(0,h)| \ge (b(h)h^{1/3})^3 c_1^{n-2} h^{(n-2)/2}/(2K_0) \ge [b(h)]^3 c_1^{n-2} h^{n/2}/(2K_0).$$

Since $\det D^2 u \ge \lambda$, Lemma 5.8 gives $|S_u(0,h)| \le C(\lambda, n)h^{n/2}$. This contradicts the preceding inequality as $h \to 0$.

Remark 4.6. In Proposition 4.5, the assumptions that $u|_{\partial\Omega}$ and $\partial\Omega$ are C^3 at 0 are in some sense optimal. Here is one example. Let Ω be a uniformly convex domain in $\mathbb{R}^2$ with $\partial\Omega \in C^\infty$,

$$\Omega \subset B_2(0) \cap \{x_2 > x_1^2\} \quad \text{and} \quad \partial\Omega \cap B_1(0) \subset \{x_2 = x_1^2\}.$$

Let $u(x) = x_1^2 x_2^{1/2} + 21 x_2^{3/2}$ in $\overline{\Omega}$. Then, $u \in C^\infty(\Omega)$ with

$$30 \le \det D^2 u = (3/2)(21 - x_1^2 x_2^{-1}) \le 32 \quad \text{in } \Omega.$$

On $\partial\Omega \cap B_1(0)$, $u = 22|x_1|^3$ which is $C^{2,1}$, and u does not separates quadratically from its tangent hyperplanes at 0. This example comes from setting $\alpha = 3$ in Proposition 6.28. We refer the reader to the scaling analysis there and related analysis in Section 10.2.

When the boundary data is C^3 and the domain boundary is uniformly convex, the constant ρ in the proof of Proposition 4.5 can be chosen for all boundary points. Thus, we obtain:

Proposition 4.7 (Quadratic separation: global version). *Let Ω be a uniformly convex domain in $\mathbb{R}^n$. Let $u \in C(\overline{\Omega})$ be a convex function satisfying, in the sense of Aleksandrov,*

$$0 < \lambda \le \det D^2 u \le \Lambda \quad \text{in } \Omega.$$

Assume that $u|_{\partial\Omega} \in C^3$ and $\partial\Omega \in C^3$. Then, on $\partial\Omega$, u separates quadratically from its tangent hyperplanes on $\partial\Omega$. This means that if $x_0 \in \partial\Omega$, then for any $p \in \partial_s u(x_0)$,

$$(4.8) \qquad \rho |x - x_0|^2 \le u(x) - u(x_0) - p \cdot (x - x_0) \le \rho^{-1} |x - x_0|^2,$$

for all $x \in \partial\Omega$, for some constant ρ depending only on n, λ, Λ, $\|u\|_{C^3(\partial\Omega)}$, the C^3 regularity of $\partial\Omega$, and the uniform convexity radius of Ω.

In the proof of Proposition 4.5, we showed that, in a suitable coordinate system, the Hessian of $Q_0^{(2)}$ is bounded below and above by positive constants depending on the boundary data. These bounds are invariant under rotations. Thus, in a principal coordinate system x' at 0, we also have

$$\rho I_{n-1} \leq D_{x'}^2 Q_0^{(2)}(0) \leq \rho^{-1} I_{n-1}.$$

Recall from (2.14) that if $u \in C^2(\overline{\Omega})$, then $D_{x'}^2 u(0) = D_{x'}^2 Q_0^{(2)}(0)$. Putting these together, we have the following remark.

Remark 4.8. If $u \in C^2(\overline{\Omega})$, in addition to the assumptions of Proposition 4.7, then in a principal coordinate system x' at any boundary point $x_0 \in \partial\Omega$,

$$\rho I_{n-1} \leq D_{x'}^2 u(x_0) \leq \rho^{-1} I_{n-1},$$

where $\rho = \rho(n, \lambda, \Lambda, M_1, M_2, K_0)$ with $M_1 := \|u\|_{C^3(\partial\Omega)}$, M_2 depending on the C^3 regularity of $\partial\Omega$, and K_0 being the uniform convexity radius of Ω.

4.3. Global Estimates up to the Second Derivatives

An important and recurring theme in this book is that of *a priori estimates*. They are estimates, in terms of given data, for all possible solutions of a nonlinear PDE, under the assumption that such solutions exist. In this section, we establish a priori global higher-order derivative estimates for the Monge–Ampère equation. We begin with C^2 estimates.

Theorem 4.9 (Global C^2 estimates). *Let Ω be a uniformly convex domain in $\mathbb{R}^n$ with boundary $\partial\Omega \in C^3$, $\varphi \in C^3(\overline{\Omega})$, $f \in C^2(\overline{\Omega})$ with $f \geq m > 0$ in $\overline{\Omega}$. Assume the convex function $u \in C^2(\overline{\Omega}) \cap C^4(\Omega)$ solves*

$$\det D^2 u = f \quad in \ \Omega, \qquad u = \varphi \quad on \ \partial\Omega.$$

Then, we have the following global C^2 estimate:

$$\|u\|_{C^2(\overline{\Omega})} \leq C(n, \Omega, m, \|f\|_{C^2(\overline{\Omega})}, \|\varphi\|_{C^3(\overline{\Omega})}).$$

Proof. Let $\bar{m} := \|f\|_{L^\infty(\Omega)}$ and $(D^2 u)^{-1} = (u^{ij})_{1 \leq i,j \leq n}$. Then $u^{ij} D_{ij} u = n$. Some of our arguments are based on the maximum principle (Theorem 2.79) applied to the linearization $L_u := u^{ij} D_{ij}$ of the operator $\log \det D^2 u$ which is a priori degenerate elliptic. We divide the proof into several steps.

Step 1. C^0 and C^1 estimates. These follow from Theorem 3.30 which gives

$$\|u\|_{L^\infty(\Omega)} + \|Du\|_{L^\infty(\Omega)} \leq C(n, \Omega, \bar{m}, \|\varphi\|_{C^2(\overline{\Omega})}).$$

Step 2. Reduction of global C^2 estimates to boundary estimates. For this, observe that the operator $\log \det r$ is concave on the space of positive definite, symmetric $n \times n$ matrices r so any pure second partial derivative $u_{\xi\xi} := (D^2 u \xi) \cdot \xi$, where ξ is any unit vector in $\mathbb{R}^n$, is a subsolution of the linearized operator of $\log \det D^2 u$. Differentiate $\log \det D^2 u = \log f$ in the direction ξ, and use Lemma 2.56 to obtain

(4.9) $$L_u(u_\xi) = u^{ij} D_{ij} u_\xi = (\log f)_\xi.$$

Here we denote $u_\xi := D_\xi u = Du \cdot \xi$ and $u_{\xi\xi} = D_\xi(D_\xi u)$. Differentiate (4.9) again in the direction ξ and use Lemmas 2.56 and 2.58 to get

$$(\log f)_{\xi\xi} = u^{ij} D_{ij} u_{\xi\xi} - u^{ik} u^{lj} D_{kl} u_\xi D_{ij} u_\xi \le u^{ij} D_{ij} u_{\xi\xi}.$$

Let $K := \| \log f \|_{C^2(\overline{\Omega})} + 1$. Then $u^{ij} D_{ij} u_{\xi\xi} > -K$. Thus

$$u^{ij} D_{ij}(u_{\xi\xi} + Ku) > -K + K u^{ij} D_{ij} u = -K + Kn > 0.$$

By the maximum principle (Theorem 2.79), the maximum of $u_{\xi\xi} + Ku$ can only happen on the boundary. It follows that

$$\max_{\overline{\Omega}} u_{\xi\xi} \le \max_{\partial\Omega}(u_{\xi\xi} + Ku) + K\|u\|_{L^\infty(\Omega)} \le \max_{\partial\Omega} u_{\xi\xi} + 2K\|u\|_{L^\infty(\Omega)}.$$

By the boundedness of u in Step 1 and Lemma 2.58(i), we have reduced the global estimates for $D^2 u$ to its boundary estimates.

We now estimate $D^2 u$ on the boundary. Consider any boundary point $x_0 \in \partial\Omega$. After a rotation of coordinates, we can assume that $x_0 = 0$ and e_n is the inner normal of $\partial\Omega$ at 0. In a principal coordinate system at $x_0 = 0$ (see Section 2.4), we have, for some $\rho > 0$ depending on Ω,

$$(4.10) \qquad \Omega \cap B_\rho(0) = \Big\{ (x', x_n) : x_n > \psi(x') = \sum_{1 \le i \le n-1} \frac{\kappa_i}{2} x_i^2 + O(|x'|^3) \Big\}$$

where $\kappa_1, \ldots, \kappa_{n-1}$ are the principal curvatures of $\partial\Omega$ at $x_0 = 0$.

Step 3. Second tangential derivative estimates. By Remark 4.8, we have

$$(4.11) \qquad\qquad \rho_1 I_{n-1} \le D_{x'}^2 u(0) \le \rho_1^{-1} I_{n-1}$$

where $\rho_1 > 0$ depends on $n, m, \bar{m}, \Omega$, and $\|\varphi\|_{C^3(\overline{\Omega})}$.

Step 4. Second mixed derivative estimates. Fix $i \in \{1, \ldots, n-1\}$. We will establish the following mixed tangential normal second derivative estimate

$$(4.12) \qquad |D_{in} u(0)| \le C = C(n, \Omega, m, \|f\|_{C^2(\overline{\Omega})}, \|\varphi\|_{C^3(\overline{\Omega})}).$$

The proof is via construction of suitable barriers. The idea is to find some expression v concerning the partial derivative $D_i u$ such that $L_u v$ is bounded and furthermore, $|v|$ is bounded from above by some strict supersolution w of the linearized operator L_u that grows at most linearly away from the boundary. One possible candidate for w is $w(x) = bx_n - 2aP_\gamma(x)$, where

$$P_\gamma(x) = \frac{1}{2}(\gamma |x'|^2 + \gamma^{1-n} x_n^2), \quad \gamma \in (0,1),$$

with $\det D^2 P_\gamma = 1$. Indeed, we have

$$L_u(bx_n - 2aP_\gamma(x)) = -a\gamma \sum_{k=1}^{n-1} u^{kk} - a\gamma^{1-n} u^{nn} - a\,\text{trace}((D^2 u)^{-1} D^2 P_\gamma).$$

By Lemma 2.57, $\text{trace}((D^2u)^{-1}D^2P_\gamma) \geq n(\det D^2u)^{-\frac{1}{n}}(\det D^2P_\gamma)^{1/n}$, so

$$(4.13) \qquad L_u(bx_n - 2aP_\gamma(x)) \leq -a\gamma \sum_{k=1}^{n} u^{kk} - an\bar{m}^{-1/n}.$$

By analyzing $\partial\Omega$ in $B_\rho(0)$ and outside $B_\rho(0)$, we find $C_0(\Omega) > 0$ such that

$$(4.14) \qquad\qquad\qquad |x|^2 \leq C_0 x_n \quad \text{on } \partial\Omega.$$

The above inequality also holds in Ω due to convexity. We may attempt to use $v = D_i u$. The issue with this choice is that we cannot control $|D_i u|$ by w on the boundary $\partial\Omega$. We can fix this issue as follows. Near the origin, $D_i + D_i\psi D_n$ is the tangential differential operator along $\partial\Omega$. Thus, $D_i(u - \varphi) + D_i\psi D_n(u - \varphi) = 0$, and

$$(4.15) \quad |D_i(u-\varphi)+\kappa_i x_i D_n(u-\varphi)| \leq C_1|x|^2 \quad \text{on } \partial\Omega\cap B_\rho, \text{ and hence on } \partial\Omega.$$

Here $C_1 = C_1(n, \Omega, \bar{m}, \|\varphi\|_{C^2(\overline{\Omega})})$. This suggests using $v = D_i(u - \varphi) + \kappa_i x_i D_n(u - \varphi)$, but the new issue is that we do not have good control on $L_u v$ as we do not have good control on $L_u(x_i D_n u)$. It turns out that if we focus on the $x_i x_n$-plane, then

$$(4.16) \qquad\qquad L_u(x_i D_n u - x_n D_i u) = (x_i D_n - x_n D_i)\log f.$$

Indeed, let (r, θ) be the polar coordinates in the $x_i x_n$-plane. Then applying (4.9) in the direction θ, we have $L_u(D_\theta u) = D_\theta \log f$, but $D_\theta = x_i D_n - x_n D_i$, so we obtain (4.16). Therefore, it is reasonable to choose

$$(4.17) \qquad\qquad v := D_i(u - \varphi) + \kappa_i[x_i D_n(u - \varphi) - x_n D_i(u - \varphi)]$$

at the expense of having an extra term $\kappa_i x_n D_i(u - \varphi)$ of order x_n, but this can be handled by increasing b in w.

Now, we establish the bound for $D_{in}u(0)$ as follows. Let v be as in (4.17). Then using (4.9), (4.16), we find

$$L_u v = D_i(\log f) + \kappa_i[(x_i D_n - x_n D_i)\log f - x_i u^{kl}D_{nkl}\varphi + x_n u^{kl}D_{ikl}\varphi]$$
$$- u^{kl}D_{ikl}\varphi,$$

which, in view of (u^{ij}) being positive definite, implies

$$(4.18) \quad |L_u v| \leq C_2\left(1 + \sum_{k=1}^{n} u^{kk}\right), \quad C_2 = C_2(\|\varphi\|_{C^3(\overline{\Omega})}, \|\log f\|_{C^1(\overline{\Omega})}, \Omega).$$

By (4.15) and the gradient bound in Step 1, we find from (4.17) that

$$(4.19) \qquad |v| \leq C_3(x_n + |x|^2), \quad \text{on } \partial\Omega, \quad C_3 = C_3(n, \Omega, \bar{m}, \|\varphi\|_{C^2(\overline{\Omega})}).$$

Let us choose

$$w(x) = bx_n - 2aP_\gamma(x) + a\tilde{\gamma}[u(x) - l(x)],$$

where l is the supporting hyperplane to the graph of u at the origin and

$$a := \frac{2(C_2 + 1)}{\min\{\gamma, n\bar{m}^{-1/n}\}}, \quad \tilde{\gamma} := \bar{m}^{-1/n}/2, \quad \text{and} \quad b = C_3 + C_0(a\gamma^{1-n} + C_3).$$

Then from (4.13) and (4.18), we obtain

$$L_u(w \pm v) \leq (C_2 - \gamma a) \sum_{k=1}^{n} u^{kk} + (C_2 - an\bar{m}^{-1/n} + an\tilde{\gamma}) < 0 \quad \text{in } \Omega,$$

while, using (4.19) and (4.14),

$$(4.20) \quad \begin{aligned} w \pm v &\geq (b - C_3)x_n - (a\gamma^{1-n} + C_3)|x|^2 \\ &= (a\gamma^{1-n} + C_3)(C_0 x_n - |x|^2) \geq 0 \quad \text{on } \partial\Omega. \end{aligned}$$

By the maximum principle (Theorem 2.79), we have

$$\min_{\overline{\Omega}}(w \pm v) = \min_{\partial\Omega}(w \pm v) \geq 0.$$

Since $w \pm v = 0$ at 0, we have $D_n w(0) \pm D_n v(0) \geq 0$. But these give

$$b + a\tilde{\gamma}D_n(u - l)(0) \pm (D_{in}u(0) - D_{in}\varphi(0)) \geq 0,$$

from which (4.12) follows.

Step 5. Second normal derivative estimates. With the second mixed derivative estimates obtained in Step 4, we can estimate $D_{nn}u(0)$ from above as follows. Using $f(0) = \det D^2 u(0)$ and expanding $\det D^2 u(0)$ with respect to the last row of $D^2 u(0)$, we obtain

$$f(0) = D_{nn}u(0) \det D^2_{x'}u(0) + P(D^2_{x'}u(0), D^2_{x'x_n}u(0)),$$

where $P(D^2_{x'}u(0), D^2_{x'x_n}u(0))$ is a polynomial of degree n depending on the second tangential and mixed derivatives of u at 0. This combined with (4.11) and (4.12) gives $0 \leq D_{nn}u(0) \leq C(n, \Omega, m, \|f\|_{C^2(\overline{\Omega})}, \|\varphi\|_{C^3(\overline{\Omega})})$, as desired.

The estimates in the above steps complete the proof of the theorem. $\square$

Remark 4.10 (Uniformly convex boundary values). In Step 4 of the proof of Theorem 4.9, we could have chosen $\gamma = 1$ and $\tilde{\gamma} = 0$. However, the introduction of these parameters is important for situations where $\partial\Omega$ is flat. Let us consider the case where $\partial\Omega$ is no longer required to be uniformly convex in a neighborhood of 0, so in (4.10), some κ_i's can be 0, and (4.14) may not hold. Suppose that on $\partial\Omega \cap B_\rho(0)$, φ has a quadratic growth:

$$\mu|x'|^2 \leq \varphi(x) \leq \mu^{-1}|x'|^2.$$

Then, a modification of Step 4 of the proof of Theorem 4.9 gives the bounds for $D_{in}u(0)$. To see this, we consider the set

$$E = \{x \in \overline{\Omega} \cap B_\rho(0) : u(x) < \mu\rho^2\}.$$

We only need to show that $w \pm v \geq 0$ on ∂E. Note that

$$w \pm v \geq (b - C_3)x_n - (a\gamma^{1-n} + C_3)|x_n|^2 - (a\gamma + C_3)|x'|^2 + a\tilde{\gamma}(u - l).$$

If $x \in E \cap \partial\Omega$, then

$$u(x) - l(x) = \varphi(x) - D_n u(0)x_n \geq \mu|x'|^2 - D_n u(0)x_n,$$

and thus, by choosing $\gamma = \mu\tilde{\gamma}/4$ and then a large so that $C_3 \leq a\gamma/2$ (hence $a\gamma + C_3 \leq (3/4)a\tilde{\gamma}\mu$) and then b large, we easily obtain $w \pm v \geq 0$ on $E \cap \partial\Omega$.

Consider now the case $x \in \partial E \cap \Omega$. Let

$$M := \sup_{B_{2\rho}(0) \cap \Omega} |Du|; \quad c_0 := \mu\rho^2/(8M).$$

We can further increase b if necessary so that if $x_n \geq c_0$, then we also have $w \pm v \geq 0$. Now, we consider the remaining case where $x \in \partial E \cap \Omega$ with $x_n \leq c_0$. In this case, we use the gradient bound for u to conclude. Let $z = (x', z_n) \in \partial\Omega$ be the intersection of the ray $x - se_n$ ($s > 0$) with the boundary $\partial\Omega$. Then, for some t between z_n and x_n, we have

$$u(x) = u(x', z_n) + D_n u(x', t)(x_n - z_n)$$
$$\geq u(x', z_n) - 2c_0 M \geq \mu|x'|^2 - 2c_0 M.$$

Thus, for b large, we have on $\partial E \cap \Omega$ where $u = \mu\rho^2$ that

$$w \pm v \geq -(a\gamma + C_3)|x'|^2 + a\tilde{\gamma}u$$
$$\geq -(3/4)a\tilde{\gamma}\mu|x'|^2 + (4/5)a\tilde{\gamma}(\mu|x'|^2 - 2c_0 M) + (1/5)a\tilde{\gamma}\mu\rho^2 \geq 0.$$

For smoother data, using the global Evans–Krylov estimates (Theorem 2.86) together with the global Schauder estimates (Theorem 2.82), we obtain global higher-order derivative estimates for the Monge–Ampère equation.

Theorem 4.11 (Global $C^{k,\alpha}$ estimates). *Let Ω be a uniformly convex domain in $\mathbb{R}^n$ with $\partial\Omega \in C^{k+2,\alpha}$, $\varphi \in C^{k+2,\alpha}(\overline{\Omega})$, and $f \in C^{k,\alpha}(\overline{\Omega})$ for some $\alpha \in (0,1)$ and integer $k \geq 2$. Assume that $m := \inf_\Omega f > 0$. Assume that the convex function $u \in C^2(\overline{\Omega}) \cap C^4(\Omega)$ solves the Dirichlet problem*

$$\det D^2 u = f \quad in \ \Omega, \qquad u = \varphi \quad on \ \partial\Omega.$$

Then $u \in C^{k+2,\alpha}(\overline{\Omega})$ with estimate

$$\|u\|_{C^{k+2,\alpha}(\overline{\Omega})} \leq C = C(n, k, \alpha, \Omega, m, \|f\|_{C^{k,\alpha}(\overline{\Omega})}, \|\varphi\|_{C^{k+2,\alpha}(\overline{\Omega})}).$$

Proof. By Theorem 4.9, we have the global C^2 estimate:

$$\|u\|_{C^2(\overline{\Omega})} \leq C_1 = C_1(n, \Omega, m, \|f\|_{C^2(\overline{\Omega})}, \|\varphi\|_{C^3(\overline{\Omega})}).$$

Let $F(D^2 u) := \log \det D^2 u = \log f$. Then, $(F_{ij}) \equiv (\frac{\partial F}{\partial r_{ij}}) = (D^2 u)^{-1}$. Therefore, thanks to Remark 2.87, F is concave and uniformly elliptic with

$$C_2^{-1} I_n \leq (F_{ij}) \leq C_2 I_n,$$

for some $C_2(m, C_1, n) > 0$. Now, we can apply Theorem 2.86 to obtain $\|u\|_{C^{2,\beta}(\overline{\Omega})} \le C_3$, where $\beta \in (0, 1)$ and C_3 are constants depending on n, C_2, m, $\|f\|_{C^2(\overline{\Omega})}$, Ω, and $\|\varphi\|_{C^3(\overline{\Omega})}$. For any unit vetor $\mathbf{e}$, $u_{\mathbf{e}} := Du \cdot \mathbf{e}$ solves the equation $F_{ij} D_{ij} u_{\mathbf{e}} = (\log f)_{\mathbf{e}}$, which is uniformly elliptic with globally $C^\beta(\overline{\Omega})$ coefficients. Thus, we can apply Theorem 2.82 (see also Remark 2.87) to get $u_{\mathbf{e}} \in C^{2,\beta}(\overline{\Omega})$, and hence $u \in C^{3,\beta}(\overline{\Omega})$. By bootstrapping, we get the conclusion of the theorem. $\qquad\square$

4.4. Existence of Classical Solutions

We are now ready to prove a fundamental theorem, due to Ivočkina [**Iv**], Krylov [**K3**], and Caffarelli, Nirenberg, and Spruck [**CNS1**], on the existence of smooth, uniformly convex solutions to the Dirichlet problem for the Monge–Ampère equation.

Theorem 4.12 (Existence and uniqueness of classical solutions). *Let Ω be a uniformly convex domain in $\mathbb{R}^n$, with $\partial\Omega \in C^{k+2,\alpha}$, $\varphi \in C^{k+2,\alpha}(\overline{\Omega})$, and $f \in C^{k,\alpha}(\overline{\Omega})$ for some $\alpha \in (0, 1)$ and integer $k \ge 2$. Assume that $m := \inf_\Omega f > 0$. Then the Dirichlet problem*

$$\begin{cases} \det D^2 u = f & in \ \Omega, \\ \quad\quad u = \varphi & on \ \partial\Omega \end{cases}$$

has a unique uniformly convex solution $u \in C^{k+2,\alpha}(\overline{\Omega})$ with estimate

$$\|u\|_{C^{k+2,\alpha}(\overline{\Omega})} \le C = C(n, k, \alpha, \Omega, m, \|f\|_{C^{k,\alpha}(\overline{\Omega})}, \|\varphi\|_{C^{k+2,\alpha}(\overline{\Omega})}).$$

Proof. By Example 3.4, any classical convex solution is also an Aleksandrov solution. Furthermore, it is unique by Corollary 3.23. It remains to prove the existence. It suffices to prove the existence result for $k = 2$ by the method of continuity. Once this is established, the existence result for $k > 2$ follows from the one for $k = 2$, taking into account Remark 2.87 and Theorem 2.82.

Assume now $k = 2$. By Lemma 2.54, we can find a uniformly convex function $u^0 \in C^{4,\alpha}(\overline{\Omega})$ such that $f^0 := \det D^2 u^0 \ge f$ in Ω and $u^0 = \varphi$ on $\partial\Omega$. For each $\sigma \in [0, 1]$, let $f_\sigma := \sigma f + (1 - \sigma) f^0$, and we would like to find a uniformly convex, $C^{4,\alpha}(\overline{\Omega})$ solution u_σ to

(4.21) $\qquad \det D^2 u_\sigma = f_\sigma \ \ in \ \Omega, \qquad u_\sigma = \varphi \ \ on \ \partial\Omega.$

Let

$I = \big\{ \sigma \in [0, 1] : \text{there is a uniformly convex function}$

$$u_\sigma \in C^{4,\alpha}(\overline{\Omega}) \text{ solving } (4.21) \big\}.$$

Note that $0 \in I$ since $u_0 := u^0$ solves (4.21) when $\sigma = 0$. So, to prove the existence result, it suffices to show that I is both open and closed in $[0, 1]$.

Step 1. I is open. For this, we use the Implicit Function Theorem. Let

$$\mathfrak{B}_1 = C_0^{4,\alpha}(\overline{\Omega}) = \{u \in C^{4,\alpha}(\overline{\Omega}) : u = 0 \text{ on } \partial\Omega\}, \ X = \mathbb{R}, \text{ and } \mathfrak{B}_2 = C^{2,\alpha}(\overline{\Omega}).$$

Consider the mapping G from $\mathfrak{B}_1 \times X$ into $\mathfrak{B}_2$ given by

$$G(w, \sigma) := \det D^2(w + \varphi) - f_\sigma.$$

Then G is continuously differentiable, and its partial Fréchet derivative $G_w(w, \sigma)$ at $w \in \mathfrak{B}_1$ is given by

$$G_w(w, \sigma)h = (\text{Cof } D^2(w + \varphi))^{ij} D_{ij} h \quad \text{for any } h \in \mathfrak{B}_1.$$

Suppose $\sigma_0 \in I$. Then, there exists a uniformly convex $C^{4,\alpha}(\overline{\Omega})$ function u_{σ_0} solving (4.21). Let $w_{\sigma_0} = u_{\sigma_0} - \varphi$. Then, clearly $G(w_{\sigma_0}, \sigma_0) = 0$, and $G_w(w_{\sigma_0}, \sigma_0) = (\text{Cof } D^2 u_{\sigma_0})^{ij} D_{ij}$ is a uniformly elliptic operator with $C^{2,\alpha}(\overline{\Omega})$ coefficients.

By Theorem 2.82 with $k = 2$, the mapping $G_w(w_{\sigma_0}, \sigma_0) : \mathfrak{B}_1 \to \mathfrak{B}_2$ is invertible. Applying the Implicit Function Theorem (Theorem 2.77), we find that for all σ close to σ_0, there is a solution $w_\sigma \in \mathfrak{B}_1$ to $G(w_\sigma, \sigma) = 0$ and $\|w_\sigma - w_{\sigma_0}\|_{C_0^{4,\alpha}(\overline{\Omega})} \leq C|\sigma - \sigma_0|$ for C independent of σ. Let $u_\sigma := w_\sigma + \varphi$. Then

$$\|u_\sigma - u_{\sigma_0}\|_{C_0^{4,\alpha}(\overline{\Omega})} \leq C|\sigma - \sigma_0|.$$

Since u_{σ_0} is uniformly convex, the function u_σ is also uniformly convex for σ sufficiently close to σ_0. This show that (4.21) has a uniformly convex $C^{4,\alpha}(\overline{\Omega})$ solution u_σ for σ sufficiently close to σ_0. Hence, I is open.

Step 2. I is closed. Let $\{\sigma_j\}_{j=1}^\infty \subset I$ be a sequence converging to σ. We need to show that $\sigma \in I$. For each j, there is a uniformly convex $C^{4,\alpha}(\overline{\Omega})$ solution u_{σ_j} to

$$(4.22) \qquad \det D^2 u_{\sigma_j} = f_{\sigma_j} \quad \text{in } \Omega, \qquad u_{\sigma_j} = \varphi \quad \text{on } \partial\Omega.$$

By the definition of f_σ, we have $f_{\sigma_j} \geq \inf_\Omega f > 0$ for each j. Furthermore

$$\|f_{\sigma_j}\|_{C^{2,\alpha}(\overline{\Omega})} \leq \|f\|_{C^{2,\alpha}(\overline{\Omega})} + \|f^0\|_{C^{2,\alpha}(\overline{\Omega})}.$$

By Theorem 4.11, we have $\|u_{\sigma_j}\|_{C^{4,\alpha}(\overline{\Omega})} \leq C$ for some constant C independent of j. Due to the Arzelà–Ascoli Theorem (Theorem 2.41), we can find a subsequence, still labeled $\{u_{\sigma_j}\}_{j=1}^\infty$, converging to a uniformly convex function $u_\sigma \in C^4(\overline{\Omega})$. Note that f_{σ_j} converges uniformly to f_σ when $j \to \infty$. Thus, letting $j \to \infty$ in (4.22), we find that u_σ solves

$$\det D^2 u_\sigma = f_\sigma \quad \text{in } \Omega, \quad u_\sigma = \varphi \quad \text{on } \partial\Omega.$$

By Theorem 4.11, we have $u_\sigma \in C^{4,\alpha}(\overline{\Omega})$. Hence $\sigma \in I$, so I is closed. This completes the proof of the theorem. $\qquad\qquad\square$

In Theorems 4.11 and 4.12, if $\partial\Omega$ is analytic and both f and φ are analytic all the way up to the boundary of Ω, then we can use the results of Friedman [**Fr**] and Morrey [**Mor**] to conclude that the function u is also analytic all the way up to the boundary of Ω.

4.5. The Role of Subsolutions

The function u^0 in the proof of Theorem 4.12 is a subsolution of $\det D^2 u = f$. In many cases, subsolutions play an important role in the derivation of global a priori estimates and in the establishment of global solvability. We mention an extension of Theorem 4.12 to Monge–Ampère-type equations in [**CNS1**].

Theorem 4.13. *Let $\alpha \in (0,1)$. Let Ω be a uniformly convex domain in $\mathbb{R}^n$ with $\partial\Omega \in C^{4,\alpha}$ and let $\varphi \in C^{4,\alpha}(\overline{\Omega})$. Let $f(x,z,p) \in C^{2,\alpha}(\overline{\Omega} \times \mathbb{R} \times \mathbb{R}^n)$ with*

$$D_z f, D_p f \in C^{2,\alpha}(\overline{\Omega} \times \mathbb{R} \times \mathbb{R}^n), \quad f \geq c_0 > 0 \quad in \ \overline{\Omega} \times \mathbb{R} \times \mathbb{R}^n.$$

Assume furthermore that one of the following conditions holds.

(a) (*Existence of a subsolution*) *There exists a convex function $\underline{u} \in C^{4,\alpha}(\overline{\Omega})$ with $\underline{u} = \varphi$ on $\partial\Omega$ such that*

$$\det D^2 \underline{u} \geq f(x, \underline{u}, D\underline{u}) \quad in \ \Omega.$$

(b) (*Monotonicity*) *$D_z f \geq 0$ in $\overline{\Omega} \times \mathbb{R} \times \mathbb{R}^n$.*

Then, the Dirichlet problem

$$(4.23) \qquad \det D^2 u = f(x, u, Du) \quad in \ \Omega, \qquad u = \varphi \quad on \ \partial\Omega$$

has a uniformly convex solution $u \in C^{4,\alpha}(\overline{\Omega})$. If (b) holds, then u is unique.

We briefly comment on the proof of and the assumptions in Theorem 4.13. Either of the additional conditions (a) and (b) can be used to obtain a priori estimates in $C^{4,\alpha}(\overline{\Omega})$ for smooth solutions of (4.23). Then, the existence follows from a degree theory argument (see, for example, Section 15.5.1). The existence of a subsolution in condition (a) can be fulfilled when f satisfies certain growth conditions such as (see Problem 2.7)

$$(4.24) \qquad 0 \leq f(x, z, p) \leq M(1 + |p|^2)^{n/2} \quad for \ x \in \overline{\Omega} \ and \ z \leq \max_{\overline{\Omega}} \varphi.$$

On the other hand, if condition (b) holds, then we can use the method of continuity as in the proof of Theorem 4.12 to obtain the existence of a unique solution. The monotonicity enters the proof to ensure the invertibility of the linearized operator in the method of continuity. The function f_σ in the proof of Theorem 4.12 is now modified to be $f_\sigma = \sigma f(x, u_\sigma, Du_\sigma) + (1 - \sigma) f^0(x)$ and the mapping $G_w(w_{\sigma_0}, \sigma_0)$ now becomes

$$G_w(w_{\sigma_0}, \sigma_0)h = (\text{Cof } D^2 u_{\sigma_0})^{ij} D_{ij} h - \sigma D_p f(x, u_{\sigma_0}, Du_{\sigma_0}) \cdot Dh$$
$$- \sigma D_z f(x, u_{\sigma_0}, Du_{\sigma_0})h.$$

For the mapping $G_w(w_{\sigma_0}, \sigma_0) : \mathfrak{B}_1 \to \mathfrak{B}_2$ to be invertible, a natural condition stemming from the Fredholm alternative is to require the coefficient of h to be nonpositive which corresponds to the monotonicity assumption $D_z f \geq 0$. In other words, we use the full scope of Theorem 2.82.

On bounded smooth domains Ω that are not necessarily convex, global $C^{2,\alpha}$ estimates for (4.23) with globally smooth right-hand side f and boundary data were first obtained by Guan and Spruck [**GS**] under the assumption that there exists a convex strict subsolution $\underline{u} \in C^2(\overline{\Omega})$ taking the boundary values φ. The strictness of the subsolution $\underline{u}$ in [**GS**] was later removed by Guan [**Gu**]. Consequently, the existence of a locally uniformly convex solution $C^\infty(\overline{\Omega})$ to (4.23) with $u \geq \underline{u}$ was obtained. Here, we indicate some details. The major changes come from the estimation of second mixed derivatives in Step 4 of the proof of Theorem 4.9. Instead of comparing $\pm v$ in (4.17) with w, we compare them with

$$\bar{w} = A[u - \underline{u} + t(\hat{\varphi} - \underline{u}) - N\text{dist}^2(\cdot, \partial\Omega)] + B|x|^2$$

for suitably large constants $A > B > 1$, where t, N are positive and $\hat{\varphi}$ is the harmonic function in Ω with boundary value φ on $\partial\Omega$.

In the two-dimensional case, the smoothness of the domains Ω in [**Gu**, **GS**] can be further relaxed. In [**LS5**], Le and Savin obtained global $C^{2,\alpha}$ estimates for the nondegenerate Monge–Ampère equation $\det D^2 u = f$ in convex polygonal domains in $\mathbb{R}^2$ provided a globally C^2, convex strict subsolution exists.

To conclude this section, we note that if $f(x, z, p) = (-z)^q$ where $q > 0$, then neither the growth condition (4.24) nor the monotonicity condition $D_z f \geq 0$ nor the nondegeneracy condition $f \geq c_0 > 0$ is satisfied. The continuity and degree theory methods do not seem to provide the existence of nonzero classical solutions to

$$\det D^2 u = (-u)^q \quad \text{in } \Omega, \quad u = 0 \quad \text{on } \partial\Omega.$$

For these equations, Tso [**Ts**] used a variational method to obtain solutions in $C^\infty(\Omega) \cap C^{0,1}(\overline{\Omega})$; see Chapter 11.

4.6. Pogorelov's Counterexamples to Interior Regularity

Without information on the boundary data, Aleksandrov solutions to the Monge–Ampère equation with smooth right-hand side can be quite singular. Put differently, there are no purely interior analogues of the global estimates in Theorem 4.12. In [**P5**], Pogorelov constructed singular solutions to the Monge–Ampère equation $\det D^2 u = g$ in a small neighborhood of the origin in $\mathbb{R}^n$, where $n \geq 3$ and g is analytic and positive. Below, we motivate these constructions using symmetry and scaling arguments.

Consider the Monge–Ampère equation

$$\det D^2 u = 1 \quad \text{in } \mathbb{R}^n = \{x = (x', x_n) \in \mathbb{R}^{n-1} \times \mathbb{R}\}.$$

It is invariant under the rescalings of u given by

$$u_\lambda(x', x_n) = \lambda^{2/n-2} u(\lambda x', x_n),$$

where $\lambda > 0$. If $\lambda = |x'|^{-1} > 0$, then the right-hand side of the above expression becomes $|x'|^{2-2/n} u(x'/|x'|, x_n)$. Thus, the scaling invariance and symmetry considerations suggest that we look for a solution of the form

$$(4.25) \qquad u(x', x_n) = |x'|^{2-2/n} f(x_n) \equiv |x'|^\alpha f(x_n) \quad \text{where } \alpha = 2 - 2/n.$$

Then, by Lemma 2.60,

$$(4.26) \qquad \det D^2 u = (2 - 2/n)^{n-1} f^{n-2} \big((\alpha - 1) f f'' - \alpha (f')^2 \big).$$

Example 4.14. By taking $f(x_n) = 1 + x_n^2$, Pogorelov showed that, in a neighborhood $B_\rho(0)$ of 0, where $\rho = 1/4$, the convex function

$$(4.27) \qquad u(x) = |x'|^{2-2/n} (1 + x_n^2) \quad (n \geq 3),$$

whose optimal regularity is $C^{1,1-2/n}$, is an Aleksandrov solution to

$$(4.28) \quad \det D^2 u = g := \left(2 - \frac{2}{n}\right)^{n-1} (x_n^2 + 1)^{n-2} \left\{ 2\left(1 - \frac{2}{n}\right) + \left(\frac{4}{n} - 6\right) x_n^2 \right\} > 0.$$

Indeed, we deduce from Lemma 2.60 that (4.28) holds classically for $x' \neq 0$. We now check, using Theorem 2.74, that (4.28) holds in the sense of Aleksandrov in $B_\rho(0)$. Note that $u \in C^1(\mathbb{R}^n) \cap C^2(\mathbb{R}^n \setminus \{x' = 0\})$. We have

$$\partial u(\{x' = 0\}) = Du(\{x' = 0\}) \subset \{x' = 0\}$$

and $|Du(\{x' = 0\})| = 0$. Therefore, for any Borel set $E \subset B_\rho(0)$, we have

$$|\partial u(E)| = |Du(E)| = |Du(E \setminus \{x' = 0\})|$$

$$= \int_{E \setminus \{x' = 0\}} \det D^2 u \, dx = \int_{E \setminus \{x' = 0\}} g \, dx = \int_E g \, dx.$$

It follows that u is an Aleksandrov solution to $\det D^2 u = g$ in $B_\rho(0)$.

Note that the graph of u contains the line segment $\{x' = 0\} \cap B_{1/4}(0)$.

Pogorelov also constructed a singular solution to $\det D^2 u = 1$, but his example has a different scaling property to the above example. In Example 4.15, we construct a singular solution of the form (4.25) to $\det D^2 u = 1$.

Example 4.15. We continue to look for u of the form (4.25) so (4.26) holds. For $n \geq 3$, by the Cauchy–Lipschitz Theorem, we can solve the ODE

$$(2 - 2/n)^{n-1} f^{n-2} \big((1 - 2/n) f f'' - (2 - 2/n)(f')^2 \big) = 1$$

with initial conditions

$$f(0) = 1, \quad f'(0) = 0$$

to find a function f that is even, smooth, convex in a neighborhood of the origin. Moreover, $f(t)$ blows up to infinity when $|t|$ approaches a finite value $r(n) > 0$. Arguing as in Example 4.14, we obtain an Aleksandrov solution to $\det D^2 u = 1$ in $B_\rho(0)$ of the form $u(x', x_n) = |x'|^{2-2/n} f(x_n)$. Clearly $u \in C^{1, 1 - \frac{2}{n}}(B_\rho(0)) \setminus C^2(B_\rho(0))$. This is a slightly different construction of Pogorelov's convex Aleksandrov solution to $\det D^2 u = 1$ that blows up at the lateral sides of a cylinder.

4.7. Application: The Classical Isoperimetric Inequality

Using the existence of classical solutions in Theorem 4.12 and the Aleksandrov maximum principle in Theorem 3.12, we can give a proof of the classical isoperimetric inequality. This proof was found by Trudinger [**Tr3**].

Theorem 4.16 (Isoperimetric inequality). *Let Ω be a bounded, open set in $\mathbb{R}^n$ with C^1 boundary. Then*

$$\frac{\mathcal{H}^{n-1}(\partial\Omega)}{|\Omega|^{(n-1)/n}} \geq \frac{\mathcal{H}^{n-1}(\partial B_1(0))}{|B_1(0)|^{(n-1)/n}} = n\omega_n^{1/n}.$$

Proof. Choose $r > 0$ such that $\Omega \subset B_r(0)$. Let $R > r$. By Theorem 2.50, we can find a sequence of functions $f_\sigma \in C^3(\overline{B_R(0)})$ such that $\sigma \leq f_\sigma \leq 2 + \sigma$ and $f_\sigma \to \chi_\Omega$ almost everywhere in $B_R(0)$ when $\sigma \to 0^+$. By Theorem 4.12, there exists a unique convex solution $u_\sigma \in C^3(\overline{B_R})$ to

$$\det D^2 u_\sigma = f_\sigma \quad \text{in } B_R(0), \quad u_\sigma = 0 \quad \text{on } \partial B_R(0).$$

Let $D(x_0) = \sup_{x \in B_R(0)} |x - x_0|$ for $x_0 \in \overline{\Omega}$. Then, $D(x_0) \leq R + r$. By Theorem 3.12,

$$|u_\sigma(x_0)|^n \leq \omega_n^{-1} D^n(x_0) |\partial u_\sigma(B_R(0))| \leq \omega_n^{-1}(R+r)^n \int_{B_R(0)} f_\sigma \, dx.$$

Thus, from Lemma 2.37(iii), we find that, for $x_0 \in \overline{\Omega}$,

$$|Du_\sigma(x_0)| \leq \frac{\sup_{\partial B_R(0)} u_\sigma - u_\sigma(x_0)}{\text{dist}(x_0, \partial B_R(0))} \leq \omega_n^{-1/n} \frac{R+r}{R-r} \left(\int_{B_R(0)} f_\sigma \, dx \right)^{1/n}.$$

Since $D^2 u_\sigma$ is positive definite, we have (see (2.15))

$$\Delta u_\sigma = \text{trace}(D^2 u_\sigma) \geq n(\det D^2 u_\sigma)^{1/n} = n f_\sigma^{1/n}.$$

Let ν be the outer unit normal to $\partial\Omega$. Then, the Divergence Theorem gives

$$\int_\Omega n f_\sigma^{1/n} \, dx \leq \int_\Omega \Delta u_\sigma \, dx = \int_{\partial\Omega} Du_\sigma \cdot \nu \, d\mathcal{H}^{n-1} \leq \mathcal{H}^{n-1}(\partial\Omega) \max_{\partial\Omega} |Du_\sigma|$$

$$\leq \omega_n^{-1/n} \mathcal{H}^{n-1}(\partial\Omega) \frac{R+r}{R-r} \left(\int_{B_R(0)} f_\sigma \, dx \right)^{1/n}.$$

Letting $\sigma \to 0$ and using the Dominated Convergence Theorem, we obtain

$$n|\Omega| \leq \omega_n^{-1/n} \mathcal{H}^{n-1}(\partial\Omega) \frac{R+r}{R-r} |\Omega|^{1/n}.$$

Letting $R \to \infty$, we obtain the isoperimetric inequality as asserted. $\square$

4.8. Application: Nonlinear Integration by Parts Inequality

An interesting application of Theorem 4.12 is the nonlinear integration by parts inequality which plays a crucial role in the proof of uniqueness of the Monge–Ampère eigenvalue and eigenfunctions on bounded convex domains that are not necessarily smooth or uniformly convex; see Chapter 11.

Proposition 4.17 (Nonlinear integration by parts inequality). *Let Ω be a bounded convex domain in $\mathbb{R}^n$. Suppose that $u, v \in C(\overline{\Omega}) \cap C^5(\Omega)$ are strictly convex functions in Ω with $u = v = 0$ on $\partial\Omega$, and suppose they satisfy*

$$\int_\Omega (\det D^2 u)^{\frac{1}{n}} (\det D^2 v)^{\frac{n-1}{n}} \, dx + \int_\Omega \det D^2 v \, dx \leq M,$$

for some constant $M > 0$. Then

(4.29)
$$\int_\Omega |u| \det D^2 v \, dx \geq \int_\Omega |v| (\det D^2 u)^{\frac{1}{n}} (\det D^2 v)^{\frac{n-1}{n}} \, dx.$$

Proof. We proceed in two steps depending on the smoothness of the data.

Step 1. We first prove (4.29) when Ω is a uniformly convex domain with C^∞ boundary and $u, v \in C^4(\overline{\Omega})$. Let $V = (V^{ij})_{1 \leq i,j \leq n} = \operatorname{Cof} D^2 v$. Then $n \det D^2 v = V^{ij} D_{ij} v$. Since V is divergence-free (see Lemma 2.56), we also have $n \det D^2 v = D_i(V^{ij} D_j v)$. Integrating by parts twice, we obtain

(4.30)
$$\int_\Omega (-u)n \det D^2 v \, dx = \int_\Omega (-u)D_i(V^{ij} D_j v) \, dx = \int_\Omega -D_{ij}u V^{ij} v \, dx.$$

From $D_{ij}u V^{ij} = \operatorname{trace}(D^2 u V)$ and Lemma 2.57, we get

$$D_{ij}u V^{ij} \geq n(\det D^2 u)^{1/n}(\det V)^{1/n} = n(\det D^2 u)^{\frac{1}{n}}(\det D^2 v)^{\frac{n-1}{n}}.$$

Recalling (4.30) and the fact that $u, v < 0$ in Ω, we obtain (4.29).

Step 2. We now prove (4.29) for the general case. Let $\{\Omega_m\}_{m=1}^\infty \subset \Omega$ be a sequence of uniformly convex domains in Ω with C^∞ boundaries $\partial\Omega_m$ that converges to Ω in the Hausdorff distance d_H; see Theorem 2.51. Since $u \in C^5(\Omega)$ is strictly convex, we have $0 < \det D^2 u \in C^3(\Omega)$. For each m, by Theorem 4.12, the Dirichlet problem

$$\det D^2 u_m = \det D^2 u \quad \text{in } \Omega_m, \qquad u_m = 0 \quad \text{on } \partial\Omega_m$$

has a unique convex solution $u_m \in C^4(\overline{\Omega_m})$. Similarly, there is a unique convex solution $v_m \in C^4(\overline{\Omega_m})$ to the Dirichlet problem

$$\det D^2 v_m = \det D^2 v \quad \text{in } \Omega_m, \qquad v_m = 0 \quad \text{on } \partial\Omega_m.$$

Applying Step 1 to the functions u_m, v_m on Ω_m, we get

$$(4.31) \quad \int_{\Omega_m} |u_m| \det D^2 v \, dx \geq \int_{\Omega_m} |v_m| (\det D^2 u)^{\frac{1}{n}} (\det D^2 v)^{\frac{n-1}{n}} \, dx := A_m.$$

Now, we let $m \to \infty$ in (4.31) to obtain (4.29). To see this, we apply the comparison principle in Theorem 3.21 to the functions v and $v_m \pm \|v\|_{L^\infty(\partial\Omega_m)}$, which have the same Monge–Ampère measure in Ω_m, to conclude

$$-\|v\|_{L^\infty(\partial\Omega_m)} + v_m \leq v \leq \|v\|_{L^\infty(\partial\Omega_m)} + v_m \quad \text{in } \Omega_m.$$

Since $v \in C(\overline{\Omega})$, $v = 0$ on $\partial\Omega$, and $d_H(\Omega_m, \Omega) \to 0$, it follows that

$$(4.32) \quad \|v - v_m\|_{L^\infty(\Omega_m)} \leq \|v\|_{L^\infty(\partial\Omega_m)} \to 0 \quad \text{when } m \to \infty.$$

From (4.32) and the bound $\int_\Omega (\det D^2 u)^{\frac{1}{n}} (\det D^2 v)^{\frac{n-1}{n}} \, dx \leq M$, we can apply the Dominated Convergence Theorem to infer that

$$(4.33) \quad A_m \to \int_\Omega |v| (\det D^2 u)^{\frac{1}{n}} (\det D^2 v)^{\frac{n-1}{n}} \, dx \quad \text{when } m \to \infty.$$

Similarly, we also have

$$(4.34) \quad \int_{\Omega_m} |u_m| \det D^2 v \, dx \to \int_\Omega |u| \det D^2 v \, dx \quad \text{when } m \to \infty.$$

Hence, using (4.33)–(4.34), we can let $m \to \infty$ in (4.31) to obtain (4.29). $\qquad \square$

In Step 1, if $n = 1$ (though our interest here is for $n \geq 2$), then we have an equality which is the standard one-dimensional integration by parts formula. Also, from the proof and that of Lemma 2.57, we see that equality occurs in Step 1 if and only if there is a positive function $\mu \in C^2(\Omega)$ such that $D^2 u = \mu D^2 v$ in Ω. It is still an open question to characterize the equality case in the general setting of Proposition 4.17. Finally, we note that the nonlinear integration by parts inequality, stated above for the Monge–Ampère operator, can be extended to k-Hessian operators.

4.9. Problems

Problem 4.1. Let Ω be a bounded convex domain in $\mathbb{R}^n$ with C^1 boundary and outer unit normal ν on $\partial\Omega$. Let $\varphi \in C^1(\overline{\Omega})$, and let $u \in C^1(\overline{\Omega})$ be a convex function such that $u = \varphi$ on $\partial\Omega$. Show that

$$Du \cdot \nu \geq \frac{\varphi - \min_{\overline{\Omega}} u}{\operatorname{diam}(\Omega)} - |D\varphi| \quad \text{on } \partial\Omega.$$

Problem 4.2. Show that in Proposition 4.4(i), in order to have the nonemptiness of $\partial_s u(x_0)$, we can replace the assumption of φ being $C^{1,1}$ at x_0 with φ being $C^{1,1}$ from below at x_0; namely, there is a quadratic polynomial P such that $P(x_0) = \varphi(x_0)$ and $P \leq \varphi$ in a neighborhood of x_0 on $\partial\Omega$.

Problem 4.3. Let Ω be a bounded domain in $\mathbb{R}^n$. Let $f(x, z, p) \in C^1(\overline{\Omega} \times \mathbb{R} \times \mathbb{R}^n)$ with $D_z f(x, z, p) \geq 0$ in $\overline{\Omega} \times \mathbb{R} \times \mathbb{R}^n$. Assume that $u, v \in C(\overline{\Omega}) \cap C^2(\Omega)$ are convex functions in Ω with $u \geq v$ on $\partial\Omega$ and

$$\det D^2 v(x) \geq f(x, v(x), Dv(x)) > 0 \quad \text{in } \Omega$$

and

$$0 < \det D^2 u(x) \leq f(x, u(x), Du(x)) \quad \text{in } \Omega.$$

Show that $u \geq v$ in Ω.

Problem 4.4. Let $\Omega = B_1(0) \cap \{x_n > 0\}$, and let $\varphi : \partial\Omega \to [1/2, 2]$ be a continuous function with $\varphi(x) = |x'|^2$ on $\partial\Omega \cap \{x_n = 0\}$. Let $u \in C(\overline{\Omega})$ be a convex function with $u = \varphi$ on $\partial\Omega$, and assume it satisfies $0 < \lambda \leq \det D^2 u \leq \Lambda$ in Ω in the sense of Aleksandrov.

(a) Let $0 < \delta < 1/2$, and define the functions

$$v(x) = \delta|x'|^2 + \Lambda\delta^{1-n}(x_n^2 - x_n), \quad w(x) = (2 + \delta)|x'|^2 + \lambda 5^{-n} x_n^2 + 3x_n.$$

Show that $v \leq u \leq w$ in Ω.

(b) Show that u separates quadratically from its tangent hyperplanes at 0.

Problem 4.5. Suppose $\varphi \equiv 0$ in Theorem 4.9. Use Problem 4.1 and the equation following (2.13) to derive the first inequality in (4.11) without using the quadratic separation.

Problem 4.6. Show that the $C^{k+2,\alpha}(\overline{\Omega})$ estimates of Theorem 4.11 are still valid for the equation

$$\det(D^2 u + A) = f \quad \text{in } \Omega, \qquad u = \varphi \quad \text{on } \partial\Omega,$$

where the matrix function $A \in C^{k,\alpha}(\overline{\Omega})$ is symmetric and the uniform convexity of u is replaced by $D^2 u + A$ being positive definite.

Problem 4.7. Let Ω be a bounded set in $\mathbb{R}^n$ with C^1 boundary. Suppose that the isoperimetric inequality is in fact an equality for Ω; that is,

$$\mathcal{H}^{n-1}(\partial\Omega) = n\omega_n^{1/n}|\Omega|^{(n-1)/n}.$$

Show that Ω is connected.

Problem 4.8. Let Ω be a bounded convex domain in $\mathbb{R}^n$ with $\partial\Omega \in C^2$. Let $u, v \in C^{1,1}(\overline{\Omega}) \cap C^4(\Omega)$ be convex functions vanishing on $\partial\Omega$. Show that

$$\int_\Omega |u| \det D^2 v \, dx \geq \int_\Omega |v|(\det D^2 u)^{\frac{n-1}{2n}} (\det D^2 v)^{\frac{n+1}{2n}} \, dx.$$

Hint: Show that if $A, B \in \mathbb{S}^n$ are nonnegative definite matrices, then

$$\text{trace}[\text{Cof}(A)B] \geq n^{1/2}(\det A)^{1/2}(\text{trace}[\text{Cof}(B)A])^{1/2}.$$

4.10. Notes

In Section 4.2, Propositions 4.5 and 4.7 are due to Caffarelli, Nirenberg, and Spruck [**CNS1**], Wang [**W4**], and Savin [**S4**]. The proof of a quantitative lower bound for μ_1 in Proposition 4.5 follows the arguments of Wang [**W4**].

We rely on Figalli [**F2**, Sections 3.1 and 3.2] for the materials in Sections 4.3, 4.4, and 4.6; see also Gilbarg–Trudinger [**GT**, Sections 17.7 and 17.8] and Han [**H**, Section 6.2]. For more on the method of continuity in fully nonlinear elliptic equations, see [**GT**, Section 17.2].

In Section 4.3, Remark 4.10 shares the same flavor with the remark proceeding Theorem 3.39: For the existence or estimates in the Monge–Ampère equation, we usually need convexity properties coming from either the domain or boundary data.

Related to the isoperimetric inequality in Theorem 4.16, see Figalli [**F2**, Section 4.6] for a proof using optimal transport.

Proposition 4.17 is from Le [**L7**].

Problem 4.6 is taken from Caffarelli–Nirenberg–Spruck [**CNS1**].

Sections and Interior First Derivative Estimates

We saw in Chapter 4 that the convex Aleksandrov solution to the Dirichlet problem for the Monge–Ampère equation $\det D^2 u = f$ in a domain Ω is smooth if the Monge–Ampère measure f, the domain boundary, and boundary values are sufficiently smooth. In many applications, the function f might depend also on u and Du, so its smoothness is not always a priori guaranteed. Thus, it is important to investigate *regularity properties of the convex Aleksandrov solution to the Dirichlet problem for the Monge–Ampère equation when the data have limited regularity.* In particular, we would like to see *if is possible to go beyond the local Lipschitz property to obtain higher-order differentiability properties.* Related to this question, we also saw from Pogorelov's Example 4.14 that there are no purely local estimates for the Monge–Ampère equation starting from dimension 3. In other words, to obtain local regularity of u around a point, we somehow need to nicely constrain the values of u on the boundary of a neighborhood of that point. But this seems to be a circular argument; as we are investigating regularity, we do not know in general how nice these boundary values are. Nevertheless, there is an interesting scenario at our disposal. As our solution is convex, it has supporting hyperplanes at any interior point x_0 of Ω. Thus, we might cut the graph of the solution by hyperplanes parallel to supporting ones and then project back onto Ω. If some projected region S is compactly contained in Ω, or even better if S can be made to be close to x_0, then the solution is an affine function on the boundary of S which gives us some hope for further local regularity. This leads us to the concept of *sections of a convex function*. To obtain regularity, it is then natural to find conditions to guarantee that sections are compactly contained, and this is the idea of *localization* in the Monge–Ampère equation. We will carry out these ideas in this chapter.

First, we introduce the notion of sections of convex functions which plays a crucial role in the study of geometry of solutions to Monge–Ampère equations. Then, we study several properties of sections including volume estimates, size estimates, engulfing property, inclusion and exclusion properties, and a localization property of intersecting sections. Using volume estimates, we prove Caffarelli's optimal estimate on the dimension of the subspace where a convex function with strictly positive Monge–Ampère measure coincides with a supporting hyperplane. We next prove two real analysis results for sections: Vitali's Covering Lemma and the Crawling of Ink-spots Lemma. They are useful in the proof of $W^{2,1+\varepsilon}$ estimates for the Monge–Ampère equation in Chapter 6 and the proof of the Harnack inequality for the linearized Monge–Ampère equation in Chapter 12.

We start the regularity theory of the Monge–Ampère equation by proving the important Localization Theorem of Caffarelli. From this together with properties of sections, we obtain Caffarelli's $C^{1,\alpha}$ regularity of strictly convex solutions. As Caffarelli's counterexamples will show, $C^{1,\alpha}$ regularity fails when solutions are not strictly convex.

Notation. For a convex function u, we always interpret the inequalities $\lambda \leq \det D^2 u \leq \Lambda$ in the sense of Aleksandrov; that is, $\lambda\, dx \leq \mu_u \leq \Lambda\, dx$. In this chapter, $\lambda \leq \Lambda$ are positive constants and $n \geq 2$.

5.1. Sections of Convex Functions

In addition to the Monge–Ampère measure, another central notion in the study of Monge–Ampère equations is that of sections (or cross sections) of convex functions, introduced and investigated by Caffarelli [**C2**, **C3**, **C4**, **C5**]. Interior sections were further developed by Caffarelli and Gutiérrez [**CG1**] and Gutiérrez and Huang [**GH2**]. Sections have the same role as Euclidean balls have in the classical theory of uniformly elliptic equations.

Definition 5.1 (Section). Let Ω be an open set in $\mathbb{R}^n$. Let $u \in C(\overline{\Omega})$ be a convex function on Ω, and let $p \in \partial u(x)$ where $x \in \Omega$. The section of u centered at x with slope p and height $h > 0$ (see Figure 5.1) is defined by

$$S_u(x, p, h) := \{y \in \overline{\Omega} : u(y) < u(x) + p \cdot (y - x) + h\}.$$

In the above definition, $S_u(x, p, h)$ depends on Ω. This dependence should be clear in concrete contexts, and *our heights of sections are always assumed to be positive*. A Euclidean ball of radius r is a section

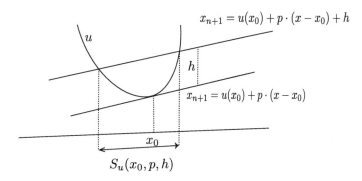

Figure 5.1. A section $S_u(x_0, p, h)$ of u with center x_0, slope p, and height h.

with height $r^2/2$ of the quadratic function $|x|^2/2$ on $\mathbb{R}^n$ whose Monge–Ampère measure is the Lebesgue measure. If $u(x) = |x|^2/2$ on $\overline{\Omega}$, then $S_u(x, Du(x), h) = B_{\sqrt{2h}}(x) \cap \overline{\Omega}$.

Remark 5.2. Definition 5.1 still makes sense for $u : \Omega \to \mathbb{R}$; in this case, we implicitly extend u to $\partial\Omega$ by restricting to $\partial\Omega$ its minimal convex extension to $\mathbb{R}^n$. This should not cause any confusion because we will mostly consider in this chapter the case of compactly contained sections. Note that if $S_u(x, p, h) \Subset \Omega$, then the convexity of u implies that $S_u(x, p, h)$ is a bounded convex domain and $u(y) = u(x) + p \cdot (y - x) + h$ on $\partial S_u(x, p, h)$.

Remark 5.3. Let $u : \Omega \to \mathbb{R}$ be convex on an open set $\Omega \subset \mathbb{R}^n$, and let $x_0 \in \Omega$.

- If $p_0 \in \partial u(x_0)$, then $S_u(x_0, p_0, h) - x_0 = S_v(0, 0, h)$, where v is defined by $v(x) := u(x + x_0) - u(x_0) - p_0 \cdot x$ for $x \in \Omega - x_0$.

- When u is differentiable at x_0, if $p \in \partial u(x_0)$, then $p = Du(x_0)$. In this case, we simply write $S_u(x_0, h)$ for $S_u(x_0, Du(x_0), h)$.

Remark 5.4. Let $u : \Omega \to \mathbb{R}$ be a convex function on an open set $\Omega \subset \mathbb{R}^n$. If $S_u(x, p, h) \Subset \Omega$, then from the convexity of u and $S_u(x, p, h)$, we find

$$\delta S_u(x, p, h) + (1 - \delta)x \subset S_u(x, p, \delta h) \quad \text{for all } 0 < \delta < 1.$$

It follows that $\delta^n |S_u(x, p, h)| \leq |S_u(x, p, \delta h)|$, so the Lebesgue measure is doubling on compactly contained sections of convex functions. If, in addition, $\lambda \leq \det D^2 u \leq \Lambda$, then the above inclusion is strict. This is a highly nontrivial fact (see Theorem 5.18), and it is the subject of this chapter.

We will investigate in what ways sections of convex solutions to the Monge–Ampère equation with suitable data are similar to Euclidean balls.

Example 5.5. Consider the convex function $u(x) = M|x|^2/2$ defined on $\mathbb{R}^n$ where $M > 0$. Then $S_u(0, 0, h) = B_{\sqrt{2h/M}}(0)$, and

$$|S_u(0, 0, h)| = \omega_n 2^{n/2} h^{n/2} M^{-n/2} = \omega_n 2^{n/2} h^{n/2} (\det D^2 u)^{-1/2}.$$

More generally, we show that, up to a structural factor of $h^{n/2}$, the volume growth of sections $S_u(x, p, h)$ that are compactly contained in the domain depends only on the bounds on the Monge–Ampère measure; moreover, these compactly contained sections are balanced around the centers.

Lemma 5.6 (Volume, balancing, and separation of compactly contained sections). *Let Ω be an open set in $\mathbb{R}^n$. Let u be a convex function satisfying $0 < \lambda \le \det D^2 u \le \Lambda$ in Ω, in the sense of Aleksandrov. Let $p \in \partial u(x)$ where $x \in \Omega$. Suppose that $S_u(x, p, h)$ is bounded, open, and convex and that $u(y) = u(x) + p \cdot (y - x) + h$ on $\partial S_u(x, p, h)$ (these hold if $S_u(x, p, h) \Subset \Omega$). Then, the following hold.*

(i) $c(n)\Lambda^{-1/2}h^{n/2} \le |S_u(x, p, h)| \le C(n)\lambda^{-1/2}h^{n/2}$, *where $c, C > 0$.*

(ii) *For all $\gamma \in (0, 1)$, we have the following separation of sections:*

$$\operatorname{dist}(S_u(x, p, \gamma h), \partial S_u(x, p, h)) \ge \frac{c(\lambda, \Lambda, n)(1 - \gamma)^n h^{n/2}}{[\operatorname{diam}(S_u(x, p, h))]^{n-1}}.$$

(iii) *There exist $\sigma(n, \lambda, \Lambda) \in (0, 1)$ and an ellipsoid $E_h = A_h^{-1} B_{h^{1/2}}(0)$ with volume $\omega_n h^{n/2}$ where $A_h \in \mathbb{M}^{n \times n}$ with $\det A_h = 1$, such that*

$$x + \sigma E_h \subset S_u(x, p, h) \subset x + \sigma^{-1} E_h.$$

(a) *If $S_u(x, p, \tau h) \Subset \Omega$ for some $\tau > 1$, then*

$$\|Du - p\|_{L^\infty(S_u(x,p,h))} \le C_* h^{1 - n/2} \quad and \quad \|A_h\| \le C_* h^{(1-n)/2},$$

for some $C_* = C_*(n, \lambda, \Lambda, \tau, \operatorname{diam}(S_u(x, p, \tau h))) > 0$.

(b) *There is $\beta(n, \lambda, \Lambda) \in (0, 1)$ such that if u is C^2 in a neighborhood of x, then for h sufficiently small, we have*

$$\begin{cases} \beta \|D^2 u(x)\|^{1/2} \le \|A_h\| \le \beta^{-1} \|D^2 u(x)\|^{1/2}, \\ \|D^2 u(x)\|^{1/2} \le C(n, \lambda, \Lambda) \|A_h^{-1}\|^{n-1}. \end{cases}$$

(iv) *There exists a constant $r = r(n, \lambda, \Lambda, h, \operatorname{diam}(S_u(x, p, h))) > 0$ such that $B_r(x) \subset S_u(x, p, h)$.*

Proof. Let $S := S_u(x, p, h)$ and $\bar{u}(z) = u(z) - [u(x) + p \cdot (z - x) + h]$. Then $\bar{u}|_{\partial S_u(x,p,h)} = 0$, and $\bar{u}$ achieves its minimum $-h$ at x. By John's Lemma (Lemma 2.63), we can find a positive definite matrix $A \in \mathbb{S}^n$ and $b \in \mathbb{R}^n$ such that

(5.1) $$B_1(b) \subset \tilde{S} := A^{-1}(S_u(x, p, h)) \subset B_n(b).$$

Let $\tilde{u}(z) = (\det A)^{-2/n}\bar{u}(Az)$ and $\tilde{x} = A^{-1}x$. Then, by Lemma 3.9, we have

$$\tilde{S} = S_{\tilde{u}}(\tilde{x}, 0, |\min_{\tilde{S}} \tilde{u}|), \quad \lambda \le \det D^2\tilde{u} \le \Lambda \quad \text{in } \tilde{S}, \quad \text{and} \quad \tilde{u} = 0 \quad \text{on } \partial\tilde{S}.$$

Now, by Lemma 3.24, we have the comparisons

$$(5.2) \qquad \frac{\Lambda^{1/n}}{2}(|z - b|^2 - n^2) \le \tilde{u}(z) \le \frac{\lambda^{1/n}}{2}(|z - b|^2 - 1) \quad \text{in } \tilde{S},$$

yielding

$$(5.3) \qquad \lambda^{1/n}/2 \le |\min_{\tilde{S}} \tilde{u}| \le n^2\Lambda^{1/n}/2.$$

In view of $|\min_{\tilde{S}} \tilde{u}| = h(\det A)^{-2/n}$, this gives

$$(5.4) \qquad c(n)\Lambda^{-1/2}h^{n/2} \le \det A \le C(n)\lambda^{-1/2}h^{n/2}.$$

On the other hand, from (5.1) and $|A^{-1}S_u(x, p, h)| = (\det A)^{-1}|S_u(x, p, h)|$, we deduce that

$$\omega_n \det A \le |S_u(x, p, h)| \le n^n\omega_n \det A,$$

and the desired volume bounds in part (i) follow from (5.4).

To prove part (ii), note that $-h \le \bar{u} \le -(1 - \gamma)h$ in $S_u(x, p, \gamma h)$. Thus, applying the Aleksandrov Theorem (Theorem 3.12) to $\bar{u}$ on $S := S_u(x, p, h)$ and then using the volume estimates in part (i), we have for any $z \in S_u(x, p, \gamma h)$,

$$(1 - \gamma)^n h^n \le |\bar{u}(z)|^n \le C(n)\text{dist}(z, \partial S)[\text{diam}(S)]^{n-1}\Lambda|S|$$
$$\le C(\lambda, \Lambda, n)\text{dist}(z, \partial S)[\text{diam}(S)]^{n-1}h^{n/2}.$$

Thus, part (ii) follows.

Note that $\tilde{S} = S_{\tilde{u}}(\tilde{x}, 0, |\min_{\tilde{S}} \tilde{u}|) \subset B_n(b)$ and $\tilde{u}$ has a minimum value at $\tilde{x}$. Therefore, (ii) and (5.3) give $\text{dist}(\tilde{x}, \partial\tilde{S}) \ge c_1(\lambda, \Lambda, n)$. This combined with $\tilde{S} \subset B_{2n}(\tilde{x})$ gives $B_{c_1}(\tilde{x}) \subset \tilde{S} := A^{-1}(S_u(x, p, h)) \subset B_{2n}(\tilde{x})$. Hence

$$x + AB_{c_1}(0) \subset S_u(x, p, h) \subset x + AB_{2n}(0).$$

Let $A_h = (\det A)^{1/n}A^{-1}$. Then, by (5.4), the ellipsoid

$$E_h := h^{1/2}(\det A)^{-1/n}AB_1(0) \equiv A_h^{-1}B_{h^{1/2}}(0)$$

satisfies the first conclusion of part (iii) for some $\sigma = \sigma(n, \lambda, \Lambda) \in (0, 1)$.

We now prove part (iii)(a). Assume $S_u(x, p, \tau h) \Subset \Omega$ where $\tau > 1$. Using part (ii) and Lemmas 2.37(iii) and 2.40(iii), we find that

$$(5.5) \qquad \max_{q \in \partial\bar{u}(\tilde{S})} \{|q|, \|D\bar{u}\|_{L^\infty(S)}\} \le \frac{\sup_{\partial S_u(x, p, \tau h)} \bar{u} - \inf_S \bar{u}}{\text{dist}(S, \partial S_u(x, p, \tau h))}$$
$$\le \bar{C}h^{1 - \frac{n}{2}},$$

where $\bar{C} = \bar{C}(\lambda, \Lambda, n, \tau, \text{diam}(S_u(x, p, \tau h)))$. This gives the asserted estimate for $\|Du - p\|_{L^\infty(S_u(x,p,h))}$, noting that $\partial \bar{u}(\overline{S_u(x, p, h)}) = \partial u(\overline{S_u(x, p, h)}) - p$.

To obtain the asserted estimate for A_h, by (5.4), it remains to prove

$$(5.6) \qquad \|A^{-1}\| \leq C(n, \lambda, \Lambda, \tau, \text{diam}(S_u(x, p, \tau h))) h^{-\frac{n}{2}}.$$

However, (5.6) follows from (5.1) and

$$(5.7) \qquad S_u(x, p, h) \supset B_{c_2 h^{\frac{n}{2}}}(x),$$

for $c_2 = 1/(4\bar{C})$. Were (5.7) not true, then there is $z \in \partial S_u(x, p, h) \cap \partial B_{c_2 h^{\frac{n}{2}}}(x)$. Let $p_z \in \partial u(z)$. Then, we obtain a contradiction from

$$h = u(z) - u(x) - p \cdot (z - x) \leq (p_z - p) \cdot (z - x) \leq 2\bar{C} h^{1 - \frac{n}{2}} (c_2 h^{\frac{n}{2}}) = h/2.$$

We next prove part (iii)(b). Assume now u is C^2 in a neighborhood of x. For simplicity, assume $x = 0$, $u(0) = 0$, $Du(0) = 0$. Then, for h small

$$S_u(x, p, h) = S_u(0, h) = \{y \in \Omega : u(y) < h\}.$$

Let $M := \frac{1}{2} D^2 u(0)$. Then M is symmetric and positive definite. From the Taylor expansion, we have

$$u(y) = (My) \cdot y + o(|y|^2) = (M^{1/2}y) \cdot (M^{1/2}y) + o(|y|^2) \quad \text{as } y \to 0.$$

Thus, for each $\delta \in (0, 1/2)$, there is $h_\delta > 0$ such that for all $h \leq h_\delta$, we have

$$\{y \in \Omega : (1 + \delta)^2 (My) \cdot y < h\} \subset S_u(0, h) \subset \{y \in \Omega : (1 - \delta)^2 (My) \cdot y < h\}.$$

It follows that

$$(1 + \delta)^{-1} M^{-1/2} B_{h^{1/2}(0)} \subset S_u(0, h) \subset (1 - \delta)^{-1} M^{-1/2} B_{h^{1/2}(0)}.$$

Because $\sigma B_{h^{1/2}}(0) \subset A_h S_u(0, h) \subset \sigma^{-1} B_{h^{1/2}}(0)$, we easily deduce that

$$\sigma(1 - \delta)\|M\|^{1/2} \leq \|A_h\| \leq \sigma^{-1}(1 + \delta)\|M\|^{1/2}.$$

This is true for all $\delta \in (0, 1/2)$ and $h \leq h_\delta$, so the estimates on $\|A_h\|$ follow.

Note that we also have $\|M^{-1/2}\| \leq (1 + \delta)\sigma^{-1}\|A_h^{-1}\|$. Let $m_1 \leq \cdots \leq m_n$ be the positive eigenvalues of $M^{1/2}$. Then each $m_i^{-1} \leq (1 + \delta)\sigma^{-1}\|A_h^{-1}\|$. Since $m_1 \ldots m_n = \det M^{1/2} \leq \Lambda^{1/2}$, we deduce that

$$\|D^2 u(x)\|^{1/2} \leq 2n m_n = \frac{2n \det M^{1/2}}{m_1 \ldots m_{n-1}} \leq 2n \Lambda^{1/2} (2\sigma^{-1}\|A_h^{-1}\|)^{n-1}.$$

Finally, we prove part (iv). Rotating coordinates, we can assume that the ellipsoid E_h in part (iii) is given by $E_h = \{x \in \mathbb{R}^n : \sum_{i=1}^n x_i^2/a_i^2 < 1\}$, where $0 < a_1 \leq \cdots \leq a_n$. Then $\omega_n h^{n/2} = |E_h| = \omega_n \prod_{i=1}^n a_i$, which gives $\prod_{i=1}^n a_i = h^{n/2}$. Since $x + \sigma E_h \subset S_u(x, p, h)$, we find that

$$a_n \leq d := \text{diam}(S_u(x, p, h))/\sigma.$$

Hence

$$a_1 = h^{n/2} \prod_{i=2}^{n} a_i^{-1} \geq h^{n/2} d^{-(n-1)} := r_1.$$

The conclusion now follows from $B_{\sigma r_1}(x) \subset x + \sigma E_h \subset S_u(x, p, h)$. $\square$

Remark 5.7. In the proof of Lemma 5.6, we deduced several properties of a compactly contained section $S_u(x, p, h)$ from the normalized function $\tilde{u}(z) = (\det A)^{-2/n}[u(Az) - u(x) - p \cdot (Az - x) - h]$ on the normalized section $\tilde{S} = A^{-1}(S_u(x, p, h))$, where $\tilde{u}$ satisfies $\lambda \leq \det D^2 \tilde{u} \leq \Lambda$ and $\tilde{u} = 0$ on $\partial \tilde{S}$. Deeper properties, depending on n, λ, Λ, of $\tilde{u}$ give corresponding properties for u. The estimate constants now depend also on $\det A$ and $\|A^{-1}\|$, as can be seen, for example, from $Du(z) = (\det A)^{2/n}(A^{-1})^t D\tilde{u}(A^{-1}z) + p$. This is the reason why we mostly study the Monge–Ampère equation on normalized convex domains.

As in the second inequality in (3.15), if $\tilde{u} \leq 0$ on $\tilde{S}$ in the proof of Lemma 5.6, then the second inequality in (5.2) still holds, and so does the second inequality in part (i) of the lemma. This is the case where the section $S_u(x, p, h)$ is only bounded but may not be compactly contained in the domain Ω. We summarize this discussion in the following useful lemma.

Lemma 5.8 (Volume estimates for bounded sections). *Let Ω be a bounded convex domain in $\mathbb{R}^n$, and let $u : \Omega \to \mathbb{R}$ be a convex function satisfying $\det D^2 u \geq \lambda > 0$ in Ω in the sense of Aleksandrov. Then for any section*

$$S_u(x, p, h) = \{y \in \overline{\Omega} : u(y) < u(x) + p \cdot (y - x) + h\}, \ x \in \overline{\Omega}, \ p \in \partial u(x),$$

we have

$$|S_u(x, p, h)| \leq C(n)\lambda^{-1/2} h^{n/2}.$$

As an application of Lemma 5.8, we prove the following theorem of Caffarelli [**C7**] concerning the size of the degeneracy set of solutions to the Monge–Ampère equation where they coincide with a supporting hyperplane.

Theorem 5.9 (Size of the degeneracy set). *Let $u : \Omega \to \mathbb{R}$ be a convex function that satisfies, in the sense of Aleksandrov, $\det D^2 u \geq \lambda > 0$ in a domain Ω in $\mathbb{R}^n$. Then, the dimension of the set where u agrees with any affine function l is smaller than $n/2$.*

Proof. Subtracting l from u, we can assume that $l = 0$. Suppose that u vanishes on a k-dimensional affine space E_k. We need to show that $k < n/2$. By a coordinate translation, we can assume that 0 is an interior point of $\{u = 0\}$ and $E_k := \{x_{k+1} = \cdots = x_n = 0\}$. In this case, any $p \in \partial u(0)$ is of the form $p = (0, \ldots, 0, p_{k+1}, \ldots, p_n)$. By subtracting $p \cdot x$ from u, we can assume $u \geq 0$ in $B_{2r} := B_{2r}(0) \subset \Omega$ and $u = 0$ on $E_k \cap B_{2r}$. By Lemma 2.40, u is Lipschitz in B_r with Lipschitz constant L. Again, by convexity, the partial derivative $D_n u(te_n)$ exists for almost every t. Since it is bounded and monotone in t for $|t| < r$, there exists a limit a_n when $t \to 0^+$. By further subtracting $a_n x_n$ from u, we can assume that $u(te_n) = o(t)$ as $t \to 0^+$.

Consider the section $S_u(0, 0, h)$ in $B_{2r}(0)$, where $h > 0$ is small. Then, $S_u(0, 0, h)$ has length $R(h)h$ in the e_n direction with $R(h) \to \infty$ as $h \to 0$. Furthermore, $S_u(0, 0, h)$ contains a ball of radius r in the subspace spanned by $\{e_1, \ldots, e_k\}$, while $S_u(0, 0, h)$ has length exceeding h/L in the $e_{k+1}, \ldots, e_{n-1}$ directions. It follows that

$$|S_u(0, 0, h)| \geq C(r, n, k) L^{-n+k+1} h^{n-k} R(h).$$

By Lemma 5.8, we have $|S_u(0, 0, h)| \leq C(n, \lambda) h^{n/2}$. Therefore,

$$C(n, \lambda) h^{n/2} \geq C(r, n, k) L^{-n+k+1} h^{n-k} R(h).$$

Since $R(h) \to \infty$ as $h \to 0$ and the above inequality holds for all $h > 0$ small, we must have $k < n/2$. $\square$

Remark 5.10. Theorem 5.9 implies that every solution to $\det D^2 u \geq 1$ in two dimensions is strictly convex. Example 5.24 will show that the estimate on the dimension of the degeneracy set in Theorem 5.9 is sharp.

We have just seen that sections of solutions to the Monge–Ampère equation that are compactly contained in the domain behave, in several ways, like Euclidean balls after suitable affine transformations. This raises an important question about whether solutions are strictly convex or not. (Recall from Definition 2.39 that a convex function u on an open set $\Omega \subset \mathbb{R}^n$ is strictly convex at $x_0 \in \Omega$ if any supporting hyperplane to u at x_0 touches its graph at only $(x_0, u(x_0))$, so $\bigcap_{h>0} S_u(x_0, p, h) = \{x_0\}$ for all $p \in \partial u(x_0)$.) More concretely, under what conditions can we conclude from $\lambda \leq \det D^2 u \leq \Lambda$ in Ω that for all $x \in \Omega$, there is a small $h(x) > 0$ such that $S_u(x, p, h(x)) \Subset \Omega$ for all $p \in \partial u(x)$? An answer to this question will be given in Theorem 5.13 whose proof is based on Caffarelli's Localization Theorem (Theorem 5.11). For its statement, we refer to Definition 2.11 for the notions of extremal points and exposed points.

5.2. Caffarelli's Localization Theorem and Strict Convexity

In this section, we prove a fundamental localization theorem of Caffarelli [**C2**] and its consequences concerning strict convexity and size estimates for sections of solutions to the Monge–Ampère equation. They pave the way for the ensuing regularity theory.

Theorem 5.11 (Caffarelli's Localization Theorem). *Let Ω be an open set in $\mathbb{R}^n$. Let u be a convex function satisfying $0 < \lambda \leq \det D^2 u \leq \Lambda$ in Ω in the sense of Aleksandrov. Let $x_0 \in \Omega$ and let $l_{x_0}(x) := u(x_0) + p_0 \cdot (x - x_0)$ be*

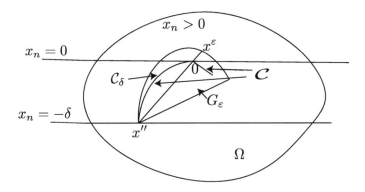

Figure 5.2. Configuration in the proof of Caffarelli's Localization Theorem.

the supporting hyperplane to the graph of the function u at $(x_0, u(x_0))$ with slope $p_0 \in \partial u(x_0)$. Then, either the contact set $\mathcal{C} = \{x \in \Omega : u(x) = l_{x_0}(x)\}$ is a single point or $\mathcal{C}$ has no extremal points in the interior of Ω.

Proof. By subtracting $l_{x_0}(x)$ from u, we can assume that $u \geq 0$ in Ω and $\mathcal{C} = \{u = 0\}$. Note that the set $\mathcal{C}$ is convex and it contains x_0. Suppose that $\mathcal{C}$ contains more than one point and that the conclusion of the theorem is false; that is, $\mathcal{C}$ contains an extremal point in the interior of Ω. Thus, by Straszewicz's Theorem (Theorem 2.12), $\mathcal{C}$ contains an exposed point in the interior of Ω. Since our argument below is local, we can now assume that Ω is bounded. By changing coordinates, we can assume that $0 \in \Omega$ is an exposed point of the set $\{u = 0\} \subset \{x_n \leq 0\}$ where $\{x_n = 0\}$ is a supporting hyperplane to $\overline{\mathcal{C}}$ that touches $\overline{\mathcal{C}}$ only at 0 and, furthermore, $\mathcal{C}_\delta := \{0 \geq x_n \geq -\delta\} \cap \mathcal{C}$ is compactly contained in Ω for some $\delta > 0$. Choose $x'' \in \{x_n = -\delta\} \cap \mathcal{C} \subset \Omega$. For $\varepsilon > 0$, consider

$$G_\varepsilon := \{x \in \Omega : v_\varepsilon(x) := u(x) - \varepsilon(x_n + \delta) \leq 0\}.$$

Then, $G_\varepsilon \subset \{x_n \geq -\delta\}$, and, as $\varepsilon \to 0$, G_ε converges in the Hausdorff distance to $\mathcal{C}_\delta$. See Figure 5.2. Hence, when $\varepsilon > 0$ is small, $G_\varepsilon \subset \Omega$ and $v_\varepsilon = 0$ on ∂G_ε. Clearly, $x'' \in \partial G_\varepsilon$, and 0 is an interior point of G_ε. On G_ε, we have $v_\varepsilon \geq -\varepsilon(d + \delta)$, where $d = \mathrm{diam}(\Omega)$. Hence,

$$v_\varepsilon(0) = -\varepsilon\delta \leq \frac{\delta}{d + \delta} \inf_{G_\varepsilon} v_\varepsilon.$$

Since 0 is an interior point on the segment connecting x'' and some point x^ε of $\partial G_\varepsilon \cap \{x_n \geq 0\}$, we can use the balancing lemma, Lemma 3.25(i), to get

$$|x_n^\varepsilon| \geq c(n, \delta, \lambda, \Lambda, d)|x_n^\varepsilon - x_n''| \geq c(n, \delta, \lambda, \Lambda, d),$$

contradicting the fact that $x_n^\varepsilon \to 0$ when $\varepsilon \to 0$. The theorem is proved. $\square$

Corollary 5.12 (Strict convexity and differentiability). *Let u be a convex function satisfying, in the sense of Aleksandrov,*

$$0 < \lambda \leq \det D^2 u \leq \Lambda \quad \text{in an open set } \Omega \subset \mathbb{R}^n.$$

(i) *If $\Omega = \mathbb{R}^n$, then u is strictly convex.*

(ii) *Assume, furthermore, u is strictly convex in Ω. Then u has a unique supporting hyperplane at each point in Ω, and thus $u \in C^1(\Omega)$ by Theorem 2.38.*

Proof. We prove part (i). Assume that u is not strictly convex. Then, it follows from Theorem 5.11 that there exist $x_0 \in \mathbb{R}^n$ and $p \in \partial u(x_0)$ such that $u(z) = u(x_0) + p \cdot (z - x_0)$ on a line L. By Lemma 2.36, $\partial u(\mathbb{R}^n)$ is contained in a hyperplane orthogonal to L. Thus, $|\partial u(\mathbb{R}^n)| = 0$. However, for any Borel set $E \subset \mathbb{R}^n$, we have $\lambda|E| \leq |\partial u(E)| \leq |\partial u(\mathbb{R}^n)| = 0$, a contradiction.

Now, we prove part (ii). Assume by contradiction that there is $x_0 \in \Omega$ such that $\partial u(x_0)$ has more than one element. By translating and rotating coordinates and subtracting an affine function from u, we can assume that $x_0 = 0$, $u(0) = 0$, $u \geq 0$ in Ω, and $\partial u(0) \supset [-\alpha e_n, \alpha e_n]$, for some $\alpha > 0$. Thus $u(x) \geq \alpha|x_n|$ in Ω. By subtracting $a x_n$ from u, where $a = \lim_{t \to 0^+} D_n u(t e_n)$, as in the proof of Theorem 5.9, we can assume further that

$$u(x) \geq \alpha x_n^-, \quad u(t e_n) = o(t) \text{ as } t \to 0^+, \quad \text{and} \quad u(t e_n) \text{ is increasing in } t > 0.$$

Here $x_n^- := \max\{0, -x_n\}$. For $0 < \sigma, \tau < \alpha$ to be chosen, consider

$$u_{\sigma,\tau}(x) := u(x) + \tau(x_n - \sigma).$$

Then $S_{\tau,\sigma} := \{x \in \overline{\Omega} : u_{\sigma,\tau}(x) < 0\} \equiv S_u(0, -\tau e_n, \tau\sigma)$. Since u is strictly convex at $0 \in \Omega$, we have $\bigcap_{\sigma > 0} S_u(0, -\tau e_n, \tau\sigma) = \{0\}$. Thus, $S_{\tau,\sigma} \Subset \Omega$ when σ and τ are small.

Note that, as $\tau < \alpha$, the function $u_{\sigma,\tau}$ attains its minimum at 0. Consider now two supporting hyperplanes $H_1 = \{x_n = T_1\}$ and $H_2 = \{x_n = -T_2\}$ to $\partial S_{\sigma,\tau}(x)$ where $T_1 > 0, T_2 > 0$. Clearly, H_1, H_2, T_1, T_2 depends on u, Ω, τ, σ, but for notational simplicity, we omit these dependences.

From $0 = u(-T_2 e_n) + \tau(-T_2 - \sigma) \geq \alpha T_2 + \tau(-T_2 - \sigma)$, we find

$$T_2 \leq \tau\sigma/(\alpha - \tau).$$

Now, choose $\tau = (2/\sigma)u(\sigma e_n/2)$ which is $o(1)$ when $\sigma \to 0^+$.

For $0 < t \leq \sigma/2$, we have

$$u_{\sigma,\tau}(t e_n) = u(t e_n) + (2/\sigma)u(\sigma e_n/2)(t - \sigma) \leq u(t e_n) - u(\sigma e_n/2) \leq 0.$$

Therefore, $T_1 \geq \sigma/2$. It follows that

$$\text{dist}(0, H_2)/\text{dist}(0, H_1) = T_2/T_1 \leq 2\tau/(\alpha - \tau) \to 0 \quad \text{as } \tau \to 0.$$

However, this contradicts Lemma 3.25(ii). Therefore, $\partial u(x)$ has one element for each $x \in \Omega$. The corollary is proved. $\square$

As a consequence of Theorem 5.11, we obtain the following important strict convexity result and its quantitative versions.

Theorem 5.13. *Let Ω be a bounded convex domain in $\mathbb{R}^n$. Let $u \in C(\overline{\Omega})$ be a convex function satisfying $0 < \lambda \leq \det D^2 u \leq \Lambda$ in the sense of Aleksandrov.*

(i) *Assume $u = L$ on $\partial\Omega$ where L is an affine function. Then u is strictly convex in Ω.*

(ii) *Assume further that $u = 0$ on $\partial\Omega$ and that $B_r(z) \subset \Omega \subset B_R(z)$, where $0 < r \leq R$ and $z \in \mathbb{R}^n$. Then, for any $x \in \Omega$ with $\text{dist}(x, \partial\Omega) \geq \alpha > 0$, there is a constant $h(\alpha, n, r, R, \lambda, \Lambda) > 0$ such that $S_u(x, p, h) \Subset \Omega$ for all $p \in \partial u(x)$.*

(iii) *Let $0 < \gamma_1 < \gamma_2 < 1$. Then, there exists $\delta_0(n, \lambda, \Lambda, \gamma_1, \gamma_2) > 0$ such that for any section $S_u(y, p, h) \Subset \Omega$, $x \in S_u(y, p, \gamma_1 h)$, and $q \in \partial u(x)$, we have $S_u(x, q, \delta_0 h) \subset S_u(y, p, \gamma_2 h)$.*

Proof. We first prove part (i). By subtracting L from u, we can assume that $u = 0$ on $\partial\Omega$. If u is not strictly convex at $x_0 \in \Omega$, then, for some supporting hyperplane l to the graph of u at $(x_0, u(x_0))$, the contact set $\mathcal{C} = \{x \in \Omega : u(x) = l(x)\}$ is not a single point. By Theorem 5.11, the set E of extremal points of $\mathcal{C}$ lies on the boundary $\partial\Omega$. Since $l = u = 0$ on E and any point in $\overline{\mathcal{C}}$ is a convex combination of points in E by Minkowski's Theorem (Theorem 2.12), we find from l being affine that $l = 0$ on $\mathcal{C}$. Therefore, $u(x_0) = l(x_0) = 0$. This combined with the convexity of u and $u = 0$ on $\partial\Omega$ allows us to deduce that $u \equiv 0$ in Ω. This contradicts $\det D^2 u \geq \lambda > 0$.

Next, we prove part (ii) by a compactness argument taking into account the strict convexity result in part (i). Assume $z = 0$. Denote $B_t = B_t(0)$ and $\Omega_\delta = \{x \in \Omega : \text{dist}(x, \partial\Omega) > \delta\}$. Assume that the conclusion is false. Then, we can find a sequence of convex functions u_k on Ω^k where $B_r \subset \Omega^k \subset B_R$, $u_k = 0$ on $\partial\Omega^k$, and $\lambda \leq \det D^2 u_k \leq \Lambda$ in Ω^k ($k = 1, 2, \ldots$), such that

$$(5.8) \qquad u_k(y_k) \leq u_k(x_k) + p_k \cdot (y_k - x_k) + h_k, \qquad p_k \in \partial u_k(x_k),$$

for sequences $x_k \in \Omega_\alpha^k$, $y_k \in \partial\Omega_{\alpha/2}^k$ and $h_k \to 0$. From Theorem 3.19, after passing to a subsequence if necessary, we may assume Ω_k converges to Ω in the Hausdorff distance where $B_r \subset \Omega \subset B_R$,

$$u_k \to u_0 \quad \text{locally uniformly on } \Omega, \quad x_k \to x_0 \in \overline{\Omega_\alpha}, \quad y_k \to y_0 \in \overline{\Omega_{\alpha/2}} \setminus \overline{\Omega_\alpha},$$

and u_0 satisfies

$$\lambda \leq \det D^2 u_0 \leq \Lambda \quad \text{in } \Omega \quad \text{and} \quad u_0 = 0 \quad \text{on } \partial\Omega.$$

On the other hand, using Lemma 2.37(iii) in conjunction with the Aleksandrov maximum principle in Theorem 3.12, we find that

$$|p_k| \leq C(\alpha, n, \Lambda, r, R) \quad \text{for all } k.$$

Thus, after passing a further subsequence, $\{p_k\}_{k=1}^{\infty}$ has a limit $p_0 \in \partial u_0(x_0)$. Moreover, we also deduce from (5.8) that

$$u_0(y_0) = u_0(x_0) + p_0 \cdot (y_0 - x_0).$$

This implies that the contact set $\{x \in \Omega : u_0(x) = u_0(x_0) + p_0 \cdot (x - x_0)\}$ is not a single point, and we have reached a contradiction to part (i).

Finally, we prove part (iii). Note that, in (ii), $\Omega = S_u(\hat{x}, 0, |\min_\Omega u|)$, where $\hat{x}$ is the minimum point of u in Ω. Moreover, by Lemma 3.24, $c_1 \leq |\min_\Omega u| \leq c_2$, where c_1, c_2 depend only on n, λ, Λ, r, and R.

Now, given $S_u(y, p, h)$, we can use John's Lemma and normalization as in the proof of Lemma 5.6 to assume that $p = 0$, $u = 0$ on $\partial S_u(y, p, h)$ while $S_u(y, p, h)$ is normalized, so $C^{-1} \leq h \leq C$ where $C = C(n, \lambda, \Lambda)$. By Lemma 5.6(ii), $\text{dist}(S_u(y, p, \gamma_1 h), \partial S_u(y, p, \gamma_2 h)) \geq \alpha(n, \lambda, \Lambda, \gamma_1, \gamma_2) > 0$. The conclusion now follows from part (ii). See also Theorem 5.30. $\qquad\square$

We can relax the zero boundary data in Theorem 5.13 to $C^{1,\beta}$ boundary data where $\beta > 1 - 2/n$. The following theorem on localization and strict convexity is also due to Caffarelli [**C2**].

Theorem 5.14 (Caffarelli's Localization and Strict Convexity Theorem). *Let Ω be a bounded convex domain in $\mathbb{R}^n$ with boundary $\partial\Omega \in C^{1,\beta}$, where $1 - 2/n < \beta \leq 1$. Let $\varphi \in C^{1,\beta}(\partial\Omega)$ and $f \in L^\infty(\Omega)$ where $0 < \lambda \leq f \leq \Lambda$. Assume that $u \in C(\overline{\Omega})$ is the convex Aleksandrov solution to*

$$\begin{cases} \det D^2 u = f & in \ \Omega, \\ \qquad\quad u = \varphi & on \ \partial\Omega. \end{cases}$$

Then, u is strictly convex in Ω.

Proof. Assume by contradiction that u is not strictly convex in Ω. Then there exist $x_0 \in \Omega$ and a supporting hyperplane $l(x) = u(x_0) + p \cdot (x - x_0)$ to the graph of u at $(x_0, u(x_0))$, where $p \in \partial u(x_0)$, such that the contact set $\mathcal{C} = \{x \in \Omega : u(x) = l(x)\}$ is not a single point. By Theorem 5.11, the set E of extremal points of $\mathcal{C}$ lies on the boundary $\partial\Omega$. Since $\overline{\mathcal{C}} = \text{conv}(E)$, E must contain at least two distinct boundary points y and z on $\partial\Omega$. Moreover, $\mathcal{C}$ contains the line segment L connecting y and z, and hence, $u = l$ on L. By subtracting l from u, we can assume that $u \geq 0$ in Ω and $u \equiv 0$ on L.

By changing coordinates, we can assume that $y = -ae_n$ and $z = ae_n$ for some $a > 0$. For each $0 < t < 1$, define the following intersection of Ω with the cylinder of radius t around the x_n-axis:

$$\Omega(t) := \{(x', x_n) \in \Omega : |x'| < t\}.$$

We continue to use φ to denote the $C^{1,\beta}(\overline{\Omega})$ function extending φ from $\partial\Omega$ to $\overline{\Omega}$. Since $\varphi = u \geq 0$ on $\partial\Omega$ and $\varphi = u = 0$ at $\pm ae_n$, we have $D_{x'}\varphi(y) = D_{x'}\varphi(z) = 0$. Now, for $x \in \partial\Omega \cap \partial\Omega(t)$ close to y, the $C^{1,\beta}$ regularity of φ and $\partial\Omega$ give

$$\varphi(x) \leq D_n\varphi(y)(x_n - y_n) + \|\varphi\|_{C^{1,\beta}(\overline{\Omega})}|x - y|^{1+\beta} \leq C_1|x' - y'|^{1+\beta} \leq C_1 t^{1+\beta}$$

for $C_1 := C_1(n, \beta, \Omega, \|\varphi\|_{C^{1,\beta}(\overline{\Omega})})$.

The above estimate implies that $\varphi \leq C_1 t^{1+\beta}$ on $\partial\Omega \cap \partial\Omega(t)$. Since $u = \varphi$ on $\partial\Omega$, the above bound and the convexity of u give $u \leq C_1 t^{1+\beta}$ on $\partial\Omega(t)$. This shows that if $h > 0$ is small, then

$$S_u(0, 0, h) \supset \Omega(t) \quad \text{for } t := (h/C_1)^{\frac{1}{1+\beta}},$$

from which we obtain the volume bounds:

$$(5.9) \qquad |S_u(0, 0, h)| \geq |\Omega(t)| \geq 2a\omega_{n-1}t^{n-1} = 2a\omega_{n-1}C_1^{-\frac{n-1}{1+\beta}}h^{\frac{n-1}{1+\beta}}.$$

Recall from Lemma 5.8 that $|S_u(0, 0, h)| \leq C(n, \lambda)h^{\frac{n}{2}}$ for all $h > 0$. Thus, we find from (5.9) that for all $h > 0$ small

$$(5.10) \qquad 2a\omega_{n-1}C_1^{-\frac{n-1}{1+\beta}}h^{\frac{n-1}{1+\beta}} \leq C(n, \lambda)h^{\frac{n}{2}}.$$

Since $\beta > 1 - 2/n$, we have $\frac{n-1}{1+\beta} < \frac{n}{2}$, and hence (5.10) cannot hold for all $h > 0$ small. This contradiction shows that u is strictly convex in Ω. $\square$

As a consequence of the balancing lemma, Lemma 3.25, we have

Lemma 5.15. *Let Ω be an open set in $\mathbb{R}^n$ that contains the origin. Assume that $u : \Omega \to \mathbb{R}$ is a convex function that satisfies $u(0) = 0, u \geq 0$, and $0 < \lambda \leq \det D^2 u \leq \Lambda$ in Ω. There exists $\theta(n, \lambda, \Lambda) \in (1/2, 1)$ such that if $S_u(0, 0, h) \Subset \Omega$, then*

$$u(\theta x) \geq u(x)/2 \quad \text{for all } x \in \partial S_u(0, 0, h).$$

It follows that $S_u(0, 0, h/2) \subset \theta S_u(0, 0, h)$.

Proof. Given $x \in \partial S_u(0, 0, h)$, let y be the intersection of the ray $\overrightarrow{x0}$ and $\partial S_u(0, 0, h)$, and let z be the intersection of the segment $[x, 0]$ and $\partial S_u(0, 0, h/2)$. Let $z = \alpha x$ and $\bar{u} := u - h$. Then $\bar{u} = 0$ on $\partial S_u(0, 0, h)$ and

$$\bar{u}(z) = -h/2 = (1/2)\min_{S_u(0,0,h)} \bar{u}.$$

Applying the balancing lemma, Lemma 3.25(i), to $\bar{u}$ on $S_u(0,0,h)$ with $\alpha = 1/2$, we find $c = c(n, \lambda, \Lambda) \in (0, 1/2)$ such that $|x - z| \geq c|x - y| > c|x|$. Hence

$$|z| = |x| - |x - z| \leq \theta|x|,$$

where $\theta = 1 - c \in (1/2, 1)$. Hence, $\alpha \leq \theta$. By the convexity of u and $u(0) = 0$, we find that $u(\theta x) \geq u(\alpha x) = u(z) = u(x)/2$. $\square$

Now, we are ready to estimate the size of compactly contained sections.

Lemma 5.16 (Size of sections). *Let Ω be an open set in $\mathbb{R}^n$. Assume that the convex function $u : \Omega \to \mathbb{R}$ satisfies $0 < \lambda \leq \det D^2 u \leq \Lambda$ in Ω, in the sense of Aleksandrov. Let $x_0 \in \Omega$ and $p_0 \in \partial u(x_0)$. Suppose that $S_u(x_0, p_0, t_0) \Subset \Omega$ satisfies $B_1(\hat{z}) \subset S_u(x_0, p_0, t_0) \subset B_n(\hat{z})$ for some $\hat{z} \in \mathbb{R}^n$. Then, there are constants $\mu \in (0, 1)$ and $C > 0$, depending on n, λ, and Λ, such that for all sections*

$$S_u(z, p, h) \Subset S_u(x_0, p_0, t_0) \quad \text{with } z \in S_u(x_0, p_0, 3t_0/4),$$

we have

$$S_u(z, p, h) \subset B_{Ch^\mu}(z).$$

If instead $B_{r_1}(\hat{z}) \subset S_u(x_0, p_0, t_0) \subset B_{r_2}(\hat{z})$, then the above inclusion still holds with the same μ, but C now depends also on r_1 and r_2.

Proof. Note that the last statement of the lemma follows from the case of $r_1 = 1$ and $r_2 = n$ via normalizing using John's Lemma (see also Remark 5.7). Thus, we only need to focus on this normalized case. From Remark 5.3, we can assume $p_0 = x_0 = 0 \in \Omega$, $u(0) = 0$, and $u \geq 0$ in Ω. Since $B_1(\hat{z}) \subset S_u(0, 0, t_0) \subset B_n(\hat{z})$, the volume estimates for sections in Lemma 5.6 show that t_0 is bounded from above and below by constants depending only on n, λ, Λ. For simplicity, let us assume that $t_0 = 1$. By Lemma 5.6(iv), there is $r_0(n, \lambda, \Lambda) > 0$ such that

$$B_{r_0}(0) \subset S_u(0, 0, 1) \subset B_{2n}(0).$$

Step 1. Reduction to $z = 0$ and $p = 0$. Consider $z \in S_u(0, 0, 3/4)$. Then, Theorem 5.13 gives $\delta = \delta(n, \lambda, \Lambda) > 0$ such that $S_u(z, p, \delta) \subset S_u(0, 0, 4/5)$ for all $p \in \partial u(z)$. Since $S_u(0, 0, 4/5) \subset B_{2n}(0)$, by Lemma 5.6, there is $r = r(n, \lambda, \Lambda) > 0$ such that $B_r(z) \subset S_u(z, p, \delta)$. Therefore $B_r(z) \subset S_u(z, p, \delta) \subset B_{2n}(z)$. Thus, from these geometric properties of $S_u(z, p, h)$, it suffices to consider the case $z = 0$ and $p = 0$.

Step 2. The case $z = 0$ and $p = 0$. Let $\theta \in (1/2, 1)$ be as in Lemma 5.15, and define $\beta > 0$ by $\theta^{1+\beta} = 2^{-1}$. Using the strict convexity result of Theorem 5.13 and a compactness argument invoking Theorem 3.19, we get

$$u|_{\partial B_{\theta r_0}(0)} \geq C^{-1}(n, \lambda, \Lambda).$$

If $x \in \Omega$, then, for any nonnegative integer m, Lemma 5.15 gives

$$u(x) \geq 2^{-m} u(\theta^{-m} x),$$

as long as $\theta^{-m} x \in \Omega$ (by then $\theta^{-m} x \in S_u(0, 0, t) \Subset \Omega$ for some $t > 0$). It suffices to consider $h \leq C^{-1}$. Now, for any $x \in S_u(0, 0, h)$ (so $|x| < \theta r_0$), let m be a positive integer such that $r_0 \geq |\theta^{-m} x| \geq \theta r_0$. Then

$$u(x) \geq \theta^{m(1+\beta)} u(\theta^{-m} x) \geq C^{-1}(n, \lambda, \Lambda) \theta^{m(1+\beta)} \geq c(n, \lambda, \Lambda) |x|^{1+\beta}.$$

It follows that $S_u(0, 0, h) \subset B_{Ch^\mu}(0)$, where $\mu = (1 + \beta)^{-1}$. □

From the proof of Lemma 5.16, we obtain the following quantitative strict convexity result.

Theorem 5.17 (Quantitative strict convexity). *Let Ω be an open set in $\mathbb{R}^n$. Assume that the convex function $u : \Omega \to \mathbb{R}$ satisfies $0 < \lambda \leq \det D^2 u \leq \Lambda$ in Ω in the sense of Aleksandrov. Let $x_0 \in \Omega$ and $p_0 \in \partial u(x_0)$. Suppose that $S_u(x_0, p_0, t_0) \Subset \Omega$ satisfies $B_1(z) \subset S_u(x_0, p_0, t_0) \subset B_n(z)$ for some $z \in \mathbb{R}^n$. Then, there are $c(n, \lambda, \Lambda) > 0$ and $\beta(n, \lambda, \Lambda) > 0$ such that*

$$u(y) \geq u(x) + p \cdot (y - x) + c|y - x|^{1+\beta}$$

for all $x, y \in S_u(x_0, p_0, t_0/2)$ and all $p \in \partial u(x)$.

5.3. Interior Hölder Gradient Estimates

As a consequence of the strict convexity result, we can now prove the following pointwise $C^{1,\alpha}$ regularity (see Definition 2.88), due to Caffarelli [**C4**].

Theorem 5.18 (Caffarelli's pointwise $C^{1,\alpha}$ regularity). *Let Ω be an open set in $\mathbb{R}^n$. Let $u : \Omega \to \mathbb{R}$ be a convex function satisfying $\lambda \leq \det D^2 u \leq \Lambda$ in Ω in the sense of Aleksandrov. Let $x_0 \in \Omega$ and $p_0 \in \partial u(x_0)$. Assume that $S := S_u(x_0, p_0, 1) \Subset \Omega$. Then, for some $\delta(n, \lambda, \Lambda) > 0$, we have*

$$(5.11) \qquad \left(\frac{1}{2} + \delta\right)[S - x_0] \subset S_u(x_0, p_0, 1/2) - x_0 \subset (1 - \delta)[S - x_0].$$

In particular, u is pointwise $C^{1,\alpha}$ at x_0 where $\alpha = \alpha(n, \lambda, \Lambda) > 0$ and

$$u(x) - u(x_0) - p_0 \cdot (x - x_0) \leq 4c_1^{-(1+\alpha)} |x - x_0|^{1+\alpha} \quad in \ S,$$

for some $c_1(n, \lambda, \Lambda, \operatorname{diam}(S)) > 0$. In fact, u is pointwise $C^{1,\alpha}$ in S.

Proof. From Remark 5.3, we can assume that $p_0 = x_0 = 0 \in \Omega$, $u(0) = 0$, and $u \geq 0$ in Ω. Thus $S = S_u(0, 0, 1)$.

By Lemma 5.15, the right inclusion of (5.11) holds for any $\delta \leq 1 - \theta$. For the left inclusion of (5.11), we note that it is invariant under affine transformation, so it suffices to prove it for $S_u(0, 0, 1)$ being normalized.

Assume by contradiction that it is false. Then, we can find sequences $\delta_k \to 0^+$, convex functions u_k on convex domains Ω^k, and points z^k ($k = 1, 2, \ldots$) with the following properties: $u_k \geq 0$, $u_k(0) = 0$,

$$\lambda \leq \det D^2 u_k \leq \Lambda \quad \text{in } \Omega^k, \quad u_k = 1 \quad \text{on } \partial\Omega^k, \quad B_1(0) \subset \Omega^k \subset B_n(0),$$

and

$$z^k \in S_{u_k}(0, 0, 1), \quad (\delta_k + 1/2)z^k \in \partial S_{u_k}(0, 0, 1/2).$$

By using a compactness argument as in the proof of Theorem 5.13, we can find a convex function u on Ω where $B_1(0) \subset \Omega \subset B_n(0)$, with

$$u(0) = 0, \quad u \geq 0, \quad \lambda \leq \det D^2 u \leq \Lambda \quad \text{on } \Omega, \quad u = 1 \quad \text{on } \partial\Omega,$$

and a point $z \in \overline{S_u(0, 0, 1)}$ such that $u(z/2) = 1/2$.

Since $u(0) = 0$, by convexity, we have $u(z/2) \leq (1/2)u(z) \leq 1/2$. Therefore, we must have $u(z) = 1$, and u is linear on the line segment connecting 0 and $z \in \partial\Omega$. This contradicts the strict convexity result in Theorem 5.13.

Now, we prove the pointwise $C^{1,\alpha}$ regularity of u in $S_u(0, 0, 1)$ using the left inclusion of (5.11). If $x \in \partial S_u(0, 0, h)$, then $u((\frac{1}{2} + \delta)x) \leq h/2$. From the convexity of u, we have

$$u\left(\frac{1}{2}x\right) \leq \frac{1}{1 + 2\delta}u\left(\left(\frac{1}{2} + \delta\right)x\right) + \frac{2\delta}{1 + 2\delta}u(0) \leq \frac{u(x)}{2(1 + 2\delta)}.$$

Let α be defined by $\frac{1}{1+2\delta} = 2^{-\alpha}$. Then for any $t \in (\frac{1}{2^{k+1}}, \frac{1}{2^k})$ where $k \in \mathbb{N}$ and for $x \in \overline{S_u(0, 0, 1)}$, we have

$$(5.12) \quad u(tx) \leq u(2^{-k}x) \leq \frac{u(x)}{[2(1 + 2\delta)]^k} = (2^{-k})^{1+\alpha}u(x) \leq (2t)^{1+\alpha}u(x).$$

Clearly, the above inequality also holds for $1/2 \leq t < 1$.

Now for any $x \in S_u(0, 0, 1)$, let $x_0 \in \partial S$ be the intersection of the ray $\overrightarrow{0x}$ and ∂S. Thus, $x = tx_0$ for some $0 < t < 1$. Since $S \Subset \Omega$, by Lemma 5.6, there exists $c_1(n, \lambda, \Lambda, \text{diam}(S)) > 0$ such that $S \supseteq B_{c_1}(0)$. It follows that $t = |x|/|x_0| \leq c_1^{-1}|x|$. Using (5.12) at x_0, we have

$$0 \leq u(x) = u(tx_0) \leq 4t^{1+\alpha}u(x_0) = 4t^{1+\alpha} \leq 4c_1^{-(1+\alpha)}|x|^{1+\alpha} \quad \text{in } S.$$

Hence $u \in C^{1,\alpha}$ at the origin.

The proof of the pointwise $C^{1,\alpha}$ regularity at other points in S is the same, so we omit it. In $S_u(0, 0, \gamma)$ where $\gamma < 1$, the pointwise $C^{1,\alpha}$ information of u depends only on n, λ, Λ, γ, and $\text{diam}(S)$. $\square$

Remark 5.19. Assume that the convex function u satisfies, in the sense of Aleksandrov,

$$0 < \lambda \leq \det D^2 u \leq \Lambda \quad \text{in an open set } \Omega \subset \mathbb{R}^n.$$

Suppose that there is a section $S_u(x, p_x, h) \Subset \Omega$ with $p_x \in \partial u(x)$ and $h > 0$. This is always the case when u is strictly convex at x because $\bigcap_{t>0} S_u(x, p_x, t) = \{x\}$. In this case, by Corollary 5.12, u is differentiable in $S_u(x, p_x, h)$ and hence at x. The following argument, which shows the differentiability of u at x, is relevant when dealing with boundary sections in Chapter 8; see Proposition 8.23.

Note that Theorem 5.18 yields

$$u(y) - u(x) - p_x \cdot (y - x) \leq C_1 |y - x|^{1+\alpha} \quad \text{for all } y \in S_u(x, p_x, h/2).$$

By Theorem 2.38, to show that u is differentiable at x, we only need to show that p_x is the only element of $\partial u(x)$. Indeed, let $p \in \partial u(x)$. Then

$$u(y) \geq u(x) + p \cdot (y - x) \quad \text{for all } y \in \Omega.$$

On the other hand, by Lemma 5.6, $B_{2r}(x) \subset S_u(x, p_x, h/2)$ for some $r > 0$ depending on n, λ, Λ, h, and $\text{diam}(S_u(x, p_x, h))$. Thus, for all $t \in (0, r]$ and all $w \in \mathbb{R}^n$ with $|w| = 1$, we apply the above estimate to $y = x + tw$ to get

$$u(x) + p \cdot (tw) \leq u(y) \leq u(x) + p_x \cdot (tw) + C_1 t^{1+\alpha}.$$

It follows that $(p - p_x) \cdot w \leq C_1 t^\alpha$ for all $t \in (0, r]$ and all $|w| = 1$. Therefore $p = p_x$ and hence u is differentiable at x.

Remark 5.20. Assume that the convex function u satisfies, in the sense of Aleksandrov, $0 < \lambda \leq \det D^2 u \leq \Lambda$ in an open set Ω in $\mathbb{R}^n$. We will simply write $S_u(x, h)$ for the section $S_u(x, p, h) \Subset \Omega$ with $p \in \partial u(x)$ because of the pointwise $C^{1,\alpha}$ estimate which gives a unique $p = Du(x)$.

As in Theorem 2.91, we can pass from pointwise $C^{1,\alpha}$ estimates in Theorem 5.18 to uniform $C^{1,\alpha}$ estimates. We give here a direct proof.

Theorem 5.21 (Interior $C^{1,\alpha}$ estimates for the Monge–Ampère equation). *Let u be a strictly convex function satisfying $0 < \lambda \leq \det D^2 u \leq \Lambda$, in the sense of Aleksandrov, in an open set Ω in $\mathbb{R}^n$. Let $\alpha(n, \lambda, \Lambda)$ be as in Theorem 5.18. If $S_u(x_0, t) \Subset \Omega$ is a normalized section, then for any $\delta \in (1/4, 1)$,*

$$|Du(y) - Du(z)| \leq C(n, \lambda, \Lambda, \delta)|y - z|^\alpha \quad \text{for all } y, z \in S_u(x_0, \delta t).$$

Proof. We can assume that $x_0 = 0$ and $u \geq 0$. From the volume estimates in Lemma 5.6, we find that t is bounded from above and below by constants depending on n, λ, Λ. Thus, for simplicity, we can assume that $t = 1$ and that $\delta = 1/2$. Fix $z \in S_u(0, 1/2)$. Then by Theorem 5.13, there is a constant $\gamma(n, \lambda, \Lambda) > 0$ such that $S_u(z, \gamma) \subset S_u(0, 3/4)$. From Lemma 5.6(iv), we find that $S_u(z, \gamma) \supset B_{c\gamma}(z)$ for some $c(n, \lambda, \Lambda) > 0$. By Lemma 5.6(iii)(a), $|Du| \leq C(n, \lambda, \Lambda)$ in $S_u(0, 3/4)$. Hence, the conclusion of the theorem clearly holds when $y \in S_u(0, 1/2)$ with $|y - z| \geq c\gamma$. Thus, it remains to consider $y \in S_u(0, 1/2)$ with $0 < 2r := |y - z| < c\gamma$.

By Theorem 5.18, there is $C = C(n, \lambda, \Lambda)$ such that for all $x \in S_u(0, 3/4)$,

$$0 \leq u(x) - u(y) - Du(y) \cdot (x - y) \leq C|x - y|^{1+\alpha},$$

$$0 \leq u(x) - u(z) - Du(z) \cdot (x - z) \leq C|x - z|^{1+\alpha}.$$

Subtracting $u(x) - u(y) - Du(y) \cdot (x - y)$ from the second estimate, we obtain

$$[u(y) - u(z) - Du(z) \cdot (y - z)] + (Du(y) - Du(z)) \cdot (x - y) \leq C|x - z|^{1+\alpha},$$

which, due to $u(y) - u(z) - Du(z) \cdot (y - z) \geq 0$, implies

$$(5.13) \quad (Du(y) - Du(z)) \cdot (x - y) \leq C|x - z|^{1+\alpha} \quad \text{for all } x \in S_u(0, 3/4).$$

Clearly, we can find $x \in \partial B_r(y)$ such that

$$(Du(y) - Du(z)) \cdot (x - y) = |Du(y) - Du(z)||x - y| = r|Du(y) - Du(z)|.$$

Note that $x \in B_{c\gamma}(y) \subset S_u(y, \gamma) \subset S_u(0, 3/4)$ and $|x - z| \leq 3r$. Then, it follows from (5.13) that

$$r|Du(y) - Du(z)| = (Du(y) - Du(z)) \cdot (x - y) \leq C|x - z|^{1+\alpha} \leq 9Cr^{1+\alpha}.$$

Therefore, recalling $|y - z| = 2r$,

$$|Du(y) - Du(z)| \leq 9Cr^{\alpha} \leq 9C|y - z|^{\alpha},$$

completing the proof of the theorem. $\qquad\qquad\qquad\qquad\qquad\qquad\square$

The proof of Theorem 5.21 invokes Theorem 5.13, which was proved by compactness arguments. A direct proof of Theorem 5.21 without using compactness arguments has been given by Forzani and Maldonado [**FM**].

As a full manifestation of the normalization idea outlined in Remark 5.7, by rescaling Theorem 5.21, we obtain the interior $C^{1,\alpha}$ estimates on all compactly contained sections.

Corollary 5.22 (Interior $C^{1,\alpha}$ estimates on arbitrary sections). *Let u be a strictly convex function satisfying $\lambda \leq \det D^2 u \leq \Lambda$, in the sense of Aleksandrov, in an open set Ω in $\mathbb{R}^n$. Let $\alpha(n, \lambda, \Lambda)$ be as in Theorem 5.18, and let $\alpha_1 = -1 + n(1 + \alpha)/2$. Assume that $S_u(x, \tau h) \Subset \Omega$ for some $\tau > 1$ and $h > 0$. Then, for all $y, z \in S_u(x, \delta h)$ where $\delta \in (1/4, 1)$, we have*

$$|Du(y) - Du(z)| \leq C(\lambda, \Lambda, n, \tau, \delta, \mathrm{diam}(S_u(x, \tau h)))h^{-\alpha_1}|y - z|^{\alpha}.$$

Proof. Let $A \in \mathbb{S}^n$, $\tilde{u}$, and $\tilde{S}$ be as in the proof of Lemma 5.6(iii) where we recall that $\tilde{S} = S_{\tilde{u}}(\tilde{x}, h(\det A)^{-2/n})$ is a normalized section.

Let $y, z \in S_u(x, \delta h)$. Then $A^{-1}y, A^{-1}z \in S_{\tilde{u}}(\tilde{x}, \delta h(\det A)^{-2/n})$. Applying the $C^{1,\alpha}$ estimate in Theorem 5.21 to $\tilde{u}$, we get

$$|D\tilde{u}(A^{-1}y) - D\tilde{u}(A^{-1}z)| \leq C_0(\lambda, \Lambda, n, \delta)|A^{-1}y - A^{-1}z|^{\alpha}.$$

Recalling the estimates (5.4) and (5.6) in the proof of Lemma 5.6 for A and

$$Du(y) - Du(z) = (\det A)^{2/n}(A^{-1})^t(D\tilde{u}(A^{-1}y) - D\tilde{u}(A^{-1}z)),$$

we obtain

$$|Du(y) - Du(z)| \leq C_0 (\det A)^{2/n} \|A^{-1}\|^{1+\alpha} |y - z|^\alpha \leq Ch^{-\alpha_1} |y - z|^\alpha,$$

where $C = C(\lambda, \Lambda, n, \tau, \delta, \mathrm{diam}(S_u(x, \tau h)))$ and $\alpha_1 = -1 + n(1 + \alpha)/2$. $\square$

Remark 5.23. We deduce from Theorem 5.21 that

$$S_u(x_0, h) \supset B_{ch^{1/(1+\alpha)}}(x_0)$$

for $h \leq t/2$ and $c(n, \lambda, \Lambda)$ small. Compared to (5.7), this will improve the norm of the map A_h in Lemma 5.6(iii) to $\|A_h\| \leq C(n, \lambda, \Lambda) h^{\frac{1}{2} - \frac{1}{1+\alpha}}$.

5.4. Counterexamples to Hölder Gradient Estimates

In this section, we construct counterexamples showing that the size of the degeneracy set of solutions to the Monge–Ampère equations in Caffarelli's Theorem 5.9 is sharp and that the strict convexity is crucial for the $C^{1,\alpha}$ regularity in Theorems 5.18 and 5.21.

Suppose now $n \geq 3$. We construct a singular, convex Aleksandrov solution to $\det D^2 u \geq \lambda > 0$ in $B_1(0)$ that vanishes on an s-dimensional subspace for all integers $s \in [1, n/2]$. We look for u of the form

(5.14) $u(x_1, \ldots, x_k, y_1, \ldots, y_s) = |x|^\alpha f(|y|)$, where $k + s = n, 1 \leq s < k$.

Then, by Lemma 2.60,

(5.15) $\det D^2 u = \alpha^k |x|^{n\alpha - 2k} f^{k-1} (f')^{s-1} |y|^{-s+1} \left((\alpha - 1) f f'' - \alpha (f')^2 \right)$.

Consider

$$\alpha = \frac{2k}{n} = 2 - \frac{2s}{n} \leq 2 - \frac{2}{n}.$$

If $s < k$, then $\alpha > 1$. We will choose f such that

$f > 0$, $f' \geq 0$, $f'' > 0$, $(f')^{s-1} |y|^{-s+1} > 0$, and $(\alpha - 1) f f'' - \alpha (f')^2 > 0$.

Example 5.24. Let $n \geq 3$, $s \in \mathbb{N} \cap [1, n/2]$, and $f(t) = t^2 + 1$. Then, we can easily check that

(5.16) $u(x, y) = |x|^{2 - \frac{2s}{n}} (|y|^2 + 1)$, $x = (x_1, \ldots, x_k)$, $y = (y_1, \ldots, y_s)$

is convex in a neighborhood $B_\rho(0)$ of 0, and, due to (5.15),

$$\det D^2 u = 2^{s-1} (2 - 2s/n)^{n-s} (|y|^2 + 1)^{n-s-1} \left[2(1 - 2s/n) + (4s/n - 6)|y|^2 \right].$$

The above computation works for $x \neq 0$. As in Example 4.14, we can check that u is an Aleksandrov solution to the above equation in $B_\rho(0)$. This example, due to Caffarelli [C7], shows that there exists a convex solution of a Monge–Ampère equation with positive analytic right-hand side that vanishes at $|x| = 0$ on an s-dimensional subspace for $1 \leq s < n/2$.

When $s = 1$, we have $k = n - 1$, and by renaming $(x_1, \ldots, x_{n-1}) = x', x_n = y_1$, we get from (5.16) Pogorelov's original example (4.27).

Remark 5.25. Let u, f, ρ, and s be as in Example 5.24. From the calculation of the Hessian in Lemma 2.60, we see that the most singular term in $D^2 u$ is comparable to $|x|^{-\frac{2s}{n}} f$ where $x \in \mathbb{R}^{n-s}$. Thus $u \in W^{2,p}(B_\rho(0))$ if and only if $-\frac{2s}{n} p > -(n-s)$, or $p < \frac{n(n-s)}{2s}$.

Note that the function u agrees with a supporting hyperplane on an s-dimensional subspace; the exponent $\frac{n(n-s)}{2s}$ is critical, due to the following result of Collins and Mooney [**CM**].

Theorem 5.26. *Let u be an Aleksandrov convex solution to $\det D^2 u \geq 1$ in $B_1(0) \subset \mathbb{R}^n$ and let $s \in [1, n/2] \cap \mathbb{N}$. If $u \in W^{2,p}(B_1(0))$ for some $p \geq \frac{n(n-s)}{2s}$, then the dimension of the set where u agrees with a supporting hyperplane is at most $s - 1$.*

Theorem 5.26 shows that if $\det D^2 u \geq 1$ in $B_1(0) \subset \mathbb{R}^n$ and $u \in W^{2,n(n-1)/2}(B_1(0))$, then u must be strictly convex. This sharpens a previous result of Urbas [**U1**] who established the strict convexity with $n(n-1)/2$ being replaced by $p > n(n-1)/2$. Moreover, Urbas also established the strict convexity of solutions to $\det D^2 u \geq 1$ in $B_1(0) \subset \mathbb{R}^n$ when $u \in C^{1,\alpha}(B_1(0))$ where $\alpha > 1 - 2/n$.

Example 5.27 (Counterexample to $C^{1,\alpha}$ regularity). Similar to the calculations in Lemma 2.60 and Example 5.24, for $u^{(\beta)}(x) = |x'| + |x'|^\beta (1 + x_n^2)$, where $x = (x', x_n)$, we have

$$\det D^2 u^{(\beta)} = \beta \left\{ \beta |x'|^{\beta-2}(1+x_n^2) + |x'|^{-1} \right\}^{n-2} |x'|^{2\beta-2} \left[2(\beta-1) - (2\beta+2)x_n^2 \right].$$

Choosing $\beta = n/2$ to cancel the singular term $|x'|^{-1}$ in the braces above, we find that the convex function

$$u(x) = |x'| + |x'|^{n/2}(1 + x_n^2),$$

found by Caffarelli (see Caffarelli–Yuan [**CaY**] for more examples), is an Aleksandrov solution to

$$\det D^2 u = \frac{n}{2} \left\{ \frac{n}{2} |x'|^{n/2-1}(1 + x_n^2) + 1 \right\}^{n-2} \left[n - 2 - (n+2)x_n^2 \right] \geq 1/2$$

in $B_{1/4}(0)$ when $n \geq 3$. The function u vanishes on the line $\{x' = 0\}$. It is only Lipschitz but not Hölder continuous. Note that when n is even, the function on the right-hand side is even analytic.

For strictly convex solutions to $\det D^2 u = f \leq \Lambda$, a positive lower bound for f is in general required for interior $C^{1,\alpha}$ regularity. Wang [**W3**] constructed counterexamples to $C^{1,\alpha}$ regularity when f takes value 0.

To conclude this section, we mention that Caffarelli and Yuan [**CaY**] recently constructed merely Lipschitz and $C^{1,\alpha}$ Aleksandrov solutions to $\det D^2 u = 1$ in $B_1(0) \subset \mathbb{R}^n$, with rational $\alpha \in (0, 1 - 2/n)$, for $n \geq 3$.

5.5. Geometry of Sections

In this section, refined geometric properties of sections will be proved, including the following: engulfing property of sections, inclusion and exclusion properties of sections, and a localization property of intersecting sections. These together with Lemma 5.6 show that sections of solutions to the Monge–Ampère equation have many properties similar to those of Euclidean balls.

The engulfing property of sections is an affine invariant version of the triangle inequality in the Monge–Ampère setting; see Figure 5.3.

Theorem 5.28 (Engulfing property of sections). *Suppose that u is a strictly convex function satisfying, in the sense of Aleksandrov,*

$$0 < \lambda \leq \det D^2 u \leq \Lambda \quad \text{in an open set } \Omega \subset \mathbb{R}^n.$$

There is a constant $\theta_0(n, \lambda, \Lambda) > 2$ with the following property:

If $S_u(y, 2h) \Subset \Omega$ and $x \in S_u(y, h)$, then $S_u(y, h) \subset S_u(x, \theta_0 h)$.

Example 5.29. Consider $u(x) = |x|^2/2$ on $\mathbb{R}^n$. If $x \in S_u(y, h) = B_{\sqrt{2h}}(y)$, then from the triangle inequality, we find that $B_{\sqrt{2h}}(y) \subset B_{2\sqrt{2h}}(x)$, or $S_u(y, h) \subset S_u(x, 4h)$. This example shows that $\theta_0(n, 1, 1) \geq 4$.

Proof of Theorem 5.28. We divide the proof into two steps.

Step 1. We first prove the engulfing property in a normalized setting. This is the case where $S_u(0, 1) \Subset \Omega$, $u(0) = 0$, $Du(0) = 0$, and $B_1(\hat{z}) \subset S_u(0, 1) \subset B_n(\hat{z})$ for some $\hat{z} \in \mathbb{R}^n$. We show that there is a $\delta(n, \lambda, \Lambda) > 0$ such that if

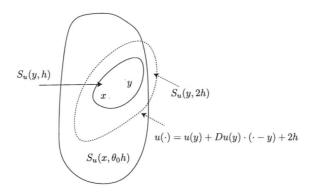

Figure 5.3. Engulfing property of sections.

$x \in S_u(0, 1/2)$, then $S_u(0, 1/2) \subset S_u(x, \delta^{-1})$. Indeed, by Lemma 5.6(iii)(a), $\|Du\|_{L^\infty(S_u(0,1/2))} \leq C(n, \lambda, \Lambda)$. Thus, if $x, z \in S_u(0, 1/2)$, then

$$u(z) - u(x) - Du(x) \cdot (z - x) \leq u(z) + |Du(x)||z - x| \leq 1/2 + 2nC.$$

Hence $S_u(0, 1/2) \subset S_u(x, \delta^{-1})$ for $\delta(n, \lambda, \Lambda) = (1/2 + 2nC)^{-1} > 0$.

Step 2. General case. Suppose $S_u(y, 2h) \Subset \Omega$ and $x \in S_u(y, h)$. We can assume that $y = 0$, and then by subtracting $u(0) + Du(0) \cdot z$ from $u(z)$, we can assume further that $u(0) = 0$ and $Du(0) = 0$. From John's Lemma (Lemma 2.63), there is a positive definite matrix $A \in \mathbb{S}^n$ and $\hat{y} \in \mathbb{R}^n$ such that

$$B_1(\hat{y}) \subset AS_u(0, 2h) \subset B_n(\hat{y}).$$

Let $\hat{u}(z) = u(A^{-1}z)/(2h)$ and $\hat{x} = Ax$. Then

$$\hat{x} \in S_{\hat{u}}(0, 1/2) = AS_u(0, h), \quad S_{\hat{u}}(0, 1) = AS_u(0, 2h).$$

By Lemma 3.9, we have

$$\lambda(\det A)^{-2}(2h)^{-n} \leq \det D^2\hat{u} \leq \Lambda(\det A)^{-2}(2h)^{-n} \quad \text{in } S_{\hat{u}}(0, 1).$$

Therefore, from Lemma 5.6 and $\omega_n \leq \det A|S_u(0, 2h)| \leq n^n\omega_n$, we find

$$C^{-1}(n, \lambda, \Lambda) \leq \det D^2\hat{u} \leq C(n, \lambda, \Lambda) \quad \text{in } S_{\hat{u}}(0, 1),$$

for some $C(n, \lambda, \Lambda) > 0$. Now, Step 1 gives $\hat{\delta} = \hat{\delta}(n, \lambda, \Lambda) > 0$ such that $S_{\hat{u}}(0, 1/2) \subset S_{\hat{u}}(\hat{x}, \hat{\delta}^{-1})$. Since $S_{\hat{u}}(0, 1/2) = AS_u(0, h)$ and $S_{\hat{u}}(\hat{x}, \hat{\delta}^{-1}) = AS_u(x, 2h\hat{\delta}^{-1})$, the result follows by choosing $\theta_0 = 2\hat{\delta}^{-1}$. $\square$

We now establish the inclusion and exclusion properties of sections. Roughly speaking, they say that for any point z contained in a compactly contained section $S = S_u(x, h)$ and $z \neq x$, there is a section S_1 centered at z that is included in S while we can also find another section S_2 centered

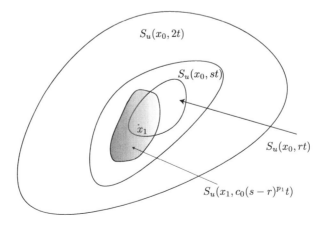

Figure 5.4. Inclusion property of sections.

at x that avoids z. The heights of these sections can be estimated based on the location of z and the structural constants. See Figure 5.4.

Theorem 5.30 (Inclusion and exclusion properties of sections). *Suppose that u is a strictly convex function satisfying, in the sense of Aleksandrov,*

$$0 < \lambda \leq \det D^2 u \leq \Lambda \quad \text{in an open set } \Omega \subset \mathbb{R}^n.$$

Then, there exist constants $c_0(n, \lambda, \Lambda) > 0$ and $p_1(n, \lambda, \Lambda) \geq 1$ with the following properties. Assume $S_u(x_0, 2t) \Subset \Omega$ and $0 < r < s \leq 1$.

(i) *If $x_1 \in S_u(x_0, rt)$, then*

$$S_u(x_1, c_0(s - r)^{p_1} t) \subset S_u(x_0, st).$$

(ii) *If $x_2 \in S_u(x_0, t) \backslash S_u(x_0, st)$, then*

$$S_u(x_2, c_0(s - r)^{p_1} t) \cap S_u(x_0, rt) = \emptyset.$$

Proof. The conclusions of the theorem are invariant under affine transformations and rescalings of the domain Ω and function u. Thus, as in the proof of Theorem 5.28, we can assume that $x_0 = 0$, $t = 1$, $S_u(0, 2) \Subset \Omega$, $u(0) = 0, Du(0) = 0$, and $B_1(\hat{z}) \subset S_u(0, 2) \subset B_n(\hat{z})$ for some $\hat{z} \in \mathbb{R}^n$.

Now, we give the proof of assertion (i) because that of assertion (ii) is similar. By Theorem 5.13(iii), there exists $c_*(n, \lambda, \Lambda) > 0$ such that $S_u(x, c_*) \subset S_u(0, 2)$ if $x \in S_u(0, 1)$. Take $c_0 \leq c_*$. Let $x_1 \in S_u(0, r)$ and $y \in S_u(x_1, c_0(s - r)^{p_1})$. Then, $u(x_1) < r$ and

$$u(y) < u(x_1) + Du(x_1) \cdot (y - x_1) + c_0(s - r)^{p_1}.$$

Because $S_u(0, 2)$ is normalized, we can deduce from Lemma 5.6(iii)(a) that

$$|Du(x_1)| \leq \|Du\|_{L^\infty(S_u(0,1))} \leq C(n, \lambda, \Lambda).$$

Combining with the estimate on the size of sections in Lemma 5.16, we find

$$u(y) < r + C|y - x_1| + c_0(s - r)^{p_1}$$
$$\leq r + C'(n, \lambda, \Lambda)(c_0(s - r)^{p_1})^\mu + c_0(s - r)^{p_1} < s$$

if $p_1 = \mu^{-1}$ and $c_0(n, \lambda, \Lambda)$ is small. Hence $S_u(x_1, c_0(s - r)^{p_1} t) \subset S_u(x_0, st)$. The theorem is proved. $\qquad\square$

Next, we prove a localization property of intersecting sections. It says that the intersection of two sections, where the center of the section with smaller height is contained in the closure of the section with larger height, contains a section of height comparable to that of the smaller section. See Figure 5.5. This property will be crucial in the proof of the $W^{2,1+\varepsilon}$ estimates for the Monge–Ampère equation in Chapter 6.

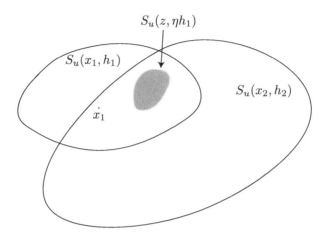

Figure 5.5. Localization property of intersecting sections.

Theorem 5.31 (Localization property of intersecting sections). *Suppose that u is a strictly convex function satisfying, in the sense of Aleksandrov,*

$$0 < \lambda \le \det D^2 u \le \Lambda \quad \text{in an open set } \Omega \subset \mathbb{R}^n.$$

There is a constant $\eta = \eta(n, \lambda, \Lambda) > 0$ with the following property: If $S_u(x_1, h_1) \Subset \Omega$ and $x_1 \in \overline{S_u(x_2, h_2)}$ where $x_2 \in \Omega$ and $h_2 \ge h_1$, then there is $z \in S_u(x_1, h_1)$ such that

$$(5.17) \qquad S_u(z, \eta h_1) \subset S_u(x_1, h_1) \cap S_u(x_2, h_2).$$

Proof. Using an affine transformation, we can assume that, for some $\hat{z} \in \mathbb{R}^n$, $B_1(\hat{z}) \subset S_u(x_1, h_1) \subset B_n(\hat{z})$ and $u = 0$ on $\partial S_u(x_1, h_1)$. Then x_1 is a minimum point of u and $Du(x_1) = 0$. We prove the theorem in this setting. Using the volume estimates for sections in Lemma 5.6, we have

$$(5.18) \qquad 0 < c_1(n, \lambda, \Lambda) \le h_1 \le C_1(n, \lambda, \Lambda).$$

By a compactness argument, we can find a small $\gamma(n, \lambda, \Lambda) > 0$ such that

$$(5.19) \qquad \text{if } y \in S_u(x_1, \gamma), \text{ then } S_u(x_1, h_1/2) \subset S_u(y, h_1).$$

Also by compactness, we can find $\delta = \delta(n, \lambda, \Lambda) > 0$ such that

$$(5.20) \qquad \text{if } y \in \Omega \setminus S_u(x_1, \gamma), \text{ then } |Du(y)| \ge \delta.$$

Case 1. $x_2 \in S_u(x_1, \gamma)$. Then by (5.19) and $h_2 \ge h_1$, we have (5.17) for $z = x_1$ and $\eta = 1/2$.

Case 2. $x_2 \notin S_u(x_1, \gamma)$. Using (5.20), we have

$$(5.21) \qquad |Du(x_2)| \ge \delta.$$

We now explain heuristically how to choose z. To fix the idea, we consider the case $u(x) = |x|^2/2$ and $x_1 \in \partial S_u(x_2, h_2) = \partial B_{\sqrt{2h_2}}(x_2)$. The segment $[x_1, x_2]$ intersects $\partial S_u(x_1, \gamma)$ at x_3. Then we can choose z to be the midpoint of $[x_1, x_3]$ having the largest η in (5.17). This z is of the form $(1 - \tau)x_1 + \tau x_2$, where $\tau > 0$ is small if h_1/h_2 is small.

Back to the proof of our theorem. There are several candidates for z, including $(1 - \tau)x_1 + \tau x_2$, $(1 - \tau)x_1 + \tau Du(x_2)$, $x_1 + \tau x_2$, $x_1 + \tau Du(x_2)$. It turns out that $z = x_1 + \tau Du(x_2)$, for an appropriate τ, does the job.

To simplify the presentation, let $\hat{x} := Du(x_2)/|Du(x_2)|$.

Using Lemma 5.6(iv) and Theorem 5.30, or by a compactness argument, we can find $r(n, \lambda, \Lambda) > 0$ and $\eta_0(n, \lambda, \Lambda) > 0$ such that if $0 < r_0 \le r$ and $z := x_1 + r_0\hat{x}$, then

$$(5.22) \qquad z \in S_u(x_1, h_1/2) \quad \text{and} \quad S_u(z, \eta_0 h_1) \subset S_u(x_1, h_1).$$

We now show the existence of $r_0 = r_0(n, \lambda, \Lambda) \in (0, r)$ such that

$$(5.23) \qquad\qquad B_{r_0/2}(z) \subset S_u(x_2, h_2).$$

For this, we use $Du(x_1) = 0$, the pointwise $C^{1,\alpha}$ of u at x_1 (see Theorem 5.18). Take any y with $|y| \le 1$. Then, by (5.21), we have

$$A := -Du(x_2) \cdot (\hat{x} + y/2) \le -|Du(x_2)| + |Du(x_2)|/2 \le -\delta/2$$

and

$$
\begin{aligned}
Du(x_2) \cdot (x_1 - x_2) &= Du(x_2) \cdot (x_1 + r_0\hat{x} + r_0 y/2 - x_2) + A r_0 \\
&\le Du(x_2) \cdot (x_1 + r_0\hat{x} + r_0 y/2 - x_2) - r_0\delta/2.
\end{aligned}
$$

Thus, by Theorem 5.18 and $Du(x_1) = 0$, we can find $C(n, \lambda, \Lambda)$ such that

$$u(x_1 + r_0\hat{x} + r_0 y/2) \le u(x_1) + C r_0^{1+\alpha}.$$

Since $x_1 \in \overline{S_u(x_2, h_2)}$, we find from the preceding estimates that

$$
\begin{aligned}
u(x_1 + r_0\hat{x} + r_0 y/2) &\le u(x_2) + Du(x_2) \cdot (x_1 - x_2) + h_2 + C r_0^{1+\alpha} \\
&\le u(x_2) + Du(x_2) \cdot (x_1 + r_0\hat{x} + r_0 y/2 - x_2) \\
&\quad + h_2 + C r_0^{1+\alpha} - \frac{r_0\delta}{2} \\
&< u(x_2) + Du(x_2) \cdot (x_1 + r_0\hat{x} + r_0 y/2 - x_2) + h_2
\end{aligned}
$$

if we choose $r_0 \in (0, r)$ such that $C r_0^{1+\alpha} - r_0\delta/2 < 0$. Hence, (5.23) holds.

From $z \in S_u(x_1, h_1/2)$ in (5.22) and the fact that $S_u(x_1, h_1)$ is normalized, we can use Lemma 5.16 to conclude that for all $h \le \eta_0 c_1 \le \eta_0 h_1$, we have

$$S_u(z, h) \subset B_{Ch^\mu}(z) \subset B_{r_0/2}(z)$$

if $h \leq \left(\frac{r_0}{2C}\right)^{\frac{1}{\mu}}$. From (5.18) and (5.23), we find that

$$S_u(z, \eta h_1) \subset S_u(x_2, h_2) \quad \text{for } \eta := \min\left\{\eta_0, C_1^{-1}\eta_0 c_1, C_1^{-1}\left(\frac{r_0}{2C}\right)^{\frac{1}{\mu}}\right\}.$$

This combined with (5.22) gives (5.17).   □

5.6. Vitali Covering and Crawling of Ink-spots Lemmas

In this section, we prove two real analysis results for sections of the Monge–Ampère equation: Vitali's Covering Lemma and the Crawling of Ink-spots Lemma. They are useful in the proof of $W^{2,1+\varepsilon}$ estimates for the Monge–Ampère equation in Chapter 6 and the proof of the Harnack inequality for the linearized Monge–Ampère equation in Chapter 12.

Lemma 5.32 (Vitali's Covering Lemma)**.** *Let $u : \Omega \to \mathbb{R}$ be a strictly convex function satisfying, in the sense of Aleksandrov, $0 < \lambda \leq \det D^2 u \leq \Lambda$ in a bounded open set Ω in $\mathbb{R}^n$. There exists $K(n, \lambda, \Lambda) > 4$ with the following properties.*

(i) *Let $\mathcal{S}$ be a collection of sections $S^x = S_u(x, h(x)) \Subset \Omega$ with $S_u(x, Kh(x)) \Subset \Omega$. Then, there exists a countable subcollection of disjoint sections $\{S_u(x_i, h(x_i))\}_{i=1}^{\infty}$ such that*

$$\bigcup_{S^x \in \mathcal{S}} S^x \subset \bigcup_{i=1}^{\infty} S_u(x_i, Kh(x_i)).$$

(ii) *Let E be a compact set in Ω. For each $x \in E$, we associate a corresponding section $S_u(x, h(x)) \Subset \Omega$. Then, there are a finite number of these sections $S_u(x_i, h(x_i)), i = 1, \ldots, m$, such that*

$$E \subset \bigcup_{i=1}^{m} S_u(x_i, h(x_i)), \quad \text{with } \left\{S_u(x_i, K^{-1}h(x_i))\right\}_{i=1}^{m} \text{ disjoint.}$$

Proof. We use the following fact for sections compactly contained in Ω. There exists $K = K(n, \lambda, \Lambda) > 4$ such that

$$\begin{cases} \text{if } S_u(x_1, Kh_1) \Subset \Omega, \; S_u(x_2, Kh_2) \Subset \Omega, \; S_u(x_1, h_1) \cap S_u(x_2, h_2) \neq \emptyset, \\ \text{and } 2h_1 \geq h_2, \text{ then } S_u(x_2, h_2) \subset S_u(x_1, Kh_1). \end{cases}$$

The proof of this fact is based on the engulfing property of sections in Theorem 5.28 and the quantitative localization property in Theorem 5.13. Suppose that $x \in S_u(x_1, h_1) \cap S_u(x_2, h_2)$ and $2h_1 \geq h_2$. Then, $S_u(x_2, h_2) \subset S_u(x, \theta_0 h_2) \subset S_u(x, 2\theta_0 h_1)$ and $x_1 \in S_u(x_1, h_1) \subset S_u(x, 2\theta_0 h_1)$. We first choose $K \geq 4\theta_0/\delta_0$ where $\delta_0 = \delta_0(n, \lambda, \Lambda, 1/4, 1/2)$ is as in Theorem 5.13 so that $S_u(x, 4\theta_0 h_1) \subset S_u(x_1, Kh_1/2)$. With this, we can apply the engulfing property to obtain $S_u(x, 2\theta_0 h_1) \subset S_u(x_1, 2\theta_0^2 h_1)$. It follows that $S_u(x_2, h_2) \subset S_u(x_1, 2\theta_0^2 h_1)$. The result follows by choosing $K = \max\{2\theta_0^2, 4\theta_0/\delta_0\}$.

Now, we prove part (i). From $S^x \Subset \Omega$ and the volume estimate in Lemma 5.6(i), we find that $d(\mathcal{S}) := \sup\{h(x) : S^x \in \mathcal{S}\} \leq C(n, \lambda, \Lambda, \Omega) < \infty$. Define

$$\mathcal{S}_i \equiv \left\{ S^x \in \mathcal{S} : \frac{d(\mathcal{S})}{2^i} < h(x) \leq \frac{d(\mathcal{S})}{2^{i-1}} \right\} \quad (i = 1, 2, \ldots)$$

and define $\mathcal{F}_i \subset \mathcal{S}_i$ as follows. Let $\mathcal{F}_1$ be a maximal disjoint collection of sections in $\mathcal{S}_1$. By the volume estimate in Lemma 5.6, $\mathcal{F}_1$ is finite. Assuming $\mathcal{F}_1, \ldots, \mathcal{F}_{i-1}$ have been selected, we choose $\mathcal{F}_i$ to be any maximal disjoint subcollection of

$$\left\{ S \in \mathcal{S}_i : S \cap S^x = \emptyset \text{ for all } S^x \in \bigcup_{j=1}^{i-1} \mathcal{F}_j \right\}.$$

Again, each $\mathcal{F}_i$ is a finite set. Let $\mathcal{F} := \bigcup_{i=1}^{\infty} \mathcal{F}_i$, and consider the countable subcollection of disjoint sections $S_u(x_i, h(x_i))$ where $S^{x_i} \in \mathcal{F}$.

We now show that this subcollection satisfies the conclusion of the lemma. Indeed, let S^x be any section in $\mathcal{S}$. Then, there is an index j such that $S^x \in \mathcal{S}_j$. By the maximality of $\mathcal{F}_j$, there is a section $S^y \in \bigcup_{i=1}^{j} \mathcal{F}_i$ with $S^x \cap S^y \neq \emptyset$. Note that $h(x) \leq 2h(y)$ because $h(y) > \frac{d(\mathcal{S})}{2^j}$ and $h(x) \leq \frac{d(\mathcal{S})}{2^{j-1}}$. Thus, by the choice of K, $S^x \subset S_u(y, Kh(y)) \subset \bigcup_{i=1}^{\infty} S_u(x_i, Kh(x_i))$.

Finally, we prove part (ii). Applying part (i) to the collection of sections $\left\{ S_u(x, h(x)/K) \right\}_{x \in E}$, we can find a countable subcollection of disjoint sections $\left\{ S_u(x_i, h(x_i)/K) \right\}_{i=1}^{\infty}$ such that

$$E \subset \bigcup_{x \in E} S_u(x, K^{-1}h(x)) \subset \bigcup_{i=1}^{\infty} S_u(x_i, h(x_i)).$$

From the compactness of E, we can choose a finite number of sections $S_u(x_i, h(x_i))$ $(i = 1, \ldots, m)$ that cover E. $\qquad\square$

Applying Vitali's Covering Lemma, we prove the following Crawling of Ink-spots Lemma which was first introduced by Krylov and Safonov [**KS2**] in the Euclidean setting. As indicated in [**KS2**], the term "Crawling of Ink-spots Lemma" was coined by E. M. Landis.

Lemma 5.33 (Crawling of Ink-spots Lemma). *Let $u : \Omega \to \mathbb{R}$ be a strictly convex function satisfying $\lambda \leq \det D^2 u \leq \Lambda$, in the sense of Aleksandrov, in a bounded open set Ω in $\mathbb{R}^n$. Assume that for some $h > 0$ and $K_1 > 4$, we have $S_u(0, K_1 h) \Subset \Omega$. Let $E \subset F \subset S_u(0, h)$ be two open sets. Assume that* *for some constant $\delta \in (0,1)$, the following two assumptions are satisfied:*

(a) *If any section $S_u(x,t) \subset S_u(0,h)$ satisfies*

$$|S_u(x,t) \cap E| > (1 - \delta)|S_u(x,t)|,$$

   then $S_u(x,t) \subset F$.

(b) *$|E| \le (1 - \delta)|S_u(0,h)|$.*

Then, $|E| \le (1 - c\delta)|F|$ for some constant $c(n, \lambda, \Lambda) > 0$ provided $K_1 = K_1(n, \lambda, \Lambda)$ is large.

Proof. Let K be as in Lemma 5.32. Because F is open, for every $x \in F$, there is a section of maximal height $S_u(x, \bar{h}(x))$ which is contained in F and contains x. Using the volume estimates in Lemma 5.6, we deduce $\bar{h}(x) \le C(n, \lambda, \Lambda)h$. By Theorem 5.13, $S_u(x, CKh) \subset S_u(0, K_1 h)$ if $K_1(n, \lambda, \Lambda)$ is large. Hence, $S_u(x, K\bar{h}(x)) \subset S_u(0, K_1 h)$. Similarly, we can require that if $S_u(x,t) \subset S_u(0,h)$, then $S_u(x, Kt) \subset S_u(0, K_1 h)$. This is our choice of K_1.

Case 1. $S_u(x, \bar{h}(x)) = S_u(0, h)$ for some $x \in F$. In this case, by assumption (b), the conclusion of the lemma is obvious.

Case 2. $S_u(x, \bar{h}(x)) \ne S_u(0, h)$ for all $x \in F$. In this case, observe that $|S_u(x, \bar{h}(x)) \cap E| \le (1 - \delta)|S_u(x, \bar{h}(x))|$, for each $x \in F$. For, otherwise, we can find a slightly larger section $\tilde{S} \subset S_u(0, h)$ containing $S_u(x, \bar{h}(x))$ such that $|\tilde{S} \cap E| > (1 - \delta)|\tilde{S}|$ and $\tilde{S} \not\subset F$, contradicting assumption (a).

Since the family of sections $\{S_u(x, \bar{h}(x))\}_{x \in F}$ covers the set F, we can use Vitali's Covering Lemma (Lemma 5.32) to find a subcollection of nonoverlapping sections $S_i := S_u(x_i, \bar{h}(x_i))$ such that $F \subset \bigcup_{i=1}^{\infty} S_u(x_i, K\bar{h}(x_i))$. The volume estimates in Lemma 5.6 then imply that, for each i,

$$|S_u(x_i, K\bar{h}(x_i))| \le C_0|S_u(x_i, \bar{h}(x_i))| = C_0|S_i|, \quad C_0 = C_0(n, \lambda, \Lambda).$$

Because $S_i \subset F$ and $|S_i \cap E| \le (1 - \delta)|S_i|$, we have $|S_i \cap (F \setminus E)| \ge \delta|S_i|$. Therefore

$$|F \setminus E| \ge \sum_{i=1}^{\infty} |S_i \cap (F \setminus E)| \ge \frac{\delta}{C_0} \sum_{i=1}^{\infty} |S_u(x_i, K\bar{h}(x_i))| \ge \frac{\delta}{C_0}|F|.$$

It follows that $|E| \le (1 - c\delta)|F|$ where $c = C_0^{-1}$. $\square$

5.7. Centered Sections

So far, we have considered sections

$$S_u(x_0, p, h) := \{x \in \Omega : u(x) < u(x_0) + p \cdot (x - x_0) + h\}$$

where $p \in \partial u(x_0)$. By an analogy with the case u being a quadratic function for which these sections are Euclidean balls centered at x_0, we call x_0 the center of $S_u(x_0, p, h)$. In general, x_0 is not the center of mass of $S_u(x_0, p, h)$.

However, the definition of $S_u(x_0, p, h)$ still makes sense for $p \in \mathbb{R}^n$, and this flexibility suggests that we look for slopes p where x_0 is in fact the center of mass for $S_u(x_0, p, h)$. In this case, we call $S_u(x_0, p, h)$ a *centered section* at x_0 with height h. The existence of centered sections was established by Caffarelli [**C5**] for any locally finite, globally defined convex function whose graph does not contain a line. Centered sections are very useful in studying the boundary regularity theory for convex potentials in optimal transport and in the proof of Chern's conjecture in Section 12.4 which is the only place we use centered sections in this book.

In many typical problems in optimal transport, convex potentials are convex Aleksandrov solutions to the second boundary value problem for the Monge–Ampère equation

$$\lambda \le \det D^2 u(x) \le \Lambda \quad \text{in } \Omega \quad \text{and} \quad Du(\Omega) = \Omega',$$

where Ω and Ω' are convex domains in $\mathbb{R}^n$; see Example 1.8. Unlike the case of Dirichlet boundary data, here we do not have much information about u on $\partial\Omega$, so it is more difficult to control the geometry of the (uncentered) sections $S_u(x_0, p, h)$ where $p \in \partial u(x_0)$ for $x_0 \in \Omega$ near the boundary. These sections can have complicated interactions with the boundary. On the other hand, the function u can be extended, for example, by setting

$$\bar{u}(x) = \sup_{y \in \Omega, p \in \partial u(y)} [u(y) + p \cdot (x - y)]$$

to a globally defined function $\bar{u}$ that satisfies

$$\lambda \chi_\Omega \le \det D^2 \bar{u} \le \Lambda \chi_\Omega \quad \text{on } \mathbb{R}^n.$$

By using centered sections of $\bar{u}$, we can view them as interior sections for which the geometric properties developed for compactly contained uncentered sections still hold. This way, the centered sections of $\bar{u}$ turn out to be quite convenient for boundary regularity; see, for example, [**C5, C6, C8, CMc, SY1**] for more details. Their existence is guaranteed by the following:

Theorem 5.34 (Existence of centered sections). *Let* $u : \mathbb{R}^n \to \mathbb{R} \cup \{+\infty\}$ *be a convex function satisfying the following conditions:*

    (a) *u is finite in a neighborhood of 0.*

    (b) *$u \ge 0$ and $u(0) = 0$.*

    (c) *The set $\partial u(\mathbb{R}^n)$ of all slopes of supporting hyperplanes to the graph of u has nonempty interior.*

Then, there is $p \in \mathbb{R}^n$ such that the section

$$S_u(0, p, 1) := \{x \in \mathbb{R}^n : u(x) < p \cdot x + 1\}$$

has center of mass at 0.

Clearly, conditions (a)–(c) in Theorem 5.34 are satisfied for the globally defined function $\bar{u}$ in the preceding example, after coordinate translations and subtractions of affine functions. Moreover, (c) implies that the graph of u does not contain a line. For, otherwise, Lemma 2.36 shows that $\partial u(\mathbb{R}^n)$ is contained in a hyperplane. We leave it to the reader to verify that (c) is equivalent to the fact that the graph of u does not contain a line.

Proof. Let $S := \partial u(\mathbb{R}^n)$. By condition (c), the interior S^0 of S is not empty. We will choose p in S^0. Observe that if $p \in S^0$, then $S_u(0, p, 1)$ is bounded. Indeed, there is $\delta > 0$ such that $B_\delta(p) \subset S^0$. For each $q \in B_\delta(p)$, there is a supporting hyperplane l_q, with slope q, to the graph of u. Choosing $q = p \pm (\delta/2)e_i$ for $1 \le i \le n$ and using $u(x) \ge l_q(x)$, we find

$$S_u(0, p, 1) \subset \left\{ x \in \mathbb{R}^n : p \cdot x + 1 \ge l_q(x) \text{ for all } q = p \pm (\delta/2)e_i, 1 \le i \le n \right\},$$

and the rightmost set is contained in a bounded rectangular box.

With $S_u(0, p, 1)$ being bounded for each $p \in S^0$, we can consider the center of mass and momentum functions on S^0. They are defined as

$$c(p) := \frac{1}{|S_u(0,p,1)|} \int_{S_u(0,p,1)} x \, dx \quad \text{and} \quad m(p) := \int_{S_u(0,p,1)} x \, dx.$$

By condition (a) and Theorem 2.26, u is pointwise Lipschitz in a neighborhood of 0. Hence, from condition (b), there is $\rho > 0$ such that $u \le 1/2$ in $B_\rho(0)$.

Step 1. We first prove the theorem for S being bounded, that is, when u is globally Lipschitz. Note that $0 \in S$. For $p \in S^0$, we have

(5.24) $S_u(0, p, 1) \supset B_{\rho_1}(0) \quad \text{where } \rho_1 := \min\left\{ \rho, \frac{1}{2\mathrm{diam}(S)} \right\}.$

Since $c(p) = 0$ if and only if $|m(p)| = 0$, we will show the existence of $p \in S^0$ with $|m(p)| = 0$. For this, we will show that

$$\lim_{p \to \partial S} |m(p)| = \infty \quad \text{and} \quad \min_{p \in S} |m(p)| = 0.$$

Step 1(a). We show that if $\{p_k\}_{k=1}^\infty \subset S^0$ converges to $p \in \partial S$, then $S_u(0, p_k, 1)$ and $c(p_k)$ are not bounded.

Indeed, if $S_u(0, p_k, 1)$ remains bounded, then $S_u(0, p, 1)$ is also bounded. In this case, the graph of u is transversal to the hyperplane $p \cdot x + 1$; for a more quantitative version of transversality, see (5.25) and the inequality preceding it. Thus, we can tilt this hyperplane a bit and still preserve the transversality. This shows that p is in the interior of S, a contradiction.

The unboundedness of $c(p_k)$ follows from that of $S_u(0, p_k, 1)$. Were this not the case, then $|c(p_k)| \le C$ for all k. By Lemma 2.66, we would have a sequence of ellipsoids E_k centered at $c(p_k)$ and contained in $S_u(0, p_k, 1)$

whose maximal lengths of principal axes tend to infinity. This implies that the graph of u contains a line, contradicting (c). Step 1(a) is proved, and we have

$$\lim_{p \to \partial S} |c(p)| = \infty, \quad \lim_{p \to \partial S} |m(p)| = \infty.$$

Step 1(b). We will show that $c(p)$ and $m(p)$ are locally Lipschitz in S^0.

Arguing as in Step 1(a), if $p \in K \subset S^0$ where K is a compact set, then $\operatorname{diam}(S_u(0, p, 1))$ and $|c(p)|$ are bounded by a constant $C_1(K, u)$. Therefore, u and $L_p(x) := p \cdot x + 1$ remain uniformly transversal along $\partial S_u(0, p, 1)$ when p varies in K. In fact, fix $p \in K$, and consider the convex function v whose graph is the cone with base $S_u(0, p, 1)$ and vertex at $(0, -1)$. Then, extend v to $\mathbb{R}^n$ by homogeneity. By Lemma 2.37(i), we have in $S_u(0, p, 1)$

$$|v(x)| \geq \frac{\operatorname{dist}(x, \partial S_u(0, p, 1))}{\operatorname{diam}(S_u(0, p, 1))} \geq \frac{\operatorname{dist}(x, \partial S_u(0, p, 1))}{C_1(K, u)}.$$

By comparing $u(x) - [p \cdot x + 1]$ with v and denoting $C_2 = C_1^{-1}$, we find that

$$u(x) - L_p(x) \leq v(x) \leq -C_2 \operatorname{dist}(x, \partial S_u(0, p, 1)) \quad \text{for all } x \in S_u(0, p, 1)$$

and

$$(5.25) \quad u(x) - L_p(x) \geq v(x) \geq C_2 \operatorname{dist}(x, \partial S_u(0, p, 1)) \quad \text{for all } x \notin S_u(0, p, 1).$$

From these uniform transversality properties, we show that for $p_1, p_2 \in K$, the following Hausdorff distance estimate holds:

$$d(p_1, p_2) := d_H(S_u(0, p_1, 1), S_u(0, p_2, 1)) \leq C_3(K, u)|p_1 - p_2|.$$

For this, it suffices to show that

$$S_u(0, p_1, 1) \subset S_u(0, p_2, 1) + B_\lambda(0),$$

where $\lambda = C_3|p_1 - p_2|$ with $C_3 := (1 + C_1)/C_2$.

Suppose otherwise that there is $x \in S_u(0, p_1, 1) \setminus S_u(0, p_2, 1)$ such that $\operatorname{dist}(x, \partial S_u(0, p_2, 1)) \geq \lambda$. Then $|x| \leq C_1$ and $p_1 \neq p_2$, so, by (5.25),

$$C_2 \lambda \leq u(x) - L_{p_2}(x) = u(x) - L_{p_1}(x) + (p_1 - p_2) \cdot x \leq (p_1 - p_2) \cdot x \leq C_1 |p_1 - p_2|.$$

We obtain a contradiction to the choice of C_3.

From the Lipschitz property of $d(p_1, p_2)$, we can show that

$$|c(p_1) - c(p_2)| + |m(p_1) - m(p_2)| \leq C_4(K, u)|p_1 - p_2|.$$

Step 1(c). For $p \in K$ where K is a compact set of S^0, we show that if $\varepsilon > 0$ is small, then the following quantitative monotonicity property holds:

$$[m(p + \varepsilon \mathbf{e}) - m(p)] \cdot \mathbf{e} \geq C_5(K, u)\varepsilon \quad \text{for all unit vectors } \mathbf{e} \in \mathbb{R}^n.$$

First, we prove the monotonicity property. Fix a unit vector $\mathbf{e}$. Consider the half-spaces

$$H_{\mathbf{e}}^+ = \{x \in \mathbb{R}^n : x \cdot \mathbf{e} > 0\} \quad \text{and} \quad H_{\mathbf{e}}^- = \{x \in \mathbb{R}^n : x \cdot \mathbf{e} < 0\}.$$

Then

$$H_{\mathbf{e}}^+ \cap S_u(0, p, 1) \subset H_{\mathbf{e}}^+ \cap S_u(0, p + \varepsilon\mathbf{e}, 1),$$

and

$$H_{\mathbf{e}}^- \cap S_u(0, p + \varepsilon\mathbf{e}, 1) \subset H_{\mathbf{e}}^- \cap S_u(0, p, 1).$$

It follows from these inclusions and writing

$$S_u(0, p, 1) = (H_{\mathbf{e}}^+ \cap S_u(0, p, 1)) \cup (H_{\mathbf{e}}^- \cap S_u(0, p, 1))$$

that

$$[m(p + \varepsilon\mathbf{e}) - m(p)] \cdot \mathbf{e} \geq g(\varepsilon) := \int_{H_{\mathbf{e}}^+ \cap [S_u(0, p+\varepsilon\mathbf{e}, 1) \setminus S_u(0, p, 1)]} x \cdot \mathbf{e} \, dx > 0.$$

Next, we show that the gain $g(\varepsilon)$ is of order ε. For this, we use the global Lipschitz property of u with Lipschitz constant $M > 0$. Observe that if $x \in H_{\mathbf{e}}^+ \cap S_u(0, p, 1)$, then

$$u(x + \varepsilon(x \cdot \mathbf{e})\mathbf{e}/M) \leq u(x) + \varepsilon(x \cdot \mathbf{e}) \leq (p + \varepsilon\mathbf{e}) \cdot x + 1,$$

so

(5.26) $$x + \varepsilon(x \cdot \mathbf{e})\mathbf{e}/M \in H_{\mathbf{e}}^+ \cap S_u(0, p + \varepsilon\mathbf{e}, 1).$$

For $p \in K$, by (5.24), we have $S_u(0, p, 1) \supset B_{\rho_1}(0)$. Let

$$G_\varepsilon := \big(S_u(0, p + \varepsilon\mathbf{e}, 1) \setminus S_u(0, p, 1)\big) \cap \{x \in \mathbb{R}^n : x \cdot \mathbf{e} \geq \rho_1/2\}.$$

Then, due to (5.26), G_ε contains

$$\big(\{x + \varepsilon(x \cdot \mathbf{e})\mathbf{e}/M : x \in S_u(0, p, 1)\} \setminus S_u(0, p, 1)\big) \cap \{x \cdot \mathbf{e} \geq \rho_1/2\};$$

hence $|G_\varepsilon| \geq C(K, u)\varepsilon$. Finally, we have

$$g(\varepsilon) \geq \int_{G_\varepsilon} x \cdot \mathbf{e} \, dx \geq \frac{\rho_1}{2}|G_\varepsilon| \geq C_5(K, u)\varepsilon.$$

Step 1(d). Finally, we are ready to show that $\min_{p \in S^0} |m(p)| = 0$.

The minimum exists due to the local Lipschitz property of m and the fact that m tends to infinity near ∂S. It suffices to show that the minimum value is 0. Were this not the case, let $p \in S^0$ be the point where we have the minimum with $m(p) \neq 0$. Consider $\mathbf{e} = m(p)/|m(p)|$. We compute

(5.27) $$|m(p - \varepsilon\mathbf{e})|^2 = |m(p - \varepsilon\mathbf{e}) \cdot \mathbf{e}|^2 + |m(p - \varepsilon\mathbf{e}) \cdot \tau|^2$$

for some unit vector τ orthogonal to $\mathbf{e}$. We estimate each term on the right-hand side as follows. Note that

$$|m(p - \varepsilon\mathbf{e}) \cdot \tau|^2 = |[m(p - \varepsilon\mathbf{e}) - m(p)] \cdot \tau|^2 \leq |m(p - \varepsilon\mathbf{e}) - m(p)|^2 \leq C_4^2 \varepsilon^2.$$

On the other hand, in view of Steps 1(b) and 1(c), we have
$$-C_4\varepsilon \le [m(p-\varepsilon\mathbf{e}) - m(p)]\cdot\mathbf{e} \le -C_5\varepsilon,$$
so for all $\varepsilon < |m(p)|/C_4$, using $m(p)\cdot\mathbf{e} = |m(p)|$, we obtain
$$0 \le |m(p)| - C_4\varepsilon \le m(p-\varepsilon\mathbf{e})\cdot\mathbf{e} \le |m(p)| - C_5\varepsilon.$$
Therefore, recalling (5.27), we find
$$|m(p-\varepsilon\mathbf{e})|^2 \le (|m(p)| - C_5\varepsilon)^2 + C_4\varepsilon^2 < |m(p)|^2$$
for $\varepsilon > 0$ small. This contradicts the minimality of $|m(p)|$.

Step 2. We prove the theorem for general u satisfying hypotheses (a)–(c). For each $M > 0$, consider the globally Lipschitz convex function
$$u_M(x) = \sup L(x)$$
where the supremum is taking over supporting hyperplanes L to the graph of u with slope $p = DL$ satisfying $|p| \le M$. For M large enough, $\partial u_M(\mathbb{R}^n)$ has nonempty interior, so we can find $p_M \in \mathbb{R}^n$ such that $S_{u_M}(0, p_M, 1)$ is a centered section of height 1 at 0. We show that for these M,
$$|p_M| \le 2/[\rho\alpha(n)],$$
where $\alpha(n)$ is as in Lemma 2.66. It suffices to consider $|p_M| > 0$. Since $u_M \ge 0$ by (b), for all $\varepsilon > 0$, we have $-(1+\varepsilon)p_M/|p_M|^2 \notin S_{u_M}(0, p_M, 1)$. From Lemma 2.66, we deduce that $\gamma(1+\varepsilon)p_M/|p_M|^2 \notin S_{u_M}(0, p_M, 1)$ if $\gamma = 2[\alpha(n)]^{-1}$. Since $\varepsilon > 0$ is arbitrary, this gives
$$u_M(\gamma p_M/|p_M|^2) \ge p_M\cdot\gamma p_M/|p_M|^2 + 1 = \gamma + 1.$$
Since $u_M \le 1/2$ in $B_\rho(0)$, we must have $\gamma p_M/|p_M|^2 \notin B_\rho(0)$, which shows that $|p_M| \le \gamma/\rho = 2/[\rho\alpha(n)]$, as asserted.

Since $|p_M| \le \frac{2}{\rho\alpha(n)}$ and $u \le 1/2$ in $B_\rho(0)$, we find
$$B_{\rho\alpha(n)/4}(0) \subset S_{u_M}(0, p_M, 1).$$
On the other hand, $\operatorname{diam}(S_{u_M}(0, p_M, 1))$ is uniformly bounded. If otherwise, then the graph of u would contain a line which contradicts condition (c).

Choose a subsequence $\{M_i\}_{i=1}^\infty$ such that p_{M_i} converges to $p \in \mathbb{R}^n$ and $\overline{S_{u_{M_i}}(0, p_{M_i}, 1)}$ converges in the Hausdorff distance to the closure $\overline{S_\infty}$ of an open convex set S_∞ containing $B_{\rho\alpha(n)/4}(0)$. Then, p is the desired slope. Indeed, we have
$$\liminf_{i\to\infty} S_{u_{M_i}}(0, p_{M_i}, 1) := \bigcup_{i=1}^\infty \bigcap_{k=i}^\infty S_{u_{M_i}}(0, p_{M_i}, 1) \subset \overline{S_\infty}.$$
Since u_{M_i} increases to u and p_{M_i} converges to p, we find
$$S_u(0, p, 1) \subset \liminf_{i\to\infty} S_{u_{M_i}}(0, p_{M_i}, 1).$$

Therefore $S_u(0, p, 1) \subset S_\infty$. It remains to show $S_\infty \subset S_u(0, p, 1)$. For this, recall that $|p_{M_i}|$ and $\operatorname{diam}(S_{u_{M_i}}(0, p_{M_i}, 1))$ are bounded. Thus, as in Step 1(b), by comparing $u_{M_i}(x) - [p_{M_i} \cdot x + 1]$ with the function v whose graph is the cone with base $S_{u_{M_i}}(0, p_{M_i}, 1)$ and vertex at $(0, -1)$, we can find a constant $C > 0$, independent of i, such that

$$u_{M_i}(x) - [p_{M_i} \cdot x + 1] \leq -C\operatorname{dist}(x, \partial S_{u_{M_i}}(0, p_{M_i}, 1)) \quad \text{in } S_{u_{M_i}}(0, p_{M_i}, 1).$$

Thus, if $x \in S_\infty$, then for large i, we have $\delta > 0$ such that

$$x \in S_{u_{M_i}}(0, p_{M_i}, 1) \quad \text{and} \quad \operatorname{dist}(x, \partial S_{u_{M_i}}(0, p_{M_i}, 1)) \geq \delta,$$

and therefore $u(x) = \lim_{i \to \infty} u_{M_i}(x) \leq p \cdot x + 1 - C\delta$. This shows that $S_\infty \subset S_u(0, p, 1)$. The proof of the theorem is complete. $\qquad\square$

By an affine transformation, we deduce from Theorem 5.34 the existence of a centered section with arbitrary center of mass and arbitrary height.

Corollary 5.35. *Let $u : \mathbb{R}^n \to \mathbb{R}$ be a convex function such that the set $\partial u(\mathbb{R}^n)$ of all slopes of supporting hyperplanes to the graph of u has nonempty interior. Then, for any $x_0 \in \mathbb{R}^n$ and any $h > 0$, there is $p \in \mathbb{R}^n$ such that x_0 is the center of mass of*

$$S_u(x_0, p, h) := \{x \in \mathbb{R}^n : u(x) < u(x_0) + p \cdot (x - x_0) + h\}.$$

Proof. Let $q \in \partial u(x_0)$. Consider $v(z) := h^{-1}[u(z + x_0) - u(x_0) - q \cdot z]$. Then v satisfies conditions (a)–(c) of Theorem 5.34, and hence, there is $p_h \in \mathbb{R}^n$ such that the center of mass of $S_v(0, p_h, 1) = \{z \in \mathbb{R}^n : v(z) < p_h \cdot z + 1\}$ is 0. Since $S_v(0, p_h, 1) = S_u(x_0, hp_h + q, h) - x_0$, if we choose $p := hp_h + q$, then x_0 is the center of mass of $S_u(x_0, p, h)$. $\qquad\square$

Remark 5.36. By a rescaling in height, Theorem 5.34 implies that for any $h > 0$, there is $p_h \in \mathbb{R}^n$ such that 0 is the center of mass of $S_u(0, p_h, h)$.

- By a different method, Caffarelli and McCann [**CMc**] proved the existence and uniqueness of the centered sections $S_u(0, p_h, h)$ for each $h > 0$. Moreover, as h varies, the continuity of $S_u(0, p_h, h)$ in the Hausdorff metric and the continuity of p_h are also established.

- Now, if we slide down the hyperplane (P), where $x_{n+1} = p_h \cdot x + h$, until it touches the graph of u at x_h, then $S_u(0, p_h, h) = S_u(x_h, p_h, \hat{h})$, where $p_h \in \partial u(x_h)$. Note that $\hat{h} \geq h$ because $\hat{h}$ is not smaller than the length h of the vertical segment from 0 to a point on (P).

- Using the result of Problem 2.11, we can show that $h \leq \hat{h} \leq C(n)h$.

We will use Theorem 5.34 together with the second bullet in Remark 5.36 in the proof of Chern's conjecture in Chapter 12.

5.8. Problems

Problem 5.1. This problem is an application to sections of the fact that a bounded convex set is balanced around its center of mass. Let $u : \mathbb{R}^n \to \mathbb{R}$ be a convex function on $\mathbb{R}^n$ with a bounded section $S_u(x_0, p, h)$ whose center of mass is x^*. Show that for all $\gamma \in (0, 1)$, there is $\theta(n, \gamma) < 1$ such that

$$x^* + \gamma[S_u(x_0, p, h) - x^*] \subset S_u(x_0, p, \theta h).$$

Hint: Take $\theta = 1 - (1 - \gamma)\alpha(n)/[1 + \alpha(n)]$, where $\alpha(n) \in (0, 1)$ is as in Lemma 2.66; see also Problem 2.11.

Problem 5.2. Let Ω be a bounded convex domain in $\mathbb{R}^n$ with boundary $\partial\Omega \in C^{1,\beta}$, where $\beta > 1 - 2/n$. Let $u \in C^{1,\beta}(\overline{\Omega})$ be a convex function satisfying $\det D^2 u \geq \lambda > 0$ in the sense of Aleksandrov. Prove that u is strictly convex in Ω. (This is a result of Urbas [**U1**].)
Hint: Suppose not. Then, we can assume that $u \equiv 0$ on a line segment $L = [-ae_n, ae_n] \Subset \Omega$ for some $a > 0$ and $u \geq 0$ in Ω. Repeat the last part of the proof of Theorem 5.14 for the domain $\Omega' := \Omega \cap \{-a \leq x_n \leq a\}$.

Problem 5.3. Let $\theta_0(n, \lambda, \Lambda)$ be the engulfing constant in Theorem 5.28.

(a) Find an explicit upper bound for $\theta_0(n, \lambda, \Lambda)$.

(b) Verify that $\theta_0(n, \lambda, \Lambda) \geq 4$. Determine Λ/λ when $\theta_0(n, \lambda, \Lambda) = 4$.

Problem 5.4. Let u be a convex function on $\mathbb{R}^n$ satisfying $\lambda \leq \det D^2 u \leq \Lambda$ in the sense of Aleksandrov. Let $\theta_0(n, \lambda, \Lambda)$ be the engulfing constant in Theorem 5.28. Show that the sections $\{S_u(x, t)\}_{x \in \mathbb{R}^n, t > 0}$ of u satisfy the separating property with the constant θ_0^2; namely, if $y \notin S_u(x, t)$, then $S_u(y, t/\theta_0^2) \cap S_u(x, t/\theta_0^2) = \emptyset$.

Problem 5.5. Let Ω be an open set in $\mathbb{R}^n$ that contains the origin. Let u be a convex function on Ω that satisfies $u(0) = 0, u \geq 0, 0 < \lambda \leq \det D^2 u \leq \Lambda$ in Ω. Suppose that for some constants $\delta \in (0, 1)$ and $\alpha > 1$, we have

$$B_\delta(0) \subset S_u(0, 0, 1) \subset S_u(0, 0, \alpha) \Subset \Omega.$$

Show that $S_u(0, 0, \alpha) \supset B_{\delta\theta^{1 - \frac{\ln \alpha}{\ln 2}}}(0)$, where θ is the constant in Lemma 5.15.

Problem 5.6. Let $\{\Omega_k\}_{k=1}^\infty$ be a sequence of convex domains in $\mathbb{R}^n$ containing $B_\delta(0)$ for some $\delta > 0$. Let $f \in L^\infty(\mathbb{R}^n)$ with $0 < \lambda \leq f \leq \Lambda$. Assume for each k, u_k is an Aleksandrov solution to $\det D^2 u_k = f$ in Ω_k which satisfies

(a) $u_k(0) = 0$, $u_k > 0$ in $\Omega_k \setminus \{0\}$, $u_k \leq 1$ in $B_\delta(0)$ and

(b) $u_k \geq \alpha_k$ on $\partial\Omega_k$ where $\alpha_k \to \infty$ when $k \to \infty$.

Show that there exists a subsequence of $\{u_k\}_{k=1}^\infty$ that converges uniformly on compact subsets of $\mathbb{R}^n$ to a solution of $\det D^2 u = f$ on $\mathbb{R}^n$.

Problem 5.7. Let $f \in L^\infty(\mathbb{R}^n)$ with $0 < \lambda \le f \le \Lambda$. For $R > 0$, let v_R be the Aleksandrov solution of
$$\det D^2 v_R = f \quad \text{in } B_R(0) \subset \mathbb{R}^n, \qquad v_R = 0 \quad \text{on } \partial B_R(0).$$
Show that $S_{v_R}(0, t_R) \subset B_R(0)$ for some $t_R > 0$ with $t_R \to \infty$ when $R \to \infty$.

Problem 5.8. Show the existence of $\gamma(\lambda, \Lambda, n)$ and $\delta(\lambda, \Lambda, n)$ as asserted in (5.19) and (5.20).

Problem 5.9. Let Ω be an open set in $\mathbb{R}^n$, and let $u : \Omega \to \mathbb{R}$ be a convex function satisfying $0 < \lambda \le \det D^2 u \le \Lambda$ in Ω in the sense of Aleksandrov. Assume that $S_u(x_i, 2t_i) \Subset \Omega$ for $i = 1, 2$, but $x_2 \notin S_u(x_1, t_1)$, while
$$S_u(x_1, (1 - \varepsilon)t_1) \cap S_u(x_2, (1 - \varepsilon)t_2) \neq \emptyset \quad \text{for some } \varepsilon \in (0, 1).$$
Show that there exist $c(n, \lambda, \Lambda) > 0$ and $p_1(n, \lambda, \Lambda) > 1$ such that $t_2 \ge c\varepsilon^{p_1} t_1$.

5.9. Notes

A large part of the materials on sections presented in this chapter (except Sections 5.4 and 5.7) is contained in Figalli [**F2**, Sections 4.1, 4.2, and 4.7] and Gutiérrez [**G2**, Chapter 3]; see also Le–Mitake–Tran [**LMT**, Section 3.2]. Lemma 5.6(iii)(b) is related to Remark 4.33 and Lemma 4.41 in Figalli [**F2**]. More general Monge–Ampère measures that satisfy a doubling condition were treated in Gutiérrez [**G2**] and Gutiérrez–Huang [**GH2**].

The proof of Theorem 5.9 follows an argument of Mooney [**Mn2**].

Related to Caffarelli's Localization Theorems in Section 5.2 and interior $C^{1,\alpha}$ estimates in Section 5.3, see also Figalli [**F2**, Sections 4.2 and 4.5], Gutiérrez [**G2**, Sections 5.2, 5.3, and 5.4], Liu–Wang [**LW**, Section 2], and Trudinger–Wang [**TW6**, Section 3.3]. Corollary 5.12(ii) is from Caffarelli [**C2**]. Theorems 5.28 and 5.30 are due to Gutiérrez and Huang [**GH2**].

In Section 5.5, Theorem 5.31, whose statement first appeared in De Philippis–Figalli–Savin [**DPFS**], is taken from Figalli [**F2**, Section 4.7].

Related to Lemma 5.32 in Section 5.6, see Savin [**S5**]. Lemma 5.33 appears in Le [**L5**] and Le–Mitake–Tran [**LMT**, Section 2.2].

Related to Problem 5.6, see Chou–Wang [**ChW1**].

Chapter 6

Interior Second Derivative Estimates

After studying fine properties of first derivatives for strictly convex solutions to the Monge–Ampère equation, we now move to the second derivatives. By Aleksandrov's Theorem, convex functions are twice differentiable almost everywhere. Thus, it is interesting to see how far we can go beyond this general fact in the Monge–Ampère equation.

In this chapter, we investigate interior second derivative estimates for strictly convex solutions to the Monge–Ampère equation with right-hand side bounded between two positive constants. With their strict convexity, we can focus our attention on compactly contained sections of these solutions and therefore, after subtracting affine functions, we can assume that the boundary data is zero. Thus, we will consider the Monge–Ampère equation

$$(6.1) \quad \det D^2 u = f \quad \text{in } \Omega \subset \mathbb{R}^n \ (n \geq 2), \quad u = 0 \quad \text{on } \partial\Omega, \quad 0 < \lambda \leq f \leq \Lambda.$$

As we saw in Chapter 5, the solution to (6.1) belongs to $C^{1,\alpha}(\Omega)$ for some positive exponent $\alpha(n, \lambda, \Lambda)$. The results presented here provide critical estimates for second-order derivatives and they are the culmination of the interior regularity theory of the Monge–Ampère equation. They include Pogorelov's second derivative estimates, Caffarelli's $W^{2,p}$ and $C^{2,\alpha}$ estimates, and De Philippis–Figalli–Savin–Schmidt $W^{2,1+\varepsilon}$ estimates.

Interior $C^{1,1}$ estimates for (6.1) were first obtained by Pogorelov when f is C^2. Combining with Calabi's C^3 estimates when f is C^3 (or Evans–Krylov $C^{2,\alpha}$ estimates for concave fully nonlinear elliptic PDEs when f is only C^2) yields $C^{3,\alpha}$ estimates. When f is C^α, Caffarelli established the interior $C^{2,\alpha}$ estimates. Interior $W^{2,p}$ estimates were also obtained by Caffarelli under the assumption that $|f - 1| \leq \varepsilon(n, p)$ locally where $\varepsilon(n, p)$ is a small positive number depending on the dimension n and $p > 1$. In particular, $u \in W^{2,p}_{\text{loc}}$ for any finite p if f is continuous.

Whenever f has large oscillation, $W^{2,p}$ estimates are not expected to hold for large values of p. Indeed, Wang showed that for any $p > 1$, there are homogenous solutions to (6.1) in two dimensions of the type

$$u(tx_1, t^\alpha x_2) = t^{1+\alpha} u(x_1, x_2) \quad \text{for } t > 0, \quad \alpha \geq 2p/(p-1),$$

which are not in $W^{2,p}$; see Proposition 6.28.

Motivated by the problem of existence of distributional solutions to the semigeostrophic equations, De Philippis and Figalli first showed that interior $W^{2,1}$ estimates hold for (6.1). In fact, they proved that $\|D^2 u\| \left|\log \|D^2 u\|\right|^k \in L^1_{\text{loc}}$ for all $k \geq 0$. After that, De Philippis, Figalli, and Savin and, independently, Schmidt obtained interior $W^{2,1+\varepsilon}$ estimates for some small $\varepsilon(n, \lambda, \Lambda) > 0$. In view of Proposition 6.28, this result is optimal. Related to this issue, we will present Mooney's counterexample to $W^{2,1}$ regularity in Proposition 6.29 when the lower bound λ degenerates to 0.

6.1. Pogorelov's Second Derivative Estimates

The following important theorem of Pogorelov [**P2**] allows us to estimate the second derivatives of solutions to the Monge–Ampère equation for smooth right-hand side.

Theorem 6.1 (Pogorelov's interior second derivative estimates). *Let Ω be a bounded convex domain in $\mathbb{R}^n$. Let $f \in C^2(\Omega) \cap C(\overline{\Omega})$ with $f > 0$ in Ω. Assume that $u \in C^4(\Omega) \cap C^2(\overline{\Omega})$ is the convex solution to*

$$\begin{cases} \det D^2 u = f & in \ \Omega, \\ u = 0 & on \ \partial\Omega. \end{cases}$$

Let ξ be a unit vector in $\mathbb{R}^n$, $D_\xi u = Du \cdot \xi$, and $D_{\xi\xi} u = (D^2 u \xi) \cdot \xi$. Let

$$h = |u| e^{\frac{1}{2}(D_\xi u)^2} D_{\xi\xi} u \quad and \quad M = \max_{\overline{\Omega}} h.$$

Then, M is attained at some interior point $x_0 \in \Omega$, and furthermore

$$M \leq e^{(D_\xi u(x_0))^2/2} \Big[n + |u(x_0)||D_\xi u(x_0)||D_\xi \log f(x_0)| $$
$$+ |D_\xi u(x_0)| + |u(x_0)||D_{\xi\xi} \log f(x_0)|^{1/2} \Big].$$

The following lemma allows us to assume that $\xi = e_1$ and that $D^2 u(x_0)$ is diagonal. Let $\mathbb{O}(n)$ denote the space of real, orthogonal $n \times n$ matrices.

Lemma 6.2. *Let Ω be a convex domain in $\mathbb{R}^n$, let $x_0 \in \Omega$, and let $\xi \in \mathbb{R}^n$ be a unit vector. Let $u \in C^2(\Omega)$ be a strictly convex function, and let $f \in C^2(\Omega)$. Then, there is a matrix $T \in \mathbb{M}^{n \times n}$ with $|\det T| = 1$ such that $\xi = Te_1$, and for $\bar{u}(x) := u(Tx)$ and $\bar{f}(x) := f(Tx)$, the Hessian matrix $D^2\bar{u}(T^{-1}x_0)$ is diagonal, while*

$$D_1\bar{f}(T^{-1}x) = D_\xi f(x), \quad D_{11}\bar{f}(T^{-1}x) = D_{\xi\xi}f(x) \quad \text{for all } x \in \Omega.$$

Proof. First, we rotate the coordinates to have ξ as the unit vector e_1 on the x_1-axis. Let $Q \in \mathbb{O}(n)$ be an orthogonal matrix with ξ being the first column. Then $Qe_1 = \xi$. Let $v(x) = u(Qx)$ and $A = D^2v(Q^tx_0) = (a_{ij})$. Then $A \in \mathbb{M}^{n \times n}$ is a symmetric positive definite matrix. Let $B \in \mathbb{M}^{n \times n}$ be the matrix, which is also viewed as a linear transformation on $\mathbb{R}^n$, given by

$$B(x_1, x_2, \ldots, x_n) = \left(x_1 - \frac{a_{12}}{a_{11}}x_2 - \frac{a_{1n}}{a_{11}}x_n, x_2, \ldots, x_n\right).$$

Then $\det B = 1$, and

$$B^tAB = \begin{pmatrix} a_{11} & 0 \\ 0 & B_1 \end{pmatrix},$$

where B_1 is an $(n-1) \times (n-1)$ symmetric positive definite matrix. Next, we can find $P \in \mathbb{O}(n-1)$ such that P^tB_1P is diagonal. Let

$$\tilde{P} = \begin{pmatrix} 1 & 0 \\ 0 & P \end{pmatrix} \in \mathbb{O}(n) \quad \text{and} \quad \bar{u}(x) = v(B\tilde{P}x).$$

Then $T := QB\tilde{P} \in \mathbb{M}^{n \times n}$ is the desired matrix. Note that $B\tilde{P}e_1 = e_1$, so $Te_1 = Q(B\tilde{P}e_1) = Qe_1 = \xi$. At $\bar{x} = T^{-1}x_0 = (B\tilde{P})^{-1}Q^tx_0$, we have

$$D^2\bar{u}(\bar{x}) = (B\tilde{P})^t D^2v(Q^tx_0)(B\tilde{P}) = (B\tilde{P})^t A(B\tilde{P})$$

$$= \tilde{P}^t \begin{pmatrix} a_{11} & 0 \\ 0 & B_1 \end{pmatrix} \tilde{P} = \begin{pmatrix} a_{11} & 0 \\ 0 & P^tB_1P \end{pmatrix},$$

which is diagonal. On the other hand, for $\bar{f}(x) := f(Tx)$, we have

$$D_{11}\bar{f}(T^{-1}x) = e_1^t D^2\bar{f}(T^{-1}x)e_1 = (Te_1)^t D^2f(x)(Te_1) = D_{\xi\xi}f(x).$$

Similarly, we have $D_1\bar{f}(T^{-1}x) = D_\xi f(x)$, and the lemma is proved. $\square$

Proof of Theorem 6.1. Because $f > 0$ in Ω, $u < 0$ and $D^2u > 0$ in Ω. Since $h = 0$ on $\partial\Omega$, h attains its maximum value on $\overline{\Omega}$ at $x_0 \in \Omega$. By Lemmas 6.2 and 3.9, we can assume that $\xi = e_1$ and that $D^2u(x_0)$ is diagonal.

In subsequent calculations, we use Lemma 2.56 and the notation below:

$$(U^{ij}) = \text{Cof } D^2u = (\det D^2u)(D^2u)^{-1} \quad \text{and} \quad (u^{ij}) = (D^2u)^{-1}.$$

Differentiating the equation $\log \det D^2u = \log f$ with respect to x_1, we obtain at x_0 using that $D^2u(x_0)$ is diagonal

$$(6.2) \qquad\qquad u^{ij}D_{1ij}u = u^{ii}D_{1ii}u = D_1(\log f)$$

and

$$D_{11}(\log f) = u^{ii} D_{11ii}u - u^{ik}u^{jl}D_{1kl}uD_{1ij}u$$
$$(6.3) \qquad\qquad\qquad = u^{ii}D_{11ii}u - u^{ii}u^{jj}[D_{1ij}u]^2.$$

By the maximality of

$$w := \log h = \log D_{11}u + \log|u| + |D_1 u|^2/2$$

at x_0, we have

$$Dw(x_0) = 0, \ \ D^2 w(x_0) \le 0.$$

It follows that, at x_0, we have

$$D_i w = \frac{D_i u}{u} + \frac{D_{11i}u}{D_{11}u} + D_1 u D_{1i}u = 0,$$

and as a consequence

$$(6.4) \qquad\qquad \frac{D_i u}{u} = -\frac{D_{11i}u}{D_{11}u} \ \ \text{for } i \ne 1.$$

Let

$$L = U^{ij}D_{ij} = (\det D^2 u)u^{ij}D_{ij} = f u^{ij}D_{ij}$$

be the linearized Monge–Ampère operator associated with u. Then, $Lw = f u^{ii}D_{ii}w = f \operatorname{trace}((D^2 u)^{-1}D^2 w) \le 0$ at x_0. Note that

$$D_{ii}w = \frac{D_{ii}u}{u} - \frac{(D_i u)^2}{u^2} + \frac{D_{11ii}u}{D_{11}u} - \frac{(D_{11i}u)^2}{(D_{11}u)^2} + (D_{1i}u)^2 + D_1 u D_{1ii}u.$$

Writing $Lw \le 0$ out and recalling that $f > 0$ in Ω, we get (still at x_0)

$$\frac{u^{ii}D_{ii}u}{u} - \frac{u^{ii}(D_i u)^2}{u^2} + \frac{u^{ii}D_{11ii}u}{D_{11}u} - \frac{u^{ii}(D_{11i}u)^2}{(D_{11}u)^2}$$
$$+ u^{ii}(D_{1i}u)^2 + u^{ii}D_1 u D_{1ii}u \le 0.$$

Clearly, the first term is n/u. The last term equals $D_1 u D_1(\log f)$ by (6.2) while the fifth term is $D_{11}u$. Thus, using (6.3) for the third term and (6.4) for the second term (and splitting the cases $i = 1$ and $i \ge 2$), we obtain

$$\left(\frac{n}{u} + D_{11}u - \frac{u^{11}(D_1 u)^2}{u^2}\right) + \left(\frac{D_{11}(\log f)}{D_{11}u} + D_1 u D_1(\log f)\right)$$
$$+ \left[\frac{u^{ii}u^{jj}(D_{1ij}u)^2}{D_{11}u} - \frac{u^{ii}(D_{11i}u)^2}{(D_{11}u)^2} - \sum_{i=2}^{n} u^{ii}\frac{(D_{11i}u)^2}{(D_{11}u)^2}\right] \le 0.$$

The bracketed term in the above inequality is nonnegative since it is

$$\sum_{j \ge 2} \frac{u^{ii}u^{jj}(D_{1ij}u)^2}{D_{11}u} - \sum_{i=2}^{n} u^{ii}\frac{(D_{11i}u)^2}{(D_{11}u)^2} = \sum_{i,j \ge 2} \frac{u^{ii}u^{jj}(D_{1ij}u)^2}{D_{11}u} \ge 0.$$

Hence, at x_0,

$$D_{11}u + \frac{n}{u} - \frac{(D_1 u)^2}{D_{11}uu^2} + D_1 u D_1(\log f) + \frac{D_{11}(\log f)}{D_{11}u} \le 0.$$

Multiplying both sides by $u^2 D_{11}ue^{(D_1 u)^2}$, we obtain

$$e^{(D_1 u)^2}u^2(D_{11}u)^2 + nuD_{11}ue^{(D_1 u)^2} - (D_1 u)^2 e^{(D_1 u)^2}$$
$$+ u^2 D_{11}ue^{(D_1 u)^2}\left(D_1 u D_1(\log f) + \frac{D_{11}(\log f)}{D_{11}u}\right) \le 0,$$

implying

$$h(x_0)^2 - \left[n + |u(x_0)||D_1 u(x_0)||D_1(\log f)(x_0)|\right]e^{[D_1 u(x_0)]^2/2}h(x_0)$$
$$- \left[(D_1 u(x_0))^2 + |u(x_0)|^2|D_{11}(\log f)(x_0)|\right]e^{[D_1 u(x_0)]^2} \le 0.$$

Using the fact that if $t^2 - bt - c \le 0$ where $b, c \ge 0$, then $t \le b + \sqrt{c}$, we obtain the desired estimate for $M = h(x_0)$. $\square$

Remark 6.3. Theorem 6.1 also applies to $f = f(x, z)$ where $x \in \mathbb{R}^n$ and $z \in \mathbb{R}$. Due to the exponent $1/2$ in $\|D_{\xi\xi}\log f(x_0, u(x_0))\|^{1/2}$, we can absorb $D_{\xi\xi}u(x_0)$ on the left-hand side. In the end, we find that

$$M \le C(|u(x_0)|, |D_\xi u(x_0)|, |f(x_0, u(x_0)|^{-1}, K(x_0)),$$

where $K(x_0)$ depends on partial derivatives of $f(x, z)$ in x and z up to second order, evaluated at $(x_0, u(x_0))$.

Combined with the Evans–Krylov Theorem, the Pogorelov estimates give the interior second and third derivative estimates for the Monge–Ampère equation with right-hand side being twice differentiable.

Lemma 6.4. *Let Ω be a bounded convex domain in $\mathbb{R}^n$. Assume that the convex function $u \in C^4(\Omega) \cap C(\overline{\Omega})$ satisfies*

$$\det D^2 u = f \quad in \ \Omega, \quad u = 0 \quad on \ \partial\Omega,$$

where $f \in C^2(\Omega) \cap C(\overline{\Omega})$ with $f \ge c_0 > 0$ in Ω. Let $\alpha \in (0, 1)$, and let $\Omega' \Subset \Omega$ be an open set. Then, there are constants C, C_α depending only on n, c_0, $\|f\|_{C^2(\Omega)}$, Ω, and $\mathrm{dist}(\Omega', \partial\Omega)$ (and C_α depending also on α), such that

$$C^{-1}I_n \le D^2 u \le CI_n \quad in \ \Omega' \quad and \quad \|u\|_{C^{3,\alpha}(\Omega')} \le C_\alpha.$$

Proof. Note that $u < 0$ in Ω. Let $M := \|u\|_{L^\infty(\Omega)}$ and $d := \mathrm{diam}(\Omega)$. Then $\Omega = S_u(x_0, M)$, where x_0 is the minimum point of u in $\overline{\Omega}$. By the Aleksandrov maximum principle (Theorem 3.12) and the volume estimate in Lemma 5.6, we have

$$0 < c_1(n, c_0, |\Omega|) \le M \le C_1(n, d, \|f\|_{L^1(\Omega)}).$$

By Lemma 2.37(i), $|u(x)| \geq \operatorname{dist}(x, \partial\Omega)M/d$ for all $x \in \Omega$. Therefore,

$$\Omega' \subset \Omega_{5\varepsilon} := \{x \in \Omega : u(x) \leq -5\varepsilon\} \quad \text{for } \varepsilon := \operatorname{dist}(\Omega', \partial\Omega)c_1/(5d) > 0.$$

Step 1. There is $C_2(n, d, \|f\|_{L^1(\Omega)}) > 0$ such that $\operatorname{dist}(\Omega_\varepsilon, \partial\Omega) \geq C_2\varepsilon^n$.

This follows from Aleksandrov's maximum principle which gives

$$\varepsilon^n \leq |u(x)|^n \leq C(n)d^{n-1}\operatorname{dist}(x, \partial\Omega)\|f\|_{L^1(\Omega)} \quad \text{for all } x \in \Omega_\varepsilon.$$

From Step 1 and the estimates on the slopes in Lemma 2.37(iii), we have

$$(6.5) \qquad |Du(x)| \leq \frac{M}{\operatorname{dist}(x, \partial\Omega)} \leq C_3(n, d, \|f\|_{L^1(\Omega)})\varepsilon^{-n} \quad \text{for all } x \in \Omega_\varepsilon.$$

Step 2. If $\xi \in \mathbb{R}^n$ with $|\xi| = 1$, then

$$(6.6) \qquad D_{\xi\xi}u \leq C_4(n, d, c_0, \|f\|_{C^2(\Omega)}, \varepsilon) \quad \text{in } \Omega_{3\varepsilon}.$$

Consider $v = u + 2\varepsilon$. Then

$$\det D^2 v = f \quad \text{in } \Omega_{2\varepsilon} \quad \text{and} \quad v = 0 \quad \text{on } \partial\Omega_{2\varepsilon}.$$

In $\Omega_{2\varepsilon} \subset \Omega_\varepsilon$, we have $|D_\xi v| = |D_\xi u| \leq C_3\varepsilon^{-n}$, due to (6.5). Thus, applying the Pogorelov estimates in Theorem 6.1 to v on $\Omega_{2\varepsilon}$ and

$$h = |v|e^{\frac{1}{2}(D_\xi v)^2}D_{\xi\xi}v,$$

we obtain at the maximum point $x_0 \in \Omega_{2\varepsilon}$ of h the estimate

$$h = |v|e^{\frac{1}{2}(D_\xi v)^2}D_{\xi\xi}v \leq h(x_0) \leq C_4(n, d, c_0, \|f\|_{C^2(\Omega)}, \varepsilon) \quad \text{on } \Omega_{2\varepsilon}.$$

Because $D_{\xi\xi}u = D_{\xi\xi}v$ and $|v| \geq |u| - 2\varepsilon \geq \varepsilon$ on $\Omega_{3\varepsilon}$, we obtain (6.6).

Step 3. We have $c_0 C_4^{1-n}I_n \leq D^2 u \leq C_4 I_n$ in $\Omega_{3\varepsilon}$.

Indeed, if $x \in \Omega_{3\varepsilon}$, then from Step 2, we have

$$D_{\xi\xi}u(x) = \sum_{1 \leq i,j \leq n} D_{ij}u(x)\xi_i\xi_j \leq C_4 \quad \text{for all } \xi = (\xi_1, \ldots, \xi_n) \quad \text{with } |\xi| = 1.$$

Let $P \in \mathbb{O}(n)$ be such that $D^2 u(x) = P^t\operatorname{diag}(\lambda_i(x))P$. Then

$$D_{\xi\xi}u(x) = (P\xi)^t\operatorname{diag}(\lambda_i(x))P\xi = \sum_{1 \leq i \leq n} \lambda_i(x)(P\xi)_i^2.$$

For each $k \in \{1, \ldots, n\}$, choosing $\xi \in \mathbb{R}^n$ such that $P\xi = e_k$ gives $\lambda_k(x) \leq C_4$. The other inequality follows from $\det D^2 u(x) = \prod_{i=1}^n \lambda_i(x) = f(x) \geq c_0$.

Step 4. For each $\alpha \in (0, 1)$, we have

$$\|u\|_{C^{3,\alpha}(\Omega_{5\varepsilon})} \leq C = C(n, \alpha, C_4, \varepsilon, \Omega, \|f\|_{C^2(\Omega)}).$$

Indeed, from Step 3, we find that the solution u to the Monge–Ampère equation $\det D^2 u = f$ satisfies the hypotheses of the Evans–Krylov Theorem (Theorem 2.85). Thus, there is $\gamma \in (0, 1)$ depending on n and C_4 and there is C_5 depending on $n, C_4, \varepsilon, \Omega$, and $\|f\|_{C^2(\Omega)}$ such that $\|u\|_{C^{2,\gamma}(\Omega_{4\varepsilon})} \leq C_5$.

Let $U = (U^{ij}) = (\det D^2 u)(D^2 u)^{-1}$. Then each partial derivative $D_k u$ $(k = 1, \ldots, n)$ satisfies the linear equation $U^{ij} D_{ij}(D_k u) = D_k f$ with uniformly elliptic, C^γ coefficients and C^1 right-hand side. By the classical Schauder theory (Theorem 2.81), we obtain interior $C^{2,\gamma}$ estimates for $D_k u$ in $\Omega_{9\varepsilon/2}$. These give the $C^{3,\gamma}$ estimates for u in $\Omega_{9\varepsilon/2}$. Now, the matrix U has $C^{1,\gamma}$ coefficients. Thus, for each $\alpha \in (0,1)$, we again apply Theorem 2.81 to $U^{ij} D_{ij}(D_k u) = D_k f$ to obtain interior $C^{2,\alpha}$ estimates for $D_k u$ in $\Omega_{5\varepsilon}$. These give the $C^{3,\alpha}$ estimates for u in $\Omega_{5\varepsilon}$.

The lemma now follows from $\Omega' \subset \Omega_{5\varepsilon}$, Step 3, and Step 4. $\qquad \square$

Now, we prove higher-order regularity, due to Pogorelov [**P2, P5**] and Cheng–Yau [**CY**], for strictly convex C^2 solutions to the Monge–Ampère equation with strictly positive and sufficiently smooth right-hand side.

Theorem 6.5. *Let Ω be an open set in $\mathbb{R}^n$, and let $u : \Omega \to \mathbb{R}$ be a strictly convex Aleksandrov solution of $\det D^2 u = f$ in Ω where $f \geq c_0 > 0$.*

(i) *If $f \in C^2(\Omega)$, then $u \in C^{3,\alpha}(\Omega)$ for all $\alpha < 1$.*

(ii) *If $f \in C^{k,\beta}(\Omega)$ where $k \geq 2$ and $\beta \in (0,1)$, then $u \in C^{k+2,\beta}(\Omega)$.*

In particular, if Ω is assumed further to be convex, the above conclusions apply to the unique convex Aleksandrov solution v to

$$\det D^2 v = f \quad in \ \Omega, \quad v = 0 \quad on \ \partial\Omega.$$

Proof. Note that the last conclusion follows from Theorem 5.13(i) and assertions (i) and (ii). We will prove assertion (i) only, since assertion (ii) is a consequence of assertion (i) and the classical Schauder theory (Theorem 2.81). The strict convexity of u and $f \in C(\Omega)$ with $f \geq c_0 > 0$ already imply that u is differentiable in Ω by Remark 5.19. Moreover, for any $x_0 \in \Omega$, there exists a small number $t > 0$ such that $S_u(x_0, 5t) \Subset \Omega$. Let

$$\bar{u}(x) := u(x) - u(x_0) - Du(x_0) \cdot (x - x_0) - 3t.$$

Then $\bar{u} = 0$ on $\partial S_u(x_0, 3t)$. Let $\{S_m\}_{m=1}^\infty$ be a sequence of uniformly convex domains with C^5 boundaries ∂S_m such that $\overline{S_m}$ converges to $\overline{S_u(x_0, 3t)}$ in the Hausdorff distance; see Theorem 2.51. Hence, there is $\tau > 0$ such that for m large, $S_u(x_0, 2t) \subset S_m \subset S_u(x_0, 4t)$ and $\text{dist}(S_u(x_0, 2t), \partial S_m) \geq \tau$.

From Theorem 2.50, we can find a sequence of $C^\infty(S_u(x_0, 4t))$ functions $\{f_m\}_{m=1}^\infty$ such that $\|f_m - f\|_{C^2(\overline{S_u(x_0, 3t)})} \to 0$ when $m \to \infty$. For each m, by Theorem 4.12, the Dirichlet problem

$$\det D^2 u_m = f_m \quad in \ S_m, \quad u_m = 0 \quad on \ \partial S_m$$

has a unique uniformly convex solution $u_m \in C^4(\overline{S_m})$. By Lemma 6.4, we have for all $\alpha \in (0,1)$,

$$\|u_m\|_{C^{3,\alpha}(S_u(x_0,t))} \leq C(\alpha, c_0, \tau, \|f_m\|_{C^2(S_u(x_0,2t))}).$$

Thanks to Theorem 3.20, when $m \to \infty$, the sequence $\{u_m\}_{m=1}^\infty$ converges uniformly to the convex Aleksandrov solution $v \in C(\overline{S_u(x_0,3t)})$ of

$$\det D^2 v = f \quad \text{in } S_u(x_0,3t), \quad v = 0 \quad \text{on } \partial S_u(x_0,3t).$$

By uniqueness, we must have $\bar{u} = v$. Therefore, the limit $\bar{u}$ of u_m also satisfies

$$\|\bar{u}\|_{C^{3,\alpha}(S_u(x_0,t))} \leq C(\alpha, c_0, \tau, \|f\|_{C^2(S_u(x_0,2t))}).$$

This shows that $u \in C^{3,\alpha}(S_u(x_0,t))$. Since x_0 is arbitrary, we conclude that $u \in C^{3,\alpha}(\Omega)$ for all $\alpha < 1$. $\qquad\qquad\square$

6.2. Second-Order Differentiability of the Convex Envelope

In this section, we discuss second-order differentiability properties of the convex envelope of a continuous function. They will be useful in establishing density estimates for the Monge–Ampère equation in Lemma 6.9.

Definition 6.6 (Convex envelope of a continuous function). Let $v \in C(\overline{\Omega})$ be a continuous function on a domain Ω in $\mathbb{R}^n$. The *convex envelope* Γ_v of v in Ω (see Figure 6.1) is defined to be

$$\Gamma_v(x) := \sup\left\{l(x) : l \leq v \text{ in } \overline{\Omega}, \ l \text{ is affine}\right\}.$$

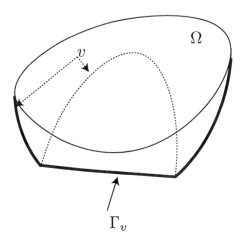

Figure 6.1. The convex envelope Γ_v of v.

Clearly, Γ_v is the largest convex function not exceeding v in $\overline{\Omega}$. Under suitable assumptions on v, we will show in Proposition 6.8 that the Hessian of Γ_v is locally bounded. For this, we need some properties of the contact set defined by

$$\mathcal{C}_v := \{x \in \overline{\Omega} : v(x) = \Gamma_v(x)\}.$$

Lemma 6.7 (Properties of the contact set). *Let Ω be a bounded convex domain in $\mathbb{R}^n$. Let $v \in C(\overline{\Omega})$ and $x_0 \in \Omega \setminus \mathcal{C}_v$. Then, there exist $(n+1)$ points $x^1, \dots, x^{n+1}$ in $\mathcal{C}_v$ such that $x_0 \in \mathrm{conv}(\{x^1, \dots, x^{n+1}\})$. Moreover, the function Γ_v is affine in $\mathrm{conv}(\{x^1, \dots, x^{n+1}\})$, and for every supporting hyperplane with slope p to Γ_v at x_0, p is also the slope of a supporting hyperplane to v at x^i for all i.*

Proof. Let p be the slope of a supporting hyperplane l_{x_0} to Γ_v at x_0, and let

$$\mathcal{C} := \{x \in \overline{\Omega} : v(x) = \Gamma_v(x_0) + p \cdot (x - x_0) \equiv l_{x_0}(x)\}.$$

Clearly $\mathcal{C}$ is compact, so by Lemma 2.6, $\mathrm{conv}(\mathcal{C})$ is convex and compact. Since $v(x) \geq \Gamma_v(x) \geq l_{x_0}(x)$ in $\overline{\Omega}$, we have $\mathcal{C} \subset \mathcal{C}_v$.

We show that $\mathcal{C} \neq \emptyset$. If otherwise, then there is $\delta > 0$ such that $v \geq l_{x_0} + \delta$ in $\overline{\Omega}$. Since $l_{x_0} + \delta$ is affine, by the definition of Γ_v, we have $\Gamma_v \geq l_{x_0} + \delta$, but this contradicts $\Gamma_v(x_0) < l_{x_0}(x_0) + \delta = \Gamma_v(x_0) + \delta$.

Now, we show that $x_0 \in \mathrm{conv}(\mathcal{C})$. Were this false, Lemma 2.9 would give an affine function l strongly separating $\mathrm{conv}(\mathcal{C})$ and x_0; that is, $l(x_0) > 0$ and $l < 0$ in $\mathrm{conv}(\mathcal{C})$. We show that there is $\gamma > 0$ such that, in $\overline{\Omega}$,

$$(6.7) \qquad\qquad v \geq l_{x_0} + \gamma l.$$

Indeed, there exist $\varepsilon > 0$ and a neighborhood $\mathcal{U}$ of $\mathrm{conv}(\mathcal{C})$ such that $l < -\varepsilon$ in $\mathcal{U}$. In $\overline{\Omega} \setminus \mathcal{U}$, $v - l_{x_0} > 0$, so we can find $\gamma > 0$ such that (6.7) holds. Since $l < 0$ in $\mathcal{U}$, (6.7) also holds there. Thus, we have (6.7) in $\overline{\Omega}$. It follows that $\Gamma_v \geq l_{x_0} + \gamma l$ in $\overline{\Omega}$. However, this does not hold at x_0, because $l(x_0) > 0$.

From $x_0 \in \mathrm{conv}(\mathcal{C})$, Carathéodory's Theorem (Theorem 2.5) gives the existence of $(n+1)$ points $x^1, \dots, x^{n+1}$ in $\mathcal{C} \subset \mathcal{C}_v$ such that $x_0 \in \mathrm{conv}(\{x^1, \dots, x^{n+1}\})$. At each x^i, we have

$$v(x^i) = \Gamma_v(x^i) = \Gamma_v(x_0) + p \cdot (x^i - x_0) \equiv l_{x_0}(x^i),$$

while $v \geq l_{x_0}$ in Ω, so p is the slope of a supporting hyperplane to v at x^i for all i. Since the convex function $\Gamma_v \geq l_{x_0}$ and the affine function l_{x_0} are equal at x^i, they must coincide in the convex hull $\mathrm{conv}(\{x^1, \dots, x^{n+1}\})$. Thus, Γ_v is affine there. $\qquad\square$

Below is our main result on the second-order differentiability of convex envelopes.

Proposition 6.8 (Regularity of convex envelopes). *Let Ω be a bounded convex domain in $\mathbb{R}^n$. Let $v \in C(\overline{\Omega}) \cap C^{1,1}(\Omega)$ be such that $v = 0$ on $\partial\Omega$ and $\inf_\Omega v < 0$. Let Γ_v be the convex envelope of v in Ω. Then:*

   (i) *$\Gamma_v \in C^{1,1}(\Omega)$.*

   (ii) *$\det D^2 \Gamma_v = 0$ almost everywhere in $\{x \in \Omega : v(x) > \Gamma_v(x)\}$.*

Proof. We first prove assertion (i). Fix $\delta_0 > 0$. Let $x_0 \in \Omega_{\delta_0} := \{x \in \Omega : \mathrm{dist}(x, \partial\Omega) > \delta_0\}$. Let p be the slope of a supporting hyperplane to the graph of Γ_v at x_0, and let $L(x) = \Gamma_v(x_0) + p \cdot (x - x_0)$. Then $v \geq \Gamma_v \geq L$.

To show that $\Gamma_v \in C^{1,1}(\Omega)$, by Theorem 2.91, it suffices to find $M > 0$ and $\delta_1 > 0$ depending only on v and δ_0, such that

$$(6.8) \qquad \Gamma_v(x_0 + z) - L(x_0 + z) \leq M|z|^2 \quad \text{for all } |z| \leq \delta_1.$$

First, consider the case $x_0 \in \mathcal{C}_v$. In this case, we have $v(x_0) = \Gamma_v(x_0)$ and $Dv(x_0) = p$. Since $v \in C^{1,1}(\Omega)$, there are $r_0 > 0$ and $M_0 > 0$ such that

$$v(x_0 + z) \leq v(x_0) + Dv(x_0) \cdot z + M_0|z|^2$$
$$= L(x_0 + z) + M_0|z|^2 \quad \text{for all } |z| \leq r_0.$$

Hence,

$$\Gamma_v(x_0 + z) - L(x_0 + z) \leq v(x_0 + z) - L(x_0 + z) \leq M_0|z|^2 \quad \text{for all } |z| \leq r_0.$$

Next, consider the case $x_0 \in \Omega_{\delta_0} \setminus \mathcal{C}_v$. By Lemma 6.7, there exist $(n+1)$ points $x^1, \ldots, x^{n+1} \in \mathcal{C}_v$ such that

$$(6.9) \qquad x_0 = \sum_{1 \leq i \leq n+1} \lambda_i x^i, \quad \lambda_i \geq 0, \quad \sum_{1 \leq i \leq n+1} \lambda_i = 1.$$

Moreover, Γ_v is equal to L in $\mathrm{conv}(\{x^1, \ldots, x^{n+1}\})$, and $Dv(x^i) = p$ for all $i = 1, \ldots, n+1$. For a positive integer m, we define

$$K_m = \mathrm{conv}(\{x \in \Omega : v(x) \leq -1/m\}).$$

Then, each K_m is a convex set which is compactly contained in Ω, and $\bigcup_{m=1}^\infty K_m = \Omega$. There is a positive integer m such that $x_0 \in K_m$.

Step 1. We show that there is $\delta_1 > 0$ depending only on n, m, $\inf_\Omega v$ such that, for some $j \in \{1, \ldots, n+1\}$, we have $\lambda_j \geq \delta_1$ and $x^j \in K_{2m}$.

Indeed, each $x \in K_m$ is a convex combination of points $z^1, \ldots, z^N$ where $\Gamma_v(z^i) \leq v(z^i) \leq -1/m$, and hence $\Gamma_v(x) \leq -1/m$ from the convexity of

Γ_v. This shows that $\Gamma_v \leq -1/m$ in K_m. Therefore, $v \leq -1/m$ in $K_m \cap \mathcal{C}_v$. Using (6.9) and the fact that Γ_v is affine in $\text{conv}(\{x^1, \ldots, x^{n+1}\})$, we find

$$-1/m \geq \Gamma_v(x_0) = \sum_{1 \leq i \leq n+1} \lambda_i \Gamma_v(x^i) = \sum_{1 \leq i \leq n+1} \lambda_i v(x^i).$$

Therefore, there must be one index $j \in \{1, \ldots, n+1\}$ such that $v(x^j) \leq -1/m$, and hence $x^j \in K_{2m}$. Now, let $\lambda := \max\{\lambda_j : x^j \in K_{2m}\}$. Then,

$$-1/m \geq \sum_{x^j \in K_{2m}} \lambda_j v(x^j) + \sum_{x^j \notin K_{2m}} \lambda_j v(x^j)$$

$$\geq \sum_{x^j \in K_{2m}} \lambda \inf_\Omega v - \frac{1}{2m} \sum_{x^j \notin K_{2m}} \lambda_j \geq (n+1)\lambda \inf_\Omega v - 1/(2m).$$

It follows that $\lambda \geq \delta_1 := (2m(n+1)|\inf_\Omega v|)^{-1}$.

Step 2. Conclusion. By relabelling, we can assume that $j = 1$ in Step 1. Then $\lambda_1 \geq \delta_1$, $v(x^1) = \Gamma_v(x^1) = L(x^1)$, and $Dv(x^1) = p$. There are $r_1 > 0$ and $M_1 > 0$ depending only on v and m such that

$$v(x^1 + z) \leq v(x^1) + Dv(x^1) \cdot z + M_1|z|^2$$
$$= L(x^1 + z) + M_1|z|^2 \quad \text{for all } |z| \leq r_1.$$

Thus, if $|z| \leq r_1\delta_1$, then $\Gamma_v(x^1 + z) \leq v(x^1 + z) \leq L(x^1 + z) + M_1|z|^2$, and we obtain from the convexity of Γ_v that

$$\Gamma_v(x_0 + z) \leq \lambda_1 \Gamma_v(x^1 + z/\lambda_1) + \sum_{2 \leq i \leq n+1} \lambda_i \Gamma_v(x^i)$$

$$\leq \lambda_1 L(x^1 + z/\lambda_1) + \sum_{2 \leq i \leq n+1} \lambda_i L(x^i) + M_1|z|^2/\lambda_1$$

$$\leq L(x_0 + z) + M_1|z|^2/\delta_1,$$

which implies (6.8). Here, we used $\Gamma_v(x^i) = L(x^i)$ for all $1 \leq i \leq n+1$. Thus, assertion (i) is proved.

Finally, we prove (ii). Since $\Gamma_v \in C^{1,1}(\Omega)$, it is twice differentiable almost everywhere. Let x_0 be a point of twice differentiability of Γ in $\Omega \setminus \mathcal{C}_v$. Since Γ_v is affine in a simplex containing x_0 as established in Lemma 6.7, one of the eigenvalues of $D^2\Gamma_v(x_0)$ must be zero, hence $\det D^2\Gamma_v(x_0) = 0$. $\quad\square$

6.3. Density Estimate

We next prove a perturbation result of Theorem 6.1 and Lemma 6.4 in the case $f \equiv 1$. It roughly says that if the Monge–Ampère measure of a strictly convex function u deviates from 1 no greater than ε, then the set where u has large second derivatives is of measure comparable to ε. It will be a key ingredient in the proof of Theorem 6.13 on $W^{2,p}$ estimates for large p.

Lemma 6.9 (Density estimate). *Assume that $\Omega \subset \mathbb{R}^n$ is a convex domain such that $B_1(0) \subset \Omega \subset B_n(0)$, $0 < \varepsilon < 1/2$, and $u \in C(\overline{\Omega}) \cap C^2(\Omega)$ is a convex function satisfying*

$$1 - \varepsilon \leq \det D^2 u \leq 1 + \varepsilon \quad in \ \Omega, \qquad u = 0 \quad on \ \partial\Omega.$$

Let $0 < \alpha < 1$, and define the set

$$\Omega_\alpha = \{x \in \Omega : u(x) < -(1-\alpha)\|u\|_{L^\infty(\Omega)}\} \equiv S_u(z_0, \alpha\|u\|_{L^\infty(\Omega)}),$$

where $z_0 \in \Omega$ is such that $u(z_0) = \min_{\overline{\Omega}} u$. For $\lambda > 0$, define

$$H_\lambda := \left\{x_0 \in \Omega : u(x) \geq u(x_0) + Du(x_0) \cdot (x - x_0) + \lambda|x - x_0|^2/2 \ in \ \Omega\right\}.$$

Then there exist $\sigma(\alpha, n) \in (0, 1)$ and $C(n) > 1$ such that

$$|\Omega_\alpha \backslash H_{\sigma(\alpha,n)}| < C(n)\varepsilon.$$

Remark 6.10. If $x_0 \in H_\sigma$, then $D^2 u(x_0) \geq \sigma I_n$. From $\det D^2 u(x_0) < 2$, we find $D^2 u(x_0) \leq 2\sigma^{-(n-1)} I_n$, so $\|D^2 u(x_0)\| \leq 2n\sigma^{-(n-1)}$. Hence,

$$\left|\Omega_\alpha \cap \{\|D^2 u\| \geq 2n\sigma^{-(n-1)}\}\right| \leq |\Omega_\alpha \backslash H_\sigma| \leq C(n)\varepsilon.$$

Proof of Lemma 6.9. Let w be the convex Aleksandrov solution to

$$\det D^2 w = 1 \quad in \ \Omega, \quad w = 0 \quad on \ \partial\Omega.$$

Then, by Theorems 5.13 and 6.5, $w \in C(\overline{\Omega}) \cap C^\infty(\Omega)$. From

$$\det D^2((1 + \varepsilon)w) = (1 + \varepsilon)^n \geq \det D^2 u$$

and $u = w = 0$ on $\partial\Omega$, we have by the comparison principle (Theorem 3.21) that $(1 + \varepsilon)w \leq u$ in Ω. Similarly, we obtain $(1 + \varepsilon)w \leq u \leq (1 - \varepsilon)w$ and

$$(1/2 + \varepsilon)w \leq u - w/2 \leq (1/2 - \varepsilon)w \quad in \ \Omega.$$

Let Γ be the convex envelope of $u - w/2$ in Ω; see Definition 6.6. Consider the contact set of Γ and $u - w/2$ in Ω:

$$\mathcal{C} := \left\{x \in \Omega : \Gamma(x) = u(x) - w(x)/2\right\}.$$

Step 1. We show that

(6.10) $$|\mathcal{C}| \geq (1 - 4n\varepsilon)|\Omega|.$$

Indeed, since w is convex, we have

(6.11) $$(1/2 + \varepsilon)w \leq \Gamma \leq (1/2 - \varepsilon)w \quad in \ \Omega.$$

Consequently, the Aleksandrov maximum principle in Theorem 3.12 gives

$$\|\Gamma - w/2\|_{L^\infty(\Omega)} \leq \varepsilon\|w\|_{L^\infty(\Omega)} \leq C_0(n)\mathrm{diam}(\Omega)\varepsilon \leq 2nC_0(n)\varepsilon \equiv \bar{\varepsilon}.$$

Since $w = 0$ on $\partial\Omega$, we deduce from (6.11) and the maximum principle that

(6.12) $$\partial((1/2 - \varepsilon)w)(\Omega) \subset \partial\Gamma(\Omega) \subset \partial((1/2 + \varepsilon)w)(\Omega).$$

By Proposition 6.8, Γ is convex with $\Gamma \in C^{1,1}(\Omega)$ and $\det D^2\Gamma = 0$ almost everywhere outside $\mathcal{C}$. Hence, by Theorem 2.74,

$$(6.13) \qquad |\partial\Gamma(\Omega)| = |D\Gamma(\Omega)| = \int_\Omega \det D^2\Gamma \, dx = \int_{\mathcal{C}} \det D^2\Gamma \, dx.$$

We now estimate $\det D^2\Gamma$ from above. For this, observe that for any $x \in \mathcal{C}$, the function $u - w/2 - \Gamma$ attains its local minimum value 0 at x. Hence,

$$D^2\Gamma(x) \leq D^2(u - w/2)(x)$$

at any point where Γ and u are twice differentiable. Therefore, this inequality holds for almost every $x \in \mathcal{C}$ by Aleksandrov's Theorem (Theorem 2.89). Thus, for almost every $x \in \mathcal{C}$, we can use the Minkowski determinant inequality in Lemma 2.59 to get

$$(\det D^2\Gamma(x))^{1/n} \leq (\det D^2 u(x))^{1/n} - (\det D^2(w/2)(x))^{1/n}$$
$$\leq (1 + \varepsilon)^{1/n} - 1/2 \leq 1/2 + \varepsilon.$$

Combining this with (6.12) and (6.13) gives

$$(1/2 - \varepsilon)^n \int_\Omega \det D^2 w \, dx \leq |\partial\Gamma(\Omega)| \leq (1/2 + \varepsilon)^n |\mathcal{C}|.$$

Now, (6.10) follows from

$$|\mathcal{C}| \geq \left(\frac{1/2 - \varepsilon}{1/2 + \varepsilon}\right)^n |\Omega| \geq (1 - 4n\varepsilon)|\Omega|.$$

Step 2. We show that, for a suitable $\sigma(\alpha, n) \in (0, 1)$, we have

$$(6.14) \qquad \mathcal{C} \cap \Omega_\alpha \subset H_\sigma \cap \Omega_\alpha.$$

Once (6.14) is established, the lemma now follows from (6.10) because

$$|\Omega_\alpha \setminus H_\sigma| \leq |\Omega_\alpha \setminus \mathcal{C}| \leq 4n\varepsilon|\Omega| \leq C(n)\varepsilon.$$

It remains to prove (6.14). From $B_1(0) \subset \Omega \equiv S_u(z_0, \|u\|_{L^\infty(\Omega)}) \subset B_n(0)$ and Lemma 5.6(ii), we have $\text{dist}(\Omega_\alpha, \partial\Omega) > c_0(\alpha, n)$. Hence, the Pogorelov estimates in Lemma 6.4 give positive constants $c_\alpha(n)$ and $C(\alpha, n)$ such that

$$(6.15) \qquad 4c_\alpha(n)I_n \leq D^2 w \leq C(\alpha, n)I_n \quad \text{in } \Omega_\alpha.$$

Let $x_0 \in \mathcal{C} \cap \Omega_\alpha$, and let l_{x_0} be a supporting hyperplane to Γ at x_0. Then, $l_{x_0}(x_0) = u(x_0) - w(x_0)/2$, $Dl_{x_0} = Du(x_0) - Dw(x_0)/2$, and

$$(6.16) \qquad u(x) \geq l_{x_0}(x) + w(x)/2 \quad \text{for all } x \in \Omega.$$

Next, we use the Taylor formula together with (6.15) to estimate

$$\bar{w}(x) := w(x) - w(x_0) - Dw(x_0) \cdot (x - x_0) \geq c_{(1+\alpha)/2}(n)|x - x_0|^2.$$

Indeed, we have

$$\bar{w}(x) = \int_0^1 t \int_0^1 (x - x_0)^t D^2 w(x_0 + \theta t (x - x_0))(x - x_0) \, d\theta dt$$

$$\geq \int_0^1 t \int_0^{1/2} (x - x_0)^t D^2 w(x_0 + \theta t (x - x_0))(x - x_0) \, d\theta dt.$$

For $t \in [0, 1]$ and $\theta \in [0, 1/2]$, we have from the convexity of $u \leq 0$ that

$$u(x_0 + t\theta(x - x_0)) \leq (1 - t\theta)u(x_0) < -(1/2 - \alpha/2)\|u\|_{L^\infty(\Omega)},$$

so $x_0 + t\theta(x - x_0) \in \Omega_{(1+\alpha)/2}$. Applying (6.15) to $\Omega_{(1+\alpha)/2}$, we find

$$\bar{w}(x) \geq \int_0^1 2t c_{(1+\alpha)/2}(n) |x - x_0|^2 \, dt = c_{(1+\alpha)/2}(n) |x - x_0|^2.$$

Combining this with (6.16), we deduce that

$$u(x) \geq l(x) + c_{(1+\alpha)/2}(n) |x - x_0|^2 / 2 \quad \text{for all } x \in \Omega,$$

where $l(x)$ is the supporting hyperplane to u at x_0 in Ω given by

$$l(x) := l_{x_0}(x) + \frac{1}{2}[w(x_0) + Dw(x_0) \cdot (x - x_0)] = u(x_0) + Du(x_0) \cdot (x - x_0).$$

Therefore $x_0 \in H_\sigma$ with $\sigma = c_{(1+\alpha)/2}(n)$, proving (6.14). $\qquad \square$

6.4. Interior Second Derivative Sobolev Estimates

In this section, we establish interior Sobolev regularity for second derivatives of convex solutions to the Monge–Ampère equation. In particular, we will prove the interior $W^{2,1+\varepsilon}$ estimates of De Philippis, Figalli, and Savin [**DPFS**] and Schmidt [**Schm**] for merely bounded right-hand side, and we will prove the interior $W^{2,p}$ estimates of Caffarelli [**C3**] for right-hand side having small oscillations. They are fundamental results for later developments and applications.

We first state these Sobolev estimates on normalized convex domains. At the end of this section, we extend them to compactly contained sections. Below is our basic $W^{2,1+\varepsilon}$ estimates.

Theorem 6.11 (De Philippis–Figalli–Savin–Schmidt interior $W^{2,1+\varepsilon}$ estimates). *Let Ω be a convex domain in $\mathbb{R}^n$ satisfying $B_1(0) \subset \Omega \subset B_n(0)$. Assume that $f \in L^\infty(\Omega)$ satisfies $0 < \lambda \leq f \leq \Lambda$. Let $u \in C(\overline{\Omega})$ be the convex Aleksandrov solution to the Monge–Ampère equation*

$$\det D^2 u = f \quad \text{in } \Omega, \quad u = 0 \quad \text{on } \partial\Omega.$$

Then, there exist positive constants ε and C depending only on n, λ, and Λ such that

$$\int_{\Omega'} \|D^2 u\|^{1+\varepsilon} \, dx \leq C, \quad \text{where } \Omega' := \{x \in \Omega : u(x) < -\|u\|_{L^\infty(\Omega)}/2\}.$$

Remark 6.12. We emphasize that no other regularities are imposed on f in Theorem 6.11. Moreover, the constant ε there depends only on n and Λ/λ. Indeed, consider $\hat{u} = \lambda^{-1/n}u$. Then $1 \le \det D^2\hat{u} \le \Lambda/\lambda$ in Ω. Moreover

$$\Omega' := \{u < -\|u\|_{L^\infty(\Omega)}/2\} = \{\hat{u} < -\|\hat{u}\|_{L^\infty(\Omega)}/2\}.$$

The remark follows by applying the conclusion of Theorem 6.11 to $\hat{u}$.

Next is our basic interior $W^{2,p}$ estimates.

Theorem 6.13 (Caffarelli's interior $W^{2,p}$ estimates). *Let Ω be a convex domain in $\mathbb{R}^n$ satisfying $B_1(0) \subset \Omega \subset B_n(0)$. Let f be a nonnegative, bounded function on Ω. Let $u \in C(\overline{\Omega})$ be the convex Aleksandrov solution to the Monge–Ampère equation*

$$\det D^2u = f \quad in\ \Omega, \quad u = 0 \quad on\ \partial\Omega.$$

Then, for any $p > 1$, there exists $\delta = \delta(p,n) = e^{-C(n)p} < 1/2$, where $C(n)$ is large, such that

$$\int_{\Omega'} \|D^2u\|^p\, dx \le C(n,p), \quad where\ \Omega' := \{x \in \Omega : u(x) < -\|u\|_{L^\infty(\Omega)}/2\},$$

provided that

$$\|f - 1\|_{L^\infty(\Omega)} \le \delta.$$

We just stated estimates for the Hessians in Ω' but they are actual $W^{2,p}$ estimates because estimates for solutions and their gradients are consequences of Theorem 3.12 and Lemma 5.6(iii).

By localizing and rescaling, we can show that Theorem 6.11 implies that $u \in W^{2,1+\varepsilon}_{\mathrm{loc}}(\Omega)$, while Theorem 6.13 implies that $u \in W^{2,p}_{\mathrm{loc}}(\Omega)$. Moreover, as a corollary of Theorem 6.13, we show that when the right-hand side is continuous, solutions belong to $W^{2,p}_{\mathrm{loc}}$ for all p.

Corollary 6.14. *Let Ω be a convex domain in $\mathbb{R}^n$, and let $u : \Omega \to \mathbb{R}$ be a strictly convex function satisfying, in the sense of Aleksandrov,*

$$\det D^2u = f \quad in\ \Omega, \quad for\ some\ function\ f > 0.$$

(i) *If $\lambda \le f \le \Lambda$, then $u \in W^{2,1+\varepsilon}_{\mathrm{loc}}(\Omega)$ where $\varepsilon = \varepsilon(n,\lambda,\Lambda) > 0$.*

(ii) *Given $p > 1$, if $\|f - 1\|_{L^\infty(\Omega)} \le \delta(p) \le 1/2$ where $\delta(p) = e^{-C(n)p}$ for a large $C(n)$, then $u \in W^{2,p}_{\mathrm{loc}}(\Omega)$.*

(iii) *If $f > 0$ is continuous in Ω, then $u \in W^{2,p}_{\mathrm{loc}}(\Omega)$ for all $p < \infty$.*

Proof. We give the proof of assertion (iii) while those for assertions (i) and (ii) are similar. Suppose that $f > 0$ is continuous in Ω. Let $x_0 \in \Omega$. By the strict convexity of u, there is $h_0 > 0$ such that $S_u(x_0, h_0) \Subset \Omega$. By subtracting an affine function, we can assume that $u(x_0) = 0$ and $Du(x_0) = 0$, so $S_u(x_0, t) = \{x \in \Omega : u(x) < t\}$ for all $t \le h_0$.

Given $p > 1$, we show that $u \in W^{2,p}$ in a small neighborhood of x_0. Indeed, let $\delta(p, n)$ be as in Theorem 6.13. The continuity of $f > 0$ in $S_u(x_0, h_0)$ implies the existence of a small $h > 0$ such that

$$(6.17) \qquad m := \frac{\sup_{S_u(x_0,h)} f}{\inf_{S_u(x_0,h)} f} < 1 + \delta(p, n).$$

We now show that $u \in W^{2,p}(S_u(x_0, h/2))$. To see this, let A_h be an affine transformation such that $B_1(0) \subset \tilde{\Omega} := A_h(S_u(x_0, h)) \subset B_n(0)$. Let

$$\tilde{u}(x) := [\inf_{S_u(x_0,h)} f]^{-1/n} (\det A_h)^{2/n} [u(A_h^{-1}x) - h], \quad x \in \tilde{\Omega}.$$

Then $\tilde{u} = 0$ on $\partial\tilde{\Omega}$, and

$$1 \leq \det D^2\tilde{u} \leq m < 1 + \delta(p, n) \quad \text{in } B_1(0) \subset \tilde{\Omega} \subset B_n(0).$$

Due to (6.17), we can apply Theorem 6.13 to conclude $\tilde{u} \in W^{2,p}(\tilde{\Omega}')$ where

$$\tilde{\Omega}' := \{x \in \tilde{\Omega} : \tilde{u}(x) < -\|\tilde{u}\|_{L^\infty(\tilde{\Omega})}/2\} = A_h(S_u(x_0, h/2)).$$

Rescaling back, we obtain $u \in W^{2,p}(S_u(x_0, h/2))$ as asserted. $\qquad \square$

The rest of this section is devoted to proving Theorems 6.11 and 6.13.

6.4.1. Reduction to a priori estimates for C^2 solutions.

It suffices to prove the estimates in Theorems 6.11 and 6.13 for C^2 solutions since the general case follows by an approximation argument, recorded below.

Proposition 6.15 (Approximation argument). *If the estimates in Theorems 6.11 and 6.13 hold for all convex solutions $u \in C^2(\Omega) \cap C(\overline{\Omega})$, then they also hold for all convex Aleksandrov solutions $u \in C(\overline{\Omega})$ as stated.*

Proof. We consider the estimates in Theorem 6.11 because those in Theorem 6.13 are similar. Let $u \in C(\overline{\Omega})$ be as in Theorem 6.11. Let $\{f_k\}_{k=1}^\infty \subset C^2(\Omega)$ satisfy $\lambda \leq f_k \leq \Lambda$ and $f_k \to f$ almost everywhere in Ω when $k \to \infty$; see Theorem 2.50. Let $u_k \in C(\overline{\Omega})$ be the Aleksandrov solution to

$$\det D^2 u_k = f_k \quad \text{in } \Omega, \quad u_k = 0 \quad \text{on } \partial\Omega.$$

By Theorem 6.5, $u_k \in C^2(\Omega)$. Assume that the estimates in Theorem 6.11 hold for all convex solutions belonging to $C^2(\Omega) \cap C(\overline{\Omega})$. Then

$$\|u_k\|_{W^{2,1+\varepsilon}(\Omega_k')} \leq C(n, \lambda, \Lambda) \quad \text{with } \Omega_k' := \{x \in \Omega : u_k(x) < -\|u_k\|_{L^\infty(\Omega)}/2\}.$$

Letting $k \to \infty$ in the above estimates and noticing that $u_k \to u$ locally uniformly by Theorem 3.19 and that Ω_k' converges to Ω' in the Hausdorff distance, we find that the estimates in Theorem 6.11 hold for u. $\qquad \square$

Remark 6.16. Concerning constants in Theorems 6.11 and 6.13:

- Since we are looking for a small $\delta < 1/2$ in Theorem 6.13, we can assume that the function f in Theorem 6.13 satisfies similar bounds as f in Theorems 6.11 where $\lambda = 1/2$ and $\Lambda = 3/2$.

- For u as in Theorems 6.11 and 6.13, Lemma 3.24 gives

$$c(\lambda, n) \leq |\min_{\Omega} u| = \|u\|_{L^\infty(\Omega)} \leq C(\Lambda, n).$$

Let $x_0 \in \Omega$ be the minimum point of u in Ω. Then

$$\Omega = S_u(x_0, \|u\|_{L^\infty(\Omega)}) \quad \text{and} \quad \Omega' = S_u(x_0, \|u\|_{L^\infty(\Omega)}/2).$$

Thus, by considering $\bar{u} := 2u/\|u\|_{L^\infty(\Omega)}$ if necessary, in the proofs of the above theorems, we can take $x_0 = 0$ and $\|u\|_{L^\infty(\Omega)} = 2$ for convenience. Note that the constant $1/2$ in Ω' can be replaced by any $\gamma \in (1/4, 1)$ where the estimate now depends also on γ.

6.4.2. Proofs of Theorems 6.11 and 6.13 for C^2 solutions. First, we record a simple fact concerning L^1 estimate for the Hessian of a C^2 convex function.

Lemma 6.17. *Let u be a C^2 convex function on a domain Ω in $\mathbb{R}^n$. Assume $E \Subset F \Subset \Omega$ where ∂E is C^1. Then*

$$\int_E \|D^2 u\| \, dx \leq \frac{\max_{\partial F} u - \min_{\partial E} u}{\text{dist}(E, \partial F)} \mathcal{H}^{n-1}(\partial E).$$

Proof. Let ν be the outer unit normal vector field on ∂E. Then, using $\|D^2 u\| \leq \Delta u$ due to u being convex and using the Divergence Theorem, we have

$$\int_E \|D^2 u\| \, dx \leq \int_E \Delta u \, dx = \int_{\partial E} Du \cdot \nu \, d\mathcal{H}^{n-1} \leq \|Du\|_{L^\infty(\partial E)} \mathcal{H}^{n-1}(\partial E).$$

The lemma follows by invoking Lemma 2.37(iii) which gives

$$\|Du\|_{L^\infty(\partial E)} \leq \frac{\max_{\partial F} u - \min_{\partial E} u}{\text{dist}(E, \partial F)}. \qquad \square$$

Obtaining higher integrability estimates for the Hessian of solutions to the Monge–Ampère equation is much more involved. They will be established in this section via suitable normalizations together with delicate measure-theoretic estimates. They are a variant of techniques introduced by De Philippis and Figalli in [**DPF1**].

In the following, let u be a C^2, convex function satisfying

(6.18) $$\lambda \leq \det D^2 u \leq \Lambda \quad \text{in a domain } \Omega \subset \mathbb{R}^n.$$

Suppose that $S_u(x_0, h) \Subset \Omega$. Then from Lemma 5.6, we can find $\sigma(n, \lambda, \Lambda) \in (0, 1)$ and a matrix $A \in \mathbb{M}^{n \times n}$ with $\det A = 1$ such that

$$(6.19) \qquad \sigma B_{\sqrt{h}}(0) \subset A(S_u(x_0, h) - x_0) \subset \sigma^{-1} B_{\sqrt{h}}(0).$$

The matrix A may not be unique. However, recall from Lemma 5.6(iii) that if u is C^2 in a neighborhood of x_0, then $\|A\|^2$ is comparable to $\|D^2 u(x_0)\|$, for all small $h > 0$. This motivates the following:

Definition 6.18 (Normalized size of a section). We say that the section $S_u(x_0, h) \Subset \Omega$ has normalized size $\boldsymbol{\alpha}$ if $\boldsymbol{\alpha} := \|A\|^2$ for some matrix $A \in \mathbb{M}^{n \times n}$ that satisfies $\det A = 1$ and (6.19).

Given a matrix $A \in \mathbb{M}^{n \times n}$ as in (6.19), we rescale u by

$$(6.20) \qquad \tilde{u}(\tilde{x}) := h^{-1} u(x), \qquad \tilde{x} = Tx := h^{-1/2} A (x - x_0).$$

Then $\tilde{u}$ solves

$$\det D^2 \tilde{u}(\tilde{x}) = \tilde{f}(\tilde{x}), \qquad \text{with} \quad \tilde{f}(\tilde{x}) := f(x), \qquad \lambda \leq \tilde{f} \leq \Lambda,$$

and the section $S_{\tilde{u}}(0, 1)$ of $\tilde{u}$ centered at 0 with height 1 satisfies

$$\sigma B_1(0) \subset S_{\tilde{u}}(0, 1) \subset \sigma^{-1} B_1(0), \qquad S_{\tilde{u}}(0, 1) = T(S_u(x_0, h)).$$

From $D^2 u(x) = A^t D^2 \tilde{u}(\tilde{x}) A$, we deduce that

$$(6.21) \qquad \|D^2 u(x)\| \leq \|A\|^2 \|D^2 \tilde{u}(\tilde{x})\|$$

and

$$(6.22) \quad n^{-1/2} \lambda_1 \|A\|^2 \leq \|D^2 u(x)\| \leq \lambda_2 \|A\|^2 \quad \text{if } \lambda_1 I_n \leq D^2 \tilde{u}(\tilde{x}) \leq \lambda_2 I_n.$$

For $\tilde{u}$ above, Lemma 6.17 gives an L^1 control of its Hessian $D^2 \tilde{u}$ in $S_{\tilde{u}}(0, 1/2)$. In the following basic lemma, we show that there is a definite fraction of this section, suitably localized, where we have L^∞ control on $D^2 \tilde{u}$. This localization will be helpful for later application of Vitali's Covering Lemma which then gives decay estimates for sets of large Hessian.

Lemma 6.19 (Bounding the Hessian in a fraction of a section). *Let u be a C^2 convex function satisfying* (6.18). *Assume that $S_u(0, 2t_0) \Subset \Omega$ satisfies*

$$(6.23) \qquad B_\sigma(0) \subset S_u(0, 2t_0) \subset B_{\sigma^{-1}}(0),$$

where $\sigma = \sigma(n, \Lambda, \Lambda)$ is as in (6.19). *Let $K = K(n, \lambda, \Lambda)$ be the constant in Vitali's Covering Lemma* (Lemma 5.32). *For $0 < t \leq 2t_0$, denote $S_t = S_u(0, t)$. Then, there is a large constant $N_0 = N_0(n, \lambda, \Lambda) > 1$ such that for any $y \in \Omega$ and $t \geq t_0$ where $0 \in \overline{S_u(y, t)} \Subset \Omega$, we have*

$$(6.24) \qquad \int_{S_{t_0}} \|D^2 u\| \, dx$$

$$\leq N_0 \left| \{x \in S_{t_0/K} : N_0^{-1} n^{1/2} I_n \leq D^2 u(x) < N_0 I_n\} \cap S_u(y, t) \right|.$$

If, in addition, $\|f - 1\|_{L^\infty(S_{2t_0})} \le \delta < 1/2$, then

$$(6.25) \quad |S_{t_0} \cap \{\|D^2 u\| \ge N_0\}|$$

$$\le N_0 \delta \left| \{x \in S_{t_0/K} : N_0^{-1} n^{1/2} I_n \le D^2 u(x) < N_0 I_n\} \cap S_u(y, t) \right|.$$

Proof. From (6.23) and the volume estimates for sections in Lemma 5.6, we find $C(n, \lambda, \Lambda)^{-1} \le t_0 \le C(n, \lambda, \Lambda)$. Assume for simplicity that $t_0 = 1$.

Assume now that $t \ge 1$ and $0 \in \overline{S_u(y, t)} \Subset \Omega$. By Theorem 5.31, there is $z \in \Omega$ and $\eta(n, \lambda, \Lambda) > 0$ such that $S_u(z, \eta/K) \subset S_{1/K} \cap S_u(y, t)$. Again, by the volume estimates for sections, we obtain

$$(6.26) \quad |S_{1/K} \cap S_u(y, t)| \ge |S_u(z, \eta/K)| \ge c_1(n, \lambda, \Lambda) > 0.$$

Because $S_u(0, 2)$ satisfies (6.23), we can deduce from Lemma 5.6(ii) that

$$\operatorname{dist}(S_u(0, 1), \partial S_u(0, 2)) \ge c(n, \lambda, \Lambda) > 0.$$

Thus, from Lemma 6.17, $S_1 \subset B_{\sigma^{-1}}(0)$, and Lemma 2.71, we have

$$(6.27) \quad \int_{S_1} \|D^2 u\| \, dx \le C(n, \lambda, \Lambda) \mathcal{H}^{n-1}(\partial S_1) \le C_1(n, \lambda, \Lambda).$$

Next, we show that there is a large constant $N(n, \lambda, \Lambda) > 0$ such that

$$(6.28) \quad \left| \{x \in S_{1/K} \cap S_u(y, t) : D^2 u(x) > 2N^{-1} I_n\} \right| \ge \frac{1}{N^{n-1}} \int_{S_1} \|D^2 u\| \, dx.$$

To prove (6.28), let us consider (see Figure 6.2)

$$E_N := \{x \in S_{1/K} \cap S_u(y, t) : D^2 u(x) > 2N^{-1} I_n\}, \quad \tilde{E} = S_{1/K} \cap S_u(y, t) \setminus E_N.$$

If $y \in \tilde{E}$, then one eigenvalue of $D^2 u(y)$ is less than or equal to $2N^{-1}$. This and $\det D^2 u(y) \ge \lambda$ imply that $\|D^2 u(y)\| \ge C_2(n, \lambda) N^{\frac{1}{n-1}}$. Therefore,

$$(6.29) \quad \|D^2 u\| \ge C_2 N^{\frac{1}{n-1}} \quad \text{in } \tilde{E}.$$

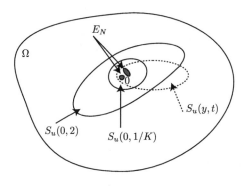

Figure 6.2. The set E_N.

Hence, using (6.27), we have

$$(6.30) \qquad C_1 \geq \int_{S_{1/K} \cap S_u(y,t)} \|D^2 u\| \, dx \geq \int_{\tilde{E}} \|D^2 u\| \, dx \geq C_2 N^{\frac{1}{n-1}} |\tilde{E}|.$$

This is a Markov-type estimate. By choosing $N_1(n, \lambda, \Lambda)$ large, we get $|\tilde{E}| \leq c_1/2$ for all $N \geq N_1$. Then, by (6.26), we have

$$(6.31) \qquad\qquad |E_N| \geq |S_{1/K} \cap S_u(y,t)| - |\tilde{E}| \geq c_1/2.$$

By (6.27), we see that (6.28) holds when N is large.

At $x \in E_N$, $D^2 u(x) > 2N^{-1} I_n$, so the inequality $\det D^2 u(x) \leq \Lambda$ gives $D^2 u(x) < C_3 N^{n-1} I_n$ for some $C_3(n, \Lambda) > n$. This shows that

$$(6.32) \qquad E_N \subset \{x \in S_{t_0/K} \cap S_u(y,t) : N_0^{-1} n^{1/2} I_n < D^2 u(x) < N_0 I_n\}$$

if we choose $N_0 \geq C_3 N^{n-1}$. Now, we find from (6.28) and (6.32) that

$$\int_{S_1} \|D^2 u\| \, dx \leq N^{n-1} |E_N|$$
$$\leq N_0 \left| \{x \in S_{t_0/K} \cap S_u(y,t) : N_0^{-1} n^{1/2} I_n < D^2 u(x) < N_0 I_n\} \right|.$$

Therefore, (6.24) holds.

It remains to prove (6.25) for $\|f - 1\|_{L^\infty(S_2)} \leq \delta < 1/2$. Here $\lambda = 1/2$, and $\Lambda = 3/2$. By the density estimate (see Lemma 6.9 with $\alpha = 1/2$ and Remark 6.10), there is $C_4(n) > 0$ such that

$$|S_1 \cap \{\|D^2 u\| \geq C_4\}| \leq C_4 \delta \leq 2\delta C_4/c_1 |E_N|,$$

where we also recall (6.31). The estimate (6.25) follows from the above inequality and from (6.32) if we choose $N_0 \geq \max\{C_4, 2C_4/c_1, C_3 N^{n-1}\}$. $\quad\square$

By rescaling Lemma 6.19, we obtain localized Hessian estimates on compactly contained sections with arbitrary heights.

Lemma 6.20. *Let u be a C^2, convex function satisfying (6.18). Let K, N_0 be as in Lemma 6.19. Let $S_u(x_0, h) \Subset \Omega$ and assume that $x_0 \in \overline{S_u(y, t/2)} \Subset \Omega$ for some $y \in \Omega$ and $t \geq h$. If $S_u(x_0, h)$ has normalized size $\boldsymbol{\alpha}$, then*

$$\int_{S_u(x_0, h/2)} \|D^2 u\| \, dx$$
$$\leq N_0 \boldsymbol{\alpha} \left| \{N_0^{-1} \boldsymbol{\alpha} < \|D^2 u\| < N_0 \boldsymbol{\alpha}\} \cap S_u(x_0, h/(2K)) \cap S_u(y, t/2) \right|.$$

If, in addition, $\|f - 1\|_{L^\infty(S_u(x_0, h))} \leq \delta < 1/2$, then

$$|S_u(x_0, h/2) \cap \{\|D^2 u\| \geq N_0 \boldsymbol{\alpha}\}|$$
$$\leq N_0 \delta \left| \{N_0^{-1} \boldsymbol{\alpha} < \|D^2 u\| < N_0 \boldsymbol{\alpha}\} \cap S_u(x_0, h/(2K)) \cap S_u(y, t/2) \right|.$$

Proof. The lemma follows by applying Lemma 6.19 with $t_0 = 1/2$ to the rescaling $\tilde{u}$ defined in (6.20). Indeed, by (6.21), $\|D^2 u(x)\| \leq \boldsymbol{\alpha}\|D^2\tilde{u}(\tilde{x})\|$, where $\tilde{x} = Tx$. Hence, integrating over $S_u(x_0, h/2)$ gives

$$|\det T| \int_{S_u(x_0,h/2)} \|D^2 u\| \, dx \leq \boldsymbol{\alpha} \int_{S_{\tilde{u}}(0,1/2)} \|D^2\tilde{u}\| \, d\tilde{x}.$$

Also, by (6.22), we obtain

$$\left\{ N_0^{-1} n^{1/2} I_n < D^2\tilde{u} < N_0 I_n \right\} \subset T\left(\left\{ N_0^{-1}\boldsymbol{\alpha} < \|D^2 u\| < N_0\boldsymbol{\alpha} \right\} \right),$$

which together with

$$S_{\tilde{u}}(0, 1/(2K)) \cap S_{\tilde{u}}(\tilde{y}, t/(2h)) = T(S_u(x_0, h/(2K)) \cap S_u(y, t/2)),$$

implies that

$$\left| \left\{ N_0^{-1} n^{1/2} I_n < D^2\tilde{u} < N_0 I_n \right\} \cap S_{\tilde{u}}(0, 1/(2K)) \cap S_{\tilde{u}}(\tilde{y}, t/(2h)) \right|$$

is bounded from above by

$$|\det T| \left| \left\{ N_0^{-1}\boldsymbol{\alpha} < \|D^2 u\| < N_0\boldsymbol{\alpha} \right\} \cap S_u(x_0, h/(2K)) \cap S_u(y, t/2) \right|.$$

The first conclusion now follows by applying Lemma 6.19 to $\tilde{u}$:

$$\int_{S_{\tilde{u}}(0,1/2)} \|D^2\tilde{u}\| \, d\tilde{x}$$
$$\leq N_0 \left| \left\{ N_0^{-1} n^{1/2} I_n < D^2\tilde{u} < N_0 I_n \right\} \cap S_{\tilde{u}}(0, 1/(2K)) \cap S_{\tilde{u}}(\tilde{y}, t/(2h)) \right|.$$

The second conclusion of the lemma follows similarly. $\qquad \square$

With the same assumptions as in the proof of Lemma 6.19 including the assumption that $t_0 = 1$, we define the following sets. For a large constant $M > 0$ to be determined later, define

$$(6.33) \qquad E_k = E_{k,M} = \{ x \in \overline{S_1} : \|D^2 u(x)\| \geq M^k \} \quad \text{for } k \in \mathbb{N}.$$

Then

$$E_k \setminus E_{k+1} = \{ x \in \overline{S_1} : M^k \leq \|D^2 u(x)\| < M^{k+1} \}.$$

These sets help estimating the L^p norm of the Hessian $D^2 u$ in S_1 as follows.

Lemma 6.21. *With $E_{k,M}$ defined above, we have for all $p > 1$*

$$\int_{S_1} \|D^2 u\|^p \, dx \leq M^p |S_1| + pM^p \sum_{k=1}^{\infty} M^{kp} |E_{k,M}|.$$

Proof. Let $\mu(t) := |S_1 \cap \{\|D^2 u\| \geq t\}|$. By the classical layer-cake formula (Lemma 2.75), we have

$$\int_{S_1} \|D^2 u\|^p \, dx = p \int_0^M t^{p-1} \mu(t) dt + \sum_{k=1}^{\infty} p \int_{M^k}^{M^{k+1}} t^{p-1} \mu(t) \, dt.$$

The result follows from the facts that $\mu(t) \leq |S_1|$ for $0 \leq t \leq M$, while $t^{p-1}\mu(t) \leq M^{(k+1)(p-1)}|E_{k,M}|$ for $M^k \leq t \leq M^{k+1}$. $\qquad\square$

In view of Lemma 6.21, to study the L^p integrability of D^2u, we need to study the decay of $|E_{k,M}|$ for k large. We start with the following decay estimate of the integral of $\|D^2u\|$ over $E_{k+1,M}$, obtained with the aid of Vitali's Covering Lemma and a variant of Widman's hole-filling technique.

Lemma 6.22. *Let u be a C^2, convex function satisfying* (6.18). *Assume that $u(0) = 0, Du(0) = 0$, and $S_2 \equiv S_u(0,2) \Subset \Omega$ satisfies for $\sigma(n,\lambda,\Lambda) \in (0,1)$*

$$B_\sigma(0) \subset S_2 \subset B_{\sigma^{-1}}(0).$$

Let $N_0(n,\lambda,\Lambda)$ be as in Lemma 6.19. Then, there is a large constant $M = M(n,\lambda,\Lambda)$ such that the sets $E_k = E_{k,M}$ defined in (6.33) *satisfy*

$$(6.34)\qquad \int_{E_{k+1}} \|D^2u\|\, dx \leq \tau \int_{E_k} \|D^2u\|\, dx \quad \text{for } k \geq 1 \quad \text{where } \tau = \frac{N_0^2}{1 + N_0^2}.$$

If, in addition, $\|f - 1\|_{L^\infty(S_2)} \leq \delta < 1/2$, then

$$(6.35)\qquad\qquad |E_{k+1}| \leq N_0^2 \delta |E_k| \quad \text{for } k \geq 1.$$

Proof. We can choose $M = N_1^2$ where $N_1(n,\lambda,\Lambda) \geq N_0$ is to be selected.

Let K be as in Lemma 6.19. For each $y \in E_{k+1}$, consider a section $S_u(y, h(y))$ of normalized size $N_0 M^k$ which is compactly contained in S_2. This is possible since for $h \to 0$, the normalized size of $S_u(u,h)$ is comparable to $\|D^2u(y)\|$ (recall Lemma 5.6(iii) and that $u \in C^2$) which is greater than $M^{k+1} > N_0 M^k$, whereas if $h = c_0 2^{-p_1}$ (c_0, p_1 being constants in Theorem 5.30), the normalized size of $S_u(y,h)$ is bounded from above by a constant $C(n,\lambda,\Lambda)$ and therefore by $N_0 M^k$ for

$$N_1 = \max\{N_0, C\}.$$

The family $\{S_u(y, h(y)/2)\}_{y \in E_{k+1}}$ covers E_{k+1}. By Vitali's Covering Lemma (Lemma 5.32), there is a sequence $\{y_i\}_{i=1}^\infty \subset E_{k+1}$ such that $E_{k+1} \subset \bigcup S_i^*$ and the sections S_i' are disjoint, where $S_i^* = S_u(y_i, h(y_i)/2)$ and $S_i' = S_u(y_i, h(y_i)/(2K))$.

Let us first prove (6.34). By Lemma 6.20,

$$\int_{S_i^*} \|D^2u\|\, dx \leq N_0^2 M^k \left|\left\{ M^k \leq \|D^2u\| < N_0^2 M^k \right\} \cap S_i' \cap S_1\right|.$$

Since $E_{k+1} \subset \bigcup_{i=1}^\infty S_i^*$ where S_i' are mutually disjoint, summing over i yields

$$\int_{E_{k+1}} \|D^2u\|\, dx \leq N_0^2 M^k |\{x \in S_1 : M^k \leq \|D^2u(x)\| < M^{k+1}\}|$$

$$(6.36)\qquad\qquad\qquad \leq N_0^2 \int_{E_k \setminus E_{k+1}} \|D^2u\|\, dx.$$

Adding $N_0^2 \int_{E_{k+1}} \|D^2 u\| \, dx$ to both sides of (6.36), we obtain (6.34).

If, in addition, $\|f - 1\|_{L^\infty(S_2)} \leq \delta < 1/2$, then by Lemma 6.20, we have

$$|S_i^* \cap \{\|D^2 u\| \geq N_0^2 M^k\}| \leq N_0^2 \delta \Big| \{M^k < \|D^2 u\| < N_0^2 M^k\} \cap S_i' \cap S_1 \Big|.$$

Since $E_{k+1} \subset \bigcup_{i=1}^\infty S_i^*$ where S_i' are mutually disjoint, summing over i gives

$$|E_{k+1}| = |E_{k+1} \cap \{\|D^2 u\| \geq N_0^2 M^k\}|$$
$$\leq N_0^2 \delta \Big| \{x \in S_1 : M^k \leq \|D^2 u(x)\| < M^{k+1}\} \Big| \leq N_0^2 \delta |E_k|.$$

This proves (6.35), completing the proof of the lemma. $\qquad\square$

Remark 6.23. The conclusions of Lemmas 6.19 and 6.20 obviously hold without localizing the right-hand side of the estimates in the sections with larger height $S_u(y, t)$. However, without this localization, we encounter an obstacle in obtaining the estimate (6.36). The reason is that the sections S_i' can escape the section S_1.

Now, we prove the crucial decay estimates for $|E_{k,M}|$.

Lemma 6.24. *Let N_0 be as in Lemma 6.19, let u, M be as in Lemma 6.22, and let $E_k = E_{k,M}$ be as in (6.33). Then, there exist $\varepsilon_0(n, \lambda, \Lambda) > 0$ and $C(n, \lambda, \Lambda) > 0$ such that*

$$(6.37) \qquad\qquad |E_k| < C M^{-k(1+\varepsilon_0)}.$$

If, in addition, $\|f - 1\|_{L^\infty(S_2)} \leq \delta < 1/2$, then for any $p_0 > 1$,

$$(6.38) \quad |E_{k,M_*}| \leq C(n) M_*^{-(k-1)(p_0+1)} \quad \text{provided that } \delta \leq (M_*^{p_0+1} N_*^2)^{-1},$$

where $M_* = M(n, 1/2, 3/2)$ and $N_* = N_0(n, 1/2, 3/2)$.

Proof. Let $\tau \in (0, 1)$ be as in Lemma 6.22. Then (6.34) and (6.27) give

$$\int_{E_k} \|D^2 u\| \, dx \leq \tau^{k-1} \int_{E_1} \|D^2 u\| \, dx \leq C_1 \tau^{k-1} \leq C \tau^k,$$

for some $C = C(n, \lambda, \Lambda)$. It follows (again, a Markov-type estimate) that

$$|E_k| = \frac{1}{M^k} \int_{E_k} M^k \, dx \leq \frac{1}{M^k} \int_{E_k} \|D^2 u\| \, dx \leq C \frac{\tau^k}{M^k} = \frac{C}{M^{k(1+\varepsilon_0)}},$$

where

$$\varepsilon_0 := \log \tau^{-1} / \log M.$$

If $\|f - 1\|_{L^\infty(S_2)} \leq \delta$, then from (6.35) we obtain for any $p_0 > 1$

$$|E_{k,M_*}| \leq C(n) (N_0^2 \delta)^{k-1} \leq C(n) M_*^{-(k-1)(p_0+1)},$$

provided that $\delta \leq (M_*^{p_0+1} N_*^2)^{-1}$ where $M_* = M(n, 1/2, 3/2)$ and $N_* = N_0(n, 1/2, 3/2)$. $\qquad\square$

We are finally ready to prove the $W^{2,1+\varepsilon}$ and $W^{2,p}$ estimates. As in Remark 6.16, we can assume that $\Omega = S_u(0,2) \equiv S_2$.

Proof of Theorem 6.11. Let M and $\varepsilon_0 > 0$ be as in Lemma 6.24. Let $E_k = E_{k,M}$ be as in (6.33). If $\varepsilon = \varepsilon_0/2$ and $p = 1 + \varepsilon$, then from (6.37) and Lemma 6.21, we have

$$\int_{S_1} \|D^2 u\|^{1+\varepsilon}\, dx \leq M^p |S_1| + pM^p \sum_{k=1}^{\infty} M^{kp}|E_k|$$

$$\leq M^p|S_1| + CpM^p \sum_{k=1}^{\infty} M^{k(p-1-\varepsilon_0)} \leq C(n,\lambda,\Lambda).$$

The theorem is proved. $\square$

Proof of Theorem 6.13. Let M_* and N_* be as in Lemma 6.24. For any $p > 1$, we choose $p_0 = p$ and $\delta = (M_*^{p_0+1} N_*^2)^{-1}$. Then $\delta = e^{-C(n)p}$ for some large $C(n) > 1$. Let E_{k,M_*} be as in (6.33).

If $\|f - 1\|_{L^\infty(\Omega)} \leq \delta$, then from (6.38) and Lemma 6.21, we find that

$$\int_{S_1} \|D^2 u\|^p\, dx \leq M_*^p|S_1| + C(n)pM_*^{p+p_0+1} \sum_{k=1}^{\infty} M_*^{k(p-p_0-1)} \leq C(n,p),$$

which proves the theorem. $\square$

In the proof of Theorem 6.13 which uses affine rescaling, we see that the condition $\|f - 1\|_{L^\infty(\Omega)} \leq \delta$ can be relaxed to its *integral average version at all scales*. Thus, we have the following theorem, due to Huang [**Hu2**].

Theorem 6.25. *Let Ω be a convex domain in $\mathbb{R}^n$ satisfying $B_1(0) \subset \Omega \subset B_n(0)$. Let f be a nonnegative, bounded function on Ω. Let $u \in C(\overline{\Omega})$ be the convex Aleksandrov solution to the Monge–Ampère equation*

$$\det D^2 u = f \quad in\ \Omega, \quad u = 0 \quad on\ \partial\Omega.$$

Then, for any $p > 1$, there exists $\delta = e^{-C(n)p}$, where $C(n)$ is large, such that

$$\int_{\Omega'} \|D^2 u\|^p\, dx \leq C(n,p), \quad where\ \Omega' := \{x \in \Omega : u(x) < -\|u\|_{L^\infty(\Omega)}/2\},$$

provided that

$$\frac{1}{|S|} \int_S |f - 1|\, dx \leq \delta \quad for\ all\ convex\ subsets\ S \subset \Omega.$$

The $W^{2,1+\varepsilon}$ regularity for Aleksandrov solutions to

$$(6.39) \qquad\qquad \lambda \leq \det D^2 \leq \Lambda \quad in\ a\ domain\ \Omega \subset \mathbb{R}^n$$

in general requires their strict convexity. This is the case for the solution in Theorem 6.11 due to the zero boundary value and the strict convexity result in Theorem 5.13. In [**Mn2**], Mooney investigates the Sobolev regularity of singular solutions to (6.39). First, he shows that any Aleksandrov solution to the Monge–Ampère inequality $\det D^2 u \geq 1$ in $B_1(0) \subset \mathbb{R}^n$ is strictly convex away from a singular set Σ (where u is not strictly convex) with $\mathcal{H}^{n-1}(\Sigma) = 0$. By constructing explicit examples, Mooney shows that the above Hausdorff dimension estimate for the singular set is optimal. As applications, Mooney uses this result to prove the remarkable $W^{2,1}$ regularity for (6.39) without requiring any strict convexity and unique continuation of singular solutions to the Monge–Ampère equation. In terms of integrability of the Hessian of singular solutions to (6.39), the L^1 regularity cannot be replaced by $L^{1+\varepsilon}$ for any $\varepsilon > 0$. However, in [**Mn3**], Mooney shows that the L^1 estimates on $D^2 u$ can be improved to $L \log^\varepsilon L$ for some $\varepsilon > 0$.

6.4.3. Rescaling and second derivative estimates.
The second derivative estimates in Theorems 6.11 and 6.13 were established for normalized convex domains. By rescaling with the help of John's Lemma as in Corollary 5.22, we can extend these estimates to all compactly contained sections.

Corollary 6.26. *Let u be a strictly convex function satisfying, in the sense of Aleksandrov, $\lambda \leq \det D^2 u \leq \Lambda$ in an open set Ω in $\mathbb{R}^n$. Let $\varepsilon(n, \lambda, \Lambda) > 0$ be as in the $W^{2,1+\varepsilon}$ estimates in Theorem 6.11, and let $\alpha_2 = n - 1 - n/[2(1 + \varepsilon)]$. If $S_u(x_0, 2h) \Subset \Omega$, then for all $\delta \in (1/4, 3/2)$, we have*

$$\|\Delta u\|_{L^{1+\varepsilon}(S_u(x_0, \delta h))} \leq C(\lambda, \Lambda, n, \delta, \operatorname{diam}(S_u(x_0, 2h))) h^{-\alpha_2}.$$

Proof. As in Remark 6.16, we prove the corollary for $\delta = 1/2$, the other cases being similar.

By John's Lemma, we can find a positive definite matrix $A \in \mathbb{S}^n$ and $b \in \mathbb{R}^n$ such that $B_1(b) \subset \tilde{S} := A^{-1}(S_u(x_0, h)) \subset B_n(b)$. Introduce the rescaled function $\tilde{u}$ on $\tilde{S}$ defined by

$$\tilde{u}(x) := (\det A)^{-2/n} \left(u(Ax) - [u(x_0) + Du(x_0) \cdot (Ax - x_0) + h] \right).$$

Recall the estimates (5.4) and (5.6) in the proof of Lemma 5.6 for A:

(6.40) $$C(n, \lambda, \Lambda)^{-1} h^{n/2} \leq \det A \leq C(n, \lambda, \Lambda) h^{n/2}$$

and

(6.41) $$\|A^{-1}\| \leq C(n, \lambda, \Lambda, \operatorname{diam}(S_u(x_0, 2h))) h^{-\frac{n}{2}}.$$

Note that $\tilde{S} = S_{\tilde{u}}(A^{-1} x_0, h(\det A)^{-2/n})$ and

$$\tilde{S}' := S_{\tilde{u}}(A^{-1} x_0, h(\det A)^{-2/n}/2) = A^{-1} S_u(x_0, h/2).$$

We have $D^2 u = (\det A)^{\frac{2}{n}} (A^{-1})^t (D^2 \tilde{u} \circ A^{-1}) A^{-1}$ on $S_u(x_0, h)$. Using the inequality $\operatorname{trace}(AB) \leq \|A\| \operatorname{trace}(B)$ for nonnegative definite matrices A, B, we thus have

$$\Delta u \leq \|A^{-1}\|^2 (\det A)^{\frac{2}{n}} \Delta \tilde{u} \circ A^{-1}.$$

By the $W^{2,1+\varepsilon}$ estimate in Theorem 6.11 applied to $\tilde{u}$ and its normalized section $\tilde{S} = A^{-1}(S_u(x_0, h))$, we have for some $\varepsilon(n, \lambda, \Lambda) > 0$

$$\|\Delta \tilde{u}\|_{L^{1+\varepsilon}(\tilde{S}')} \leq C(n, \lambda, \Lambda).$$

Using (6.40) and (6.41), we find that

$$\|\Delta u\|_{L^{1+\varepsilon}(S_u(x_0, h/2))} \leq \|A^{-1}\|^2 (\det A)^{\frac{2}{n}} \|\Delta \tilde{u} \circ A^{-1}\|_{L^{1+\varepsilon}(S_u(x_0, h/2))}$$

$$= \|A^{-1}\|^2 (\det A)^{\frac{2}{n} + \frac{1}{1+\varepsilon}} \|\Delta \tilde{u}\|_{L^{1+\varepsilon}(\tilde{S}')}$$

$$\leq C h^{-n} h^{1 + \frac{n}{2}\frac{1}{1+\varepsilon}} \leq C h^{-\alpha_2},$$

for $C(\lambda, \Lambda, n, \text{diam}(S_u(x_0, 2h))) > 0$ and $\alpha_2 = n - 1 - n/[2(1+\varepsilon)] > 0$. $\quad\square$

Remark 6.27. The estimate in Corollary 6.26 appears to be counterintuitive and contradicts Theorem 6.11, because it seems to indicate that the $L^{1+\varepsilon}$ norm of $D^2 u$ on $S_u(x_0, h)$ blows up as $h \to 0$. This estimate is sharp and the appearance of the negative power $h^{-\alpha_2}$ is due to the affine invariance of the Monge–Ampère equation. We can look at the following example. Let us consider for $h > 0$ small the convex function

$$u(x) = \frac{h x_1^2}{2} + \cdots + \frac{h x_{n-1}^2}{2} + \frac{x_n^2}{2 h^{n-1}} \quad \text{in } \mathbb{R}^n.$$

Then $\det D^2 u = 1$. The section $S_u(0, h) \Subset B_2(0)$ is equivalent to an ellipsoid with length of major axes comparable to $1, \ldots, 1, h^{n/2}$. Since $h^{-(n-1)} \leq \|D^2 u\| \leq 2 h^{-(n-1)}$ and $\Delta u \geq \|D^2 u\|$, we find that for all $\varepsilon > 0$

$$\|\Delta u\|_{L^{1+\varepsilon}(S_u(x_0, h))} \geq c(n) h^{-\alpha} \quad \text{where } \alpha = n - 1 - \frac{n}{2}\frac{1}{1+\varepsilon}.$$

The negative power in $h^{-\alpha}$ reflects the degeneracy of the Monge–Ampère equation on $\mathbb{R}^n$. Note also that when h is small, any section $S_u(0, t)$ where t is fixed is not normalized.

6.5. Counterexamples of Wang and Mooney

Wang's counterexample to the interior $W^{2,p}$ estimates. First, we discuss Wang's counterexample to the interior $W^{2,p}$ estimates in [**W3**].

Proposition 6.28 (Counterexample to the interior $W^{2,p}$ estimates). *For $\alpha > 2$, consider the following function in $\mathbb{R}^2$:*

$$u(x) = \begin{cases} |x_1|^\alpha + \dfrac{\alpha^2 - 1}{\alpha(\alpha - 2)} x_2^2 |x_1|^{2-\alpha} & \text{if } |x_2| \leq |x_1|^{\alpha-1}, \\[2ex] \dfrac{1}{2\alpha} x_1^2 |x_2|^{\frac{\alpha-2}{\alpha-1}} + \dfrac{4\alpha - 5}{2(\alpha - 2)} |x_2|^{\frac{\alpha}{\alpha-1}} & \text{if } |x_2| \geq |x_1|^{\alpha-1}. \end{cases}$$

Then

$$0 < \lambda(\alpha) \leq \det D^2 u \leq \Lambda(\alpha) \quad \text{on } \mathbb{R}^2,$$

in the sense of Aleksandrov. Moreover,

$$u \notin W^{2,p}_{\text{loc}} \quad \text{for } p \geq \frac{\alpha}{\alpha - 2} \quad \left(\text{or, equivalently, } \alpha \geq \frac{2p}{p-1}\right).$$

This example shows that the results in Theorems 6.11 and 6.13 and Corollary 6.14 are optimal: The constant $\varepsilon > 0$ in Theorem 6.11 clearly depends on n, λ, and Λ, and it can be as small as we wish provided Λ/λ is large; moreover, for the higher integrability of the Hessian in Theorem 6.13 and Corollary 6.14, in general we need the small oscillation of the Monge–Ampère measure. We also note that in Corollary 6.14(iii), p cannot be taken to be ∞ in general. This is due to another example of Wang: There exists a convex function, with unbounded Hessian, around the origin in $\mathbb{R}^2$ with strictly positive, continuous Monge–Ampère measure; see Problem 6.9.

Now, we briefly explain Wang's construction. We look for a convex function u on $\mathbb{R}^2$ with Monge–Ampère measure bounded between two positive constants and such that it has the scaling

$$u(tx_1, t^\beta x_2) = t^\alpha u(x_1, x_2) \quad \text{for all } t > 0,$$

where $\beta, \alpha > 0$. Since taking the Hessian determinant on both sides gives

$$t^{2+2\beta}(\det D^2 u)(tx_1, t^\beta x_2) = t^{2\alpha} \det D^2 u(x_1, x_2)$$

and $\det D^2 u$ is comparable to 1, the exponents β and α must satisfy $\alpha = \beta + 1$. Thus $u(tx_1, t^{\alpha-1}x_2) = t^\alpha u(x_1, x_2)$. To simplify the discussion, we consider $x_1 > 0, x_2 > 0$. For $x_1 \neq 0$ and $t = x_1^{-1}$, we have

$$u(x_1, x_2) = t^{-\alpha}u(tx_1, t^{\alpha-1}x_2) = x_1^\alpha u(1, x_1^{1-\alpha}x_2).$$

Let $s = x_1^{1-\alpha}x_2$ and $f(s) = u(1, s)$. Then

$$(6.42) \quad \det D^2 u(x_1, x_2) = (\alpha - 1)f''(s)[\alpha f(s) - (\alpha - 2)sf'(s)] - [f'(s)]^2.$$

We construct a C^1 function $f(s)$ of the form $f(s) = 1 + At^2$ in the region $0 \leq s \leq 1$. In this region, by (6.42), we have

$$(6.43) \quad \det D^2 u = \alpha(\alpha - 1)2A - (\alpha - 2)(\alpha - 3)2A^2 s^2.$$

In the region $x_2 \leq x_1^{\alpha-1}$, u has the form $x_1^\alpha(1 + Ax_2^2 x_1^{-2(\alpha-1)})$. By symmetry, in the region $x_2 \geq x_1^{\alpha-1}$, or $x_1 \leq x_2^{\frac{1}{\alpha-1}}$, we look for u having the form

$$x_2^{\frac{1}{\alpha-1}+1}(B + BDx_1^2 x_2^{\frac{-2}{\alpha-1}}) = Bx_2^\gamma(1 + D\rho^2)$$

where

$$\gamma = \frac{\alpha}{\alpha - 1}, \quad \rho = x_1 x_2^{-\frac{1}{\alpha-1}} = s^{-\frac{1}{\alpha-1}}.$$

In the region $0 \leq \rho \leq 1$, by (6.42), we have

$$(6.44) \quad \det D^2 u = B^2[\gamma(\gamma - 1)2D - (\gamma - 2)(\gamma - 3)2D^2\rho^2].$$

Note that $u(x_1, x_2) = x_1^\alpha f(s)$ where

$$f(s) = \begin{cases} 1 + As^2 & \text{if } 0 \le s \le 1, \\ BDs^{\frac{\alpha-2}{\alpha-1}} + Bs^{\frac{\alpha}{\alpha-1}} & \text{if } s \ge 1. \end{cases}$$

The continuity of u and Du across $x_2 = x_1^{\alpha-1}$, or $s = \rho = 1$, requires

$$1 + A = B + BD, \quad 2A = BD\frac{\alpha-2}{\alpha-1} + B\frac{\alpha}{\alpha-1}.$$

Solving this system yields

$$B = \frac{1+A}{1+D} = \frac{2A}{D\frac{\alpha-2}{\alpha-1} + \frac{\alpha}{\alpha-1}}, \quad D = \frac{\alpha - (\alpha-2)A}{\alpha A - (\alpha-2)}.$$

We will restrict ourselves to $\alpha > 2$ and

(6.45) $$\frac{\alpha-2}{\alpha} < A < \frac{\alpha}{\alpha-2}.$$

For $\det D^2 u$ to be bounded between two positive constants, we need from (6.43) and (6.44) that

$$\alpha(\alpha-1) - (\alpha-2)(\alpha-3)A > 0$$

and

$$\gamma(\gamma-1) - (\gamma-2)(\gamma-3)D = \frac{\alpha - D(\alpha-2)(2\alpha-3)}{(\alpha-1)^2} > 0.$$

The last inequality and the relation between A and D force

$$A > \frac{2\alpha(\alpha-1)(\alpha-2)}{\alpha^2 + (\alpha-2)^2(2\alpha-3)}.$$

Choosing one appropriate value of A leads to corresponding values of D and B. We can take

$$A = \frac{\alpha^2-1}{\alpha(\alpha-2)}, \quad D = \frac{\alpha-2}{\alpha(4\alpha-5)}, \quad B = \frac{4\alpha-5}{2(\alpha-2)}.$$

This leads to the function found by Wang.

Next we study the integrability of $D^2 u$ on the first quadrant. In this domain, if $x_2 \ge x_1^{\alpha-1}$, then, we have

$$D_{22}u(x) = \frac{1}{2(\alpha-1)^2} x_2^{\frac{2-\alpha}{\alpha-1}} \left(\frac{\alpha(4\alpha-5)}{\alpha-2} - \frac{\alpha-2}{\alpha} x_1^2 x_2^{-\frac{2}{\alpha-1}} \right) \ge c(\alpha) x_2^{\frac{2-\alpha}{\alpha-1}}$$

for some $c(\alpha) > 0$. For any $r > 0$ and $p > 0$, we have

$$\int_{B_r(0)} |D_{22}u|^p \, dx \ge c(\alpha, p) \int_0^r \left(\int_{x_1^{\alpha-1}}^r x_2^{\frac{p(2-\alpha)}{\alpha-1}} \, dx_2 \right) dx_1$$

$$= \frac{c(\alpha, p)}{\frac{(2-\alpha)p}{\alpha-1} + 1} \int_0^r \left(r^{\frac{(2-\alpha)p}{\alpha-1}+1} - x_1^{p(2-\alpha)+\alpha-1} \right) dx_1 = +\infty$$

if $p(2-\alpha) + \alpha - 1 \le -1$, or $p \ge \alpha/(\alpha-2)$.

Mooney's counterexample to the interior $W^{2,1}$ regularity. As alluded to at the beginning of this chapter, the $W^{2,1}$ regularity of solutions to $0 < \lambda \leq \det D^2 u \leq \Lambda$ was used by Ambrosio, Colombo, De Philippis, and Figalli [**ACDF1, ACDF2**] to establish the global existence of weak solutions to the semigeostrophic equations when the initial density is supported in the whole space. In physically interesting cases, the initial density is compactly supported. This naturally leads to an open question raised in De Philippis–Figalli [**DPF2**] and Figalli [**F3**] about the $W^{2,1}$ regularity of convex solutions to the Dirichlet problem for the degenerate Monge–Ampère equation

$$(6.46) \qquad \det D^2 u \leq 1 \quad \text{in } \Omega, \quad u = 0 \quad \text{on } \partial\Omega,$$

where Ω is a bounded convex domain in $\mathbb{R}^n (n \geq 2)$. In [**Mn4**], Mooney answers this open problem negatively. He shows that in all dimensions $n \geq 2$, there exist solutions to (6.46) that are not $W^{2,1}$. More precisely, in dimensions $n \geq 3$, Mooney constructs examples of solutions to (6.46) that have a Lipschitz singularity on a hyperplane. In two dimensions, Mooney delicately constructs a solution to (6.46) whose second derivatives have nontrivial Cantor part. We refer the reader to [**Mn4**] for details when $n = 2$.

Below, we present Mooney's construction of nonnegative Aleksandrov solutions to $\det D^2 u \leq 1$ in $B_4(0)$ in dimensions $n \geq 3$ that have a Lipschitz singularity on a hyperplane. For each such u, by subtracting a suitable constant from u on a suitable sublevel set Ω, we obtain a solution to (6.46) that is not in $W^{2,1}_{\text{loc}}$.

Proposition 6.29 (Counterexample to the interior $W^{2,1}$ regularity in dimensions $n \geq 3$). *Let $n \geq 3$. Then, for any $\alpha \geq n/(n-2)$, there exists a nonnegative convex Aleksandrov solution to $\det D^2 u \leq 1$ in $B_4(0) \subset \mathbb{R}^n$ that is a positive multiple of $|x_n|$ in $\{|x_n| \geq \big(\max\{|x'| - 1, 0\}\big)^\alpha\}$.*

Proof. For $x = (x', x_n) \in \mathbb{R}^n$, we let $r = |x'|$. Let $h(r) := \max\{r - 1, 0\}$. We will construct a convex function $w(x) = v(r, x_n)$ in $\{|x_n| < h(r)^\alpha\}$, with a suitable $\alpha > 1$, that glues to a multiple of $|x_n|$ across the boundary. For this, we look for $v(r, x_n)$ with the scaling

$$v(1 + tr, t^\alpha x_n) = t^\alpha v(1 + r, x_n) \quad \text{for all } t > 0,$$

so that $D_n w$ is invariant under the above rescaling. Let

$$v(r, x_n) = \begin{cases} h(r)^\alpha + h(r)^{-\alpha} x_n^2 & \text{if } |x_n| < h(r)^\alpha, \\ 2|x_n| & \text{if } |x_n| \geq h(r)^\alpha. \end{cases}$$

Then Dw is continuous on $\partial\{|x_n| < h(r)^\alpha\}\backslash\{r = 1, x_n = 0\}$. Moreover,

$$\partial w(\{r = 1, x_n = 0\}) = [-2e_n, 2e_n]$$

which has measure zero. Thus, as in Examples 4.14 and 5.24, to show that $\det D^2 w \leq C$ in the sense of Aleksandrov, we only need to verify it pointwise in each of the regions separated by $\{|x_n| = h(r)^\alpha\}$.

In the region $\{|x_n| \geq h(r)^\alpha\}$, we have $\det D^2 w = 0$. Consider now the region $\{|x_n| < h(r)^\alpha\}$. As in Lemma 2.60, in a suitable coordinate system, the Hessian of w is

$$D^2 w = \begin{pmatrix} 2h(r)^{-\alpha} & 0 & \cdots & 0 & -2\alpha x_n h(r)^{-\alpha-1} \\ 0 & M_1/r & 0 & 0 & 0 \\ \vdots & 0 & \cdots & \cdots & 0 \\ 0 & \cdots & \cdots & M_1/r & \cdots \\ -2\alpha x_n h(r)^{-\alpha-1} & 0 & \cdots & 0 & M_2 \end{pmatrix},$$

where

$$M_1 := \alpha h(r)^{\alpha-1}\Big(1 - \frac{x_n^2}{h(r)^{2\alpha}}\Big), \quad M_2 := \alpha h(r)^{\alpha-2}\Big(\alpha - 1 + (\alpha+1)\frac{x_n^2}{h(r)^{2\alpha}}\Big).$$

As such, in $\{|x_n| < h(r)^\alpha\}$, we have

$$\det D^2 w = 2\alpha^{n-1}(\alpha-1)r^{2-n}h(r)^{\alpha(n-2)-n}\Big[1 - \Big(\frac{x_n}{h(r)^\alpha}\Big)^2\Big]^{n-1}.$$

Thus, for $\alpha \geq n/(n-2)$, $\det D^2 w \leq C(n)$ in $B_4(0)$. Now set $u = w/C$ to obtain u satisfying the conclusion of the proposition. $\square$

6.6. Interior Second Derivative Hölder Estimates

In this section, we prove Caffarelli's interior $C^{2,\alpha}$ estimates in [**C3**] for the Monge–Ampère equation when the right-hand side is C^α and bounded between two positive constants.

6.6.1. Pointwise $C^{2,\alpha}$ estimate at an interior point.
Key to the proof of interior $C^{2,\alpha}$ estimates for the Monge–Ampère equation is the following pointwise $C^{2,\alpha}$ estimate.

Theorem 6.30 (Pointwise $C^{2,\alpha}$ estimates for the Monge–Ampère equation).
Let Ω be a bounded convex domain in $\mathbb{R}^n$, satisfying $B_{R^{-1}}(\hat{z}) \subset \Omega \subset B_R(\hat{z})$ for some $\hat{z} \in \mathbb{R}^n$ and $R > 1$. Let $f \in L^\infty(\Omega)$ with $0 < \lambda \leq f \leq \Lambda$. Let $u \in C(\overline{\Omega})$ be the convex Aleksandrov solution to the Monge–Ampère equation

$$\det D^2 u = f \quad in \ \Omega, \quad u = a \ given \ constant \ on \ \partial\Omega.$$

Let $z \in \Omega$ be the minimum point of u in Ω. Assume that for some $\alpha \in (0,1)$ and $M \geq 0$, we have

$$|f(x) - f(z)| \leq M|x - z|^\alpha \quad for \ all \ x \in \Omega.$$

Then, there is a convex, homogeneous quadratic polynomial $\mathcal{P}_z$ with

$$\det D^2 \mathcal{P}_z = f(z), \quad \|D^2 \mathcal{P}_z\| \leq C,$$

where the constant C depends only on $R, n, \lambda, \Lambda, \alpha$, and M, such that

$$|u(x) - u(z) - P_z(x-z)| \leq C|x-z|^{2+\alpha} \quad in \ \Omega.$$

The rest of this section is devoted to the proof of Theorem 6.30 which uses the perturbation arguments of Caffarelli. The presentation below follows the arguments for the global $C^{2,\alpha}$ estimates in Savin [**S4**].

Simplifications. By Theorem 5.21, $u \in C^{1,\beta}(\Omega)$ for some $\beta(n, \lambda, \Lambda) \in (0,1)$. Clearly $Du(z) = 0$. By changing coordinates and replacing u by $u - u(z)$, we can assume $z = 0$ and $u(0) = 0$. Now, $u = \bar{h}$ on $\partial\Omega$ and $\Omega = S_u(0, \bar{h})$. By considering $[f(0)]^{-1/n}u$ instead of u and dividing f, M, λ, Λ by $f(0)$, we can assume that $f(0) = 1$. Since $B_{R^{-1}}(\hat{z}) \subset \Omega = S_u(0, \bar{h}) \subset B_R(\hat{z})$, the volume estimates in Lemma 5.6 give $t_1 \leq \bar{h} \leq t_1^{-1}$, where t_1 depends only on n, λ, Λ, and R. In what follows, we only use a fixed t_1, and in view of John's Lemma, we can also assume $R = n$. Summarizing, we assume

(6.47) $\quad z = Du(0) = 0, R = n, u(0) = 0, f(0) = 1, |f(x) - 1| \leq M|x|^\alpha$ in Ω.

By Lemma 5.6, each compactly contained section $S_u(0, h)$ has volume comparable to $\omega_n h^{n/2}$. Thus, there is a linear transformation A_h with $\det A_h = 1$ that makes the section $S_u(0, h)$ comparable to $B_{h^{1/2}}(0)$.

Reduction to M being small. In the first step, we show that, for an arbitrarily small constant $\sigma > 0$, if h is sufficiently small, then the corresponding rescaling u_h of u using A_h satisfies the hypotheses of Theorem 6.30 in which the constant M is replaced by σ.

Lemma 6.31. *Let u be as in Theorem 6.30 where (6.47) is enforced. Given any $\sigma > 0$, there exist small positive constants $h = h_0(M, \sigma, \alpha, n, \lambda, \Lambda)$, $\kappa(n, \lambda, \Lambda)$, $\mu(n, \lambda, \Lambda)$ and rescalings of u and f*

$$u_h(x) := h^{-1}u(h^{1/2}A_h^{-1}x), \quad f_h(x) := f(h^{1/2}A_h^{-1}x)$$

where $A_h \in \mathbb{S}^n$ is positive definite with

$$\det A_h = 1, \quad \|A_h^{-1}\| \leq \kappa^{-1}h^{\mu - 1/2}, \quad \|A_h\| \leq \kappa^{-1}h^{-1/2},$$

such that

(i) $u_h(0) = 0$, $Du_h(0) = 0$, $B_\kappa(0) \subset S_{u_h}(0, 1) \Subset B_{\kappa^{-1}}(0)$, *and*

(ii) $\det D^2 u_h = f_h$ *and* $|f_h(x) - 1| \leq \sigma|x|^\alpha$ *in* $S_{u_h}(0, 2)$.

Proof. Observe that $\Omega = S_u(0, \bar{h})$, where $2t_0 \leq \bar{h} \leq (2t_0)^{-1}$ for some $t_0(n, \lambda, \Lambda) > 0$. From Lemmas 2.40 and 5.6(ii), we have the Lipschitz bound $|Du| \leq C_0(n, \lambda, \Lambda)$ in $S_u(0, t_0)$. This together with Lemma 5.16 gives

$$B_{ch}(0) \subset S_u(0, h) \subset B_{Ch^\mu}(0) \quad \text{for } h \leq t_0,$$

where $0 < \mu, c(\leq t_0)$, and C depend only on n, λ, and Λ. By Lemma 5.6(iii), for all $h \leq t_0$, there is an ellipsoid $E_h = A_h^{-1} B_{h^{1/2}}(0)$ where $A_h \in \mathbb{S}^n$ is positive definite with $\det A_h = 1$ such that

$$\kappa_0 E_h \subset S_u(0, h) \Subset \kappa_0^{-1} E_h, \quad \kappa_0 = \kappa_0(n, \lambda, \Lambda).$$

From the above inclusions for $S_u(0, h)$, we find

$$\|h^{1/2} A_h^{-1}\| \leq C\kappa_0^{-1} h^\mu \quad \text{and} \quad \|A_h\| \leq c^{-1}\kappa_0^{-1} h^{-1/2}.$$

Let u_h and f_h be as in the statement of the lemma, and let $\kappa = \min\{c\kappa_0, \kappa_0/C\}$. Then, from $S_{u_h}(0,1) = h^{-1/2} A_h S_u(0, h)$, all the conclusions of the lemma have been verified, except property (ii) which we now prove. Indeed, if $x \in S_{u_h}(0, 2) = h^{-1/2} A_h S_u(0, 2h)$, then $\det D^2 u_h(x) = f_h(x) = f(h^{1/2} A_h^{-1} x)$ and

$$|f_h(x) - 1| \leq M|h^{1/2} A_h^{-1} x|^\alpha \leq M(C\kappa_0^{-1} h^\mu)^\alpha |x|^\alpha \leq \sigma |x|^\alpha$$

if $h = h_0(M, \sigma, \alpha, n, \lambda, \Lambda)$ is sufficiently small. $\qquad\square$

Approximation of u_h by a convex, quadratic polynomial. In the next lemma, we show that if σ is sufficiently small, then u_h can be well-approximated by a quadratic polynomial near the origin. This can be accomplished with the aid of Pogorelov estimates.

Lemma 6.32. *For any $\delta_0 \in (0, 1)$, $\varepsilon_0 \in (0, 1)$, there exist positive constants $\mu_0(\varepsilon_0, n, \lambda, \Lambda)$ and $\sigma_0(\delta_0, \varepsilon_0, \alpha, n, \lambda, \Lambda)$ such that for any function u_h satisfying properties (i)–(ii) of Lemma 6.31 with $\sigma \leq \sigma_0$, we can find rescalings*

$$\tilde{u}(x) := \mu_0^{-2} u_h(\mu_0 x), \quad \tilde{f}(x) := f_h(\mu_0 x) \quad \text{in } \tilde{\Omega} := \mu_0^{-1} S_{u_h}(0, 1) \supset B_1(0)$$

that satisfy $\tilde{u}(0) = 0$, $D\tilde{u}(0) = 0$, and

$$\det D^2 \tilde{u} = \tilde{f}, \quad |\tilde{f}(x) - 1| \leq \delta_0 \varepsilon_0 |x|^\alpha, \quad |\tilde{u} - P_0| \leq \varepsilon_0 \quad \text{in } B_1(0),$$

for some convex, homogeneous quadratic polynomial P_0 with

$$\det D^2 P_0 = 1, \quad \|D^2 P_0\| \leq C_0(n, \lambda, \Lambda).$$

Proof. We use a compactness argument. Assume by contradiction that the statement is false for a sequence $\{u_m\}_{m=1}^\infty$ satisfying properties (i)–(ii) of Lemma 6.31 with $\sigma_m \to 0$. Note that $u_m = 1$ on $\partial S_{u_m}(0, 1)$. Then, by Theorem 3.19, we may assume after passing to a subsequence if necessary that

    (a) u_m converges locally uniformly to a convex function u_∞ defined on a set $\Omega_\infty = S_{u_\infty}(0, 1)$ where $B_\kappa(0) \subset \Omega_\infty \Subset B_{\kappa^{-1}}(0)$,

    (b) $\overline{S_{u_m}(0, 1)}$ converges to $\overline{S_{u_\infty}(0, 1)}$ in the Hausdorff distance, and

    (c) $u_\infty(0) = 0, Du_\infty(0) = 0$, and $\det D^2 u_\infty = 1$ in Ω_∞.

From Lemma 6.4 and Theorem 6.5, we can find $c_0(n, \lambda, \Lambda) \leq \kappa$ such that
$$|u_\infty - P_\infty| \leq c_0^{-1}|x|^3 \quad \text{in } B_{c_0}(0),$$
where P_∞ is a convex, homogeneous quadratic polynomial such that
$$\det D^2 P_\infty = 1, \quad \|D^2 P_\infty\| \leq c_0^{-1}.$$
Choose $\mu_0 = \mu_0(\varepsilon_0, n, \lambda, \Lambda) = c_0\varepsilon_0/32$. Then,
$$|u_\infty - P_\infty| \leq \varepsilon_0\mu_0^2/4 \quad \text{in } B_{2\mu_0}(0),$$
which together with (a) implies that for all large m
$$|u_m - P_\infty| \leq \varepsilon_0\mu_0^2/2 \quad \text{in } B_{\mu_0}(0) \subset S_{u_m}(0, 1).$$
Then, for all large m, the functions
$$\tilde{u}_m(x) := \mu_0^{-2}u_m(\mu_0 x), \quad \tilde{f}_m(x) := f_m(\mu_0 x)$$
on $\tilde{\Omega}_m := \mu_0^{-1}S_{u_m}(0, 1)$ satisfy in $B_1(0) \subset \tilde{\Omega}_m$
$$|\tilde{u}_m - P_\infty| \leq \varepsilon_0/2$$
and
$$\det D^2\tilde{u}_m = \tilde{f}_m, \quad \text{with } |\tilde{f}_m(x) - 1| \leq \sigma_m(\mu_0|x|)^\alpha \leq \delta_0\varepsilon_0|x|^\alpha.$$
Then $\tilde{u}_m$, $P_m := P_\infty$ satisfy the conclusion of the lemma for all large m, and we reached a contradiction. □

Second proof of Lemma 6.32 without using compactness. We note that $|f_h - 1| \leq \sigma\kappa^{-\alpha}$ in $S_{u_h}(0, 1)$. We assume $\sigma < \kappa^\alpha/2$. Let v be the convex Aleksandrov solution to
$$\det D^2 v = 1 \quad \text{in } S_{u_h}(0, 1), \quad v = 1 \quad \text{on } \partial S_{u_h}(0, 1).$$
By the comparison principle in Theorem 3.21, we have
$$(1 + \sigma\kappa^{-\alpha})(v - 1) \leq u_h - 1 \leq (1 - \sigma\kappa^{-\alpha})(v - 1) \quad \text{in } S_{u_h}(0, 1),$$
so
$$(6.48) \qquad |u_h - v| \leq \sigma\kappa^{-\alpha}(1 - v) \leq \bar{C}(n, \kappa)\sigma\kappa^{-\alpha} \quad \text{in } S_{u_h}(0, 1),$$
where the Aleksandrov maximum principle was used in the last inequality. Since $u_h(0) = 0$, we have
$$(6.49) \qquad\qquad |v(0)| \leq \bar{C}\sigma\kappa^{-\alpha}.$$
Let z be the minimum point of v in $S_{u_h}(0, 1)$. Then, due to (6.48) and $v(z) - u_h(z) \leq v(z) \leq v(0) - u_h(0)$, it follows that
$$(6.50) \qquad\qquad |v(z)| \leq \bar{C}\sigma\kappa^{-\alpha} \leq 1/2$$
if we further reduce σ.

Since $v = 1$ on $\partial S_{u_h}(0,1)$, the Aleksandrov maximum principle gives

$$\text{dist}(z, \partial S_{u_h}(0,1)) \geq c_0(\kappa, n).$$

Let Ω' be the convex hull of $\{x \in \Omega : \text{dist}(x, \partial S_{u_h}(0,1)) \geq c_0\}$ and $B_{\kappa/2}(0)$. From Lemma 6.4 and Theorem 6.5, we can find $C_1 = C_1(\kappa, n)$ such that

$$(6.51) \qquad \|v\|_{C^3} \leq C_1 \quad \text{and} \quad D^2 v \geq C_1^{-1} I_n \quad \text{in } \Omega'.$$

Using Taylor's expansion and $Dv(z) = 0$, we can find $\zeta \in [z, 0]$ such that

$$v(0) = v(z) + (D^2 v(\zeta)z) \cdot z/2 \geq v(z) + (C_1^{-1}/2)|z|^2.$$

It follows from (6.49)–(6.51) that $|z| \leq 2(C_1 \bar{C} \sigma \kappa^{-\alpha})^{1/2}$, and hence

$$|Dv(0)| = |Dv(0) - Dv(z)| \leq \|D^2 v\|_{L^\infty(\Omega')}|z| \leq C_2(n, \kappa, \alpha)\sigma^{1/2}.$$

From (6.51), we have

$$|v(x) - v(0) - Dv(0) \cdot x - (D^2 v(0)x) \cdot x/2| \leq C_1 |x|^3 \quad \text{in } B_{\kappa/2}(0).$$

Therefore, defining the convex, homogeneous quadratic polynomial

$$P_0(x) := (D^2 v(0)x) \cdot x/2,$$

we have $\det D^2 P_0 = 1$, $\|D^2 P_0\| \leq C_1(n, \kappa) = C_0(n, \lambda, \Lambda)$, and

$$|u_h(x) - P_0(x)| \leq |u_h(x) - v(x)| + C_1|x|^3 + |v(0)| + |Dv(0)||x|$$
$$\leq 2\bar{C}\sigma\kappa^{-\alpha} + C_1|x|^3 + C_2\sigma^{1/2}|x| \quad \text{in } B_{\kappa/2}(0).$$

Now, for $\mu_0 \leq \kappa/2$, let

$$\tilde{u}(x) := \mu_0^{-2} u_h(\mu_0 x), \quad \tilde{f}(x) := f_h(\mu_0 x) \quad \text{in } \tilde{\Omega} := \mu_0^{-1} S_{u_h}(0,1) \supset B_1(0).$$

Then in $B_1(0)$, we have

$$|\tilde{u}(x) - P_0(x)| \leq 2\bar{C}\mu_0^{-2}\sigma\kappa^{-\alpha} + C_1\mu_0 + C_2\mu_0^{-1}\sigma^{1/2}$$

and

$$|\tilde{f}(x) - 1| = |f_h(\mu_0 x) - 1| \leq \sigma|\mu_0 x|^\alpha = \sigma\mu_0^\alpha|x|^\alpha.$$

Thus, given $\delta_0 \in (0,1)$ and $\varepsilon_0 \in (0,1)$, we first choose $\mu_0(\varepsilon_0, n, \lambda, \Lambda)$ small such that $C_1\mu_0 \leq \varepsilon_0/2$ and then choose $\sigma_0(\delta_0, \varepsilon_0, n, \alpha, \lambda, \Lambda)$ small such that

$$2\bar{C}\mu_0^{-2}\sigma_0\kappa^{-\alpha} + C_2\mu_0^{-1}\sigma_0^{1/2} \leq \varepsilon_0/2 \quad \text{and} \quad \sigma_0\mu_0^\alpha \leq \delta_0\varepsilon_0.$$

The lemma is proved. $\qquad\qquad\qquad\qquad\qquad\qquad\qquad\qquad\qquad\qquad\qquad\square$

The perturbation arguments for pointwise $C^{2,\alpha}$ estimates. For u as in Theorem 6.30 where (6.47) is enforced, from Lemmas 6.31 and 6.32, we see that given any $\delta_0, \varepsilon_0 \in (0,1)$, there exists a positive definite matrix

$$T := \mu_0 h_0^{1/2} A_{h_0}^{-1} \in \mathbb{S}^n \quad \text{with } \|T^{-1}\| + \|T\| \le C_*(M, \delta_0, \varepsilon_0, \alpha, n, \lambda, \Lambda),$$

such that the rescalings

$$\tilde{u}(x) := (\det T)^{-2/n} u(Tx), \quad \tilde{f}(x) := f(Tx) \quad \text{in } \tilde{\Omega} := T^{-1}\Omega \supset B_1(0)$$

satisfy

$$\det D^2 \tilde{u} = \tilde{f}, \quad |\tilde{f} - 1| \le \delta_0 \varepsilon_0 |x|^\alpha, \quad \text{and } |\tilde{u} - P_0| \le \varepsilon_0 \qquad \text{in } B_1(0),$$

for some convex, homogeneous quadratic polynomial P_0 with

$$\det D^2 P_0 = 1, \quad \|D^2 P_0\| \le C_0(n, \lambda, \Lambda).$$

By choosing δ_0, ε_0 appropriately small, depending on $\alpha, n, \lambda, \Lambda$, we will show that $\tilde{u}$ is well-approximated of order $2 + \alpha$ by quadratic polynomials $b_m \cdot x + P_m(x)$ in each ball $B_{r_0^m}(0)$ for some small $r_0 > 0$. This proves $\tilde{u} \in C^{2,\alpha}(0)$ and hence Theorem 6.30. The key lies in the following lemma.

Lemma 6.33 (Perturbation arguments). *Let Ω be a convex domain in $\mathbb{R}^n$ such that $\Omega \supset B_1(0)$. Let $u \in C(\Omega)$ be a strictly convex function that satisfies $u(0) = 0$, $Du(0) = 0$, and in $B_1(0)$,*

$$\det D^2 u = f, \quad |f - 1| \le \delta_0 \varepsilon_0 |x|^\alpha, \quad \text{and} \quad |u - P_0| \le \varepsilon_0,$$

where $\varepsilon_0, \delta_0 > 0$ and P_0 is a convex, homogeneous quadratic polynomial with

$$\det D^2 P_0 = 1, \quad \|D^2 P_0\| \le C_0.$$

Then, there exist ε_0, δ_0, r_0 small, depending only on n, α, and C_0, such that for all $m \in \mathbb{N}_0$, we can find a linear function $l_m(x) = b_m \cdot x$ and a convex, homogeneous quadratic polynomial P_m satisfying

(6.52) $$|u - l_m - P_m| \le \varepsilon_0 r_0^{m(2+\alpha)} \quad \text{in } B_{r_0^m}(0)$$

and

(6.53) $$r_0^m |b_{m+1} - b_m| + r_0^{2m} \|D^2 P_{m+1} - D^2 P_m\| \le C_\infty(n, \alpha, C_0) \varepsilon_0 r_0^{m(2+\alpha)},$$

with

$$\det D^2 P_m = 1, \quad |b_m| \le 1, \quad \|D^2 P_m\| \le 2C_0.$$

Proof. We prove the lemma by induction on m. For $m = 0$, we take $b_0 = 0$, while for $m = -1$, we take $b_{-1} = 0$ and $P_{-1}(x) = |x|^2/2$. Suppose that the conclusion of the lemma holds for $m \ge 0$. We prove it for $m + 1$. Let

$$r := r_0^m, \quad \varepsilon := \varepsilon_0 r^\alpha = \varepsilon_0 r_0^{m\alpha}, \quad v(x) := r^{-2}(u - l_m)(rx).$$

Then, in $B_1(0) \subset r^{-1}\Omega$, we have $v(0) = 0$,

(6.54) $$|v - P_m| \le \varepsilon, \quad |\det D^2 v(x) - 1| = |f(rx) - 1| \le \delta_0 \varepsilon_0 r^\alpha = \delta_0 \varepsilon.$$

Let $\lambda_1 \leq \cdots \leq \lambda_n$ be the positive eigenvalues of $D^2 P_m$. Since their product is 1 and $\lambda_n \leq 2C_0$, we find $\lambda_1 \geq (2C_0)^{1-n}$. In $B_1(0)$, we have

$$(2C_0)^{1-n}|x|^2/2 - \varepsilon \leq P_m(x) - \varepsilon \leq v(x) \leq \varepsilon + P_m(x) \leq \varepsilon + C_0|x|^2.$$

Thus, we can find $\varepsilon_0 \leq \varepsilon_1(n, C_0), \tau(n, C_0)$, and $t(n, C_0)$, all positive and small, such that

$$B_{2\tau}(0) \subset S_t := \{x \in B_1(0) : v(x) < t\} \subset B_{1/2}(0).$$

Since v is equal to the constant t on ∂S_t, by Theorem 3.38, there exists a unique convex Aleksandrov solution $w \in C(\overline{S_t})$ to

$$\det D^2 w = 1 \quad \text{in } S_t, \quad w = v \quad \text{on } \partial S_t.$$

It can be estimated that for some $C_2 = C_2(n)$,

(6.55) $$|v - w| \leq C_2 \delta_0 \varepsilon \quad \text{in } S_t.$$

Indeed, we have $v = w = t$ on ∂S_t. Using

$$\det D^2[(1 - \delta_0\varepsilon)(w - t)] = (1 - \delta_0\varepsilon)^n \leq \det D^2 v \leq \det D^2[(1 + \delta_0\varepsilon)(w - t)],$$

we obtain from the comparison principle (Theorem 3.21) that

$$(1 + \delta_0\varepsilon)(w - t) \leq v - t \leq (1 - \delta_0\varepsilon)(w - t) \quad \text{in } S_t.$$

Therefore, combining with the Aleksandrov maximum principle, we obtain

$$|v - w| \leq \delta_0\varepsilon|w - t| \leq C_2\delta_0\varepsilon \quad \text{in } S_t.$$

From (6.54), (6.55) we can conclude that if δ_0 is sufficiently small, then

$$|w - P_m| \leq 2\varepsilon \quad \text{in } B_{2\tau}(0).$$

Since w is a constant on ∂S_t, it is strictly convex in S_t by Theorem 5.13. From Lemma 6.4 and Theorem 6.5, we deduce that

$$D^2 w \geq c_0 I_n \quad \text{in } B_{c_0}(0) \quad \text{and} \quad \|D^2 w\|_{C^{1,1/2}(B_{c_0}(0))} \leq c_0^{-1},$$

for some small constant $c_0 = c_0(n, \tau)$. In $B_{c_0}(0)$, we have

$$0 = \det D^2 w - \det D^2 P_m$$

$$= \int_0^1 \frac{d}{dt} \det \left((1-t)D^2 P_m + t D^2 w\right) dt = A^{ij} D_{ij}(w - P_m),$$

where

$$A = (A^{ij}) = \int_0^1 \text{Cof} \left((1-t)D^2 P_m + t D^2 w\right) dt.$$

It follows that

$$c_1(n, C_0) I_n \leq A \leq c_1^{-1}(n, C_0) I_n, \quad \|A\|_{C^{1,1/2}(B_{c_0}(0))} \leq C(C_0, n).$$

By the classical Schauder estimates (Theorem 2.81) for the uniformly elliptic equation $A^{ij}D_{ij}(w - P_m) = 0$, we can find $C_3 = C_3(C_0, n)$ such that

$$\|w - P_m\|_{C^{2,1}(B_{c_0/2}(0))} \leq C_3\|w - P_m\|_{L^\infty(B_{c_0}(0))} \leq 2C_3\varepsilon.$$

Therefore, the second-order Taylor expansion $\tilde{a}_m + \tilde{b}_m \cdot x + \tilde{P}_m(x)$ of $w - P_m$ at 0, where $\tilde{P}_m$ is a homogeneous quadratic polynomial, satisfies

$$(6.56) \quad |w(x) - P_m(x) - \tilde{a}_m - \tilde{b}_m \cdot x - \tilde{P}_m(x)| \leq 2C_3\varepsilon|x|^3 \quad \text{in } B_{c_0/2}(0),$$

with

$$\max\{|\tilde{a}_m|, |\tilde{b}_m|, \|D^2\tilde{P}_m\|\} \leq 2C_3\varepsilon.$$

Since $\tilde{a}_m + \tilde{b}_m \cdot x + P_m(x) + \tilde{P}_m(x)$ is the quadratic expansion for w at 0, we also have that $D^2 P_m + D^2\tilde{P}_m \in \mathbb{S}^n$ is positive definite and

$$\det D^2(P_m + \tilde{P}_m) = 1.$$

We define

$$P_{m+1}(x) := P_m(x) + \tilde{P}_m(x), \quad l_{m+1}(x) := b_{m+1}x, \quad b_{m+1} := b_m + r\tilde{b}_m.$$

Note that

$$
\begin{aligned}
(6.57) \qquad & r_0^m|b_{m+1} - b_m| + r_0^{2m}\|D^2 P_{m+1} - D^2 P_m\| \\
& = r_0^{2m}(|\tilde{b}_m| + \|D^2\tilde{P}_m\|) \\
& \leq 4C_3\varepsilon r_0^{2m} = 4C_3\varepsilon_0 r_0^{m(2+\alpha)}.
\end{aligned}
$$

From (6.55), (6.56), and $\tilde{a}_m = w(0) = w(0) - v(0)$, we find

$$
\begin{aligned}
|v - \tilde{b}_m \cdot x - P_{m+1}| &\leq (2C_3 r_0^3 + C_2\delta_0)\varepsilon + |\tilde{a}_m| \\
&\leq (2C_3 r_0^3 + 2C_2\delta_0)\varepsilon \qquad \text{in } B_{r_0}(0).
\end{aligned}
$$

Thus, by first choosing $r_0 = r_0(C_3, \alpha)$ small, and then choosing δ_0 small depending on r_0, C_2, we obtain

$$|v - \tilde{b}_m \cdot x - P_{m+1}| \leq \varepsilon r_0^{2+\alpha} \quad \text{in } B_{r_0}(0).$$

It follows from the definitions of v and ε that

$$|u - b_{m+1} \cdot x - P_{m+1}| \leq \varepsilon r^2 r_0^{2+\alpha} = \varepsilon_0 r_0^{(m+1)(2+\alpha)} \quad \text{in } B_{rr_0}(0) = B_{r_0^{m+1}}(0).$$

From (6.57), $b_0 = 0$, and $\|D^2 P_0\| \leq C_0$, we can choose ε_0 small so that

$$|b_m| \leq 1, \quad \|D^2 P_m\| \leq 2C_0 \quad \text{for all } m.$$

With the above choices of ε_0, δ_0, and r_0, the induction hypotheses hold for $m + 1$, completing the proof of the lemma. $\qquad \square$

With all ingredients in place, we can now complete the proof of the pointwise $C^{2,\alpha}$ estimates in Theorem 6.30.

Proof of Theorem 6.30. Recall that (6.47) is enforced. As discussed after the proof of Lemma 6.32, given $\delta_0 \in (0,1)$ and $\varepsilon_0 \in (0,1)$, there is a rescaling $\tilde{u}$ of u satisfying the hypotheses of Lemma 6.33 with $C_0 = C_0(n, \lambda, \Lambda)$. By choosing ε_0, δ_0, r_0 small, depending on n, α, and C_0 (and hence on λ, Λ), we find for each $m \in \mathbb{N}_0$ a vector $b_m \in \mathbb{R}^n$ and a convex, homogeneous quadratic polynomial P_m as in Lemma 6.33. Denote $B_t = B_t(0)$.

Upon letting $m \to \infty$, b_m has a limit $\tilde{b} \in \mathbb{R}^n$ with $|\tilde{b}| \le 1$, and P_m has a limit $\tilde{P}$ with $\det D^2\tilde{P} = 1$ and $\|D^2\tilde{P}\| \le 2C_0$. We show that

$$(6.58) \qquad |\tilde{u}(x) - \tilde{b} \cdot x - \tilde{P}(x)| \le \tilde{C}|x|^{2+\alpha} \quad \text{in } B_1, \quad \tilde{C} = \tilde{C}(\alpha, n, \lambda, \Lambda).$$

Indeed, from (6.52) and (6.53), we infer that

$$\|\tilde{u} - \tilde{b} \cdot x - \tilde{P}\|_{L^\infty(B_{r_0^m})} \le \|\tilde{u} - b_m \cdot x - P_m\|_{L^\infty(B_{r_0^m})}$$
$$+ \sum_{i=m}^{\infty} \|(b_{i+1} - b_i) \cdot x + P_{i+1} - P_i\|_{L^\infty(B_{r_0^m})}$$
$$\le \varepsilon_0 r_0^{m(2+\alpha)} + \sum_{i=m}^{\infty} 2C_\infty \varepsilon_0 r_0^{\alpha i + 2m} \le \bar{C}\varepsilon_0 r_0^{m(2+\alpha)}.$$

For any $x \in B_1$, let $m \in \mathbb{N}_0$ be such that $r_0^{m+1} < |x| \le r_0^m$. Then, the above estimates imply (6.58), as claimed with $\tilde{C} := \bar{C}\varepsilon_0 r_0^{-(2+\alpha)}$.

Since $D\tilde{u}(0) = 0$, we obtain $\tilde{b} = 0$ and therefore

$$|(\det T)^{-2/n} u(Ty) - \tilde{P}(y)| \le \tilde{C}|y|^{2+\alpha} \quad \text{in } B_1,$$

where we recall $\|T^{-1}\| + \|T\| \le C_*(n, M, \alpha, \lambda, \Lambda)$. Rescaling back, we find

$$|u(x) - \mathcal{P}_0(x)| \le \tilde{C}(\det T)^{2/n}|T^{-1}x|^{2+\alpha} \quad \text{in } TB_1 \subset \Omega,$$

where $\mathcal{P}_0(x) = (\det T)^{2/n}\tilde{P}(T^{-1}x)$ is a convex, homogeneous quadratic polynomial with

$$\det D^2\mathcal{P}_0 = 1 = f(0), \quad \|D^2\mathcal{P}_0\| \le C(n, M, \alpha, \lambda, \Lambda).$$

Since $\|T^{-1}\| + \|T\| \le C_*$, we have $TB_1 \supset B_{1/C_*}$, and hence

$$|u(x) - \mathcal{P}_0(x)| \le C(n, M, \alpha, \lambda, \Lambda)|x|^{2+\alpha} \quad \text{in } B_{1/C_*}.$$

Since $|u|$ and $\|D^2\mathcal{P}_0\|$ are bounded in Ω by a constant depending only on $n, M, \alpha, \lambda, \Lambda$, the above inequality also holds in Ω (for a larger C). This completes the proof of the theorem. $\qquad\square$

6.6.2. Interior $C^{2,\alpha}$ estimates. We are now ready to state and prove the interior $C^{2,\alpha}$ estimates for the Monge–Ampère equation.

Theorem 6.34 (Caffarelli's interior $C^{2,\alpha}$ estimates for the Monge–Ampère equation). *Let Ω be a convex domain in $\mathbb{R}^n$ satisfying $B_{R^{-1}}(0) \subset \Omega \subset B_R(0)$ for some $R \geq 1$. Let $u \in C(\overline{\Omega})$ be the convex Aleksandrov solution to*

$$\det D^2 u = f \quad in \ \Omega, \qquad u = 0 \quad on \ \partial\Omega,$$

where $0 < \lambda \leq f \leq \Lambda$ and $f \in C^\alpha(\Omega)$ for some $\alpha \in (0,1)$. Then,

$$\|u\|_{C^{2,\alpha}(\overline{\Omega'})} \leq C(n, \lambda, \Lambda, R, \alpha, \|f\|_{C^\alpha(\Omega)}) \quad where \ \Omega' := \{u < -\|u\|_{L^\infty(\Omega)}/2\}.$$

Proof. Via normalization using John's Lemma, we can assume $R = n$ for simplicity. Let x_0 be the minimum point of u in Ω, and let $h := \|u\|_{L^\infty(\Omega)}/4$. Then $\Omega = S_u(x_0, 4h)$, $\Omega' = S_u(x_0, 2h)$, and the volume estimates in Lemma 5.6 give $0 < c(n, \lambda, \Lambda) \leq h \leq C(n, \lambda, \Lambda)$. For any $z \in \overline{S_u(x_0, 2h)}$, there is $\gamma(n, \lambda, \Lambda) > 0$ such that $S_u(z, 2\gamma) \subset S_u(x_0, 3h)$. Let $M := \|f\|_{C^\alpha(\Omega)}$. Then

$$|f(x) - f(z)| \leq M|x - z|^\alpha \quad in \ S_u(z, 2\gamma).$$

Applying Theorem 6.30 to $u(x) - [u(z) + Du(z) \cdot (x - z)]$, we find a quadratic polynomial $\mathcal{P}_z(x)$,

$$\mathcal{P}_z(x) = u(z) + Du(z) \cdot (x - z) + \frac{1}{2}(A_z(x - z)) \cdot (x - z),$$

where $A_z \in \mathbb{S}^n$, and a positive constant $C_0(n, \lambda, \Lambda, \alpha, M)$ such that

$$\|D^2 \mathcal{P}_z\| \leq C_0 \quad and \quad |u(x) - \mathcal{P}_z(x)| \leq C_0|x - z|^{2+\alpha} \quad for \ all \ x \in S_u(z, \gamma).$$

This holds for all $z \in \overline{\Omega'}$ so, by Theorem 2.91, we have $\|u\|_{C^{2,\alpha}(\overline{\Omega'})} \leq C$ where $C = C(n, \lambda, \Lambda, \alpha, M)$. The theorem is proved. $\square$

In the proof of Theorem 6.30, we also have

$$\|D^2\tilde{P} - D^2 P_0\| \leq \sum_{m=0}^{\infty} \|D^2 P_{m+1} - D^2 P_m\| \leq C\varepsilon_0 \sum_{m=0}^{\infty} r_0^{m\alpha} = C(n, \alpha, C_0)\varepsilon_0.$$

Hence, upon applying (6.58) to the function u in Lemma 6.33, we find that $\|D^2 u(0) - D^2 P_0\| \leq C\varepsilon_0$. We obtain this Hessian closeness estimate from u being close to P_0 and $\det D^2 u$ being close to $\det D^2 P_0$ in a pointwise C^α sense at 0. When this pointwise condition is upgraded to C^α closeness in a region, we obtain corresponding closeness results for $D^2 u$ to $\det D^2 P_0$ in C^α. This is the content of the following perturbation theorem.

Theorem 6.35 (Perturbation theorem for interior $C^{2,\alpha}$ estimates). *Let Ω be a bounded convex domain containing $B_1(0)$ in $\mathbb{R}^n$. Assume that $u \in C(\Omega)$ is a convex Aleksandrov solution to*

$$\det D^2 u = f \quad in \ \Omega, \quad where \ 0 < \lambda \leq f \leq \Lambda.$$

Assume that in $B_1(0)$, we have for $\varepsilon \leq \varepsilon_0 < 1$,

$$\|f - \det D^2 P\|_{C^\alpha(B_1(0))} \leq \varepsilon \quad and \quad |u - P| \leq \varepsilon,$$

for some convex, homogenous quadratic polynomial $P(x)$ with

$$\det D^2 P \geq M^{-1}, \quad \|D^2 P\| \leq M.$$

There exist $\varepsilon_0(n, M) > 0$ and $r(n, \lambda, \Lambda, M) > 0$ such that if $\varepsilon \leq \varepsilon_0$, then

$$\|u - P\|_{C^{2,\alpha}(B_r(0))} \leq C(n, \lambda, \Lambda, \alpha, M)\varepsilon.$$

Proof. The proof follows from the interior $C^{2,\alpha}$ estimates in Theorem 6.34 and Schauder estimates. By an orthogonal transformation of coordinates, we can assume that $P(x) = \sum_{i=1}^{n} \lambda_i x_i^2 / 2$. From the assumptions on P, we have $M^{-n} \leq \lambda_i \leq M$ for all $i = 1, \ldots, n$. From $|u - P| \leq \varepsilon$ in $B_1(0)$, we can find $r_0 > 0, \varepsilon_0 > 0$, and $t_0 > 0$ depending on n and M such that if $\varepsilon \leq \varepsilon_0$,

$$B_{r_0}(0) \subset \Omega := \{x \in B_1(0) : u(x) < t_0\} \subset B_{3/4}(0).$$

Therefore, from Lemma 5.6, we can find $r(n, \lambda, \Lambda, M) > 0$ such that

$$B_{2r}(0) \subset \Omega' := \{x \in \Omega : u(x) - t_0 < -\|u - t_0\|_{L^\infty(\Omega)}/2\} \subset B_{3/4}(0).$$

Clearly, $\|f\|_{C^\alpha(B_{3/4}(0))} \leq C_1(n, \lambda, \Lambda, M)$. Applying Theorem 6.34, we obtain

(6.59) $$\|u\|_{C^{2,\alpha}(\Omega')} \leq C_2(n, \alpha, \lambda, \Lambda, M).$$

Since $f = \det D^2 u$, we have

$$f - \det D^2 P = \int_0^1 \frac{d}{dt} \det\left((1-t)D^2 P + tD^2 u\right) dt = A^{ij} D_{ij}(u - P),$$

where

$$A = (A^{ij}) = \int_0^1 \operatorname{Cof}\left((1-t)D^2 P + tD^2 u\right) dt.$$

From (6.59), we deduce that for some $C_3(n, M, \alpha, \lambda, \Lambda) > 0$,

$$C_3^{-1} I_n \leq A \leq C_3 I_n \quad \text{in } B_{2r}(0), \quad \|A\|_{C^{0,\alpha}(B_{2r}(0))} \leq C_3.$$

By the interior Schauder estimates in Theorem 2.81, we obtain

$$\|u - P\|_{C^{2,\alpha}(B_r(0))} \leq C(\|u - P\|_{L^\infty(B_{2r}(0))} + \|f - \det D^2 P\|_{C^\alpha(B_{2r}(0))})$$
$$\leq C(n, \alpha, \lambda, \Lambda, M)\varepsilon.$$

This completes the proof of the theorem. $\square$

6.6.3. Interior smoothness for degenerate equations. By combining the interior $C^{2,\alpha}$ estimates for the Monge–Ampère equation with the Schauder theory, we can obtain interior regularity and uniqueness results for certain degenerate Monge–Ampère equations with zero boundary condition that are relevant to the study of the Monge–Ampère eigenvalue problem in Chapter 11. They include equations of the form $\det D^2 u = |u|^p$ having zero boundary values so they are degenerate near the boundary.

Proposition 6.36 (Interior smoothness for degenerate equations). *Let Ω be a bounded convex domain in $\mathbb{R}^n$, $0 \le p < \infty$, and $\delta \in [0, \infty)$. Assume that $u \in C(\overline{\Omega})$ is a nonzero convex Aleksandrov solution to the Dirichlet problem*

$$\det D^2 u = (|u| + \delta)^p \quad in \ \Omega, \quad u = 0 \quad on \ \partial\Omega.$$

Then u is strictly convex in Ω and $u \in C^\infty(\Omega)$.

Proof. Note that $u < 0$ in Ω. Let $h := \|u\|_{L^\infty(\Omega)} > 0$. For $\varepsilon \in (0, h)$, let $\Omega_\varepsilon = \{x \in \Omega : u(x) < -\varepsilon\}$. Then, Ω_ε is convex, because $u \in C(\overline{\Omega})$ is convex. Since

$$\varepsilon^p \le \det D^2 u = (|u| + \delta)^p \le (h + \delta)^p \quad in \ \Omega_\varepsilon, \quad u = -\varepsilon \quad on \ \partial\Omega_\varepsilon,$$

the function u is strictly convex in Ω_ε by Theorem 5.13. Thus, by the $C^{1,\alpha}$ estimates in Theorems 5.18 and 5.21, $u \in C^{1,\alpha}(\Omega_\varepsilon)$ for some $\alpha \in (0, 1)$ depending only on $n, p, \varepsilon, \delta$, and h. Now, the right-hand side $(|u| + \delta)^p = (-u + \delta)^p$ is in $C^{1,\alpha}(\Omega_\varepsilon)$. Thus, using Caffarelli's $C^{2,\alpha}$ estimates in Theorem 6.34, we have $u \in C^{2,\alpha}(\Omega_\varepsilon)$. In the interior of Ω_ε, the equation $\det D^2 u = (|u| + \delta)^p$ now becomes uniformly elliptic with $C^{2,\alpha}$ right-hand side. By a simple bootstrap argument using the Schauder estimates (Theorem 2.81) as in Step 4 of the proof of Lemma 6.4, we have $u \in C^\infty(\Omega_\varepsilon)$. Since $\varepsilon \in (0, h)$ is arbitrary, we conclude $u \in C^\infty(\Omega)$ and u is strictly convex in Ω. $\square$

For equations of the form $\det D^2 u = (-u)^p$ where $p > 0$, we cannot directly use the maximum principle to prove uniqueness for nonzero solutions; compare with Problem 4.3. In fact, this uniqueness issue is wide open for $p > n$; see Huang [**Hua1**] for some results in this direction. However, when p is in the subcritical range $0 < p < n$, uniqueness can be obtained.

Proposition 6.37 (Uniqueness for degenerate equations). *Let $p \in [0, n)$, $\delta \in [0, \infty)$, and let Ω be a bounded, convex domain in $\mathbb{R}^n$. Then there is at most one nonzero convex Aleksandrov solution $u \in C(\overline{\Omega})$ to*

$$\det D^2 u = (|u| + \delta)^p \quad in \ \Omega, \quad u = 0 \quad on \ \partial\Omega.$$

Proof. Suppose we have two nonzero convex solutions u and v to the given equation with $u - v$ being positive somewhere in Ω. By Proposition 6.36, $u, v \in C^\infty(\Omega)$ and u, v are strictly convex in Ω. By translation of coordinates, we can assume that $0 \in \Omega$ satisfies $u(0) - v(0) = \max_{\overline{\Omega}}(u - v) = \tilde{\delta} > 0$. For $1 < \gamma \le 2$, consider for $x \in \Omega$,

$$u_\gamma(x) = u(x/\gamma) \quad and \quad \eta_\gamma(x) = v(x)/u_\gamma(x).$$

If $\operatorname{dist}(x, \partial\Omega) \to 0$, then $\eta_\gamma(x) \to 0$. Note that $u, v < 0$ in Ω and

$$\eta_\gamma(0) = v(0)/u(0) = [u(0) - \tilde{\delta}]/u(0) \ge 1 + \varepsilon \quad for \ some \ \varepsilon > 0.$$

Therefore, the function η_γ attains its maximum value at $x_\gamma \in \Omega$ with $\eta_\gamma(x_\gamma) \geq 1 + \varepsilon$. At $x = x_\gamma$, we have $D\eta_\gamma(x_\gamma) = 0$, $D^2\eta_\gamma(x_\gamma) \leq 0$, and $u_\gamma(x_\gamma) \leq 0$, so we can compute $D^2 v = \eta_\gamma D^2 u_\gamma + D^2 \eta_\gamma u_\gamma \geq \eta_\gamma D^2 u_\gamma$. Hence,

$$(|v(x_\gamma)| + \delta)^p = \det D^2 v(x_\gamma) \geq \eta_\gamma^n \det D^2 u_\gamma(x_\gamma)$$
$$= \eta_\gamma^n \gamma^{-2n}(|u(x_\gamma/\gamma)| + \delta)^p.$$

Using $\eta_\gamma(x_\gamma) \geq 1 + \varepsilon$, we find

$$|v(x_\gamma)| + \delta = \eta(x_\gamma)|u(x_\gamma/\gamma)| + \delta \leq \eta(x_\gamma)(|u(x_\gamma/\gamma)| + \delta),$$

and we deduce from the preceding estimates that

$$1 \geq [\eta_\gamma(x_\gamma)]^{n-p} \gamma^{-2n} \geq (1 + \varepsilon)^{n-p} \gamma^{-2n}.$$

Letting $\gamma \searrow 1$, using $p < n$ and $\varepsilon > 0$, we obtain a contradiction. $\qquad \square$

6.7. Further Remarks on Pogorelov-type Estimates

In previous sections, we saw the crucial role of Pogorelov's second derivative estimates in the interior regularity theory of the Monge–Ampère equation, including interior $W^{2,p}$ and $C^{2,\alpha}$ estimates. In this section, we make further remarks pertaining to the Pogorelov estimates in Theorem 6.1.

When $f \in C^3$ in Lemma 6.4, as an alternative to the Evans–Krylov estimates, we can use Calabi's interior C^3 estimates for the Monge–Ampère equation [**Cal1**] to obtain similar $C^{3,\alpha}$ estimates.

Theorem 6.38 (Calabi's interior C^3 estimates). *Let Ω be a bounded convex domain in $\mathbb{R}^n$. Assume that $u \in C^5(\Omega)$ is a convex function satisfying $\det D^2 u = f$ in Ω with $\|D^2 u\| \leq M$ in Ω, where $f \in C^3(\Omega)$ satisfies $0 < \lambda \leq f \leq \Lambda$. Then, for any open set $\Omega' \Subset \Omega$, we have*

$$\|D^3 u\|_{L^\infty(\Omega')} \leq C = C(M, n, \lambda, \Lambda, \|f\|_{C^3(\Omega)}, \Omega', \mathrm{dist}(\Omega', \partial\Omega)).$$

6.7.1. Choice of the auxiliary function. To estimate $|u|\|D^2 u\|$, instead of using the function h as in Theorem 6.1, Trudinger and Wang [**TW5**, **TW6**] chose the auxiliary function

$$h = |u|\left(1 - |Du|^2/(4M)\right)^{-1/8} D_{\xi\xi} u, \quad \text{where } M := \sup_\Omega |Du|^2,$$

and they obtained, after performing similar calculations,

$$(6.60) \qquad \sup_\Omega |u|\|D^2 u\| \leq C(n, \sup_\Omega |u|, \|\log f\|_{C^2(\Omega)})(1 + \sup_\Omega |Du|^2).$$

Thus, $\sup_\Omega \|D^2 u\|$ can be estimated from above by a linear function of $\sup_\Omega |Du|^2$. We refer the reader to [**TW5**] for an application of this linear dependence to global $C^{2,\alpha}$ estimates for the Monge–Ampère equation with global C^α Monge–Ampère measure under optimal boundary conditions.

Consider now an auxiliary function of the form $h = |u|^\beta \eta(|Du|^2) D_{\xi\xi}u$, where $\beta > 0$ and η is a positive function. Let x_0 be the maximum point of h in $\overline{\Omega}$. Then, the method of the proof of Theorem 6.1 gives an interior bound for $\|D^2 u(x)\|$ in terms of n, $\|u\|_{C^1(\overline{\Omega})}$, $\|\log f\|_{C^2(\Omega)}$, and $\text{dist}(x, \partial\Omega)$ if in the inequality $u^{ij}(x_0) D_{ij} \log h(x_0) \leq 0$, we can at least bound the expression T_h collecting all third-order derivatives from below by some (possibly negative) quantities involving $u(x_0)$, $Du(x_0)$, and $D^2 u(x_0)$. For the auxiliary function in the statement of Theorem 6.1 for the Monge–Ampère equation, T_h is nonnegative because

$$D_{11} u T_h = u^{ii} u^{jj} (D_{1ij}u)^2 - \frac{u^{ii}(D_{11i}u)^2}{D_{11}u} - \sum_{i=2}^{n} u^{ii} \frac{(D_{11i}u)^2}{D_{11}u}$$

$$= \sum_{i,j \geq 2} u^{ii} u^{jj} (D_{1ij}u)^2 \geq 0.$$

Below, we will briefly mention some examples in applying the above ideas to related fully elliptic nonlinear equations.

6.7.2. Hessian equations. Let $1 \leq k \leq n$. Consider the k-Hessian equation

$$\sum_{1 \leq i_1 < \cdots < i_k \leq n} \lambda_{i_1}(D^2 u) \ldots \lambda_{i_k}(D^2 u) = f \quad \text{on } \Omega,$$

on a bounded domain Ω in $\mathbb{R}^n$, where the eigenvalues $\lambda_{i_1}(D^2 u), \ldots, \lambda_{i_n}(D^2 u)$ of $D^2 u$ are now assumed to lie on a suitable cone Γ_k in $\mathbb{R}^n$ that contains $\Gamma_n := \{x \in \mathbb{R}^n : x_1 > 0, \ldots, x_n > 0\}$. When $k = n$, we have the Monge–Ampère equation. Though the algebra for the maximum principle type argument concerning $|u|^\beta \eta(|Du|^2) D_{\xi\xi}u$ is more involved, Chou and Wang [**ChW2**] were able to extend the above method and obtained interior second derivative estimates for Hessian equations with zero boundary data.

6.7.3. Complex Monge–Ampère equation. Let us try to mimic the proof of Theorem 6.1 for the complex Monge–Ampère equation

(6.61) $$\det u_{z_j \bar{z}_k} = f \quad \text{in } \Omega \subset \mathbb{C}^n, \quad u = 0 \quad \text{on } \partial\Omega,$$

where Ω is a bounded pseudo-convex domain in $\mathbb{C}^n$ $(n \geq 2)$, $f \in C^2(\Omega) \cap C(\overline{\Omega})$ with $f > 0$ in Ω, and $u \in C^4(\Omega) \cap C^2(\overline{\Omega})$ is a real-valued plurisubharmonic function; that is,

$$\sum_{j,k=1}^{n} \frac{\partial^2}{\partial z_j \partial \bar{z}_k} u(z) \xi_j \bar{\xi}_k \geq 0 \quad \text{for all } z \in \Omega \text{ and all } \xi = (\xi_1, \ldots, \xi_n) \in \mathbb{C}^n.$$

Here, for $z_j = x_j + \sqrt{-1} y_j$, we use the notation $\bar{z}_j = x_j - \sqrt{-1} y_j$,

$$u_{z_j} = \frac{\partial u}{\partial z_j} = \frac{1}{2}\left(\frac{\partial u}{\partial x_j} - \sqrt{-1}\frac{\partial u}{\partial y_j}\right) \quad \text{and} \quad u_{\bar{z}_j} = \frac{\partial u}{\partial \bar{z}_j} = \frac{1}{2}\left(\frac{\partial u}{\partial x_j} + \sqrt{-1}\frac{\partial u}{\partial y_j}\right).$$

Now, at the maximum point x_0 of any auxiliary function h that involves $u_{z_j \bar{z}_k} \xi_j \bar{\xi}_k$ and derivatives up to the first order of u, we have

$$u^{j\bar{k}}(x_0)(\log h)_{z_j \bar{z}_k}(x_0) \le 0,$$

where $(u^{j\bar{k}}) = (u_{z_j \bar{z}_k})^{-1}$ is the inverse of the complex Hessian matrix $(u_{z_j \bar{z}_k})$. The term involving all third-order derivatives is of the form

$$\tilde{T}_h = \sum_{j,k=1}^{n} \frac{|u_{z_1 \bar{z}_k z_j}|^2}{u_{z_k \bar{z}_k} u_{z_j \bar{z}_j}} - \sum_{j=1}^{n} \frac{|u_{z_1 \bar{z}_1 z_j}|^2}{u_{z_1 \bar{z}_1} u_{z_j \bar{z}_j}} - \sum_{j=2}^{n} \frac{|u_{z_1 \bar{z}_1 z_j}|^2}{u_{z_j \bar{z}_j} u_{z_1 \bar{z}_1}}.$$

If, for all $j, k = 1, \ldots, n$, we have

(6.62) $$|u_{z_j \bar{z}_j z_k}| = |u_{z_j \bar{z}_k z_j}|,$$

then $\tilde{T}_h = \sum_{2 \le j, k \le n} \frac{|u_{z_1 \bar{z}_k z_j}|^2}{u_{z_k \bar{z}_k} u_{z_j \bar{z}_j}} \ge 0$, and we will obtain a Pogorelov-type estimate for (6.61).

Unfortunately, the equality (6.62) involving mixed third-order derivatives is false; take, for example $u(z_1, z_2) = |z_1|^2 \text{Re}(z_2)$ in $\mathbb{C}^2$. Even until now, it is not known if there is a complex analogue of the Pogorelov estimates in Theorem 6.1. We refer to Błocki [**Bl**] for an account of previous attempts in this direction.

6.7.4. Optimal transportation with nonquadratic cost and the Ma–Trudinger–Wang condition.
In Example 1.8, we considered the Monge–Kantorovich optimal transportation problem between $(\Omega, f_1 \, dx)$ and $(\Omega', f_2 \, dy)$ with quadratic cost $c(x, y) = |x - y|^2$ (which is actually equivalent to the cost function $c(x, y) = -x \cdot y$ from the formulation of the problem) with the optimal map $T_u = Du$ where the convex potential function u solves a Monge–Ampère equation of the form $\det D^2 u = k(x, Du)$. For more general nonquadratic cost $c(x, y)$, the potential function solves a Monge–Ampère-type equation

(6.63) $$\det(D^2 u - \mathcal{A}(x, Du(x))) = f(x, Du(x)),$$

where $\mathcal{A} = (\mathcal{A}_{ij}(x, p))_{1 \le i, j \le n}$, the optimal map T_u, f, and the potential function u are determined by the following relations:

$$\mathcal{A}(x, Du(x)) = -D_x^2 c(x, T_u(x)), \quad Du(x) = -D_x c(x, T_u(x)),$$

and

$$f(x, Du(x)) = |\det D_{xy}^2 c(x, T_u(x))| f_1(x) / f_2 \circ T_u(x).$$

Let us carry out the Pogorelov method to establish interior second derivative estimates for u. In doing so, we need to differentiate (6.63) twice and in the expression $\text{trace}([D^2 u - \mathcal{A}(x_0, Du)]^{-1} D^2 \log h(x_0)) \le 0$ the term $\hat{T}_h$

collecting all third-order derivatives of u involves coefficients that are fourth-order derivatives of the cost function c. To handle this term, Ma–Trudinger–Wang [**MTW**] introduced a condition involving c (or equivalently, involving $\mathcal{A}$), called (A3), that allows estimating $\tilde{T}_h$ from below by some negative quantities involving $u(x_0)$, $Du(x_0)$, and $D^2u(x_0)$. This condition requires

$$(A3) \quad D^2_{p_k p_l} \mathcal{A}_{ij}(x,p)\xi_i\xi_j\eta_k\eta_l \geq c_0|\xi|^2|\eta|^2 \quad \text{for all } \xi, \eta \in \mathbb{R}^n, \text{with } \xi \cdot \eta = 0,$$

for some positive constant c_0. With (A3), they were able to establish interior regularity for (6.63). It turns out that (A3) is also necessary for the interior regularity of the optimal transportation problem, as shown by Loeper [**Lo3**]. The condition (A3) is now the well-known *MTW condition* (named after the originators) in optimal transportation. We refer the reader to Trudinger–Wang [**TW6**, Section 3.8] and Figalli [**F2**, Section 5.3] for further results and subsequent developments.

6.8. Problems

Problem 6.1. Let Ω, u, f be as in Theorem 6.1. Prove the estimate (6.60).

Problem 6.2. Let Ω be a bounded convex domain in $\mathbb{R}^n$, and let $f, g \in L^1(\Omega)$ be nonnegative functions. Let $u, v \in C(\overline{\Omega}) \cap W^{2,n}_{loc}(\Omega)$ be convex Aleksandrov solutions of $\det D^2u = f$ and $\det D^2v = g$ in Ω. Show that

$$\max_{\overline{\Omega}}(u-v) \leq \max_{\partial\Omega}(u-v) + C(n)\text{diam}(\Omega)\| \max\{g^{1/n} - f^{1/n}, 0\}\|_{L^n(\Omega)}.$$

Problem 6.3. Let Ω be a convex domain in $\mathbb{R}^n$ $(n \geq 2)$ satisfying $B_1(0) \subset \Omega \subset B_n(0)$. Let $\varepsilon \in (0, 1/2)$. Assume that $w \in C(\overline{\Omega})$ is a convex function that satisfies, in the sense of Aleksandrov,

$$1 - \varepsilon \leq \det D^2 w \leq 1 + \varepsilon \quad \text{in } \Omega, \quad w = 0 \quad \text{on } \partial\Omega.$$

By Theorem 6.13, we know that for some $p = p(n, \varepsilon) > 1$,

$$\|D^2w\|_{L^p(B_{1/2}(0))} \leq C(n, \varepsilon).$$

(a) Show that we can take $p > c(n)\log\frac{1}{\varepsilon}$ for some $c(n) > 0$.

(b) ($\star$) When $n = 2$, can we take $p > \frac{c}{\varepsilon}$ with some constant $c > 0$ independent of ε? (*As of this writing, this is still an open problem.*)

Problem 6.4. Let Ω be a convex domain in $\mathbb{R}^n$ satisfying $B_1(0) \subset \Omega \subset B_n(0)$. Let $u \in C^2(\Omega)$ be a convex function satisfying $0 < \lambda \leq \det D^2u \leq \Lambda$ in Ω with $u = 0$ on $\partial\Omega$. Let $\Omega_{1/2} := \{x \in \Omega : u(x) < -\|u\|_{L^\infty(\Omega)}/2\}$.

(a) Show that when $n = 2$, there exist constants $\varepsilon, C > 0$ depending only on λ and Λ such that $\|(D^2u)^{-1}\|_{L^{1+\varepsilon}(\Omega_{1/2})} \leq C$.

(b) ($\star$) For $n \geq 3$, do there exist constants $\varepsilon, C > 0$ depending only on n, λ, and Λ such that $\|(D^2 u)^{-1}\|_{L^{1+\varepsilon}(\Omega_{1/2})} \leq C$? (*As of this writing, this is still an open problem.*)

Problem 6.5. Let P be a strictly convex quadratic polynomial on $\mathbb{R}^n$. Suppose that u is a convex function on $B_r(z)$ satisfying

$$\|u - P\|_{L^\infty(B_r(z))} < \varepsilon \quad \text{and} \quad |\det D^2 u - \det D^2 P| \leq \varepsilon \quad \text{in } B_r(z).$$

Let $p > n^2$. Show that for $\varepsilon > 0$ small, depending only on r, n, p, and the eigenvalues of $D^2 P$, u is strictly convex in $B_{r/2}(z)$ and $u \in W^{2,p}(B_{r/2}(z))$; hence $u \in C^{1,1-1/n}(\overline{B_{r/2}(z)})$ by the Sobolev Embedding Theorem.
Hint: For the strict convexity, use arguments similar to those for the inclusions preceding (6.59).

Problem 6.6 (C^2 estimates for Dini continuous right-hand side). Let Ω be a convex domain in $\mathbb{R}^n$ satisfying $B_1(0) \subset \Omega \subset B_n(0)$ and let $f \in C(\Omega)$ with $0 < \lambda \leq f \leq \Lambda$. Let $u \in C(\overline{\Omega})$ be the convex Aleksandrov solution to the Dirichlet problem

$$\det D^2 u = f \quad \text{in } \Omega, \quad u = 0 \quad \text{on } \partial\Omega.$$

Assume that f is Dini continuous; that is, its modulus of continuity

$$\omega_f(r) := \sup_{x,y \in \Omega; |x-y| < r} |f(x) - f(y)|$$

satisfies $\int_0^1 \frac{\omega_f(r)}{r}\, dr < \infty$. Show that, for $\Omega' := \{u < -\|u\|_{L^\infty(\Omega)}/2\}$, we have

$$\|u\|_{C^2(\overline{\Omega'})} \leq C(n, \lambda, \Lambda, \omega_f).$$

Problem 6.7. Let Ω be a bounded convex domain in $\mathbb{R}^n$. Let $u_0 \in C(\Omega)$ be a nonzero convex function on Ω. Let $\{a_k\}_{k=0}^\infty \subset [\lambda, \infty)$ where $\lambda > 0$. For $k \geq 0$, we recursively define u_{k+1} as the convex Aleksandrov solution to the Dirichlet problem

$$\det D^2 u_{k+1} = a_k |u_k|^n \quad \text{in } \Omega, \quad u_{k+1} = 0 \quad \text{on } \partial\Omega.$$

Prove that u_{k+1} is strictly convex in Ω and $u_{k+1} \in C^{2k, \frac{1}{n}}(\Omega)$ for all $k \geq 1$.

Problem 6.8. This is a continuation of the example in Proposition 6.28.

(a) Consider the function u in Proposition 6.28 with $\alpha > 2$. Find the largest exponent $\beta = \beta(\alpha) \in (0, 1)$ so that $u \in C^{1,\beta}(\mathbb{R}^2)$.

(b) Show that for any $\beta \in (0, 1)$, there are positive constants $\lambda(\beta) < \Lambda(\beta)$ and a convex function u satisfying $\lambda(\beta) \leq \det D^2 u \leq \Lambda(\beta)$ in $\mathbb{R}^2$ in the sense of Aleksandrov, but $u \notin C^{1,\beta}(\mathbb{R}^2)$.

Problem 6.9. Consider the following function in $B_{1/100}(0)$ in $\mathbb{R}^2$:

$$u(x) = x_1^2 / \log|\log r^2| + x_2^2 \log|\log r^2|, \quad r = |x| = (x_1^2 + x_2^2)^{1/2}.$$

Show that the Monge–Ampere measure μ_u of u has positive continuous density f in $B_{1/100}(0)$ and $f(x) = 4 + O(\log|\log r^2|/\log r^2)$ when $r \to 0$.

Problem 6.10. Let $p \in [n, \infty]$. Study the validity of the conclusion of Theorem 6.30 when the condition $|f(x) - f(z)| \leq M|x - z|^\alpha$ for all $x \in \Omega$ is replaced by

$$\|f - f(z)\|_{L^p(B_r(z))} \leq M r^{1+\alpha} \quad \text{for all } B_r(z) \subset \Omega.$$

6.9. Notes

Pogorelov's second derivative estimates in Section 6.1 can be found in many expository works, including Pogorelov [**P5**, Chapter 5], Gilbarg–Trudinger [**GT**, Section 17.6], Gutiérrez [**G2**, Sections 4.1 and 4.2], Figalli [**F2**, Section 3.3], and Han [**H**, Section 6.3]. The two-dimensional case appeared in [**P1**, Chapter 8]. In two dimensions, Heinz [**He**] obtained the interior second derivative estimates for $\det D^2 u = f$ with f being Lipschitz; the interior $C^{2,\alpha}$ estimates then follow from applying the Hölder gradient estimates for uniformly elliptic equations in two dimensions to the gradient (see [**GT**, Section 12.2]). In Theorem 6.5, if f is analytic, then a classical result of Hopf [**Hop**] allows us to conclude that the solution u is also analytic in Ω. A simple proof of the analyticity of smooth solutions to fully nonlinear, uniformly elliptic, analytic equations can be found in Hashimoto [**Has**].

In Section 6.2, in the proofs of Lemma 6.7 and Proposition 6.8 regarding the second-order differentiability of convex envelopes, we follow the arguments in Gutiérrez [**G2**, Section 6.6]; see also De Philippis–Figalli [**DPF3**].

The proof of Lemma 6.9 follows Gutiérrez [**G2**, Section 6.1].

For materials presented in Section 6.4, see also Figalli [**F2**, Section 4.8]. The proof of Theorem 6.11 follows the original argument of De Philippis, Figalli, and Savin [**DPFS**] who also obtained interior $W^{2,1+\varepsilon}$ estimates when the function f is replaced by a more general measure μ of the form

$$\mu = \sum_{i=1}^{N} f_i |P_i|^{\beta_i}, \quad \lambda \leq f_i \leq \Lambda, \quad P_i \text{ polynomial}, \quad \text{and} \quad \beta_i \geq 0.$$

That both Theorems 6.11 and 6.13 follow from a common approach (while invoking Lemma 6.9 for the latter) was from Figalli [**F1**]. Related to Section 6.4, see Gutiérrez [**G2**, Chapter 6] for a different approach to $W^{2,p}$ estimates, and see Gutiérrez–Huang [**GH3**] for $W^{2,p}$ estimates for the parabolic Monge–Ampère equation.

Related to Section 6.6, different proofs of the interior $C^{2,\alpha}$ estimates for the Monge–Ampère equation can be found in Jian–Wang [**JW1**] (where estimates similar to (2.27) were obtained), Figalli [**F2**, Section 4.10], and Gutiérrez [**G2**, Chapter 8]. The proof of the pointwise $C^{2,\alpha}$ estimates in Theorem 6.30 presented here is similar to the one in Han [**H**, Section 8.5]. A new proof of the interior $C^{2,\alpha}$ estimates using Green's function can be found in Wang–Wu [**WW**]. When the right-hand side is Dini continuous, Wang [**W1**] established the C^2 estimates; see Problem 6.6. Proposition 6.36 is from Le [**L7**]. Proposition 6.37 is from Tso [**Ts**].

Problem 6.2 is based on Huang [**Hu2**] and Jian–Wang [**JW1**]. The example in Problem 6.9 is from Wang [**W3**].

Chapter 7

Viscosity Solutions and Liouville-type Theorems

This chapter explores various implications of the interior regularity results for Aleksandrov solutions to the Monge–Ampère equation. These include relating several notions of solutions, establishing Liouville-type theorems, and uniting the Monge–Ampère equation to other well-known partial differential equations.

We will introduce the concept of viscosity solutions to the Monge–Ampère equation and discuss the following topics: the relationship between Aleksandrov solutions and viscosity solutions, viscosity solutions with boundary discontinuities and their compactness. As a consequence of the interior $C^{2,\alpha}$ regularity estimates, we prove the classical theorem of Jörgens, Calabi, and Pogorelov asserting that global Aleksandrov or viscosity solutions of $\det D^2 u = 1$ on the whole space must be quadratic polynomials. Also presented are an alternative proof of this theorem in two dimensions and its connections with the following equations: Laplace, minimal surface, and incompressible Euler. We conclude with a brief introduction to the Legendre–Lewy rotation which explains and generalizes these connections.

7.1. Viscosity Solutions of the Monge–Ampère Equation

In this section, we define the concept of *viscosity solutions* of the Monge–Ampère equation and discuss its relations with Aleksandrov solutions. As mentioned in Section 3.1, this concept is based on the maximum principle.

Basic idea. Consider the nonlinear PDE

$$(7.1) \qquad F(D^2u(x), Du(x), u(x), x) = 0 \quad \text{in } \Omega \subset \mathbb{R}^n$$

for a given continuous operator $F(r, p, z, x) : \mathbb{S}^n \times \mathbb{R}^n \times \mathbb{R} \times \Omega \to \mathbb{R}$ and an unknown $u : \Omega \to \mathbb{R}$ where Ω is an open set in $\mathbb{R}^n$. Assume that F is *elliptic*, that is, F is nondecreasing in $r \in \mathbb{S}^n$.

Suppose that $v \in C^2(\Omega)$ is a subsolution of (7.1), that is,

$$(7.2) \qquad F(D^2v(x), Dv(x), v(x), x) \geq 0 \quad \text{in } \Omega.$$

If $\varphi \in C^2(\Omega)$ and $v - \varphi$ has a local maximum at x_0, then $Dv(x_0) = D\varphi(x_0)$ and $D^2v(x_0) \leq D^2\varphi(x_0)$, so it follows from the ellipticity of F that

$$F(D^2\varphi(x_0), D\varphi(x_0), v(x_0), x_0) \geq F(D^2v(x_0), Dv(x_0), v(x_0), x_0).$$

Therefore,

$$(7.3) \qquad F(D^2\varphi(x_0), D\varphi(x_0), v(x_0), x_0) \geq 0.$$

We have just moved the derivatives of v in (7.2) to the test function φ in (7.3). This is usually referred to as *differentiation by parts*. Moreover, for (7.3) and the local maximality of $v - \varphi$ to make sense at x_0, only the continuity of v is needed. This is the basic idea for the theory of viscosity solutions to (7.1) which is a powerful tool for proving existence, uniqueness, and stability results for nonlinear elliptic equations. Here, for simplicity, we restrict ourselves to continuous viscosity solutions.

Definition 7.1. Consider the elliptic PDE (7.1). A continuous function $u \in C(\Omega)$ is called

(a) a viscosity subsolution of (7.1) when the following condition holds: If $\varphi \in C^2(\Omega)$ and $u - \varphi$ has a local maximum at $x_0 \in \Omega$, then

$$F(D^2\varphi(x_0), D\varphi(x_0), u(x_0), x_0) \geq 0.$$

We write $F(D^2u(x), Du(x), u(x), x) \geq 0$ in the viscosity sense.

(b) a viscosity supersolution of (7.1) when the following condition holds: If $\varphi \in C^2(\Omega)$ and $u - \varphi$ has a local minimum at $x_0 \in \Omega$, then

$$F(D^2\varphi(x_0), D\varphi(x_0), u(x_0), x_0) \leq 0.$$

We write $F(D^2u(x), Du(x), u(x), x) \leq 0$ in the viscosity sense.

(c) a viscosity solution of (7.1) if it is both a viscosity subsolution and a viscosity supersolution.

Using viscosity solutions, we can give a local characterization for a continuous function $u : \mathbb{R}^n \to \mathbb{R}$ to be convex. Recall that, in the case u is twice differentiable, the convexity of u is equivalent to its Hessian D^2u being nonnegative definite on $\mathbb{R}^n$. The last condition can be rephrased as

$\lambda_{\min}(D^2u) \geq 0$, where $\lambda_{\min}(D^2u)$ is the smallest eigenvalue of D^2u, which is an elliptic operator (see Problem 7.1). This equivalence turns out to hold for merely continuous functions u where the inequality $\lambda_{\min}(D^2u) \geq 0$ is interpreted in the viscosity sense, as observed by Oberman [**Ob**].

Theorem 7.2. *A continuous function* $u : \mathbb{R}^n \to \mathbb{R}$ *is convex if and only if* $\lambda_{\min}(D^2u) \geq 0$ *in the viscosity sense.*

Proof. Assume that the continuous function $u : \mathbb{R}^n \to \mathbb{R}$ is convex. We show that $\lambda_{\min}(D^2u) \geq 0$ in the viscosity sense. Let $\varphi \in C^2(\mathbb{R}^n)$ and $x_0 \in \mathbb{R}^n$ satisfy that $u - \varphi$ has a local maximum at x_0. Then

$$(7.4) \qquad \partial u(x_0) = \{D\varphi(x_0)\} \quad \text{and} \quad D^2\varphi(x_0) \geq 0,$$

where the last inequality clearly implies $\lambda_{\min}(D^2\varphi(x_0)) \geq 0$ as desired.

To see (7.4), let $p \in \partial u(x_0)$ and $l(x) := u(x_0) + p \cdot (x - x_0)$. Then $l - u$ has a local maximum at x_0. Therefore, $l - \varphi = (l - u) + (u - \varphi)$ has a local maximum at x_0. Thus, $D(l - \varphi)(x_0) = 0$ and $D^2(l - \varphi)(x_0) \leq 0$ from which we find $D\varphi(x_0) = Dl(x_0) = p$ and $D^2\varphi(x_0) \geq D^2l(x_0) = 0$. See Figure 7.1.

Now, assume that $\lambda_{\min}(D^2u) \geq 0$ in the viscosity sense. We show that u is convex. If otherwise, then there exist $y, z \in \mathbb{R}^n$ and $0 < \theta < 1$ such that

$$(7.5) \qquad u(\theta y + (1 - \theta)z) > \theta u(y) + (1 - \theta)u(z).$$

We can choose a coordinate system so that $y = (0, y_n)$ and $z = (0, z_n)$ where $z_n > y_n > 0$. Let

$$w = \theta y + (1 - \theta)z = (0, \theta y_n + (1 - \theta)z_n).$$

Let $a(x_n - w_n)^2/2 + b(x_n - w_n) + u(w)$ be the quadratic function which is equal to u at y, z, and w. From (7.5), we have $a < 0$. For $\delta \in (0, -a)$,

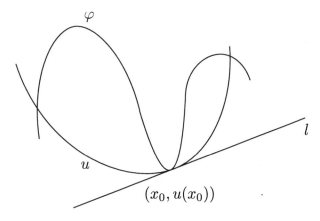

Figure 7.1. If $u - \varphi$ has a local maximum at x_0, then $l - \varphi$ has a local maximum at x_0.

consider

$$q(x) = (a + \delta)(x_n - w_n)^2/2 + b(x_n - w_n) + u(w)$$

Then $u - q = 0$ at w while $u - q < 0$ at y and z.

Consider first the case $n = 1$. Then there is $\tilde{w} \in (y, z)$ such that $u - q$ has a local maximum at $\tilde{w}$. We obtain a contradiction since $\lambda_{\min}(D^2 q(\tilde{w})) = a + \delta < 0$.

Now, consider the case $n > 1$. Observe that there is $\varepsilon \in (0, (z_n - y_n)/4)$ such that $q > u$ in $B_\varepsilon(y)$ and $B_\varepsilon(z)$. Let K be the convex hull of $\overline{B_\varepsilon(y)}$ and $\overline{B_\varepsilon(z)}$, and let $M = \max_K u$. Consider the function

$$\varphi(x) = q(x) + M|x'|^2/\varepsilon^2.$$

Then $\varphi(w) = u(w)$, while $\varphi \geq u$ on ∂K. Indeed, if $x \in \partial K$ and $|x'| < \varepsilon$, then $x \in \partial B_\varepsilon(y) \cup \partial B_\varepsilon(z)$, so $\varphi(x) \geq q(x) \geq u(x)$, while if $x \in \partial K$ and $|x'| = \varepsilon$, then $\varphi(x) \geq M \geq u(x)$. Note that $\lambda_{\min}(D^2\varphi) = a + \delta < 0$, and w is an interior point of K.

If w is a local maximum of $u - \varphi$, then from $\lambda_{\min}(D^2\varphi(w)) = a + \delta < 0$, we obtain a contradiction. If not, then from $\varphi(w) = u(w)$ and $\varphi \geq u$ on ∂K, we infer that $u - \varphi$ has a local maximum at some point $\tilde{w}$ in the interior of K. Since $\lambda_{\min}(D^2\varphi(\tilde{w})) = a + \delta < 0$, we also obtain a contradiction. $\square$

Remark 7.3. Theorem 7.2 suggests defining convex functions on a domain Ω, not necessarily convex, as continuous functions on Ω that are viscosity solutions of $\lambda_{\min}(D^2 u) \geq 0$. This definition needs only local information and does not requiring extending functions to the whole space.

Since the Monge–Ampère operator is not elliptic in general, to define viscosity solutions of the Monge–Ampère equation with convex solutions, we need to slightly modify Definition 7.1. This is to make sure that when we apply the maximum principle, the Hessians of the smooth solutions (if any) and the test functions should be nonnegative definite matrices where the Monge–Ampère operator is in fact elliptic.

Definition 7.4. Let $f \in C(\Omega)$ with $f \geq 0$ in an open set Ω in $\mathbb{R}^n$. Let $u \in C(\Omega)$ be a convex function. We say that:

(a) u is a viscosity supersolution of $\det D^2 u = f$, and we will write

$$\det D^2 u \leq f \quad \text{in the viscosity sense}$$

if for any convex function $\varphi \in C^2(\Omega)$ such that $u - \varphi$ has a local minimum at $x_0 \in \Omega$, then $\det D^2\varphi(x_0) \leq f(x_0)$.

(b) u is a viscosity subsolution of $\det D^2 u = f$, and we will write

$$\det D^2 u \geq f \quad \text{in the viscosity sense}$$

if for any $\varphi \in C^2(\Omega)$ such that $u - \varphi$ has a local maximum at $x_0 \in \Omega$, then $\det D^2 \varphi(x_0) \geq f(x_0)$.

(c) u is a viscosity solution of $\det D^2 u = f$ if u is both a viscosity supersolution and a viscosity subsolution of $\det D^2 u = f$.

Note that the test functions are only required to be convex in the definition of viscosity supersolutions for the Monge–Ampère equation, and this requirement is necessary for classical solutions to be viscosity supersolutions as shown in the following example.

Example 7.5. Let $n \geq 2$ be an even integer. The convex function $u \equiv 0$ is the classical solution of the Monge–Ampère equation $\det D^2 u = 0$ on $\mathbb{R}^n$. Let $\varphi(x) = -|x|^2$. Then $u(x) - \varphi(x) = |x|^2$ has a local minimum at $x_0 = 0$. We have $\det D^2 \varphi(0) = 2^n > 0$, so, without requiring the test functions to be convex, u cannot be a viscosity supersolution of $\det D^2 u = 0$.

On the other hand, at any touching point with viscosity subsolutions for the Monge–Ampère equation, test functions have a sort of pointwise convexity.

Remark 7.6. Let $u \in C(\Omega)$ be a convex function. Assume that $\varphi \in C^2(\Omega)$ and $x_0 \in \Omega$ satisfy that $u - \varphi$ has a local maximum at x_0. Then, as in the first part of the proof of Theorem 7.2, we have

$$\partial u(x_0) = \{D\varphi(x_0)\} \quad \text{and} \quad D^2 \varphi(x_0) \geq 0.$$

We show that Aleksandrov solutions to the Monge–Ampère equations with continuous right-hand side are also viscosity solutions.

Proposition 7.7 (Aleksandrov solutions are also viscosity solutions). *Let $f \in C(\Omega)$ with $f \geq 0$ in an open set Ω in $\mathbb{R}^n$. Let $u \in C(\Omega)$ be a convex function.*

(i) *If $\mu_u \geq f \, dx$, then $\det D^2 u \geq f$ in the viscosity sense.*

(ii) *If $\mu_u \leq f \, dx$, then $\det D^2 u \leq f$ in the viscosity sense.*

Proof. We prove part (i). Assume $\mu_u \geq f \, dx$. Let $\varphi \in C^2(\Omega)$, and let $x_0 \in \Omega$ be such that $u - \varphi$ has a local maximum at x_0. We need to show that $\det D^2 \varphi(x_0) \geq f(x_0)$. For $\delta > 0$ small, consider

$$\varphi_\delta(x) = \varphi(x) + \delta |x - x_0|^2 - \varphi(x_0) + u(x_0).$$

Then $u(x_0) = \varphi_\delta(x_0)$ and $u(x) - \varphi_\delta(x) < 0$ for all $0 < |x - x_0| \leq \delta$. From Remark 7.6, we have $D^2 \varphi(x_0) \geq 0$. Thus, we can choose $\delta > 0$ small so that φ_δ is strictly convex in $B_\delta(x_0) \subset \Omega$.

For any $0 < \varepsilon < (1/2) \min_{\delta/2 \le |x-x_0| \le \delta} \{|u(x) - \varphi_\delta(x)|\}$, consider the set

$$K_\varepsilon := \{x \in B_\delta(x_0) : u(x) + \varepsilon > \varphi_\delta(x)\}.$$

By the definition of ε, we have $x_0 \in K_\varepsilon \subset B_{\delta/2}(x_0)$ and $u + \varepsilon = \varphi_\delta$ on ∂K_ε. By the maximum principle in Lemma 3.11, $\partial(u + \varepsilon)(K_\varepsilon) \subset \partial\varphi_\delta(K_\varepsilon)$. Therefore, using $\mu_u \ge f\,dx$ and invoking Theorem 2.74, we have

$$\int_{K_\varepsilon} f\,dx \le |\partial(u + \varepsilon)(K_\varepsilon)| \le |\partial\varphi_\delta(K_\varepsilon)| = \int_{K_\varepsilon} \det D^2\varphi_\delta\,dx.$$

Since f is continuous at x_0, by dividing the above inequalities by $|K_\varepsilon|$ and letting $\varepsilon \to 0^+$, we get $\det D^2\varphi_\delta(x_0) \ge f(x_0)$. Now, let $\delta \to 0^+$ to obtain $\det D^2\varphi(x_0) \ge f(x_0)$, as desired.

Next, we prove part (ii). Assume $\mu_u \le f\,dx$. Let $\varphi \in C^2(\Omega)$ be convex such that $u - \varphi$ has a local minimum at $x_0 \in \Omega$. We need to show that $\det D^2\varphi(x_0) \le f(x_0)$. Note that $D^2\varphi(x_0) \ge 0$ by the convexity of φ. If one of the eigenvalues of $D^2\varphi(x_0)$ is zero, then we are done. Assume now $D^2\varphi(x_0) > 0$. Then, we can find $\delta > 0$ such that $D^2\varphi(x_0) > 2\delta I_n$ in $B_\delta(x_0)$. In this case,

$$\underline{\varphi}_\delta(x) = \varphi(x) - \delta|x - x_0|^2 - \varphi(x_0) + u(x_0)$$

is convex in $B_\delta(x_0)$. Now, as in part (i), we compare $\underline{\varphi}_\delta$ with $u - \varepsilon$ from below for a suitable $\varepsilon > 0$. The rest of the proof is the same. $\qquad \square$

A converse version of Proposition 7.7 will be given in Proposition 7.11 below. For this, we need several properties of viscosity solutions.

Similar to Lemma 3.10, we observe the following superadditive and multiplicative properties.

Lemma 7.8. *Let Ω be an open set in $\mathbb{R}^n$, $f \in C(\Omega)$ with $f \ge 0$ in Ω. Let $u \in C(\Omega)$ be a convex, viscosity subsolution to $\det D^2u = f$.*

(i) *Let $g \in C(\Omega)$ with $g \ge 0$ in Ω. Assume that $v \in C^2(\Omega)$ is a convex subsolution to $\det D^2v = g$. Then $u + v$ is a viscosity subsolution to $\det D^2(u + v) = f + g$.*

(ii) *Let $A \in \mathbb{S}^n$ be positive definite and $w(x) = u(Ax)$ on $\Omega' := A^{-1}\Omega$. Then w is a viscosity subsolution to $\det D^2w = (\det A)^2 f \circ A$.*

Proof. We prove part (i). Let $\varphi \in C^2(\Omega)$, and let $x_0 \in \Omega$ be such that $u + v - \varphi = u - (\varphi - v)$ has a local maximum at x_0. We show that $\det D^2\varphi(x_0) \ge f(x_0) + g(x_0)$. Indeed, since u is a viscosity subsolution, we have $\det D^2(\varphi - v)(x_0) \ge f(x_0)$. By Remark 7.6, $D^2(\varphi - v)(x_0) \ge 0$, so (2.17) gives

$$\det D^2\varphi(x_0) \ge \det D^2(\varphi - v)(x_0) + \det D^2v(x_0) \ge f(x_0) + g(x_0).$$

Now, we prove part (ii). Consider $z_0 \in \Omega'$ so $x_0 := Az_0 \in \Omega$, and let $\varphi \in C^2(\Omega')$ be such that $w - \varphi$ has a local maximum at z_0. We show that

$$(7.6) \qquad \det D^2\varphi(z_0) \geq (\det A)^2 f(Az_0) = (\det A)^2 f(x_0).$$

To see this, let $\tilde{\varphi}(x) := \varphi(A^{-1}x)$. Then $\tilde{\varphi} \in C^2(\Omega)$ and $u - \tilde{\varphi}$ has a local maximum at x_0. This gives $\det D^2\tilde{\varphi}(x_0) \geq f(x_0)$ which implies (7.6), due to $\det D^2\varphi(z_0) = (\det A)^2 \det D^2\tilde{\varphi}(x_0)$. The lemma is proved. $\qquad\square$

Next, we will prove the comparison principle for viscosity solutions. We start with the case when one of the two solutions is smooth.

Lemma 7.9 (Comparison principle with smooth functions). *Let Ω be a bounded open set in $\mathbb{R}^n$. Let $f, g \in C(\Omega)$ with $f, g \geq 0$ in Ω.*

(i) *Let $u \in C(\overline{\Omega})$ be a convex, viscosity subsolution to $\det D^2 u = f$. Assume that the convex function $v \in C^2(\Omega) \cap C(\overline{\Omega})$ satisfies $\det D^2 v \leq g$ in Ω. If $f \geq g$ in Ω, then*

$$\max_{\overline{\Omega}}(u - v) = \max_{\partial\Omega}(u - v).$$

(ii) *Let $u \in C(\overline{\Omega})$ be a convex, viscosity supersolution to $\det D^2 u = f$. Assume that the convex function $v \in C^2(\Omega) \cap C(\overline{\Omega})$ satisfies $\det D^2 v \geq g$ in Ω. If $f \leq g$ in Ω, then*

$$\min_{\overline{\Omega}}(u - v) = \min_{\partial\Omega}(u - v).$$

Proof. The proofs of assertions (i) and (ii) are similar. Let us prove (i), for example. We first make the strict inequality in $\det D^2 u \geq f$ by adding $\varepsilon|x|^2$ to u. By Lemma 7.8, for each $\varepsilon > 0$, the convex function $u_\varepsilon := u + \varepsilon|x|^2$ satisfies $\det D^2 u_\varepsilon \geq f + (2\varepsilon)^n$ in the viscosity sense. We will prove that

$$\max_{\overline{\Omega}}(u_\varepsilon - v) = \max_{\partial\Omega}(u_\varepsilon - v)$$

and then let $\varepsilon \to 0^+$ to conclude. Suppose otherwise that $\max_{\overline{\Omega}}(u_\varepsilon - v) > \max_{\partial\Omega}(u_\varepsilon - v)$. Then $u_\varepsilon - v$ must have a local maximum at some $x_0 \in \Omega$. By the definition of viscosity subsolutions, we must have $\det D^2 v(x_0) \geq f(x_0) + (2\varepsilon)^n$, but this contradicts the fact that $\det D^2 v(x_0) \leq g(x_0) \leq f(x_0)$. $\qquad\square$

Now, we can prove the full version of the comparison principle.

Theorem 7.10 (Comparison principle for viscosity solutions). *Let Ω be a bounded convex domain in $\mathbb{R}^n$. Assume that $f \in C(\Omega)$ with $f \geq 0$. Suppose that the convex functions $u, v \in C(\overline{\Omega})$ satisfy*

$$\det D^2 u \geq f \geq \det D^2 v \quad in \ \Omega$$

in the viscosity sense. If $u \leq v$ on $\partial\Omega$, then $u \leq v$ in Ω.

Proof. By proving the assertion for v and $u - \delta$ for $\delta > 0$ and then letting $\delta \to 0$, we can assume that $u < v$ on $\partial\Omega$. For a sufficiently small $\varepsilon > 0$, we still have $u + \varepsilon|x|^2 < v$ on $\partial\Omega$ while, by Lemma 7.8, we also have $\det D^2(u + \varepsilon|x|^2) \geq f + (2\varepsilon)^n$ in the viscosity sense. These inequalities still hold in uniformly convex domains $K \subset \Omega$ with smooth boundaries ∂K, where $\mathrm{dist}(K, \partial\Omega)$ is small. These domains exist by Theorem 2.51. Thus, it suffices to prove the lemma under the strengthened conditions that Ω is uniformly convex with C^∞ boundary $\partial\Omega$, and for some $\delta > 0$,

$$(7.7) \qquad \det D^2 u \geq f + \delta \geq f \geq \det D^2 v \quad \text{in } \Omega \quad \text{and} \quad u < v \quad \text{on } \partial\Omega.$$

Choose smooth functions $\varphi, g \in C^\infty(\overline{\Omega})$ such that

$$(7.8) \qquad u < \varphi < v \quad \text{on } \partial\Omega \quad \text{and} \quad f < g < f + \delta \quad \text{in } \Omega.$$

Let $w \in C(\overline{\Omega})$ be the unique Aleksandrov solution to

$$\det D^2 w = g \quad \text{in } \Omega, \quad w = \varphi \quad \text{on } \partial\Omega.$$

Then, by Theorem 4.12, $w \in C^\infty(\Omega)$. In view of (7.7) and (7.8), we obtain from Lemma 7.9 that $u < w < v$ in Ω. The theorem is proved. $\qquad\square$

The following proposition says that viscosity solutions to the Monge–Ampère equation with continuous and strictly positive right-hand side are also Aleksandrov solutions.

Proposition 7.11. *Let Ω be an open set in $\mathbb{R}^n$. Let $u \in C(\Omega)$ be convex.*

  (i) *If u is a viscosity supersolution to $\det D^2 u \leq \Lambda$ in Ω where $\Lambda > 0$, then u satisfies $\det D^2 u \leq \Lambda$ in the sense of Aleksandrov.*

  (ii) *Assume Ω is bounded. Let $f \in C(\overline{\Omega})$ with $f > 0$ in $\overline{\Omega}$. If u is a viscosity solution to $\det D^2 u = f$ in Ω, then u is also an Aleksandrov solution to $\det D^2 u = f$ in Ω.*

Proof. We first prove part (i). It suffices to show that if $B := B_r(x_0) \subset \Omega$, then $|\partial u(B)| \leq \Lambda|B|$. Indeed, let $w \in C(\overline{B})$ be the Aleksandrov solution to

$$\det D^2 w = \Lambda \quad \text{in } B, \quad w = u \quad \text{on } \partial B$$

which exists by Theorem 3.37. By Proposition 7.7, w is also a viscosity solution. By Theorem 7.10, we have $u \geq w$ in B. Since $w = u$ on ∂B, by the maximum principle in Lemma 3.11, we have $\partial u(B) \subset \partial w(B)$. Therefore, $|\partial u(B)| \leq |\partial w(B)| = \Lambda|B|$, as asserted.

For the reader's convenience, we explain why μ_u is absolutely continuous with respect to the Lebesgue measure. Let $A \subset \Omega$ be a set of Lebesgue measure zero. We show that $\mu_u(A) = 0$. It suffices to show that $\mu_u(A_k) = 0$ for each $k \in \mathbb{N}$, where

$$A_k := A \cap \{x \in \Omega : \mathrm{dist}(x, \partial\Omega) > 1/k\}.$$

Hence, we can assume that $\delta := \text{dist}(A, \partial\Omega)/4 > 0$. Given any $\varepsilon > 0$, by the characterization of the Lebesgue measure in (2.20), we can find cubes $Q_i \Subset \Omega$ with $\text{diam}(Q_i) \le \delta$ and $\{x \in \Omega : \text{dist}(x, \partial Q_i) \le \delta\} \Subset \Omega$ such that

$$A \subset \bigcup_{i=1}^{\infty} Q_i \quad \text{and} \quad \sum_{i=1}^{\infty} |Q_i| < \varepsilon.$$

For each i and $x_0 \in Q_i$ (say $\text{diam}(Q_i) = \delta$) we have $Q_i \subset B_\delta(x_0) \Subset \Omega$ and

$$|\partial u(Q_i)| \le |\partial u(B_\delta(x_0))| \le \Lambda |B_\delta(x_0)| \le 2^n \omega_n \Lambda |Q_i|.$$

Then

$$|\partial u(A)| \le \sum_{i=1}^{\infty} |\partial u(Q_i)| \le 2^n \omega_n \Lambda \sum_{i=1}^{\infty} |Q_i| < 2^n \omega_n \Lambda \varepsilon.$$

This proves the absolute continuity of μ_u with respect to the Lebesgue measure. Hence, there is a function $g \in L^1_{\text{loc}}(\Omega)$ such that

$$\mu_u(E) = \int_E g \, dx \quad \text{for all Borel sets } E \subset \Omega.$$

The inequality $\mu_u(B_r(x_0)) \le \Lambda |B_r(x_0)|$ for all $B_r(x_0) \subset \Omega$ implies that $g(x) \le \Lambda$ at all Lebesgue points $x \in \Omega$ of g. Thus $g \le \Lambda$ almost everywhere in Ω. Hence $\mu_u(E) \le \Lambda |E|$ for all Borel sets $E \subset \Omega$.

Finally, we prove part (ii). Since $f \in C(\overline{\Omega})$ and $f > 0$ in $\overline{\Omega}$, there are $0 < \lambda \le \Lambda$ such that $\lambda \le f \le \Lambda$. Fix $x_0 \in \Omega$. For any $0 < \varepsilon < \lambda/2$, there is $0 < \delta = \delta(\varepsilon, x_0, f, \Omega) < \text{dist}(x_0, \partial\Omega)$ such that

$$f(x_0) - \varepsilon \le f \le f(x_0) + \varepsilon \quad \text{in } B_\delta(x_0).$$

As in part (i), we can show that for any $0 < r \le \delta$

$$(f(x_0) - \varepsilon)|B_r(x_0)| \le |\partial u(B_r(x_0))| = \mu_u(B_r(x_0)) \le (f(x_0) + \varepsilon)|B_r(x_0)|.$$

This shows that the Monge–Ampère measure μ_u of u has a density F which satisfies for almost all $x \in \Omega$ and for all $0 < \varepsilon < \lambda/2$

$$f(x_0) - \varepsilon \le F(x_0) \le f(x_0) + \varepsilon.$$

Therefore, $F = f$, and u is an Aleksandrov solution to $\det D^2 u = f$ in Ω. $\square$

Similar to the weak continuity of the Monge–Ampère measure in Theorem 3.7, we have the following *stability* property of viscosity solutions.

Theorem 7.12 (Stability property of viscosity solutions). *Let $\{\Omega_k\}_{k=1}^{\infty}$ be a sequence of bounded convex domains in $\mathbb{R}^n$ that converges, in the Hausdorff distance, to a bounded convex domain Ω in $\mathbb{R}^n$. Let $f_k \in C(\Omega_k)$, $f_k \ge 0$, and let convex function $u_k \in C(\Omega_k)$ satisfy $\det D^2 u_k \ge (\le) f_k$ in Ω_k in the viscosity sense. Suppose that $\{u_k\}_{k=1}^{\infty}$ converges uniformly on compact* *subsets of Ω to a convex function u and that $\{f_k\}_{k=1}^{\infty}$ converges uniformly on compact subsets of Ω to $f \in C(\Omega)$. Then $\det D^2 u \geq (\leq) f$ in Ω in the viscosity sense.*

Proof. We provide the proof for viscosity subsolutions; the proof for viscosity supersolutions is similar. Suppose $\varphi \in C^2(\Omega)$ and suppose $u - \varphi$ has a local maximum at $x_0 \in \Omega$. We show that $\det D^2 \varphi(x_0) \geq f(x_0)$. Indeed, by adding $u(x_0) - \varphi(x_0)$ to φ and using local maximality, we can find a ball $B_r := B_r(x_0) \Subset \Omega$ such that $u(x) \leq \varphi(x)$ in B_r, while $u(x_0) = \varphi(x_0)$. Note that $B_r \subset \Omega_k$ when $k \geq k_0$. For $0 < \varepsilon < 1$, consider $\varphi_\varepsilon(x) = \varphi(x) + \varepsilon|x - x_0|^2$. Then $u - \varphi_\varepsilon$ has a strict local maximum at x_0, and $\max_{\partial B_{\varepsilon r}}(u - \varphi_\varepsilon) < (u - \varphi_\varepsilon)(x_0)$. Since u_k converges uniformly to u in $\overline{B_{\varepsilon r}}$, we have $\max_{\partial B_{\varepsilon r}}(u_k - \varphi_\varepsilon) < (u_k - \varphi_\varepsilon)(x_0)$ for all $k \geq k_1(\varepsilon) \geq k_0$. Hence $u_k - \varphi_\varepsilon$ has a local maximum at some point $x_\varepsilon \in B_{\varepsilon r} \subset \Omega_k$. Since u_k satisfies $\det D^2 u_k \geq f_k$ in the viscosity sense in Ω_k, we have

$$f_k(x_\varepsilon) \leq \det D^2 \varphi_\varepsilon(x_\varepsilon) = \det(D^2 \varphi(x_\varepsilon) + 2\varepsilon I_n)$$

for $k \geq k_1(\varepsilon)$. Now, letting $\varepsilon \to 0$ and recalling $|x_\varepsilon - x_0| \leq \varepsilon r$ and the uniform convergence of f_k to f in $\overline{B_r}$, we obtain $\det D^2 \varphi(x_0) \geq f(x_0)$. $\square$

7.2. Viscosity Solutions with Boundary Discontinuities

In solving the inhomogeneous Dirichlet problem (Section 3.7) and in establishing the interior regularity (Sections 5.2 and 5.3), we employed compactness arguments for the Monge–Ampère equation with either affine boundary values or on strictly convex domains. Now, we consider the compactness problem for more general boundary data. Our study naturally leads to a discussion of viscosity solutions to the Monge–Ampère equation with boundary discontinuities and their compactness. They will be used in the proof of Savin's Boundary Localization Theorem (Theorem 8.5).

A compactness problem. The basic compactness problem is this: Given a sequence of convex viscosity solutions $u_k : \Omega_k \to \mathbb{R}$ $(k = 1, 2, \ldots)$ to the Monge–Ampère equation

$$\det D^2 u_k = f_k \quad \text{on } \Omega_k$$

on bounded, convex domains Ω_k in $\mathbb{R}^n$, with boundary values

$$u_k = \varphi_k \quad \text{on } \partial\Omega_k,$$

(say, φ_k is equal to the restriction to $\partial\Omega_k$ of a convex extension of u_k to $\mathbb{R}^n$), study the limit, if possible, of the quadruples $(\Omega_k, u_k, \varphi_k, f_k)$. More concretely, suppose that in some sense to be specified, we have

(7.9) $\qquad\qquad \Omega_k \to \Omega, \quad f_k \to f, \quad u_k \to u, \quad \varphi_k \to \varphi.$

Can we then conclude that

$$\det D^2 u = f \quad \text{in } \Omega, \quad u = \varphi \quad \text{on } \partial\Omega?$$

That u satisfies the Monge–Ampère equation $\det D^2 u = f$ in Ω usually follows from the stability property of viscosity solutions in Theorem 7.12. So the main question is whether $u = \varphi$ on $\partial\Omega$.

Note carefully that this question has two ingredients: (a) assigning the boundary values of u on $\partial\Omega$ for a convex function u originally defined on Ω and (b) verifying the equality between these values and φ. For (a), it is natural to extend u to $\partial\Omega$ by restricting to $\partial\Omega$ its minimal convex extension to $\mathbb{R}^n$ (see also Remark 2.22). This process gives the minimal boundary values for u. Interestingly, when Ω is convex, we can also express them in terms of the convex epigraph $\mathrm{epi}(u,\Omega)$ of u over Ω. This is especially relevant to our compactness problem as it involves convergence of functions defined on different domains and their boundaries. Now, the compactness problem can be rephrased in terms of convergences in the Hausdorff distance of convex sets, including $\mathrm{epi}(u_k, \overline{\Omega}_k)$ and $\mathrm{epi}(\varphi_k, \partial\Omega_k)$.

We can expect the answer to the compactness problem to be affirmative when $\partial\Omega$ is strictly convex for a large class of boundary functions; see Theorem 3.35. However, when $\partial\Omega$ is not strictly convex, the equation $u = \varphi$ may not hold in the usual pointwise sense; even more seriously, φ may be nonconvex on $\partial\Omega$; compare with Remark 3.34. Thus, for the compactness problem to have a positive answer, we may need to understand the expression $u = \varphi$ on $\partial\Omega$ in a sense different from the pointwise one.

As will be seen later, $\varphi \geq u$ on $\partial\Omega$ in the pointwise sense; we saw this in the very special case of Theorem 3.35 where we obtain $g \geq u$ on $\partial\Omega$ from the inequality (3.27). Since u is convex on $\partial\Omega$, it is not greater than the convex envelope φ^c of φ on $\partial\Omega$. So, the best hope is to have $u = \varphi^c$ on $\partial\Omega$, and this is actually what we will obtain. If this holds, we will say that $u = \varphi$ on $\partial\Omega$ in the epigraphical sense. An interesting point to notice is that when $\partial\Omega$ has a flat part Γ which lies on an $(n-1)$-dimensional hyperplane, there are many choices of $\varphi \geq u$ on $\partial\Omega$ for $u = \varphi$ in the epigraphical sense.

For the convergences in the Hausdorff distance alluded to above, we only need the closedness of the eipigraphs so the convex functions $u_k : \overline{\Omega}_k \to \mathbb{R} \cup \{+\infty\}$ and the boundary data $\varphi_k : \partial\Omega_k \to \mathbb{R} \cup \{+\infty\}$ are only required to be lower semicontinuous. When $\varphi : \partial\Omega \to \mathbb{R}$ is not continuous, the Monge–Ampère equation with boundary data $u = \varphi$ on $\partial\Omega$ in the epigraphical sense has boundary discontinuities.

In what follows, we will make precise the above discussions.

Recall that the epigraph of a function $f : E \to \mathbb{R} \cup \{+\infty\}$ over a set $E \subset \mathbb{R}^n$ is defined by

$$\mathrm{epi}(f, E) := \{(x, x_{n+1}) \in E \times \mathbb{R} : x_{n+1} \geq f(x)\}.$$

Definition 7.13. Let $E \subset \mathbb{R}^n$ be a nonempty set. A function $f : E \to \mathbb{R} \cup \{+\infty\}$ is said to be *lower semicontinuous* if for all $x_0 \in E$, we have

$$f(x_0) \leq \liminf_{x \to x_0,\ x \in E} f(x).$$

An equivalent definition of lower semicontinuity for f is that its epigraph $\mathrm{epi}(f, E)$ is closed in $E \times \mathbb{R}$.

Definition 7.14 (Boundary values of a convex function). Consider a convex function $u \in C(\Omega)$ on a bounded and convex domain Ω in $\mathbb{R}^n$. We define the boundary values of u on $\partial\Omega$ to be equal to $\varphi : \partial\Omega \to \mathbb{R} \cup \{+\infty\}$ in the *epigraphical sense* and write

$$u|_{\partial\Omega,\mathrm{epi}} = \varphi$$

if the epigraph of φ is given by the closure of $\mathrm{epi}(u, \Omega)$ restricted to $\partial\Omega \times \mathbb{R}$; that is,

$$\mathrm{epi}(\varphi, \partial\Omega) := \overline{\mathrm{epi}(u, \Omega)} \cap (\partial\Omega \times \mathbb{R}).$$

Remark 7.15. Let $u \in C(\Omega)$ be a convex function on a bounded, convex domain Ω in $\mathbb{R}^n$. We have the following comments on the notation $u|_{\partial\Omega,\mathrm{epi}}$.

- In [**Ba5**, Section 10.3], $u|_{\partial\Omega,\mathrm{epi}}$ is called the *border* of u. It can be verified that $\inf_\Omega u > -\infty$ (Problem 2.2), so the boundary values of u cannot be $-\infty$. Moreover, $u|_{\partial\Omega,\mathrm{epi}}$ gives the usual pointwise boundary values of a convex function $u \in C(\overline{\Omega})$; see Lemma 7.16.

- With the values of u on $\partial\Omega$ understood as $u|_{\partial\Omega,\mathrm{epi}}$, we have

$$\mathrm{epi}(u, \overline{\Omega}) = \overline{\mathrm{epi}(u, \Omega)}.$$

  In Definition 7.14, if $z \in \partial\Omega$ does not belong to the projection of $\overline{\mathrm{epi}(u, \Omega)}$ onto $\mathbb{R}^n$, then $\varphi(z) = +\infty$.

- From the definition of $u|_{\partial\Omega,\mathrm{epi}} = \varphi$, we see that $\mathrm{epi}(\varphi, \partial\Omega)$ is closed; hence, φ is lower semicontinuous. Moreover, φ is convex in the following sense: $\mathrm{epi}(\varphi, \partial\Omega)$ is also the restriction to $\partial\Omega \times \mathbb{R}$ of the convex hull generated by $\mathrm{epi}(\varphi, \partial\Omega)$.

Now, we relate the boundary values in the epigraphical sense, convex extensions and pointwise values of a convex function u on a convex domain Ω. All statements remain unchanged after subtracting the minimum value of u, so we can assume that u has a minimum value 0 at $0 \in \Omega$.

Lemma 7.16 (Minimal convex extension). *Let $u \in C(\Omega)$ be a convex function on a bounded convex domain Ω in $\mathbb{R}^n$ with $0 \in \Omega$. Assume $u(0) = 0$ and $u \geq 0$ in Ω. Define the boundary values of u on $\partial\Omega$ in the epigraphical sense as in Definition 7.14. Then:*

(i) *$u : \overline{\Omega} \to \mathbb{R} \cup \{+\infty\}$ is the minimal convex extension of u from Ω to $\overline{\Omega}$.*

(ii) *$u(x) = \lim_{t \to 1^-} u(tx)$ for $x \in \partial\Omega$.*

Proof. We prove part (i). Let $v : \overline{\Omega} \to \mathbb{R} \cup \{+\infty\}$ be a convex function such that $v = u$ in Ω. We will show that $\mathrm{epi}(v, \overline{\Omega}) \subset \mathrm{epi}(u, \overline{\Omega})$, so $u \leq v$ on $\partial\Omega$. Indeed, if $(x, \alpha) \in \mathrm{epi}(v, \overline{\Omega})$, then $v(x) \leq \alpha$, and for any $0 \leq t < 1$, by convexity, we have

$$(7.10) \qquad u(tx) = v(tx) \leq tv(x) + (1-t)v(0) = tv(x) \leq t\alpha.$$

Therefore, $(tx, t\alpha) \in \mathrm{epi}(u, \Omega)$. Letting $t \to 1^-$, we find $(x, \alpha) \in \overline{\mathrm{epi}(u, \Omega)}$. From $\mathrm{epi}(u, \overline{\Omega}) = \overline{\mathrm{epi}(u, \Omega)}$, we deduce $(x, \alpha) \in \mathrm{epi}(u, \overline{\Omega})$ and this proves (i).

Now, we prove part (ii). If $w : \overline{\Omega} \to \mathbb{R} \cup \{+\infty\}$ is defined by $w(x) = u(x)$ for $x \in \Omega$, while $w(x) = \lim_{t \to 1^-} u(tx)$ for $x \in \partial\Omega$, then w is convex, and using (7.10), we infer that it is in fact the minimal convex extension of u to $\overline{\Omega}$. Hence, we obtain the conclusion of part (ii) from part (i). $\qquad\square$

Definition 7.17 (Convex envelope of a function defined on the boundary). Let $\varphi : \partial\Omega \to \mathbb{R} \cup \{+\infty\}$ be lower semicontinuous. We define φ^c to be the convex envelope of φ on $\partial\Omega$; that is, $\mathrm{epi}(\varphi^c, \partial\Omega)$ is the restriction to $\partial\Omega \times \mathbb{R}$ of the convex hull $\mathcal{C}_\varphi$ generated by $\mathrm{epi}(\varphi, \partial\Omega)$. See Figure 7.2.

Note that $\varphi^c \leq \varphi$. We always have $(\varphi^c)^c = \varphi^c$. If Ω is strictly convex, then for a large class of boundary functions, we have $\varphi^c = \varphi$, as shown next.

Lemma 7.18. *Let Ω be a bounded, strictly convex domain in $\mathbb{R}^n$, and let $\varphi : \partial\Omega \to \mathbb{R}$ be lower semicontinuous and bounded. Then $\varphi^c = \varphi$.*

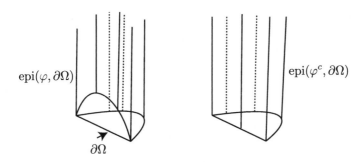

Figure 7.2. Epigraphs of φ and φ^c on $\partial\Omega$.

Proof. Since $\varphi^c \leq \varphi$, it remains to show that $\varphi^c \geq \varphi$. Let $x_0 \in \partial\Omega$ be an arbitrary point. We show that $\varphi^c(x_0) \geq \varphi(x_0)$. The proof is inspired by (3.24). We can assume that $x_0 = 0$ and $\Omega \subset \{x \in \mathbb{R}^n : x_n > 0\}$. Let $\varepsilon > 0$. From the lower semicontinuity of φ, there exists $\delta(\varepsilon) > 0$ such that

$$\varphi(x) > \varphi(0) - \varepsilon \quad \text{for all } x \in \partial\Omega \cap B_\delta(0).$$

Since Ω is strictly convex, there exists $\eta(\varepsilon) > 0$ such that $x_n \geq \eta$ for all $x \in \partial\Omega \setminus B_\delta(0)$. It can be verified that for $C_1 = 2\|\varphi\|_{L^\infty(\partial\Omega)}/\eta$, we have

$$l(x) := \varphi(0) - \varepsilon - C_1 x_n < \varphi(x) \quad \text{for all } x \in \partial\Omega.$$

Thus, the hyperplane $x_{n+1} = l(x)$ lies strictly below the convex hull $\mathcal{C}_\varphi$ of epi$(\varphi, \partial\Omega)$. From the definition of φ^c, we find $\varphi^c \geq l$. In particular, we have $\varphi^c(0) \geq l(0) = \varphi(0) - \varepsilon$ for all $\varepsilon > 0$, so $\varphi^c(0) \geq \varphi(0)$ as desired. $\qquad\square$

Example 7.19. We give here some examples of convex envelope.

    (a) Let Ω be a uniformly convex domain in $\mathbb{R}^n$ with $0 \in \partial\Omega$. Let $\varphi(x) = 1$ if $x \in \partial\Omega \setminus \{0\}$, and let $\varphi(0) = 0$. Then

$$\text{epi}\,(\varphi, \partial\Omega) = \partial\Omega \setminus \{0\} \times [1, \infty) \cup \{0\} \times [0, \infty),$$

    and $\varphi^c \equiv \varphi$ on $\partial\Omega$.

    (b) Let $\Omega \subset \mathbb{R}^2$ be a triangle with vertices z_1, z_2, z_3. Let

$$\varphi(x) = \begin{cases} 1 & \text{if } x \in \partial\Omega \setminus \{z_1, z_2, z_3\}, \\ 0 & \text{if } x \in \{z_1, z_2, z_3\}. \end{cases}$$

    Then $\varphi^c \equiv 0$ on $\partial\Omega$.

    We introduce the following extension of Definition 7.14 to boundary functions that are not necessarily convex. When we discuss compactness results for viscosity solutions, it will be a very convenient notion as the boundary data is sort of preserved in the limit.

Definition 7.20. Let $\varphi : \partial\Omega \to \mathbb{R} \cup \{+\infty\}$ be a lower semicontinuous function. If a convex function $u : \Omega \to \mathbb{R}$ satisfies $u|_{\partial\Omega,\text{epi}} = \varphi^c$, then we write

$$u \overset{\text{epi}}{=} \varphi \quad \text{on } \partial\Omega.$$

Remark 7.21. Some remarks on the notation in Definition 7.20 are in order.

    • Whenever φ^c and φ do not coincide, we can think of the graph of u as having a vertical part on $\partial\Omega$ between φ^c and φ.

    • If a convex function $u : \Omega \to \mathbb{R}$ satisfies $u \overset{\text{epi}}{=} \varphi$ on $\partial\Omega$, then, on the domain $S_h := \{x \in \Omega : u(x) < h\}$, the boundary values of u

are given by $u \overset{\text{epi}}{=} \varphi_h$ on ∂S_h, where

$$\varphi_h = \begin{cases} \varphi & \text{on } \partial\Omega \cap \{\varphi \le h\} \subset \partial S_h, \\ h & \text{on the remaining part of } \partial S_h. \end{cases}$$

This observation will be useful in the study of boundary sections in Chapter 8.

Using the Hausdorff distance d_H (see Definition 2.13), we define the notion of convergence for functions defined on different domains.

Definition 7.22 (Hausdorff convergence of epigraphs). Let $\Omega, \Omega_k \subset \mathbb{R}^n$ ($k = 1, 2, \ldots$) be convex domains.

(a) Let $u_k : \Omega_k \to \mathbb{R}$ be a sequence of convex functions. We say that u_k converges to $u : \Omega \to \mathbb{R}$ and we write $u_k \to u$ in the sense of Hausdorff convergence of epigraphs if, in the Hausdorff distance d_H, the closures of their epigraphs converge; that is,

$$d_H\big(\overline{\text{epi}(u_k, \Omega_k)}, \overline{\text{epi}(u, \Omega)}\big) \to 0 \quad \text{when } k \to \infty.$$

In particular it follows that $\overline{\Omega_k} \to \overline{\Omega}$ in the Hausdorff distance and u_k converges uniformly to u on compact subsets of Ω.

(b) Let $\varphi_k : \partial\Omega_k \to \mathbb{R} \cup \{+\infty\}$ be a sequence of lower semicontinuous functions. We say that φ_k converges to $\varphi : \partial\Omega \to \mathbb{R} \cup \{+\infty\}$, and we write $\varphi_k \to \varphi$ in the sense of Hausdorff convergence of epigraphs if the epigraphs $\text{epi}(\varphi_k, \partial\Omega_k)$ converge in d_H to $\text{epi}(\varphi, \partial\Omega)$.

Remark 7.23. It follows from Theorem 2.18 that if Ω_k, u_k are uniformly bounded, then we can always extract a convergent subsequence $u_{k_m} \to u$. Similarly, if Ω_k, φ_k are uniformly bounded, then we can extract a convergent subsequence $\varphi_{k_m} \to \varphi$. The convergences here are in the sense of Hausdorff convergence of epigraphs.

Remark 7.24. Suppose that $u_k : \Omega_k \to \mathbb{R}$ are convex functions and $u_k|_{\partial\Omega_k, \text{epi}} = \varphi_k$. Then, the functions φ_k are convex but the limit function φ in Definition 7.22 is not necessarily so. Here is an example. Consider uniformly convex domains $\Omega_k \subset \mathbb{R}^2$ that converge to $\Omega = \{x \in \mathbb{R}^2 : |x| < 1, x_2 > 0\}$. We can find convex functions φ_k on $\partial\Omega_k$ that converges to a nonconvex function φ on $\partial\Omega$ where $\varphi(0) = 0$ and $\varphi(x) = 1$ for $x \in \partial\Omega \setminus \{0\}$.

With the above definitions, we now proceed to address the compactness problem raised at the beginning of this section. For this aim, some preparatory results are needed.

First, we show that viscosity subsolutions of the Monge–Ampère equation with continuous and bounded right-hand side can be approximated globally by subsolutions that are continuous up to the boundary.

Lemma 7.25. *Let $f \in C(\Omega) \cap L^\infty(\Omega)$ with $f \geq 0$ on a bounded convex domain Ω in $\mathbb{R}^n$. Assume that $u \in C(\Omega)$ is a convex viscosity subsolution to $\det D^2 u = f$ in Ω. Define the values of u on $\partial\Omega$ to be $u|_{\partial\Omega,\text{epi}}$ as in Definition 7.14. Then, there exists a sequence of convex functions $\{u_k\}_{k=1}^\infty \subset C(\overline{\Omega})$ satisfying $\det D^2 u_k \geq f$ in Ω in the viscosity sense, and*

$$\lim_{k \to \infty} u_k = u \quad \text{in } \overline{\Omega}.$$

Proof. By translating coordinates and subtracting a supporting hyperplane to the graph of u, we can assume that $0 \in \Omega$, $u(0) = 0$, and $u \geq 0$ in Ω. Then, by convexity, the functions $v_\varepsilon \in C(\overline{\Omega})$ defined by

$$v_\varepsilon(x) = u((1 - \varepsilon)x) \quad (\varepsilon > 0)$$

form an increasing family of continuous functions. By Lemma 7.16(ii),

$$\lim_{\varepsilon \to 0} v_\varepsilon = u \quad \text{in } \overline{\Omega}.$$

By Lemma 7.8, v_ε satisfies, in the viscosity sense,

$$\det D^2 v_\varepsilon(x) \geq (1 - \varepsilon)^{2n} f((1 - \varepsilon)x).$$

The above right-hand side is not yet f, so we modify v_ε as follows. Let

$$f_\varepsilon(x) = |f(x) - (1 - \varepsilon)^{2n} f((1 - \varepsilon)x)|,$$

and let $g_\varepsilon \in C^2(\Omega)$ be such that

(7.11) $f_\varepsilon + \varepsilon \leq g_\varepsilon$ in Ω and $\displaystyle\int_\Omega g_\varepsilon \, dx \to 0$ when $\varepsilon \to 0$.

Here is a construction of g_ε. For $\eta > 0$, let $\Omega_\eta = \{x \in \Omega : \text{dist}(x, \partial\Omega) > \eta\}$. We first construct $g_\varepsilon^1 \in C^2(\Omega)$ such that $g_\varepsilon^1 \geq 0$ and

$$f_\varepsilon + \varepsilon \leq g_\varepsilon^1 \leq f_\varepsilon + 2\varepsilon \quad \text{in } \Omega_{4\varepsilon}, \qquad g_\varepsilon^1 = 2\|f\|_{L^\infty(\Omega)} + \varepsilon \quad \text{in } \Omega \backslash \Omega_{2\varepsilon}.$$

Next, we construct $g_\varepsilon^2 \in C^2(\Omega)$ such that $g_\varepsilon^2 \geq 0$ and

$$f_\varepsilon + \varepsilon \leq g_\varepsilon^2 \leq f_\varepsilon + 2\varepsilon \quad \text{in } \Omega_{2\varepsilon}, \qquad g_\varepsilon^2 = 2\|f\|_{L^\infty(\Omega)} + \varepsilon \quad \text{in } \Omega \backslash \Omega_\varepsilon.$$

Then the function $g_\varepsilon := g_\varepsilon^1 + g_\varepsilon^2$ satisfies $f_\varepsilon + \varepsilon \leq g_\varepsilon$ in Ω. Using the Dominated Convergence Theorem, we find $\int_\Omega g_\varepsilon \, dx \to 0$ when $\varepsilon \to 0$.

Let $w_\varepsilon \in C(\overline{\Omega})$ be the Aleksandrov solution of

$$\det D^2 w_\varepsilon = g_\varepsilon \quad \text{in } \Omega, \qquad w_\varepsilon = 0 \quad \text{on } \partial\Omega.$$

Since $g_\varepsilon \in C^2(\Omega)$, $w_\varepsilon \in C^3(\Omega)$ by Theorem 6.5. From Lemma 7.8 and

$$(1 - \varepsilon)^{2n} f((1 - \varepsilon)x) + g_\varepsilon(x) \geq (1 - \varepsilon)^{2n} f((1 - \varepsilon)x) + f_\varepsilon(x) + \varepsilon > f(x),$$

we deduce that $\det D^2(v_\varepsilon + w_\varepsilon) \geq f$ in the viscosity sense. By (7.11) and the Aleksandrov maximum principle, w_ε converges uniformly to 0 in $\overline{\Omega}$. Thus $v_\varepsilon + w_\varepsilon \leq u$, and $v_\varepsilon + w_\varepsilon$ converges to u in $\overline{\Omega}$. The lemma is proved by setting $u_k := v_{1/k} + w_{1/k}$ for $k \in \mathbb{N}$.   □

Now, we extend the comparison principle in Theorem 7.10 to viscosity solutions with boundary discontinuities. One immediate consequence is that viscosity solutions of $\det D^2 u = f$ are determined uniquely by their boundary values $u|_{\partial\Omega,\text{epi}}$.

Theorem 7.26 (Comparison principle for viscosity solutions with boundary discontinuities). *Let Ω be a bounded convex domain in $\mathbb{R}^n$, and let $u, v : \Omega \to \mathbb{R}$ be convex functions. Let $f \geq 0$ be a bounded and continuous function on Ω.*

    (i) *Assume that $\det D^2 u \geq f \geq \det D^2 v$ in the viscosity sense and that*

$$u|_{\partial\Omega,\text{epi}} \leq v|_{\partial\Omega,\text{epi}}.$$

    Then $u \leq v$ in Ω.

    (ii) (General boundary data version) *Let $\varphi, \psi : \partial\Omega \to \mathbb{R}$ be lower semicontinuous functions such that $\varphi \leq \psi$ on $\partial\Omega$. If*

$$\det D^2 u \geq f \geq \det D^2 v \quad \text{in the viscosity sense in } \Omega$$

    and

$$u \overset{\text{epi}}{=} \varphi, \quad v \overset{\text{epi}}{=} \psi \quad \text{on } \partial\Omega,$$

    then $u \leq v$ in Ω.

Proof. It suffices to prove assertion (i) since assertion (ii) follows from assertion (i), Definition 7.20, and $\varphi^c \leq \psi^c$. By proving (i) first for u and $v + \varepsilon$ where $\varepsilon > 0$ and then letting $\varepsilon \to 0$, we can assume that $u|_{\partial\Omega,\text{epi}} < v|_{\partial\Omega,\text{epi}}$.

By Lemma 7.25, u can be approximated globally on $\overline{\Omega}$ by viscosity subsolutions of $\det D^2 u \geq f$ that are continuous on $\overline{\Omega}$. Thus, it suffices to prove assertion (i) in the case where u is continuous on $\overline{\Omega}$ and $u < v|_{\partial\Omega,\text{epi}}$ on $\partial\Omega$. Then, since v is lower semicontinuous on $\overline{\Omega}$ and it is the minimal convex extension of its interior values (see Lemma 7.16), we deduce that $u < v$ in a small neighborhood of $\partial\Omega$. Thus, for $\delta > 0$ small, we have $u < v$ on $\partial\Omega_\delta$, where $\Omega_\delta := \{x \in \Omega : \text{dist}(x, \partial\Omega) > \delta\}$. The comparison principle in Theorem 7.10 applied to u and v in $\overline{\Omega_\delta}$ gives $u \leq v$ in Ω_δ. This holds for all $\delta > 0$ small, so $u \leq v$ in Ω. $\square$

Next, we identify the boundary values of the limit solution of a sequence of viscosity solutions that converges in the sense of Hausdorff convergence of epigraphs. Recall Definition 7.20 for φ^c.

Theorem 7.27 (Boundary values of the limit viscosity solution). *Let Ω_k be convex domains in $\mathbb{R}^n$, $f_k \in C(\Omega_k)$ with $f_k \geq 0$ in Ω_k ($k = 1, 2, \ldots$). For each $k \in \mathbb{N}$, let $u_k : \Omega_k \to \mathbb{R}$ be a convex function such that*

$$\det D^2 u_k = f_k \quad \text{in the viscosity sense in } \Omega_k \quad \text{and} \quad u_k|_{\partial\Omega_k,\text{epi}} = \varphi_k.$$

Assume that, as $k \to \infty$,

(7.12) $$u_k \to u, \quad \varphi_k \to \varphi,$$

in the sense of Hausdorff convergence of epigraphs, and

$\{f_k\}_{k=1}^{\infty}$ *is uniformly bounded and $f_k \to f$ uniformly on compact sets of Ω.*
Then, in the viscosity sense,

$$\det D^2 u = f \quad in \ \Omega, \quad u|_{\partial\Omega,\mathrm{epi}} = \varphi^c \quad on \ \partial\Omega.$$

The conclusion still holds if $0 < \lambda \leq f_k \leq \Lambda$. In this case, we replace
$\det D^2 u = f$ *by $\lambda \leq \det D^2 u \leq \Lambda$.*

Proof. The stability property of viscosity solutions in Theorem 7.12 tells us that $\det D^2 u = f$ in Ω in the viscosity sense. From (7.12) and Definition 7.22, we have

(7.13) $$\mathrm{diam}(\Omega_k) + \|f_k\|_{L^\infty(\Omega)} \leq M$$

for some constant M; furthermore,

$$\overline{\mathrm{epi}(u_k, \Omega_k)} \to \overline{\mathrm{epi}(u, \Omega)}, \quad \mathrm{epi}(\varphi_k, \partial\Omega_k) \to \mathrm{epi}(\varphi, \partial\Omega),$$

and

$$\mathrm{epi}(\varphi_k, \partial\Omega_k) \subset \overline{\mathrm{epi}(u_k, \Omega_k)},$$

which imply that $\mathrm{epi}(\varphi, \partial\Omega) \subset \overline{\mathrm{epi}(u, \Omega)}$. This gives $u|_{\partial\Omega,\mathrm{epi}} \leq \varphi^c$.

Let $\mathcal{C}_\varphi$ be the convex hull of $\mathrm{epi}(\varphi, \partial\Omega)$. Then

$$\mathrm{epi}(\varphi^c, \partial\Omega) = \mathcal{C}_\varphi \cap (\partial\Omega \times \mathbb{R}) \subset \mathcal{C}_\varphi \subset \overline{\mathrm{epi}(u, \Omega)}.$$

By Definition 7.14, in order to prove that $u|_{\partial\Omega,\mathrm{epi}} = \varphi^c$, that is,

$$\mathrm{epi}(\varphi^c, \partial\Omega) = \overline{\mathrm{epi}(u, \Omega)} \cap (\partial\Omega \times \mathbb{R}),$$

it remains to show that

(7.14) $$\overline{\mathrm{epi}(u, \Omega)} \cap (\partial\Omega \times \mathbb{R}) \subset \mathcal{C}_\varphi.$$

Indeed, consider a hyperplane (H) $x_{n+1} = l(x)$ which lies strictly below $\mathcal{C}_\varphi$. Then, for all large k, (H) is strictly below $\mathrm{epi}(\varphi_k, \partial\Omega_k)$; hence $u_k - l \geq 0$ on $\partial\Omega_k$. By Remark 3.28 and (7.13), we have

$$u_k - l \geq -C(n, M)\mathrm{dist}^{1/n}(\cdot, \partial\Omega_k) \quad in \ \Omega_k.$$

Letting $k \to \infty$, we obtain

$$u - l \geq -C(n, M)\mathrm{dist}^{1/n}(\cdot, \partial\Omega) \quad in \ \Omega.$$

Thus, no point on $\partial\Omega \times \mathbb{R}$ that is below (H) belongs to $\overline{\mathrm{epi}(u, \Omega)}$. Therefore, (7.14) holds. $\qquad\square$

A consequence of Theorem 7.27 is the unique solvability of the Dirichlet problem for the Monge–Ampère equation with bounded, lower semicontinuous boundary data and bounded, continuous right-hand side.

Theorem 7.28. *Let Ω be a bounded convex domain in $\mathbb{R}^n$, and let $f \geq 0$ be a bounded and continuous function on Ω. Let $\varphi : \partial\Omega \to \mathbb{R}$ be bounded and lower semicontinuous. Then, there is a unique convex viscosity solution to the Dirichlet problem*

$$(7.15) \qquad \det D^2 u = f \quad \text{in } \Omega, \qquad u \overset{\text{epi}}{=} \varphi \quad \text{on } \partial\Omega.$$

Proof. We use an approximation argument, invoking Theorems 2.50 and 2.51. We first approximate φ by a sequence of Lipschitz continuous functions on $\partial\Omega$ that increasingly converges pointwise to φ; see also Problem 7.4. Let $\{\Omega_k\}_{k=1}^{\infty} \subset \Omega$ be uniformly convex domains converging to Ω in the Hausdorff distance. From the assumptions and the above approximation of φ, we can find sequences $\{f_k\}_{k=1}^{\infty}$, $\{\varphi_k\}_{k=1}^{\infty}$ of continuous, uniformly bounded functions defined on $\overline{\Omega_k}$ and $\partial\Omega_k$, respectively, such that $f_k > 0$ in $\overline{\Omega_k}$, $f_k \to f$ uniformly on compact sets of Ω, and $\text{epi}(\varphi_k, \partial\Omega_k) \to \text{epi}(\varphi, \partial\Omega)$ in the Hausdorff distance.

For each k, by Theorem 3.37, there is a unique Aleksandrov solution $u_k \in C(\overline{\Omega_k})$ to

$$\det D^2 u_k = f_k \quad \text{in } \Omega_k, \qquad u_k = \varphi_k \quad \text{on } \partial\Omega_k.$$

By Proposition 7.7, u_k is also a viscosity solution to the above problem. In view of Corollary 3.14 and Lemma 2.40, the functions u_k are uniformly Lipschitz in each compact subset K of Ω. Using the Arzelà–Ascoli Theorem (Theorem 2.41) and Theorem 7.27, we see that u_k must converge to a convex viscosity solution u to (7.15), which is unique by Theorem 7.26. $\qquad\square$

We have the following compactness result for viscosity solutions with right-hand sides bounded between two positive constants; see also Problem 7.7 for the case of continuous, uniformly bounded right-hand sides.

Theorem 7.29 (Compactness of viscosity solutions with boundary discontinuities and bounded boundary data). *Let $\{\Omega_k\}_{k=1}^{\infty}$ be a sequence of uniformly bounded convex domains in $\mathbb{R}^n$. Consider uniformly bounded, lower semicontinuous functions $\varphi_k : \partial\Omega_k \to \mathbb{R}$. Assume that the convex functions $u_k : \Omega_k \to \mathbb{R}$ satisfy*

$$0 < \lambda \leq \det D^2 u_k \leq \Lambda \quad \text{in the viscosity sense in } \Omega_k, \qquad u_k \overset{\text{epi}}{=} \varphi_k \quad \text{on } \partial\Omega_k.$$

Then there exist a subsequence $\{k_m\}_{m=1}^{\infty}$, a convex function $u : \Omega \to \mathbb{R}$, and a lower semicontinuous function $\varphi : \partial\Omega \to \mathbb{R}$ such that

$$u_{k_m} \to u, \quad \varphi_{k_m} \to \varphi$$

in the sense of Hausdorff convergence of epigraphs, with

$$\lambda \leq \det D^2 u \leq \Lambda \quad in \ \Omega, \qquad u \overset{\mathrm{epi}}{=} \varphi \quad on \ \partial\Omega.$$

Note that the functions φ_k in Theorem 7.29 are not necessarily convex, while φ_k in Theorem 7.27 are convex.

Proof. In this proof, convergences of functions on different domains are understood in the sense of Hausdorff convergence of epigraphs. Note that $u_k|_{\partial\Omega_k,\mathrm{epi}} = \varphi_k^c$. Since Ω_k and φ_k are uniformly bounded, using the convexity of u_k and Remark 3.28, we find that $\{u_k\}_{k=1}^\infty$ is uniformly bounded from above and below. Thus, up to extracting a subsequence, we can assume that $u_k \to u$, $\varphi_k \to \varphi$, and $\varphi_k^c \to \psi$, where $u : \Omega \to \mathbb{R}$ is convex and $\varphi, \psi : \partial\Omega \to \mathbb{R}$ are lower semicontinuous.

By Theorem 7.27, the limit function u of u_k satisfies

$$\lambda \leq \det D^2 u \leq \Lambda \quad in \ \Omega \quad and \quad u|_{\partial\Omega,\mathrm{epi}} = \psi^c.$$

Recall that $\varphi_k^c \leq \varphi_k$. Thus, from $\varphi_k \to \varphi$, we find $\psi \leq \varphi$ which implies

$$\psi^c \leq \varphi^c.$$

Moreover, we have $\mathrm{epi}(\varphi_k, \partial\Omega_k) \to \mathrm{epi}(\varphi, \partial\Omega)$. Thus, for any $\varepsilon > 0$ and any hyperplane K below $\mathrm{epi}(\varphi^c, \partial\Omega)$ by a distance ε, we find that $\mathrm{epi}(\varphi_k, \partial\Omega_k)$ is above K for any large k. Thus the convergence $\varphi_k^c \to \psi$ implies that $\mathrm{epi}(\psi, \partial\Omega)$ is above K. Hence $\psi \geq \varphi^c$. Therefore $\varphi^c \leq \psi^c$. Consequently, $\varphi^c = \psi^c$ and this completes the proof of the theorem. $\qquad\square$

Remark 7.30. From now on, for a convex function $u : \Omega \to \mathbb{R}$ on a bounded convex domain Ω in $\mathbb{R}^n$ and a lower semicontinuous function $\varphi : \partial\Omega \to \mathbb{R} \cup \{+\infty\}$, we simply write $u = \varphi$ on $\partial\Omega$ when $u|_{\partial\Omega,\mathrm{epi}} = \varphi^c$. We believe this will not cause confusions, since all the boundary values in this book are interpreted in the usual pointwise sense, except the following: Definition 8.9, the compactness argument in Section 8.3 in the proof of the Boundary Localization Theorem (Theorem 8.5), Theorem 10.7, and Lemma 10.11.

7.3. Liouville-type Theorems

In this section, we show that entire convex solutions of the Monge–Ampère equation $\det D^2 u = 1$ in $\mathbb{R}^n$ must be quadratic polynomials. In the smooth setting or more generally for Aleksandrov solutions, this is the classical theorem of Jörgens [**J**] (in two dimensions), Calabi [**Cal1**] (for dimensions up to 5), and Pogorelov [**P3**] who proved the result in all dimensions. For viscosity solutions, this classification is due to Caffarelli.

Theorem 7.31 (Jörgens–Calabi–Pogorelov Theorem). *Let* $u : \mathbb{R}^n \to \mathbb{R}$ *be an Aleksandrov convex solution of* $\det D^2 u = 1$ *in* $\mathbb{R}^n$. *Then* u *is a quadratic polynomial.*

Proof. By Corollary 5.12, u is strictly convex. By Theorem 6.5, $u \in C^\infty(\mathbb{R}^n)$. We will prove that $D^2 u = D^2 u(0)$ in $\mathbb{R}^n$ in the following steps.

Step 1. Normalization. We can normalize u at the origin so that

$$(7.16) \qquad u(0) = 0, \quad Du(0) = 0, \quad D^2 u(0) = I_n.$$

Indeed, let $v(x) = u(x) - u(0) - Du(0) \cdot x$. Then

$$v(0) = 0, \quad Dv(0) = 0, \quad v \geq 0 \quad \text{in } \mathbb{R}^n, \quad \det D^2 v = 1 \quad \text{in } \mathbb{R}^n.$$

Let $\lambda_1, \ldots, \lambda_n$ be the positive eigenvalues of $D^2 v(0)$. Then, there exists an $n \times n$ orthogonal matrix P such that $P^t D^2 v(0) P = \text{diag}(\lambda_1, \ldots, \lambda_n)$. Let $w(x) = v(PQx)$ where $Q = \text{diag}(\lambda_1^{-1/2}, \ldots, \lambda_n^{-1/2})$. Then

$$D^2 w(0) = (PQ)^t D^2 v(0) PQ = Q^t (P^t D^2 v(0) P) Q = I_n.$$

Clearly, $w(0) = 0$, $Dw(0) = (PQ)^t Dv(0) = 0$, and

$$\det D^2 w(x) = (\det(PQ))^2 \det D^2 v(PQx) = (\det Q)^2 = (\det D^2 v(0))^{-1} = 1.$$

Now, we prove the theorem for u satisfying (7.16).

Step 2. Boundedness of sections. We show that the section $S_u(0, M) = \{x \in \mathbb{R}^n : u(x) < M\}$ is bounded for each $M > 0$.

Indeed, observe that if $|x| \leq 1$ and $C = \sup_{B_1(0)} \|D^3 u\| + 1$, then by Taylor's formula and (7.16)

$$u(x) \geq u(0) + Du(0) \cdot x + (1/2) D^2 u(0) x \cdot x - C |x|^3 = |x|^2 / 2 - C |x|^3.$$

Thus, $u(x) \geq r^2 / 4 > 0$ if $|x| = r := 1/(4C)$.

Now, consider $x \in S_u(0, M)$ with $|x| \geq r$. Then, by convexity

$$\frac{1}{4} r^2 \leq u\left(\frac{r}{|x|} x + \left(1 - \frac{r}{|x|}\right) 0\right) \leq \frac{r}{|x|} u(x) < \frac{r}{|x|} M.$$

It follows that $|x| \leq 4M/r$, proving the boundedness of $S_u(0, M)$.

It follows from $S_u(0, M) \Subset \mathbb{R}^n$ and Lemma 5.6(iii) that there is an $n \times n$ matrix a_M with $\det a_M = 1$ such that

$$B_{C(n)^{-1} M^{1/2}}(0) \subset a_M(S_u(0, M)) \subset B_{C(n) M^{1/2}}(0).$$

Step 3. Uniform Hessian bound. We next show that

$$\max\{\|a_M\|, \|a_M^{-1}\|, \sup_{\mathbb{R}^n} \|D^2 u\|\} \leq C_1(n),$$

for some constant $C_1(n) > 0$. Indeed, let $\Omega := M^{-1/2} a_M(S_u(0, M))$ and

$$w(x) := M^{-1} u(a_M^{-1}(M^{1/2} x)) - 1, \quad \text{for } x \in \Omega.$$

Then $B_{1/C(n)}(0) \subset \Omega \subset B_{C(n)}(0)$ and

$$\det D^2 w = 1 \quad \text{in } \Omega, \quad \text{and } w = 0 \quad \text{on } \partial\Omega.$$

By Lemma 6.4, there exists $C_* = C_*(n) > 0$ such that

$$\|D^2 w\| \le C_* \quad \text{in } B_{1/C_*}(0).$$

In particular, $\|D^2 w(0)\| \le C_*(n)$. Since $D^2 u(0) = I_n$, we have

$$D^2 w(0) = (a_M^{-1})^t D^2 u(0) a_M^{-1} = (a_M^{-1})^t (a_M^{-1}).$$

It follows that

$$\|a_M^{-1}\| = \left[\text{trace}((a_M^{-1})^t (a_M^{-1})) \right]^{1/2} \le C_2(n).$$

Since $\det(a_M^{-1}) = 1$, we get from $a_M = \text{Cof } a_M^{-1}$ that $\|a_M\| \le C_3(n)$. Note that if $|y| \le M^{1/2}/(C_* C_3(n))$, then $y = a_M^{-1}(M^{1/2} x)$ where $x = M^{-1/2} a_M y \in B_{1/C_*}(0)$ with $\|D^2 w(x)\| \le C_*$. It follows that

$$\|D^2 u(y)\| = \|a_M^t D^2 w(x) a_M\| \le C_4(n).$$

Since M can be chosen arbitrarily large, we obtain the desired estimates for $\|D^2 u\|$, a_M, a_M^{-1} as stated where $C_1(n) = \max\{C_2(n), C_3(n), C_4(n)\}$.

Step 4. Conclusion. By Step 3 and (7.16), $u(x) \le C_1(n)|x|^2$ for all $|x| \ge 1$.

For any $\bar{x} \in \mathbb{R}^n$, we show that $D^2 u(\bar{x}) = D^2 u(0) = I_n$. Indeed, for $R > 2(|\bar{x}| + 1)$, consider

$$w(y) = R^{-2} u(Ry), \quad \text{where } y \in B_1(0).$$

Then

$$|w| \le C_1(n), \quad \det D^2 w = 1, \quad \|D^2 w\| \le C_1(n) \quad \text{in } B_1(0).$$

Let $\bar{y} = \bar{x}/R$. Then $|\bar{y}| \le 1/2$. By the Evans–Krylov interior $C^{2,\alpha}$ estimates in Theorem 2.85, there exist $\alpha(n) \in (0,1)$ and $C_5(n) > 0$ such that

$$\|D^2 w(\bar{y}) - D^2 w(0)\| \le C_5 |\bar{y}|^\alpha.$$

This translates to

$$\|D^2 u(\bar{x}) - D^2 u(0)\| \le C_5 \left(|\bar{x}|/R \right)^\alpha.$$

Sending $R \to \infty$, we have $D^2 u(\bar{x}) = D^2 u(0)$. This completes the proof. $\square$

It suffices to assume in Theorem 7.31 that u is a convex viscosity solution to $\det D^2 u = 1$ in $\mathbb{R}^n$. The following extension is due to Caffarelli.

Theorem 7.32. *Let $u \in C(\mathbb{R}^n)$ be a convex viscosity solution of $\det D^2 u = 1$ on $\mathbb{R}^n$. Then u is a quadratic polynomial.*

Proof. By Proposition 7.11, u is an Aleksandrov solution of $\det D^2 u = 1$. Hence, by Theorem 7.31, u is a quadratic polynomial. $\square$

Theorem 7.31 can be viewed as a Liouville-type result for the Monge–Ampère equation. Due to the convexity of the solutions u, Theorem 7.31 is quite different from the Liouville-type results for harmonic functions (those satisfying $\Delta u = 0$ on $\mathbb{R}^n$), as it excludes all polynomials of degree greater than 2. On the other hand, in $\mathbb{R}^n$, the space of harmonic, homogeneous polynomials of degree d has dimension $\binom{n+d-1}{n-1} - \binom{n+d-3}{n-1}$. However, in two dimensions, we can see in Section 7.4 how to go from Liouville's Theorem for harmonic functions to Jörgens's Theorem for the Monge–Ampère equation to Bernstein's Theorem in minimal surface.

The Liouville-type result in Theorem 7.31 has been extended to half-spaces and planar quadrants. Due to the constancy of the right-hand sides in the following two theorems, the notion of solutions can be understood either in the sense of Aleksandrov or in the viscosity sense.

On half-spaces, Savin [**S6**] proved the following Liouville-type theorem.

Theorem 7.33 (Liouville-type theorem for the Monge–Ampère equation on a half-space). *Let $\mathbb{R}^n_+ = \{x \in \mathbb{R}^n : x_n > 0\}$. Assume the convex function $u \in C(\overline{\mathbb{R}^n_+})$ satisfies*

$$\det D^2 u = 1 \quad in \ \mathbb{R}^n_+ \quad and \quad u(x', 0) = |x'|^2/2.$$

If there exists $\varepsilon \in (0,1)$ such that $u(x) = O(|x|^{3-\varepsilon})$ as $|x| \to \infty$, then

$$u(Ax) = bx_n + |x|^2/2,$$

for some constant $b \in \mathbb{R}$ and some linear map $A : \mathbb{R}^n \to \mathbb{R}^n$ of the form $Ax = x + \tau x_n$, where $\tau \in \mathbb{R}^n$ with $\tau \cdot e_n = 0$.

Without the subcubic growth condition in Theorem 7.33, there is another nonquadratic solution

$$\frac{x_1^2}{2(1 + x_n)} + \frac{1}{2}(x_2^2 + \cdots + x_n^2) + \frac{x_n^3}{6}.$$

In the first quadrant in the plane, Le and Savin [**LS5**] proved the following Liouville-type theorem.

Theorem 7.34 (Liouville-type theorem for the Monge–Ampère equation in the planar quadrant). *Let $Q := \{x = (x_1, x_2) \in \mathbb{R}^2 : x_1, x_2 > 0\}$. Let $c > 0$. Assume that $u \in C(\overline{Q})$ is a convex solution to*

(7.17) $$\det D^2 u = c \quad in \ Q \quad and \quad u(x) = |x|^2/2 \quad on \ \partial Q.$$

If $u \geq 0$ in Q, then it is necessary that $c \leq 1$. Moreover:

  (i) *If $c = 1$ and $u \geq 0$ in Q, then the only solution is $u(x) = |x|^2/2$.*

  (ii) *If $c < 1$, then either*

$$u = P_c^{\pm} \equiv \frac{x_1^2}{2} + \frac{x_2^2}{2} \pm \sqrt{1-c}\, x_1 x_2$$

or

$$u(x) = \lambda^2 \, \overline{P}_c(x/\lambda) \quad or \quad u(x) = \lambda^2 \, \underline{P}_c(x/\lambda),$$

for some $\lambda > 0$, where $\overline{P}_c, \underline{P}_c$ are particular solutions of (7.17) satisfying

$$\underline{P}_c < P_c^- < \overline{P}_c < P_c^+ \quad in \ Q \qquad and \qquad \overline{P}_c(1,1) = 1, \quad \underline{P}_c(1,1) = 0.$$

In Theorem 7.34, near the origin, $\overline{P}_c(x) = P_c^+(x) + O(|x|^{2+\alpha})$ for some $\alpha(c) \in (0,1)$ while $\underline{P}_c$ has a conical singularity where $\|D^2\underline{P}_c\|$ is unbounded. At infinity, we have

$$\overline{P}_c(x) - P_c^-(x) = O(|x|^{2-\alpha}), \quad \underline{P}_c(x) - P_c^-(x) = O(|x|^{2-\alpha}).$$

7.4. Application: Connections between Four Important PDEs in Two Dimensions

In this section, we will describe interesting connections between four important PDEs in two dimensions. We go from Liouville's Theorem concerning the Laplace equation to Jörgens's Theorem concerning the Monge–Ampère equation (a two-dimensional case of Theorem 7.31) to Bernstein's Theorem concerning the minimal surface equation. In retrospect, the classification of global solutions to the Monge-Ampère equation was motivated by that of the minimal surface equation. We also show the equivalence between the Monge–Ampère equation and the stationary incompressible Euler equations in the class of homogenous solutions. To illustrate the main ideas, we assume all solutions in this section are reasonably smooth. We recall:

Theorem 7.35 (Liouville's Theorem). *A bounded harmonic function on Euclidean space is a constant.*

There are many proofs of this classical theorem, including applying (2.22) with $k = 1$ and then letting $r \to \infty$. There is an elementary proof due to Nelson [**Nel**] that uses no notation at all; it uses, however, the mean value property of harmonic functions.

7.4.1. Jörgens's Theorem. The following theorem is due to Jörgens [**J**].

Theorem 7.36 (Jörgens's Theorem). *Let $u : \mathbb{R}^2 \to \mathbb{R}$ be a convex solution of $\det D^2u = 1$ on $\mathbb{R}^2$. Then u is a quadratic polynomial.*

Proof. The elementary proof we give here is due to Nitsche [**Ni**]. It is convenient to write $u = u(z)$ where $z = (x,y) \in \mathbb{R}^2$, $u_x = \partial u/\partial x$, $u_y = \partial u/\partial y$, $u_{xy} = (u_x)_y$, etc. Then $\det D^2u = u_{xx}u_{yy} - u_{xy}^2 = 1$.

Let $z_1 = (x_1, y_1)$, $z_2 = (x_2, y_2)$. Then, the convexity of u and the monotonicity of the normal mapping (see Remark 2.34) give

(7.18) $(z_2 - z_1) \cdot (Du(z_2) - Du(z_1)) \geq 0.$

Let us now consider the transformation $\gamma = \mathbf{F}(z)$, due to Lewy [**Lw**] from $z = (x, y) \in \mathbb{R}^2$ to $\gamma = (\alpha, \beta) \in \mathbb{R}^2$, where

$$\mathbf{F}(z) = (\alpha(x, y), \beta(x, y)), \ \alpha(x, y) = x + u_x, \ \beta(x, y) = y + u_y, \ \gamma = (\alpha, \beta).$$

Then $\mathbf{F}$ is a diffeomorphism from $\mathbb{R}^2$ to $\mathbb{R}^2$. We can explain this fact using the Inverse Function Theorem as follows. Let $\mathbf{F}(z_j) = \gamma_j = (\alpha_j, \beta_j) \in \mathbb{R}^2$ for $j = 1, 2$. First, since $Du(z_j) = \gamma_j - z_j$, we rewrite (7.18) in the form

$$|z_2 - z_1|^2 \leq (z_2 - z_1) \cdot (\gamma_2 - \gamma_1) \leq |z_2 - z_1||\gamma_2 - \gamma_1|,$$

where we used the Cauchy–Schwartz inequality in the last estimate. Hence, we obtain $|z_2 - z_1| \leq |\gamma_2 - \gamma_1|$. Thus, $\mathbf{F}$ is one-to-one and $\mathbf{F}(\mathbb{R}^2)$ is closed.

We compute the Jacobian

$$J_{\mathbf{F}}(x, y) = \begin{pmatrix} \alpha_x & \alpha_y \\ \beta_x & \beta_y \end{pmatrix} = \begin{pmatrix} 1 + u_{xx} & u_{xy} \\ u_{xy} & 1 + u_{yy} \end{pmatrix}$$

and its determinant while recalling $\det D^2 u = 1$

$$|J_{\mathbf{F}}(x, y)| = 1 + \Delta u + \det D^2 u = 2 + \Delta u.$$

By the convexity of u, we have $u_{xx} \geq 0$ and $u_{yy} \geq 0$ and thus $\Delta u = u_{xx} + u_{yy} \geq 0$. Thus $J_{\mathbf{F}}$ is invertible. We now use the Inverse Function Theorem to conclude that $\mathbf{F}$ is an invertible function near p for all $p \in \mathbb{R}^2$. That is, an inverse function to $\mathbf{F}$ exists in some neighborhood of $\mathbf{F}(p)$. Thus $\mathbf{F}(\mathbb{R}^2)$ is open. This combined with the closedness of $\mathbf{F}(\mathbb{R}^2)$ shows that $\mathbf{F}(\mathbb{R}^2) = \mathbb{R}^2$ and $\mathbf{F}$ is a diffeomorphism from $\mathbb{R}^2$ to $\mathbb{R}^2$.

We now compute the inverse of $J_{\mathbf{F}}$:

$$J_{\mathbf{F}}^{-1}(x, y) = \begin{pmatrix} x_\alpha & x_\beta \\ y_\alpha & y_\beta \end{pmatrix} = \frac{1}{2 + \Delta u} \begin{pmatrix} 1 + u_{yy} & -u_{xy} \\ -u_{xy} & 1 + u_{xx} \end{pmatrix}.$$

Denote $M = x - u_x$ and $N = -y + u_y$. Let us complexify $\mathbb{R}^2$ and set

$$\gamma = \alpha + \sqrt{-1}\beta, \ w(\gamma) = x - u_x - \sqrt{-1}(y - u_y) \equiv M + \sqrt{-1}N.$$

We show that w is analytic, so M_α, N_α are harmonic. Indeed, compute

$$\frac{\partial M}{\partial \alpha} = \frac{\partial M}{\partial x}\frac{\partial x}{\partial \alpha} + \frac{\partial M}{\partial y}\frac{\partial y}{\partial \alpha} = \frac{(1 - u_{xx})(1 + u_{yy}) - u_{xy}(-u_{xy})}{\Delta u + 2} = \frac{-u_{xx} + u_{yy}}{\Delta u + 2}$$

and

$$\frac{\partial N}{\partial \beta} = \frac{\partial N}{\partial x}\frac{\partial x}{\partial \beta} + \frac{\partial N}{\partial y}\frac{\partial y}{\partial \beta} = \frac{u_{yx}(-u_{xy}) + (-1 + u_{yy})(1 + u_{xx})}{\Delta u + 2} = \frac{-u_{xx} + u_{yy}}{\Delta u + 2}.$$

Similar calculations lead to the Cauchy–Riemann equations

$$\frac{\partial M}{\partial \alpha} = \frac{\partial N}{\partial \beta} = \frac{-u_{xx} + u_{yy}}{\Delta u + 2} \quad \text{and} \quad \frac{\partial M}{\partial \beta} = -\frac{\partial N}{\partial \alpha} = \frac{-2u_{xy}}{\Delta u + 2}.$$

Thus, w is analytic in γ and so is $w' = M_\alpha + \sqrt{-1}N_\alpha$. We compute

$$|w'(\gamma)|^2 = \frac{(u_{xx} - u_{yy})^2 + 4(u_{xx}u_{yy} - 1)}{(\Delta u + 2)^2} = \frac{(\Delta u)^2 - 4}{(\Delta u + 2)^2} = \frac{\Delta u - 2}{\Delta u + 2} < 1.$$

Thus, w' is a constant by Liouville's Theorem in complex analysis. We can also use Theorem 7.35 because the real and imaginary parts of w' are harmonic functions. We observe, using $u_{xx}u_{yy} - u_{xy}^2 = 1$, that

$$|1 - w'|^2 = \left(1 - \frac{-u_{xx} + u_{yy}}{\Delta u + 2}\right)^2 + \left(\frac{2u_{xy}}{\Delta u + 2}\right)^2 = \frac{(2 + 2u_{xx})^2 + 4u_{xy}^2}{(\Delta u + 2)^2} = \frac{4u_{xx}}{\Delta u + 2}.$$

Similarly, $1 - |w'|^2 = \frac{4}{\Delta u + 2}$, and thus $u_{xx} = \frac{|1 - w'|^2}{1 - |w'|^2}$ is a constant. In the same vein, we find that

$$u_{yy} = \frac{|1 + w'|^2}{1 - |w'|^2}, \quad u_{xy} = \frac{\sqrt{-1}(\bar{w}' - w')}{1 - |w'|^2}$$

are also constants. Thus, all second partial derivatives of u are constant and hence u must be a quadratic polynomial. $\qquad\square$

Remark 7.37. The map $\mathbf{F}$ in the above proof is nothing but the gradient of $v(x, y) := u(x, y) + (x^2 + y^2)/2$. Since $\mathbf{F} = Dv$ is one-to-one and onto, the Legendre transform $v^*(x^*, y^*)$ of v is defined on the whole plane $\mathbb{R}^2$, with

$$D^2 v^*(x^*, y^*) = (D^2 v(x, y))^{-1} = (D^2 u + I_2)^{-1}, \quad \text{where } (x^*, y^*) = Dv(x, y).$$

Since $\det D^2 u = 1$, we have $\Delta v^*(x^*, y^*) = 1$. Applying the Liouville Theorem to each entry of $D^2 v^*$, which lies in $[-1, 1]$, we find that $D^2 v^*$ is a constant matrix, and so is $D^2 u$. Hence, u must be a quadratic polynomial.

Remark 7.38. We can also prove Jörgens's Theorem using the partial Legendre transform as follows. Consider the partial Legendre transform in the x_1-variable:

$$\begin{cases} (y_1, y_2) & = \big(u_{x_1}(x_1, x_2), x_2\big), \\ u^\star(y_1, y_2) & = x_1 u_{x_1}(x_1, x_2) - u(x_1, x_2), \end{cases}$$

where we denote $u_{x_1} = \partial u/\partial x_1$, $u_{x_2} = \partial u/\partial x_2$, etc. We find that

$$u^\star(y_1, y_2) = \sup_{x_1 \in \mathbb{R}} (x_1 y_1 - u(x_1, y_2)).$$

Furthermore, $u^\star_{y_1} = x_1$, $u^\star_{y_2} = -u_{x_2}$, and

$$u^\star_{y_1 y_1} = \frac{1}{u_{x_1 x_1}}, \quad u^\star_{y_2 y_2} = -\frac{\det D^2 u}{u_{x_1 x_1}}, \quad u^\star_{y_1 y_2} = -\frac{u_{x_1 x_2}}{u_{x_1 x_1}}.$$

It follows that
$$u^\star_{y_1 y_1} \det D^2 u + u^\star_{y_2 y_2} = 0.$$
Assume now $\det D^2 u = 1$. Then u^* is harmonic and so is $u^\star_{y_1 y_1}$ with $u^\star_{y_1 y_1} \geq 0$. By Liouville's Theorem for harmonic functions bounded from below, we deduce that $u^\star_{y_1 y_1}$ is a constant. Hence $u_{x_1 x_1}$ is a constant. Similarly, $u_{x_2 x_2}$ is a constant and so is $u_{x_1 x_2}$. Therefore, u must be a quadratic polynomial.

The partial Legendre transform can also be used to prove the following Liouville Theorem for degenerate Monge–Ampère equations; see Jin–Xiong [**JX**].

Theorem 7.39. *Let $u(x_1, x_2)$ be an Aleksandrov solution to*
$$\det D^2 u(x_1, x_2) = |x_2|^\alpha \quad in \ \mathbb{R}^2,$$
where $\alpha > -1$. Then
$$u(x_1, x_2) = \frac{1}{2a}x_1^2 + bx_1 x_2 + \frac{ab^2}{2}x_2^2 + \frac{a}{(\alpha + 2)(\alpha + 1)}|x_2|^{\alpha + 2} + l(x_1, x_2)$$
for some constants $a > 0$, $b \in \mathbb{R}$ and a linear function $l(x_1, x_2)$.

7.4.2. Minimal surface equation and Bernstein's Theorem. The minimal surface equation on $\mathbb{R}^n$ is
$$\operatorname{div}\left(\frac{Du}{\sqrt{1 + |Du|^2}}\right) = 0.$$
The following theorem is due to Bernstein [**Bn**].

Theorem 7.40 (Bernstein's Theorem). *Let u be a solution of the minimal surface equation in the plane; that is, $u(x, y)$ satisfies*
$$(7.19) \qquad \frac{\partial}{\partial x}\left(\frac{u_x}{\sqrt{1 + u_x^2 + u_y^2}}\right) + \frac{\partial}{\partial y}\left(\frac{u_y}{\sqrt{1 + u_x^2 + u_y^2}}\right) = 0 \quad in \ \mathbb{R}^2,$$
where $u_x = \partial u/\partial x$, $u_y = \partial u/\partial y$. Then u is an affine function, so its graph is a plane. Consequently, a complete minimal surface in the three-dimensional Euclidean space, which is a graph in some direction, must be a plane.

Proof. This proof, using Jörgens's Theorem, is due to E. Heinz; see [**J**, p. 133]. Consider the following functions:
$$v = \sqrt{1 + |Du|^2}, \quad f = (1 + u_x^2)/v, \quad g = u_x u_y/v, \quad h = (1 + u_y^2)/v.$$
Then, by direct calculations, we have
$$\frac{\partial f}{\partial y} - \frac{\partial g}{\partial x} = \frac{u_y}{v^3}\left[(1 + u_y^2)u_{xx} - 2u_x u_y u_{xy} + (1 + u_x^2)u_{yy}\right] = 0,$$

because the minimal surface equation (7.19) can be rewritten as

$$[(1 + u_y^2)u_{xx} - 2u_xu_yu_{xy} + (1 + u_x^2)u_{yy}](1 + u_x^2 + u_y^2)^{-3/2} = 0.$$

Therefore, the vector field (f, g) is conservative. Hence, there is a function $\rho : \mathbb{R}^2 \to \mathbb{R}$ such that $\rho_x = f, \rho_y = g$. Similarly, the vector field (g, h) is conservative, and there is a function $\alpha : \mathbb{R}^2 \to \mathbb{R}$ such that $\alpha_x = g, \alpha_y = h$. Since $\rho_y = \alpha_x = g$, the vector field (ρ, α) is conservative. Hence, there is a function $\varphi : \mathbb{R}^2 \to \mathbb{R}$ such that $D\varphi = (\rho, \alpha)$. With this φ, we have

$$\frac{\partial^2 \varphi}{\partial x^2} = \varphi_{xx} = \rho_x = f, \quad \frac{\partial^2 \varphi}{\partial x \partial y} = \varphi_{xy} = \rho_y = g, \quad \frac{\partial^2 \varphi}{\partial y^2} = \varphi_{yy} = \alpha_y = h.$$

From the above relations and the definition of f, g, h, we find that φ satisfies

$$(7.20) \qquad\qquad \varphi_{xx}\varphi_{yy} - \varphi_{xy}^2 = fh - g^2 = 1.$$

In $\mathbb{R}^2$, a function φ satisfying the Monge–Ampère equation (7.20) is either concave or convex. By changing the sign of φ if necessary, we can assume that φ is convex. Therefore, by Jörgens's Theorem (Theorem 7.36), φ is a quadratic polynomial and thus f, g, h are all constants. Thus $f/h = (1 + u_x^2)/(1 + u_y^2)$ is a constant. This together with $f = (1 + u_x^2)(1 + u_x^2 + u_y^2)^{-1/2}$ being a constant shows that u_x^2 and u_y^2 are constants. Therefore, u_x and u_y are constants. Hence u is an affine function, and the theorem is proved. $\square$

7.4.3. Incompressible Euler equations in two dimensions.

We consider the incompressible Navier–Stokes equations in two dimensions

$$(7.21) \quad \begin{cases} \partial_t \mathbf{u}(x, t) + (\mathbf{u} \cdot D)\mathbf{u} - \nu\Delta\mathbf{u} + Dp - \mathbf{f} = 0 & \text{in } \mathbb{R}^2 \times (0, \infty), \\ \text{div } \mathbf{u} = 0 & \text{in } \mathbb{R}^2 \times (0, \infty), \\ \mathbf{u}(x, 0) = \mathbf{u}_0(x) & \text{in } \mathbb{R}^2. \end{cases}$$

Here, $\mathbf{u}(x, t) = (u^1(x, t), u^2(x, t))$ is the velocity vector field, $\nu \geq 0$ is the viscosity, p is the pressure, Dp is the gradient of p, $\mathbf{f}$ is the force, and $\mathbf{u}_0$ is a divergence-free initial velocity. In terms of the scalar functions u^1, u^2, the expression $(\mathbf{u} \cdot D)\mathbf{u}$ can be written as

$$(\mathbf{u} \cdot D)\mathbf{u} = \left(\sum_{i=1}^2 u^i \frac{\partial}{\partial x_i}\right)\mathbf{u} = \left(\sum_{i=1}^2 u^i \frac{\partial u^1}{\partial x_i}, \sum_{i=1}^2 u^i \frac{\partial u^2}{\partial x_i}\right).$$

When $\nu = 0$, (7.21) become the incompressible Euler equations describing the motion of an ideal incompressible fluid.

The incompressibility condition $\text{div } \mathbf{u} = 0$ implies the existence of a stream function $\Psi : \mathbb{R}^2 \times [0, \infty) \to \mathbb{R}$ such that $\mathbf{u} = D^\perp\Psi \equiv (-D_2\Psi, D_1\Psi)$. Taking the divergence of the first equation in (7.21), we obtain

$$(7.22) \qquad\qquad \text{div}[(\mathbf{u} \cdot D)\mathbf{u}] + \Delta p - \text{div } \mathbf{f} = 0.$$

A calculation reveals that the divergence term is equal to $-2 \det D^2 \Psi$. As a consequence, we obtain a Monge–Ampère equation from the incompressible Navier–Stokes equations.

Lemma 7.41. *Consider the incompressible Navier–Stokes equations* (7.21). *If* $\mathbf{u} = D^\perp \Psi$, *then* $\det D^2 \Psi = (\Delta p - \operatorname{div} \mathbf{f})/2$.

We now consider the stationary incompressible Euler equations, that is (7.21) with $\nu = 0$ and no time-dependence. In general, the above procedure does not establish the equivalence between the Monge–Ampère equation and the stationary incompressible Euler equations. They do, however, become equivalent in the homogeneous case, as observed by Luo and Shvydkoy in [**LSh2**]. We briefly explain this equivalence.

Euler equations with homogenous solutions. Consider the stationary incompressible Euler equations with no force

$$(7.23) \qquad \begin{cases} (\mathbf{u} \cdot D)\mathbf{u} + Dp = 0 & \text{in } \mathbb{R}^2, \\ \operatorname{div} \mathbf{u} = 0 & \text{in } \mathbb{R}^2. \end{cases}$$

We denote by Ψ the stream function; that is, $\mathbf{u} = D^\perp \Psi$. Let (r, θ) be the polar coordinates. If we are looking for homogenous solutions of (7.23) with locally finite energy, then as explained in [**LSh1**, Section 2], we can focus on solutions of the form

$$(7.24) \qquad \Psi(r, \theta) = r^\lambda \psi(\theta), \quad p = r^{2(\lambda-1)} P(\theta) \quad \text{where } \lambda > 0.$$

Plugging this ansatz into (7.23), we find that P must be a constant and furthermore, ψ solves

$$(7.25) \quad 2(\lambda - 1)P = -(\lambda - 1)(\psi')^2 + \lambda^2 \psi^2 + \lambda \psi'' \psi, \quad \psi(0) = \psi(2\pi) \text{ if } \lambda \neq 1.$$

For completeness, we indicate how to obtain (7.25) from the ansatz (7.24) with P being a constant. We will use the following fact in polar coordinates:

$$(7.26) \quad \det D^2 v = r^{2(\gamma-2)} \det \begin{pmatrix} \gamma(\gamma-1)h & (\gamma-1)h' \\ (\gamma-1)h' & h'' + \gamma h \end{pmatrix} \text{ if } v = r^\gamma h(\theta).$$

Now, for the ansatz (7.24), using (7.26), we have

$$\det D^2 \Psi = \det D^2(r^\lambda \psi(\theta)) = r^{2(\lambda-2)}(\lambda - 1)[\lambda \psi \psi'' + \lambda^2 \psi^2 - (\lambda - 1)(\psi')^2]$$

and

$$\Delta p = \Delta(r^{2(\lambda-1)} P) = 4(\lambda - 1)^2 r^{2(\lambda-2)} P.$$

Thus, from Lemma 7.41 with $f = 0$, we obtain (7.25) if $\lambda \neq 1$.

Monge–Ampère equation with homogenous solutions. In [**DS**], Daskalopoulos and Savin studied the Monge–Ampère equation

(7.27) $\det D^2 w = |x|^\alpha$ in $\mathbb{R}^2$,   where $\alpha > -2$.

Homogenous solutions to (7.27) must be of the form

$$w = r^\beta g(\theta) \quad \text{where } \beta = 2 + \alpha/2.$$

We calculate using (7.26)

$$\det D^2 w = r^{2(\beta - 2)}(\beta - 1)[\beta g g'' + \beta^2 g^2 - (\beta - 1)(g')^2].$$

Using (7.27) and $\beta = 2 + \alpha/2$, we deduce that

(7.28) $\beta g g'' + \beta^2 g^2 - (\beta - 1)(g')^2 = 1/(\beta - 1).$

From (7.25) and (7.28) and in the special case of $\lambda = \beta$, we see that (7.23)
and (7.27) are equivalent in the class of homogenous solutions.

Remark 7.42. The classification of homogenous stationary solutions to the
two-dimensional Euler equations (7.23) with the ansatz (7.24) in [**LSh1**] left
open the ranges $3/4 < \lambda < 1$ and $1 < \lambda < 4/3$. From the equivalence between the Monge–Ampère equation and the stationary incompressible Euler
equations in the homogeneous case, Luo and Shvydkoy [**LSh2**] were able
to use results on the Monge–Ampère equation in [**DS**] to complete their
analysis in [**LSh1**] and thus the classification of all homogenous stationary
solutions to the two-dimensional incompressible Euler equations.

7.5. Legendre–Lewy Rotation

In Section 7.4, we used the Lewy transformation to convert the Monge–
Ampère equation $\det D^2 u = 1$ to a harmonic function $\Delta w = 0$ in two
dimensions. In this section, we look more in depth into this interesting
transformation, which we call the *Legendre–Lewy rotation*; this is due to its
connection with the Legendre transform as discussed in Remark 7.37. The
idea is to rotate the *Lagrangian graph* $\{(x, Du(x)) : x \in \mathbb{R}^n\} \subset \mathbb{R}^n \times \mathbb{R}^n$ of
a function $u : \mathbb{R}^n \to \mathbb{R}$. In two-dimensional coordinates, *clockwise* rotating
a point $(x_1, x_2) \in \mathbb{R}^2$ by an angle α gives us a new point $(\bar{x}_1, \bar{x}_2)$, where

$$\bar{x}_1 = \cos \alpha \, x_1 + \sin \alpha \, x_2, \quad \bar{x}_2 = -\sin \alpha \, x_1 + \cos \alpha \, x_2.$$

Applying this rotation to the Lagrangian graph $M = \{(x, Du(x)) : x \in \mathbb{R}^n\}$
of a function $u : \mathbb{R}^n \to \mathbb{R}$ gives us a subset $\bar{M} = \{(\bar{x}, \bar{y}) \in \mathbb{R}^n \times \mathbb{R}^n\}$, where

$$\bar{x} = \cos \alpha \, x + \sin \alpha \, Du(x), \quad \bar{y} = -\sin \alpha \, x + \cos \alpha \, Du(x).$$

Observe that, for u convex, we have

(1) $\bar{M}$ is still a graph over the whole $\bar{x}$-space $\mathbb{R}^n$. The proof uses
    similar ideas to $|z_1 - z_2| \le |\gamma_2 - \gamma_1|$ in the proof of Theorem 7.36.

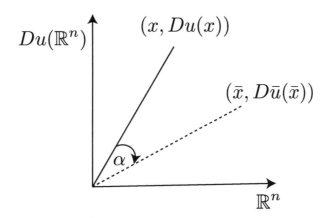

Figure 7.3. Legendre–Lewy rotation of the Lagrangian graph $\{(x, Du(x))\}$.

(2) $\bar{M}$ is also a Lagrangian graph. That is, there is a function $\bar{u}$: $\mathbb{R}^n \to \mathbb{R}$ such that $\bar{M} = \{(\bar{x}, D\bar{u}(\bar{x})) : \bar{x} \in \mathbb{R}^n\}$. For a proof of this fact, see Harvey–Lawson [**HLa**, Lemma 2.2, p. 91].

Definition 7.43 (Legendre–Lewy rotation). Consider the Lagrangian graph $M = \{(x, Du(x)) : x \in \mathbb{R}^n\}$ of a convex function $u : \mathbb{R}^n \to \mathbb{R}$. The Legendre–Lewy rotation of M by an angle α, in the clockwise direction, gives us a new Lagrangian graph $\bar{M} = \{(\bar{x}, D\bar{u}(\bar{x})) : \bar{x} \in \mathbb{R}^n\}$, where

(7.29)
$$\begin{cases} \bar{x} = \cos\alpha \; x + \sin\alpha \; Du(x), \\ D\bar{u}(\bar{x}) = -\sin\alpha \; x + \cos\alpha \; Du(x). \end{cases}$$

See Figure 7.3. The case of $\alpha = \pi/4$ is the original Lewy rotation [**Lw**]. The general case originates from Yuan [**Yu2**].

An important property of the Legendre–Lewy rotation is that it gives better uniform controls on the Hessian of the new function for a suitably chosen rotation angle. This can be deduced from the following simple result.

Lemma 7.44 (Hessian matrix under the Legendre–Lewy rotation). *We have the following relation between the Hessians (taken with respect to $\bar{x}$ and x, respectively):*

(7.30) $D^2\bar{u}(\bar{x}) = [\cos\alpha \, I_n + \sin\alpha \, D^2u(x)]^{-1}(-\sin\alpha \, I_n + \cos\alpha \, D^2u(x)).$

Proof. By the chain rule, we have
$$D^2_{\bar{x}}\bar{u}(\bar{x}) = D_x[D_{\bar{x}}\bar{u}(\bar{x})]D_{\bar{x}}x = D_x[D_{\bar{x}}\bar{u}(\bar{x})](D_x\bar{x})^{-1}.$$
Substituting (7.29) into the last expression, we obtain (7.30). $\square$

Now, we derive some implications of (7.30). Let us denote by $\lambda_i = \lambda_i(D^2(x))$ $(i = 1, \ldots, n)$ the eigenvalues of $D^2u(x)$ and let us denote by $\bar{\lambda}_i$

$(i = 1, \ldots, n)$ the eigenvalues of $D^2\bar{u}(\bar{x})$. Denote the Lagrangian angles by

$$\theta_j = \arctan \lambda_j, \quad \bar{\theta}_j = \arctan \bar{\lambda}_j.$$

Then (7.30) gives

$$\bar{\lambda}_i = \frac{(\cos\alpha)\lambda_i - \sin\alpha}{\cos\alpha + (\sin\alpha)\lambda_i} = \frac{\tan\theta_i - \tan\alpha}{1 + \tan\alpha\tan\theta_i} = \tan(\theta_i - \alpha).$$

Since $\bar{\lambda}_i = \tan\bar{\theta}_i$, we find that

$$\bar{\theta}_i = \theta_i - \alpha.$$

Thus the *Lagrangian angles are reduced by* α. This simple observation is significant in the study of global solutions to *special Lagrangian equations*

$$\sum_{i=1}^{n} \arctan\lambda_i(D^2 u) \equiv \sum_{i=1}^{n} \theta_i = \text{constant}.$$

For a convex, C^2 function u, we have $\theta_j \in (0, \pi/2)$ but no controls on the eigenvalues of $D^2 u$; however, when $0 < \alpha < \pi/2$, we have $-\alpha < \bar{\theta}_j < \pi/2 - \alpha$, so the Hessian matrix $D^2\bar{u}(\bar{x})$ is bounded with

(7.31) $-\tan\alpha\, I_n \leq D^2\bar{u}(\bar{x}) \leq \cot\alpha\, I_n.$

Example 7.45 (Monge–Ampère and special Lagrangian equations). Let us take another look at the Monge–Ampère equation $\det D^2 u = 1$ in $\mathbb{R}^2$ with u being convex. In this case, we have $1 = \lambda_1\lambda_2 = \tan\theta_1\tan\theta_2$, so the Lagrangian angles $\theta_1 > 0$ and $\theta_2 > 0$ satisfy

$$\theta_1 + \theta_2 = \pi/2.$$

This is a special Lagrangian equation. By a Legendre–Lewy rotation of angle $\alpha := \pi/4$ in the clockwise direction, we have a new function $\bar{u}$ on $\mathbb{R}^2$ with the Lagrangian angles $\bar{\theta}_1$ and $\bar{\theta}_2$ satisfying $\bar{\theta}_1 = \theta_1 - \pi/4$ and $\bar{\theta}_2 = \theta_2 - \pi/4$, so $\bar{\theta}_1 + \bar{\theta}_2 = 0$. Therefore

$$0 = \tan(\bar{\theta}_1 + \bar{\theta}_2) = (\bar{\lambda}_1 + \bar{\lambda}_2)/(1 - \bar{\lambda}_1\bar{\lambda}_2).$$

It follows that $\Delta\bar{u} = 0$, so $\bar{u}$ is harmonic. So is each entry of $D^2\bar{u}$ which is bounded, thanks to (7.31). By Liouville's Theorem for harmonic functions, we conclude that $D^2\bar{u}$ must be a constant matrix. From (7.30), we find that $D^2 u$ is a constant matrix. Hence u is a quadratic polynomial.

We now relate to the variables γ and quantities M and N introduced by Nitsche in his proof of Jörgens's Theorem. Using the variables $z = (x, y) \in \mathbb{R}^2$, we see that the gradient graph $\{(z, Du(z)) : z \in \mathbb{R}^2\}$ is transformed to the gradient graph $\{(\bar{z}, D\bar{u}(\bar{z})) : \bar{z} \in \mathbb{R}^2\}$ where, by (7.29),

$$
\begin{cases}
\bar{z} = \dfrac{1}{\sqrt{2}}(z + Du(z)) = \dfrac{1}{\sqrt{2}}(x + u_x, y + u_y), \\[2mm]
D\bar{u}(\bar{z}) = \dfrac{1}{\sqrt{2}}(-z + Du(z)) = \dfrac{1}{\sqrt{2}}(-x + u_x, -y + u_y).
\end{cases}
$$

We have

$$
\gamma = (x + u_x, y + u_y) = \sqrt{2}\bar{z}, \quad (-M, N) = \sqrt{2}D\bar{u}(\bar{z}).
$$

Example 7.46 (Legendre–Lewy rotation and Legendre transform). Let us consider the Legendre–Lewy rotation by $\alpha := \pi/2$. Then $\bar{\theta}_j = \theta_j - \pi/2$ and

$$
\bar{\lambda}_j = \tan\bar{\theta}_j = -1/\tan\theta_j = -1/\lambda_j.
$$

Hence $D^2\bar{u}(\bar{x}) = -(D^2u(x))^{-1}$. Let $u^*(\bar{x}) = -\bar{u}(\bar{x})$. Then

$$
D^2u^*(\bar{x}) = (D^2u(x))^{-1}, \quad \lambda_j^* = 1/\lambda_j.
$$

From the formula of the rotation, we have

$$
\bar{x} = Du(x), \quad Du^*(\bar{x}) = -D\bar{u}(\bar{x}) = x.
$$

This is the Legendre transform.

We refer to Yuan [**Yu1, Yu2**] and Shankar–Yuan [**ShYu**] for important applications of the Legendre–Lewy rotation in classifying global solutions to special Lagrangian equations and for the sigma-2 equation.

7.6. Problems

Problem 7.1. Show that $\lambda_{\min}(D^2u)$, the smallest eigenvalue of D^2u, is an elliptic operator.

Problem 7.2. Using the definition of viscosity solutions, verify that the Pogorelov example $u(x) = |x'|^{2-2/n}(1 + x_n^2)$, where $n \geq 3$, is a viscosity solution of

$$
\det D^2u = \left(2 - \frac{2}{n}\right)^{n-1}(|x_n|^2 + 1)^{n-2}\left\{2 - \frac{4}{n} + \left(\frac{4}{n} - 6\right)|x_n|^2\right\}
$$

in a small ball $B_\rho(0)$ in $\mathbb{R}^n$.

Problem 7.3. Let Ω be a bounded convex domain in $\mathbb{R}^n$ with $0 \in \Omega$. Let $\{u_k\}_{k=1}^{\infty} \subset C(\overline{\Omega})$ be a sequence of convex functions satisfying $u_k(0) = 0$, $u_k \geq 0$ in Ω, and $\int_{\partial\Omega} u_k \, d\mathcal{H}^{n-1} \leq 1$.

    (a) Show that there is a subsequence $\{u_{k_j}\}_{j=1}^{\infty}$ and a convex function $u : \Omega \to \mathbb{R}$ such that $u_{k_j} \to u$ on compact subsets of Ω.

(b) Define the boundary values of u to be $u|_{\partial\Omega,\text{epi}}$ as in Definition 7.14. Show that

$$\lim_{j\to\infty} \int_\Omega u_{k_j}\,dx = \int_\Omega u\,dx \quad \text{and} \quad \int_{\partial\Omega} u\,d\mathcal{H}^{n-1} \leq \liminf_{j\to\infty} \int_{\partial\Omega} u_{k_j}\,d\mathcal{H}^{n-1}.$$

Problem 7.4. Let $E \subset \mathbb{R}^n$, and let $\varphi : E \to \mathbb{R}$ be bounded and lower semicontinuous. For each $k \in \mathbb{N}$, define the function $\varphi_k : E \to \mathbb{R}$ by

$$\varphi_k(x) = \inf\big\{\varphi(x) + k|x - y| : y \in E\big\}.$$

Show that each φ_k is Lipschitz continuous and that for each $x \in E$, the sequence $\{\varphi_k(x)\}_{k=1}^\infty$ converges increasingly to $\varphi(x)$.

Problem 7.5. Let $u_k : \Omega_k \to \mathbb{R}$ ($k \in \mathbb{N}$) be a sequence of convex functions on convex domains Ω_k in $\mathbb{R}^n$. Assume that u_k converges to $u : \Omega \to \mathbb{R}$ in the sense of Hausdorff convergence of epigraphs. Define the boundary values of u_k and u to be $u_k|_{\partial\Omega_k,\text{epi}}$ and $u|_{\partial\Omega,\text{epi}}$, respectively, as in Definition 7.14. Show that $\{u_k\}_{k=1}^\infty$ Γ-*converges* to u; that is, the following hold:

(a) (*Liminf inequality*) For any sequence $\{x_k\}_{k=1}^\infty$ with $x_k \in \overline{\Omega_k}$ that converges to $x \in \overline{\Omega}$, we have $u(x) \leq \liminf_{k\to\infty} u_k(x_k)$.

(b) (*Recovery sequence*) For any $x \in \overline{\Omega}$, there is a sequence $\{x_k\}_{k=1}^\infty$, where $x_k \in \overline{\Omega_k}$, converging to x with $u(x) = \lim_{k\to\infty} u_k(x_k)$.

Problem 7.6. Let $\Omega \subset \mathbb{R}^n$ be a uniformly convex domain with C^1 boundary. Let $u \in C(\Omega) \cap L^\infty(\Omega)$ be a convex function satisfying $\det D^2 u \leq 1$ in Ω in the sense of Aleksandrov. Define the boundary values of u to be $u|_{\partial\Omega,\text{epi}}$. Show that $\partial_s u(x) \neq \emptyset$ for $\mathcal{H}^{n-1}$ almost every $x \in \partial\Omega$.

Problem 7.7. In Theorem 7.29, suppose we replace $\lambda \leq \det D^2 u_k \leq \Lambda$ by $\det D^2 u_k = f_k$ where $f_k \geq 0$ are continuous, uniformly bounded and $f_k \to f$ uniformly on compact sets of Ω. Show that the conclusion of Theorem 7.29 still holds where $\lambda \leq \det D^2 u \leq \Lambda$ is now replaced by $\det D^2 u = f$.

Problem 7.8 (Partial Legendre transform for the Monge–Ampère equation). Let $u \in C^2(\mathbb{R}^n)$ be a strictly convex function. Consider the partial Legendre transformation

$$y_i = D_i u(x) \quad (i \leq n-1), \quad y_n = x_n, \quad u^*(y) = x' \cdot D_{x'} u - u(x).$$

The function u^* is obtained by taking the Legendre transform of the function u on each slice $x_n = \text{constant}$.

(a) Show that

$$D_j u^* = x_j \quad \text{if } j < n, \quad D_n u^* = -D_n u.$$

(b) Consider the $n \times n$ transformation matrices

$$(Y_{ij}) = \left(\frac{\partial y_i}{\partial x_j}\right) \quad \text{and} \quad (Y_{ij})^{-1} = (X_{ij}) = \left(\frac{\partial x_i}{\partial y_j}\right).$$

Let (U^{ij}) be the cofactor matrix of $D^2 u$. Show that

$$Y_{ij} = D_{ij} u \quad \text{if } i \le n-1, \quad Y_{nj} = \delta_{jn},$$

and

$$(X_{ij}) = \left(\begin{array}{c|c} (D^2_{x'} u)^{-1} & U^{ni} / \det D^2_{x'} u \\ \hline 0 & 1 \end{array} \right).$$

(c) Show that $(u^*)^* = u$ and u^* satisfies

$$(\det D^2 u) \det D^2_{y'} u^* + D_{nn} u^* = 0.$$

Problem 7.9. Suppose $u \in C^2(\mathbb{R}^2)$ is a function satisfying

$$\det D^2 u = -1 \quad \text{with } D_{11} u \ne 0 \quad \text{in } \mathbb{R}^2.$$

Use the partial Legendre transform to find a general formula for u.

Problem 7.10. Verify Lemma 7.41 and formula (7.26).

Problem 7.11. Let $\mathbb{R}^2_+ = \{x = (x_1, x_2) \in \mathbb{R}^2 : x_2 > 0\}$. Assume the convex function $u \in C(\overline{\mathbb{R}^2_+})$ satisfies

$$\det D^2 u = 1 \quad \text{in } \mathbb{R}^2_+ \quad \text{and} \quad u(x_1, 0) = x_1^2.$$

Use the partial Legendre transform to find all possible solutions u.

7.7. Notes

In Section 7.1, see Caffarelli–Cabré [**CC**] and Crandall–Ishii–Lions [**CIL**] for viscosity solutions of second-order partial differential equations. For viscosity solutions to the Monge–Ampère equation, see also Gutiérrez [**G2**, Sections 1.3 and 1.7], Ishii–Lions [**IL**], and Urbas [**U2**]. The theory of viscosity solutions was first set forth for first-order equations by Crandall and Lions; see Evans [**E2**, Chapter 10] for an introduction to this subject.

Section 7.2 is based on Savin [**S3, S4**]. For an introduction to Γ-convergence, see Dal Maso [**DM**]. In the calculus of variations, Gamma-convergence is also called epigraphical convergence.

The proof of Theorem 7.31 here essentially follows Caffarelli–Li [**CL**]; see also Pogorelov [**P5**, Chapter 6], Gutiérrez [**G2**, Chapter 4], Figalli [**F2**, Section 4.3], Han [**H**, Section 6.4], and Trudinger–Wang [**TW1**]. For a parabolic Monge–Ampère version of the Jörgens–Calabi–Pogorelov Theorem, see Gutiérrez–Huang [**GH1**]. For Liouville-type theorems for certain singular Monge-Ampère equations on half-spaces, see Huang–Tan–Wang [**HTW**] and Jian–Wang [**JW2**]. These results have interesting applications to the regularity of the free boundary of the Monge-Ampère obstacle problem and the affine hyperbolic spheres.

It is more challenging to establish Liouville-type theorems for the complex Monge–Ampère equation. In fact, the classification of smooth plurisubharmonic solutions to $\det(u_{z_j \bar{z}_k}) = 1$ in $\mathbb{C}^n$ where $n \geq 2$ is still in its infancy, even if one imposes the quadratic growth on u at infinity! However, if one imposes further that the metric $\sqrt{-1} u_{z_j \bar{z}_k} dz_j \wedge d\bar{z}_k$ is complete, then u is a quadratic polynomial; see Li–Sheng [**LiSh**].

Related to Section 7.4, see also Trudinger's survey [**Tr4**] which discussed the relationship between the theorems of Liouville, Jörgens, and Bernstein. Though interchangeably connected in two dimensions, the higher-dimensional analogues of Bernstein's and Jörgens's Theorems are quite different. Contrary to the Jörgens–Calabi–Pogorelov Theorem for the Monge–Ampère equation, the higher-dimensional Bernstein Theorem for the minimal surface equation is only valid up to dimensions 7 due to works of Bernstein [**Bn**], De Giorgi [**DG2**], Almgren [**Alm**], and Simons [**Sms**], while Bombieri, De Giorgi, and Giusti [**BDGG**] found a counterexample in dimension 8.

Problem 7.7 is based on Savin [**S4**]. Problem 7.8 is taken from Le–Savin [**LS4**]. Problem 7.11 is based on Savin [**S6**] and Figalli [**F2**, Section 5.1].

Chapter 8

Boundary Localization

So far, we have a fairly complete picture of basic interior regularity results for the Monge–Ampère equation together with relevant technical tools developed along the way. Now, we might wonder *if these results and tools are also possible near the boundary.* This chapter will discuss a fundamental tool for the analysis of Monge–Ampère equations near the boundary. The main focus is *boundary localization* properties for the Monge–Ampère equation

$$\lambda \le \det D^2 u \le \Lambda \quad \text{in } \Omega \subset \mathbb{R}^n \quad (n \ge 2),$$

under suitable assumptions on the domain Ω and the boundary data of u on $\partial\Omega$. In particular, we will prove Savin's Boundary Localization Theorem (Theorem 8.5) which provides precise quantitative information on the shape of boundary sections. This theorem will play a very important role in the boundary regularity theory for the Monge–Ampère equation in Chapter 10 and for the linearized Monge–Ampère equation in Part 2.

The concept of solutions to the Monge–Ampère equation in this chapter is understood in the sense of Aleksandrov. By Proposition 7.7, they are also viscosity solutions. All the proofs in this chapter mostly use the comparison principle for the Monge–Ampère equation so the results here are also valid for viscosity solutions. For Aleksandrov solutions, the comparison principle comes from Theorem 3.21 while for viscosity solutions, it comes from Theorem 7.26. We implicitly use these conventions in this chapter.

Recall that Caffarelli's Localization Theorem (Theorem 5.14) allows us to localize interior sections of solutions to the Monge–Ampère equation $\lambda \le \det D^2 u \le \Lambda$ in a domain Ω in $\mathbb{R}^n$ with appropriate boundary data. By this theorem and Lemma 5.6, for each section $S_u(x_0, p, h) \Subset \Omega$ with $x_0 \in \Omega, p \in \partial u(x_0)$, there is an ellipsoid $E_h = A_h^{-1} B_{h^{1/2}}(0)$ where $A_h \in \mathbb{S}^n$ is positive definite with $\det A_h = 1$ and $\sigma(n, \Lambda, \Lambda) > 0$ such that

$$(8.1) \qquad \sigma E_h \subset S_u(x_0, p, h) - x_0 \subset \sigma^{-1} E_h.$$

We already saw in Chapters 5 and 6 the critical role of this localization property in the study of interior $C^{1,\alpha}$, $W^{2,1+\varepsilon}$, $W^{2,p}$, and $C^{2,\alpha}$ regularity of the Monge–Ampère equation.

In this chapter, we extend Caffarelli's localization properties for interior sections to localization properties for sections with centers at the boundary. The version developed here will be used to extend the above-mentioned interior regularity results to those at the boundary, including the global $C^{2,\alpha}$ estimates which then allow global higher-order regularity results from the Schauder theory. As we saw in Chapter 4, a sufficient condition for these estimates to hold when the Monge–Ampère measure is bounded between two positive constants is the second tangential derivatives being bounded from below. In a less regular setting, this is more or less the quadratic separation of solutions from their tangent hyperplanes at the boundary, which is our key assumption in what follows. As will be seen in Section 10.2, this quadratic separation condition is also necessary for the validity of global second derivative estimates.

8.1. Preliminary Localization Property of Boundary Sections

In this section, we establish a boundary version of (8.1) for solutions to the Monge–Ampère equation that separate quadratically from their tangent hyperplanes at the boundary. We recall the notion of special subdifferential $\partial_s u(x_0)$ in Definition 4.1, and the section of u at x_0 with slope p and height $h > 0$ is denoted by

$$S_u(x_0, p, h) = \{x \in \overline{\Omega} : u(x) < u(x_0) + p \cdot (x - x_0) + h\}.$$

Theorem 8.1 (A preliminary localization property of boundary sections). *Let Ω be a bounded convex domain in $\mathbb{R}^n$, $x_0 \in \partial\Omega$, and let L be a supporting hyperplane to $\partial\Omega$ at x_0. Assume that for some $\rho \in (0, 1)$, $\Omega \subset B_{1/\rho}(0)$ and Ω satisfies at x_0 the interior ball condition with radius ρ; that is, there exists $B_\rho(y) \subset \Omega$ with $x_0 \in \partial B_\rho(y)$. Let $u \in C(\overline{\Omega})$ be a convex function satisfying*

$$0 < \lambda \leq \det D^2 u \leq \Lambda \quad in \ \Omega.$$

Assume that u separates quadratically from a tangent hyperplane with slope p at x_0; that is, there exist $p \in \partial_s u(x_0)$ and $\mu > 0$ such that

$$\mu|x - x_0|^2 \leq u(x) - u(x_0) - p \cdot (x - x_0) \leq \mu^{-1}|x - x_0|^2$$
$$\textit{for all } x \in \partial\Omega \cap \{x \in \mathbb{R}^n : \mathrm{dist}(x, L) \leq \rho\}.$$

Here, $p \in \partial_s u(x_0)$ means that $x_{n+1} = u(x_0) + p \cdot (x - x_0)$ is a supporting hyperplane for the graph of u at $(x_0, u(x_0))$, but for any $\varepsilon > 0$,

$$x_{n+1} = u(x_0) + (p - \varepsilon \nu_{x_0}) \cdot (x - x_0)$$

is not a supporting hyperplane where ν_{x_0} is the outer unit normal to $\partial \Omega$ at x_0.

Then, there exist positive constants $\kappa(n, \lambda, \Lambda, \mu)$ and $c(n, \lambda, \Lambda, \mu, \rho)$ such that, for each $h < c$, there is an ellipsoid E_h of volume $\omega_n h^{n/2}$ satisfying

$$\kappa E_h \cap \overline{\Omega} \subset S_u(x_0, p, h) - x_0 \subset \kappa^{-1} E_h \cap \overline{\Omega}.$$

The quadratic separation from tangent hyperplanes on the boundary for solutions to the Monge–Ampère equation is a crucial assumption in Theorem 8.1. By Proposition 4.5, this is the case when $u|_{\partial \Omega}$ and $\partial \Omega$ are pointwise C^3 at x_0 and $\partial \Omega$ is uniformly convex at x_0; see also Problem 4.4. Recall that the pointwise C^3 condition of $\partial \Omega$ at x_0 also guarantees the interior ball condition there.

The hypotheses of Theorem 8.1 imply that u is in fact differentiable at x_0, and then $\partial u(x_0) = \{Du(x_0)\}$, where $Du(x_0)$ is the classical gradient. In fact, we will show in Proposition 8.23 that u is pointwise $C^{1,\alpha}$ at x_0 for all $\alpha < 1$; see also Problem 8.1.

Observe from Lemma 2.67 and (8.1) that an interior section $S_u(x_0, p, h)$ is balanced around its center of mass and around x_0. For a boundary section, the latter no longer holds. Nevertheless, its center of mass, around which it is balanced, continues to be valuable geometric information in the analysis, especially in determining the ellipsoid E_h.

The rest of this section will be devoted to proving Theorem 8.1.

8.1.1. Simplified setting.
By translating and rotating coordinates, we can rewrite the hypotheses of Theorem 8.1 in the following simplified setting.

Denote the upper half of $B_r(0)$ by $B_r^+(0) = B_r(0) \cap \mathbb{R}_+^n$. Let $e_n = (0, \ldots, 0, 1) \in \mathbb{R}^n$. Let Ω be a bounded convex domain in $\mathbb{R}^n$ with

$$(8.2) \qquad\qquad B_\rho(\rho e_n) \subset \Omega \subset B_{1/\rho}^+(0),$$

for some small $\rho > 0$. Let $u \in C(\overline{\Omega})$ be a convex function satisfying

$$(8.3) \qquad\qquad 0 < \lambda \leq \det D^2 u \leq \Lambda \quad \text{in } \Omega.$$

We extend u to be ∞ outside $\overline{\Omega}$.

Subtracting an affine function from u, we can assume $0 \in \partial_s u(0)$, so

$$(8.4) \qquad\qquad x_{n+1} = 0 \text{ is the tangent hyperplane to } u \text{ at } 0;$$

see Definition 4.1. This means that $u(0) = 0$, $u \geq 0$, and any hyperplane $x_{n+1} = \varepsilon x_n$, $\varepsilon > 0$, is not a supporting hyperplane for u.

Assume that for some $\mu > 0$,

(8.5) $\mu|x|^2 \leq u(x) \leq \mu^{-1}|x|^2$ on $\partial\Omega \cap \{x_n \leq \rho\}$.

The section $S_u(x_0, p, h)$ now becomes the section of u at 0 with slope 0 and height h, but for simplicity, we will *omit the slope* 0 *in writing*

$$S_u(x_0, p, h) = S_u(0, h) := \{x \in \overline{\Omega} : u(x) < h\}.$$

We will show that $S_u(0, h)$ is equivalent to a half-ellipsoid centered at 0, with volume comparable to $h^{n/2}$. While the upper bound on the volume of $S_u(0, h)$ follows easily from Lemma 5.8, the lower bound is more involved. The proof of this fact for interior sections, see (5.2) in the proof of Lemma 5.6, suggests that we should compare u with a quadratic polynomial in $S_u(0, h)$. This is possible with the help of the quadratic separation at the boundary. In Proposition 8.4 which gives more quantitative information than Theorem 8.1, we confirm this expectation by showing that the volume of $S_u(0, h)$ is proportional to $h^{n/2}$, and after a sliding along $x_n = 0$ (of controlled norm) we may assume that the center of mass of $S_u(0, h)$ lies on the x_n-axis.

8.1.2. Center of mass of boundary sections. We begin the proof of Theorem 8.1 by showing that the center of mass of $S_u(0, h)$ is of distance at least $h^{n/(n+1)}$ from the hyperplane $x_n = 0$. The choice of the exponent $\alpha = n/(n+1)$ will be apparent in (8.8).

Lemma 8.2 (Lower bound on the altitude of the center of mass). *Assume that Ω and u satisfy (8.2)–(8.5). Then, there are a large constant $C_0 = C_0(\mu, \Lambda, n, \rho) > 1$ and a small constant $c_1 = c_1(\mu, \Lambda, n, \rho) > 0$ such that*

(8.6) $S_u(0, h) \cap \{x_n \leq \rho\} \subset \{x_n > C_0^{-1}(\mu|x'|^2 - h)\}$

and the center of mass x_h^* of $S_u(0, h)$ satisfies for $h \leq c_1$ the estimate

(8.7) $x_h^* \cdot e_n \geq c_1 h^{n/(n+1)}.$

Proof. The proof will be accomplished by constructing suitable quadratic lower bounds for u in $\Omega \cap \{x_n \leq \rho\}$. For (8.6), consider

$$v(x) := \mu|x'|^2 + \frac{\Lambda}{\mu^{n-1}}x_n^2 - C_0 x_n, \quad \text{where } C_0 = \mu\rho^{-3} + \frac{\Lambda\rho}{\mu^{n-1}}.$$

Then

$$v \leq u \quad \text{on } \partial\Omega \cap \{x_n \leq \rho\}, \quad v \leq 0 \leq u \quad \text{on } \Omega \cap \{x_n = \rho\},$$

and

$$\det D^2 v = 2^n \Lambda > \Lambda \geq \det D^2 u \quad \text{in } \Omega \cap \{x_n \leq \rho\}.$$

By the comparison principle, $v \leq u$ in $\Omega \cap \{x_n \leq \rho\}$, so (8.6) follows from

$$S_u(0, h) \cap \{x_n \leq \rho\} \subset \{x \in \overline{\Omega} : v(x) < h\} \subset \{x_n > C_0^{-1}(\mu|x'|^2 - h)\}.$$

We now prove (8.7) by contradiction. Denote

$$S_h := S_u(0, h) \quad \text{and} \quad \alpha = n/(n+1).$$

Assume that (8.7) is not true. Then, $x_h^* \cdot e_n \leq c_1 h^\alpha$ for all $h \leq c_1$, for c_1 sufficiently small with $c_1 \leq \min\{\alpha(n)/2, \rho^{1/\alpha}\}$, where $\alpha(n)$ is as in John's Lemma (Lemma 2.66). From (8.6) and Lemma 2.67, we obtain for $h \leq c_1$,

$$S_h \subset \{x_n \leq 2\alpha^{-1}(n)c_1 h^\alpha \leq h^\alpha\} \cap \{|x'| \leq C_1 h^{\alpha/2}\}, \quad \text{where } C_1 = (C_0 + 1)/\mu.$$

We claim that if $\varepsilon > 0$ and $c_1 = c_1(\mu, \Lambda, n, \rho)$ are sufficiently small, then

$$w(x) := \varepsilon x_n + \frac{h}{2} \left(\frac{|x'|}{C_1 h^{\alpha/2}} \right)^2 + \Lambda C_1^{2(n-1)} h \left(\frac{x_n}{h^\alpha} \right)^2$$

is a lower barrier for u in S_h. Indeed, if ε and c_1 are sufficiently small, then

$$w \leq \frac{h}{4} + \frac{h}{2} + \Lambda C_1^{2(n-1)} (2\alpha^{-1}(n)c_1)^2 h < h \quad \text{in } S_h,$$

and for all $h \leq c_1$, using $x_n \leq |x'|^2/\rho$ on $S_h \cap \partial\Omega$, we get

$$w \leq \varepsilon x_n + \frac{h^{1-\alpha}}{C_1^2} |x'|^2 + \Lambda C_1^{2(n-1)} (2\alpha^{-1}(n)c_1) h \frac{x_n}{h^\alpha} \leq \mu |x'|^2 \leq u \quad \text{on } S_h \cap \partial\Omega.$$

Note that

(8.8) $\det D^2 w = 2\Lambda h^{(1-\alpha)(n-1)+1-2\alpha} = 2\Lambda > \det D^2 u.$

Therefore, the comparison principle gives $w \leq u$ in S_h, as claimed. It follows that $\varepsilon x_n \leq u$ in S_h and hence also in Ω by convexity. This contradicts (8.4). Thus (8.7) is proved.   □

8.1.3. Volume of boundary sections. Next, we introduce the notion of a *sliding map* which will be useful in studying the geometry of boundary sections.

Definition 8.3 (Sliding along a hyperplane). Let L be a hyperplane in $\mathbb{R}^n$ given by the equation $\nu \cdot (x - x_0) = 0$, where ν is a unit normal vector of L and $x_0 \in L$. Any linear transformation $A : \mathbb{R}^n \to \mathbb{R}^n$ of the form

$$Ax = x - [\nu \cdot (x - x_0)]\tau, \quad \text{where } \tau \in \mathbb{R}^n \quad \text{with } \tau \cdot \nu = 0,$$

is called a *sliding along* the hyperplane L. See Figure 8.1.

In particular, a *sliding along* the hyperplane $x_n = 0$ is of the form

$$Ax = x - \tau x_n, \quad \text{where } \tau \in \mathbb{R}^n \quad \text{with } \tau \cdot e_n = 0.$$

In this case, $\tau \in \text{span}\{e_1, \ldots, e_{n-1}\}$, and the map A is also called a *sliding along the $x_1, \ldots, x_{n-1}$ directions*; see Definition 8.14.

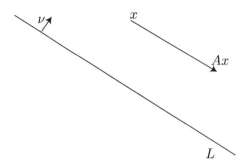

Figure 8.1. Sliding along the hyperplane L.

If $Ax = x - [\nu \cdot (x - x_0)]\tau$ is a sliding along L, then so is $A^{-1}x = x + [\nu \cdot (x - x_0)]\tau$. Notice that the sliding map A along the hyperplane L is the identity map when it is restricted to L, and it becomes a translation of vector $-s\tau$ when it is restricted to the hyperplane $\nu \cdot (x - x_0) = s$.

In the following proposition, we show that the volume of $S_u(0, h)$ is comparable to $h^{n/2}$, and by sliding along the hyperplane $x_n = 0$ by an affine transformation with norm not larger than $h^{-\frac{n}{2(n+1)}}$, we may assume that the center of mass of $S_u(0, h)$ lies on the x_n-axis.

Proposition 8.4. *Assume that Ω and u satisfy (8.2)–(8.5). Then, there are positive constants $\kappa_0(\mu, \lambda, \Lambda, n)$ and $c(\mu, \lambda, \Lambda, n, \rho)$ with the following properties. For all $h \leq c$, there exists a sliding along $x_n = 0$ given by*

$$A_h(x) = x - \tau_h x_n \quad \text{with } \tau_h = (\tau_1, \ldots, \tau_{n-1}, 0) \quad \text{and} \quad |\tau_h| \leq c^{-1} h^{-\frac{n}{2(n+1)}},$$

such that the rescalings of Ω, u, and $S_u(0, h)$ defined by

$$\tilde{\Omega} := A_h \Omega, \; \tilde{u}(x) := u(A_h^{-1} x), \; S_{\tilde{u}}(0, h) := A_h S_u(0, h) = \{x \in \overline{\tilde{\Omega}} : \tilde{u}(x) < h\}$$

satisfy the following.

(i) *The center of mass $\tilde{x}_h^*$ of $S_{\tilde{u}}(0, h)$ lies on the x_n-axis; that is, $\tilde{x}_h^* = d_h e_n$ where*

$$d_h \geq c h^{\frac{n}{n+1}}.$$

(ii) *$\tilde{G}_h := \partial S_{\tilde{u}}(0, h) \cap \{x \in \overline{\tilde{\Omega}} : \tilde{u}(x) < h\}$ is a graph of a function g_h; that is, $\tilde{G}_h = \{(x', g_h(x'))\}$, where g_h satisfies*

$$g_h(x') \leq 2|x'|^2/\rho \quad \text{and} \quad (\mu/2)|x'|^2 \leq \tilde{u}(x) \leq 2\mu^{-1}|x'|^2 \quad \text{on } \tilde{G}_h.$$

(iii) *After rotating the $x_1, \ldots, x_{n-1}$ coordinates, we have*

$$(8.9) \qquad \kappa_0 D_h B_1(0) \subset S_{\tilde{u}}(0, h) - \tilde{x}_h^* \subset \kappa_0^{-1} D_h B_1(0),$$

where $D_h := \operatorname{diag}(d_1, \ldots, d_n)$ is a diagonal matrix that satisfies
- *$d_i \geq \kappa_0 h^{1/2}$ for $i = 1, \ldots, n-1$ and*
- *$d_n \geq c h^{\frac{n}{n+1}}$, $\kappa_0 d_h \leq d_n \leq \kappa_0^{-1} d_h$, and $\prod_{i=1}^{n} d_i = h^{n/2}$.*

Moreover, we have the volume estimates

$$\kappa_0 h^{n/2} \leq |S_{\tilde{u}}(0, h)| = |S_u(0, h)| \leq \kappa_0^{-1} h^{n/2}.$$

Proof. We prove part (i). Denote $S_h := S_u(0, h)$, $\alpha := n/(n+1)$, and denote the center of mass of S_h by $x_h^* = (x_h^{*\prime}, x_h^* \cdot e_n)$. We choose the map A_h by defining

$$A_h x = x - \tau_h x_n \quad \text{where} \quad \tau_h = (\tau_h', 0) \equiv \frac{(x_h^{*\prime}, 0)}{x_h^* \cdot e_n},$$

and we set

$$\tilde{u}(x) := u(A_h^{-1} x) \quad \text{for } x \in \tilde{\Omega} := A_h \Omega.$$

The center of mass of

$$\tilde{S}_h := S_{\tilde{u}}(0, h) = A_h S_h$$

is $\tilde{x}_h^* = A_h x_h^*$, and it lies on the x_n-axis from the definition of A_h. Thus, $\tilde{x}_h^* = d_h e_n$ where $d_h = \tilde{x}_h^* \cdot e_n = x_h^* \cdot e_n \geq c_1 h^{\frac{n}{n+1}}$ by (8.7). This proves (i).

Moreover, since $x_h^* \in S_h$, we see from (8.6) and (8.7) that

$$|\tau_h| \leq \frac{C(n, \mu, \Lambda, \rho)(x_h^* \cdot e_n)^{1/2}}{(x_h^* \cdot e_n)} \leq C_1(n, \mu, \Lambda, \rho) h^{-\alpha/2} \quad \text{for } h \leq c_1(\mu, \Lambda, n, \rho).$$

Next, we prove part (ii). Note that, from (8.2) and (8.5), we have

$$\partial S_h \cap \partial \Omega \subset \{x_n \leq |x'|^2/\rho\} \cap \{|x'| < C(\mu) h^{1/2}\}.$$

Therefore, if $h \leq c_1$, then, on $\partial S_h \cap \partial \Omega$,

$$(8.10) \qquad |A_h x - x| = |\tau_h'| x_n \leq C_1 h^{-\alpha/2} |x'|^2/\rho \leq C_2(n, \mu, \Lambda, \rho) h^{\frac{1-\alpha}{2}} |x'|.$$

Clearly, $\tilde{G}_h = \partial S_{\tilde{u}}(0, h) \cap \{x \in \overline{\tilde{\Omega}} : \tilde{u}(x) < h\} \subset A_h(\partial S_h \cap \partial \Omega)$ is a graph of a function g_h; that is, $\tilde{G}_h = \{(y', g_h(y'))\}$, where

$$y' = x' - \tau_h' x_n \quad \text{and} \quad g_h(y') = x_n \quad \text{for } (x', x_n) \in \partial S_h \cap \partial \Omega.$$

Reducing c_1 if necessary, we find from (8.10) that if $h \leq c_1$, then on $\tilde{G}_h$,

$$|y'|^2 \geq |x'|^2 - 2|\tau_h'| x_n |x'| \geq |x'|^2 - 2C_2 h^{\frac{1-\alpha}{2}} |x'|^2 \geq (3/4)|x'|^2$$

and

$$|y'|^2 \le |x'|^2 + 2|\tau_h'||x_n||x'| + |\tau_h'|^2 x_n^2$$
$$\le (1 + 2C_2 h^{\frac{1-\alpha}{2}} + C_2^2 h^{1-\alpha})|x'|^2 \le 2|x'|^2.$$

Thus, on $\tilde{G}_h$, we have

$$g_h(y') = x_n \le |x'|^2/\rho \le 2|y'|^2/\rho,$$

and, from (8.5), we obtain

$$(\mu/2)|y'|^2 \le \tilde{u}(y) = u(x) \le \mu^{-1}(|x'|^2 + 4|y'|^4/\rho^2) \le 2\mu^{-1}|y'|^2$$

if $h \le c_1$ is small (so $|y'|$ is also small). Therefore, part (ii) is proved.

We now prove part (iii).

Step 1. Choosing D_h. Consider the cross section $\tilde{S}_h' := \tilde{S}_h \cap \{x_n = d_h\}$ of $\tilde{S}_h$ with the hyperplane $x_n = d_h$. By Lemma 2.67, after a rotation of the first $n-1$ coordinates, we might assume that $\tilde{S}_h'$ is equivalent to an ellipsoid with principal axes of lengths $d_1 \le \cdots \le d_{n-1}$; that is,

$$\left\{\sum_{i=1}^{n-1}\left(\frac{x_i}{d_i}\right)^2 \le 1\right\} \cap \{x_n = d_h\} \subset \tilde{S}_h' \subset \left\{\sum_{i=1}^{n-1}\left(\frac{x_i}{d_i}\right)^2 \le C(n)\right\} \cap \{x_n = d_h\}.$$

It follows that

$$(8.11) \quad \left\{\sum_{i=1}^{n-1}\left(\frac{x_i}{d_i}\right)^2 \le 1\right\} \cap \{0 \le x_n \le d_h\} \subset \tilde{S}_h$$

$$\subset \left\{\sum_{i=1}^{n-1}\left(\frac{x_i}{d_i}\right)^2 \le C(n)\right\} \cap \{0 \le x_n \le C(n)d_h\}.$$

Consequently,

$$\text{diag}\,(d_1,\ldots,d_h)B_1(0) \subset S_{\tilde{u}}(0,h) - \tilde{x}_h^* \subset C(n)\text{diag}\,(d_1,\ldots,d_h)B_1(0).$$

Let $d_n = h^{n/2}\prod_{i=1}^{n-1} d_i^{-1}$ and $D_h := \text{diag}\,(d_1,\ldots,d_n)$.

Step 2. Properties of D_h. By Lemma 5.8, $|\tilde{S}_h| \le C(\lambda,n)h^{n/2}$. Thus

$$(8.12) \qquad\qquad d_h \prod_{i=1}^{n-1} d_i \le K(\lambda,n)h^{n/2}.$$

Since $\tilde{u} \le 2\mu^{-1}|x'|^2$ on $\tilde{G}_h = \{(x', g_h(x'))\}$, we see from (8.11) that

$$d_i \ge [C(n)]^{-1/2}(\mu h/2)^{1/2} \equiv \kappa_1 h^{1/2} \quad (i = 1,\ldots,n-1),$$

for some $\kappa_1(n,\mu) > 0$.

By (8.12) and part (i), we have $d_n \geq d_h/K \geq (c_1/K)h^\alpha$.
We show that there exists $\kappa_2 = \kappa_2(n, \mu, \Lambda) > 0$ such that for all small h,

$$(8.13) \qquad d_h \prod_{i=1}^{n-1} d_i \geq \kappa_2 h^{n/2}.$$

Granted (8.13), then from (8.11) and $\prod_{i=1}^{n} d_i = h^{n/2}$, it is clear that $|S_h| \geq k_0 h^{n/2}$, while $\kappa_0 d_h \leq d_n \leq \kappa_0^{-1} d_h$, and hence (8.9) is confirmed, for some $\kappa_0(n, \lambda, \Lambda, \mu) > 0$. Thus, part (iii) is proved.

To prove (8.13), we will use the quadratic polynomial

$$w(x) = \varepsilon x_n + \kappa h \Big[\sum_{i=1}^{n-1} \Big(\frac{x_i}{d_i}\Big)^2 + \Big(\frac{x_n}{d_h}\Big)^2 \Big]$$

as a lower barrier for $\tilde{u}$ on $\partial \tilde{S}_h$ for suitably chosen ε and κ. On $\tilde{G}_h$, we have $x_n \leq 2|x'|^2/\rho$ and $x_n \leq C(n)d_h$. Therefore, using $d_h \geq c_1 h^\alpha$, we find on $\tilde{G}_h$,

$$w \leq \varepsilon x_n + \frac{\kappa}{\kappa_1^2}|x'|^2 + \kappa h C(n)\frac{x_n}{d_h} \leq \Big(\frac{2\varepsilon}{\rho} + \frac{\kappa}{\kappa_1^2} + \frac{2\kappa h^{1-\alpha}C(n)}{c_1\rho}\Big)|x'|^2.$$

We choose κ sufficiently small depending on μ and n, so that for all $h < c(n, \lambda, \Lambda, \mu, \rho) \leq c_1^{1/(1-\alpha)}$ and $\varepsilon \leq \varepsilon_0(\mu, h, \rho)$, we have

$$w \leq (\mu/2)|x'|^2 \leq \tilde{u} \quad \text{on } \tilde{G}_h,$$

while

$$w \leq \varepsilon/\rho + \kappa[C(n)]^2 h \leq h \quad \text{on } \partial \tilde{S}_h \setminus \tilde{G}_h.$$

In other words, $w \leq \tilde{u}$ on $\partial \tilde{S}_h$. We claim that (8.13) holds for $\kappa_2 = \Lambda^{-1/2}(2\kappa)^{n/2}$. Suppose otherwise. Then $d_h \prod_{i=1}^{n-1} d_i < \Lambda^{-1/2}(2\kappa)^{n/2}h^{n/2}$, so

$$\det D^2 w = (2\kappa h)^n d_h^{-2} \prod_{i=1}^{n-1} d_i^{-2} > \Lambda \geq \det D^2 \tilde{u}.$$

Thus, by the comparison principle, $w \leq \tilde{u}$ in $\tilde{S}_h$. Because $\tilde{u}$ is obtained from u by a sliding along $x_n = 0$, $x_{n+1} = 0$ is still the tangent hyperplane of $\tilde{u}$ at 0. We reach a contradiction since $\tilde{u} \geq w \geq \varepsilon x_n$, and (8.13) is proved with the above value of κ_2. The proof of the proposition is complete. $\qquad \square$

Proof of Theorem 8.1. We can assume that Ω and u satisfy (8.2)–(8.5) and we use the notation of Proposition 8.4. From (8.9), we have

$$S_{\tilde{u}}(0, h) \subset 2\kappa_0^{-1} D_h B_1(0) \cap A_h \overline{\Omega}.$$

From (8.11) and $\kappa_0 d_h \leq d_n \leq \kappa_0^{-1} d_h$ in Proposition 8.4(iii), we find that

$$(8.14) \qquad (\kappa_0/2) D_h B_1(0) \cap A_h \overline{\Omega} \subset S_{\tilde{u}}(0, h).$$

Hence, for $\kappa := \kappa_0/2$, we have

$$\kappa D_h B_1(0) \cap A_h \overline{\Omega} \subset A_h S_u(0, h) = S_{\tilde{u}}(0, h) \subset \kappa^{-1} D_h B_1(0) \cap A_h \overline{\Omega}.$$

The ellipsoid $E_h := A_h^{-1} D_h B_1(0)$ satisfies the conclusion of the theorem. $\qquad \square$

8.2. Savin's Boundary Localization Theorem

This section and Sections 8.3 and 8.4 will be devoted to proving a striking result, due to Savin [**S3, S4**], which gives precise information on τ_h and d_n in Proposition 8.4. This result significantly improves the lower bound for d_n from $d_n \geq c(n, \mu, \Lambda, \rho) h^{\frac{n}{n+1}}$ to $d_n \geq c(n, \mu, \lambda, \Lambda) h^{\frac{1}{2}}$, and the upper bound for τ_h from $|\tau_h| \leq C(n, \mu, \Lambda, \rho) h^{-\frac{n}{2(n+1)}}$ to $|\tau_h| \leq C(n, \mu, \lambda, \Lambda) |\log h|$.

Theorem 8.5 (Savin's Boundary Localization Theorem). *Let Ω be a convex domain in $\mathbb{R}^n$, and let $u : \overline{\Omega} \to \mathbb{R}$ be a continuous and convex function. Assume that Ω and u satisfy the following hypotheses for some positive constants $\rho, \lambda, \Lambda,$ and μ:*

(a) *$B_\rho(\rho e_n) \subset \Omega \subset \{x_n \geq 0\} \cap B_{1/\rho}(0)$,*

(b) *$\lambda \leq \det D^2 u \leq \Lambda$ in Ω,*

(c) *$x_{n+1} = 0$ is the tangent hyperplane to u at 0; that is, $u \geq 0$, $u(0) = 0$, and any hyperplane $x_{n+1} = \varepsilon x_n$ (where $\varepsilon > 0$) is not a supporting hyperplane for u, and*

(d) *$\mu |x|^2 \leq u(x) \leq \mu^{-1} |x|^2$ on $\partial \Omega \cap \{x_n \leq \rho\}$.*

Consider the sections $S_u(0, h) = \{x \in \overline{\Omega} : u(x) < h\}$ of u with heights $h > 0$ and center at 0. Then, for each $h < c(\rho, \mu, \lambda, \Lambda, n)$, there exists an ellipsoid E_h of volume $\omega_n h^{n/2}$ such that

$$\kappa E_h \cap \overline{\Omega} \subset S_u(0, h) \subset \kappa^{-1} E_h \cap \overline{\Omega},$$

where $\kappa = \kappa(\mu, \lambda, \Lambda, n) > 0$. Moreover, E_h is obtained from the ball $B_{h^{1/2}}(0)$ by a linear transformation A_h^{-1} which is sliding along the hyperplane $x_n = 0$:

$$E_h = A_h^{-1} B_{h^{1/2}}(0), \quad \text{where } A_h(x) = x - \tau_h x_n, \quad \tau_h = (\tau_1, \ldots, \tau_{n-1}, 0),$$

with

$$|\tau_h| \leq \kappa^{-1} |\log h|.$$

More generally, if u and Ω satisfy the hypotheses of Theorem 8.1 at $x_0 \in \partial \Omega$, then the ellipsoid E_h there can be chosen to be $A_h^{-1} B_{h^{1/2}}(x_0)$ where $A_h \in \mathbb{M}^{n \times n}$ with $\det A_h = 1$, $\|A_h\| \leq \kappa^{-1} |\log h|$, and A_h is a sliding along the supporting hyperplane L to $\partial \Omega$ at x_0.

The last statement of Theorem 8.5 is a consequence of the first statement concerning $S_u(0, h)$, so we will focus on this case. The ellipsoid E_h, or equivalently the linear map A_h, provides quantitative information about

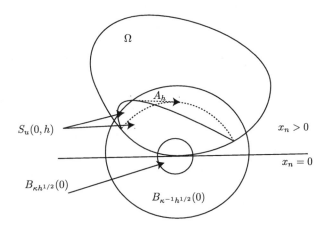

Figure 8.2. Geometry of the Boundary Localization Theorem.

the behavior of the second derivatives of u near the origin; see Figure 8.2. Heuristically, Theorem 8.5 implies that in $S_u(0, h)$, the second tangential derivatives of u are bounded from above and below, the second mixed derivatives are bounded by $C|\log h|$, and by using $\det D^2 u \le \Lambda$, the second normal derivative of u is bounded by $C|\log h|^2$. One way to see these is to apply Lemma 5.6(iii)(b) to the function $\tilde{u}(x) := u(A_h^{-1}x)$ whose section of height h at the origin is equivalent to a half-ball of radius $h^{1/2}$ for all h small. Thus, roughly speaking, the Hessian of $\tilde{u}$ is bounded near the origin, and this gives the heuristic bounds on the Hessian of u near the origin. Strikingly, all these are inferred just from bounds on the Monge–Ampère measure and the quadratic separation at the boundary of the solution. Moreover, to transform a boundary section $S_u(0, h)$ into a shape comparable to the upper half-ball $B_{h^{1/2}}(0) \cap \mathbb{R}^n_+$, we only need a linear transformation of norm $C|\log h|$ which is much smaller than the typical norm of order $h^{\frac{1}{2} - \frac{1}{1+\alpha}}$ where $\alpha(n, \lambda, \Lambda) \in (0, 1)$ for interior sections; see Remark 5.23. This is another striking feature of Theorem 8.5.

We now prove Theorem 8.5 by compactness arguments, continuing using the notation of Proposition 8.4. The most important part is to establish the correct lower bound on d_n. The upper bound $|\tau_h| \le C|\log h|$ for τ_h then follows as a consequence.

8.2.1. Normalizing altitude of boundary sections.

The key property of d_n in Proposition 8.4 will be established in Proposition 8.21 which asserts that d_n is comparable to $h^{1/2}$ when h is small. In view of assertion (iii) in Proposition 8.4 and the fact that d_n is comparable to the altitude (in the x_n direction) of $S_{\tilde{u}}(0, h)$ (and also $S_u(0, h)$), this is equivalent to showing that

$h^{-1/2} \sup_{S_u(0,h)} x_n$ is bounded from below. The latter quantity, denoted by $b_{u,\Omega}(h)$, has many invariant properties under affine transformations.

Definition 8.6 (Normalized altitude of boundary sections $b_{u,\Omega}$). Let Ω be a bounded convex domain in $\mathbb{R}^n_+$ with $0 \in \partial\Omega$. For any convex function $u : \overline{\Omega} \to \mathbb{R}$ satisfying $u(0) = 0$ and $u \geq 0$, we define for $h > 0$

$$b_{u,\Omega}(h) = h^{-1/2} \sup_{S_u(0,h)} x_n.$$

Remark 8.7 (Invariant properties of $b_{u,\Omega}(h)$). Let Ω, u, and $b_{u,\Omega}$ be as in Definition 8.6. Then, the quantity $b_{u,\Omega}(h)$ has the following properties:

(a) If $h_1 \leq h_2$, then

$$\left(\frac{h_1}{h_2}\right)^{\frac{1}{2}} \leq \frac{b_{u,\Omega}(h_1)}{b_{u,\Omega}(h_2)} \leq \left(\frac{h_2}{h_1}\right)^{\frac{1}{2}}.$$

(b) If a linear transformation A is a sliding along the hyperplane $x_n = 0$ and if $\tilde{u}(x) = u(Ax)$, then $b_{\tilde{u},A^{-1}\Omega}(h) = b_{u,\Omega}(h)$.

(c) If A is a linear transformation which leaves the hyperplane $x_n = 0$ invariant (such as $Ax = (A'x', ax_n)$) and if $\tilde{u}(x) = u(Ax)$, then the quotients $b_{u,\Omega}(h_1)/b_{u,\Omega}(h_2)$ remain unchanged; that is,

$$\frac{b_{\tilde{u},A^{-1}\Omega}(h_1)}{b_{\tilde{u},A^{-1}\Omega}(h_2)} = \frac{b_{u,\Omega}(h_1)}{b_{u,\Omega}(h_2)}.$$

(d) If $\beta > 0$ and $\tilde{u}(x) = \beta u(x)$, then $b_{\tilde{u},\Omega}(\beta h) = \beta^{-1/2} b_{u,\Omega}(h)$.

Proof. We only need to verify property (a). Note that the first inequality comes from convexity of u, while the second inequality comes from the inclusion $S_u(0,h_1) \subset S_u(0,h_2)$. □

We restate Proposition 8.4 in a normalized setting as follows.

Proposition 8.8. *Assume that Ω and u satisfy (8.2)–(8.5). We normalize the function $\tilde{u}$, the section $S_{\tilde{u}}(0,h)$, and the entries d_i of the matrix D_h in Proposition 8.4 by*

$$v(x) = h^{-1}\tilde{u}(D_h x) \quad on \ \Omega_v := \{v < 1\} = D_h^{-1} S_{\tilde{u}}(0,h) \quad and \quad a_i = h^{-1/2}d_i.$$

Then $\prod_{i=1}^n a_i = 1$ with

$$a_i \geq \kappa_0 \quad for \ i = 1, \ldots, n-1 \quad and \quad \kappa_0 \leq a_n^{-1} b_{u,\Omega}(h) = b_{v,\Omega_v}(1) \leq \kappa_0^{-1}.$$

The function $v \in C(\overline{\Omega}_v)$ is convex and it satisfies

$$v(0) = 0, \ v \geq 0, \quad and \quad \lambda \leq \det D^2 v \leq \Lambda \quad in \ \Omega_v.$$

Moreover,

$$0 \in \partial\Omega_v, \quad x^* + \kappa_0 B_1(0) \subset \Omega_v \subset \kappa_0^{-1} B_1^+(0),$$

for some x^*, and the section $S_v(0, t)$ of v centered at 0 with height t satisfies

$$\kappa_0 t^{n/2} \leq |S_v(0, t)| \leq \kappa_0^{-1} t^{n/2} \quad \text{for all } t \leq 1.$$

The set $G := \partial\Omega_v \cap \{v < 1\}$ is a graph $\{(x', g(x'))\}$ with

(8.15) $$G \subset \{x_n \leq \sigma\}, \quad \sigma := C(n, \mu, \lambda, \Lambda, \rho) h^{\frac{1}{n+1}}$$

and on G, we have

$$\frac{1}{2}\mu \sum_{i=1}^{n-1} a_i^2 x_i^2 \leq v \leq 2\mu^{-1} \sum_{i=1}^{n-1} a_i^2 x_i^2.$$

Furthermore, $v = 1$ on $\partial\Omega_v \backslash G$. The constant κ_0 depends on n, λ, Λ, and μ.

All assertions in Proposition 8.8 are straightforward, except perhaps (8.15) which can be proved as follows. If $x \in G = D_h^{-1}\tilde{G}_h = \{(x', g(x'))\}$, then $D_h x \in \tilde{G}_h$, and therefore, Proposition 8.4(ii) gives

$$d_n g(x') \leq (2/\rho) \sum_{i=1}^{n-1} d_i^2 x_i^2$$

and

$$x_n d_n = g_h(x') \leq 4(\mu\rho)^{-1}\tilde{u}(D_h x) \leq 4(\mu\rho)^{-1} h.$$

Hence, using $d_n \geq ch^{\frac{n}{n+1}}$ from Proposition 8.4(iii), we obtain (8.15).

Since $S_v(0, t) = D_h^{-1}S_{\tilde{u}}(0, th)$, we have $b_{v,\Omega_v}(t) = d_n^{-1}h^{1/2}b_{\tilde{u},\tilde{\Omega}}(th)$, and thus, by Remark 8.7(b),

(8.16) $$\frac{b_{v,\Omega_v}(t)}{b_{v,\Omega_v}(1)} = \frac{b_{u,\Omega}(th)}{b_{u,\Omega}(h)}.$$

8.2.2. Key property of the normalized altitude of boundary sections. To study deeper properties of $b_{u,\Omega}$, or equivalently b_{v,Ω_v}, we introduce a class $\mathcal{D}_\sigma^\mu$ of solutions to the Monge–Ampère equation that have the same properties as v and Ω_v in Proposition 8.8.

Fix $0 < \lambda \leq \Lambda < \infty$.

Let π_n be the projection map from $\mathbb{R}^n$ to $\mathbb{R}^{n-1}$ defined as follows:

$$\pi_n(E) = \{x' \in \mathbb{R}^{n-1} : \text{there is } t \in \mathbb{R} \text{ such that } (x', t) \in E\} \text{ for all } E \subset \mathbb{R}^n.$$

We will study $\mathcal{D}_\sigma^\mu$ using compactness methods, so for this purpose, an appropriate notion of boundary values of a convex function defined on an open bounded set is that of Definition 7.20.

Recall that $u \stackrel{\text{epi}}{=} \varphi$ on $\partial\Omega$ means $u|_{\partial\Omega,\text{epi}} = \varphi^c$. Moreover, we will often use Remark 7.21 on boundary values of a convex function in a section.

Definition 8.9 (Class $\mathcal{D}_\sigma^\mu(a_1, \ldots, a_{n-1})$). Fix small, positive constants μ and σ ($\mu, \sigma \le 1$) and fix an increasing sequence

$$a_1 \le \cdots \le a_{n-1} \quad \text{with } a_1 \ge \mu.$$

We introduce the class $\mathcal{D}_\sigma^\mu(a_1, \ldots, a_{n-1})$ consisting of pairs of function u and domain $\Omega \subset \mathbb{R}^n$ satisfying the following conditions:

(a) $u : \Omega \to \mathbb{R}$ is convex, where

(8.17)
$$\begin{cases} 0 \in \partial\Omega, \quad B_\mu(x^*) \subset \Omega \subset B_{1/\mu}^+(0) \quad \text{for some } x^* \in \mathbb{R}^n, \\ u(0) = 0, \quad 0 \le u \le 1, \text{ and } \lambda \le \det D^2 u \le \Lambda \quad \text{in } \Omega. \end{cases}$$

(b) For all $h \le 1$, we have

(8.18)
$$\mu h^{n/2} \le |S_u(0, h)| \le \mu^{-1} h^{n/2}.$$

(c) The boundary $\partial\Omega$ has a closed subset $G \subset \{x_n \le \sigma\} \cap \partial\Omega$ which is a graph in the e_n direction with projection $\pi_n(G) \subset \mathbb{R}^{n-1}$ along the e_n direction satisfying

$$\left\{ \mu^{-1} \sum_{i=1}^{n-1} a_i^2 x_i^2 \le 1 \right\} \subset \pi_n(G) \subset \left\{ \mu \sum_{i=1}^{n-1} a_i^2 x_i^2 \le 1 \right\}.$$

On $\partial\Omega$, $u \overset{\text{epi}}{=} \varphi$ in the epigraphical sense of Definition 7.20, where φ satisfies the following: $\varphi = 1$ on $\partial\Omega \setminus G$ and

$$\mu \sum_{i=1}^{n-1} a_i^2 x_i^2 \le \varphi(x) \le \min\left\{ 1, \mu^{-1} \sum_{i=1}^{n-1} a_i^2 x_i^2 \right\} \qquad \text{on } G.$$

In Proposition 8.8, by relabeling the $x_1, \ldots, x_{n-1}$ variables, we can assume that $a_1 \le \cdots \le a_{n-1}$. Then, $(v, \Omega_v) \in \mathcal{D}_\sigma^{\bar\mu}(a_1, \ldots, a_{n-1})$ where $\bar\mu = \min\{\kappa_0, \mu/2\}$ depends only on n, λ, Λ, and μ. Furthermore, the set G in Definition 8.9 is the closure of the set G in Proposition 8.8.

The key property of the normalized altitude of boundary sections in the class $\mathcal{D}_\sigma^\mu(a_1, \ldots, a_{n-1})$ is given by the following proposition.

Proposition 8.10. *For any $M_0 > 0$, there exists $C_*(n, \lambda, \Lambda, \mu, M_0) > 0$ such that if $(u, \Omega) \in \mathcal{D}_\sigma^\mu(a_1, \ldots, a_{n-1})$ with $a_{n-1} \ge C_*$ and $\sigma \le C_*^{-1}$, then, for some $h \in [C_*^{-1}, 1]$, we have*

$$b_{u,\Omega}(h) = h^{-1/2}\Big(\sup_{S_u(0,h)} x_n \Big) \ge M_0.$$

The proof of Proposition 8.10 will be given in Section 8.3 below. It will be used to complete the proof of Theorem 8.5 in Section 8.4.

8.3. Normalized Altitude of Boundary Sections

In this section, we prove Proposition 8.10 by compactness using Proposition 8.13 below. For this purpose, we introduce the limiting solutions from the class $\mathcal{D}_\sigma^\mu(a_1, \ldots, a_{n-1})$, defined in Definition 8.9, when $a_{k+1} \to \infty$ and $\sigma \to 0$. Recall that $0 < \lambda \leq \Lambda$ are fixed, and we always assume $\mu \in (0, 1)$.

Definition 8.11 (Class $\mathcal{D}_0^\mu(a_1, \ldots, a_k, \infty, \ldots, \infty)$). Fix $k \in \{0, 1, \ldots, n-2\}$. For $\mu \leq a_1 \leq \cdots \leq a_k$, we denote by $\mathcal{D}_0^\mu(a_1, \ldots, a_k, \infty, \ldots, \infty)$ the class of pairs of convex function u (on Ω) and domain Ω in $\mathbb{R}^n$ satisfying the following conditions:

    (a) u and Ω satisfy properties (8.17) and (8.18).

    (b) The boundary $\partial\Omega$ has a closed subset G such that

$$G \subset \{x = (x_1, \ldots, x_n) \in \mathbb{R}^n : x_i = 0, \text{ for all } i > k\} \cap \partial\Omega.$$

    (c) If we restrict to the space $\mathbb{R}^k$ generated by the first k coordinates, then

$$\left\{ \mu^{-1} \sum_{i=1}^k a_i^2 x_i^2 \leq 1 \right\} \subset G \subset \left\{ \mu \sum_{i=1}^k a_i^2 x_i^2 \leq 1 \right\}.$$

    (d) $u \overset{\text{epi}}{=} \varphi$ on $\partial\Omega$ in the sense of Definition 7.20 with $\varphi = 1$ on $\partial\Omega \setminus G$ and

$$\mu \sum_{i=1}^k a_i^2 x_i^2 \leq \varphi \leq \min\left\{ 1, \mu^{-1} \sum_{i=1}^k a_i^2 x_i^2 \right\} \qquad \text{on } G.$$

Remark 8.12. If $a_k \leq K < \infty$ and $(u, \Omega) \in \mathcal{D}_0^\mu(a_1, \ldots, a_k, \infty, \ldots, \infty)$, then (see also Problem 8.2)

$$(u, \Omega) \in \mathcal{D}_0^{\tilde\mu}(\underbrace{1, \ldots, 1}_{k \text{ times}}, \infty, \ldots, \infty), \quad \text{where } \tilde\mu := \min\{\mu^3, \mu K^{-2}\}.$$

Suppose we have a sequence of function-domain pairs

$$(u_m, \Omega_m) \in \mathcal{D}_{\sigma_m}^\mu(a_1^m, \ldots, a_{n-1}^m)$$

with

$$\sigma_m \to 0 \quad \text{and} \quad a_{k+1}^m \to \infty$$

for some fixed $0 \leq k \leq n - 2$. Then, by the compactness result in Theorem 7.29, we can extract a subsequence converging, in the sense of Hausdorff convergence of epigraphs in Definition 7.22, to a function-domain pair (u, Ω) with $(u, \Omega) \in \mathcal{D}_0^\mu(a_1, \ldots, a_l, \infty, \ldots, \infty)$, for some $l \leq k$ and $a_1 \leq \cdots \leq a_l$.

    The following proposition gives the main property for function-domain pairs in the class $\mathcal{D}_0^\mu(a_1, \ldots, a_k, \infty, \ldots, \infty)$.

Proposition 8.13. *For any $M > 0$ and $0 \leq k \leq n - 2$, there exists $c_k > 0$ depending on $M, \mu, \lambda, \Lambda, n, k$ such that if*

$$(8.19) \qquad\qquad (u, \Omega) \in \mathcal{D}_0^\mu(a_1, \ldots, a_k, \infty, \ldots, \infty),$$

then, for some $h \in [c_k, 1]$, we have

$$b_{u,\Omega}(h) = h^{-1/2} \Big(\sup_{S_u(0,h)} x_n \Big) \geq M.$$

Proof of Proposition 8.10 assuming Proposition 8.13. Fix $M_0 > 0$. Assume otherwise that the conclusion of Proposition 8.10 does not hold. Then for each positive integer m, there exists a function-domain pair

$$(u_m, \Omega_m) \in \mathcal{D}_{\sigma_m}^\mu(a_1^m, \ldots, a_{n-1}^m)$$

satisfying $a_{n-1}^m \geq m$, $\sigma_m \leq 1/m$, and

$$b_{u_m,\Omega_m}(h) < M_0 \quad \text{for all } h \in [1/m, 1].$$

Using the compactness result in Theorem 7.29, we can extract a subsequence of $\{(u_m, \Omega_m)\}_{m=1}^\infty$ that converges, in the sense of Hausdorff convergence of epigraphs in Definition 7.22, to a limiting function-domain pair (u, Ω) satisfying (8.19) for some $k \leq n-2$ for which $b_{u,\Omega}(h) \leq M_0$ for all $h > 0$ (see also Problem 8.4). This contradicts Proposition 8.13 with $M := 2M_0$. $\qquad\square$

The rest of this section is devoted to the proof of Proposition 8.13 which goes by induction on k. We introduce some conventions and notation. The proofs here rely on the results in Section 7.2. Thus, by the comparison principle, we mean the one in Theorem 7.26 for viscosity solutions with boundary discontinuities. When $k = 0$, solutions in the class $\mathcal{D}_0^\mu(\infty, \ldots, \infty)$ have boundary discontinuities. Moreover, to simplify, we write $u = \varphi$ on $\partial\Omega$ to mean that $u \overset{\text{epi}}{=} \varphi$ on $\partial\Omega$.

Write $\mathbb{R}^n = \mathbb{R}^k \times \mathbb{R}^{n-1-k} \times \mathbb{R}$, and for $x = (x', x_n) \in \mathbb{R}^n$,

$$(8.20) \quad x' = (y, z), \ y = (x_1, \ldots, x_k) \in \mathbb{R}^k, \ z = (x_{k+1}, \ldots, x_{n-1}) \in \mathbb{R}^{n-1-k}.$$

The following definition extends Definition 8.3 from $k = n - 1$ to $k \leq n - 2$.

Definition 8.14. Let $1 \leq k \leq n - 2$. We say that a linear transformation $T : \mathbb{R}^n \to \mathbb{R}^n$ is a *sliding along the $x_1, \ldots, x_k$ directions (or, y direction, for short)* if there exist $\tau_1, \ldots, \tau_{n-k} \in \text{span}\{e_1, \ldots, e_k\}$ such that

$$Tx := x + \tau_1 z_1 + \tau_2 z_2 + \cdots + \tau_{n-k-1} z_{n-k-1} + \tau_{n-k} x_n.$$

Clearly, T is the identity map when restricted to the subspace $\mathbb{R}^k \times \{0\} \times \{0\}$, and it leaves the (z, x_n) components invariant. Observe that if T is a sliding along the y direction, then so is T^{-1} and $\det T = 1$.

We will use the following linear algebra fact about transformations T above.

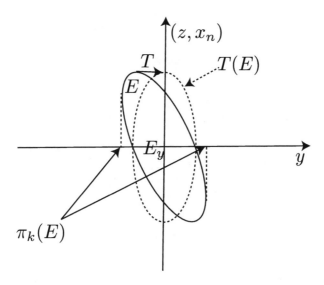

Figure 8.3. Geometry in the proof of Lemma 8.15.

Lemma 8.15. *Fix* $1 \leq k \leq n-2$. *Let* $E \subset \mathbb{R}^n$ *be an ellipsoid with center at* 0. *Then there exists a sliding* T *along the* y *direction (in* $\mathbb{R}^k$*) such that*

$$\frac{1}{2}(E_y \times E') \subset TE \subset E_y \times E',$$

where E_y *is an ellipsoid in the* y *variable and* E' *is an ellipsoid in the* (z, x_n) *variable.*

Proof. Consider the ellipsoid $E_y = E \cap \{(y, 0, 0) : y \in \mathbb{R}^k\}$. Let π_k be the restriction map defined by

$$\pi_k(K) = \{(y, 0, 0) : \quad \text{there is } (z, x_n) \in \mathbb{R}^{n-k} \quad \text{such that } (y, z, x_n) \in K\}.$$

Then $E_y \subset \pi_k(E)$. The ellipsoid E can be tilted in the y direction in the sense that the set $\pi_k(E)$ might strictly contain E_y. However, by a suitable sliding T along the y direction, we have

$$\pi_k(TE) = E_y.$$

See Figure 8.3. The lemma follows by choosing $E' = TE \cap \{(0, z, x_n) : (z, x_n) \in \mathbb{R}^{n-k-1} \times \mathbb{R}\}$. $\qquad\square$

We explain how to obtain the base case $k = 0$ in Proposition 8.13. Suppose that the conclusion of the proposition is false in this case. Then, we have $(u, \Omega) \in \mathcal{D}_0^\mu(\infty, \dots, \infty)$ with $b_{u,\Omega}(h) \leq M$ for all h. Moreover, $u(0) = 0$, while $u = 1$ on $\partial\Omega\backslash\{0\}$. By constructing a barrier, we show that $S_u(0, h)$ is trapped in a cone with height $Mh^{1/2}$ and opening bounded by a structural constant (independent of M). This is the content of Lemma 8.16. With this initial localization, we further localize to show that $S_u(0, h)$

is trapped in another cone with smaller opening. After a finite number of steps, we eventually show that $S_u(0, h)$ is trapped in a convex set with height of order h and width of order $h^{1/2}$, and this contradicts the lower bound on the volume growth of $S_u(0, h)$. Proposition 8.17 accomplishes this task.

In the following lemma, we localize the sections $S_u(0, h)$ for (u, Ω) in the class $\mathcal{D}_0^\mu(\infty, \ldots, \infty)$ with uniformly bounded $b_{u,\Omega}$.

Lemma 8.16. *Assume that* $(u, \Omega) \in \mathcal{D}_0^\mu(\infty, \ldots, \infty)$ *with* $b_{u,\Omega}(h) \leq M$ *for all* $h \in (0, 1]$. *Then, there are* $\delta(\mu) > 0$ *and* $N(n, \Lambda, \mu) > 0$, *such that*

$$u(x', x_n) \geq \delta|x'| - N x_n \quad in \ \Omega$$

and

$$S_u(0, h) \subset C(n, \Lambda, \mu, M)h^{1/2}B_1^+(0) \quad for \ all \ h \leq 1.$$

Proof. Since $(u, \Omega) \in \mathcal{D}_0^\mu(\infty, \ldots, \infty)$, we have then $G = \{0\}$ and $u = \varphi = 1$ on $\partial\Omega \setminus \{0\}$ in Definition 8.11. Let

$$v(x) := \delta(|x'| + |x'|^2) + \frac{\Lambda}{\delta^{n-1}}x_n^2 - N x_n, \quad \text{where } \delta := \frac{\mu^2}{2}, \quad N := \frac{\Lambda}{\mu\delta^{n-1}}.$$

Then $v \leq 1$ in $B_{1/\mu}^+(0)$, $v \leq \varphi$ on $\partial\Omega$, and

$$\det D^2 v \geq 2^n \Lambda > \det D^2 u \quad in \ \Omega.$$

Now, by the comparison principle (Theorem 7.26), $v \leq u$ in Ω. Hence,

$$u \geq \delta|x'| - N x_n \quad in \ \Omega$$

(this implies that $S_u(0, h)$ is trapped in a cone around the x_n-axis with opening $2\arctan(N/\delta)$) and we obtain

$$S_u(0, h) \subset \{|x'| \leq (N + 1)\delta^{-1}(x_n + h)\}.$$

Since $b_{u,\Omega}(h) \leq M$, we obtain from Definition 8.6 that $x_n \leq M h^{1/2}$ in $S_u(0, h)$. Combining this with the above inclusion, we conclude that

$$S_u(0, h) \subset C(n, \Lambda, \mu, M)h^{1/2}B_1^+(0) \quad for \ all \ h \leq 1.$$

The lemma is proved. □

The key step in the proof of Proposition 8.13 is the following proposition which asserts that the scenario in Lemma 8.16 does not happen.

Proposition 8.17. *Let* $0 \leq k \leq n - 2$. *Assume that* $u : \Omega \to \mathbb{R}$ *is a convex function on a domain* Ω *in* $\mathbb{R}^n$, *and assume that* u *satisfies*

$$u(x) \geq \gamma(|z| - K x_n) \quad in \ \Omega$$

for some $\gamma, K > 0$, where the notation $x = (y, z, x_n) \in \mathbb{R}^k \times \mathbb{R}^{n-1-k} \times \mathbb{R}$ in (8.20) is used. Assume furthermore that for each section $S_u(0, h)$ of u with height $h \in (0, 1)$, there exists a sliding T_h along the y direction such that

$$T_h S_u(0, h) \subset C_0 h^{1/2} B_1^+(0),$$

for some constant C_0. Then

$$(u, \Omega) \notin \mathcal{D}_0^\mu(\underbrace{1, \ldots, 1}_{k \ times}, \infty, \ldots, \infty) \quad \textit{for all } \mu \in (0, 1].$$

Proof. Assume by contradiction that $(u, \Omega) \in \mathcal{D}_0^\mu(\underbrace{1, \ldots, 1}_{k \text{ times}}, \infty, \ldots, \infty)$ for some $\mu \in (0, 1]$ and that (u, Ω) satisfies the hypotheses of the proposition with $K \leq K_0$ for some $K_0 > 0$.

We claim that for some $\gamma' \in (0, \gamma/2)$ and $K' = K - \eta$, where $\eta > 0$ depends on $\mu, C_0, \Lambda, n, K_0$ and is chosen so that $K/\eta \notin \mathbb{N}$, we have

(8.21) $$u(x) \geq \gamma'(|z| - K' x_n) \quad \text{in } \Omega.$$

Assume (8.21) holds. Then, as $K' \leq K_0$, we can apply (8.21) a finite number of times (as long as $K' > 0$) and obtain

$$u \geq \gamma_0(|z| + x_n),$$

for some small $\gamma_0 > 0$. Thus, $S_u(0, h) \subset \{x_n \leq \gamma_0^{-1} h\}$. Now, recalling that T_h is a sliding along the y direction, we obtain $T_h S_u(0, h) \subset \{x_n \leq \gamma_0^{-1} h\}$. This together with the hypothesis $T_h S_u(0, h) \subset C_0 h^{1/2} B_1^+(0)$ gives

$$|S_u(0, h)| = |T_h S_u(0, h)| = O(h^{(n+1)/2}) \quad \text{as } h \to 0,$$

which contradicts the lower bound on the volume of $S_u(0, h)$ in (8.18).

Thus, to complete the proof of the proposition, it remains to prove (8.21). We achieve this by rescaling u using T_h to obtain a new function w with section of height 1 at the origin being bounded and the corresponding γ being large (if h is small) from which the lower bound for w can be improved by δx_n. For this, we recall the definition of the class $\mathcal{D}_0^\mu(\underbrace{1, \ldots, 1}_{k \text{ times}}, \infty, \ldots, \infty)$ which (u, Ω) belongs to. From Definition 8.11, there exists a closed set

$$G_h \subset \partial S_u(0, h) \cap \{z = 0, x_n = 0\}$$

such that on the subspace $\{(y, 0, 0) : y \in \mathbb{R}^k\}$, we have

$$\{\mu^{-1}|y|^2 \leq h\} \subset G_h \subset \{\mu|y|^2 \leq h\},$$

and the boundary values φ_h of u on ∂S_h, in the epigraphical sense of Definition 7.20, satisfy (see Remark 7.21)

$$\varphi_h = h \quad \text{on } \partial S_u(0, h) \setminus G_h \quad \text{and} \quad \mu|y|^2 \leq \varphi_h \leq \min\{h, \mu^{-1}|y|^2\} \quad \text{on } G_h.$$

For $h > 0$ small, to be chosen later, we define the rescaling w of u by

$$w(x) := h^{-1} u(h^{1/2} T_h^{-1} x).$$

Then

$$S_w(0, 1) := \Omega_w = h^{-1/2} T_h S_u(0, h) \subset B_{C_0}^+(0),$$

and

$$G_w := h^{-1/2} G_h \subset \partial \Omega_w \cap \{z = 0, x_n = 0\}.$$

Moreover, the boundary values φ_w of w on $\partial \Omega_w$ satisfy

$$\varphi_w = 1 \quad \text{on } \partial \Omega_w \setminus G_w \quad \text{and} \quad \mu |y|^2 \le \varphi_w \le \min\{1, \mu^{-1} |y|^2\}.$$

The hypothesis $u(x) \ge \gamma(|z| - K x_n)$ in Ω now becomes

(8.22)
$$w(x) \ge \frac{\gamma}{h^{1/2}} (|z| - K x_n) \quad \text{in } \Omega_w.$$

Pick an arbitrary unit vector $\mathbf{e}$ in the z-hyperplane. For simplicity, we take $\mathbf{e}$ to be along the positive z_1-axis so $z_1 = z \cdot \mathbf{e}$. Consider the function

$$v(x) := \delta(|x|^2 + x_n) + \frac{\Lambda}{\delta^{n-1}} (z_1 - K x_n)^2 + N(z_1 - K x_n),$$

where

$$\delta := \min\left\{ \frac{\mu}{2}, \frac{1}{2(C_0^2 + C_0)} \right\} \quad \text{and} \quad N := \frac{2\Lambda(1 + K_0) C_0}{\delta^{n-1}}.$$

Then

$$\det D^2 v = 2^n (\delta^n + K^2 \Lambda + \Lambda) > \Lambda.$$

We show that if h is small, then v is a lower barrier for w in Ω_w. Since $\det D^2 w \le \Lambda$ in Ω_w, from the comparison principle, it suffices to show that $\varphi_w \ge v$ on $\partial \Omega_w$. We will consider different parts of the boundary $\partial \Omega_w$.

Note that the function $g(t) := (\Lambda/\delta^{n-1}) t^2 + Nt$ is increasing in the interval $|t| \le (1 + K_0) C_0$ and there is $t_0 = t_0(\delta, \Lambda, N) > 0$ such that

$$g(t) = \frac{\Lambda}{\delta^{n-1}} t^2 + Nt \le \frac{1}{2} \quad \text{on } [0, t_0].$$

Case 1. On the part of $\partial \Omega_w$ where $z_1 \le K x_n$. Then, $g(z_1 - K x_n) \le 0$ and

$$v(x) = \delta(|x|^2 + x_n) + g(z_1 - K x_n) \le \delta(|x|^2 + x_n) \le \frac{1}{2} \varphi_w,$$

where the last inequality can be easily checked as follows.

If $x = (y, z, x_n) \in G_w$, then $x_n = 0$ and $z = 0$; hence

$$v(x) \le \delta |y|^2 \le \frac{\mu}{2} |y|^2 \le \frac{1}{2} \varphi_w.$$

If $x \notin G_w$, then

$$v(x) \le \delta(C_0^2 + C_0) \le \frac{1}{2} = \frac{1}{2} \varphi_w.$$

Case 2. On the part of $\partial \Omega_w$ where $z_1 > Kx_n$. On this part, $x \notin G_w$. We use (8.22), and obtain, provided that $h \leq (t_0\gamma)^2$,

$$1 = \varphi_w \geq t_0^{-1}(|z| - Kx_n) \geq t_0^{-1}(z_1 - Kx_n).$$

Then $z_1 - Kx_n \leq t_0$ and hence $g(z_1 - Kx_n) \leq 1/2$. Therefore

$$v(x) = \delta(|x|^2 + x_n) + g(z_1 - Kx_n) < \delta(|x|^2 + x_n) + 1/2 \leq 1 = \varphi_w.$$

In all cases, if h is small, then $\varphi_w \geq v$ on $\partial\Omega_w$ and, therefore, v is a lower barrier for w in Ω_w. It follows that

$$w \geq N(z \cdot \mathbf{e} - Kx_n) + \delta x_n \quad \text{in } \Omega_w$$

for all unit vectors $\mathbf{e}$ in the z-hyperplane, and we obtain

$$w \geq N(|z| - (K - \eta)x_n) \quad \text{in } \Omega_w \quad \text{where } \eta := \delta/N.$$

We can slightly reduce η if necessary so that $K/\eta \notin \mathbb{N}$. Rescaling back and letting

$$\gamma' = Nh^{1/2} \leq Nt_0\gamma \leq \gamma/2,$$

where $g(t_0) \leq 1/2$ was used, we get

$$u \geq \gamma'(|z| - (K - \eta)x_n) \quad \text{in } S_u(0, h).$$

Since u is convex and $u(0) = 0$, the above inequality also holds in Ω. Therefore (8.21) is proved, completing the proof of the proposition. $\quad\square$

Proof of Proposition 8.13. We prove the proposition by induction on k. The base case $k = 0$ is a consequence of Lemma 8.16 and Proposition 8.17. We assume that the proposition holds for all nonnegative integers up to $k - 1$, $1 \leq k \leq n - 2$, and we prove it for k. In this proof we denote by c, C positive *structural* constants that depend on $M, \mu, \lambda, \Lambda, n$, and k.

Let $(\bar{u}, \bar{\Omega}) \in \mathcal{D}_0^\mu(a_1, \ldots, a_k, \infty, \ldots, \infty)$. By the induction hypothesis and a compactness argument using Theorem 7.29, we can find a positive constant $C_k(\mu, M, \lambda, \Lambda, n)$ such that if $a_k \geq C_k$, then $b_{\bar{u}, \bar{\Omega}}(h) \geq M$ for some $h \in [C_k^{-1}, 1]$. Thus, we only need to consider the case when $a_k < C_k$.

Assume by contradiction that the conclusion of the proposition for $\bar{u}$ and $\bar{\Omega}$ does not hold for any $c_k > 0$. Then, again by compactness, we can find a limiting solution u on a domain $\Omega \subset B_{1/\tilde{\mu}}^+$ such that

$$(8.23) \qquad (u, \Omega) \in \mathcal{D}_0^{\tilde{\mu}}(\underbrace{1, \ldots, 1}_{k \text{ times}}, \infty, \ldots, \infty),$$

with $\tilde{\mu} \in (0, 1]$ depending on μ and C_k (see Remark 8.12), and

$$(8.24) \qquad b_{u,\Omega}(h) \leq M \quad \text{for all } h \in (0, 1].$$

We show that such a pair of function and domain (u, Ω) does not exist.

Recall the notation in (8.20). Let $S_h := S_u(0, h)$ be the section of u centered at the origin with height h.

Step 1. Rescaling u using a sliding along the y direction. As in the proof of Lemma 8.16, we consider the function

$$v(x) := \delta|x'|^2 + \delta|z| + \frac{\Lambda}{\delta^{n-1}}x_n^2 - Nx_n, \quad \text{where } \delta := \frac{\tilde{\mu}^2}{2} \text{ and } N := \frac{\Lambda}{\delta^{n-1}\tilde{\mu}}.$$

Then, $\det D^2 v \geq 2^n \Lambda > \det D^2 u$ in Ω, while the boundary values of u satisfy $\varphi \geq v$ on $\partial\Omega$. Thus, by the comparison principle, we have $u \geq v$ in Ω. Hence

$$(8.25) \qquad\qquad u(x) \geq \delta|z| - Nx_n \quad \text{in } \Omega.$$

It follows from (8.24)–(8.25) that

$$(8.26) \qquad\qquad S_h \subset \{|z| < \delta^{-1}(Nx_n + h)\} \cap \{x_n \leq Mh^{1/2}\}.$$

Let x_h^* be the center of mass of S_h. By John's Lemma (Lemma 2.66), there is an ellipsoid $E_h = A_h B_1(0)$ centered at the origin, of the same volume as S_h such that for some $c(n) = \alpha(n) > 0$ and $C(n) = [\alpha(n)]^{-1}$,

$$(8.27) \qquad\qquad c(n)E_h \subset S_h - x_h^* \subset C(n)E_h, \quad |E_h| = |S_h|.$$

For this ellipsoid E_h, we can use Lemma 8.15 and its proof to find T_h, a sliding along the y direction as in Definition 8.14, such that

$$(8.28) \qquad\qquad T_h E_h = |E_h|^{1/n} AB_1(0),$$

where $A \in \mathbb{M}^{n \times n}$ with $\det A = 1$ and T_h leaves invariant the subspaces $\{(y, 0, 0) : y \in \mathbb{R}^k\}$ and $\{(0, z, x_n) : (z, x_n) \in \mathbb{R}^{n-k-1} \times \mathbb{R}\}$. By choosing appropriate coordinate systems in the y and z variables, we can assume that

$$A(y, z, x_n) = (A_1 y, A_2(z, x_n))$$

where

$$A_1 = \text{diag}(\beta_1, \ldots, \beta_k) \quad \text{with } 0 < \beta_1 \leq \cdots \leq \beta_k$$

and

$$A_2 = \begin{pmatrix} \gamma_{k+1} & 0 & \cdots & 0 & \tau_1 \\ 0 & \gamma_{k+2} & \cdots & 0 & \tau_2 \\ \vdots & \vdots & \ddots & \vdots & \vdots \\ 0 & 0 & \cdots & \gamma_{n-1} & \tau_{n-k-1} \\ 0 & 0 & \cdots & 0 & \theta_n \end{pmatrix} \quad \text{with } \gamma_j > 0 \quad \text{and} \quad \theta_n > 0.$$

Observe that

$$(8.29) \qquad\qquad \theta_n \prod_{i=1}^{k} \beta_i \prod_{j=k+1}^{n-1} \gamma_j = \det A = 1.$$

Rescale u by
$$\tilde{u}(x) := u(T_h^{-1}x) \quad \text{where } x \in \tilde{\Omega} := T_h\Omega.$$
Then, the section $\tilde{S}_h := S_{\tilde{u}}(0, h)$ of $\tilde{u}$ centered at 0 with height h satisfies $\tilde{S}_h = T_h S_h$ and (8.26).

Step 2. Structural upper bound for θ_n, γ_j, $|\tau_l|$ and structural lower bound for β_i. Since $(u, \Omega) \in \mathcal{D}_0^{\tilde{\mu}}(\underbrace{1, \ldots, 1}_{k \text{ times}}, \infty, \ldots, \infty)$, there exists $\tilde{G}_h = G_h$ with

$$\tilde{G}_h \subset \{(y, z, x_n) \in \mathbb{R}^n : z = 0, x_n = 0\} \cap \partial\tilde{S}_h$$

such that on the subspace $\mathbb{R}^k \times \{0\} = \{(y, 0, 0) : y \in \mathbb{R}^k\}$

$$\{\tilde{\mu}^{-1}|y|^2 \leq h\} \subset \tilde{G}_h \subset \{\tilde{\mu}|y|^2 \leq h\}$$

and the boundary values $\tilde{\varphi}_h$ of $\tilde{u}$ on $\partial\tilde{S}_h$ satisfy

$$\tilde{\varphi}_h = h \quad \text{on } \partial\tilde{S}_h \setminus \tilde{G}_h \quad \text{and} \quad \tilde{\mu}|y|^2 \leq \tilde{\varphi}_h \leq \min\{h, \tilde{\mu}^{-1}|y|^2\} \quad \text{on } \tilde{G}_h.$$

Moreover, we have
$$\tilde{\mu}h^{n/2} \leq |S_h| \leq \tilde{\mu}^{-1}h^{n/2}.$$
Since $|E_h| = |S_h|$, it follows from (8.28) that

$$(8.30) \qquad \tilde{\mu}^{1/n}h^{1/2}AB_1(0) \subset T_h E_h \subset \tilde{\mu}^{-1/n}h^{1/2}AB_1(0).$$

Because $0 \in \partial S_h$, the second inclusion in (8.27) implies that $-x_h^* \in C(n)E_h$. Hence $x_h^* \in C(n)E_h$ and

$$(8.31) \qquad x_h^* + c(n)E_h \subset S_h \subset 2C(n)E_h.$$

In view of $\tilde{S}_h = T_h S_h$, we obtain

$$(8.32) \qquad \tilde{x}_h^* + c_1 h^{1/2}AB_1(0) \subset \tilde{S}_h \subset C_1 h^{1/2}AB_1(0),$$

where $\tilde{x}_h^*$ is the center of mass of $\tilde{S}_h$, c_1 and C_1 depend on $\tilde{\mu}$ and n.

Note that $\tilde{x}_h^*$ has positive x_n component. Thus, (8.32) gives

$$(8.33) \qquad C_1^{-1}b_{u,\Omega}(h) \leq \theta_n \leq c_1^{-1}b_{u,\Omega}(h).$$

Since S_h satisfies (8.26) and $\tilde{S}_h = T_h S_h$ where T_h is a sliding along the y direction, we find that

$$\tilde{S}_h \subset \{(y, z, x_n) \in \mathbb{R}^n : |(z, x_n)| \leq C(n, M, \delta, N)h^{1/2}\}.$$

This together with the inclusion (8.32) gives $\|A_2\| \leq C$. It follows that

$$\theta_n + \max_{k+1 \leq j \leq n-1} \gamma_j + \max_{1 \leq l \leq n-k-1} |\tau_l| \leq C.$$

Note that $\tilde{G}_h$ contains a ball of radius $(\tilde{\mu}h)^{1/2}$ in $\mathbb{R}^k \times \{0\}$. Thus, using $\tilde{G}_h \subset \tilde{S}_h$ together with the second inclusion of (8.32), we find that

$$\beta_i \geq \bar{c} = \tilde{\mu}^{1/2}C_1^{-1} > 0, \quad \text{for all } i = 1, \ldots, k.$$

Step 3. Structural lower bounds for $b_{u,\Omega}$ and γ_j and completion of the proof. We will use the induction hypothesis to show that there exist positive constants $c_\star$ and $c'_\star$ depending on $\tilde{\mu}, M, \lambda, \Lambda, n, k$ such that

$$(8.34) \qquad\qquad b_{u,\Omega}(h) \geq c_\star \quad \text{for all small } h$$

and

$$(8.35) \qquad\qquad \gamma_j \geq c'_\star \quad \text{for all } j = k+1, \ldots, n-1.$$

We first show that these lower bounds complete the proof of the proposition. Indeed, due to (8.33), (8.34) gives a lower bound for θ_n. In view of (8.29), this together with (8.35) implies that β_k is bounded from above. Therefore

$$\|A\| \leq C(\tilde{\mu}, M, \lambda, \Lambda, n, k).$$

Due to the second inclusion in (8.32), this will imply that

$$(8.36) \qquad\qquad \tilde{S}_h = T_h S_h \subset C_0 h^{n/2} B_1^+(0),$$

for structural constant C_0. Since assertions (8.36), (8.23), (8.25) contradict Proposition 8.17, the proof of Proposition 8.13 is then finished.

It remains to prove (8.34) and (8.35). We define the rescaling w of $\tilde{u}$ by

$$w(x) := h^{-1}\tilde{u}(h^{1/2}Ax), \quad x \in \Omega_w := h^{-1/2}A^{-1}\tilde{S}_h \equiv S_w(0,1).$$

Then (8.32) gives $B_{c_1}(x^*) \subset \Omega_w \subset B_{C_1}^+(0)$ for some $x^* \in \mathbb{R}^n$. Let $G_w := h^{-1/2}A^{-1}\tilde{G}_h$. Then $w = \varphi_w$ on $\partial\Omega_w$ where $\varphi_w = 1$ on $\partial\Omega_w \setminus G_w$, and

$$\tilde{\mu}\sum_{i=1}^{k}\beta_i^2 x_i^2 \leq \varphi_w \leq \min\left\{1, \tilde{\mu}^{-1}\sum_{i=1}^{k}\beta_i^2 x_i^2\right\} \quad \text{on } G_w.$$

Therefore, for some small $\bar{\mu}$ depending on $\tilde{\mu}, M, \lambda, \Lambda, n, k$, we have

$$(w, \Omega_w) \in \mathcal{D}_0^{\bar{\mu}}(\beta_1, \ldots, \beta_k, \infty, \ldots, \infty).$$

Recalling (8.29) and (8.33), we have

$$(8.37) \qquad b_{u,\Omega}(h)\prod_{i=1}^{k}\beta_i \prod_{j=k+1}^{n-1}\gamma_j \geq c_1(\tilde{\mu}, n)\theta_n \prod_{i=1}^{k}\beta_i \prod_{j=k+1}^{n-1}\gamma_j = c_1.$$

If $b_{u,\Omega}(h) \leq \hat{c}_\star$ for some small $\hat{c}_\star = \hat{c}_\star(\tilde{\mu}, M, \lambda, \Lambda, n, k)$, then from $\gamma_j \leq C$ and (8.37), we deduce that $\beta_k \geq C_k(\bar{\mu}, \bar{M}, \lambda, \Lambda, n) > 8$, where $\bar{M} := 2\bar{\mu}^{-1}$. By the induction hypothesis

$$b_{w,\Omega_w}(\tilde{h}) \geq \bar{M} \geq 2b_{w,\Omega_w}(1)$$

for some $C_k^{-1} \leq \tilde{h} < 1$. This combined with Remark 8.7(c) gives

$$\frac{b_{u,\Omega}(h\tilde{h})}{b_{u,\Omega}(h)} = \frac{b_{\tilde{u},\tilde{\Omega}}(h\tilde{h})}{b_{\tilde{u},\tilde{\Omega}}(h)} = \frac{b_{w,\Omega_w}(\tilde{h})}{b_{w,\Omega_w}(1)} \geq 2,$$

which implies $b_{u,\Omega}(h\tilde{h}) \geq 2b_{u,\Omega}(h)$. By Remark 8.7(a), the value $\tilde{h}$ satisfies

$$2 < \frac{b_{u,\Omega}(h\tilde{h})}{b_{u,\Omega}(h)} \leq \frac{1}{\tilde{h}^{1/2}},$$

so $\tilde{h} \leq 1/4$. Therefore, using Remark 8.7(a) and $b_{u,\Omega}(1) \geq \tilde{\mu}^n/(n\omega_{n-1}) > 0$, we can apply Lemma 8.18 below to $L = 1$ and $\alpha := \min\{C_k^{-1}, \hat{c}_\star, 1/8\}$ and obtain the lower bound (8.34) with $c_\star := \alpha^2$.

To prove (8.35), we use the same argument as in the proof of (8.34). Indeed, (8.37) implies that if some γ_j is smaller than a small value $c'_\star$, then

$$\beta_k \geq C_k(\bar{\mu}, \bar{M}_1, \lambda, \Lambda, n) \quad \text{where } \bar{M}_1 := \frac{2M}{\bar{\mu}c_\star}.$$

By the induction hypothesis and $b_{w,\Omega_w}(1) \leq \bar{\mu}^{-1}$, we have

$$b_{w,\Omega_w}(\tilde{h}) \geq \bar{M}_1 \geq \frac{2M}{c_\star}b_{w,\Omega_w}(1),$$

for some $C_k^{-1} < \tilde{h} \leq 1$. Thus,

$$\frac{b_{u,\Omega}(h\tilde{h})}{b_{u,\Omega}(h)} = \frac{b_{w,\Omega_w}(\tilde{h})}{b_{w,\Omega_w}(1)} \geq \frac{2M}{c_\star}.$$

In view of (8.34), this gives $b_{u,\Omega}(h\tilde{h}) \geq 2M$, which contradicts (8.24). Hence, (8.35) must hold. The proof of our proposition is complete. $\qquad\square$

In Step 3 of the proof of Proposition 8.13 as well as the proof of Proposition 8.21 in the next section, we use the following simple but useful result.

Lemma 8.18. *Let $L \in (0,1]$, and let $\alpha \in (0,1/8]$. Let $b : (0,L] \to (0,\infty)$ be a function satisfying the following conditions:*

    (a) *$b(x_1)/b(x_2) \geq (x_1/x_2)^{1/2}$ for all $x_1 \leq x_2 \in (0,L]$.*

    (b) *If $x \in (0,L]$ and $b(x) \leq \alpha$, then there is $t \in [\alpha, 1-\alpha]$ such that $b(tx) \geq 2b(x)$.*

Then, for all $x \in (0,\bar{h}]$ where $\bar{h} = \bar{h}(\alpha, L, b(L)) > 0$, we have $b(x) \geq \alpha^2$.

Proof. Let $h_1 := L$. Due to condition (b), we can define a sequence $\{h_k\}_{k=1}^\infty$ inductively as follows: Once h_k is determined for $k \geq 1$, we set

$$h_{k+1} := \begin{cases} \alpha h_k & \text{if } b(h_k) > \alpha, \\ t_k h_k & \text{if } b(h_k) \leq \alpha \text{ and } t_k \in [\alpha, 1-\alpha] \text{ satisfies } b(t_k h_k) \geq 2b(h_k). \end{cases}$$

First, we show that

$$b(h_{k_1}) \geq \alpha^{3/2} \quad \text{for some } k_1 \leq k_0 := \min\left\{k \in \mathbb{N} : k > 1 + \log_2(\alpha^{3/2}/b(L))\right\}.$$

If, otherwise, then $b(h_k) < \alpha^{3/2} < \alpha$ for all $k \leq k_0$. Thus, by construction, $b(h_k) \geq 2b(h_{k-1})$ for all $k \leq k_0$. We now obtain a contradiction from

$$b(h_{k_0}) \geq 2^{k_0-1}b(h_1) = 2^{k_0-1}b(L) > \alpha^{3/2}.$$

Next, we prove by induction that $b(h_k) \geq \alpha^{3/2}$ for all $k \geq k_1$. The case $k = k_1$ is obvious by the choice of k_1. Suppose that the inequality $b(h_k) \geq \alpha^{3/2}$ holds for $k \geq k_1$. We prove it for $k + 1$. If $\alpha^{3/2} \leq b(h_k) \leq \alpha$, then by the choice of h_{k+1}, we have $b(h_{k+1}) \geq 2b(h_k) \geq 2\alpha^{3/2}$. If $b(h_k) > \alpha$, then, by the choice of $h_{k+1} = \alpha h_k$ and condition (a), we have

$$b(h_{k+1}) \geq b(h_k)\big(h_{k+1}/h_k\big)^{1/2} \geq b(h_k)\alpha^{1/2} \geq \alpha^{3/2}.$$

Note that the choice of the sequence $\{h_k\}$ implies that $h_{k+1}/h_k \geq \alpha$ for all $k \geq 1$, and it converges to 0. Now, for any $x \in (0, h_{k_1}]$, there is an integer $k \geq k_1$ such that $h_{k+1} < x \leq h_k$. Then, by condition (a) again, we have

$$b(x) \geq b(h_k)\big(x/h_k\big)^{1/2} \geq b(h_k)\alpha^{1/2} \geq \alpha^2.$$

Since $h_{k_0} \geq \alpha^{k_0-1}L$, we can set $\bar{h} = \alpha^{k_0-1}L$ to complete the proof. $\square$

8.4. Proof of the Boundary Localization Theorem

In this section, we will complete the proof of Theorem 8.5. First, we record a consequence of Proposition 8.10.

Lemma 8.19. *In the setting of Proposition 8.8, if σ, a_n are sufficiently small depending on n, μ, λ, Λ, then the function v satisfies*

$$b_{v,\Omega_v}(t) \geq 2b_{v,\Omega_v}(1)$$

for some $1 > t \geq c_0 = c_0(n, \mu, \lambda, \Lambda) > 0$.

Proof. By relabeling the $x_1, \ldots, x_{n-1}$ variables, we can assume that $a_1 \leq \cdots \leq a_{n-1}$. Then, $(v, \Omega_v) \in \mathcal{D}_\sigma^{\bar{\mu}}(a_1, \ldots, a_{n-1})$, where $\bar{\mu} = \min\{\kappa_0, \mu/2\}$ depends only on n, λ, Λ, and μ. Since the product of $a_1, \ldots, a_n$ is 1 and $a_1 \geq \bar{\mu}$, we see that a_{n-1} must be large if a_n is small. Thus, provided that σ, a_n are sufficiently small, the pair of function v and domain Ω_v satisfies the hypotheses of Proposition 8.10. By choosing

$$M_0 = 2\bar{\mu}^{-1} \geq 2b_{v,\Omega_v}(1)$$

in Proposition 8.10 with μ replaced by $\bar{\mu}$, we can find $t \in [c_0, 1)$ where $c_0(n, \mu, \lambda, \Lambda) > 0$, such that $b_{v,\Omega_v}(t) \geq M_0 \geq 2b_{v,\Omega_v}(1)$. $\square$

The crucial property of $b_{u,\Omega}$ is contained in the following lemma which states that if the normalized altitude $b_{u,\Omega}(h)$ of the boundary section $S_u(0, h)$ is less than a critical value c_0, then we can find a smaller section at height still comparable to h with twice the normalized altitude.

Lemma 8.20. *Assume that Ω and u satisfy* (8.2)–(8.5). *There exist $c_0 = c_0(n, \mu, \lambda, \Lambda) > 0$, $\hat{c} = \hat{c}(n, \mu, \lambda, \Lambda, \rho) > 0$ such that if $h \le \hat{c}$ and $b_{u,\Omega}(h) \le c_0$, then, for some $t \in [c_0, 1/4]$, we have*

$$b_{u,\Omega}(th) > 2b_{u,\Omega}(h).$$

Proof. If $h \le \hat{c} = \hat{c}(n, \mu, \lambda, \Lambda, \rho)$ is sufficiently small, then, the function v in Proposition 8.8 satisfies (8.15) with σ small. Moreover, if $b_{u,\Omega}(h) \le c_1$ is small depending only on n, λ, Λ, and μ, then a_n in Proposition 8.8 is also small. Thus, by Lemma 8.19, the function v satisfies

$$b_{v,\Omega_v}(t) \ge 2b_{v,\Omega_v}(1)$$

for some $1 > t \ge \bar{c}_0(n, \mu, \lambda, \Lambda) > 0$. By Remark 8.7(a), we have $t \le 1/4$. Hence, by (8.16), we find $b_{u,\Omega}(th) > 2b_{u,\Omega}(h)$. The lemma is proved with $c_0 = \min\{c_1, \bar{c}_0\}$.   □

We are now in a position to state and prove a crucial property of the quantity d_n in Proposition 8.4.

Proposition 8.21 (Altitude of boundary sections). *Assume that Ω and u satisfy* (8.2)–(8.5). *Let d_n be as in Proposition 8.4. Then, there exist positive constants $\bar{\kappa}(n, \mu, \lambda, \Lambda)$ and $c(n, \mu, \lambda, \Lambda, \rho)$ such that for all $h \le c$, we have*

$$d_n \ge \bar{\kappa} h^{1/2}.$$

Proof. By Lemma 8.20 and Remark 8.7(a), the function $b(t) := b_{u,\Omega}(t)$ satisfies the hypotheses of Lemma 8.18 with $L = \hat{c}$ and $\alpha = c_0$. Thus, Lemma 8.18 gives $b(h) \ge c_0^2$ for all $h \le c(n, \mu, \lambda, \Lambda, \rho)$. Due to (8.9) and Remark 8.7(b),

$$c(n, \kappa_0)d_n \le b(h)h^{1/2} \le C(n, \kappa_0)d_n.$$

These combined with $b(h) \ge c_0^2$ gives the conclusion of the proposition.   □

Completion of the proof of the Boundary Localization Theorem (Theorem 8.5). Let A_h, $\tilde{u}$, τ_h, and d_i $(i = 1, \ldots, n)$ be defined as in Proposition 8.4. We will show that the ellipsoid E_h defined by

$$E_h := A_h^{-1} B_{h^{1/2}}(0)$$

satisfies the conclusion of the theorem.

From Propositions 8.4 and 8.21, we deduce that all d_i $(i = 1, \ldots, n)$ are bounded from below by $\bar{\kappa}_0 h^{1/2}$ for some $\bar{\kappa}_0 = \bar{\kappa}_0(n, \mu, \lambda, \Lambda) > 0$. Since their product is $h^{n/2}$, by reducing $\bar{\kappa}_0$ if necessary, we find that

$$\bar{\kappa}_0 h^{1/2} \le d_i \le \bar{\kappa}_0^{-1} h^{1/2} \quad \text{for all } i = 1, \ldots, n$$

for all $h \le c(n, \mu, \lambda, \Lambda, \rho)$. From (8.9), we obtain

$$S_{\tilde{u}}(0, h) \subset C(n)\bar{\kappa}_0^{-1} h^{1/2} B_1(0).$$

From (8.14), we find that for $\kappa_1 := \min\{\kappa_0\bar{\kappa}_0/2, \bar{\kappa}_0/C(n)\}$,

$$\kappa_1 h^{1/2} B_1(0) \cap A_h \overline{\Omega} \subset A_h S_u(0,h) \subset \kappa_1^{-1} h^{1/2} B_1(0).$$

Thus the ellipsoid E_h defined above satisfies

$$\kappa_1 E_h \cap \overline{\Omega} \subset S_u(0,h) \subset \kappa_1^{-1} E_h.$$

Clearly,

$$\kappa_1 E_{h/2} \cap \overline{\Omega} \subset S_u(0,h/2) \subset S_u(0,h) \subset \kappa_1^{-1} E_h;$$

hence, we easily obtain $A_h A_{h/2}^{-1} B_1(0) \subset 2\kappa_1^{-2} B_1(0)$. This inclusion implies $|\tau_h - \tau_{h/2}| \leq 2\kappa_1^{-2}$, which gives the desired bound

$$|\tau_h| \leq \kappa_2^{-1} |\log h|$$

for all small h, where κ_2 depends only on n, μ, λ, and Λ. Upon choosing $\kappa = \min\{\kappa_1, \kappa_2\}$, the proof of the theorem is complete. $\qquad \square$

Remark 8.22. Compactness plays a crucial role in the proof of the Boundary Localization Theorem, Theorem 8.5. It would be very interesting to find a new proof of Theorem 8.5 without using compactness.

8.5. Pointwise Hölder Gradient Estimates at the Boundary

We now show that the hypotheses of Theorem 8.1 imply that u is always differentiable at x_0, and then $\partial u(x_0) = \{Du(x_0)\}$, where $Du(x_0)$ is defined in the usual classical sense. In fact, u is pointwise $C^{1,\alpha}$ at x_0 for all $\alpha < 1$.

Proposition 8.23 (Pointwise $C^{1,\alpha}$ estimates at the boundary). *Assume that u and Ω satisfy the hypotheses of Theorem 8.1 at a point $x_0 \in \partial\Omega$. Then u is differentiable at x_0, and for $x \in \overline{\Omega} \cap B_r(x_0)$ where $r \leq \delta(n, \lambda, \Lambda, \rho, \mu)$, we have*

$$(8.38) \qquad \begin{aligned} |x - x_0|^3 &\leq u(x) - u(x_0) - Du(x_0) \cdot (x - x_0) \\ &\leq C\,|x - x_0|^2\,(\log|x - x_0|)^2. \end{aligned}$$

Moreover, if u and Ω satisfy the hypotheses of Theorem 8.1 also at another point $z_0 \in \partial\Omega \cap B_r(x_0)$, then

$$|Du(z_0) - Du(x_0)| \leq C(n, \rho, \mu, \lambda, \Lambda) r |\log r|^2.$$

Proof. We can assume that $x_0 = 0 \in \partial\Omega$ and that Ω and u satisfy (8.2)–(8.5). For all $h \leq c_0(n, \lambda, \Lambda, \rho, \mu)$, by Theorem 8.5, $S_u(0,h)$ satisfies

$$\kappa E_h \cap \overline{\Omega} \subset S_u(0,h) \subset \kappa^{-1} E_h \cap \overline{\Omega}, \quad \text{where } E_h = A_h^{-1} B_{h^{1/2}}(0)$$

where $A_h \in \mathbb{M}^{n \times n}$ is a matrix having the following properties:

$$\det A_h = 1, \quad A_h x = x - \tau_h x_n, \quad \tau_h \cdot e_n = 0, \quad \|A_h^{-1}\| + \|A_h\| \leq \kappa^{-1} |\log h|.$$

It follows that, for c and C depending only on $\rho, \mu, \lambda, \Lambda$, and n, we have

$$(8.39) \qquad \overline{\Omega} \cap B^+_{ch^{1/2}/|\log h|}(0) \subset S_u(0,h) \subset B^+_{Ch^{1/2}|\log h|}(0),$$

or $|u| \le h$ in $\overline{\Omega} \cap B^+_{ch^{1/2}/|\log h|}(0)$. Thus, for all x close to the origin

$$(8.40) \qquad |u(x)| \le C|x|^2 (\log|x|)^2,$$

which shows that u is differentiable at 0. The other inclusion of (8.39) gives a lower bound for u near the origin

$$(8.41) \qquad u(x) \ge c|x|^2 (\log|x|)^{-2} \ge |x|^3.$$

Therefore, (8.38) is proved.

Suppose now the hypotheses of Theorem 8.1 are satisfied at $z_0 \in \partial\Omega \cap B_r(0)$. We need to show that for $\hat{z} := Du(z_0)$,

$$|\hat{z}| = |Du(z_0)| \le Cr|\log r|^2.$$

For this, we use (8.38) for z_0 at 0 and for 0 and z_0 at all points x in a ball

$$B := \overline{B_{c(n,\rho)r}(y)} \subset \Omega \cap B_r(0) \cap B_r(z_0).$$

For $x \in B$, we have

$$Du(z_0) \cdot x \le u(x) + [-u(z_0) + Du(z_0) \cdot z_0] - |x - z_0|^3$$
$$\le C|x|^2 (\log|x|)^2 + [C|z_0|^2 (\log|z_0|)^2 - u(0)] \le Cr^2 |\log r|^2.$$

The lower bound for $Du(z_0) \cdot x$ is obtained similarly, and we have

$$|\hat{z} \cdot x| = |Du(z_0) \cdot x| \le C_0(n, \rho, \mu, \lambda, \Lambda) r^2 |\log r|^2 \quad \text{for all } x \in B.$$

We use the above inequality at y and $\hat{y} := y + cr\hat{z}/|\hat{z}|$ (if $\hat{z} \ne 0$) to get

$$cr|\hat{z}| = \hat{z} \cdot \hat{y} - \hat{z} \cdot y \le 2C_0 r^2 |\log r|^2.$$

Therefore $|\hat{z}| \le (2C_0/c)r|\log r|^2$, as desired. $\qquad \square$

We record here a useful consequence of Propositions 4.7 and 8.23.

Proposition 8.24. *Let Ω be a uniformly convex domain in $\mathbb{R}^n$ ($n \ge 2$). Assume that the convex function $u \in C(\overline{\Omega})$ satisfies $0 < \lambda \le \det D^2 u \le \Lambda$ in Ω, in the sense of Aleksandrov. Assume that $u|_{\partial\Omega}$ and $\partial\Omega$ are of class C^3. Then $u \in C^1(\overline{\Omega})$. Moreover, u separates quadratically from its tangent hyperplanes on $\partial\Omega$, and for all $x_0, x \in \partial\Omega$, we have*

$$\rho|x - x_0|^2 \le u(x) - u(x_0) - Du(x_0) \cdot (x - x_0) \le \rho^{-1}|x - x_0|^2,$$

for some small constant $\rho > 0$ depending only on $n, \lambda, \Lambda, \|u\|_{C^3(\partial\Omega)}$, the C^3 regularity of $\partial\Omega$, and the uniform convexity radius of Ω.

8.6. Boundary Localization for Degenerate and Singular Equations

In this section, we mention several boundary localization theorems for degenerate or singular Monge–Ampère equations $\det D^2 u = f$, where f is comparable to some (positive or negative) power of the distance to the boundary. Theorem 8.5 corresponds to the power being zero.

8.6.1. Degenerate Monge–Ampère equations.

For degenerate Monge–Ampère equations $\det D^2 u = g[\mathrm{dist}(\cdot, \partial\Omega)]^\alpha$ where $\alpha > 0$ and g is a continuous strictly positive function, Savin [**S6**] established a remarkable boundary localization theorem which is the basic ingredient in the global regularity of the Monge–Ampère eigenfunctions (see Section 11.5). His result states that under natural conditions, the sections $S_u(0, h)$ of the solution u at $0 \in \partial\Omega$ are equivalent, up to a sliding along $x_n = 0$, to the sections $\mathcal{E}_h := \{|x'|^2 + x_n^{2+\alpha} < h\}$ of the function $|x'|^2 + x_n^{2+\alpha}$ which is a solution to $\det D^2 v = 2^{n-1}(2 + \alpha)(1 + \alpha)[\mathrm{dist}(\cdot, \partial\mathbb{R}_+^n)]^\alpha$ in the upper half-space.

Theorem 8.25 (Boundary Localization Theorem for degenerate equations).
Let $\alpha > 0$. Let Ω be a bounded convex domain in $\mathbb{R}^n$ with $0 \in \partial\Omega$ and which satisfies the inclusions $B_\rho(\rho\, e_n) \subset \Omega \subset B_{1/\rho}(0) \cap \{x_n > 0\}$ for some $\rho > 0$. Let $u \in C(\overline{\Omega})$ be a convex function satisfying the following conditions:

(a) *The Monge–Ampère measure of u is close to $[\mathrm{dist}(\cdot, \partial\Omega)]^\alpha$ near 0; that is,*

$$(1 - \varepsilon_0)[\mathrm{dist}(\cdot, \partial\Omega)]^\alpha \leq \det D^2 u \leq (1 + \varepsilon_0)[\mathrm{dist}(\cdot, \partial\Omega)]^\alpha \text{ in } B_\rho(0) \cap \Omega,$$

and $\det D^2 u \leq 1/\rho'$ in $\{x_n < \rho\} \cap \Omega$.

(b) *$u \geq 0$, $u(0) = 0$, and any hyperplane $x_{n+1} = \varepsilon x_n$ where $\varepsilon > 0$ is not a supporting hyperplane for u.*

(c) *$u \geq \rho'$ on $\partial\Omega \cap \{x_n \leq \rho\} \setminus B_{\rho/2}(0)$, while for some $\varepsilon_0 \in (0, 1/4)$,*

$$(1 - \varepsilon_0)\hat{\varphi}(x') \leq u(x) \leq (1 + \varepsilon_0)\hat{\varphi}(x') \quad \text{for all } x \in \partial\Omega \cap B_{\rho/2}(0),$$

where $\hat{\varphi}(x')$ is a function of $(n - 1)$ variables satisfying

$$\mu I_{n-1} \leq D_{x'}^2 \hat{\varphi} \leq \mu^{-1}\, I_{n-1}.$$

If ε_0 is sufficiently small, depending only on n, α, and μ, then

$$\kappa\, A\mathcal{E}_h \cap \overline{\Omega} \subset S_u(0, h) \subset \kappa^{-1}\, A\mathcal{E}_h \cap \overline{\Omega} \qquad \text{for all } h < c,$$

where $\mathcal{E}_h := \{|x'|^2 + x_n^{2+\alpha} < h\}$ and A is a sliding along $x_n = 0$; that is,

$$Ax = x + \tau x_n, \quad \text{where } \tau = (\tau_1, \ldots, \tau_{n-1}, 0), \quad \text{with } |\tau| \leq C.$$

The constant κ above depends only on n, α, and μ, and c, C depend on n, α, μ, ρ, and ρ'.

8.6.2. Singular Monge–Ampère equations. For certain classes of singular Monge–Ampère equations such as $\det D^2 u = g(x)[\text{dist}(x, \partial\Omega)]^{-\alpha}$, where $\alpha \in (0, 2)$ and g is bounded between two positive constants, Savin and Zhang [**SZ2**] established very interesting boundary localization theorems. For example, when $\alpha \in (0, 1)$, they proved the following theorem.

Theorem 8.26 (Boundary Localization Theorem for singular equations). *Let Ω be a bounded convex domain in $\mathbb{R}^n$ with C^2 boundary $\partial\Omega$. Let $0 < \lambda \leq \Lambda$. Assume $u \in C(\overline{\Omega})$ is a convex function satisfying*

$$\lambda[\text{dist}(\cdot, \partial\Omega)]^{-\alpha} \leq \det D^2 u \leq \Lambda[\text{dist}(\cdot, \partial\Omega)]^{-\alpha} \quad \text{in } \Omega$$

for some $\alpha \in (0, 1)$, and on $\partial\Omega$, u separates quadratically from its tangent hyperplane; namely, there exists $\mu > 0$ such that for all $x_0, x \in \partial\Omega$,

$$\mu|x - x_0|^2 \leq u(x) - u(x_0) - Du(x_0) \cdot (x - x_0) \leq \mu^{-1}|x - x_0|^2.$$

Then, there is a constant $c > 0$ depending only on n, λ, Λ, α, μ, $\text{diam}(\Omega)$, and the C^2 regularity of $\partial\Omega$, such that for each $x_0 \in \partial\Omega$ and $h \leq c$,

$$\mathcal{E}_{ch}(x_0) \cap \overline{\Omega} \subset S_u(x_0, h) \subset \mathcal{E}_{c^{-1}h}(x_0),$$

where

$$\mathcal{E}_h(x_0) := \left\{ x \in \mathbb{R}^n : |(x - x_0)_\tau|^2 + |(x - x_0) \cdot \nu_{x_0}|^{2-\alpha} < h \right\},$$

with ν_{x_0} denoting the outer unit normal to $\partial\Omega$ at x_0 and $(x - x_0)_\tau$ being the projection of $x - x_0$ onto the supporting hyperplane of $\partial\Omega$ at x_0.

We note that the proof of Theorem 8.26 is more or less along the lines of that of Theorem 8.5, while that of Theorem 8.25 is much more intricate.

8.7. Problems

Problem 8.1. Let Ω and u satisfy (8.2), (8.4), (8.5), and furthermore, $\det D^2 u \leq \Lambda$ in Ω. Show that u is pointwise $C^{1,1/n}$ at 0; that is,

$$u(x) \leq C|x|^{(n+1)/n} \quad \text{in } \Omega \cap B_c(0)$$

where C and c are positive constants depending only on n, ρ, μ, and Λ.

Problem 8.2. Verify the claim in Remark 8.12.

Problem 8.3. Show that the conclusion of Theorem 8.5 continues to hold if instead of requiring $u : \overline{\Omega} \to \mathbb{R}$ to be continuous up to the boundary of $\overline{\Omega}$, we require that (8.5) holds for the boundary values of u understood as $u|_{\partial\Omega,\text{epi}}$ in Definition 7.20.

Problem 8.4. Let $\{\Omega_m\}_{m=1}^\infty \subset B_M(0) \cap \mathbb{R}^n_+$ be convex domains with $0 \in \partial\Omega_m$. Let $u_m : \Omega_m \to \mathbb{R}$ be convex with $u_m(0) = 0$, $u_m \geq 0$, and $b_{u_m,\Omega_m}(h) \leq K$ for all $h \leq c$. Assume that $\{u_m\}_{m=1}^\infty$ converges to $u : \Omega \to \mathbb{R}$

in the sense of Hausdorff convergence of epigraphs. Show that $b_{u,\Omega}(h) \leq K$ for all $h \leq c$.

Problem 8.5. Let Ω be a uniformly convex domain in $\mathbb{R}^n$ with smooth boundary. Let $\lambda > 0$ and $q > 0$. Assume that $u \in C(\overline{\Omega})$ is a nonzero convex Aleksandrov solution of

$$\det D^2 u \ = \lambda |u|^q \quad \text{in } \Omega, \qquad u = 0 \quad \text{on } \partial\Omega.$$

(a) Show that on $\partial\Omega$, the function u separates quadratically from its tangent hyperplanes at each $x_0 \in \partial\Omega$. That is, for any $p \in \partial_s u(x_0)$,

$$\rho|x - x_0|^2 \leq u(x) - u(x_0) - p \cdot (x - x_0) \leq \rho^{-1}|x - x_0|^2,$$

for all $x \in \partial\Omega$, where $\rho(n, \lambda, q, \|u\|_{L^\infty(\Omega)}, \Omega) > 0$.

(b) Show that u is pointwise $C^{1,1/n}$ at the boundary.

Problem 8.6. Let Ω and u be as in the statement of Theorem 8.25. Let x_h^* be the center of mass of the section $S_u(0, h)$. Prove that there exists a small constant $c_0(n, \alpha, \mu, \rho, \rho') > 0$ such that if $h \leq c_0$, then

$$c_0 h^n \leq |S_u(0, h)|^2 (x_h^* \cdot e_n)^\alpha \leq c_0^{-1} h^n.$$

8.8. Notes

This chapter is based on Savin [**S3, S4**].

The volume estimates for the boundary sections in Proposition 8.4, under a slightly different boundary condition for the solutions, were also established by Trudinger and Wang [**TW5**].

Proposition 8.23 is from Le–Savin [**LS1**].

For the degenerate equations in Problem 8.5, better regularity results are available. For these, we need to apply Theorem 8.25 and its improvements; see Section 11.5. Problem 8.6 is taken from Savin [**S6**].

Chapter 9

Geometry of Boundary Sections

This chapter can be viewed as a boundary counterpart of the geometry of compactly contained sections of the Monge–Ampère equation in Chapter 5. Using the Boundary Localization Theorem (Theorem 8.5), we will investigate several important geometric properties of boundary sections and maximal interior sections of Aleksandrov solutions to the Monge–Ampère equation. *They will provide crucial tools in the investigation of boundary regularity for the Monge–Ampère equation in Chapter 10 and for the linearized Monge–Ampère equation in Chapters 13 and 14.*

We will establish the following properties of boundary sections: dichotomy of sections, volume estimates, engulfing and separating properties, inclusion and exclusion properties, and a chain property. We prove Besicovitch's Covering Lemma and employ it to prove a covering theorem and a strong-type (p, p) estimate for the maximal function with respect to boundary sections. Moreover, we will introduce a quasi-distance induced by boundary sections and show that the structure of the Monge–Ampère equation gives rise to a space of homogeneous type. We will also establish global $C^{1,\alpha}$ estimates for the Monge–Ampère equation.

Many results in this chapter hold under the local hypotheses of the Boundary Localization Theorem (Theorem 8.5). However, since we are mostly concerned with the geometry of sections having different centers on the boundary, we state many results under a more global quadratic separation condition (9.4) or (9.26) below. Moreover, when the hypotheses of Theorem 8.1 are satisfied at $x_0 \in \partial\Omega$, $p \in \partial_s u(x_0)$ is unique and it is equal

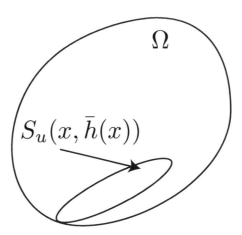

Figure 9.1. A maximal interior section of u.

to the classical gradient $Du(x_0)$ by Proposition 8.23. Thus, we will use this gradient when stating the quadratic separation condition.

If $u \in C(\overline{\Omega})$ is a convex function, then the section of u centered at $x \in \overline{\Omega}$, where u is differentiable, with height $h > 0$, is defined by

$$S_u(x, h) := \big\{ y \in \overline{\Omega} : \quad u(y) < u(x) + Du(x) \cdot (y - x) + h \big\}.$$

Our conditions on u in this chapter always guarantee u being differentiable at x when we speak of the sections $S_u(x, h)$. Moreover, for $x \in \Omega$, we denote by $\bar{h}(x)$ the maximal height of all sections of u centered at x and contained in Ω; that is,

$$\bar{h}(x) := \sup\{h > 0 : \quad S_u(x, h) \subset \Omega\}.$$

In this case, $S_u(x, \bar{h}(x))$ is called the *maximal interior section of u with center $x \in \Omega$*, and it is tangent to the boundary. See Figure 9.1.

9.1. Maximal Interior Sections and Rescalings

In global analysis of Monge–Ampère equations, maximal interior sections are helpful devices in *connecting interior results to those at the boundary*. Thus, we need to understand their geometry. Moreover, the boundary analysis becomes more manageable once boundary sections behave geometrically like half-balls with some boundary portions sufficiently flat. This can be done via rescalings using the Boundary Localization Theorem. We take up these issues in this section, and our analysis is local in nature.

Structural hypotheses for local analysis. Let Ω be a bounded convex domain in $\mathbb{R}^n$ ($n \geq 2$ throughout this chapter) satisfying

(9.1) $B_\rho(\rho e_n) \subset \Omega \subset \{x_n \geq 0\} \cap B_{1/\rho}(0),$

for some small $\rho > 0$. Assume that, at each $y \in \partial\Omega \cap B_\rho(0)$, Ω satisfies the interior ball condition with radius ρ; that is,

(9.2)   there is a ball $B_\rho(\hat{y}) \subset \Omega$ such that $y \in \partial B_\rho(\hat{y})$.

Assume that $u \in C(\overline{\Omega})$ is a strictly convex Aleksandrov solution to

(9.3) $0 < \lambda \leq \det D^2 u \leq \Lambda$ in Ω.

We assume that on $\partial\Omega \cap B_\rho(0)$, u separates quadratically from its tangent hyperplanes on $\partial\Omega$; that is, if $x_0 \in \partial\Omega \cap B_\rho(0)$, then, for all $x \in \partial\Omega$,

(9.4) $\rho \, |x - x_0|^2 \leq u(x) - u(x_0) - Du(x_0) \cdot (x - x_0) \leq \rho^{-1} \, |x - x_0|^2$.

Since u is strictly convex, by Theorem 5.21, $u \in C^{1,\gamma}(\Omega)$ for some $\gamma = \gamma(n, \lambda, \Lambda) \in (0, 1)$. The strict convexity of u will follow from Theorem 5.14 and the conditions (9.24)–(9.26) for many results in later sections.

Let us focus on a boundary point $x_0 \in \partial\Omega \cap B_\rho(0)$. By changing coordinates and subtracting an affine function, we can assume that $x_0 = 0$ and that u and Ω satisfy the hypotheses of the Boundary Localization Theorem (Theorem 8.5) at the origin with $\mu = \rho$. This means that

(9.5) $\begin{cases} \Omega \text{ is convex,}\quad B_\rho(\rho e_n) \subset \Omega \subset \{x_n \geq 0\} \cap B_{1/\rho}(0), \\ u \in C(\overline{\Omega}) \text{ is convex,}\ \ \lambda \leq \det D^2 u \leq \Lambda\ \ \text{in } \Omega,\ \ \ u \geq 0, u(0) = 0, \\ \text{for any } \varepsilon > 0, x_{n+1} = \varepsilon x_n \text{ is not a supporting hyperplane for } u, \\ \rho |x|^2 \leq u(x) \leq \rho^{-1}|x|^2\ \ \text{on } \partial\Omega \cap \{x_n \leq \rho\}. \end{cases}$

To simplify, we use the following constant κ:

(9.6) $\kappa(n, \rho, \lambda, \Lambda)$ is the minimum of the constants c and κ

in Theorems 8.1, and 8.5.

By Theorem 8.5, we know that for all $h \leq \kappa$, $S_u(0, h)$ satisfies

(9.7) $\kappa E_h \cap \overline{\Omega} \subset S_u(0, h) \subset \kappa^{-1} E_h$, $E_h = A_h^{-1} B_{h^{1/2}}(0)$,

with A_h being a linear transformation on $\mathbb{R}^n$ with the following properties:

(9.8) $\det A_h = 1$, $A_h x = x - \tau_h x_n$, $\tau_h \cdot e_n = 0$, $\|A_h^{-1}\| + \|A_h\| \leq \kappa^{-1}|\log h|$.

We recall the following consequences for $h \leq \kappa$:

- (Altitude estimate for section) For (u, Ω) satisfying (9.5),

(9.9) $x_n \leq \kappa^{-1} h^{1/2}$ for all $x \in S_u(0, h)$.

- (Volume estimates for section) For all $x_0 \in \partial\Omega \cap B_\rho(0)$,

(9.10) $C^{-1} h^{n/2} \leq |S_u(x_0, h)| \leq C h^{n/2}$.

- (Size estimates for section) For all $x_0 \in \partial\Omega \cap B_\rho(0)$,

$$(9.11) \qquad \overline{\Omega} \cap B_{ch^{1/2}/|\log h|}(x_0) \subset S_u(x_0, h) \subset B_{Ch^{1/2}|\log h|}(x_0) \cap \overline{\Omega}.$$

Here c and C depend on $n, \rho, \lambda, \Lambda$.

In view of (9.7)–(9.8), the altitude of $S_u(0, h)$ is comparable to $h^{1/2}$, thus matching the upper bound in (9.9). More generally, the following lemma shows that for any boundary section $S_u(x, t)$, there is a point on $\partial S_u(x, t)$ whose distance to the supporting hyperplane to the domain at the boundary point x is comparable to $t^{1/2}$. It justifies the definition of the *apex* of sections in Remark 13.20 in connection with the boundary Harnack inequality for the linearized Monge–Ampère equation.

Lemma 9.1 (Apex of sections). *Assume that Ω and u satisfy (9.1)–(9.4). Let $x_0 \in \partial\Omega \cap B_{\rho/2}(0)$ and $t \le c_1$, where $c_1 = c_1(n, \rho, \lambda, \Lambda)$ is small. Then, for some constant $C(n, \rho, \lambda, \Lambda) > 0$, the following hold.*

- (i) $\mathrm{dist}(y, \partial\Omega) \le Ct^{1/2}$ *for all $y \in S_u(x_0, t)$.*
- (ii) *There exists $y \in \partial S_u(x_0, t) \cap \Omega$ such that $\mathrm{dist}(y, \partial\Omega) \ge C^{-1}t^{1/2}$.*

Proof. We can assume that $x_0 = 0$ and that (u, Ω) satisfies (9.5).

If $y \in S_u(0, t)$, then by (9.9), $\mathrm{dist}(y, \partial\Omega) \le y_n \le \kappa^{-1}t^{1/2}$. This proves (i).

Now, we prove (ii). Applying (9.7)–(9.8) to $S_u(0, t)$, we infer the existence of $y \in \partial S_u(0, t)$ such that $y_n \ge \kappa t^{1/2}$. From $B_\rho(\rho e_n) \subset \Omega$, we find that $x_n \le \rho^{-1}|x'|^2$ on $\partial\Omega \cap B_\rho(0)$. Moreover, by (9.11), $S_u(0, t) \subset B_{Ct^{1/2}|\log t|}(0)$. Thus, if $x \in \partial\Omega$ satisfies $|y - x| = \mathrm{dist}(y, \partial\Omega)$, then from $\mathrm{dist}(y, \partial\Omega) \le |y|$, we have $|x| \le 2|y| \le 2Ct^{1/2}|\log t|$, and hence $x_n \le 4\rho^{-1}C^2 t|\log t|^2$. Therefore

$$\mathrm{dist}(y, \partial\Omega) = |y - x| \ge y_n - x_n \ge y_n - 4\rho^{-1}C^2 t|\log t|^2 \ge (\kappa/2)t^{1/2}$$

if $t \le c_1(n, \rho, \lambda, \Lambda)$ where c_1 is small. Assertion (ii) is proved. $\qquad\square$

In the following proposition, we summarize key geometric properties of maximal interior sections.

Proposition 9.2 (Shape of maximal interior sections). *Let u and Ω satisfy (9.5). Assume that for some $y \in \Omega$, the maximal interior section $S_u(y, \bar h(y)) \subset \Omega$ is tangent to $\partial\Omega$ at 0; that is, $\partial S_u(y, \bar h(y)) \cap \partial\Omega = \{0\}$. If $h := \bar h(y) \le c$ where $c = c(n, \rho, \lambda, \Lambda) > 0$ is small, then the following hold.*

- (i) *There exists a small positive constant $\kappa_0(\rho, \lambda, \Lambda, n) < \kappa$ such that*

$$\begin{cases} Du(y) = ae_n \quad \textit{for some } a \in [\kappa_0 h^{1/2}, \kappa_0^{-1}h^{1/2}], \\ \kappa_0 E_h \subset S_u(y, h) - y \subset \kappa_0^{-1}E_h, \textit{ and} \\ \kappa_0 h^{1/2} \le \mathrm{dist}(y, \partial\Omega) \le \kappa_0^{-1}h^{1/2}, \end{cases}$$

  with E_h and κ the ellipsoid and constant defined in (9.7)–(9.8).

(ii) *If $h/2 < t \le c$, then $S_u(y, 2t) \subset S_u(0, (2\kappa_0^{-4} + 4)t)$.*

(iii) *If $t \le h/2$, then, there are positive constants $\bar{\mu}(n, \lambda, \Lambda) \le 1/8$ and $C(n, \rho, \lambda, \Lambda)$ such that $S_u(y, t) \subset B_{Ct^{\bar{\mu}}}(y)$.*

Proof. We begin with the proof of assertion (i). Because the section

$$S_u(y, h) = \{x \in \overline{\Omega} : u(x) < u(y) + Du(y) \cdot (x - y) + h\} \subset \Omega$$

is tangent to $\partial\Omega$ at 0, we must have

$$u(0) = u(y) + Du(y) \cdot (0 - y) + h \quad \text{and} \quad Du(0) - Du(y) = -ae_n$$

for some $a \in \mathbb{R}$. Since $u(0) = 0$ and $Du(0) = 0$, we have

$$Du(y) = ae_n, \quad u(y) + h = ay_n, \quad \text{and} \quad S_u(y, h) = \{x \in \overline{\Omega} : u(x) < ax_n\}.$$

The same arguments show that

(9.12) $$S_u(y, t) = \{x \in \overline{\Omega} : u(x) < ax_n + t - h\} \quad \text{for all } t > 0.$$

For $t > 0$, we denote

$$S_t' := \{x \in \overline{\Omega} : u(x) < tx_n\},$$

and clearly $S_{t_1}' \subset S_{t_2}'$ if $t_1 \le t_2$.

To prove assertion (i), we show that $S_{ch^{1/2}}'$ has the shape of the ellipsoid E_h in (9.7)–(9.8) for all small h. This will follow from (9.13) and $|S_{\kappa h^{1/2}}'| \ge c|E_h|$. Consider $h \le \kappa$. From (9.7)–(9.8), we know

$$S_u(0, h) := \{x \in \overline{\Omega} : u(x) < h\} \subset \kappa^{-1} E_h \subset \{x_n \le \kappa^{-1} h^{1/2}\}.$$

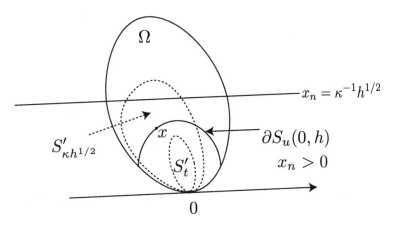

Figure 9.2. If $S_{\kappa h^{1/2}}' \subset S_u(0, h) \cap \Omega$ does not hold, then the point x will lead to a contradiction.

Step 1. We show that

(9.13) $$S'_{\kappa h^{1/2}} \subset S_u(0, h) \cap \Omega.$$

Indeed, if (9.13) does not hold, then from $u(0) = 0$ and the convexity of u, we can find $x \in S'_{\kappa h^{1/2}} \cap \partial S_u(0, h) \cap \Omega$. Thus $u(x) = h$, and $x_n \le \kappa^{-1} h^{1/2}$ by (9.9). However, this contradicts $u(x) < \kappa h^{1/2} x_n$ from x being in $S'_{\kappa h^{1/2}}$. Hence, (9.13) holds. See Figure 9.2.

Step 2. We now show that $|S'_{\kappa h^{1/2}}| \ge c_1 h^{n/2}$ for $c_1(n, \rho, \lambda, \Lambda) > 0$.

Let $\theta = \kappa^4/4$. Applying (9.7)–(9.8) to $S_u(0, \theta h)$, we deduce the existence of $z \in \partial S_u(0, \theta h)$ such that $z_n \ge \kappa(\theta h)^{1/2}$. Let $\tilde{u}(x) := u(x) - \kappa h^{1/2} x_n$. We evaluate $\tilde{u}$ at z and find

$$\tilde{u}(z) \le \theta h - \kappa h^{1/2} \kappa(\theta h)^{1/2} = -\theta h.$$

Let x_0 be the minimum point of $\tilde{u}$ in $\overline{S'_{\kappa h^{1/2}}}$. Since $\tilde{u} = 0$ on $\partial S'_{\kappa h^{1/2}}$, we can find $\tilde{h} \ge \theta h$ such that

$$S'_{\kappa h^{1/2}} = S_{\tilde{u}}(x_0, \tilde{h}) \quad \text{and} \quad \tilde{u}(x) = \tilde{u}(x_0) + D\tilde{u}(x_0) \cdot (x - x_0) + \tilde{h} \quad \text{on } \partial S'_{\kappa h^{1/2}}.$$

Since $\det D^2 \tilde{u} \le \Lambda$, we can use Lemma 5.6 to estimate

$$|S'_{\kappa h^{1/2}}| = |S_{\tilde{u}}(x_0, \tilde{h})| \ge c(n, \Lambda) \tilde{h}^{n/2} \ge c_1(n, \rho, \lambda, \Lambda) h^{n/2}.$$

Step 3. Completion of the proof of assertion (i). By the volume estimates for sections in Lemma 5.6, we have

$$c(\Lambda, n) h^{n/2} \le |S_u(y, h)| = |S'_a| \le C(\lambda, n) h^{n/2}.$$

Thus, if $\alpha = \alpha(n, \rho, \lambda, \Lambda)$ is small, then from (9.13) and (9.7), we obtain

$$|S'_{\kappa(\alpha h)^{1/2}}| \le \kappa^{-n} |E_{\alpha h}| = \kappa^{-n} (\alpha h)^{n/2} \omega_n \le c(\Lambda, n) h^{n/2} \le |S'_a|.$$

It follows that

$$a \ge \kappa \alpha^{1/2} h^{1/2}.$$

Similarly, if $\beta = \beta(n, \rho, \lambda, \Lambda)$ is large, then from Step 2, we have

$$|S'_{\kappa(\beta h)^{1/2}}| \ge c_1(\beta h)^{n/2} \ge C(\lambda, n) h^{n/2} \ge |S'_a|.$$

Therefore

$$a \le \kappa \beta^{1/2} h^{1/2}.$$

Now, we have

$$S_u(y, h) = S'_a \subset S'_{\kappa(\beta h)^{1/2}} \subset S_u(0, \beta h) \cap \Omega \subset \kappa^{-1} E_{\beta h} \subset C E_h$$

for some $C = C(n, \rho, \lambda, \Lambda)$. Hence $y \in C E_h$. Therefore

$$\text{dist}(y, \partial\Omega) \le C\kappa^{-1} h^{1/2} \quad \text{and} \quad S_u(y, h) - y \subset 2C E_h = A_h^{-1} B_{2Ch^{1/2}}(0).$$

The convex set $A_h(S_u(y, h) - y) \subset B_{2Ch^{1/2}}(0)$ contains the origin and has volume at least $c(\Lambda, n)h^{n/2}$, so it must contain a ball $B_{ch^{1/2}}(0)$. From $B_{ch^{1/2}}(0) \subset A_h(S_u(y, h) - y) \subset B_{2Ch^{1/2}}(0)$, we infer that

$$cE_h \subset S_u(y, h) - y \subset 2CE_h.$$

Since

$$y_n = (u(y) + h)/a \geq h/a \geq (\kappa\beta^{1/2})^{-1}h^{1/2},$$

we obtain $\text{dist}(y, \partial\Omega) \geq c_0(n, \rho, \Lambda, \Lambda)h^{1/2}$ if $h \leq c$ where $c = c(n, \rho, \lambda, \Lambda)$ is small, as in the proof of Lemma 9.1. Thus, assertion (i) is proved.

Next, we prove assertion (ii). We will prove a slightly stronger statement that if $h/2 < t \leq c$, then

$$(9.14) \qquad S_u(y, 2t) \subset S_u(0, t^*) \quad \text{with } t^* := \kappa_0^{-4}h + 2(2t - h),$$

where $c > 0$ is small enough such that

$$\kappa_0^{-4}h + 2(2t - h) \leq 2\kappa_0^{-4}c + 4c \leq \kappa.$$

Suppose otherwise that (9.14) is not true. Then, using the convexity of $S_u(y, 2t)$ and $S_u(0, t^*)$ and the fact that their closures both contain 0, we can find a point $x \in \partial S_u(0, t^*) \cap S_u(y, 2t) \cap \Omega$. Recalling (9.12), we have

$$\kappa_0^{-4}h + 2(2t - h) = t^* = u(x) < ax_n + 2t - h \leq \kappa_0^{-1}h^{1/2}x_n + 2t - h,$$

where we used assertion (i) in the last inequality. Hence, $\kappa_0^{-4}h + 2t - h < \kappa_0^{-1}h^{1/2}x_n$. Thus, applying the altitude estimate (9.9) to $S_u(0, \kappa_0^{-4}h + 2(2t - h))$ while also recalling $\kappa^{-1} \leq \kappa_0^{-1}$, we obtain

$$\kappa_0^{-4}h + 2t - h < \kappa_0^{-1}h^{1/2}\big[\kappa_0^{-1}\big(\kappa_0^{-4}h + 2(2t - h)\big)^{1/2}\big].$$

Squaring, we get $(2t - h)^2 < 0$, a contradiction. Hence (9.14) is valid.

Finally, we prove assertion (iii). By assertion (i), the rescaling $\tilde{u}$ of u

$$\tilde{u}(x) := h^{-1}\Big[u(y + h^{1/2}A_h^{-1}x) - u(y) - Du(y) \cdot (h^{1/2}A_h^{-1}x)\Big]$$

satisfies $\lambda \leq \det D^2\tilde{u} \leq \Lambda$ in the sense of Aleksandrov, and

$$(9.15) \enspace B_{\kappa_0}(0) \subset S_{\tilde{u}}(0, 1) \subset B_{\kappa_0^{-1}}(0), \qquad S_{\tilde{u}}(0, 1) = h^{-1/2}A_h(S_u(y, h) - y),$$

where $S_{\tilde{u}}(0, 1)$ represents the section of $\tilde{u}$ at the origin with height 1.

Now, Lemma 5.16 gives $\mu(n, \lambda, \Lambda) \in (0, 1)$ and $C_*(n, \lambda, \Lambda, \rho)$ such that

$$h^{-1/2}A_h(S_u(y, t) - y) = S_{\tilde{u}}(0, t/h) \subset B_{C_*(t/h)^\mu}(0).$$

Using (9.7)–(9.8) and $t \leq h/2 \leq c/2$, we can take $\bar{\mu} = \min\{\mu, 1/8\}$ and obtain

$$S_u(y, t) - y \subset h^{1/2} A_h^{-1} B_{C_*(t/h)^\mu}(0) \subset \kappa^{-1} h^{1/4} B_{C_*(t/h)^\mu}(0) \subset B_{Ct^{\bar{\mu}}}(0)$$

for some $C(n, \rho, \lambda, \Lambda) > 0$. Hence assertion (iii) is proved, and the proof of the proposition is complete. $\qquad\square$

In boundary analysis of Monge–Ampère equations, the rescaling below, using the Boundary Localization Theorem, will be repeatedly employed. Its usefulness consists in *preserving the geometry of the Boundary Localization Theorem and improving the flatness of the boundary.*

Lemma 9.3 (Properties of the rescaled functions using the Boundary Localization Theorem). *Assume that Ω and u satisfy (9.1)–(9.4), with $u(0) = 0$ and $Du(0) = 0$. Let A_h be as in (9.7)–(9.8). Define the rescaled convex function u_h on the rescaled domain Ω_h by*

$$u_h(x) := h^{-1} u(h^{1/2} A_h^{-1} x), \quad \Omega_h := h^{-1/2} A_h \Omega.$$

Then, the function $u_h \in C(\overline{\Omega_h})$ satisfies, in the sense of Aleksandrov,

$$\lambda \leq \det D^2 u_h \leq \Lambda,$$

and the section of u_h centered at the origin with height 1 satisfies

$$S_{u_h}(0, 1) = h^{-1/2} A_h S_u(0, h) \quad and \quad B_\kappa(0) \cap \overline{\Omega_h} \subset S_{u_h}(0, 1) \subset B_{\kappa^{-1}}(0).$$

Moreover, there are constants c and C depending on $\rho, n, \lambda, \Lambda$ such that the following assertions hold for $h \leq c$.

   (i) *For any $x, x_0 \in \partial\Omega_h \cap B_{2/\kappa}(0)$, we have*

$$(9.16) \quad \frac{\rho}{4} |x - x_0|^2 \leq u_h(x) - u_h(x_0) - Du_h(x_0) \cdot (x - x_0) \leq 4\rho^{-1} |x - x_0|^2.$$

   (ii) *$\partial\Omega_h \cap B_{2/\kappa}(0)$ is a graph G_h in the e_n direction whose $C^{1,1}$ norm is bounded by $Ch^{1/2}$.*

   (iii) *If $r \leq c$, then u_h and $S_{u_h}(0, 1)$ satisfy the hypotheses of Theorem 8.1 at all $x_0 \in \partial\Omega_h \cap B_r(0)$, with $S_{u_h}(x_0, 1/2) \subset S_{u_h}(0, 1)$.*

   (iv) *If $r \leq c$, then $|Du_h| \leq Cr|\log r|^2$ in $\overline{\Omega_h} \cap B_r(0)$.*

   (v) *If $\delta \leq c$ where $\delta(\rho, n, \lambda, \Lambda)$ is small, then for any $x \in S_{u_h}(0, \delta) \cap \Omega_h$, the maximal interior section $S_{u_h}(x, \bar{h}(x))$ of u_h centered at x is tangent to $\partial\Omega_h$ at $z \in \partial\Omega_h \cap B_{c/2}(0)$ and $\bar{h}(x) \leq c/2$.*

Proof. Thanks to (9.7), and (9.8), the first two assertions hold. We only need to prove assertions (i)–(v). Let us start with assertions (i) and (ii).

Denote $B_r = B_r(0)$ and $T := h^{1/2} A_h^{-1}$. For $x, x_0 \in \partial\Omega_h \cap B_{2/\kappa}$, let

$$X = Tx, \quad X_0 = Tx_0.$$

Then, using (9.8), we have $X, X_0 \in \partial\Omega \cap B_{Ch^{1/2}|\log h|}$ where $C = C(\rho, n, \lambda, \Lambda)$. Let ν_0 be the outer unit normal to Ω at X_0. Then, since Ω satisfies at X_0 the interior ball condition with radius ρ, we have

$$(9.17) \qquad |(X - X_0) \cdot \nu_0| \leq |X - X_0|^2/\rho.$$

First, we show that

$$(9.18) \qquad \frac{|x - x_0|}{2} \leq \frac{|X - X_0|}{h^{1/2}} \leq 2|x - x_0|.$$

Indeed, this is equivalent to

$$1/2 \leq |A_h Y|/|Y| \leq 2, \quad \text{where } Y := X - X_0.$$

Since Ω has an interior tangent ball of radius ρ at each point on $\partial\Omega \cap B_\rho$, $\partial\Omega$ is $C^{1,1}$ in a neighborhood of the origin. Therefore,

$$(9.19) \qquad |\nu_0 + e_n| \leq C(\rho)|X_0| \leq Ch^{1/2}|\log h|.$$

It follows from $|Y| \leq Ch^{1/2}|\log h|$ and (9.17) (namely $|Y \cdot \nu_0| \leq |Y|^2/\rho$) that

$$|Y_n| \leq |Y \cdot (e_n + \nu_0)| + |Y \cdot \nu_0| \leq Ch^{1/2}|\log h|(|Y_n| + |Y'|).$$

Thus, if $h \leq c$ where $c(\rho, n, \lambda, \Lambda)$ is small, then $|Y_n| \leq Ch^{1/2}|\log h||Y'|$, and

$$|A_h Y - Y| = |\tau_h Y_n| \leq Ch^{1/2}|\log h|^2|Y'| \leq |Y|/2,$$

and (9.18) is proved.

With (9.18), we obtain assertion (i) from (9.4) and the equality

$$u_h(x) - u_h(x_0) - Du_h(x_0) \cdot (x - x_0) = \frac{u(X) - u(X_0) - Du(X_0) \cdot (X - X_0)}{h}.$$

Next, we prove assertion (ii). Clearly, $\partial\Omega_h \cap B_{2/\kappa}$ is a graph G_h in the e_n direction. Due to (9.18), (9.17) implies $|(x - x_0) \cdot T^t\nu_0| \leq Ch|x - x_0|^2$, or

$$|(x - x_0) \cdot \tilde{\nu}_0| \leq C\frac{h}{|T^t\nu_0|}|x - x_0|^2, \quad \text{where } \tilde{\nu}_0 := T^t\nu_0/|T^t\nu_0|.$$

From the formula for A_h, we see that

$$A_h^{-1}z = z + \tau_h z_n \quad \text{and} \quad (A_h^{-1})^t z = z + (\tau_h \cdot z)e_n.$$

Therefore, $(A_h^{-1})^t e_n = e_n$, and for $h \leq c$, using (9.19) and (9.8), we have

$$|(A_h^{-1})^t\nu_0| \geq |(A_h^{-1})^t e_n| - |(A_h^{-1})^t(\nu_0 + e_n)| \geq 1 - Ch^{1/2}|\log h|^2 \geq 1/2$$

and

$$|T^t\nu_0| = h^{1/2}|(A_h^{-1})^t\nu_0| \geq h^{1/2}/2,$$

which shows

$$|(x - x_0) \cdot \tilde{\nu}_0| \leq Ch^{1/2}|x - x_0|^2.$$

This implies assertion (ii) about the $C^{1,1}$ norm of G_h.

We now prove assertion (iii). From assertions (i)–(ii) together with $B_\kappa(0) \cap \overline{\Omega}_h \subset S_{u_h}(0,1) \subset B_{\kappa^{-1}}(0)$, we see that $(u_h, S_{u_h}(0,1))$ satisfies (9.5) where ρ there is now replaced by a small constant $\tilde{\rho}(\rho, n, \lambda, \Lambda) > 0$.

We consider $x_0 \in \partial \Omega_h \cap B_r$ where $r \leq c(n, \rho, \lambda, \Lambda)$ and show that u_h and $S_{u_h}(0,1)$ satisfy the hypotheses of Theorem 8.1 also at x_0, with

(9.20) $$S_{u_h}(x_0, 1/2) \subset S_{u_h}(0,1) \subset B_{1/\kappa}.$$

Indeed, by the volume estimates (9.10),

$$|S_{u_h}(x_0, 1/2)| = |h^{-1/2} A_h S_u(X_0, h/2)| = h^{-n/2} |S_u(X_0, h/2)| \geq 2^{-n} C^{-1}.$$

If (9.20) holds, then the convexity of $S_{u_h}(x_0, 1/2)$ implies that $S_{u_h}(x_0, 1/2)$ contains $S_{u_h}(x_0, 1/2) \cap B_{2\hat{\rho}}(x_0)$ for some $\hat{\rho}(n, \lambda, \Lambda, \rho) > 0$, so $S_{u_h}(0,1)$ has an interior ball of radius $\hat{\rho}$ that is tangent to $\partial S_{u_h}(0,1)$ at x_0. Thus, in view of assertions (i) and (ii), it remains to prove (9.20) if r is small.

From (8.38), $u_h(0) = 0$, and $Du_h(0) = 0$ (since $Du(0) = 0$), we have

(9.21) $$|u_h| \leq Cr^2 |\log r|^2 \quad \text{in } \overline{\Omega}_h \cap B_{2r},$$

which, by the convexity of u_h, gives

$$D_n u_h(x_0) \leq Cr |\log r|^2.$$

We next show that

$$|D_{x'} u_h(x_0)| \leq Cr.$$

Indeed, by (9.16) applied at x and 0, we have

$$(\rho/4)|x|^2 \leq u_h(x) \leq 4\rho^{-1} |x|^2 \quad \text{on } \partial \Omega_h \cap B_r,$$

and therefore, applying (9.16) again at x and x_0, we find

$$|u_h(x_0) + Du_h(x_0) \cdot (x - x_0)| \leq Cr^2 \quad \text{on } \partial \Omega_h \cap B_r.$$

Using the above estimate at $x \in \partial \Omega_h \cap B_r$ and 0, we obtain

$$|Du_h(x_0) \cdot x| \leq Cr^2 \quad \text{on } \partial \Omega \cap B_r.$$

Since $x_n \geq 0$ on $\partial \Omega_h$, we see that if $D_n u_h(x_0) \geq 0$, then,

$$D_{x'} u_h(x_0) \cdot x' \leq Cr^2 \quad \text{if } |x'| \leq r/2,$$

which gives $|D_{x'} u_h(x_0)| \leq Cr$.

Similarly, if $D_n u_h(x_0) \leq 0$, then, using $D_{x'} u_h(x_0) \cdot x' \geq -Cr^2$ for $|x'| \leq r/2$, we also obtain $|D_{x'} u_h(x_0)| \leq Cr$, as asserted.

If the first inclusion in (9.20) does not hold, then we can find a point $z \in S_{u_h}(x_0, 1/2) \cap \partial S_{u_h}(0,1)$. Thus, $z \in B_{1/\kappa} \setminus B_\kappa$, so $z_n \geq (x_0)_n$. Using the upper bounds on $D_n u_h(x_0)$ and $|D_{x'} u_h(x_0)|$, we obtain

$$u_h(z) \leq u_h(x_0) + Du_h(x_0) \cdot (z - x_0) + 1/2 \leq Cr^2 + Cr |\log r|^2 + 1/2 < 1,$$

provided r is small. This contradicts $u_h(z) = 1$. Therefore, (9.20) holds.

We prove assertion (iv). Since u_h and $S_{u_h}(0,1)$ satisfy the hypotheses of Theorem 8.1 at all points $x_0 \in \partial\Omega_h \cap B_r(0)$, by Proposition 8.23, we have

$$(9.22) \qquad |Du_h(x_0)| \leq Cr|\log r|^2 \quad \text{for all } x_0 \in \partial\Omega_h \cap B_r(0).$$

Now, the claimed estimate at any interior point $\bar{x} \in \Omega_h \cap B_r(0)$ follows by convexity. Take, for example, $\bar{x}$ on the x_n-axis. Then, the cone

$$K_{\bar{x}} := B_r(\bar{x}) \cap \{x \in \mathbb{R}^n : x_n - \bar{x}_n \geq |x' - \bar{x}'|\} \subset B_{2r} \cap \Omega_h \subset S_{u_h}(0,1).$$

Thus, for any unit vector $\gamma \in \mathbb{R}^n$ with $\gamma_n \geq |\gamma'|$, we have $\bar{x} + r\gamma \in B_{2r} \cap \Omega_h$. By the convexity of u_h and $0 \leq u_h \leq Cr^2|\log r|^2$ in $B_{2r} \cap \Omega_h$, we have

$$D_\gamma u_h(\bar{x}) := Du_h(\bar{x}) \cdot \gamma \leq [u_h(\bar{x} + r\gamma) - u_h(\bar{x})]/r \leq Cr|\log r|^2,$$

which gives an upper bound for $D_n u_h(\bar{x})$ and $|D_i u_h(\bar{x})|$ for $1 \leq i \leq n-1$. We use the convexity again to get the lower bound $D_n u_h(\bar{x}) \geq D_n u_h(0) = 0$.

Finally, we prove assertion (v). Let $x \in S_{u_h}(0,\delta)$ for $\delta(n,\rho,\lambda,\Lambda)$ small. Then, $x \in \overline{\Omega_h} \cap B_{C\delta^{1/2}|\log\delta|}$. Since $S_{u_h}(x, \bar{h}(x))$ is balanced around x by Lemma 5.6(iii), the tangent point $z \in \partial S_{u_h}(x, \bar{h}(x)) \cap \partial\Omega_h$ must be on $\partial\Omega_h \cap B_{2/\kappa}$. Furthermore,

$$0 = u_h(0) \geq u_h(x) + Du_h(x) \cdot (0 - x) + \bar{h}(x),$$

and

$$u_h(z) \leq u_h(x) + Du_h(x) \cdot (z - x) + \bar{h}(x).$$

These together with assertion (i) give

$$\rho|z|^2/4 \leq u_h(z) \leq Du_h(x) \cdot z \leq |Du_h(x)||z|.$$

By assertion (iv), $|Du_h(x)| \leq C\delta^{1/2}|\log\delta|^3$, so

$$|z| \leq C\delta^{1/2}|\log\delta|^3/\rho \quad \text{and} \quad \bar{h}(x) \leq Du_h(x) \cdot x \leq C\delta|\log\delta|^5.$$

Therefore, $z \in \partial\Omega_h \cap B_{c/2}$, and $\bar{h}(x) \leq c/2$ if δ is small. $\qquad\square$

Another geometric property of the rescaling in Lemma 9.3 is that the distance to the boundary, when appropriately rescaled, is essentially preserved. This will be useful in studying fine regularity properties of the linearized Monge–Ampère equation in Chapter 13.

Lemma 9.4 (Distance comparison under rescaling using the Boundary Localization Theorem). *Assume that Ω and u satisfy (9.1)–(9.4), with $u(0) = 0$ and $Du(0) = 0$. Let A_h be as in (9.7)–(9.8), and let $\Omega_h := h^{-1/2}A_h\Omega$. Then, there exists $C(n,\rho,\lambda,\Lambda) > 0$, such that for all $x \in \Omega_h \cap B^+_{\kappa^{-1}}(0)$ where $h \leq C^{-1}$, we have*

$$1 - Ch^{1/2}|\log h|^2 \leq \frac{h^{-1/2}\operatorname{dist}(h^{1/2}A_h^{-1}x, \partial\Omega)}{\operatorname{dist}(x, \partial\Omega_h)} \leq 1 + Ch^{1/2}|\log h|^2.$$

Proof. Let $X = h^{1/2}A_h^{-1}x$. Let ξ_x and ξ_X be the inner unit normal vectors to $\partial\Omega_h$ and $\partial\Omega$ at x and X, respectively. We compare $h^{-1/2}\text{dist}(X, \partial\Omega)$ with $\text{dist}(x, \partial\Omega_h)$ by computing the directional derivative of $h^{-1/2}\text{dist}(X, \partial\Omega)$ along ξ_x. We have

$$D_x(h^{-1/2}\text{dist}(X, \partial\Omega)) \cdot \xi_x = h^{-1/2}(h^{1/2}A_h^{-1})^t D_X \text{dist}(X, \partial\Omega) \cdot \xi_x$$
$$= \xi_X \cdot (A_h^{-1}\xi_x).$$

Since $\partial\Omega$ is $C^{1,1}$ at 0 and $|X| \leq Ch^{1/2}|\log h|$, we get $|\xi_X - e_n| \leq Ch^{1/2}|\log h|$. Moreover, the $C^{1,1}$ bound of $\partial\Omega_h$ from Lemma 9.3(ii) gives

$$|\xi_x - e_n| \leq Ch^{1/2} \quad \text{if } h \leq c(n, \lambda, \Lambda, \rho).$$

Thus, for $h \leq c$, we have

$$|\xi_X \cdot (A_h^{-1}\xi_x) - e_n \cdot (A_h^{-1}e_n)| \leq Ch^{1/2}|\log h|\|A_h^{-1}\| \leq Ch^{1/2}|\log h|^2.$$

Using $e_n \cdot (A_h^{-1}e_n) = 1$, we obtain

$$|D_x(h^{-1/2}\text{dist}(X, \partial\Omega)) \cdot \xi_x - 1| \leq Ch^{1/2}|\log h|^2$$

if $h \leq c(n, \lambda, \Lambda, \rho)$. The asserted distance ratio estimates then follow. $\qquad\square$

9.2. Global Hölder Gradient Estimates

Extending the interior $C^{1,\alpha}$ estimates in Theorem 5.21 to the boundary, we establish the following global $C^{1,\alpha}$ estimates, due to Le and Savin [**LS1**], for the Monge–Ampère equation with right-hand side bounded between two positive constants.

Theorem 9.5 (Global $C^{1,\alpha}$ estimates for the Monge–Ampère equation). *Assume that the convex domain Ω in $\mathbb{R}^n$ satisfies $\Omega \subset B_{1/\rho}(0)$ and it satisfies the interior ball condition with radius ρ at each point on $\partial\Omega$. Assume that the convex function $u \in C(\overline{\Omega})$ satisfies, in the sense of Aleksandrov,*

$$\lambda \leq \det D^2 u \leq \Lambda \quad \text{in } \Omega$$

and that u separates quadratically from its tangent hyperplanes on $\partial\Omega$; namely,

$$\rho |x - x_0|^2 \leq u(x) - u(x_0) - Du(x_0) \cdot (x - x_0) \leq \rho^{-1} |x - x_0|^2,$$

for all $x, x_0 \in \partial\Omega$. Then, there exists $\alpha(n, \lambda, \Lambda) \in (0, 1)$ such that

$$[Du]_{C^\alpha(\overline{\Omega})} := \sup_{x \neq y \in \overline{\Omega}} \frac{|Du(x) - Du(y)|}{|x - y|^\alpha} \leq C = C(n, \lambda, \Lambda, \rho).$$

Note that if u satisfies the hypotheses of the theorem, then so does $u(x) + z \cdot x$ for any $z \in \mathbb{R}^n$, with the same structural constants. Thus Du is not bounded by any structural constant, but its oscillation is. This is the reason why we are only asserting the bound on the C^α seminorm.

Proof. By Proposition 8.23 and covering $\partial\Omega$ by finitely many small balls, we deduce that $u \in C^{1,\beta}(\partial\Omega)$ for all $\beta \in (0,1)$ and $\mathrm{osc}_{\partial\Omega}|Du| \le C(n,\lambda,\Lambda,\rho)$. Thus, by Theorem 5.14, u is strictly convex in Ω. By Theorem 5.21, $u \in C^{1,\alpha}(\Omega)$ for some $\alpha(n,\lambda,\Lambda) \in (0,1)$.

Step 1. Oscillation estimate for Du in a maximal interior section. For $y \in \Omega$, let $y_0 \in \partial\Omega$ satisfy $|y - y_0| = r := \mathrm{dist}(y,\partial\Omega)$. Let $S_u(y,\bar{h})$ be the maximal interior section of u centered at y. We show that if r is small, then for some $C_1(n,\lambda,\Lambda,\rho) > 0$ and $\alpha_1 := (1+\alpha)/2$,

$$|Du(z_1) - Du(z_2)| \le C_1|z_1 - z_2|^\alpha \text{ in } S_u(y,\bar{h}/2) \text{ and } |Du(y) - Du(y_0)| \le r^{\alpha_1}.$$

Indeed, if $r \le c_1(n,\lambda,\Lambda,\rho)$ where c_1 is small, then Lemma 9.3 gives $\bar{h} \le c$, and by Proposition 9.2 applied at the point $x_0 \in \partial S_u(y,\bar{h}) \cap \partial\Omega$, we have

$$c\bar{h}^{1/2} \le r \le C\bar{h}^{1/2}, \quad |Du(y) - Du(x_0)| \le C\bar{h}^{1/2}, \quad cE \subset S_u(y,\bar{h}) - y \subset CE,$$

for some c small and C depending on n,ρ,λ,Λ, where

$$E := A_{\bar{h}}^{-1}B_{\bar{h}^{1/2}}(0), \quad \text{with } \|A_{\bar{h}}\| + \|A_{\bar{h}}^{-1}\| \le C|\log\bar{h}|.$$

Consider the rescaling $\tilde{u}$ of u given by

$$\tilde{u}(\tilde{x}) := \bar{h}^{-1}[u(y + \bar{h}^{1/2}A_{\bar{h}}^{-1}\tilde{x}) - u(y) - Du(y) \cdot (\bar{h}^{1/2}A_{\bar{h}}^{-1}\tilde{x})].$$

Then $\tilde{u}$ satisfies

$$\lambda \le \det D^2\tilde{u} \le \Lambda \quad \text{in } S_{\tilde{u}}(0,1),$$

and the section $\tilde{S}_1 := S_{\tilde{u}}(0,1)$ of $\tilde{u}$ at the origin with height 1 satisfies

$$B_c(0) \subset \tilde{S}_1 \subset B_C(0), \qquad \tilde{S}_1 = \bar{h}^{-1/2}A_{\bar{h}}(S_u(y,\bar{h}) - y).$$

The interior $C^{1,\alpha}$ estimate in Theorem 5.21 (see also Corollary 5.22) gives

$$(9.23) \quad |D\tilde{u}(\tilde{z}_1) - D\tilde{u}(\tilde{z}_2)| \le C_1(n,\lambda,\Lambda,\rho)|\tilde{z}_1 - \tilde{z}_2|^\alpha \quad \text{for all } \tilde{z}_1,\tilde{z}_2 \in \tilde{S}_{1/2}.$$

For $\tilde{z} \in \tilde{S}_1$, let $z = y + \bar{h}^{1/2}A_{\bar{h}}^{-1}\tilde{z}$. Rescaling back the estimate (9.23) and using $\tilde{z}_1 - \tilde{z}_2 = \bar{h}^{-1/2}A_{\bar{h}}(z_1 - z_2)$, we find, for all $z_1,z_2 \in S_u(y,\bar{h}/2)$,

$$\begin{aligned}
|Du(z_1) - Du(z_2)| &= \left|\bar{h}^{1/2}A_{\bar{h}}^t(D\tilde{u}(\tilde{z}_1) - D\tilde{u}(\tilde{z}_2))\right| \\
&\le C_1\bar{h}^{1/2}\|A_{\bar{h}}\||\tilde{z}_1 - \tilde{z}_2|^\alpha \\
&\le C_1\bar{h}^{(1-\alpha)/2}\|A_{\bar{h}}\|^{1+\alpha}|z_1 - z_2|^\alpha \\
&\le C_1\bar{h}^{(1-\alpha)/2}(C|\log\bar{h}|)^{1+\alpha}|z_1 - z_2|^\alpha \le C_1|z_1 - z_2|^\alpha
\end{aligned}$$

if c_1 is small.

From $|y - y_0| = r$, we have

$$|x_0 - y_0| \le |x_0 - y| + |y - y_0| \le C\bar{h}^{1/2}|\log\bar{h}| + r \le r(\log r)^2$$

if c_1 is small. By Proposition 8.23, we then find

$$|Du(y) - Du(y_0)| \leq |Du(y) - Du(x_0)| + |Du(x_0) - Du(y_0)|$$
$$\leq C\bar{h}^{1/2} + C|x_0 - y_0|(\log|x_0 - y_0|)^2 \leq r^{\alpha_1}.$$

Step 2. Oscillation estimate for Du near the boundary. Let $x, y \in \Omega$ with $\max\{\operatorname{dist}(x, \partial\Omega), \operatorname{dist}(y, \partial\Omega), |x - y|\} \leq c_1$ small. We show that for $\alpha_2 := (\alpha_1 + \alpha)/2$,

$$|Du(x) - Du(y)| \leq \max\{C_1|x - y|^\alpha, |x - y|^{\alpha_2}\}.$$

Indeed, let $x_0, y_0 \in \partial\Omega$ be such that $|x - x_0| = \operatorname{dist}(x, \partial\Omega) := r_x$ and $|y - y_0| = \operatorname{dist}(y, \partial\Omega) := r_y$. We can assume $r_y \leq r_x \leq c_1(n, \lambda, \Lambda, \rho)$.

Note that $(1/2)(S_u(y, \bar{h}(y)) - y) \subset S_u(y, \bar{h}(y)/2) - y$. Thus

$$S_u(y, \bar{h}(y)/2) \supset B_{\bar{h}(y)^{1/2}/(2C|\log \bar{h}(y)|)}(y) \supset B_{cr_y^{1+\gamma}}(y),$$

for c_1 and c depending on n, ρ, λ, and Λ, where $\gamma := (\alpha_1 - \alpha_2)/(2\alpha_2)$.

If $|y - x| \leq cr_x^{1+\gamma}$, then $y \in S_u(x, \bar{h}(x)/2)$ and hence, Step 1 gives

$$|Du(x) - Du(y)| \leq C_1|x - y|^\alpha.$$

Consider now the case $|y - x| \geq cr_x^{1+\gamma}$. Then,

$$|x_0 - y_0| \leq |x_0 - x| + |x - y| + |y - y_0| \leq 2r_x + |x - y| \leq C|x - y|^{1/(1+\gamma)}.$$

Thus, from

$$|Du(x) - Du(y)| \leq |Du(x) - Du(x_0)|$$
$$+ |Du(x_0) - Du(y_0)| + |Du(y_0) - Du(y)|,$$

Step 1, and Proposition 8.23 and noting that $\alpha_1/(1 + \gamma) > \alpha_2$, we have

$$|Du(x) - Du(y)| \leq r_x^{\alpha_1} + C|x_0 - y_0|(\log|x_0 - y_0|)^2 + r_y^{\alpha_1}$$
$$\leq 2r_x^{\alpha_1} + C|x - y|^{1/(1+\gamma)}(\log|x - y|)^2 \leq |x - y|^{\alpha_2}.$$

Step 3. Conclusion. Since $\operatorname{osc}_{\partial\Omega}|Du| \leq C(n, \lambda, \Lambda, \rho)$, by the convexity of u, we also have $\operatorname{osc}_\Omega|Du| \leq C(n, \lambda, \Lambda, \rho)$. Combining this with Step 2 and Theorem 5.21, we easily obtain the conclusion of the theorem. $\square$

If Ω is uniformly convex and if $\partial\Omega$ and $u|_{\partial\Omega}$ are C^3, then by combining Proposition 4.4 (which gives the global Lipschitz estimates) and Proposition 4.7 (which gives the quadratic separation of u from its tangent hyperplanes on $\partial\Omega$), we obtain from Theorem 9.5 the following global $C^{1,\alpha}$ regularity.

Theorem 9.6 (Global $C^{1,\alpha}$ regularity for the Monge–Ampère equation). *Let Ω be a uniformly convex domain in $\mathbb{R}^n$, with boundary $\partial\Omega \in C^3$, and let $\varphi \in C^3(\overline{\Omega})$. Let $u \in C(\overline{\Omega})$ be a convex function satisfying*

$$0 < \lambda \leq \det D^2 u \leq \Lambda \quad \text{in } \Omega, \qquad u = \varphi \quad \text{on } \partial\Omega.$$

Then $u \in C^{1,\alpha}(\overline{\Omega})$, where $\alpha = \alpha(n, \lambda, \Lambda) \in (0, 1)$, with the estimate

$$\|u\|_{C^{1,\alpha}(\overline{\Omega})} \le C = C(n, \lambda, \Lambda, \Omega, \varphi).$$

The above global $C^{1,\alpha}$ regularity constitutes a crucial step in the solvability of singular Abreu equations in higher dimensions [**KLWZ, L14**]. The two-dimensional case of these equations will be discussed in depth in Section 15.5.

In the remaining sections of this chapter, we will use the following structural hypotheses in our geometric investigation of boundary sections.

Structural hypotheses for global analysis. We assume that the convex domain Ω in $\mathbb{R}^n$ satisfies

(9.24) $\Omega \subset B_{1/\rho}(0)$, and for each $y \in \partial\Omega$, there is an interior ball

$$B_\rho(\hat{y}) \subset \Omega \text{ such that } y \in \partial B_\rho(\hat{y}).$$

Let $u : \overline{\Omega} \to \mathbb{R}$ be a continuous and convex function satisfying

(9.25) $0 < \lambda \le \det D^2 u \le \Lambda$ in Ω, in the sense of Aleksandrov.

Assume further that on $\partial\Omega$, u separates quadratically from its tangent hyperplanes; namely, for all $x, x_0 \in \partial\Omega$,

(9.26) $\rho |x - x_0|^2 \le u(x) - u(x_0) - Du(x_0) \cdot (x - x_0) \le \rho^{-1} |x - x_0|^2.$

By Proposition 8.23, u is $C^{1,\beta}$ on the boundary $\partial\Omega$ for all $\beta \in (0, 1)$. Consequently, by Theorem 5.14, u is strictly convex in Ω. Hence, by Theorem 5.21, $u \in C^{1,\alpha}(\Omega)$ for some $\alpha = \alpha(n, \lambda, \Lambda) \in (0, 1)$. Moreover, for each $y \in \Omega$, there is a maximal interior section $S_u(y, \bar{h}(y))$ centered at y.

A direct consequence of Theorem 9.5 is the following lemma.

Lemma 9.7. *Assume that Ω and u satisfy (9.24)–(9.26). Then, there exists $M(n, \rho, \lambda, \Lambda) > 0$ such that $S_u(x_0, M) \supset \overline{\Omega}$, for all $x_0 \in \overline{\Omega}$.*

Proof. By Theorem 9.5, $[Du]_{C^\alpha(\overline{\Omega})} \le C$ for some constants $\alpha(n, \lambda, \Lambda) \in (0, 1)$ and $C(n, \rho, \lambda, \Lambda) > 0$. Now, let $x_0 \in \overline{\Omega}$. Then, for any $x \in \overline{\Omega}$, we have

$$u(x) - u(x_0) - Du(x_0) \cdot (x - x_0) \le [Du]_{C^\alpha(\overline{\Omega})} |x - x_0|^{1+\alpha}$$
$$\le C(2\rho^{-1})^{1+\alpha} =: M.$$

Thus, $\overline{\Omega} \subset S_u(x_0, M)$. $\qquad\square$

9.3. Dichotomy, Volume Growth, and Engulfing Properties

We first prove a dichotomy for sections of solutions to the Monge–Ampère equation satisfying (9.24)–(9.26): Any section is either an interior section or is included in a boundary section with comparable height. This allows us to focus our attention on only interior sections and boundary sections.

Proposition 9.8 (Dichotomy for sections). *Assume that Ω and u satisfy (9.24)–(9.26). Let $S_u(x_0, t_0)$ be a section of u with $x_0 \in \overline{\Omega}$ and $t_0 > 0$. Then one of the following assertions is true:*

(i) *$S_u(x_0, 2t_0)$ is an interior section; that is, $S_u(x_0, 2t_0) \subset \Omega$.*

(ii) *$S_u(x_0, 2t_0)$ is included in a boundary section of comparable height; that is, there exist $z \in \partial\Omega$ and $\bar{c}(n, \rho, \lambda, \Lambda) > 1$ such that*

$$S_u(x_0, 2t_0) \subset S_u(z, \bar{c}t_0).$$

Proof. It suffices to consider the case $x_0 \in \Omega$. Let $S_u(x_0, \bar{h})$ be the maximal interior section with center x_0 where $\bar{h} = \bar{h}(x_0)$. If $t_0 \leq \bar{h}/2$, then assertion (i) is true. We now consider the case $\bar{h}/2 < t_0$ and show that assertion (ii) is valid with $\bar{c} := \max\{2\kappa_0^{-4} + 4, M/c\}$, where $M(n, \rho, \lambda, \Lambda)$ is the constant given by Lemma 9.7 and κ_0 is the constant in Proposition 9.2.

Without loss of generality, we assume that $\Omega \subset \mathbb{R}^n_+$, $\partial S_u(x_0, \bar{h})$ is tangent to $\partial\Omega$ at 0, $u(0) = 0$, and $Du(0) = 0$. If $\bar{h}/2 < t_0 \leq c(n, \rho, \lambda, \Lambda)$, then, by Proposition 9.2(ii),

$$S_u(x_0, 2t_0) \subset S_u(0, (2\kappa_0^{-4} + 4)t_0) \subset S_u(0, \bar{c}t_0).$$

Therefore, assertion (ii) is valid in this case.

In the case $c < t_0$, by using Lemma 9.7, we obtain assertion (ii) from

$$S_u(x_0, 2t_0) \subset \overline{\Omega} \subset S_u(0, M) \subset S_u(0, \bar{c}t_0).$$

This completes the proof of the proposition. $\square$

As an application of the dichotomy of sections, we obtain the following volume growth of sections.

Corollary 9.9 (Volume growth and doubling of sections). *Assume that Ω and u satisfy (9.24)–(9.26). Then the following hold.*

(i) *There exist constants c_0, c_1, C_2 depending only on ρ, λ, Λ, and n such that for any section $S_u(x, t)$ with $x \in \overline{\Omega}$ and $t \leq c_0$, we have*

(9.27) $$c_1 t^{n/2} \leq |S_u(x, t)| \leq C_2 t^{n/2}.$$

(ii) *For all $x, y \in \overline{\Omega}$ and $t > 0$, $K \geq 2$, we have*

$$|S_u(y, Kt)| \leq C(\rho, \lambda, \Lambda, n, K) |S_u(x, t)|.$$

Proof. Let $c_0 := \kappa/\bar{c}$, where κ is as in (9.6) and $\bar{c}$ is as in Proposition 9.8.

First, we prove assertion (i). Let $S_u(x_0, t)$ be a section with $t \leq c_0$. If $x_0 \in \partial\Omega$, then the volume estimates in (9.10) give the desired result. Now, consider $x_0 \in \Omega$, and let $S_u(x_0, \bar{h})$ be the maximal interior section with center x_0 where $\bar{h} = \bar{h}(x_0)$.

If $t \leq \bar{h}/2$, then the result follows from the volume growth of interior sections in Lemma 5.6.

If $\bar{h}/2 < t \leq c_0$, then Proposition 9.8 gives $S_u(x_0, 2t) \subset S_u(z, \bar{c}t)$ for some $z \in \partial\Omega$. Hence, by invoking the volume estimates in (9.10), we obtain the second inequality in (9.27). To prove the first inequality in (9.27), we first note that if $t < 2\bar{h}$, then, by Lemma 5.6,

$$|S_u(x_0, t)| \geq |S_u(x_0, \bar{h}/2)| \geq c(n, \lambda, \Lambda)\bar{h}^{n/2} \geq 2^{-n/2}ct^{n/2}.$$

Consider now $t \geq 2\bar{h}$. Without loss of generality, we assume that $\Omega \subset \mathbb{R}^n_+$, $\partial S_u(x_0, \bar{h})$ is tangent to $\partial\Omega$ at 0, $u(0) = 0$, and $Du(0) = 0$. Then, as in (9.12), we get for some positive number a

$$S_u(x_0, t) = \{x \in \overline{\Omega} : u(x) < ax_n + t - \bar{h}\}.$$

Using this, together with $\bar{h} \leq t/2$ and $x_n \geq 0$ in Ω, we obtain

$$S_u(x_0, t) \supset \{x \in \overline{\Omega} : u(x) < t/2\} = S_u(0, t/2).$$

By the volume estimates in (9.10), the first inequality in (9.27) follows.

Next, we prove assertion (ii.) If $Kt \leq c_0$, then (ii) easily follows from assertion (i). Now assume $Kt > c_0$. Then, from $\Omega \subset B_{1/\rho}(0)$, we have

$$|S_u(y, Kt)| \leq |B_{1/\rho}(0)| = C\, c_1 (c_0/K)^{\frac{n}{2}} \leq C\, |S_u(x, c_0/K)| \leq C\, |S_u(x, t)|,$$

where $C = C(\rho, \lambda, \Lambda, n, K)$, and we used assertion (i) in the second inequality. The proof is thus complete. $\qquad\square$

Next, we prove the engulfing and separating properties of sections.

Theorem 9.10 (Engulfing and separating properties of sections). *Assume that Ω and u satisfy (9.24)–(9.26). There exists $\theta_*(n, \rho, \lambda, \Lambda) > 0$ such that:*

  (i) *The sections $\{S_u(x, t)\}$ of u satisfy the engulfing property with the constant θ_*; that is, if $y \in S_u(x, t)$ with $x \in \overline{\Omega}$ and $t > 0$, then $S_u(x, t) \subset S_u(y, \theta_* t)$.*

  (ii) *The sections $\{S_u(x, t)\}$ of u satisfy the separating property with the constant θ_*^2; namely, if $y \notin S_u(x, t)$, then $S_u(y, \frac{t}{\theta_*^2}) \cap S_u(x, \frac{t}{\theta_*^2}) = \emptyset$.*

Proof. We first prove assertion (i) with the help of Proposition 9.8.

Step 1. The case $x \in \partial\Omega$. Without loss of generality, we assume that x is the origin and that u and Ω satisfy (9.5). We show that there exists $\theta_0(n, \rho, \lambda, \Lambda) > 0$ such that if $X \in S_u(0, t)$ with $t > 0$, then

$$(9.28) \qquad\qquad S_u(0, t) \subset S_u(X, \theta_0 t).$$

Let $t \leq c_0$ with $c_0 \leq \kappa$ to be chosen later, where $\kappa = \kappa(\rho, \lambda, \Lambda, n)$ is as in (9.6). Let us consider $h \in [t, \kappa]$. Let A_h be as in (9.7)–(9.8). Let

$$u_h(z) := h^{-1} u(h^{1/2} A_h^{-1} z) \quad \text{for } z \in \Omega_h := h^{-1/2} A_h \Omega.$$

For $X, Y \in S_u(0, t)$, let $x := h^{-1/2} A_h X$ and $y := h^{-1/2} A_h Y$. Then

$$S_{u_h}(0, 1) = h^{-1/2} A_h S_u(0, h), \quad x, y \in S_{u_h}(0, t/h).$$

By Lemma 9.3(iii), u_h satisfies the hypotheses of the Boundary Localization Theorem (Theorem 8.5) in $S_{u_h}(0, 1)$. Hence, by (9.11), there is $C(n, \rho, \lambda, \Lambda) > 0$ such that

$$|x|, |y| \leq C \left(\frac{t}{h}\right)^{1/2} \left| \log \left(\frac{t}{h}\right) \right|.$$

Let c be as in Lemma 9.3, and let $M_1 > 1$ be such that $C M_1^{-1/2} \log M_1 \leq c$.

Now, take $c_0 := \kappa/M_1$ and $h = t M_1 \in [t, \kappa]$. Then, $x, y \in B_c(0)$. By Lemma 9.3(iv), we have $|Du_h(x)| \leq c |\log c|^2$. Thus,

$$\frac{u(Y) - u(X) - Du(X) \cdot (Y - X)}{h} = u_h(y) - u_h(x) - Du_h(x) \cdot (y - x)$$
$$\leq t/h + 2c^2 |\log c|^2,$$

which shows $Y \in S_u\big(X, (1 + 2M_1 c^2 |\log c|^2) t\big)$. Hence, for any $X \in S_u(0, t)$ with $t \leq c_0$, we have $S_u(0, t) \subset S_u\big(X, (1 + 2M_1 c^2 |\log c|^2) t\big)$. In the case $X \in S_u(0, t)$ with $t > c_0$, then by using Lemma 9.7, we obtain

$$S_u(0, t) \subset \overline{\Omega} \subset S_u(X, M) \subset S_u(X, Mt/c_0).$$

Therefore, by taking $\theta_0 := \max\{1 + 2M_1 c^2 |\log c|^2, M/c_0\}$, we obtain (9.28).

Step 2. General case. To simplify, we denote $S(z, t) := S_u(z, t)$. By Step 1, it remains to consider the case $x \in \Omega$. Let $S(x, \bar{h})$ be the maximal interior section of u with center x where $\bar{h} = \bar{h}(x)$ and let $y \in S(x, t)$.

If $t < \bar{h}/2$, then $S(x, 2t)$ is an interior section and the result follows from the engulfing properties of interior sections in Theorem 5.28; namely, $S(x, t) \subset S(y, \theta t)$ for some $\theta(\lambda, \Lambda, n) > 0$.

It remains to consider the case $\bar{h}(x)/2 \leq t$. Then, by Proposition 9.8, $S(x, 2t) \subset S(z, \bar{c}t)$ for some $z \in \partial\Omega$. Since $y \in S(z, \bar{c}t)$, by Step 1, we have $S(z, \bar{c}t) \subset S(y, \theta_0 \bar{c}t)$. Therefore the result follows with $\theta_* := \max\{\theta, \theta_0 \bar{c}\}$.

Now, we prove assertion (ii). Suppose that $y \notin S_u(x,t)$. If, otherwise, there is $z \in S_u(y, \frac{t}{\theta_*^2}) \cap S_u(x, \frac{t}{\theta_*^2})$, then, by assertion (i),

$$S_u\left(x, \frac{t}{\theta_*^2}\right) \cup S_u\left(y, \frac{t}{\theta_*^2}\right) \subset S_u\left(z, \frac{t}{\theta_*}\right).$$

Hence, $x, y \in S_u(z, \frac{t}{\theta_*})$. Again, by assertion (i), we have $S_u(z, \frac{t}{\theta_*}) \subset S_u(x,t)$, so $y \in S_u(x,t)$, which is a contradiction. □

9.4. Global Inclusion, Exclusion, and Chain Properties

In this section, we prove the global inclusion and exclusion properties and establish a chain property for sections of solutions to the Monge–Ampère equations in Theorem 9.12 and Theorem 9.13, respectively.

We start with estimates for the size of a section in terms of its height, essentially removing the height restriction in Proposition 9.2.

Lemma 9.11 (Size of boundary sections). *Assume that Ω and u satisfy (9.24)–(9.26). Let M be as in Lemma 9.7, and let $\bar{\mu}$ be as in Proposition 9.2. Then, there is $\bar{C}(n, \rho, \lambda, \Lambda) > 0$ such that for all $t \leq M$ and $y \in \bar{\Omega}$,*

$$S_u(y,t) \subset B_{\bar{C}t^{\bar{\mu}}}(y).$$

Proof. We write $S(z,h)$ for $S_u(z,h)$. We first prove the lemma for the case where y is a boundary point which can be assumed to be $0 \in \partial\Omega$. Furthermore, we can assume that u and Ω satisfy (9.5). Thus, if $t \leq \kappa(\leq 1)$, then from (9.11), we have

$$S(0,t) \subset B_{Ct^{1/2}|\log t|}(0) \subset B_{Ct^{1/4}}(0)$$

for some $C(n, \rho, \lambda, \Lambda) > 0$. Hence, we can find $C_1(n, \rho, \lambda, \Lambda) > 0$ such that

$$(9.29) \qquad S(0,t) \subset B_{C_1 t^{\bar{\mu}}}(0) \quad \text{for all } t \leq M.$$

Next, we prove the lemma for $y \in \Omega$. Let $S(y, \bar{h}(y))$ be the maximal interior section centered at y. Let $c(n, \rho, \lambda, \Lambda)$ be the constant in Proposition 9.2. We will treat separately whether y is away from the boundary or not.

Case 1. $\bar{h}(y) > c/2$. In this case, $S(y, c/2) \Subset \Omega$, and we can use Lemma 5.16 to find $\mu(n, \lambda, \Lambda) \in [\bar{\mu}, 1)$ and $C_0(\rho, n, \lambda, \Lambda) > 0$ such that $S_u(y,t) \subset B_{C_0 t^{\mu}}(y)$ for all $t \leq c/4$. By increasing C_0 if necessary, we obtain

$$(9.30) \qquad S_u(y,t) \subset B_{C_0 t^{\bar{\mu}}}(y) \quad \text{for all } t \leq M.$$

Case 2. $\bar{h}(y) \leq c/2$. Consider the section $S(y,t)$ with $t \leq M$. Then, either there exists $z \in \partial\Omega$ such that $z \in S(y, 2t)$ or $S(y, 2t) \subset \Omega$. In the first case, by Theorem 9.10, $S(y, 2t) \subset S(z, 2\theta_* t)$. Thus, as in (9.29), we find

$$S(y,t) \subset S(z, 2\theta_* t) \subset B_{\hat{C}_1 t^{\bar{\mu}}}(z) \quad \text{where } \hat{C}_1 := (2\theta_*)^{\bar{\mu}} C_1.$$

It follows that $|y - z| \leq \hat{C}_1 t^{\bar{\mu}}$ and therefore

$$(9.31) \qquad S(y, t) \subset B_{2\hat{C}_1 t^{\bar{\mu}}}(y) \quad \text{for all } t \leq M.$$

In the remaining case $S(y, 2t) \subset \Omega$, we have $2t \leq \bar{h}(y) \leq c/2$ by the definition of $\bar{h}(y)$. Using Proposition 9.2(iii), we obtain

$$(9.32) \qquad S(y, t) \subset B_{C_2 t^{\bar{\mu}}}(y).$$

The lemma now follows from (9.29)–(9.32) with $\bar{C} = C_0 + 2\hat{C}_1 + C_2$. $\qquad \square$

We now prove global inclusion and exclusion properties for sections.

Theorem 9.12 (Global inclusion and exclusion properties for sections). *Assume that Ω and u satisfy (9.24)–(9.26). Let $M(n, \rho, \lambda, \Lambda)$ and $\bar{\mu}(n, \lambda, \Lambda) \in (0, 1/8]$ be as in Lemmas 9.7 and 9.11, respectively. Let $p_1 = \bar{\mu}^{-1}$. Then, there exists $c_0(n, \rho, \lambda, \Lambda) > 0$ such that, for any $x_0 \in \overline{\Omega}$, the following hold.*

(i) *If $0 < r < s \leq 3, s - r \leq 1, 0 < t \leq M$, and $x_1 \in S_u(x_0, rt)$, then*

$$S_u(x_1, c_0(s - r)^{p_1} t) \subset S_u(x_0, st).$$

(ii) *If $0 < r < s < 1, 0 < t \leq M$, and $x_1 \in S_u(x_0, t) \backslash S_u(x_0, st)$, then*

$$S_u(x_1, c_0(s - r)^{p_1} t) \cap S_u(x_0, rt) = \emptyset.$$

Proof. We first prove assertion (i). Let $0 < r < s \leq 3, s - r \leq 1$, and $0 < t \leq M$. Let c, δ be as in Lemma 9.3 and let θ_* be as in Theorem 9.10.

Step 1. We first consider the case $t \leq c\delta/(4\theta_*)$. If $S_u(x_0, 4t) \subset \Omega$, then assertion (i) follows from the interior result in Theorem 5.30. Suppose now $S_u(x_0, 4t) \cap \partial\Omega \neq \emptyset$. Without loss of generality, we can assume that $0 \in S_u(x_0, 4t) \cap \partial\Omega$ and that u and Ω satisfy (9.5). By Theorem 9.10, we have

$$(9.33) \qquad S_u(x_0, 4t) \subset S_u(0, 4\theta_* t).$$

Let $h = 4\theta_* \delta^{-1} t \leq c$, and let A_h be as in (9.7)–(9.8). Consider the rescaled function

$$u_h(x) := h^{-1} u(h^{1/2} A_h^{-1} x) \quad \text{where } x \in \Omega_h := h^{-1/2} A_h \Omega.$$

Denote $x_{i,h} = h^{-1/2} A_h x_i$ for $i = 0, 1$. Let $\bar{t} = \delta/(4\theta_*)$. Then, $t/h = \bar{t}$. Since $x_1 \in S_u(x_0, rt)$, we have $x_{1,h} \in S_{u_h}(x_{0,h}, r\bar{t})$,

$$h^{-1/2} A_h S_u(x_1, c_0(s - r)^{p_1} t) = S_{u_h}(x_{1,h}, c_0(s - r)^{p_1} \bar{t}),$$

and

$$h^{-1/2} A_h S_u(x_0, st) = S_{u_h}(x_{0,h}, s\bar{t}).$$

Now, assertion (i) is equivalent to showing that for some $c_0(n, \rho, \lambda, \Lambda) > 0$,

$$(9.34) \qquad S_{u_h}(x_{1,h}, c_0(s - r)^{p_1} \bar{t}) \subset S_{u_h}(x_{0,h}, s\bar{t}).$$

Suppose that $y \in S_{u_h}(x_{1,h}, c_0(s-r)^{p_1}\bar{t})$ and $x_{1,h} \in S_{u_h}(x_{0,h}, r\bar{t})$. Then

$$
\begin{aligned}
u_h(y) &< u_h(x_{1,h}) + Du_h(x_{1,h}) \cdot (y - x_{1,h}) + c_0(s-r)^{p_1}\bar{t} \\
&< u_h(x_{0,h}) + Du_h(x_{0,h}) \cdot (x_{1,h} - x_{0,h}) + r\bar{t} \\
&\quad + Du_h(x_{1,h}) \cdot (y - x_{1,h}) + c_0(s-r)^{p_1}\bar{t} \\
&= u_h(x_{0,h}) + Du_h(x_{0,h}) \cdot (y - x_{0,h}) \\
&\quad + [Du_h(x_{1,h}) - Du_h(x_{0,h})] \cdot (y - x_{1,h}) + c_0(s-r)^{p_1}\bar{t} + r\bar{t}.
\end{aligned}
$$
(9.35)

We note that $x_{1,h} \in S_{u_h}(x_{0,h}, r\bar{t}) \subset S_{u_h}(0, \delta)$, by (9.33). By Lemma 9.3(iv),

$$
(9.36) \qquad |Du_h(x_{0,h})| + |Du_h(x_{1,h})| \leq C, \quad \text{for } C = C(n, \rho, \lambda, \Lambda).
$$

On the other hand, from $y \in S_{u_h}(x_{1,h}, c_0(s-r)^{p_1}\bar{t})$, we can estimate

$$
(9.37) \qquad |y - x_{1,h}| \leq C(n, \rho, \lambda, \Lambda)(c_0(s-r)^{p_1}\bar{t})^{\bar{\mu}}.
$$

Indeed, if $x_{1,h} \in \Omega_h$, then by Lemma 9.3(v), $S_{u_h}(x_{1,h}, \bar{h}(x_{1,h}))$—the maximal interior section of u_h centered at $x_{1,h}$—is tangent to $\partial\Omega_h$ at $z \in \partial\Omega_h \cap B_{c/2}(0)$ and $\bar{h}(x_{1,h}) \leq c/2$. Thus, the proof of Lemma 9.11 applies, with Case 1 there skipped, and we obtain (9.37).

It follows from (9.35)–(9.37) and $0 < s - r \leq 1$ that

$$
\begin{aligned}
u_h(y) - u_h(x_{0,h}) &- Du_h(x_{0,h}) \cdot (y - x_{0,h}) \\
&< C(c_0(s-r)^{p_1}\bar{t})^{\bar{\mu}} + c_0(s-r)^{p_1}\bar{t} + r\bar{t} < s\bar{t},
\end{aligned}
$$

if we choose c_0 small, depending on $n, \rho, \lambda, \Lambda$. Here, we recall $p_1 = \bar{\mu}^{-1}$. Therefore $S_{u_h}(x_{1,h}, c_0(s-r)^{p_1}\bar{t}) \subset S_{u_h}(x_{0,h}, s\bar{t})$, proving (9.34), as claimed.

Step 2. Finally, we consider the case $M \geq t \geq c\delta/(4\theta_*)$. Suppose that $z \in S_u(x_1, c_0(s-r)^{p_1}t)$ and $x_1 \in S_u(x_0, rt)$.

Computing as in (9.35), we show that if c_0 is small, then

$$
\begin{aligned}
(9.38) \quad u(z) - u(x_0) &- Du(x_0) \cdot (z - x_0) \\
&< [Du(x_1) - Du(x_0)] \cdot (z - x_1) + c_0(s-r)^{p_1}t + rt < st,
\end{aligned}
$$

which implies that $S_u(x_1, c_0(s-r)^{p_1}t) \subset S_u(x_0, st)$, as asserted in the theorem. Indeed, by Lemma 9.11, we have

$$
|z - x_1| \leq \bar{C}(c_0(s-r)^{p_1}t)^{\bar{\mu}} \quad \text{and} \quad |x_1 - x_0| \leq \bar{C}(rt)^{\bar{\mu}}.
$$

It follows from the global $C^{1,\alpha}$ estimate in Theorem 9.5 that for some $\alpha(n, \lambda, \Lambda) \in (0, 1)$,

$$
\begin{aligned}
|[Du(x_1) - Du(x_0)] \cdot (z - x_1)| &\leq C_\alpha |x_1 - x_0|^\alpha |z - x_1| \\
&\leq C(rt)^{\alpha\bar{\mu}}(c_0(s-r)^{p_1}t)^{\bar{\mu}},
\end{aligned}
$$
(9.39)

where $C = C(n, \rho, \lambda, \Lambda)$. If $c\delta/(4\theta_*) \leq t \leq M$, then (9.38) follows from

$$C(rt)^{\alpha\bar{\mu}}(c_0(s-r)^{p_1}t)^{\bar{\mu}} + c_0(s-r)^{p_1}t + rt < st$$

if we choose c_0 small.

Note that the proof of assertion (ii) is quite similar to that of assertion (i). We include here for example the proof for the case $M \geq t \geq c\delta/(4\theta_*)$. From the convexity of u and $x_1 \in S_u(x_0, t) \backslash S_u(x_0, st)$, we have for all $z \in \overline{\Omega}$

$$\begin{aligned} u(z) &\geq u(x_1) + Du(x_1) \cdot (z - x_1) \\ &\geq u(x_0) + Du(x_0) \cdot (x_1 - x_0) + st + Du(x_1) \cdot (z - x_1) \\ &= u(x_0) + Du(x_0) \cdot (z - x_0) + [Du(x_1) - Du(x_0)] \cdot (z - x_1) + st. \end{aligned}$$

We will show that, for all $z \in S_u(x_1, c_0(s-r)^{p_1}t)$,

(9.40)
$$\begin{aligned} u(z) &- u(x_0) - Du(x_0) \cdot (z - x_0) \\ &> [Du(x_1) - Du(x_0)] \cdot (z - x_1) + st > rt, \end{aligned}$$

which implies that $S_u(x_1, c_0(s-r)^{p_1}t) \cap S_u(x_0, rt) = \emptyset$.

Suppose now $z \in S_u(x_1, c_0(s-r)^{p_1}t)$. Then, Lemma 9.11 gives

$$|z - x_1| \leq \bar{C}(c_0(s-r)^{p_1}t)^{\bar{\mu}} \quad \text{and} \quad |x_1 - x_0| \leq \bar{C}t^{\bar{\mu}}.$$

As in (9.39), we have

$$|[Du(x_1) - Du(x_0)] \cdot (z - x_1)| \leq Ct^{\alpha\bar{\mu}}(c_0(s-r)^{p_1}t)^{\bar{\mu}},$$

and thus

$$u(z) - u(x_0) - Du(x_0) \cdot (z - x_0) \geq st - Ct^{\alpha\bar{\mu}}(c_0(s-r)^{p_1}t)^{\bar{\mu}} \geq rt,$$

proving (9.40), provided that

$$Ct^{\alpha\bar{\mu}}(c_0(s-r)^{p_1}t)^{\bar{\mu}} < (s-r)t.$$

Since $p_1 = \bar{\mu}^{-1}$, $M \geq t \geq c\delta/(4\theta_*)$, and $0 < s - r \leq 1$, this can be achieved by choosing c_0 small. The proof of the theorem is complete. $\qquad\square$

We now establish a chain property for sections of the Monge–Ampère equation. It roughly says that any pair of interior points x and y, where x is close to the boundary and the distance of y to the boundary is structurally large compared to that of x, can be connected by a chain of finitely many (and structurally bounded) interior sections. See Figure 9.3.

Theorem 9.13 (Chain property for sections). *Assume u and Ω satisfy the hypotheses of the Boundary Localization Theorem (Theorem 8.5) with $\rho = \mu$ at all points on $\partial\Omega \cap B_\rho(z)$ where $z \in \partial\Omega$. Let $\kappa_0(n, \rho, \lambda, \Lambda)$ be the constant in Proposition 9.2. Let $\tau \in (0, 1)$. Assume that $x \in \Omega$ with $\mathrm{dist}(x, \partial\Omega) \leq c_0(n, \rho, \lambda, \Lambda)$ where c_0 is small and that it satisfies $\partial\Omega \cap \partial S_u(x, \bar{h}(x)) = \{z\}$.*

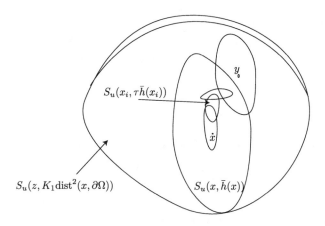

Figure 9.3. A chain of interior sections connecting x and y.

Then, the following hold:

(i) *There exist positive constants $K, \bar{K}$ depending on $\tau, n, \rho, \lambda, \Lambda$, and there exists a sequence $x_0 = x, x_1, \ldots, x_m$ in $S_u(z, \bar{K} \operatorname{dist}^2(x, \partial\Omega))$ with $m \leq K$ such that*

$$x_i \in S_u(x_{i+1}, \tau\bar{h}(x_{i+1})) \quad \text{for all } i = 0, 1, \ldots, m - 1$$

and

$$\operatorname{dist}(x_m, \partial\Omega) \geq 2\kappa_0^{-2} \operatorname{dist}(x, \partial\Omega).$$

(ii) *Conversely, for given positive constants K_1 and K_2 and any $y \in S_u(z, K_1 \operatorname{dist}^2(x, \partial\Omega))$ with $\operatorname{dist}(y, \partial\Omega) \geq K_2 \operatorname{dist}(x, \partial\Omega)$, there exists a sequence $x_0 = x, x_1, \ldots, x_m = y$ in $S_u(z, K_1 \operatorname{dist}^2(x, \partial\Omega))$ with m depending on $K_1, K_2, \tau, n, \rho, \lambda, \Lambda$ such that*

$$x_i \in S_u(x_{i+1}, \tau\bar{h}(x_{i+1})) \quad \text{for all } i = 0, 1, \ldots, m - 1.$$

Proof. We give here the proof of assertion (i) while that of assertion (ii) follows similarly. For simplicity, we can assume $z = 0$ and that (u, Ω) satisfies (9.5). Let c be as in Proposition 9.2, and let $c_0 \leq c$. Then

$$\kappa_0 \bar{h}^{1/2}(x) \leq \operatorname{dist}(x, \partial\Omega) \leq \kappa_0^{-1}\bar{h}^{1/2}(x).$$

Case 1. $c_0 c \leq \operatorname{dist}(x, \partial\Omega) \leq c_0$.

By Lemma 9.1, there is $y \in S_u(0, c)$ such that $\operatorname{dist}(y, \partial\Omega) \geq c_3(\rho, n, \lambda, \Lambda)$. Thus if $c_0 = c_0(n, \rho, \lambda, \Lambda)$ is small, then for $K_1 := 6\kappa_0^{-2}$, we have

(9.41) $$\operatorname{dist}(y, \partial\Omega) \geq K_1 \operatorname{dist}(x, \partial\Omega).$$

Let

$$D = \{y \in \overline{S_u(0, c)} : \operatorname{dist}(y, \partial\Omega) \geq K_1 \operatorname{dist}(x, \partial\Omega)\} \subset S_u(0, \bar{K}\operatorname{dist}^2(x, \partial\Omega)/4)$$

where $\bar{K} = 4(cc_0^2)^{-1}$. With y as in (9.41), we find $D \neq \emptyset$. Since $D \subset \bigcup_{y \in D} S_u(y, \tau \bar{h}(y))$, by Lemma 5.32, we can find a covering

$$(9.42) \qquad D \subset \bigcup_{i=1}^{K} S_u(y_i, \tau \bar{h}(y_i))$$

such that $\{S_u(y_i, \delta \tau \bar{h}(y_i))\}$ is a disjoint family of sections where $\delta = \delta(n, \lambda, \Lambda)$.

Now, with y as in (9.41), we can find a sequence $x_0 = x, x_1, \ldots, x_m = y$ in $S_u(0, \bar{K}\text{dist}^2(x, \partial\Omega)/4)$ with $m \leq K$ such that $x_i \in S_u(x_{i+1}, \tau \bar{h}(x_{i+1}))$ for all $i = 0, 1, \ldots, m-1$. Assertion (i) is proved once we verify that K is bounded by a constant depending on $\tau, n, \rho, \lambda, \Lambda$. However, this follows from the volume estimate. Indeed, by Lemma 5.8, $|D| \leq C(n, \rho, \lambda, \Lambda)$. On the other hand, from Proposition 9.2, we have for each $i = 1, \ldots, K$,

$$\bar{h}(y_i) \geq \kappa_0^2 \text{dist}^2(y_i, \partial\Omega) \geq \kappa_0^2 \text{dist}^2(x, \partial\Omega) \geq (\kappa_0 cc_0)^2.$$

Hence, by (9.42) and the lower bound on volume of sections in Lemma 5.6, we obtain the desired bound for K.

Case 2. $\text{dist}(x, \partial\Omega) \leq cc_0$.

Let $h = \text{dist}^2(x, \partial\Omega)/(c_0^2 c) \leq c$, and let A_h be as in (9.7)–(9.8). Consider

$$u_h(x) := h^{-1} u(h^{1/2} A_h^{-1} x) \quad \text{where } x \in \Omega_h := h^{-1/2} A_h \Omega.$$

By Lemma 9.3, u_h and $S_{u_h}(0, 1)$ satisfy the hypotheses of the Boundary Localization Theorem (Theorem 8.5) at all points $x_0 \in \partial\Omega_h \cap B_c(0)$. By Lemma 9.4, we have for all $y_h = h^{-1/2} A_h y \in \Omega_h \cap B_{k-1}^+(0)$,

$$(9.43) \qquad 1 - Ch^{1/2}|\log h|^2 \leq \frac{h^{-1/2}\text{dist}(y, \partial\Omega)}{\text{dist}(y_h, \partial\Omega_h)} \leq 1 + Ch^{1/2}|\log h|^2.$$

If $h \leq c(n, \rho, \lambda, \Lambda)$ is small, we have

$$(9.44) \qquad Ch^{1/2}|\log h|^2 \leq 1/2.$$

This and $\text{dist}(x, \partial\Omega) = c_0(ch)^{1/2}$ imply that for $x_h = h^{-1/2} A_h x$,

$$\frac{1}{2} c_0 c^{1/2} \leq \text{dist}(x_h, \partial\Omega_h) \leq 2c_0 c^{1/2}.$$

As in Case 1, we can find a sequence

$$\{x_{0,h} = x_h, x_{1,h}, \ldots, x_{m,h}\} \subset S_{u_h}(0, \bar{K}\text{dist}^2(x_h, \partial\Omega_h)/4),$$

with $m \leq K(\tau, n, \rho, \lambda, \Lambda)$ such that

$$x_{i,h} \in S_{u_h}(x_{i+1,h}, \tau \bar{h}(x_{i+1,h})) \quad \text{for all } i = 0, 1, \ldots, m-1$$

and

$$\text{dist}(x_{m,h}, \partial\Omega_h) \geq K_1 \text{dist}(x_h, \partial\Omega_h).$$

It follows that for $x_i = h^{1/2} A_h^{-1} x_{i,h}$, we have $x_i \in S_u(x_{i+1}, \tau \bar{h}(x_{i+1}))$ for all $i = 0, 1, \ldots, m - 1$. Recalling (9.43) and (9.44), we find

$$\operatorname{dist}(x_m, \partial\Omega) \geq (K_1/3)\operatorname{dist}(x, \partial\Omega) = 2\kappa_0^{-2}\operatorname{dist}(x, \partial\Omega).$$

Furthermore, for all $i = 0, 1, \ldots, m$, we have from (9.43) and (9.44)

$$x_i \in S_u(0, h\bar{K}\operatorname{dist}^2(x_h, \partial\Omega_h)/4) \subset S_u(0, \bar{K}\operatorname{dist}^2(x, \partial\Omega)).$$

The proof of assertion (i) is complete. $\qquad\square$

9.5. Besicovitch's Covering Lemma, Maximal Function, and Quasi-Distance

This section is concerned with various real analysis topics related to boundary sections of solutions to the Monge–Ampère equation. They include Besicovitch's Covering Lemma, covering theorem, global strong-type (p, p) estimate for the maximal function, and quasi-distance induced by sections.

We begin with the following covering lemma.

Lemma 9.14 (Besicovitch's Covering Lemma for boundary sections). *Assume that Ω and u satisfy (9.24)–(9.26). Let M be as in Lemma 9.7, and let θ_* be the engulfing constant in Theorem 9.10. Let $E \subset \overline{\Omega}$, and let $\mathcal{F}$ be a family of sections $\{S_u(x, t_x)\}_{x \in E}$ where each $t_x \leq M$. Then, there exists a countable subfamily $\{S_u(x_k, t_k)\}_{k=1}^{\infty}$ of $\mathcal{F}$ with the following properties:*

(i) $E \subset \bigcup_{k=1}^{\infty} S_u(x_k, t_k)$,

(ii) $x_k \notin \bigcup_{j<k} S_u(x_j, t_j)$ *for all* $k \geq 2$,

(iii) *the family* $\{S_u(x_k, t_k/(2\theta_*^2))\}_{k=1}^{\infty}$ *is disjoint, and*

(iv) *there exists a constant* $K(\rho, \lambda, \Lambda, n) > 0$ *such that*

$$\sum_{k=1}^{\infty} \chi_{S_u(x_k, (1-\varepsilon)t_k)} \leq K \log \frac{1}{\varepsilon} \quad \textit{for all } 0 < \varepsilon < 1.$$

Proof. To simplify, we denote $S(z, t) := S_u(z, t)$. Without loss of generality, we assume that $M = \sup\{t : S(x, t) \in \mathcal{F}\}$. Consider

$$\mathcal{F}_0 := \{S(x, t) \in \mathcal{F} : M/2 < t \leq M\} \quad \text{and} \quad E_0 := \{x : S(x, t) \in \mathcal{F}_0\}.$$

Pick $S(x_1, t_1) \in \mathcal{F}_0$ such that $t_1 > 3M/4$. If $E_0 \subset S(x_1, t_1)$, then we stop. Otherwise, the set

$$\left\{ t : S(x, t) \in \mathcal{F}_0 \text{ and } x \in E_0 \setminus S(x_1, t_1) \right\} \neq \emptyset,$$

and let M_1 be its supremum. Pick t_2 in this set such that $t_2 > 3M_1/4$, and let $S(x_2, t_2)$ be the corresponding section. Then, $x_2 \notin S(x_1, t_1)$, and

$$t_1 > 3M/4 \geq 3M_1/4 \geq 3t_2/4.$$

If $E_0 \subset S(x_1, t_1) \cup S(x_2, t_2)$, then we stop. Otherwise, we continue the process. In this way, we have constructed a (possibly infinite) family of sections denoted by

$$\mathcal{F}_0' = \{S(x_i^0, t_i^0)\}_{i=1}^{\infty} \quad \text{with } x_j^0 \in E_0 \setminus \bigcup_{i<j} S(x_i^0, t_i^0).$$

We next consider the family

$$\mathcal{F}_1 := \{S(x, t) \in \mathcal{F} : M/4 < t \leq M/2\}$$

and let

$$E_1 := \left\{x : S(x, t) \in \mathcal{F}_1 \text{ and } x \notin \bigcup_{i=1}^{\infty} S(x_i^0, t_i^0)\right\}.$$

We repeat the above construction for E_1 and obtain a family of sections denoted by

$$\mathcal{F}_1' = \{S(x_i^1, t_i^1)\}_{i=1}^{\infty} \quad \text{with } x_j^1 \in E_1 \setminus \bigcup_{i<j} S(x_i^1, t_i^1).$$

We continue this process. In the kth stage, we consider the family

$$\mathcal{F}_k := \left\{S(x, t) \in \mathcal{F} : \frac{M}{2^{k+1}} < t \leq \frac{M}{2^k}\right\}$$

and let

$$E_k := \left\{x : S(x, t) \in \mathcal{F}_k \text{ and } x \notin \bigcup \text{sections previously selected}\right\}.$$

Thus, we obtain a family of sections denoted by

$$\mathcal{F}_k' = \{S(x_i^k, t_i^k)\}_{i=1}^{\infty} \quad \text{with } x_j^k \in E_k \setminus \bigcup_{i<j} S(x_i^k, t_i^k).$$

We now show that the collection of all sections in all generations $\mathcal{F}_k'$, $k \geq 0$, is the family that satisfies the conclusion of the lemma.

Step 1. The overlapping in each generation $\mathcal{F}_k'$ is at most $\bar{\kappa}(\rho, \lambda, \Lambda, n)$.

To show this, let us suppose that $z \in S(x_{j_1}^k, t_{j_1}^k) \cap \cdots \cap S(x_{j_N}^k, t_{j_N}^k)$, with $S(x_{j_i}^k, t_{j_i}^k) \in \mathcal{F}_k'$ and $j_1 < \cdots < j_N$. For simplicity, we set $x_{j_i}^k = x_i$, $t_{j_i}^k = t_i$, and let $t = \max\{t_i : 1 \leq i \leq N\}$. Then, due to $\frac{M}{2^{k+1}} < t_i \leq \frac{M}{2^k}$, we have $t_i > t/2$, and the engulfing property of sections in Theorem 9.10 gives

(9.45) $$\bigcup_{i=1}^{N} S(x_i, t_i) \subset S(z, \theta_* t).$$

For any $l > i$, as $x_l \notin S(x_i, t_i)$, we obtain from the separating property of sections in Theorem 9.10 that $S(x_l, \frac{t_i}{\theta_*^2}) \cap S(x_i, \frac{t_i}{\theta_*^2}) = \emptyset$. Thus, recalling

$t_i > t/2$, we have

(9.46) $S\left(x_l, \dfrac{t_l}{2\theta_*^2}\right) \cap S\left(x_i, \dfrac{t_i}{2\theta_*^2}\right) = \emptyset, \quad S\left(x_l, \dfrac{t}{2\theta_*^2}\right) \cap S\left(x_i, \dfrac{t}{2\theta_*^2}\right) = \emptyset.$

Combining this fact with (9.45), we obtain

(9.47)
$$\sum_{i=1}^{N} \left| S\left(x_i, \frac{t}{2\theta_*^2}\right) \right| \le |S(z, \theta_* t)|.$$

Applying Corollary 9.9 with $K = 2\theta_*^3$, we can find $\bar{\kappa}(n, \lambda, \Lambda, \rho) > 0$ such that

$$|S(z, \theta_* t)| \le \bar{\kappa} \left| S\left(x_i, \frac{t}{2\theta_*^2}\right) \right| \quad \text{for all } i = 1, \ldots, N.$$

Therefore, from (9.47), we obtain $N \le \bar{\kappa}$.

Step 2. The family $\mathcal{F}'_k = \{S(x_i^k, t_i^k)\}_{i=1}^{\infty}$ is actually finite.

Indeed, by Step 1, we have $\sum_i \chi_{S(x_i^k, t_i^k)} \le \bar{\kappa}$. Integrating over Ω, we get

$$\sum_i |S(x_i^k, t_i^k)| \le \bar{\kappa} |\Omega|.$$

Recall that $M/2^{k+1} < t_i^k$. Therefore, for $t := \min\{M/2^{k+1}, c_0\}$, we have from the above inequality and Corollary 9.9 that

$$\sum_i c_1 t^{n/2} \le \sum_i |S(x_i^k, t)| \le \bar{\kappa} |\Omega|.$$

Thus, the number of terms in the sum is finite, and Step 2 is proved.

From Step 2 and our construction, we get $E_k \subset \bigcup_{i=1}^{\infty} S(x_i^k, t_i^k)$, and thus property (i) is valid. Since each generation $\mathcal{F}'_k$ has a finite number of sections, by relabeling the indices of all sections of all generations $\mathcal{F}'_k$, we obtain property (ii).

Next, we prove property (iii). Let $x_i \ne x_j$. If $S(x_i, t_i)$ and $S(x_j, t_j)$ belong to the same generation, then $S(x_i, t_i/(2\theta_*^2)) \cap S(x_j, t_j/(2\theta_*^2)) = \emptyset$ by (9.46). Now, suppose $S(x_i, t_i) \in \mathcal{F}'_k$ and $S(x_j, t_j) \in \mathcal{F}'_{k+m}$ for some $m \ge 1$. Then, by construction, $x_j \notin S(x_i, t_i)$, so $S(x_j, t_i/\theta_*^2) \cap S(x_i, t_i/\theta_*^2) = \emptyset$ by the separating property in Theorem 9.10. Since $t_i > t_j$, this gives $S(x_j, t_j/\theta_*^2) \cap S(x_i, t_i/\theta_*^2) = \emptyset$ and so property (iii) is proved.

For property (iv), we estimate the overlapping of sections belonging to different generations. Let $0 < \varepsilon < 1$ and

(9.48)
$$z \in \bigcap_i S\big(x_{r_i}^{m_i}, (1 - \varepsilon) t_{r_i}^{m_i}\big),$$

where $m_1 < m_2 < \cdots$ are positive integers (the generations) and the heights $t_{r_i}^{m_i}$ satisfy $M 2^{-(m_i+1)} < t_{r_i}^{m_i} \le M 2^{-m_i}$. To simplify notation, set $x_i = x_{r_i}^{m_i}$, $t_i = t_{r_i}^{m_i}$, and $\alpha := 2\theta_*^2$. Our aim is to show that the number of sections in

(9.48) is not more than $C \log \frac{1}{\varepsilon}$. We only need to consider $\varepsilon < 1 - \frac{1}{\alpha}$ since otherwise the sections are disjoint by property (iii).

Step 3. We show that if $j > i$, then

$$(9.49) \qquad\qquad m_j - m_i \le C(\rho, \lambda, \Lambda, n) \log \frac{1}{\varepsilon}.$$

In particular, the number of members in (9.48) is at most $C \log \frac{1}{\varepsilon}$, which together with Step 1 gives property (iv).

To prove (9.49), we first note from $z \in S(x_j, (1 - \varepsilon)t_j)$ and the engulfing property of sections in Theorem 9.10 that

$$(9.50) \qquad\qquad S(x_j, (1 - \varepsilon)t_j) \subset S(z, \theta_*(1 - \varepsilon)t_j).$$

On the other hand, since $z \in S(x_i, (1 - \varepsilon)t_i)$, by the inclusion property of sections in Theorem 9.12 (with $r = 1 - \varepsilon, s = 1$), we can find $\hat{c}_0 > 0$ and $p_1 > 1$ depending on $n, \rho, \lambda, \Lambda$ such that $S(z, \hat{c}_0 \varepsilon^{p_1} t_i) \subset S(x_i, t_i)$.

By construction, $x_j \notin S(x_i, t_i)$, so (9.50) gives $\theta_*(1 - \varepsilon)t_j \ge \hat{c}_0 \varepsilon^{p_1} t_i$. Hence,

$$\left(\frac{1}{\varepsilon}\right)^{p_1} \ge \frac{\hat{c}_0}{\theta_*} \frac{t_i}{t_j} > \frac{\hat{c}_0}{2\theta_*} 2^{m_j - m_i}$$

which implies (9.49). Thus the proof of the lemma is complete. $\qquad\square$

Using Lemma 9.14, we can prove the following covering theorem.

Theorem 9.15 (Covering theorem for boundary sections). *Assume that Ω and u satisfy* (9.24)–(9.26). *Let $\mathcal{O} \subset \Omega$ be open and let $\varepsilon \in (0, 1)$. Suppose that for each $x \in \mathcal{O}$, a section $S_u(x, t_x)$ is given with*

$$\frac{|S_u(x, t_x) \cap \mathcal{O}|}{|S_u(x, t_x)|} = \varepsilon.$$

Then, if $\sup \{t_x : x \in \mathcal{O}\} < +\infty$, there exists a countable subfamily of sections $\{S_u(x_k, t_k)\}_{k=1}^{\infty} \subset \{S_u(x, t_x)\}_{x \in \mathcal{O}}$ satisfying

(i) $\mathcal{O} \subset \bigcup_{k=1}^{\infty} S_u(x_k, t_k)$ *and*

(ii) $|\mathcal{O}| \le C(n, \rho, \lambda, \Lambda) \sqrt{\varepsilon} \left| \bigcup_{k=1}^{\infty} S_u(x_k, t_k) \right|.$

Proof. Let $0 < \delta < 1/2$, to be chosen later. By applying Lemma 9.14 to the family $\mathcal{F} := \{S_u(x, t_x)\}_{x \in \mathcal{O}}$, we can find a countable subfamily, denoted by $\{S_u(x_k, t_k)\}_{k=1}^{\infty}$, such that $\mathcal{O} \subset \bigcup_{k=1}^{\infty} S_u(x_k, t_k)$ (so property (i) is valid), and

$$(9.51) \qquad\qquad \sum_{k=1}^{\infty} \chi_{S_u(x_k, (1-\delta)t_k)} \le K(n, \rho, \lambda, \Lambda) \log \frac{1}{\delta}.$$

To simplify, we write S_k for $S_u(x_k, t_k)$. Then, by our assumptions,

$$|\mathcal{O}| = \left| \mathcal{O} \cap \bigcup_{k=1}^{\infty} S_k \right| = \lim_{N \to \infty} \left| \mathcal{O} \cap \bigcup_{k=1}^{N} S_k \right|$$

$$\leq \limsup_{N \to \infty} \sum_{k=1}^{N} |\mathcal{O} \cap S_k| = \varepsilon \limsup_{N \to \infty} \sum_{k=1}^{N} |S_k|.$$

Moreover, by the doubling property in Corollary 9.9, we get

$$|S_k| \leq C \left| S_u(x_k, t_k/2) \right| \leq C \left| S_u\big(x_k, (1-\delta)t_k\big) \right|,$$

where $C = C(n, \rho, \lambda, \Lambda)$. Therefore,

$$(9.52) \qquad |\mathcal{O}| \leq C\varepsilon \limsup_{N \to \infty} \sum_{k=1}^{N} \left| S_u\big(x_k, (1-\delta)t_k\big) \right|.$$

Denote the overlapping function for the family $\{ S_u\big(x_k, (1-\delta)t_k\big) \}_{k=1}^{N}$ by

$$n_N^{\delta}(x) := \begin{cases} \#\{ k : x \in S_u\big(x_k, (1-\delta)t_k\big) \} & \text{if } x \in \bigcup_{k=1}^{N} S_u\big(x_k, (1-\delta)t_k\big), \\ 1 & \text{if } x \notin \bigcup_{k=1}^{N} S_u\big(x_k, (1-\delta)t_k\big). \end{cases}$$

Then $n_N^{\delta} \leq K \log \frac{1}{\delta}$ by (9.51), and

$$\sum_{k=1}^{N} \chi_{S_u\big(x_k, (1-\delta)t_k\big)} = n_N^{\delta} \chi_{\bigcup_{k=1}^{N} S_u\big(x_k, (1-\delta)t_k\big)}.$$

Hence, integrating over Ω, we obtain

$$\sum_{k=1}^{N} \left| S_u\big(x_k, (1-\delta)t_k\big) \right| \leq K \log \frac{1}{\delta} \left| \bigcup_{k=1}^{N} S_u\big(x_k, (1-\delta)t_k\big) \right|.$$

We infer from this and (9.52) that

$$|\mathcal{O}| \leq CK\varepsilon \log \frac{1}{\delta} \left| \bigcup_{k=1}^{\infty} S_u\big(x_k, (1-\delta)t_k\big) \right| \quad \text{for all } 0 < \delta < 1/2.$$

By choosing $\delta > 0$ such that $\log \frac{1}{\delta} = 2\varepsilon^{-1/2}$, we obtain property (ii). □

Now, we establish some global strong-type (p, p) estimates for the maximal function with respect to sections.

Theorem 9.16 (Global strong-type (p, p) estimates for the maximal function with respect to sections). *Assume that Ω and u satisfy* (9.24)–(9.26). *For $f \in L^1(\Omega)$, define the maximal function*

$$\mathcal{M}(f)(x) := \sup_{t > 0} \frac{1}{|S_u(x, t)|} \int_{S_u(x,t)} |f| \, dy \quad \text{for all } x \in \Omega.$$

Then, the following hold:

(i) (Weak-type $(1,1)$ estimate) *For all $\beta > 0$, we have*

$$\left|\{x \in \Omega : \mathcal{M}(f)(x) > \beta\}\right| \leq \frac{C(n, \rho, \lambda, \Lambda)}{\beta} \int_{\Omega} |f| \, dy.$$

(ii) (Strong-type (p,p) estimate) *For any $1 < p < \infty$, we have*

$$\|\mathcal{M}(f)\|_{L^p(\Omega)} \leq C(p, \rho, \lambda, \Lambda, n) \|f\|_{L^p(\Omega)}.$$

When $u(x) = |x|^2/2$, $\mathcal{M}$ is the classical Hardy-Littlewood maximal operator.

Proof. The proof is based on Besicovitch's Covering Lemma. Let $E_\beta := \{x \in \Omega : \mathcal{M}(f)(x) > \beta\}$ and let M be the constant in Lemma 9.7. Then, clearly

$$\mathcal{M}(f)(x) = \sup_{t \leq M} \frac{1}{|S_u(x,t)|} \int_{S_u(x,t)} |f| \, dy \quad \text{for all } x \in \Omega.$$

Therefore for each $x \in E_\beta$, we can find $t_x \leq M$ satisfying

(9.53) $$\frac{1}{|S_u(x,t_x)|} \int_{S_u(x,t_x)} |f| \, dy \geq \beta.$$

Consider the family $\{S_u(x, 2t_x)\}_{x \in E_\beta}$. Then by Lemma 9.14, there exists a countable subfamily $\{S_u(x_k, 2t_k)\}_{k=1}^{\infty}$ such that $E_\beta \subset \bigcup_k S_u(x_k, 2t_k)$ and

$$\sum_{k=1}^{\infty} \chi_{S_u(x_k, (1-\varepsilon)2t_k)} \leq K(n, \rho, \lambda, \Lambda) \log \frac{1}{\varepsilon}$$

for every $0 < \varepsilon < 1/2$. We can fix $\varepsilon = 1/4$ for example.

From the doubling property in Corollary 9.9 and (9.53), we obtain

$$|E_\beta| \leq \sum_k |S_u(x_k, 2t_k)| \leq C \sum_k |S_u(x_k, t_k)|$$

$$\leq \frac{C}{\beta} \sum_k \int_{S_u(x_k, (1-\varepsilon)2t_k)} |f| \, dy$$

$$= \frac{C}{\beta} \int_{\Omega} \sum_k \chi_{S_u(x_k, (1-\varepsilon)2t_k)} |f| \, dy \leq \frac{CK \log \frac{1}{\varepsilon}}{\beta} \int_{\Omega} |f| \, dy.$$

Thus we have proved the weak-type $(1,1)$ estimate in assertion (i).

Now we prove assertion (ii) for $f \in L^p(\Omega)$ where $1 < p < \infty$. Define

$$f_1(x) := \begin{cases} f(x) & \text{if } |f(x)| \geq \beta/2, \\ 0 & \text{otherwise.} \end{cases}$$

Then $f_1 \in L^1(\Omega)$,

$$|f| \le |f_1| + \beta/2, \quad \mathcal{M}(f) \le \mathcal{M}(f_1) + \beta/2,$$

and therefore

$$E_\beta := \{x \in \Omega : \mathcal{M}(f)(x) > \beta\} \subset \{x \in \Omega : \mathcal{M}(f_1)(x) > \beta/2\},$$

which, thanks to assertion (i), gives

$$|E_\beta| \le \frac{2C \log \frac{1}{\varepsilon}}{\beta} \|f_1\|_{L^1(\Omega)} = \frac{2C \log \frac{1}{\varepsilon}}{\beta} \int_{\{y \in \Omega : |f(y)| > \beta/2\}} |f|\, dy.$$

By the layer-cake formula and interchanging the orders of integration, we have

$$\int_\Omega |\mathcal{M}(f)|^p\, dx = p \int_0^\infty \beta^{p-1} |E_\beta|\, d\beta$$

$$\le p \int_0^\infty \beta^{p-1} \left(\frac{2C \log \frac{1}{\varepsilon}}{\beta} \int_{\{|f(y)| > \beta/2\}} |f(y)|\, dy \right) d\beta$$

$$= 2pC \log \frac{1}{\varepsilon} \int_\Omega |f(y)| \left(\int_0^{2|f(y)|} \beta^{p-2}\, d\beta \right) dy$$

$$= \frac{2^p pC \log \frac{1}{\varepsilon}}{p-1} \|f\|_{L^p(\Omega)}^p.$$

Thus, the strong-type (p, p) estimate in assertion (ii) is proved. $\qquad\square$

Remark 9.17. Theorem 9.16 asserts that the maximal operator $\mathcal{M}$ maps $L^p(\Omega)$ to $L^p(\Omega)$ when $1 < p < \infty$. It would be very interesting to investigate where $\mathcal{M}$ maps $W^{1,p}(\Omega)$ to $W^{1,p}(\Omega)$. This is indeed the case of the classical Hardy-Littlewood maximal operator corresponding to $u(x) = |x|^2/2$, as showed by Kinnunen and Lindqvist [**KL**].

Since the engulfing property of sections of the Monge–Ampère equation is an affine invariant version of the triangle inequality, it is of interest to look deeper into the metric aspects of sections. In fact, quasi-distances generated by solutions of the Monge–Ampère equation will be convenient tools in proving the boundary Harnack inequality for the linearized Monge–Ampère equation in Section 13.5.

The following lemma is concerned with metric properties of sections.

Lemma 9.18 (Metric properties of sections). *Assume that Ω and u satisfy (9.24)–(9.26). Let θ_* be as in Theorem 9.10. Then, the sections of u have the following properties:*

(i) $\bigcap_{r>0} S_u(x, r) = \{x\}$ *for every* $x \in \overline{\Omega}$.

(ii) $\bigcup_{r>0} S_u(x, r) = \overline{\Omega}$ *for every* $x \in \overline{\Omega}$.

(iii) *For each $x \in \overline{\Omega}$, the map $r \mapsto S_u(x, r)$ is nondecreasing in r.*

(iv) *If $y \in S_u(x, r)$, then*

$$S_u(x, r) \subset S_u(y, \theta_*^2 r) \quad and \quad S_u(y, r) \subset S_u(x, \theta_*^2 r).$$

Proof. Clearly, property (ii) follows from Lemma 9.7 and property (iii) is obvious, while property (iv) is a consequence of the engulfing property in Theorem 9.10. To verify property (i), it suffices to show that $\bigcap_{r>0} S_u(x, r) \subset \{x\}$. If x is a boundary point of $\overline{\Omega}$, then, by (9.11), we have

$$\bigcap_{r>0} S_u(x, r) \subset \bigcap_{0<r<1/2} B_{Cr^{1/2}|\log r|}(x) = \{x\}.$$

If x is an interior point of Ω, then property (i) follows from u being strictly convex in Ω as remarked before the statement of Lemma 9.7. $\qquad \square$

In the following theorem, we introduce a quasi-distance d induced by sections of solutions u to the Monge–Ampère equation in Ω. Moreover, if Ω and u satisfy (9.24)–(9.26), then $(\overline{\Omega}, d, |\cdot|)$ is a space of homogeneous type where $|\cdot|$ denotes the n-dimensional Lebesgue measure restricted to $\overline{\Omega}$.

Theorem 9.19 (Quasi-metric space). *Assume that the convex domain Ω and the convex function u satisfy (9.24)–(9.26). Let $\theta_* > 1$ be the engulfing constant in Theorem 9.10. We define a function $d : \overline{\Omega} \times \overline{\Omega} \longrightarrow [0, \infty)$ by*

$$d(x, y) := \inf \{r > 0 : x \in S_u(y, r) \text{ and } y \in S_u(x, r)\} \quad for \ all \ x, y \in \overline{\Omega}.$$

The induced d-ball with center $x \in \overline{\Omega}$ and radius $r > 0$ is defined by

$$B_d(x, r) := \{y \in \overline{\Omega} : d(x, y) < r\}.$$

Then, the following hold.

(i) *$d(x, y) = d(y, x)$ for all $x, y \in \overline{\Omega}$.*

(ii) *$d(x, y) = 0$ if and only if $x = y$.*

(iii) *$d(x, y) \leq \theta_*^2 \left[d(x, z) + d(z, y) \right]$ for all $x, y, z \in \overline{\Omega}$.*

(iv) *$S_u(x, r/(2\theta_*^2)) \subset B_d(x, r) \subset S_u(x, r)$ for all $x \in \overline{\Omega}$ and $r > 0$.*

Proof. From Lemma 9.18(ii), we deduce that d is defined on all of $\overline{\Omega} \times \overline{\Omega}$. Note that the symmetry result in property (i) is a direct consequence of the definition of d.

Let us prove property (ii). It is clear that $d(x, x) = 0$. Assume now $d(x, y) = 0$. Then, we have $y \in S_u(x, \frac{1}{k})$ for all positive integers k. Using Lemma 9.18(i), we get $x = y$.

We now prove property (iii). Let $r = d(x, z)$ and $s = d(z, y)$. Then for any $\varepsilon > 0$, we have $y \in S_u(z, s + \varepsilon) \subset S_u(z, \theta_*(r + s + \varepsilon))$ and $z \in S_u(x, r + \varepsilon)$. Thus, by Theorem 9.10, we have

$$S_u(x, r + \varepsilon) \subset S_u(z, \theta_*(r + \varepsilon)) \subset S_u(z, \theta_*(r + s + \varepsilon)).$$

Again, by Theorem 9.10, we have

$$S_u(z, \theta_*(r + s + \varepsilon)) \subset S_u(x, \theta_*^2(r + s + \varepsilon)).$$

It follows that $y \in S_u(x, \theta_*^2(r + s + \varepsilon))$. This and symmetry give $d(x, y) \leq \theta_*^2(r + s + \varepsilon)$ and property (iii) follows by letting $\varepsilon \to 0$.

Finally, we verify property (iv). The second inclusion is a rephrasing of the definition of d. For the first inclusion, let $y \in S_u(x, \frac{r}{2\theta_*^2})$ where $x \in \overline{\Omega}$ and $r > 0$. We need to show that $y \in B_d(x, r)$. Indeed, by Lemma 9.18(iv), we have $x \in S_u(x, \frac{r}{2\theta_*^2}) \subset S_u(y, \frac{r}{2})$. Hence $d(x, y) \leq \frac{r}{2} < r$, so $y \in B_d(x, r)$. $\square$

As a consequence of Corollary 9.9 and Theorem 9.19, we obtain the following doubling property for d-balls: For all $x \in \overline{\Omega}$ and $r > 0$, we have

$$|B_d(x, 2r)| \leq |S_u(x, 2r)| \leq C\,|S(x, r/(2\theta_*^2))| \leq C\,|B_d(x, r)|,$$

where C depends only on ρ, λ, Λ, and n. Thus, $(\overline{\Omega}, d, |\cdot|)$ is a doubling quasi-metric space and hence it is a space of homogeneous type; see Coifman-Weiss [**CWs**]. This allows us to place the Monge–Ampère setting in a more general context where many real analytic problems have been studied; see Di Fazio-Gutiérrez-Lanconelli [**DFGL**].

9.6. Problems

Problem 9.1. Assume the hypotheses of Theorem 9.5 are satisfied. Assume furthermore that $\det D^2 u = f$ in Ω where $f \in C(\overline{\Omega})$. Let $\alpha \in (0, 1)$. Show that there is a constant C depending only on $n, \alpha, \rho, \lambda, \Lambda$, and the modulus of continuity of f in $\overline{\Omega}$ such that

$$[Du]_{C^\alpha(\overline{\Omega})} \leq C.$$

Problem 9.2. Assume the hypotheses of Theorem 9.5 are satisfied and furthermore $\det D^2 u \in C^{k,\beta}(\overline{\Omega})$ where $k \in \mathbb{N}_0$ and $\beta \in (0, 1)$. Show that

$$\|D^{k+2}u\| \leq K|\log \operatorname{dist}(\cdot, \partial\Omega)|^{k+2}\operatorname{dist}^{-k}(\cdot, \partial\Omega) \quad \text{near } \partial\Omega,$$

where K is a constant depending on $n, k, \beta, \rho, \lambda, \Lambda$, and $\|\det D^2 u\|_{C^{k,\beta}(\overline{\Omega})}$.

Problem 9.3. Assume that the sections $\{S_u(x, t)\}_{x \in \overline{\Omega}, t > 0}$ of a convex function u defined on a convex domain Ω satisfy the separating property with the constant θ. Show that the sections $\{S_u(x, t)\}_{x \in \overline{\Omega}, t > 0}$ satisfy the engulfing property with the constant θ^2. This together with Theorem 9.10(ii) implies that the engulfing property of sections is equivalent to the separating property of sections.

Problem 9.4. Give details of the proof of Theorem 9.12(ii) for the case $t \leq c\delta/(4\theta_*)$.

Problem 9.5. Give details of the proof of Theorem 9.13(ii).

Problem 9.6. Assume that the convex domain Ω and the convex function u satisfy (9.24)–(9.26). Prove a boundary version of Theorem 5.31 for sections of u.

9.7. Notes

The results in this chapter are boundary versions of those established by Caffarelli and Gutiérrez [**CG1**, **CG2**], Gutiérrez and Huang [**GH2**], and Aimar, Forzani, and Toledano [**AFT**] for interior sections (see also Figalli [**F2**, Sections 4.1, 4.2, 4.5, and 4.7] and Gutiérrez [**G2**, Sections 3.2, 3.3, and 6.5]). The results in this chapter are crucial for the boundary analysis of the linearized Monge–Ampère equation in Chapters 13 and 14. They also provide key tools for the global $W^{2,p}$ and $W^{1,p}$ estimates for the linearized Monge–Ampère equation in Le–Nguyen [**LN2**, **LN3**]. These are boundary versions of interior estimates obtained in Gutiérrez–Nguyen [**GN1**, **GN2**].

Section 9.1 is based on Savin [**S5**] and Le–Savin [**LS1**].

In Section 9.2, the global Hölder gradient estimates for solutions to the Monge–Ampère equation in Theorems 9.5 and 9.6 are due to Le and Savin [**LS1**]. Savin and Zhang [**SZ1**] established these estimates under optimal conditions when the domain is uniformly convex or has flat boundary. When the domain is uniformly convex, it suffices to have $C^{2,\beta}$ boundary and boundary data for the global Hölder gradient estimates when the Monge–Ampère measure is bounded between two positive constants. Recently, Caffarelli, Tang, and Wang [**CTW**] established these estimates when the Monge–Ampère measure is doubling and bounded from above, which allows for degeneracy and the case of zero right-hand side.

Sections 9.3 and 9.5 are from Le–Nguyen [**LN1**].

Section 9.4 is from Le [**L4**].

Related to Problem 9.1, see Theorem 10.2. Problem 9.2 is based on Le–Savin [**LS1**] and Savin [**S5**]. Problem 9.3 is taken from Le–Nguyen [**LN1**].

Chapter 10

Boundary Second Derivative Estimates

This chapter is concerned with fundamental boundary second derivative estimates for Aleksandrov solutions to the Monge–Ampère equation with Monge–Ampère measure bounded between two positive constants under suitable assumptions on the domain in $\mathbb{R}^n$ ($n \geq 2$) and the boundary data. It is the culmination of tools and techniques developed in previous chapters. We will use Savin's Boundary Localization Theorem to prove the global $W^{2,1+\varepsilon}$, $W^{2,p}$, and $C^{2,\alpha}$ estimates, together with the pointwise $C^{2,\alpha}$ estimates at the boundary. They are boundary versions of the interior regularity results in Chapter 6. We will present Wang's examples to show that the conditions imposed on the boundary are in fact optimal.

10.1. Global Second Derivative Sobolev Estimates

In this section, we prove Savin's global $W^{2,1+\varepsilon}$ and $W^{2,p}$ estimates for the Monge–Ampère equation [**S5**]. The technique here combines corresponding interior estimates and a covering argument using maximal interior sections with the aid of the Boundary Localization Theorem (Theorem 8.5).

Theorem 10.1 (Global $W^{2,1+\varepsilon}$ estimates). *Let Ω be a uniformly convex domain in $\mathbb{R}^n$ ($n \geq 2$). Let $u \in C(\overline{\Omega})$ be a convex function satisfying*

$$0 < \lambda \leq \det D^2 u \leq \Lambda \quad in \; \Omega,$$

in the sense of Aleksandrov. Assume that $u|_{\partial\Omega}$ and $\partial\Omega$ are of class C^3. Then, there exist constants $\varepsilon = \varepsilon(n, \lambda, \Lambda) > 0$ and K depending only on

n, λ, Λ, $\|u\|_{C^3(\partial\Omega)}$, *the uniform convexity radius of* $\partial\Omega$, *and the* C^3 *regularity of* $\partial\Omega$ *such that*

$$\|D^2 u\|_{L^{1+\varepsilon}(\Omega)} \leq K.$$

Proof. By Proposition 4.7, u and Ω satisfy the hypotheses of the Boundary Localization Theorem (Theorem 8.5) at every boundary point on $\partial\Omega$ with small positive constants $\mu = \rho$ depending only on n, λ, Λ, $\|u\|_{C^3(\partial\Omega)}$, the C^3 regularity of $\partial\Omega$, and the uniform convexity radius of $\partial\Omega$. By Proposition 8.24, $u \in C^1(\overline{\Omega})$. We divide the proof into two steps.

Step 1. L^p estimate for the Hessian in maximal interior sections. Consider the maximal interior section $S_u(y, h)$ of u centered at $y \in \Omega$ with height

$$h := \bar{h}(y) = \sup\{t > 0 : S_u(y, t) \subset \Omega\}.$$

If $h \leq c$ where $c = c(n, \lambda, \Lambda, \rho)$ is small, then Proposition 9.2 gives

$$\kappa_0 E_h \subset S_u(y, h) - y \subset \kappa_0^{-1} E_h,$$

where E_h is an ellipsoid in $\mathbb{R}^n$ given by

$$E_h := h^{1/2} A_h^{-1} B_1(0) \quad \text{with } \det A_h = 1, \quad \|A_h\| + \|A_h^{-1}\| \leq C |\log h|.$$

Here κ_0 and C depend on $n, \lambda, \Lambda, \rho$. We use the following rescalings:

$$u_h(x) := h^{-1}\big[u(y + h^{1/2} A_h^{-1} x) - u(y) - Du(y) \cdot (h^{1/2} A_h^{-1} x) - h\big],$$

for $x \in \Omega_h := h^{-1/2} A_h(\Omega - y)$. Then

$$B_{\kappa_0}(0) \subset S_{u_h}(0, 1) \equiv h^{-1/2} A_h\big(S_u(y, h) - y\big) \subset B_{\kappa_0^{-1}}(0).$$

We have

$$\lambda \leq \det D^2 u_h \leq \Lambda \quad \text{in } S_{u_h}(0, 1), \qquad u_h = 0 \quad \text{on } \partial S_{u_h}(0, 1).$$

By the interior $W^{2,p}$ estimates in Theorem 6.11 (see also Corollary 6.26),

$$\int_{S_{u_h}(0,1/2)} |D^2 u_h|^p \, dx \leq C, \quad p = 1 + \varepsilon,$$

where $\varepsilon(n, \lambda, \Lambda) > 0$ and $C(n, \lambda, \Lambda, \kappa_0) > 0$. Since

$$D^2 u(y + h^{1/2} A_h^{-1} x) = (A_h)^t \, D^2 u_h(x) \, A_h,$$

we obtain

$$\int_{S_u(y,h/2)} |D^2 u(z)|^p \, dz = h^{\frac{n}{2}} \int_{S_{u_h}(0,1/2)} |A_h^t \, D^2 u_h(x) \, A_h|^p \, dx$$

(10.1)
$$\leq C \, h^{\frac{n}{2}} |\log h|^{2p} \int_{S_{u_h}(0,1/2)} |D^2 u_h(x)|^p \, dx$$

$$\leq C \, h^{\frac{n}{2}} |\log h|^{2p}.$$

From Proposition 9.2, we find that if $y \in \Omega$ with $\bar{h}(y) \leq c$ small, then

$$S_u(y, \bar{h}(y)) \subset y + \kappa_0^{-1} E_h \subset D_{C\bar{h}(y)^{1/2}} := \{x \in \overline{\Omega} : \mathrm{dist}(x, \partial\Omega) \leq C\bar{h}(y)^{1/2}\},$$

where $C = 2\kappa_0^{-2}$.

Step 2. A covering argument. By Lemma 5.32, there exists a covering $\bigcup_{i=1}^{\infty} S_u(y_i, \bar{h}(y_i)/2)$ of Ω where the sections $S_u(y_i, \delta\bar{h}(y_i))$ are disjoint for some $\delta = \delta(n, \lambda, \Lambda) < 1/2$. We have

$$(10.2) \qquad \int_{\Omega} |D^2 u|^p \, dx \leq \sum_{i=1}^{\infty} \int_{S_u(y_i, \bar{h}(y_i)/2)} |D^2 u|^p \, dx.$$

We will estimate the sum in (10.2), depending on the heights $\bar{h}(y_i)$. Note that, by Lemma 5.6, there exists a constant $c_0 = c_0(n, \lambda, \Lambda) > 0$ such that

$$(10.3) \qquad |S_u(y_i, \delta\bar{h}(y_i))| \geq c_0 \bar{h}(y_i)^{n/2}.$$

Step 2(a). Sections with large height. In view of (10.3), the number of sections $S_u(y_i, \bar{h}(y_i))$ with $\bar{h}(y_i) \geq c$ is bounded by $K(n, \lambda, \Lambda, \rho)$. Together with the interior $W^{2,p}$ estimates in Theorem 6.11, we infer that

$$\sum_{\bar{h}(y_i) \geq c} \int_{S_u(y_i, \bar{h}(y_i)/2)} |D^2 u|^p \, dx \leq C(n, \lambda, \Lambda, \rho).$$

Step 2(b). Sections with small height. Now, for $d \leq c$, we consider the family $\mathcal{F}_d$ of indices i for sections $S_u(y_i, \bar{h}(y_i)/2)$ such that $d/2 < \bar{h}(y_i) \leq d$. Let M_d be the number of indices in $\mathcal{F}_d$. Since $S_u(y_i, \delta\bar{h}(y_i)) \subset D_{Cd^{1/2}}$ are disjoint for $i \in \mathcal{F}_d$, we find from (10.3) that

$$M_d c_0 (d/2)^{n/2} \leq \sum_{i \in \mathcal{F}_d} |S_u(y_i, \delta\bar{h}(y_i))| \leq |D_{Cd^{1/2}}| \leq C_* d^{1/2}$$

where $C_* > 0$ depends only on n, ρ, and the C^2 regularity of $\partial\Omega$. Therefore

$$(10.4) \qquad M_d \leq C_b(\rho, n, \lambda, \Lambda, \Omega) d^{\frac{1}{2} - \frac{n}{2}}.$$

It follows from (10.1) and (10.4) that

$$\sum_{i \in \mathcal{F}_d} \int_{S_u(y_i, \bar{h}(y_i)/2)} |D^2 u|^p \, dx \leq C M_d d^{\frac{n}{2}} |\log d|^{2p} \leq C d^{\frac{1}{2}} |\log d|^{2p}.$$

Adding these inequalities for $d = c2^{-k}$ where $k = 0, 1, 2, \ldots$, we obtain

$$\sum_{\bar{h}(y_i) \leq c} \int_{S_u(y_i, \bar{h}(y_i)/2)} |D^2 u|^p \, dx = \sum_{k=0}^{\infty} \sum_{i \in \mathcal{F}_{c2^{-k}}} \int_{S_u(y_i, \bar{h}(y_i)/2)} |D^2 u|^p \, dx$$

$$\leq \sum_{k=0}^{\infty} C(c2^{-k})^{\frac{1}{2}} |\log(c2^{-k})|^{2p} \leq C,$$

where $C = C(n, \lambda, \Lambda, \rho)$. Combining this with Step 2(a) and recalling (10.2), we obtain the desired global L^p estimate for D^2u. $\square$

When the Monge–Ampère measure of the function u in Theorem 10.1 has small oscillations, the exponent ε there can be taken to be large. This is the content of our next theorem concerning the global $W^{2,p}$ estimates for the Monge–Ampère equation. We omit its proof since it is similar to that of Theorem 10.1 where Theorem 6.11 is now replaced by the interior $W^{2,p}$ estimates in Theorem 6.13 and Corollary 6.14.

Theorem 10.2 (Global $W^{2,p}$ estimates). *Let Ω be a uniformly convex domain in $\mathbb{R}^n$ $(n \geq 2)$. Let $u \in C(\overline{\Omega})$ be a convex Aleksandrov solution of*

$$\det D^2 u = f \geq 0 \quad in\ \Omega.$$

Assume that $u|_{\partial\Omega}$ and $\partial\Omega$ are of class C^3.

(i) *Let $p > 1$. Then, there exist $\delta = e^{-C(n)p}$ and a constant K depending only on $n, p, \|u\|_{C^3(\partial\Omega)}$, the C^3 regularity of $\partial\Omega$, and the uniform convexity radius of $\partial\Omega$ such that*

$$\|D^2 u\|_{L^p(\Omega)} \leq K,$$

provided that

$$\|f - 1\|_{L^\infty(\Omega)} \leq \delta.$$

(ii) *Assume that $f \in C(\overline{\Omega})$ and $0 < \lambda \leq f \leq \Lambda$. Then for all $p \in (1, \infty)$, there exists a constant M depending only on n, p, $\|u\|_{C^3(\partial\Omega)}$, the C^3 regularity of $\partial\Omega$, the uniform convexity radius of $\partial\Omega$, and the modulus of continuity of f in $\overline{\Omega}$ such that*

$$\|D^2 u\|_{L^p(\Omega)} \leq M.$$

We will use the global $W^{2,1+\varepsilon}$ estimates in establishing global integrability estimates for the gradient of the Green's function of the linearized Monge–Ampère operator in Section 14.5 and in solving fourth-order singular Abreu equations in Section 15.5.

10.2. Wang's Counterexamples

In Theorems 10.1 and 10.2, when establishing global $W^{2,1+\varepsilon}$ and $W^{2,p}$ estimates for the Monge–Ampère equation, we imposed the C^3 regularity on the domain boundary and boundary data. In this section, we present Wang's counterexamples [**W4**] to show that these regularity assumptions are in some sense optimal.

Note that the convex function $u(x_1, x_2) = x_1^2 x_2^{1/2} + 21 x_2^{3/2}$ on a smooth, uniformly convex domain $\Omega \subset \{x_2 > x_1^2\}$ with boundary portion $\{x_2 = x_1^2\}$

near the origin in $\mathbb{R}^2$ is a $C^\infty(\Omega)$ counterexample to global $W^{2,3}(\Omega)$ regularity when the boundary values are only in $C^{2,1}$. This example in Remark 4.6, where $30 \le \det D^2 u \le 32$, comes from setting $\alpha = 3$ in Proposition 6.28 which makes the boundary curve $x_2 = |x_1|^{\alpha-1}$ smooth and uniformly convex near the origin. We can use this function as a sort of subsolution to construction counterexamples with constant Monge–Ampère measures. The following lemma, due to Wang [**W4**], is key to the constructions.

Lemma 10.3. *Let Ω be a uniformly convex domain with C^2 boundary in $\mathbb{R}^2$ and $0 \in \partial\Omega$. Assume that for some $K \ge 1$,*

$$\Omega \subset \{x_2 > x_1^2/K\}, \quad \Omega \cap \{x_2 \le 1/K\} \subset \{x_2 < Kx_1^2\},$$

and $\varphi \in C(\overline{\Omega})$ satisfies

$$K^{-1}(|x_1|^3 + x_2^2) \le \varphi(x_1, x_2) \le K(|x_1|^3 + x_2^2) \quad \text{on } \partial\Omega.$$

Then, there exists $\varepsilon_0(K, \operatorname{diam}(\Omega)) > 0$ such that for all $\varepsilon \in (0, \varepsilon_0)$, the convex Aleksandrov solution $u \in C(\overline{\Omega})$ to the Dirichlet problem

$$(10.5) \qquad \det D^2 u = \varepsilon^2 \quad \text{in } \Omega, \qquad u = \varphi \quad \text{on } \partial\Omega$$

does not lie in $W^{2,3}(\Omega)$.

Proof. By Remark 5.10 and Theorem 6.5, the convex Aleksandrov solution $u \in C(\overline{\Omega})$ to (10.5) belongs to $C^\infty(\Omega)$.

Let $w(x_1, x_2) = \varepsilon_0(x_1^2 x_2^{1/2} + (K+1)x_2^{3/2})$. Then

$$\det D^2 w = \frac{3}{2}\varepsilon_0^2(K + 1 - x_1^2 x_2^{-1}) \ge \varepsilon_0^2 \quad \text{in } \Omega.$$

If $\varepsilon_0(K, \operatorname{diam}(\Omega)) > 0$ is small, then $w \le \varphi \le K(K^{3/2}x_2^{3/2} + x_2^2)$ on $\partial\Omega$. By the comparison principle in Theorem 3.21, we have for $\varepsilon \in (0, \varepsilon_0)$,

$$0 < w(x_1, x_2) \le u(x_1, x_2) \le K(K^{3/2}x_2^{3/2} + x_2^2) \quad \text{in } \Omega.$$

We prove the lemma by showing that $\int_\Omega (D_{22}u)^3 \, dx = +\infty$. For this, since $D_{11}u D_{22}u = (D_{12}u)^2 + \varepsilon^2 \ge \varepsilon^2$, it suffices to show $\int_\Omega (D_{11}u)^{-3} \, dx = +\infty$.

For any $t \in (0, K^{-1})$, let $(\psi^\pm(t), t)$ be the intersection points of $\partial\Omega$ with $\{x_2 = t\}$ so that $\psi^-(t) < 0 < \psi^+(t)$. Then

$$\sqrt{t/K} < |\psi^\pm(t)| < \sqrt{Kt}.$$

For any $x_2 \in (0, K^{-1})$, there exists $x_2^+ \in (\frac{1}{2}\psi^+(x_2), \psi^+(x_2))$ such that

$$D_1 u(x_2^+, x_2) = \frac{u(\psi^+(x_2), x_2) - u(\frac{1}{2}\psi^+(x_2), x_2)}{\psi^+(x_2) - \frac{1}{2}\psi^+(x_2)} > 0;$$

hence

$$|D_1 u(x_2^+, x_2)| \le \frac{2K^{5/2}x_2^{3/2}}{\sqrt{x_2/K}/2} = \alpha(K)x_2, \quad \text{where } \alpha(K) = 4K^3.$$

Similarly, there exists $x_2^- \in (\psi^-(x_2), \frac{1}{2}\psi^-(x_2))$ such that $D_1 u(x_2^-, x_2) < 0$ and

$$|D_1 u(x_2^-, x_2)| = \left| \frac{u(\psi^-(x_2), x_2) - u(\frac{1}{2}\psi^-(x_2), x_2)}{\psi^-(x_2) - \frac{1}{2}\psi^-(x_2)} \right| \leq \alpha(K) x_2.$$

By the convexity of u, $x_2^\pm$ are unique. Moreover,

$$x_2^+ - x_2^- \geq \frac{1}{2}(\psi^+(x_2) - \psi^-(x_2)) \geq \sqrt{x_2/K}.$$

The key point in the proof is to show that, for a fixed portion of the interval $[x_2^-, x_2^+]$, the second partial derivative $D_{11}u(\cdot, x_2)$ is proportional to $x_2^{1/2}$. This is a quantitative version of the failure of the quadratic separation property. To see this, let

$$S_{x_2} = \{ x_1 \in (x_2^-, x_2^+) : D_{11}u(x_1, x_2) < 4\alpha(K)\sqrt{K}x_2^{1/2} \}$$

and

$$L_{x_2} = (x_2^-, x_2^+) \setminus S_{x_2}.$$

Then

$$2\alpha(K)x_2 \geq D_1 u(x_2^+, x_2) - D_1 u(x_2^-, x_2) \geq \int_{L_{x_2}} D_{11}u(x_1, x_2)\, dx_1$$

$$\geq 4\alpha(K)\sqrt{K}x_2^{1/2}|L_{x_2}|.$$

Hence, $|L_{x_2}| \leq \sqrt{x_2/K}/2$, from which we deduce that

$$|S_{x_2}| \geq x_2^+ - x_2^- - |L_{x_2}| \geq \sqrt{x_2/K}/2.$$

Therefore,

$$\int_\Omega (D_{11}u)^{-3}\, dx \geq \int_0^{K^{-1}} dx_2 \int_{S_{x_2}} (D_{11}u)^{-3}\, dx_1$$

$$\geq \int_0^{K^{-1}} (4\alpha(K)\sqrt{K}x_2^{1/2})^{-3}|S_{x_2}|\, dx_2$$

$$\geq \int_0^{K^{-1}} \beta(K)x_2^{-1}\, dx_2 = +\infty,$$

where $\beta(K) > 0$. The lemma is proved. $\qquad\square$

From Lemma 10.3, we can construct explicit examples showing the failure of global $W^{2,p}$ estimates, at least for $p \geq 3$, for the Monge–Ampère equation with constant Monge–Ampère measure when either the boundary data or the domain boundary fails to be C^3. Below are Wang's examples.

Example 10.4. Let $\Omega = B_1(e_2) \subset \mathbb{R}^2$ and $\varphi(x_1, x_2) = |x_1|^3 + x_2^2$. Then, the solution to (10.5) for some small $\varepsilon > 0$ does not belong to $W^{2,3}(\Omega)$. In this example, $\partial\Omega \in C^\infty$ and $\varphi \in C^{2,1}(\overline{\Omega})$.

Example 10.5. Let Ω be a uniformly convex domain with C^2 boundary in $\mathbb{R}^2$ such that $B_1(e_2) \subset \Omega \subset B_2(e_2)$, and near the origin, $\partial\Omega$ is given by $x_2 = \frac{1}{2}x_1^2 - |x_1|^3$. Let $\varphi(x_1, x_2) = \frac{1}{2}x_1^2 + x_2^2 - x_2 + x_2^4$. Then, the solution to (10.5) for some small $\varepsilon > 0$ does not belong to $W^{2,3}(\Omega)$. In this example, $\partial\Omega \in C^{2,1}$ and $\varphi \in C^\infty(\overline{\Omega})$.

In the next three sections, we will prove boundary Pogorelov estimates (Section 10.3), pointwise $C^{2,\alpha}$ estimates at the boundary (Section 10.4), and global $C^{2,\alpha}$ estimates (Section 10.5) for the Monge–Ampère equation

$$\det D^2 u = f \quad \text{in } \Omega, \quad u = \varphi \quad \text{on } \partial\Omega,$$

where $f \in C(\overline{\Omega})$ and $f > 0$ in $\overline{\Omega}$, under sharp conditions on the right-hand side and boundary data. These important results were obtained by Savin [**S4**] and Trudinger and Wang [**TW5**].

10.3. Boundary Pogorelov Second Derivative Estimates

In the proof of the interior pointwise $C^{2,\alpha}$ estimates in Theorem 6.30, we saw the crucial role of the higher-order regularity estimates for the Monge–Ampère equation with constant right-hand side in Lemma 6.4 and Theorem 6.5 whose proofs are based on approximations and the Pogorelov second derivative estimates in Theorem 6.1. In this section, we will prove boundary versions of these Pogorelov-type estimates for domains lying in a half-space, say $\mathbb{R}^n_+$, and that the boundary data on $\{x_n = 0\}$ is a quadratic polynomial. They will be used in the proof of pointwise $C^{2,\alpha}$ estimates at the boundary for the Monge–Ampère equation in Section 10.4.

We could choose to prove boundary Pogorelov estimates for domains Ω with boundary having a flat part on $\{x_n = 0\}$. However, in practice, we need solutions to be sufficiently smooth up to the boundary to apply it. The smoothness is not given a priori but could be obtained via approximations as in the interior case of Theorem 6.5. Therefore, we will prove a version of boundary Pogorelov estimates for domains which are uniformly convex and sufficiently close to Ω. In these domains, by Theorem 4.12, Aleksandrov solutions with sufficiently smooth boundary data are smooth up to the boundary. Moreover, implicit in Lemma 6.4 is the fact that the domain Ω is a section of a convex solution to a Monge–Ampère equation. In Proposition 10.6, our domain is almost a boundary section while in Theorem 10.7 on the boundary Pogorelov estimates, our domain is actually a boundary section.

Proposition 10.6 (Boundary Pogorelov estimates)**.** *Assume that the convex domain Ω in $\mathbb{R}^n$ satisfies, for some $\kappa \in (0, 1/9)$ and small $\varepsilon > 0$,*

$$\begin{cases} B_{9\kappa}(0) \cap B_{\varepsilon^{-1}}(\varepsilon^{-1}e_n) \subset \Omega \subset B_{\kappa^{-1}/9}(0) \cap B_{\varepsilon^{-1}}(\varepsilon^{-1}e_n) \\ \text{and } B_1(0) \cap \partial B_{\varepsilon^{-1}}(\varepsilon^{-1}e_n) \subset \partial\Omega. \end{cases}$$

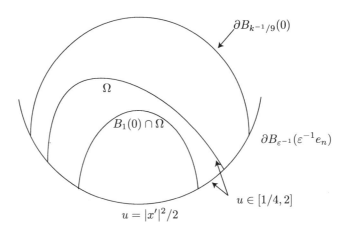

Figure 10.1. Boundary values of u in Proposition 10.6.

Assume the convex function $u \in C^4(\overline{\Omega})$ solves the Monge–Ampère equation

$$\det D^2 u = 1 \quad in \ \Omega, \quad u \geq -\varepsilon \quad in \ \Omega$$

and its boundary values satisfy (see Figure 10.1)

$$\begin{cases} u = \dfrac{1}{2}|x'|^2 & on \ B_1(0) \cap \partial B_{\varepsilon^{-1}}(\varepsilon^{-1}e_n) \subset \partial\Omega, \\ u \in [1/4, 2] & on \ the \ rest \ of \ \partial\Omega. \end{cases}$$

Then, for all ε and κ_0 small, depending only on n and κ, we have

$$\|u\|_{C^{3,1}(\{u < \kappa_0^2\})} \leq C(\kappa, n).$$

Proof. We divide the proof into several steps. Denote $B_r = B_r(0)$. Let ν_x be the outer unit normal to $\partial\Omega$ at $x \in \partial\Omega$. If $x \in \partial B_\kappa \cap \partial B_{\varepsilon^{-1}}(\varepsilon^{-1}e_n)$, then $|x'| = \kappa_\varepsilon := \kappa\sqrt{1 - \varepsilon^2\kappa^2/4}$ and $x_n = \varepsilon\kappa^2/2$.

Step 1. We show that $Du \cdot \nu \leq C(\kappa, n)$ on $\partial\Omega \cap \{x : |x'| \leq 2\kappa_\varepsilon\}$.

For each $x_0 \in \partial\Omega \cap \{x : |x'| \leq 2\kappa_\varepsilon\}$, consider the barrier

$$w_{x_0}(x) = \frac{1}{2}|x_0'|^2 + x_0' \cdot (x' - x_0') + \delta\{|x - x_0|^2 - |(x - x_0) \cdot \nu_{x_0}|^2\}$$
$$+ \delta^{1-n}\{|(x - x_0) \cdot \nu_{x_0}|^2 + 2\kappa^{-1}(x - x_0) \cdot \nu_{x_0}\},$$

where $\delta > 0$ is small, to be chosen later, so that $w_{x_0} \leq 1/4$ in $B_{\kappa^{-1}/9} \cap \Omega$.

Since $0 \geq (x - x_0) \cdot \nu_{x_0} \geq -2\kappa^{-1}$ in Ω, we have for $x \in B_{\kappa^{-1}/9} \cap \Omega$,

$$w_{x_0}(x) \leq x_0' \cdot x' - \frac{1}{2}|x_0|^2 + \delta|x - x_0|^2 \leq |x_0'|\kappa^{-1}/9 - \frac{1}{2}|x_0'|^2 + \delta\kappa^{-2}$$
$$\leq 2/9 - 2\kappa_\varepsilon^2 + \delta\kappa^{-2} \leq 1/4$$

if $\delta \leq \kappa^2/36$. To fix the idea, we take $\delta = \kappa^2/36$.

Using an orthogonal coordinate frame at x_0 containing ν_{x_0}, we compute
$$\det D^2 w_{x_0} = 2^n.$$
We now show that if $\varepsilon \leq \varepsilon_0 \leq \kappa$ where $\varepsilon_0 = \varepsilon_0(n, \kappa)$ is small, then
$$w_{x_0} \leq u \quad \text{on } B_1 \cap \partial B_{\varepsilon^{-1}}(\varepsilon^{-1} e_n) \subset \partial\Omega.$$
Indeed, for $x \in B_1 \cap \partial B_{\varepsilon^{-1}}(\varepsilon^{-1} e_n) \subset \partial\Omega$, we have $u(x) = |x'|^2/2$ and $|x_n - (x_0)_n| \leq C(\kappa)\varepsilon|x' - x_0'|$, so
$$\begin{aligned} w_{x_0}(x) &\leq \frac{1}{2}|x_0'|^2 + x_0' \cdot (x' - x_0') + \delta|x - x_0|^2 \\ &= u(x) - (1/2 - \delta)|x' - x_0'|^2 + \delta|x_n - (x_0)_n|^2 \\ &\leq u(x) - \delta|x' - x_0'|^2 + \delta C(\kappa)\varepsilon^2|x' - x_0'|^2 \leq u(x) \end{aligned}$$
if ε is small.

Summing up, we have $w_{x_0}(x_0) = u(x_0)$, $\det D^2 w_{x_0} > \det D^2 u$ in Ω,
$$w_{x_0} \leq u \qquad \text{on } B_1 \cap \partial B_{\varepsilon^{-1}}(\varepsilon^{-1} e_n) \subset \partial\Omega,$$
$$w_{x_0} \leq 1/4 \leq u \quad \text{on } \partial\Omega \setminus (B_1 \cap \partial B_{\varepsilon^{-1}}(\varepsilon^{-1} e_n)).$$
Thus, by the comparison principle in Theorem 3.21,
$$(10.6) \quad u(x) \geq w_{x_0}(x) \geq u(x_0) + x_0' \cdot (x' - x_0') + 2\kappa^{-1}\delta^{1-n}(x - x_0) \cdot \nu_{x_0} \quad \text{in } \Omega.$$
By taking $x = x_0 - t\nu_{x_0}$ in (10.6), subtracting $u(x_0)$ from both sides, dividing by $t > 0$, and then letting $t \to 0^+$, we obtain the assertion in Step 1 from
$$(10.7) \qquad Du(x_0) \cdot \nu_{x_0} \leq x_0' \cdot \nu_{x_0}' + 2\kappa^{-1}\delta^{1-n} \leq C(\kappa, n).$$

Since $u = |x'|^2/2$ on $\partial\Omega \cap B_1$, we have bounds on the tangential derivatives of u on $\partial\Omega \cap B_1$. Combining them with (10.7), we also have a lower bound (depending on κ and n) for $D_n u$ on $\partial\Omega \cap \{x : |x'| \leq 2\kappa_\varepsilon\}$.

Step 2. Letting $A := \{u < \kappa_\varepsilon^2/2\}$ and $c_0 := \kappa^{2n+1}/36^n$, we show that
$$|Du| \leq C(\kappa, n) \quad \text{on } A' := \partial A \cap \{x_n \leq c_0\}.$$
For any $x \in \Omega$ with $x' \neq 0$, take
$$x_0 = (\kappa_\varepsilon x'/|x'|, \varepsilon\kappa^2/2) \subset \partial B_{\varepsilon^{-1}}(\varepsilon^{-1} e_n) \cap \partial B_\kappa \subset \partial\Omega.$$
Then, $\nu_{x_0} = (\varepsilon x_0', \varepsilon(x_0)_n - 1)$, and, recalling $(x_0)_n = \varepsilon\kappa^2/2$,
$$\begin{aligned} &x_0' \cdot (x' - x_0') + 2\kappa^{-1}\delta^{1-n}(x - x_0) \cdot \nu_{x_0} \\ &= \kappa_\varepsilon(1 + 2\varepsilon\kappa^{-1}\delta^{1-n})(|x'| - \kappa_\varepsilon) - \kappa^{-1}\delta^{1-n}(2 - \varepsilon^2\kappa^2)(x_n - \varepsilon\kappa^2/2). \end{aligned}$$
Thus, with the above choice of x_0, we have from (10.6),
$$\begin{aligned} u(x) &\geq \kappa_\varepsilon^2/2 + \kappa_\varepsilon(1 + 2\varepsilon\kappa^{-1}\delta^{1-n})(|x'| - \kappa_\varepsilon) \\ &\quad - \kappa^{-1}\delta^{1-n}(2 - \varepsilon^2\kappa^2)(x_n - \varepsilon\kappa^2/2). \end{aligned}$$

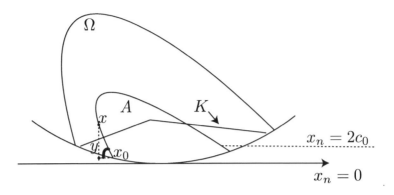

Figure 10.2. Geometry in Step 2 in the proof of Proposition 10.6.

In particular, from $u(x) < \kappa_\varepsilon^2/2$ in A, we find that $A \subset \tilde{A}$, where

$$(10.8) \quad \tilde{A} = \left\{ (2 - \varepsilon^2\kappa^2)(x_n - \varepsilon\kappa^2/2) \geq \kappa_\varepsilon\kappa\delta^{n-1}(1 + 2\varepsilon\kappa^{-1}\delta^{1-n})(|x'| - \kappa_\varepsilon) \right\}.$$

Thus, if $\varepsilon \leq \kappa$, then

$$(10.9) \qquad x \in \overline{\tilde{A}} \cap \{x_n \leq 2c_0\} \text{ implies that } |x'| \leq 3\kappa_\varepsilon/2.$$

Step 2(a). A lower bound for $D_n u$ on $\partial A \cap \{x_n \leq 2c_0\}$.

Consider $x \in \partial A \cap \{x_n \leq 2c_0\}$. Then, $|x'| \leq 2\kappa_\varepsilon$ by (10.9) and $A \subset \tilde{A}$. Let $y = (y', y_n) \equiv (x', y_n) \in \partial\Omega$. Clearly, $y_n \leq x_n$ (see Figure 10.2). Thus, by the convexity of u and Step 1, we have

$$D_n u(x) \geq D_n u(y) \geq -C(\kappa, n).$$

Step 2(b). An upper bound for $D_n u$ and $|D_i u|$ for $1 \leq i \leq n-1$ on A'.

To see this, let K be the intersection of Ω with the cone generated by $2\kappa e_n$ and $B_1(0) \cap \{x_n = 0\}$. Then, $\partial K \cap (\partial\tilde{A} \setminus \partial\Omega)$ is the sphere $|x'| = r \geq \kappa_\varepsilon$ in the hyperplane $x_n = \tilde{c}$, where $\tilde{c} \geq 2c_0$. For otherwise, by (10.9), $|x'| \leq 3\kappa_\varepsilon/2$, while using the equation for ∂K, we have

$$2c_0 \geq x_n = 2\kappa(1 - |x'|) \geq 2\kappa(1 - 3\kappa_\varepsilon/2) \geq \kappa,$$

which contradicts the choice of c_0.

Now, let $\bar{x} \in A'$. Then, there is $c_1(\kappa, n) > 0$ such that the cone

$$K_{\bar{x}} := B_{c_1}(\bar{x}) \cap \{x \in \mathbb{R}^n : x_n - \bar{x}_n \geq |x' - \bar{x}'|\} \subset \overline{\Omega}$$

does not intersect K. Thus, $\bar{x} + c_1\gamma \in \Omega$ for any unit direction $\gamma \in \mathbb{R}^n$ with $\gamma_n \geq |\gamma'|$. By the convexity of u together with $-\varepsilon \leq u \leq 2$ in Ω, we have

$$D_\gamma u(\bar{x}) := Du(\bar{x}) \cdot \gamma \leq [u(\bar{x} + c_1\gamma) - u(\bar{x})]/c_1 \leq 3/c_1,$$

which gives an upper bound (depending on κ and n) for $D_n u$ and $|D_i u|$ for $1 \leq i \leq n-1$ on A'. This completes the proof of Step 2.

Step 3. We show $|Du| \leq C(\kappa, n)$ in $\{u < 8\kappa_0^2\}$ for some small $\kappa_0(\kappa, n) > 0$.

We use volume estimate and convexity. For any $h > 0$ small, $S_h := \{x \in \overline{\Omega} : u(x) < h\}$ is a section of u centered at the minimum point $x_0 \in S_h$ with height $h - u(x_0) \leq h + \varepsilon$. By the volume estimate in Lemma 5.8, we have $|S_h| \leq C(n)(h + \varepsilon)^{n/2}$. By our assumption, $\overline{S_h}$ contains an $(n-1)$-dimensional ball with boundary $\{|x'| \leq \sqrt{2h}\} \cap \partial\Omega$. Since S_h is a convex set in the upper half-space, we deduce that

$$x_n \leq \frac{C(n)(h + \varepsilon)^{n/2}}{(\sqrt{2h})^{n-1}} \leq c_0/2 \quad \text{in } S_h$$

if $h \leq 8\kappa_0^2$ where $\kappa_0(\kappa, n)$ is small and $\varepsilon \leq h$.

From the proofs of Steps 2(a) and 2(b), we find that $|Du| \leq C(\kappa, n)$ on $\partial S_{8\kappa_0^2}$. By the convexity of u, we have $|Du| \leq C(\kappa, n)$ in $S_{8\kappa_0^2} = \{u < 8\kappa_0^2\}$.

Step 4. We show that

$$\|D^2 u\| \leq C(\kappa, n) \quad \text{on } E := B_1 \cap \partial B_{\varepsilon^{-1}}(\varepsilon^{-1}e_n) \cap \{|x'| \leq \kappa_\varepsilon/2\}.$$

Note that the principal curvatures of $\partial\Omega$ at $x \in B_1 \cap \partial B_{\varepsilon^{-1}}(\varepsilon^{-1}e_n)$ are $\kappa_1 = \cdots = \kappa_{n-1} = \varepsilon$. From $u = \varphi(x) = |x'|^2/2$ on $B_1 \cap \partial B_{\varepsilon^{-1}}(\varepsilon^{-1}e_n) \subset \partial\Omega$, we have, with respect to a principal coordinate system $\{\tau_1, \ldots, \tau_{n-1}, \nu\}$ at any point $x \in B_1 \cap \partial B_{\varepsilon^{-1}}(\varepsilon^{-1}e_n)$ (see (2.13))

$$D^2_{\tau_i \tau_j}(u - \varphi) = D_\nu(u - \varphi)\kappa_i \delta_{ij} = D_\nu(u - \varphi)\varepsilon\delta_{ij} \quad \text{for all } i, j = 1, \ldots, n-1.$$

It follows that, for ε small, the tangential Hessian $D^2_\tau u$ of u satisfies

$$(1/2)I_{n-1} \leq D^2_\tau u \leq 2I_{n-1}.$$

Since $\det D^2 u = 1$, it suffices to prove that $|D^2_{\tau_i \nu} u| := |(D^2 u \tau_i) \cdot \nu|$ is bounded in E for each $i = 1, \ldots, n-1$. Clearly, this follows from Remark 4.10.

Step 5. We show that $\|D^2 u\| \leq C(\kappa, n)$ in $\{u < 6\kappa_0^2\}$.

For this, we apply the Pogorelov second derivative estimates in the set $F := \{x \in \overline{\Omega} : u(x) < 8\kappa_0^2\}$. Fix $i \in \{1, \ldots, n\}$. Consider the function

$$h(x) = (8\kappa_0^2 - u)e^{\frac{1}{2}(D_i u)^2} D_{ii} u.$$

It assumes a maximum value on $\overline{F}$ at some $P \in \overline{F}$ which can only be on $\partial F \cap \partial\Omega$ or in the interior $\text{int}(F)$ of F.

If $P \in \partial F \cap \partial\Omega$, then by Steps 3 and 4, we have $h(P) \leq C(n, \kappa)$. Therefore, for all $x \in \{u < 6\kappa_0^2\}$, we obtain

$$(2\kappa_0^2)D_{ii} u(x) \leq h(x) \leq h(P) \leq C(n, \kappa),$$

which implies an upper bound for $D_{ii} u(x)$.

If $P \in \text{int}(F)$, then the Pogorelov estimates in Theorem 6.1 give

$$h(P) \leq (n + |D_i u(P)|)e^{(D_i u(P))^2/2}.$$

From Step 3, we have $h(P) \leq C(n, \kappa)$ from which we also deduce an upper bound for $D_{ii}u(x)$ if $x \in \{u < 6\kappa_0^2\}$.

Step 6. Conclusion. From Step 5, we find that the Monge–Ampère equation $\det D^2 u = 1$ is uniformly elliptic in $\{u < 6\kappa_0^2\}$. By the Evans–Krylov Theorem (Theorem 2.86), $\|u\|_{C^{2,\alpha}(\{u<4\kappa_0^2\})} \leq C(n, \kappa)$ for some $\alpha = \alpha(n, \kappa) \in (0, 1)$.

For any unit vetor $\mathbf{e}$, $u_{\mathbf{e}} := Du \cdot \mathbf{e}$ solves the equation $U^{ij} D_{ij} u_{\mathbf{e}} = 0$ which is uniformly elliptic with global C^α coefficients $(U^{ij}) = (D^2 u)^{-1}$ in $\{u < 4\kappa_0^2\}$. Thus, we can apply the global Schauder estimates in Theorem 2.82 to get $u_{\mathbf{e}} \in C^{2,\alpha}$ and hence $u \in C^{3,\alpha}$ in $\{u < 3\kappa_0^2\}$. From this, we see that $u_{\mathbf{e}}$ satisfies $U^{ij} D_{ij} u_{\mathbf{e}} = 0$ with global $C^{1,\alpha}$ coefficients in $\{u < 3\kappa_0^2\}$. Again, from Theorem 2.82, we deduce the $C^{3,\alpha}$ bound for $u_{\mathbf{e}}$ in $\{u < 2\kappa_0^2\}$, which easily gives the desired $C^{3,1}$ bound for u in $\{u < \kappa_0^2\}$. $\qquad \square$

From Proposition 10.6, we obtain the following boundary $C^{3,1}$ estimates in a half-space. We use the notation $B_r^+(0) = B_r(0) \cap \mathbb{R}_+^n$.

Theorem 10.7 (Boundary Pogorelov estimates in a half-space). *Let Ω be a bounded convex domain in $\mathbb{R}^n$ that satisfies, for some $\kappa \in (0, 1/2)$,*

$$B_{2\kappa}^+(0) \subset \Omega \subset B_{\kappa^{-1}}^+(0).$$

Let $u : \Omega \to [0, \infty)$ be a convex function satisfying the Monge–Ampère equation $\det D^2 u = 1$ in Ω in the sense of Aleksandrov. Assume that the boundary values of u, in the epigraphical sense of Definition 7.20, are given by

$$\begin{cases} u = p(x') & \text{on } \{p(x') \leq 1\} \cap \{x_n = 0\} \subset \partial\Omega, \\ u = 1 & \text{on the rest of } \partial\Omega, \end{cases}$$

where p is a quadratic polynomial that satisfies, for some constant $\rho > 0$,

$$\rho|x'|^2 \leq p(x') \leq \rho^{-1}|x'|^2.$$

Then, there exists $c_0 = c_0(n, \kappa, \rho) \in (0, 1)$ such that

(10.10) $$\|u\|_{C^{3,1}(B_{c_0}^+(0))} \leq c_0^{-1}.$$

Proof. Let

$$A = \begin{pmatrix} (D^2 p)^{-1/2} & 0 \\ 0 & (\det D^2 p)^{1/2} \end{pmatrix}.$$

Then, $C^{-1} I_n \leq A \leq C I_n$ for some $C(\rho, n) > 0$. By considering $\tilde{u}(x) = u(Ax)$ and $\tilde{p}(x') = p[(D^2 p)^{-1/2} x']$, where $\det D^2 \tilde{u} = 1$ and $\tilde{p}(x') = |x'|^2/2$, we can assume that $p(x') = |x'|^2/2$.

First, we approximate u on $\partial\Omega$ by a sequence of smooth functions φ_m on $\partial\Omega_m \in C^\infty$ ($m \in \mathbb{N}$), with $\Omega_m \subset \Omega$ being uniformly convex and $0 \in \partial\Omega_m$, so that φ_m and Ω_m satisfy the boundary conditions of Proposition 10.6 with $\varepsilon_m \searrow 0$.

Next, we let $u_m \in C(\overline{\Omega_m})$ be the convex Aleksandrov solution to

$$\det D^2 u_m = 1 \quad \text{in } \Omega_m, \quad u_m = \varphi_m \quad \text{on } \partial\Omega_m.$$

By Theorem 4.12, u_m is smooth up to the boundary of Ω_m. By Theorems 7.29 and 7.26, u_m converges to u in the sense of Hausdorff convergence of epigraphs. Thus, u_m converges uniformly to u on compact subsets of Ω. Moreover, since $u \geq 0$, we deduce that $\inf_{\Omega_m} u_m \to 0$ when $m \to \infty$. Thus, for m large, Proposition 10.6 can be applied for u_m to get

$$\|u_m\|_{C^{3,1}(\{u_m < \kappa_0^2\})} \leq C(\kappa, \rho, n),$$

where $\kappa_0 = \kappa_0(\kappa, \rho, n) \in (0, 1)$. From the convexity of u_m and u, together with their boundary data, we can find $c_0(n, \kappa, \rho) > 0$ small (for example, $c_0 = \min\{\kappa\kappa_0^2, \sqrt{\rho}\kappa_0\}/4$) such that

$$B_{2c_0}^+(0) \cap \Omega_m \subset \{u_m < \kappa_0^2\} \quad \text{for large } m \quad \text{and} \quad B_{2c_0}^+(0) \subset \{u < \kappa_0^2\}.$$

Letting $m \to \infty$ in the above $C^{3,1}$ estimates, we obtain (10.10). $\qquad\square$

10.4. Pointwise Second Derivative Hölder Estimates at the Boundary

In this section, we prove a boundary version of the interior pointwise $C^{2,\alpha}$ estimates in Theorem 6.30. The setting here is similar to that of the Localization Theorem (Theorem 8.5). For the rest of this chapter, the concept of solutions can be understood as either in the sense of Aleksandrov or in the viscosity sense, so we will not specify it in the statements that follow.

Let Ω be a bounded convex domain in the upper half-space $\mathbb{R}_+^n$ satisfying

$$(10.11) \qquad B_\rho(\rho e_n) \subset \Omega \subset \{x_n \geq 0\} \cap B_{1/\rho}(0),$$

for some small $\rho > 0$, so Ω contains an interior ball of radius ρ that is tangent to $\partial\Omega$ at the origin. Let $u \in C(\overline{\Omega})$ be a convex function satisfying

$$(10.12) \qquad \det D^2 u = f, \qquad 0 < \lambda \leq f \leq \Lambda \quad \text{in } \Omega,$$

and, in the sense of Definition 4.1,

$$(10.13) \qquad x_{n+1} = 0 \text{ is a tangent hyperplane to the graph of } u \text{ at } 0.$$

We also assume that on $\partial\Omega$, in a neighborhood of $\{x_n = 0\}$, u separates quadratically from the tangent hyperplane $\{x_{n+1} = 0\}$,

$$(10.14) \qquad \rho|x|^2 \leq u(x) \leq \rho^{-1}|x|^2 \quad \text{on } \partial\Omega \cap \{x_n \leq \rho\}.$$

The following theorem, due to Savin [**S4**], shows that if at the origin, f is pointwise C^α, $\partial\Omega$ and $u|_{\partial\Omega}$ are pointwise $C^{2,\alpha}$, then u is pointwise $C^{2,\alpha}$.

Theorem 10.8 (Savin's pointwise $C^{2,\alpha}$ estimates at the boundary for the Monge–Ampère equation). *Let Ω and u satisfy* (10.11)–(10.14). *Suppose that for some $\alpha \in (0,1)$, f is C^α at the origin, $\partial\Omega$ and $u|_{\partial\Omega}$ are $C^{2,\alpha}$ at the origin. Then, u is $C^{2,\alpha}$ at the origin.*

More precisely, assume that, for some $M > 0$ and quadratic polynomials $p(x')$ and $q(x')$, the following estimates hold:

$$|f(x) - f(0)| \le M|x|^\alpha \qquad in \ \Omega \cap B_\rho(0),$$

$$|x_n - q(x')| + |u(x) - p(x')| \le M|x'|^{2+\alpha} \qquad on \ \partial\Omega \cap B_\rho(0).$$

Then, there exists a convex, homogeneous quadratic polynomial $\mathcal{P}_0$ with

$$\det D^2\mathcal{P}_0 = f(0), \quad \|D^2\mathcal{P}_0\| \le C,$$

such that

$$|u(x) - \mathcal{P}_0(x)| \le C|x|^{2+\alpha} \quad in \ \Omega \cap B_\rho(0),$$

where C depends on M, ρ, λ, Λ, n, and α.

Remark 10.9. In view of (10.11) and (10.14), p and q are homogenous quadratic polynomials. Moreover, $\|D^2q\| \le n^2\rho^{-1}$, $\|D^2p\| \le n^2\rho^{-1}$, and $D^2p \ge 2\rho I_{n-1}$.

The proof of Theorem 10.8 has three steps. We already implemented this strategy in the proof of the interior pointwise $C^{2,\alpha}$ estimates in Theorem 6.30. First, in Lemma 10.10, we use the Boundary Localization Theorem (Theorem 8.5) and rescaling to reduce the theorem to the case when M is arbitrarily small. Second, in Lemma 10.11, we use the Pogorelov estimates in the upper half-space to further reduce the theorem to the case when u is arbitrarily close to a convex quadratic polynomial. Finally, in Lemma 10.16, we use a perturbation argument to show that u is well-approximated by convex quadratic polynomials at all scales.

By considering $[f(0)]^{-1/n}u$ and $[f(0)]^{-1}f$ instead of u and f, we can assume that

$$f(0) = 1.$$

Notation. We denote $B_r := B_r(0)$ and $B_r^+ = B_r(0) \cap \mathbb{R}_+^n$. If $P(x)$ is a polynomial in $x \in \mathbb{R}^n$, then for simplicity, we write $P(x')$ for $P(x', 0)$.

10.4.1. Reduction to M being small. Rescaling as in Lemma 9.3 using the Boundary Localization Theorem, we show in the following lemma that if h is sufficiently small, the corresponding rescaling u_h satisfies the hypotheses of u in which the constant M is replaced by an arbitrarily small constant σ.

Lemma 10.10. *Let u and Ω be as in Theorem 10.8. Let $\sigma > 0$. Then, there exist small constants $h_0(M, \sigma, \rho, \lambda, \Lambda, n, \alpha)$, $\kappa = \kappa(\rho, \lambda, \Lambda, n) > 0$ and rescalings of u, f, q, and Ω:*

$$u_h = h^{-1} u \circ (h^{1/2} A_h^{-1}), \ f_h = f \circ (h^{1/2} A_h^{-1}), \ q_h = h^{1/2} q, \ \Omega_h = h^{-1/2} A_h \Omega,$$

where A_h is a linear transformation on $\mathbb{R}^n$ with

$$\det A_h = 1, \quad \|A_h^{-1}\| + \|A_h\| \le \kappa^{-1} |\log h|,$$

such that when $h \le h_0$, the following properties hold.

(i) *$B_\kappa(0) \cap \overline{\Omega}_h \subset S_{u_h}(0,1) := \{x \in \overline{\Omega}_h : u_h(x) < 1\} \subset B_{\kappa^{-1}}(0) \cap \mathbb{R}_+^n$.*

(ii) *$\det D^2 u_h = f_h, \quad |f_h(x) - 1| \le \sigma |x|^\alpha \quad$ in $\Omega_h \cap B_{\kappa^{-1}}(0)$.*

(iii) *On $\partial \Omega_h \cap B_{\kappa^{-1}}(0)$, we have*

$$x_n \le \sigma |x'|^2, \quad |q_h(x')| \le \sigma, \quad |x_n - q_h(x')| + |u_h(x) - p(x')| \le \sigma |x'|^{2+\alpha}.$$

Proof. Recall that $S_u(0, h) = \{x \in \overline{\Omega} : u(x) < h\}$. By the Boundary Localization Theorem (Theorem 8.5), for all $h \le c(\rho, \lambda, \Lambda, n)$, we have

$$\kappa E_h \cap \overline{\Omega} \subset S_u(0, h) \subset \kappa^{-1} E_h \cap \overline{\Omega},$$

for some $\kappa(\rho, \lambda, \Lambda, n) > 0$, where

$$E_h = A_h^{-1} B_{h^{1/2}}, \ A_h x = x - \tau_h x_n, \ \tau_h \cdot e_n = 0, \ \|A_h^{-1}\| + \|A_h\| \le \kappa^{-1} |\log h|.$$

Note that $\tau_h = (\tau_h', 0)$. Let u_h, f_h, q_h, and Ω_h be as in the statement of the lemma. Then $\det D^2 u_h = f_h$ in Ω_h, and

$$S_{u_h}(0, 1) := \{x \in \overline{\Omega}_h : u_h(x) < 1\} = h^{-1/2} A_h S_u(0, h).$$

Hence $B_\kappa \cap \overline{\Omega}_h \subset S_{u_h}(0, 1) \subset B_{\kappa^{-1}} \cap \mathbb{R}_+^n$, which proves property (i).

For any $x \in \overline{\Omega}_h \cap B_{\kappa^{-1}}$, let $y := h^{1/2} A_h^{-1} x \in \overline{\Omega}$. Then

$$h^{1/2} x_n = y_n, \quad h^{1/2} x' = y' - \tau_h' y_n, \quad \text{and} \quad h^{-1/2} y' = x' + \tau_h' x_n.$$

Moreover, if $h \le h_1(\rho, \kappa)$ is small, then $y \in \overline{\Omega} \cap B_\rho \cap B_{h^{1/4}}$, since

$$|y| \le \kappa^{-1} h^{1/2} |\log h| |x| \le \kappa^{-2} h^{1/2} |\log h| < \min\{\rho, h^{1/4}\}.$$

Therefore, when $h \le h_1$ and $x \in \Omega_h \cap B_{\kappa^{-1}}$, we have

$$|f_h(x) - 1| = |f(h^{1/2} A_h^{-1} x) - 1| \le M |h^{1/2} A_h^{-1} x|^\alpha$$
$$\le M(h^{1/2} \kappa^{-1} |\log h|)^\alpha |x|^\alpha \le \sigma |x|^\alpha$$

if $h \le h_0 \le h_1$ where $h_0(M, \sigma, \rho, \lambda, \Lambda, n, \alpha)$ is small. This proves (ii).

To prove property (iii), we consider $x \in \partial \Omega_h \cap B_{\kappa^{-1}}$ with $h \le h_0$. Then, $y = h^{1/2} A_h^{-1} x \in \partial \Omega \cap B_\rho \cap B_{h^{1/4}}$. Since Ω has an interior tangent ball of radius ρ to $\partial \Omega$ at the origin, we have $|y_n| \le \rho^{-1} |y'|^2$. Therefore,

$$|\tau_h y_n| \le \kappa^{-1} \rho^{-1} |\log h| |y'|^2 \le \kappa^{-1} \rho^{-1} h^{1/4} |\log h| |y'| \le |y'|/2$$

if h_0 is small. Thus, from $h^{1/2}x' = y' - \tau'_h y_n$, we find

$$\frac{1}{2}|y'| \leq |h^{1/2}x'| \leq \frac{3}{2}|y'|,$$

and, if h_0 is small,

$$0 \leq x_n = h^{-1/2}y_n \leq h^{-1/2}\rho^{-1}|y'|^2 \leq 4\rho^{-1}h^{1/2}|x'|^2 \leq \sigma|x'|^2.$$

Now, we estimate $q_h(x')$, $|x_n - h^{1/2}q(x')|$, and $|u_h - p(x')|$ on $\partial\Omega_h \cap B_{\kappa^{-1}}(0)$. Recall that q is a quadratic polynomial with $\|D^2q\| \leq n^2\rho^{-1}$. Thus,

$$|q_h(x')| = h^{1/2}|q(x')| \leq h^{1/2}\|D^2q\|\|x'\|^2 \leq n^2 h^{1/2}\rho^{-1}\kappa^{-2} \leq \sigma$$

if h_0 is small. Moreover, using $h^{-1/2}y' = x' + \tau'_h x_n$ and $x_n \leq \sigma|x'|^2$,

$$|x_n - h^{1/2}q(x')| \leq h^{-1/2}|y_n - q(y')| + h^{1/2}|q(h^{-1/2}y') - q(x')|$$
$$\leq Mh^{-1/2}|y'|^{2+\alpha} + C(n,\rho)h^{1/2}\left(|x'||\tau_h x_n| + |\tau_h x_n|^2\right).$$

Since p is a quadratic polynomial with $\|D^2p\| \leq n^2\rho^{-1}$, we also have

$$|u_h(x) - p(x')| \leq h^{-1}|u(y) - p(y')| + |p(h^{-1/2}y') - p(x')|$$
$$\leq Mh^{-1}|y'|^{2+\alpha} + C(n,\rho)\left(|x'||\tau_h x_n| + |\tau_h x_n|^2\right).$$

Therefore, from $|y'| \leq 2h^{1/2}|x'|$, $0 \leq x_n \leq C(\rho)h^{1/2}|x'|^2$, we obtain

$$|x_n - h^{1/2}q(x')| + |u_h(x) - p(x')|$$
$$\leq 16Mh^{\alpha/2}|x'|^{2+\alpha} + C(n,\rho,\kappa)\left(h^{1/2}|\log h||x'|^3 + h|\log h|^2|x'|^4\right)$$
$$\leq \sigma|x'|^{2+\alpha}$$

if $h \leq h_0(\leq 1)$ is small. This finishes the proof of the lemma. $\qquad\square$

10.4.2. Approximation of u_h by quadratic polynomials. We next show by a compactness argument that if σ is sufficiently small, then u_h in Lemma 10.10 can be well-approximated by convex quadratic polynomials near the origin.

Lemma 10.11. *For any $\delta_0 \in (0,1)$ and $\varepsilon_0 \in (0,1)$, there exist positive constants $\sigma_0(\delta_0, \varepsilon_0, \alpha, n, \lambda, \Lambda, \rho)$ and $\mu_0(\varepsilon_0, n, \lambda, \Lambda, \rho)$ such that for any convex function u_h satisfying properties* (i)–(iii) *of Lemma 10.10 with $\sigma \leq \sigma_0$, we can find a rescaling*

$$\tilde{u}(x) := \mu_0^{-2}(u_h - l_h)(\mu_0 x) \quad in \ \tilde{\Omega} := \mu_0^{-1}S_{u_h}(0,1) \supset \{te_n : 0 < t \leq 1\},$$

where

$$l_h(x) = a_h x_n, \quad |a_h| \leq C_0 = C_0(\alpha, n, \lambda, \Lambda, \rho),$$

with the following properties:

(i) in $\tilde{\Omega} \cap B_1(0)$, we have

$$\det D^2 \tilde{u} = \tilde{f}, \qquad |\tilde{f}(x) - 1| \le \delta_0 \varepsilon_0 |x|^\alpha, \qquad |\tilde{u} - P_0| \le \varepsilon_0,$$

for some convex, homogeneous quadratic polynomial P_0 with

$$\det D^2 P_0 = 1, \quad \|D^2 P_0\| \le C_0;$$

(ii) on $\partial\tilde{\Omega} \cap B_1(0)$, we have $x_n \le \delta_0 \varepsilon_0 |x'|^2$ and

$$|x_n - \tilde{q}_0(x')| + |\tilde{u}(x) - \tilde{p}_0(x')| \le \delta_0 \varepsilon_0 |x'|^{2+\alpha}, \qquad |\tilde{q}_0(x')| \le \delta_0 \varepsilon_0,$$

for some quadratic polynomials $\tilde{p}_0(x')$ and $\tilde{q}_0(x')$ with

$$\tilde{p}_0(x') = P_0(x'), \qquad (\rho/2)|x'|^2 \le \tilde{p}_0(x') \le 2\rho^{-1}|x'|^2.$$

Proof. Assume by contradiction that the conclusion of the lemma is false for a sequence of convex functions u_m on convex domains Ω_m ($m \in \mathbb{N}$) satisfying properties (i)–(iii) of Lemma 10.10 with $\sigma_m \to 0$, where we also recall $\|D^2 p_m\| \le n^2 \rho^{-1}$ and $D^2 p_m \ge 2\rho I_{n-1}$. Then, up to extracting a subsequence if necessary, we can assume that

$$(10.15) \qquad p_m(x') \to p_\infty(x'), \quad q_m(x') \to 0 \quad \text{uniformly on } B_{\kappa^{-1}}^+,$$

and $(u_m, S_{u_m}(0,1))$ converges in the sense of Hausdorff convergence of epigraphs in Definition 7.22 to $(u_\infty, S_{u_\infty}(0,1))$. By Theorem 7.29, u_∞ satisfies

$$B_\kappa^+ \subset \Omega_\infty = S_{u_\infty}(0,1) \subset B_{\kappa^{-1}}^+, \qquad \det D^2 u_\infty = 1,$$

and its boundary values in the epigraphical sense of Definition 7.20 satisfy

$$\begin{cases} u_\infty = p_\infty(x') & \text{on } \{p_\infty(x') < 1\} \cap \{x_n = 0\} \subset \partial\Omega_\infty, \\ u_\infty = 1 & \text{on the rest of } \partial\Omega_\infty. \end{cases}$$

From the boundary Pogorelov estimates in Theorem 10.7, we can find a positive constant $c_0(n, \lambda, \Lambda, \rho)$, a linear function $l_\infty(x) := a_\infty x_n$, and a convex, homogeneous quadratic polynomial P_∞ such that

$$|u_\infty(x) - l_\infty(x) - P_\infty(x)| \le c_0^{-1} |x|^3 \quad \text{in } B_{c_0}^+,$$

where

$$|a_\infty| \le c_0^{-1}, \quad P_\infty(x') = p_\infty(x'), \quad \det D^2 P_\infty = 1, \quad \text{and} \quad \|D^2 P_\infty\| \le c_0^{-1}.$$

Choose $\mu_0 = \mu_0(\varepsilon_0, n, \lambda, \Lambda, \rho) \le \kappa/2$ small such that

$$c_0^{-1} \mu_0 \le \varepsilon_0/32.$$

Then

$$|u_\infty - l_\infty - P_\infty| \le \frac{1}{4} \varepsilon_0 \mu_0^2 \quad \text{in } B_{2\mu_0}^+.$$

This together with the convergence $u_m \to u_\infty$ implies that for all large m

$$|u_m - l_\infty - P_\infty| \leq \frac{1}{2}\varepsilon_0\mu_0^2 \quad \text{in } S_{u_m}(0,1) \cap B_{\mu_0}^+.$$

Now, for all large m, the functions

$$\tilde{u}_m(x) := \mu_0^{-2}(u_m - l_\infty)(\mu_0 x), \ \tilde{f}_m(x) := f_m(\mu_0 x) \quad \text{on } \tilde{\Omega}_m := \mu_0^{-1}S_{u_m}(0,1)$$

satisfy

$$\det D^2\tilde{u}_m = \tilde{f}_m \quad \text{in } \tilde{\Omega}_m \cap B_1,$$

together with the following estimates:

$$|\tilde{u}_m - P_\infty| \leq \varepsilon_0/2, \quad |\tilde{f}_m(x) - 1| \leq \sigma_m(\mu_0|x|)^\alpha \leq \delta_0\varepsilon_0|x|^\alpha \quad \text{in } \tilde{\Omega}_m \cap B_1.$$

Next, we will modify (P_∞, p_m, q_m) into $(P_m, \tilde{p}_m, \tilde{q}_m)$ so that $\tilde{u}_m, P_m, \tilde{p}_m, \tilde{q}_m$ satisfy the conclusion of the lemma on $\tilde{\Omega}_m \supset \{te_n : 0 < t \leq 1\}$ for all large m, and we reach a contradiction.

Step 1. We modify $\tilde{q}_m$ into

$$\tilde{q}_m := \mu_0 q_m.$$

On $\partial\tilde{\Omega}_m \cap B_1$, we use property (iii) of Lemma 10.10 for (u_m, Ω_m) to estimate x_n, $|\tilde{q}_m(x')|$, and $|x_n - \tilde{q}_m(x')|$ as follows. We have

$$x_n = \mu_0^{-1}(\mu_0 x_n) \leq \mu_0^{-1}\sigma_m|\mu_0 x'|^2 \leq \delta_0\varepsilon_0|x'|^2,$$

$$|\tilde{q}_m(x')| = \mu_0^{-1}|q_m(\mu_0 x')| \leq \mu_0^{-1}\sigma_m \leq \delta_0\varepsilon_0,$$

$$|x_n - \tilde{q}_m(x')| = \mu_0^{-1}|\mu_0 x_n - q_m(\mu_0 x')| \leq \mu_0^{-1}\sigma_m|\mu_0 x'|^{2+\alpha} \leq \delta_0\varepsilon_0|x'|^{2+\alpha}.$$

Step 2. We modify p_m into

$$\tilde{p}_m := p_m - a_\infty q_m.$$

This modification guarantees that $|\tilde{u}_m(x) - \tilde{p}_m(x')|$ is small compared to $|x'|^{2+\alpha}$. Indeed, we have

$$|\tilde{u}_m(x) - \tilde{p}_m(x')| = \mu_0^{-2}|(u_m - l_\infty)(\mu_0 x) - p_m(\mu_0 x') + a_\infty q_m(\mu_0 x')|$$

$$\leq \mu_0^{-2}\Big(|u_m(\mu_0 x) - p_m(\mu_0 x')| + |a_\infty||\mu_0 x_n - q_m(\mu_0 x')|\Big)$$

$$\leq \sigma_m\mu_0^\alpha(1 + |a_\infty|)|x'|^{2+\alpha} \leq \delta_0\varepsilon_0|x'|^{2+\alpha}.$$

In view of (10.15), we find

$$\tilde{p}_m \to p_\infty, \quad \tilde{q}_m \to 0 \quad \text{uniformly in } B_1.$$

Note that $\rho|x'|^2 \leq p_m(x') \leq \rho^{-1}|x'|^2$ so for large m, we have

$$(\rho/2)|x'|^2 \leq \tilde{p}_m(x') \leq 2\rho^{-1}|x'|^2.$$

Step 3. We modify P_∞ into

$$P_m(x) = \tilde{p}_m(x') + \sum_{i=1}^{n-1} D_{in}P_\infty x_i x_n + (1/2)(D_{nn}P_\infty + \gamma_m)x_n^2,$$

where $\gamma_m \in \mathbb{R}$ is chosen so that $\det D^2 P_m = 1$. Note that this is a linear equation in γ_m. Moreover, since $\tilde{p}_m(x') \to p_\infty(x')$ uniformly in B_1, $P_\infty(x') = p_\infty(x')$, and $\det D^2 P_\infty = 1$, we infer the existence of γ_m for m large and that $\gamma_m \to 0$ when $m \to \infty$. It follows that

$$P_m(x') = \tilde{p}_m(x'), \quad \det D^2 P_m = 1, \quad P_m \to P_\infty \quad \text{uniformly in } B_1.$$

Then, for m large enough, we have $|P_\infty - P_m| \le \varepsilon_0/2$ in B_1, and hence

$$|\tilde{u}_m - P_m| \le |\tilde{u}_m - P_\infty| + |P_\infty - P_m| \le \varepsilon_0 \quad \text{in } \tilde{\Omega}_m \cap B_1.$$

Now, $\tilde{u}_m, P_m, \tilde{p}_m, \tilde{q}_m$ satisfy the conclusion of the lemma on $\tilde{\Omega}_m$ for all large m, and the proof of the lemma is complete. $\square$

Remark 10.12. We infer from Lemmas 10.10 and 10.11 that for $\delta_0 > 0$ and $\varepsilon_0 > 0$, there exist a linear transformation $\tilde{A} := \mu_0 h_0^{1/2} A_{h_0}^{-1} \in \mathbb{M}^{n \times n}$ and a linear function $l(x) := a_0 x_n$ with

$$\max\left\{|a_0|, \|\tilde{A}^{-1}\|, \|\tilde{A}\|\right\} \le C(M, \delta_0, \varepsilon_0, n, \lambda, \Lambda, \rho, \alpha),$$

such that the rescalings of u and f in Theorem 10.8,

$$\tilde{u}(x) := (\det \tilde{A})^{-2/n}(u - l)(\tilde{A}x), \quad \tilde{f}(x) := f(\tilde{A}x)$$

defined in a domain $\tilde{\Omega}$ in $\mathbb{R}_+^n$ containing $\{te_n : 0 < t \le 1\}$, satisfy:

(i) in $\tilde{\Omega} \cap B_1(0)$, we have

$$\det D^2\tilde{u} = \tilde{f}, \quad |\tilde{f}(x) - 1| \le \delta_0\varepsilon_0|x|^\alpha, \quad \text{and} \quad |\tilde{u} - P_0| \le \varepsilon_0,$$

where P_0 is a convex, homogeneous quadratic polynomial with

$$\det D^2 P_0 = 1 \quad \text{and} \quad \|D^2 P_0\| \le C_0(\alpha, n, \lambda, \Lambda, \rho);$$

(ii) on $\partial\tilde{\Omega} \cap B_1(0)$, we have $0 \le x_n \le \delta_0\varepsilon_0|x'|^2$ and

$$|x_n - \tilde{q}(x')| + |\tilde{u}(x) - \tilde{p}(x')| \le \delta_0\varepsilon_0|x'|^{2+\alpha} \quad \text{and} \quad |\tilde{q}(x')| \le \delta_0\varepsilon_0,$$

where $\tilde{p}$ and $\tilde{q}$ are quadratic polynomials with

$$(\rho/2)|x'|^2 \le \tilde{p}(x') = P_0(x') \le 2\rho^{-1}|x'|^2.$$

Remark 10.13. For the purpose of approximating $(\tilde{u} - P_0)/\varepsilon_0$, we can use affine transformations, at the cost of increasing ε_0 by a factor of $C_1(\rho, C_0, n)$, to assume that P_0 in Remark 10.12 is the quadratic polynomial $|x|^2/2$. The details are as follows.

(a) **Transformation of $\tilde{p}(x')$ to $|x'|^2/2$.** Let

$$A = \begin{pmatrix} (D^2\tilde{p})^{-1/2} & 0 \\ 0 & (\det D^2\tilde{p})^{1/2} \end{pmatrix}.$$

Then, $C^{-1}I_n \leq A \leq CI_n$ for some $C = C(\rho, n, C_0) > 0$. Let $\hat{\Omega} = A^{-1}\tilde{\Omega}$. Define the following functions on $\hat{\Omega}$:

$$\hat{u} = \tilde{u} \circ A, \quad \hat{f} = \tilde{f} \circ A, \quad \hat{P} = P_0 \circ A, \quad \hat{q} = (\det D^2\tilde{p})^{-1/2}\tilde{q}.$$

It can be verified easily that $\hat{\Omega}, \hat{u}, \hat{P}, \hat{q}$ satisfy properties (i) and (ii) in Remark 10.12, where $\hat{p}(x') = \hat{P}(x') = |x'|^2/2$ and ε_0 is replaced by $C_1(\rho, C_0, n)\varepsilon_0$.

(b) **Transformation of $\hat{P}(x)$ to $|x|^2/2$ via sliding.** Note that

$$\hat{P}(x) = \frac{1}{2}|x'|^2 + (\tau' \cdot x')x_n + \frac{1}{2}a_{nn}x_n^2 = \frac{1}{2}|x' + x_n\tau'|^2 + \frac{1}{2}|x_n|^2,$$

where $\tau' \in \mathbb{R}^{n-1}$ with $\det D^2\hat{P} = a_{nn} - |\tau'|^2 = 1$. Consider the sliding map

$$\check{A}(x) = (x' - x_n\tau', x_n).$$

Now, we define $\check{\Omega} = \check{A}^{-1}\hat{\Omega}$ and the following functions on $\check{\Omega}$:

$$\check{u} = \hat{u} \circ \check{A}, \quad \check{f} = \hat{f} \circ \check{A}, \quad \check{P} = \hat{P} \circ \check{A}, \quad \check{q} = \hat{q}.$$

Note that $\check{P}(x) = |x|^2/2$. As in the proof of Lemma 10.10, we can show that they satisfy properties (i) and (ii) in Remark 10.12, where ε_0 is replaced by $C_*(\rho, C_0, n)\varepsilon_0$.

10.4.3. Approximation by solutions of the linearized equation. In this section, we show that if δ_0 and ε_0 are small, then in the setting of Remark 10.12, $(\tilde{u} - P_0)/\varepsilon_0$ is well-approximated by solutions of the linearized operator of P_0. Due to Remark 10.13, we can take P_0 to be the quadratic polynomial $|x|^2/2$ to simplify the arguments. See Figure 10.3.

Proposition 10.14 (Approximation by harmonic functions). *Let Ω be a convex domain in $\mathbb{R}^n_+$ containing $\{te_n : 0 < t \leq 1\}$. Denote $P(x) = |x|^2/2$. Let $u \in C(\overline{\Omega})$ be a convex function satisfying*

$$\det D^2u = f, \quad |f - 1| \leq \delta\varepsilon, \quad and \quad |u - P| \leq \varepsilon \quad in \ \Omega \cap B_1(0).$$

Assume that on $\partial\Omega \cap B_1(0)$, we have

$$x_n \leq \delta\varepsilon, \quad |u(x) - |x'|^2/2| \leq \delta\varepsilon.$$

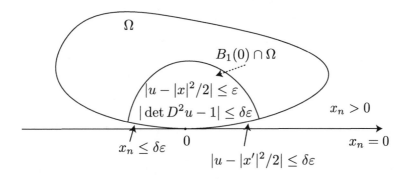

Figure 10.3. Setting of Proposition 10.14.

Then, for any small $\eta > 0$, we can find a solution to

(10.16) $\Delta w = 0$ *in* $\mathbb{R}^n_+$, $w = 0$ *on* $\{x_n = 0\}$, *with* $|w| \leq 1$ *in* $B_1(0) \cap \mathbb{R}^n_+$,

such that

$$\left| (u - P)/\varepsilon - w \right| \leq \eta \quad in \; B_{1/2}(0) \cap \overline{\Omega},$$

provided that ε and δ are chosen sufficiently small, depending only on n and η. Consequently, there are $a_0 \in \mathbb{R}, a' \in \mathbb{R}^{n-1}$ with $|a_0| \leq C(n)$ and $|a'| \leq C(n)$ such that

(10.17) $|u - P - \varepsilon(a_0 + a' \cdot x')x_n| \leq \varepsilon(\eta + C(n)|x|^3) \quad in \; B_{1/2}(0) \cap \overline{\Omega}.$

Proof. Let

$$v_\varepsilon = (u - P)/\varepsilon.$$

Step 1. We first show that v_ε separates almost "linearly" away from $\{x_n = 0\}$. More precisely, we show that there is constant $C = C(n) > 0$ such that

$$|v_\varepsilon(x', x_n)| \leq 2\delta + Cx_n \quad in \; B_{1/2}(0) \cap \overline{\Omega}.$$

Indeed, let $x_0 \in B_{1/2}(0) \cap \{x_n = 0\}$. We consider the function

$$\bar{w}(x) = \frac{1}{2}|x'|^2 + 16\varepsilon|x' - x'_0|^2 + \frac{1 - \delta\varepsilon}{(1 + 32\varepsilon)^{n-1}}\frac{x_n^2}{2} + C\varepsilon x_n + 2\delta\varepsilon,$$

for some $C(n)$ large to be determined. Then,

$$\det D^2\bar{w} = 1 - \delta\varepsilon \leq \det D^2 u \quad in \; \Omega \cap B_1(0).$$

On $\partial B_1(0) \cap \Omega$, we have $16|x' - x'_0|^2 + 4x_n \geq 1$; hence, for $C = C(n)$ large,

$$\bar{w}(x) \geq \frac{1}{2}|x|^2 + 16\varepsilon|x' - x'_0|^2 + \frac{C}{2}\varepsilon x_n \geq P(x) + \varepsilon \geq u.$$

On $\partial\Omega \cap B_1(0)$, we have

$$\bar{w}(x) \geq |x'|^2/2 + 2\delta\varepsilon \geq u.$$

By the comparison principle (Theorem 3.21), we obtain $u \leq \bar{w}$ in $\Omega \cap B_1(0)$ which implies that $u(x_0', x_n) \leq \bar{w}(x_0', x_n)$ when $x_n \geq 0$, so

$$v_\varepsilon(x_0', x_n) \leq 2\delta + C x_n.$$

To show that

$$v_\varepsilon(x_0', x_n) \geq -2\delta - C x_n,$$

we argue similarly by considering the following lower barrier for u:

$$\underline{w}(x) = \frac{1}{2}|x'|^2 - 16\varepsilon|x' - x_0'|^2 + \frac{1+\delta\varepsilon}{(1-32\varepsilon)^{n-1}}\frac{x_n^2}{2} - C\varepsilon x_n - 2\delta\varepsilon.$$

Step 2. We show that v_ε is well-approximated by harmonic functions.

By our hypothesis, $|v_\varepsilon| \leq 1$ in $B_1(0) \cap \Omega$. It suffices to show that, for a sequence of $\varepsilon, \delta \to 0$, the corresponding v_ε's converge uniformly in $B_{1/2}(0) \cap \overline{\Omega}$ to a solution of (10.16) along a subsequence. From Step 1, we find that v_ε grows at most linearly away from $\{x_n = 0\}$. It remains to prove the uniform convergence of v_ε's on compact subsets of $B_1^+ := B_1(0) \cap \mathbb{R}_+^n$.

Fix a ball $B_{4r}(z) \subset B_1^+$. Note that, for $\delta \leq 1$ and ε small, u belongs to $W^{2,n^2}(B_{2r}(z))$ and $C^{1,1-1/n}(\overline{B_{2r}(z)})$; see Problem 6.5. Let u_0 be the convex solution to the Dirichlet problem

$$\det D^2 u_0 = 1 \quad \text{in } B_{2r}(z), \qquad u_0 = u \quad \text{on } \partial B_{2r}(z).$$

We claim that

$$|u - u_0| \leq 4r^2\delta\varepsilon \quad \text{in } B_{2r}(z).$$

To see this, we first note that u_0 is strictly convex by Theorem 5.14, and hence it is smooth by Theorem 6.5. Now, we use the comparison principle in Theorem 3.21 and the matrix inequality

$$\det(A + \lambda I_n) \geq \det A + n\lambda(\det A)^{\frac{n-1}{n}} \quad \text{if } A \in \mathbb{S}^n, A \geq 0, \quad \text{and} \quad \lambda \geq 0$$

to obtain in $B_{2r}(z)$

$$u + \delta\varepsilon(|x - z|^2 - (2r)^2) \leq u_0 \quad \text{and} \quad u_0 + \delta\varepsilon(|x - z|^2 - (2r)^2) \leq u,$$

from which the claim follows.

Now, if we denote

$$v_0 := (u_0 - P)/\varepsilon,$$

then

$$|v_\varepsilon - v_0| = |u - u_0|/\varepsilon \leq 4r^2\delta \quad \text{in } B_{2r}(z),$$

and hence, $v_\varepsilon - v_0 \to 0$ uniformly in $\overline{B_r(z)}$ as $\delta \to 0$.

Next, we show that, as $\varepsilon \to 0$, the corresponding v_0's converge uniformly, up to extracting a subsequence, in $\overline{B_r(z)}$, to a solution of (10.16). Note that

$$0 = \frac{1}{\varepsilon}(\det D^2 u_0 - \det D^2 P) = \text{trace}(A_\varepsilon D^2 v_0),$$

where, using Cof M to denote the cofactor matrix of the matrix M,

$$A_\varepsilon = \int_0^1 \mathrm{Cof}\,(D^2P + t(D^2u_0 - D^2P))\,dt.$$

When $\varepsilon \to 0$, we have $u \to P$, and therefore $D^2u_0 \to D^2P = I_n$ uniformly in $\overline{B_r(z)}$. This shows that $A_\varepsilon \to I_n$ uniformly in $\overline{B_r(z)}$, and thus v_0's must converge to a harmonic function w satisfying (10.16). The bound $|w| \leq 1$ in B_1^+ follows from the corresponding bound for v_ε and the convergence $v_\varepsilon - v_0 \to 0$.

Step 3. We now establish (10.17). Consider a solution w to (10.16). Then w is smooth. Since $w = 0$ on $\{x_n = 0\}$, $w(0) = 0$, $D_iw(0) = 0$ for $i < n$, and $D_{ij}w(0) = 0$ for $i, j < n$. Therefore, $\Delta w(0) = 0$ gives $D_{nn}w(0) = 0$. Thus the quadratic expansion of w around 0 has the form

$$w(x) = D_nw(0)x_n + \sum_{i=1}^{n-1} D_{in}w(0)x_ix_n + O(|x|^3) \quad \text{as } |x| \to 0.$$

It follows that

$$|w(x) - (a_0 + a' \cdot x')x_n| \leq C(n)|x|^3 \quad \text{in } B_1(0),$$

where $a' = (a_1, \ldots, a_{n-1})$,

$$|a_0| = |D_nw(0)| \leq C(n), \quad \text{and} \quad |a_i| = |D_{in}w(0)| \leq C(n).$$

From Step 2, we obtain

$$|v_\varepsilon(x) - (a_0 + a' \cdot x')x_n| \leq \eta + C(n)|x|^3 \quad \text{in } B_{1/2}(0) \cap \overline{\Omega},$$

provided that ε and δ are sufficiently small, depending on n and η. This proves (10.17), completing the proof of the proposition. $\qquad\square$

Remark 10.15. In Proposition 10.14, if we let

$$\tilde{P}(x) := \varepsilon(a' \cdot x')x_n + \varepsilon^2|a'|^2x_n^2/2,$$

then

$$\det D^2(P + \tilde{P}) = 1, \quad \|D^2\tilde{P}\| \leq C(n)\varepsilon,$$

and, from (10.17), we obtain

$$|u - P - \varepsilon a_0 x_n - \tilde{P}| \leq \varepsilon(\eta + C(n)\varepsilon + C(n)|x|^3) \quad \text{in } B_{1/2}(0) \cap \overline{\Omega}.$$

10.4.4. The perturbation arguments for pointwise $C^{2,\alpha}$ estimates. By choosing δ_0, ε_0 appropriately small, depending on $n, \lambda, \Lambda, \rho, \alpha$, we will show that the rescaled function $\tilde{u}$ (as in Remark 10.12) can be approximated to order $2 + \alpha$ by convex quadratic polynomials $l_m + P_m$ in each ball $B_{r_0^m}(0)$ for some small $r_0 > 0$. This proves $\tilde{u} \in C^{2,\alpha}(0)$ and Theorem 10.8.

Lemma 10.16 (Perturbation arguments at the boundary). *Let $\tilde{\Omega}$ be a convex domain in $\mathbb{R}^n_+$ containing $\{te_n : 0 < t \le 1\}$ and let $\tilde{u} \in C(\overline{\tilde{\Omega}})$ be a convex function satisfying the following conditions:*

(a) *in $\tilde{\Omega} \cap B_1(0)$, we have*
$$\det D^2 \tilde{u} = \tilde{f}, \quad |\tilde{f}(x) - 1| \le \delta_0 \varepsilon_0 |x|^\alpha, \quad and \quad |\tilde{u} - P_0| \le \varepsilon_0,$$
where P_0 is a convex, homogeneous quadratic polynomial with
$$\det D^2 P_0 = 1 \quad and \quad \|D^2 P_0\| \le C_0;$$

(b) *on $\partial\tilde{\Omega} \cap B_1(0)$, we have*
$$|x_n - \tilde{q}(x')| \le \delta_0 \varepsilon_0 |x'|^{2+\alpha}, \quad |\tilde{q}(x')| \le \delta_0 \varepsilon_0, \quad |\tilde{u} - \tilde{p}(x')| \le \delta_0 \varepsilon_0 |x'|^{2+\alpha},$$
where $\tilde{p}$ and $\tilde{q}$ are quadratic polynomials with
$$(\rho/2)|x'|^2 \le \tilde{p}(x') = P_0(x') \le 2\rho^{-1}|x'|^2.$$

There exist ε_0, δ_0, r_0 small and C large, depending only on C_0, n, α, ρ, such that for each $m \in \mathbb{N}_0$, we can find a linear function $l_m(x) = \gamma_m x_n$ and a convex, homogeneous quadratic polynomial P_m with the following properties:

(i) $\det D^2 P_m = 1$, $P_m(x') = \tilde{p}(x') - \gamma_m \tilde{q}(x')$;

(ii) $|\tilde{u} - l_m - P_m| \le \varepsilon_0 r_0^{m(2+\alpha)}$ *in $B_{r_0^m}(0) \cap \tilde{\Omega}$;*

(iii) $r_0^m |\gamma_{m+1} - \gamma_m| + r_0^{2m}\|D^2 P_{m+1} - D^2 P_m\| \le C\varepsilon_0 r_0^{m(2+\alpha)}$. *Furthermore, $|\gamma_m| \le 1$, and $\|D^2 P_m\| \le 2C_0$.*

Proof. We prove the lemma by induction on m. Let us denote $B_t := B_t(0)$. For $m = 0$, we take $\gamma_0 = 0$, while for $m = -1$, we take $\gamma_{-1} = 0$ and $P_{-1} = P_0$. Assume the conclusion of the lemma holds for $m \ge 0$. We now prove it for $m + 1$. Let
$$r := r_0^m, \quad \varepsilon := \varepsilon_0 r^\alpha = \varepsilon_0 r_0^{m\alpha}, \quad and \quad \Omega_r := r^{-1}\tilde{\Omega} \supset \{te_n : 0 < t \le 1\}.$$
Define
$$v(x) := \frac{(\tilde{u} - l_m)(rx)}{r^2} \quad on \ \Omega_r.$$
Then, in $\Omega_r \cap B_1$, we have from property (ii),

(10.18) $|v - P_m| \le \varepsilon$,

and, by condition (a),
$$|\det D^2 v(x) - 1| = |\tilde{f}(rx) - 1| \le \delta_0 \varepsilon_0 |rx|^\alpha = \delta_0 \varepsilon |x|^\alpha \le \delta_0 \varepsilon.$$
On $\partial\Omega_r \cap B_1$, using condition (b), we have

(10.19) $|x_n/r - \tilde{q}(x')| = r^{-2}|rx_n - \tilde{q}(rx')| \le \delta_0 \varepsilon |x'|^{2+\alpha} \le \delta_0 \varepsilon$.

It follows from $\alpha \in (0,1)$ and condition (b) that

(10.20) $|x_n| \le r|\tilde{q}(x')| + \delta_0 \varepsilon \le \delta_0 \varepsilon_0 r + \delta_0 \varepsilon \le 2\delta_0 \varepsilon$ on $\partial\Omega_r \cap B_1$.

Since $\|D^2 P_m\| \leq 2C_0$, we have

$$|P_m(x) - P_m(x')| \leq 2nC_0|x||x_n|.$$

Therefore, using $P_m(x') = \tilde{p}(x') - \gamma_m \tilde{q}(x')$, we have in $B_1 \cap \Omega_r$

$$|v(x) - P_m(x)| \leq r^{-2}|\tilde{u}(rx) - P_m(rx') - r\gamma_m x_n| + |P_m(x) - P_m(x')|$$
$$\leq r^{-2}|\tilde{u}(rx) - \tilde{p}(rx')| + |\gamma_m||x_n/r - \tilde{q}(x')| + 2nC_0|x_n|.$$

By condition (b), (10.19), and (10.20), we can find $C_1(n, C_0) > 1$ such that

$$(10.21) \qquad |v - P_m| \leq C_1 \delta_0 \varepsilon \quad \text{in } \partial\Omega_r \cap B_1.$$

Step 1. Construction of l_{m+1} and P_{m+1}. For any $\eta > 0$, there exist δ_0 and ε_0 small, so that, by Remark 10.13, Proposition 10.14 together with Remark 10.15, we have

$$|v - P_m - \tilde{\gamma}_m x_n - \tilde{P}_m| \leq \varepsilon(2\eta + C_2|x|^3) \quad \text{in } \Omega_r \cap B_{1/2},$$

where $C_2 = C_2(n, \rho, C_0)$ and

$$(10.22) \quad \det D^2(P_m + \tilde{P}_m) = 1, \ \tilde{P}_m(x') = 0, \ |\tilde{\gamma}_m| \leq C_2\varepsilon, \ \|D^2\tilde{P}_m\| \leq C_2\varepsilon.$$

Now, let $\eta = C_2 r_0^3/2$ where $r_0 < 1/2$ is to be chosen. Then, since $\Omega_r \subset \mathbb{R}_+^n$,

$$(10.23) \qquad |v - P_m - \tilde{\gamma}_m x_n - \tilde{P}_m| \leq 2\varepsilon C_2 r_0^3 \quad \text{in } \Omega_r \cap B_{r_0}^+.$$

This inequality suggests that, when rescaling back to $\tilde{u}$, a natural choice for l_{m+1} would be

$$l_{m+1}(x) := \gamma_{m+1} x_n \quad \text{where } \gamma_{m+1} = \gamma_m + r\tilde{\gamma}_m.$$

On the other hand, due to (10.22), $\hat{P}_m := P_m + \tilde{P}_m$ is a candidate for P_{m+1}. However, for γ_{m+1} above, $\hat{P}_m$ does not satisfy property (i) because

$$\hat{P}_m(x') = P_m(x') = \tilde{p}(x') - \gamma_m \tilde{q}(x') = \tilde{p}(x') - \gamma_{m+1}\tilde{q}(x') + r\tilde{\gamma}_m \tilde{q}(x').$$

We correct this mismatch by choosing $\eta_m \in \mathbb{R}$ such that the convex, homogeneous quadratic polynomial

$$P_{m+1}(x) := P_m(x) + \tilde{P}_m(x) - r\tilde{\gamma}_m \tilde{q}(x') + \eta_m x_n^2$$

satisfies $\det D^2 P_{m+1} = 1$.

Step 2. Choice of ε_0, δ_0, and r_0. We show that for ε_0, δ_0, and r_0 small, P_{m+1} satisfies properties (i)–(iii). Indeed, from

$$(2C_0)^{1-n} I_n \leq D^2 P_m \leq 2C_0 I_n, \quad \|D^2\tilde{P}_m\| \leq 2C_2\varepsilon, \ |\tilde{\gamma}_m| \leq C_2\varepsilon, \ |\tilde{q}| \leq \delta_0\varepsilon_0,$$

we deduce

$$(10.24) \qquad \|D^2 P_{m+1} - D^2(P_m + \tilde{P}_m)\| \leq C_3\delta_0\varepsilon,$$

for some $C_3 = C_3(n, \rho, C_0, C_2)$. Therefore

$$(10.25) \qquad |\gamma_{m+1} - \gamma_m| + \|D^2 P_{m+1} - D^2 P_m\| \leq C_4\varepsilon = C_4\varepsilon_0 r_0^{m\alpha}.$$

Combining (10.23) with (10.24) and denoting $\tilde{l}_m(x) = \tilde{\gamma}_m x_n$, we get

$$|v - \tilde{l}_m - P_{m+1}| \leq |v - \tilde{l}_m - P_m - \tilde{P}_m| + |P_{m+1} - (P_m + \tilde{P}_m)|$$
$$\leq (2C_2 r_0^3 + C_4 \delta_0)\varepsilon \quad \text{in } \Omega_r \cap B_{r_0}^+.$$

Thus, by first choosing r_0 small (depending on C_2 and α) and then choosing δ_0 small, depending on r_0 and C_4, we obtain

$$|v - \tilde{l}_m - P_{m+1}| \leq \varepsilon r_0^{2+\alpha} \quad \text{in } \Omega_r \cap B_{r_0}^+.$$

Hence, from the definition of v, we find

$$|\tilde{u} - l_{m+1} - P_{m+1}| \leq \varepsilon r^2 r_0^{2+\alpha} = \varepsilon_0 (rr_0)^\alpha \quad \text{in } \tilde{\Omega} \cap B_{rr_0}^+.$$

We are left with choosing ε_0. From (10.25) together with $\gamma_0 = 0$ and $\|D^2 P_0\| \leq C_0$, we can choose ε_0 small so that

$$|\gamma_m| \leq 1, \quad \|D^2 P_m\| \leq 2C_0 \quad \text{for all } m.$$

With the above choices of ε_0, δ_0, and r_0, the induction hypotheses hold for $m + 1$. The lemma is proved. $\qquad \square$

With all ingredients in place, we can now complete the proof of the pointwise $C^{2,\alpha}$ estimates at the boundary in Theorem 10.8.

Proof of Theorem 10.8. By Remark 10.12, given $\delta_0 \in (0,1)$ and $\varepsilon_0 \in (0,1)$, there is a rescaling $\tilde{u}$ of u satisfying the hypotheses of Lemma 10.16 with $C_0 = C_0(\alpha, n, \lambda, \Lambda, \rho)$.

By choosing $\varepsilon_0, \delta_0, r_0$ small, depending on n, α, and C_0 (and, hence, on $\alpha, n, \lambda, \Lambda, \rho$), we find for each $m \in \mathbb{N}_0$ a constant γ_m and a convex, homogeneous quadratic polynomial P_m as in Lemma 10.16.

Upon letting $m \to \infty$, γ_m has a limit γ with $|\gamma| \leq 1$ and P_m has a limit P with $\det D^2 P = 1$ and $\|D^2 P\| \leq 2C_0$. As in the proof of (6.58), we have

$$(10.26) \qquad |\tilde{u}(x) - \gamma x_n - P(x)| \leq C(\alpha, n, \lambda, \Lambda, \rho)|x|^{2+\alpha} \quad \text{in } \tilde{\Omega} \cap B_1(0).$$

As in the last paragraph in the proof of Theorem 6.30, by rescaling back and recalling (10.13), we can find a convex, homogeneous quadratic polynomial $\mathcal{P}_0$ with

$$\det D^2 \mathcal{P}_0 = 1 = f(0), \quad \|D^2 \mathcal{P}_0\| \leq C^*,$$

and

$$|u(x) - \mathcal{P}_0(x)| \leq C^* |x|^{2+\alpha} \quad \text{in } \Omega \cap B_\rho(0),$$

where C^* depends only on $n, \alpha, \rho, \lambda, \Lambda$, and M. This proves the theorem. $\qquad \square$

10.5. Global Second Derivative Hölder Estimates

Combining the pointwise $C^{2,\alpha}$ estimates at the boundary in Theorem 10.8 with Caffarelli's interior $C^{2,\alpha}$ estimates in Theorem 6.34, we obtain the following theorem of Savin [**S4**] on global $C^{2,\alpha}$ estimates.

Theorem 10.17 (Savin's global $C^{2,\alpha}$ estimates). *Let Ω be a bounded, convex domain in $\mathbb{R}^n$ ($n \geq 2$), and let $u : \overline{\Omega} \to \mathbb{R}$ be a convex, Lipschitz continuous function, satisfying*

$$\det D^2 u = f, \qquad 0 < \lambda \leq f \leq \Lambda \quad in \ \Omega.$$

Assume that

$$\partial\Omega \in C^{2,\alpha}, \quad u|_{\partial\Omega} \in C^{2,\alpha}, \qquad f \in C^{\alpha}(\overline{\Omega}),$$

for some $\alpha \in (0,1)$, and assume that there exists a constant $\rho_0 > 0$ such that

$$u(y) - u(x) - p(x) \cdot (y - x) \geq \rho_0 |y - x|^2 \qquad for \ all \ x, y \in \partial\Omega,$$

where $p(x) \in \partial_s u(x)$ is understood in the sense of Definition 4.1. Then $u \in C^{2,\alpha}(\overline{\Omega})$ with estimate

$$\|u\|_{C^{2,\alpha}(\overline{\Omega})} \leq C,$$

where C depends only on ρ_0, λ, Λ, n, α, $\|f\|_{C^{\alpha}(\overline{\Omega})}$, $\|u\|_{C^{0,1}(\overline{\Omega})}$, $\|u\|_{C^{2,\alpha}(\partial\Omega)}$, and the $C^{2,\alpha}$ regularity of $\partial\Omega$. Moreover, $p(x) = Du(x)$ on $\partial\Omega$.

Proof. Since $u|_{\partial\Omega}, \partial\Omega \in C^{2,\alpha}$, by Theorem 5.14, u is strictly convex in Ω. By Caffarelli's interior $C^{2,\alpha}$ estimates in Theorem 6.34, $u \in C^{2,\alpha}(\Omega)$ with

$$(10.27) \qquad \|u\|_{C^{2,\alpha}(\Omega')} \leq C_1 \quad in \ \Omega' \Subset \Omega,$$

where C_1 depends on $n, \lambda, \Lambda, \alpha, \|f\|_{C^{\alpha}(\overline{\Omega})}, \|u\|_{C^{0,1}(\overline{\Omega})}, \|u\|_{C^{2,\alpha}(\partial\Omega)}$, the $C^{2,\alpha}$ regularity of $\partial\Omega$, and $\mathrm{dist}(\Omega', \partial\Omega)$.

Step 1. We prove $C^{2,\alpha}$ estimates near the boundary. From the hypotheses of the theorem, we can find a constant $\rho \in (0, \rho_0)$ depending on ρ_0, $\|u\|_{C^{0,1}(\overline{\Omega})}$, the $C^{2,\alpha}$ regularity of $\partial\Omega$, and $\|u\|_{C^{2,\alpha}(\partial\Omega)}$ such that $\Omega \subset B_{1/\rho}(0)$,

$$\rho|y - x|^2 \leq u(y) - u(x) - p(x) \cdot (y - x) \leq \rho^{-1}|y - x|^2 \qquad for \ all \ x, y \in \partial\Omega,$$

and at each $y \in \partial\Omega$, there is a ball $B_\rho(\hat{y}) \subset \Omega$ such that $y \in \partial B_\rho(\hat{y})$. By Proposition 8.23, $u \in C^1(\overline{\Omega})$, and therefore $p(x) = Du(x)$ on $\partial\Omega$.

From the hypotheses, we can find $M > 0$ depending on n, α, $\|f\|_{C^{\alpha}(\overline{\Omega})}$, the $C^{2,\alpha}$ regularity of $\partial\Omega$, and $\|u\|_{C^{2,\alpha}(\partial\Omega)}$ such that the hypotheses of Theorem 10.8 are satisfied at each boundary point $z \in \partial\Omega$ (after subtracting the tangent hyperplane to u at z and rotating coordinates).

Take $z \in \Omega$, and let $z_0 \in \partial\Omega$ be such that $|z - z_0| = \mathrm{dist}(z, \partial\Omega) = \delta$, where $\delta \leq \delta_0 (\leq \rho/4)$ is to be determined.

By Theorem 10.8, there exist $M_0 = M_0(M, \rho, \lambda, \Lambda, n, \alpha)$, an affine function $l_{z_0}(x)$, and a convex, homogeneous quadratic polynomial P_{z_0} with

$$(10.28) \qquad \det D^2 P_{z_0} = f(z_0), \quad \|D^2 P_{z_0}\| \leq M_0,$$

such that

$$(10.29) \quad |u(x) - l_{z_0}(x) - P_{z_0}(x - z_0)| \leq M_0 |x - z_0|^{2+\alpha} \quad \text{in } \Omega \cap B_\rho(z_0).$$

We now turn (10.29) into a uniform $C^{2,\alpha}$ estimate for u around z. Let $r(n, \lambda, \Lambda, M_0), \varepsilon_0(n, M_0)$ be the constants in Theorem 6.35. Consider

$$\tilde{u}(x) := \delta^{-2}(u - l_{z_0})(z + \delta x) + P_{z_0}(x) - P_{z_0}(x + (z - z_0)/\delta) \quad \text{for } x \in B_1(0).$$

Since $P_{z_0}(x) - P_{z_0}(x + (z - z_0)/\delta)$ is affine in x, we have

$$\det D^2 \tilde{u}(x) = \det D^2 u(z + \delta x) = f(z + \delta x) := \tilde{f}(x) \quad \text{in } B_1(0).$$

For $x \in B_1(0)$, $|z - z_0 + \delta x| \leq 2\delta \leq \rho/2$, and therefore, (10.29) gives

$$|\tilde{u}(x) - P_{z_0}(x)| = \delta^{-2}|(u - l_{z_0})(z + \delta x) - P_{z_0}(z + \delta x - z_0)| \leq 2^{2+\alpha} M_0 \delta^\alpha.$$

Furthermore

$$\|\tilde{f} - \det D^2 P_{z_0}\|_{C^\alpha(B_1(0))} = \|f(z + \delta \cdot) - f(z_0)\|_{C^\alpha(B_1(0))} \leq M(2\delta)^\alpha.$$

Thus, for δ_0 satisfying

$$\max\{M(2\delta_0)^\alpha, 2^{2+\alpha} M_0 \delta_0^\alpha\} \leq \varepsilon_0,$$

we obtain from Theorem 6.35 that

$$\|\tilde{u} - P_{z_0}\|_{C^{2,\alpha}(B_r(0))} \leq C_0(n, \lambda, \Lambda, \alpha, M)\delta^\alpha.$$

In particular, this implies that there is $C_2(n, \lambda, \Lambda, \alpha, M) > 0$ such that

$$(10.30) \quad \|D^2 u\|_{C^\alpha(B_{r|z-z_0|}(z))} \leq C_2 \quad \text{and} \quad \|D^2 u(z) - D^2 P_{z_0}\| \leq C_2 |z - z_0|^\alpha.$$

Step 2. We prove that $D^2 u$ is C^α at the boundary by showing that

$$\|D^2 P_z - D^2 P_{\bar{z}}\| \leq C_3(n, \lambda, \Lambda, \alpha, M)|z - \bar{z}|^\alpha \quad \text{for } z, \bar{z} \in \partial\Omega.$$

The proof is similar to that of Proposition 8.23. To simplify notation, we assume $\bar{z} = 0 \in \partial\Omega$. Due to (10.28), it suffices to consider $r := |z| \leq \rho/2$. Then $\Omega \cap B_\rho(0) \cap B_\rho(z) \supset \Omega \cap B_r(0)$. By writing (10.29) at $x \in \Omega \cap B_r(0)$ for 0 and z and then subtracting, we find

$$|l_z(x) - l_0(x) + P_z(x - z) - P_0(x)| \leq M_0(|x|^{2+\alpha} + |x - z|^{2+\alpha})$$
$$\leq 9 M_0 r^{2+\alpha} \quad \text{for } x \in \Omega \cap B_r(0).$$

Note that the expression in the left-hand side is a quadratic polynomial in x, with homogeneous quadratic term $(Rx) \cdot x$ where $R := (1/2)(D^2 P_z - D^2 P_0)$. By Lemma 10.18 below, we have $\|R\| \leq C(\rho, n) M_0 r^\alpha$, as desired.

Step 3. Global $C^{2,\alpha}$ estimates. From (10.27) and the second inequality in (10.30), we deduce $\|u\|_{C^{1,1}(\overline{\Omega})} \leq C_4$. As in the proof of Theorem 9.5 (see Step 2 there), by combining (10.27), (10.30), and Step 2, we obtain $\|D^2 u\|_{C^\alpha(\overline{\Omega})} \leq C_5$ where C_4 and C_5 depend on $n, \lambda, \Lambda, \alpha, M$. Therefore

$$\|u\|_{C^{2,\alpha}(\overline{\Omega})} \leq C(n, \lambda, \Lambda, \alpha, M).$$

The proof of the theorem is complete. $\qquad\square$

In the proof of Theorem 10.17, we used the following simple lemma.

Lemma 10.18. *Let Ω be a bounded convex domain in $\mathbb{R}^n$ that contains a ball $B_\rho(z) \subset \Omega$ that is tangent to $\partial\Omega$ at 0. Let $a \in \mathbb{R}$, $\mathbf{b} \in \mathbb{R}^n$, and $R = (r_{ij})_{1 \leq i,j \leq n} \in \mathbb{S}^n$ be such that*

$$|a + \mathbf{b} \cdot x + (Rx) \cdot x| \leq Kr^{2+\alpha} \quad \text{in } \Omega \cap B_r(0),$$

where $\alpha \in (0,1)$ and $r \leq \rho/2$. Then, there is a constant $C(\rho, n)$ such that

$$\|R\| \leq CKr^\alpha.$$

Proof. For simplicity, we assume that $K = 1$. We need to show that

(10.31) $$|r_{ij}| \leq C(\rho, n)r^\alpha \quad \text{for all } 1 \leq i, j \leq n.$$

First, with $x = 0$, we have $|a| \leq r^{2+\alpha}$. Thus

(10.32) $$|\mathbf{b} \cdot x + (Rx) \cdot x| \leq 2r^{2+\alpha} \quad \text{in } \Omega \cap B_r(0).$$

We will use a *scaling argument*. Applying (10.32) to $x/2$ where $x \in \Omega \cap B_r(0)$ and then multiplying the resulting inequality by 2, we obtain

(10.33) $$|\mathbf{b} \cdot x + (Rx) \cdot x/2| \leq 4r^{2+\alpha} \quad \text{in } \Omega \cap B_r(0).$$

It follows from (10.32) and (10.33) that

(10.34) $$|(Rx) \cdot x| \leq 12r^{2+\alpha} \quad \text{in } \Omega \cap B_r(0).$$

By rotating coordinates, we might assume that the inner unit normal to $\partial\Omega$ at 0 is e_n. Furthermore, there is $\delta(\rho, n) > 0$ such that

$$B := \{x \in \mathbb{R}^n : |x'| \leq \delta r, x_n = r/2\} \subset \Omega \cap B_r(0).$$

Letting $x' = 0$ and $x_n = r/2$ in (10.34) yields $|r_{nn}| \leq 48r^\alpha$, so (10.34) gives

(10.35) $$\left| \sum_{1 \leq i,j < n} r_{ij} x_i x_j + \sum_{1 \leq i \leq n-1} r_{in} x_i x_n \right| \leq 60r^{2+\alpha} \quad \text{in } B.$$

Using a scaling argument as above, we obtain

$$\left| \sum_{1 \leq i,j < n} r_{ij} x_i x_j \right| \leq 360r^{2+\alpha} \quad \text{in } B.$$

This implies $|r_{ij}| \leq 360\delta^{-2}r^\alpha$ for $1 \leq i, j < n$. Thus, from (10.35) and $x_n = r/2$ in B, we find $|r_{in}| \leq \bar{C}(\rho, n)r^\alpha$, completing the proof of (10.31). $\qquad\square$

Remark 10.19. In Theorem 10.17, if we assume $f \in C^\alpha(\overline{D_\delta})$ instead of $f \in C^\alpha(\overline{\Omega})$ where $D_\delta := \{x \in \overline{\Omega} : \text{dist}(x, \partial\Omega) \leq \delta\}$, then, from its proof, we find $u \in C^{2,\alpha}(\overline{D_{\delta/2}})$, with estimate

$$\|u\|_{C^{2,\alpha}(\overline{D_{\delta/2}})} \leq C,$$

where C now depends on δ, $\|f\|_{C^\alpha(\overline{D_\delta})}$, ρ_0, λ, Λ, n, α, $\|u\|_{C^{0,1}(\overline{\Omega})}$, $\|u\|_{C^{2,\alpha}(\partial\Omega)}$, and the $C^{2,\alpha}$ regularity of $\partial\Omega$.

By Theorem 3.30, the global Lipschitz bound in Theorem 10.17 is guaranteed from the C^2 character of $u|_{\partial\Omega}$ if Ω is uniformly convex. If, in addition, $\partial\Omega$ and $u|_{\partial\Omega}$ are C^3, then Proposition 4.7 gives the quadratic separation of u from its tangent hyperplanes on $\partial\Omega$. Thus, from Theorem 10.17, we have the following theorem on global $C^{2,\alpha}$ estimates for the Monge–Ampère equation established by Trudinger and Wang [**TW5**] when the Monge–Ampère measure is only assumed to be globally C^α. It is a global version of Theorem 6.34.

Theorem 10.20 (Trudinger–Wang global $C^{2,\alpha}$ estimates)**.** *Let Ω be a uniformly convex domain in $\mathbb{R}^n$ with $\partial\Omega \in C^3$, $\varphi \in C^3(\overline{\Omega})$, and $f \in C^\alpha(\overline{\Omega})$, for some $\alpha \in (0,1)$, satisfying $\inf_\Omega f > 0$. Then, the Dirichlet problem*

$$\begin{cases} \det D^2 u = f & in \ \Omega, \\ \qquad u = \varphi & on \ \partial\Omega \end{cases}$$

has a unique uniformly convex solution $u \in C^{2,\alpha}(\overline{\Omega})$ with estimate

$$\|u\|_{C^{2,\alpha}(\overline{\Omega})} \leq C$$

where C is a constant depending on n, α, $\inf_\Omega f$, $\|f\|_{C^\alpha(\overline{\Omega})}$, $\|\varphi\|_{C^3(\overline{\Omega})}$, the uniform convexity radius of $\partial\Omega$, and the C^3 regularity of $\partial\Omega$.

We refer to Section 15.5 for an application of the global $C^{2,\alpha}$ estimates in Theorem 10.20 to the solvability of fourth-order singular Abreu equations.

10.6. Problems

Problem 10.1. Prove that the global $W^{2,1+\varepsilon}$ and $W^{2,p}$ estimates in Theorems 10.1 and 10.2 continue to hold when the Lebesgue measure dx (see, for example, the left-hand side of (10.2)) is replaced by the weighted Lebesgue measure $\text{dist}^{-\gamma}(\cdot, \partial\Omega)\, dx$ where $0 < \gamma < 1$. Of course, the constants K and M in the estimates now also depend on γ.

Problem 10.2. Let Ω be a uniformly convex domain in $\mathbb{R}^n$ $(n \geq 2)$ with boundary $\partial\Omega \in C^3$. Let $p > n$. For each $k \in \mathbb{N}$, let $u_k \in C(\overline{\Omega})$ be the convex solution to

$$\det D^2 u_k = f_k \geq 0 \quad in \ \Omega, \qquad u_k = 0 \quad on \ \partial\Omega$$

where $\|f_k - 1\|_{L^\infty(\Omega)} \le e^{-C(n)p}$. Show that if $C(n)$ is sufficiently large, then, up to a subsequence, $\{u_k\}_{k=1}^\infty$ converges uniformly to a convex function $u \in C(\overline{\Omega})$ and $\{\det D^2 u_k\}_{k=1}^\infty$ converges weakly to $\det D^2 u$ in $L^{p/n}(\Omega)$.

Problem 10.3. Let Ω be an open triangle and let $u \in C^\infty(\Omega) \cap C(\overline{\Omega})$ satisfy

$$\det D^2 u = 1 \quad \text{in } \Omega, \quad u = 0 \quad \text{on } \partial\Omega.$$

Show that $D^2 u \notin L^1(\Omega)$.
Hint: Use John's Lemma and Example 3.32. Compare with Problem 2.14.

Problem 10.4. Assume that the convex function $u \in C(\overline{\mathbb{R}_+^n})$ satisfies

$$\det D^2 u = 1 \text{ in } \mathbb{R}_+^n, \quad \rho|x|^2 \le u(x) \le \rho^{-1}|x|^2 \text{ in } \mathbb{R}_+^n, \quad \text{and} \quad u(x', 0) = |x'|^2/2,$$

for some $\rho \in (0, 1/2)$. Show that u is a quadratic polynomial.

Problem 10.5. Extend the result of Lemma 10.18 to polynomials P of degree $m \ge 3$ satisfying $|P(x)| \le Kr^{m+\alpha}$ in $\Omega \cap B_r(0)$.

Problem 10.6. Fix $p > 1$. Is it possible to construct counterexamples to global $W^{2,p}$ estimates for the Monge–Ampère equation with strictly positive, continuous Monge–Ampère measure when either the C^3 regularity on the domain boundary or boundary data fails?

Problem 10.7. Let Ω be a bounded convex domain in the upper half-space $\mathbb{R}_+^n$ containing $\{te_n : 0 < t \le 1\}$. Let $v \in C(\overline{\Omega})$ be a convex function and let P be a convex quadratic polynomial. Assume that, for $\delta, \varepsilon > 0$ and $K, M > 1$,

  (a) $x_n \le \delta\varepsilon$ and $|v - P| \le K\delta\varepsilon$ on $\partial\Omega \cap B_1(0)$;
  (b) $\det D^2 P = 1$, $\|D^2 P\| \le M$;
  (c) $|\det D^2 v - 1| \le \delta\varepsilon$ and $|v - P| \le \varepsilon$ in $\Omega \cap B_1(0)$.

Consider the convex solution w to the Monge–Ampère equation

$$\begin{cases} \det D^2 w = 1 & \text{in } B_{1/8}^+(0), \\ \qquad w = v & \text{on } \partial B_{1/8}^+(0) \cap \Omega, \\ \qquad w = P & \text{on } \partial B_{1/8}^+(0) \setminus \Omega, \end{cases}$$

where $B_r^+(0) := B_r(0) \cap \mathbb{R}_+^n$. Show that

$$|w - P| \le 2K\delta\varepsilon \quad \text{in } B_{1/8}^+(0) \setminus \Omega \quad \text{and} \quad |w - v| \le 4K\delta\varepsilon \quad \text{in } \Omega \cap B_{1/8}^+,$$

provided that $\delta \le \bar{\delta}(n, K, M)$ and $\varepsilon \le \bar{\varepsilon}(n, K, M)$.

Problem 10.8. Let $Q := \{(x_1, x_2) \in \mathbb{R}^2 : x_1, x_2 > 0\}$ be the first quadrant in the plane. Let $u \in C(\overline{Q})$ be a nonnegative convex function satisfying

$$\det D^2 u = m \quad \text{in } Q \quad \text{and} \quad u(x) = |x|^2/2 \quad \text{on } \partial Q,$$

for some constant $m > 0$. Show that there is a constant $C(m)$ such that

$$\|u\|_{C^3} \le C(m) \quad \text{in } (B_2(0) \setminus B_1(0)) \cap Q.$$

Problem 10.9. Let Ω be a uniformly convex domain in $\mathbb{R}^n$ ($n \ge 2$) with C^3 boundary. Assume that the convex function $u \in C^2(\Omega) \cap C(\overline{\Omega})$ satisfies

$$\begin{cases} \det D^2 u = \big(f|Du| + g|u|\big)^m & \text{in } \Omega, \\ \quad\quad u = 0 & \text{on } \partial\Omega, \end{cases}$$

where m is a positive integer, f and g are strictly positive and smooth functions on $\overline{\Omega}$. Show that $u \in C^{2,\alpha}(\overline{\Omega})$ for all $\alpha \in (0, 1)$.

10.7. Notes

This chapter is based on Savin [**S4, S5**].

The conclusions of Theorems 10.1 and 10.2 still hold if the conditions on $\partial\Omega$ and $u|_{\partial\Omega}$ there (uniform convexity and C^3 regularity) are replaced by $\partial\Omega$ and $u|_{\partial\Omega}$ being $C^{1,1}$ and u separating quadratically from its tangent hyperplanes on $\partial\Omega$. Proposition 10.6 and Theorem 10.7 are based on Savin [**S4**]. A boundary Pogorelov estimate similar to that of Proposition 10.6 was also proved by Trudinger and Wang [**TW5**].

The proof of Lemma 10.16 is based on the argument of Le and Savin [**LS4**]. The original proof of Lemma 10.16 in [**S4**] used an approximation lemma (see Problem 10.7) and the boundary Pogorelov estimates.

The conclusion of Problem 10.4 also holds for convex solutions growing subcubically at infity; see Theorem 7.33. Problem 10.8 is based on Le–Savin [**LS5**].

Chapter 11

Monge–Ampère Eigenvalue and Variational Method

So far, we have studied the Monge–Ampère equation using either the measure-theoretic method or the maximum principle method. However, as mentioned at the end of Section 4.5, there is a large class of Monge–Ampère equations such as $\det D^2 u = (-u)^p$, for positive p, on a bounded domain with zero boundary values, where the comparison principle is not applicable, while it is not convenient to use the measure-theoretic method because the right-hand side depends on the solution itself. For this class of equations, the *variational method* is very useful. The case p being the dimension is the Monge–Ampère eigenvalue problem.

The main focus of this chapter is the Monge–Ampère eigenvalue and eigenfunctions. Similar to the variational characterization of the first Laplace eigenvalue using the Rayleigh quotient, the Monge–Ampère eigenvalue also has a variational characterization using a Monge–Ampère functional. We will study some basic properties of this Monge–Ampère functional. The existence of the Monge–Ampère eigenfunctions is established using compactness and solutions to certain degenerate Monge–Ampère equations. In turn, these solutions are found using a parabolic Monge–Ampère flow.

To motivate what will follow, we recall two basic features of the spectral theory of the Laplace operator on a bounded domain Ω in $\mathbb{R}^n$ with smooth boundary $\partial\Omega$. First of all, the Laplace eigenvalue problem

$$(11.1) \qquad \Delta w = -\lambda w \quad \text{in } \Omega, \qquad w = 0 \quad \text{on } \partial\Omega$$

has an increasing sequence of eigenvalues

$$\lambda_1 < \lambda_2 \leq \lambda_3 \leq \cdots \to +\infty,$$

where $\lambda_1 > 0$, and corresponding nontrivial eigenfunctions $w_k \in C^\infty(\overline{\Omega})$. Second, the first eigenvalue λ_1 of the Laplace operator has a variational characterization via the Rayleigh quotient:

$$(11.2) \qquad \lambda_1 = \inf \left\{ \frac{\int_\Omega |Du|^2 \, dx}{\int_\Omega |u|^2 \, dx} : u \in W_0^{1,2}(\Omega) \right\}.$$

We will study the Monge–Ampère analogue of the Laplace eigenvalue problem. We would like to replace the Laplace operator Δw in (11.1) by certain expressions involving the Monge–Ampère operator $\det D^2 w$. Due to the linear scaling of (11.1), that is, if w is a solution, then so is αw for all $\alpha \in \mathbb{R}$, a reasonable replacement will be $(\det D^2 w)^{1/n}$. We will focus on the Monge–Ampère operator in the elliptic setting so we will assume functions involved to be convex. Thus, the Monge–Ampère eigenvalue problem can be stated as follows: Find $\gamma \in \mathbb{R}$ and a nonzero convex function u solving

$$(\det D^2 u)^{\frac{1}{n}} = -\gamma u \quad \text{in } \Omega, \qquad u = 0 \quad \text{on } \partial\Omega.$$

In this setting, Ω is a sublevel set of a strictly convex function u, so it is convex. Furthermore, $u < 0$ in Ω, so $-\gamma u = \gamma|u|$, and from the equation for u, we find that $\gamma > 0$. Thus, we can recast the above problem as follows.

Formulation of the Monge–Ampère eigenvalue problem. Given a bounded, convex domain Ω in $\mathbb{R}^n$, find constants $\lambda > 0$ and nonzero convex functions u solving

$$(11.3) \qquad \begin{cases} \det D^2 u = \lambda |u|^n & \text{in } \Omega, \\ \quad u = 0 & \text{on } \partial\Omega. \end{cases}$$

In one dimension, the Laplace operator coincides with the Monge–Ampère operator; however, there is an important difference between the eigenvalue problems (11.1) and (11.3) due to the requirement of *convexity of the eigenfunctions* for the later problem. To illustrate, consider $\Omega = (0, \pi) \subset \mathbb{R}$. In this case, the Laplace eigenvalues are $\lambda_k = k^2$ and the corresponding eigenfunctions are $w_k = c \sin(kx)$ where $k = 1, 2 \ldots$. These are solutions of (11.1). For each $k \geq 2$, w_k is convex on a proper subset of Ω and it is concave on the complement. Only when $k = 1$ can we choose $w_1 = -c \sin x$, where $c > 0$, to be convex. See Figure 11.1. Thus, (11.3) has only one eigenvalue $\lambda = \lambda_1 = 1$. This turns out to be true in all dimensions. As will be seen in Theorem 11.17, there is only one eigenvalue for the Monge–Ampère operator on any bounded, convex domain Ω. Thus, we can speak of the Monge–Ampère eigenvalue $\lambda[\Omega]$.

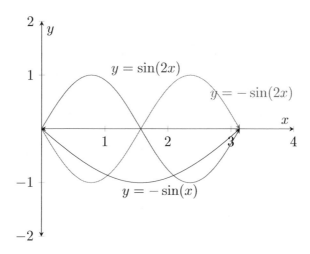

Figure 11.1. Graphs of $\sin x$, $-\sin(2x)$, and $\sin(2x)$.

Next, we would like to obtain an analogue of the variational characterization (11.2) for the Monge–Ampère eigenvalue. Again, the idea is to replace the Laplace operator by certain expressions involving the Monge–Ampère operator. Thus, we need to rewrite (11.2) so as to have the appearance of the Laplace operator on its right-hand side. In general, this could be problematic as u is only required to be $W_0^{1,2}(\Omega)$. However, to fix the idea, let us assume that u is smooth. Then, an integration by parts gives the following expression for the Rayleigh quotient:

$$(11.4) \qquad \frac{\int_\Omega |Du|^2 \, dx}{\int_\Omega |u|^2 \, dx} = \frac{\int_\Omega -u\Delta u \, dx}{\int_\Omega |u|^2 \, dx}.$$

Now, the Laplace operator appears in (11.4) and we can try to replace it by the Monge–Ampère operator for a variational characterization of $\lambda[\Omega]$. Note that (11.4) is invariant under the linear scaling $u \mapsto \alpha u$ when $\alpha > 0$. Thus, in analogy to (11.2), we expect that

$$\lambda[\Omega] = \inf\left\{ \frac{\int_\Omega -u \det D^2 u \, dx}{\int_\Omega |u|^{n+1} \, dx} : u \in C(\overline{\Omega}) \text{ convex}, u = 0 \text{ on } \partial\Omega \right\}.$$

Theorem 11.17 confirms this expectation. In so doing, it is important to investigate the expression $\int_\Omega -u \det D^2 u \, dx$, called the Monge–Ampère energy of a convex function u, with no additional regularity. This is the subject of the next section. For a general convex function u, it is appropriate to replace $\det D^2 u$ by its Monge–Ampère measure μ_u (see Definition 3.1).

11.1. The Monge–Ampère Energy

We assume in this chapter that Ω is a bounded convex domain in $\mathbb{R}^n$ ($n \geq 2$).

For a convex function $u \in C(\overline{\Omega})$, we define its Monge–Ampère energy by

$$(11.5) \qquad I(u) = I(u; \Omega) = \int_{\Omega} (-u) \, d\mu_u.$$

Our analysis is mostly for the case $u = 0$ on $\partial\Omega$. Then, $u \leq 0$ in Ω, and the Monotone Convergence Theorem gives

$$\lim_{\varepsilon \to 0} \int_{\left\{ x \in \Omega : u(x) < -\varepsilon \right\}} (-u) \, d\mu_u = I(u) = \int_{\Omega} |u| \, d\mu_u.$$

In this section, we study basic properties of the Monge–Ampère energy including: continuity, monotonicity, energy estimates, and approximation with strictly convex, smooth functions. In $I(u)$, the convex function u is integrated against the Monge–Ampère measure μ_u, so it is not surprising to see many similarities between $I(u)$ and the Monge–Ampère mass $\mu_u(\Omega)$ of u.

First, we show certain continuity property of $I(\cdot; \Omega)$ along a sequence of convex functions converging uniformly on the whole domain.

Proposition 11.1. *Let Ω be a bounded convex domain in $\mathbb{R}^n$, and let $I(\cdot) = I(\cdot; \Omega)$ be as in (11.5). Let $\{u_k\}_{k=1}^{\infty}, \{f_k\}_{k=1}^{\infty} \subset C(\overline{\Omega})$ be sequences of functions converging uniformly on $\overline{\Omega}$ to $u, f \in C(\overline{\Omega})$, respectively. Furthermore, assume that $f = 0$ on $\partial\Omega$, u_k and u are convex with $\sup_{k \geq 1} \mu_{u_k}(\Omega) < \infty$.*

  (i) *Then $\lim_{k \to \infty} \int_{\Omega} f_k \, d\mu_k = \int_{\Omega} f \, d\mu_u$.*
      In particular, if $u = 0$ on $\partial\Omega$, then $\lim_{k \to \infty} I(u_k) = I(u)$.

  (ii) *Assume that $u_k, u \leq 0$ in Ω. Then $I(u) \leq \liminf_{k \to \infty} I(u_k)$.*

Proof. Assume $\sup_{k \geq 1} \mu_{u_k}(\Omega) \leq E$ for a constant E. Let $\varphi \in C_c(\Omega)$. Then,

$$(11.6) \qquad \lim_{k \to \infty} \int_{\Omega} \varphi f_k \, d\mu_{u_k} = \int_{\Omega} \varphi f \, d\mu_u.$$

Indeed, write

$$\int_{\Omega} \varphi f_k \, d\mu_{u_k} - \int_{\Omega} \varphi f \, d\mu_u = \int_{\Omega} \varphi f \, d\mu_{u_k} - \int_{\Omega} \varphi f \, d\mu_u - \int_{\Omega} \varphi(f - f_k) \, d\mu_{u_k}.$$

Then the uniform convergence of u_k to u on Ω and Theorem 3.7 give

$$\int_{\Omega} \varphi f \, d\mu_{u_k} - \int_{\Omega} \varphi f \, d\mu_u \to 0 \quad \text{as } k \to \infty.$$

Moreover, by Example 3.5, we also have

$$\left| \int_\Omega \varphi(f - f_k) \, d\mu_{u_k} \right| \leq \|\varphi\|_{L^\infty(\Omega)} \|f - f_k\|_{L^\infty(\mathrm{spt}(\varphi))} \mu_{u_k}(\mathrm{spt}(\varphi))$$

$$\leq \|\varphi\|_{L^\infty(\Omega)} \|f - f_k\|_{L^\infty(\mathrm{spt}(\varphi))} \omega_n [\mathrm{dist}(\mathrm{spt}(\varphi), \partial\Omega)]^{-n} (\underset{\Omega}{\mathrm{osc}}\, u_k)^n \to 0$$

as $k \to \infty$. Therefore, (11.6) holds. Here, we did not use $\sup_{k\geq 1} \mu_{u_k}(\Omega) \leq E$.

Fix $\varepsilon > 0$. Let $\Omega_\varepsilon \Subset \Omega$ be an open set to be determined. Let $\varphi_\varepsilon \in C_c(\Omega)$ be such that $0 \leq \varphi_\varepsilon \leq 1$ in Ω and $\varphi_\varepsilon \equiv 1$ on $\overline{\Omega_\varepsilon}$. Then, for any $k \geq 1$,

$$\int_\Omega f \, d\mu_u - \int_\Omega f_k \, d\mu_{u_k} = \int_\Omega \varphi_\varepsilon f \, d\mu_u - \int_\Omega \varphi_\varepsilon f_k \, d\mu_{u_k}$$

$$(11.7) \qquad\qquad + \int_{\Omega\setminus\overline{\Omega_\varepsilon}} (1 - \varphi_\varepsilon) f \, d\mu_u - \int_{\Omega\setminus\overline{\Omega_\varepsilon}} (1 - \varphi_\varepsilon) f_k \, d\mu_{u_k}.$$

We prove assertion (i). Since μ_{u_k} converges weakly* to μ_u and recalling $\sup_{k\geq 1} \mu_{u_k}(\Omega) \leq E$, we have

$$(11.8) \qquad\qquad \mu_u(\Omega) \leq E.$$

We choose $\Omega_\varepsilon := \{x \in \Omega : |f(x)| > \varepsilon\}$. Then $|f| \leq \varepsilon$ in $\Omega \setminus \overline{\Omega_\varepsilon}$. From the uniform convergence of f_k to f in $\overline{\Omega}$, there is $C > 1$ independent of k and ε such that $|f_k| \leq C\varepsilon$ in $\Omega \setminus \overline{\Omega_\varepsilon}$. Now, (11.8) together with $\mu_{u_k}(\Omega) \leq E$ gives

$$\left| \int_{\Omega\setminus\overline{\Omega_\varepsilon}} (1 - \varphi_\varepsilon) f \, d\mu_u \right| + \left| \int_{\Omega\setminus\overline{\Omega_\varepsilon}} (1 - \varphi_\varepsilon) f_k \, d\mu_{u_k} \right| \leq 2CE\varepsilon.$$

Therefore, by (11.7),

$$\left| \int_\Omega f \, d\mu_u - \int_\Omega f_k \, d\mu_{u_k} \right| \leq \left| \int_\Omega \varphi_\varepsilon f \, d\mu_u - \int_\Omega \varphi_\varepsilon f_k \, d\mu_{u_k} \right| + 2CE\varepsilon.$$

Consequently, letting $k \to \infty$ and recalling (11.6), we have

$$\limsup_{k\to\infty} \left| \int_\Omega f \, d\mu_u - \int_\Omega f_k \, d\mu_{u_k} \right| \leq 2CE\varepsilon.$$

Since this holds for all $\varepsilon > 0$, we conclude that $\lim_{k\to\infty} \int_\Omega f_k \, d\mu_{u_k} = \int_\Omega f \, d\mu_u$.

Finally, we prove assertion (ii). Assume $u_k, u \leq 0$ in Ω. Then, by (11.7) applied to $f_k = u_k$ and $f = u$, we have

$$I(u_k) - I(u) \geq \int_\Omega \varphi_\varepsilon u \, d\mu_u - \int_\Omega \varphi_\varepsilon u_k \, d\mu_{u_k} + \int_{\Omega\setminus\overline{\Omega_\varepsilon}} (1 - \varphi_\varepsilon) u \, d\mu_u.$$

Therefore, invoking (11.6) again, we find

$$\liminf_{k\to\infty} I(u_k) \geq I(u) + \int_{\Omega\setminus\overline{\Omega_\varepsilon}} (1 - \varphi_\varepsilon) u \, d\mu_u.$$

Choose $\Omega_\varepsilon = \{x \in \Omega : \mathrm{dist}(x, \partial\Omega) > \varepsilon\}$. Then, upon letting $\varepsilon \to 0$, the Dominated Convergence Theorem and (11.8) give $\liminf_{k\to\infty} I(u_k) \geq I(u)$. $\square$

Refining Proposition 11.1, we establish the lower semicontinuity of $I(\cdot; \Omega)$ along locally uniformly converging sequences of convex functions.

Proposition 11.2 (Lower semicontinuity of the Monge–Ampère energy). *Let Ω be a bounded convex domain in $\mathbb{R}^n$. Let $\{u_k\}_{k=1}^{\infty} \subset C(\Omega)$ be a uniformly bounded sequence of nonpositive convex functions that converges uniformly on compact subsets of Ω to a convex function u on Ω. Then*

$$I(u; \Omega) \leq \liminf_{k \to \infty} I(u_k; \Omega).$$

Proof. Let K be any compact subset of Ω. Choose an open convex set $\mathcal{O}$ such that $K \subset \mathcal{O} \subset \overline{\mathcal{O}} \Subset \Omega$. Let $M := \sup_k \|u_k\|_{L^{\infty}(\Omega)} < \infty$. From Lemma 2.37(iii), we have $\partial u_k(\mathcal{O}) \subset B_r(0)$ where $r := 2M[\text{dist}(\overline{\mathcal{O}}, \partial\Omega)]^{-1}$, and thus $\mu_{u_k}(\mathcal{O}) \leq |B_r(0)| < \infty$. By Proposition 11.1(ii), we have

$$I(u; \mathcal{O}) \leq \lim_{k \to \infty} I(u_k; \mathcal{O}) \leq \liminf_{k \to \infty} I(u_k; \Omega).$$

It follows that

$$I(u; K) \leq \liminf_{k \to \infty} I(u_k; \Omega).$$

Taking the supremum over K completes the proof of the proposition. $\square$

As observed by Krylov [**K1**], the Monge–Ampère energy enjoys corresponding monotonicity properties of the Monge–Ampère measure.

Proposition 11.3 (Monotonicity of the Monge–Ampère energy). *Let Ω be a bounded convex domain in $\mathbb{R}^n$. Let $u, v \in C^{0,1}(\overline{\Omega})$ be convex functions on $\overline{\Omega}$ with $u = v = 0$ on $\partial\Omega$. Let $w(t) = (1-t)u + tv$ for $t \in [0, 1]$. Then*

$$\frac{d}{dt} \int_{\Omega} w(t) \, d\mu_{w(t)} = (n+1) \int_{\Omega} \partial_t w(t) \, d\mu_{w(t)}.$$

As a consequence, if $u \geq v$ in Ω, then $I(u; \Omega) \leq I(v; \Omega)$.

Proof. We divide the proof into several steps.

Step 1. $u, v \in C^4(\overline{\Omega})$, $u = v = 0$ on $\partial\Omega$, and $\partial\Omega \in C^4$. In this case, letting $W(t) = (W^{ij}(t))_{1 \leq i,j \leq n}$ be the cofactor matrix of $D^2 w(t)$, we compute

$$\frac{d}{dt} \int_{\Omega} w(t) \det D^2 w(t) \, dx = \int_{\Omega} \partial_t w(t) \det D^2 w(t) \, dx$$

$$+ \int_{\Omega} w(t) W^{ij}(t) D_{ij}(\partial_t w(t)) \, dx$$

$$= (n+1) \int_{\Omega} \partial_t w(t) \det D^2 w(t) \, dx.$$

In the last line, we use integration by parts twice together with the fact that $W(t)$ is divergence-free and $W^{ij}(t)D_{ij}w(t) = n \det D^2 w(t)$ to get

$$\int_\Omega w(t)W^{ij}(t)D_{ij}(\partial_t w(t))\,dx = \int_\Omega D_{ij}(w(t)W^{ij}(t))\partial_t w(t)\,dx$$

$$= n\int_\Omega \partial_t w(t) \det D^2 w(t)\,dx.$$

Step 2. General case. The Lispchitz properties of u and v allow us to extend u and v to convex, Lipschitz functions on a bounded convex set $\hat\Omega \ni \Omega$. Then

$$(11.9) \qquad \mu_u(\hat\Omega) + \mu_v(\hat\Omega) + \mu_{w(t)}(\hat\Omega) \leq E < \infty.$$

By Theorem 2.51, there is a sequence of uniformly convex domains $\{\Omega^k\}_{k=1}^\infty$ with $\partial\Omega^k \in C^\infty$ such that $\Omega^k \subset \Omega^{k+1} \subset \Omega$ and Ω^k converges to Ω in the Hausdorff distance. We can find convex functions $u^k, v^k \in C^5(\hat\Omega)$ such that $u^k = v^k = 0$ on $\partial\Omega^k$ and $u^k \to u$, $v^k \to v$ uniformly on $\overline\Omega$. To see this, let $u^k \in C^5(\overline{\Omega^k})$ be the convex solution of the equation (see Theorem 4.12)

$$\det D^2 u^k = \det D^2(u * \varphi_{1/k}) + 1/k \quad \text{in } \Omega^k, \quad u^k = 0 \quad \text{on } \partial\Omega^k,$$

and then extend u^k into $\hat\Omega$ as in the proof of Lemma 2.52. Here $\varphi_{1/k}$ is the standard mollifier (Definition 2.49). We can use Theorem 3.20 to get the uniform convergence of u^k to u on $\overline\Omega$.

Let $w^k(t) = (1-t)u^k + tv^k$. Since $w^k(t)$ converges uniformly to $w(t)$ on Ω, $\mu_{w^k(t)}$ converges weakly* to $\mu_{w(t)}$ in Ω for each $t \in [0,1]$; see Theorem 3.7. Thus, from (11.9), we have $\mu_{w^k(t)}(\Omega) \leq E$.

Applying Proposition 11.1(i) to $f^k := w^k\chi_{\Omega^k}$ and $f = w$, we have

$$(11.10) \qquad \lim_{k\to\infty} \int_{\Omega^k} w^k(t)\,d\mu_{w^k(t)} = \int_\Omega w(t)\,d\mu_{w(t)} \quad \text{for each } t \in [0,1].$$

Applying Proposition 11.1(i) to $f^k := \partial_t w^k \chi_{\Omega^k}$ and $f = \partial_t w$, we find

$$(11.11) \qquad \lim_{k\to\infty} \int_{\Omega^k} \partial_t w^k(t)\,d\mu_{w^k(t)} = \int_\Omega \partial_t w(t)\,d\mu_{w(t)} \quad \text{for each } t \in [0,1].$$

Since $\int_{\Omega^k} w^k(t)\,d\mu_{w^k(t)} = \int_{\Omega^k} w^k(t) \det D^2 w^k(t)\,dx$ is a polynomial in t of degree no greater than $n+1$, we can exchange the limit and differentiating with respect to t in this expression; that is,

$$\frac{d}{dt}\left(\lim_{k\to\infty} \int_{\Omega^k} w^k(t)\,d\mu_{w^k(t)}\right) = \lim_{k\to\infty} \frac{d}{dt}\int_{\Omega^k} w^k(t)\,d\mu_{w^k(t)}.$$

Combining the above equation with (11.10), we get

$$(11.12) \qquad \frac{d}{dt}\int_\Omega w(t)\,d\mu_{w(t)} = \lim_{k\to\infty} \frac{d}{dt}\int_{\Omega^k} w^k(t)\,d\mu_{w^k(t)}.$$

Applying the result in Step 1 to (11.12) and then recalling (11.11), we find

$$\frac{d}{dt}\int_\Omega w(t)\,d\mu_{w(t)} = \lim_{k\to\infty}(n+1)\int_{\Omega^k}\partial_t w^k(t)\,d\mu_{w^k(t)} = (n+1)\int_\Omega \partial_t w(t)\,d\mu_{w(t)}.$$

If $u \geq v$ in Ω, then $\partial_t w = v - u \leq 0$ on Ω, and hence

$$I(u) - I(v) = \int_0^1\left(\frac{d}{dt}\int_\Omega w(t)\,d\mu_{w(t)}\right)dt = (n+1)\int_0^1\int_\Omega \partial_t w(t)\,d\mu_{w(t)} \leq 0.$$

The proposition is proved. $\square$

The following energy estimate, due to Krylov [**K1**], resembles the Aleksandrov maximum principle in Theorem 3.12.

Theorem 11.4 (Energy estimate). *Let Ω be a bounded convex domain in $\mathbb{R}^n$, and let $D(x) := \max_{y\in\partial\Omega}|x - y|$. Let $u \in C(\overline{\Omega})$ be a convex function with $u = 0$ on $\partial\Omega$. Then for all $x_0 \in \Omega$, we have*

$$|u(x_0)|^{n+1} \leq \min\left\{n\omega_{n-1}^{-1}D^{n-1}(x_0)\,\mathrm{dist}(x_0,\partial\Omega)I(u;\Omega),\ \omega_n^{-1}D^n(x_0)I(u;\Omega)\right\}.$$

Proof. We divide the proof into two steps.

Step 1. $u \in C^{0,1}(\overline{\Omega})$. Let v be the convex function whose graph is the cone with vertex $(x_0, u(x_0))$ and the base Ω, with $v = 0$ on $\partial\Omega$. Then $v \geq u$ in Ω, since u is convex. Moreover, $v \in C^{0,1}(\overline{\Omega})$, and

$$I(v;\Omega) = |v(x_0)||\partial v(\Omega)| = |u(x_0)||\partial v(\Omega)|.$$

From Theorem 3.12, we have

$$|\partial v(\Omega)| \geq \max\left\{\omega_n\left(\frac{|u(x_0)|}{D(x_0)}\right)^n, \frac{\omega_{n-1}}{n}\left(\frac{|u(x_0)|}{D(x_0)}\right)^{n-1}\frac{|u(x_0)|}{\mathrm{dist}(x_0,\partial\Omega)}\right\}.$$

By Proposition 11.3, $I(u;\Omega) \geq I(v;\Omega)$. Thus, the desired estimate follows.

Step 2. General case. It suffices to consider the case $u < 0$ in Ω. Let $x_0 \in \Omega$. Take $\varepsilon \in (0, |u(x_0)|/2)$. Then $x_0 \in \Omega_\varepsilon := \{x \in \Omega : u(x) < -\varepsilon\}$. Moreover, by Lemma 2.40, we have $u \in C^{0,1}(\overline{\Omega_\varepsilon})$. Using Step 1 for $\bar{u} = u+\varepsilon$ in Ω_ε, noting that $|\bar{u}(x)| = |u(x)| - \varepsilon \leq |u(x)|$ in Ω_ε, so $I(\bar{u};\Omega_\varepsilon) \leq I(u;\Omega)$, and then letting $\varepsilon \to 0$, we obtain the desired estimate. $\square$

From the lower semicontinuity and energy estimates for the Monge–Ampère energy, we obtain the following basic existence theorem on minimizers of the Monge–Ampère energy with a constraint.

Theorem 11.5 (Minimizers of the Monge–Ampère energy with constraint). *Let Ω be a bounded convex domain in $\mathbb{R}^n$. Denote*

$$K(\Omega) = \left\{v\ \text{convex}, v \in C(\overline{\Omega}),\ \|v\|_{L^{n+1}(\Omega)} = 1,\ v = 0\ \text{on}\ \partial\Omega\right\}.$$

Then, there exists a convex function $u \in K(\Omega)$ such that

$$I(u;\Omega) = \inf_{v\in K(\Omega)} I(v;\Omega).$$

Proof. For simplicity, denote $I(u) = I(u; \Omega)$. Let $\{u_k\}_{k=1}^{\infty} \subset K(\Omega)$ be a minimizing sequence for the energy functional I so that

$$\lim_{k \to \infty} I(u_k) = \inf_{v \in K(\Omega)} I(v) =: \lambda[\Omega].$$

By taking a convex function $v \in K(\Omega)$ whose graph is the cone with vertex at $(x_0, v(x_0)) \in \Omega \times \mathbb{R}$, we see that $\lambda[\Omega] < \infty$. From Theorem 11.4 and Lemma 2.37(iv), we find that $\{u_k\}_{k=1}^{\infty}$ is uniformly bounded in $C^{\frac{1}{n+1}}(\overline{\Omega})$. Thus, by the Arzelà–Ascoli Theorem (Theorem 2.41), there exists a subsequence $\{u_{k_i}\}_{i=1}^{\infty}$ that converges uniformly on $\overline{\Omega}$ to a convex function $u \in C^{\frac{1}{n+1}}(\overline{\Omega})$ with $u = 0$ on $\partial \Omega$. It follows that $\|u\|_{L^{n+1}(\Omega)} = \lim_{i \to \infty} \|u_{k_i}\|_{L^{n+1}(\Omega)} = 1$, so $u \in K(\Omega)$. As a result, $I(u) \geq \lambda[\Omega]$. On the other hand, by Proposition 11.2,

$$I(u) \leq \liminf_{i \to \infty} I(u_{k_i}) = \lambda[\Omega].$$

Therefore, $I(u) = \lambda[\Omega]$, and u satisfies the conclusion of the theorem. $\square$

The constant $\lambda[\Omega]$ is called the Monge–Ampère eigenvalue of Ω. We will prove in Theorem 11.17 that there exists a minimizer $u \in C(\overline{\Omega})$, called the Monge–Ampère eigenfunction of Ω, satisfying $u \in C^{\infty}(\Omega)$ and

$$\det D^2 u = \lambda[\Omega]|u|^n \quad \text{in } \Omega.$$

For this purpose, we establish various useful uniform estimates for $I(\cdot; \Omega)$ and for the degenerate Monge–Ampère equations $\det D^2 u = M|u|^p$.

Lemma 11.6 (Uniform estimates for degenerate Monge–Ampère equations). *Let Ω be a bounded convex domain in $\mathbb{R}^n$. Let $0 \leq p < \infty$.*

(i) *Let $u \in C(\overline{\Omega})$ be a convex function in Ω with $u = 0$ on $\partial \Omega$. Then*

$$c(n, p)|\Omega|\|u\|_{L^{\infty}(\Omega)}^p \leq \int_{\Omega} |u|^p \, dx \leq |\Omega|\|u\|_{L^{\infty}(\Omega)}^p$$

and

$$I(u; \Omega) = \int_{\Omega} |u| \, d\mu_u \geq c(n)|\Omega|^{-1}\|u\|_{L^{\infty}(\Omega)}^{n+1}.$$

(ii) *There exists a nonzero convex function $u \in C(\overline{\Omega}) \cap C^{\infty}(\Omega)$ with $u = 0$ on $\partial \Omega$ such that*

$$I(u; \Omega) \leq C(n)|\Omega|^{-1}\|u\|_{L^{\infty}(\Omega)}^{n+1}.$$

If Ω is uniformly convex with $C^{4,\alpha}$ boundary $\partial \Omega$ where $\alpha \in (0, 1)$, then we can choose u to satisfy additionally $u \in C^{4,\alpha}(\overline{\Omega})$.

(iii) *Let $M > 0$ be a positive constant. Assume that $u \in C(\overline{\Omega})$ is a nonzero convex Aleksandrov solution to the Dirichlet problem*

$$\det D^2 u = M|u|^p \quad \text{in } \Omega, \qquad u = 0 \quad \text{on } \partial\Omega.$$

Then

$$c(n,p)|\Omega|^{-2} \leq M\|u\|_{L^\infty(\Omega)}^{p-n} \leq C(n,p)|\Omega|^{-2}.$$

Proof. Under the affine transformations $\Omega \mapsto T(\Omega), u(x) \mapsto u(T^{-1}x)$, where $Tx = Ax + b : \mathbb{R}^n \to \mathbb{R}^n$ for $A \in \mathbb{M}^{n\times n}$ with $\det A = 1$, the equation $\det D^2 u = M|u|^p$, the quantities $I(u; \Omega), \|u\|_{L^p(\Omega)}, \|u\|_{L^\infty(\Omega)}$, and $|\Omega|$ are unchanged. Thus, by John's Lemma, Lemma 2.63, we can assume that

$$B_R(0) \subset \Omega \subset B_{nR}(0) \quad \text{for some } R > 0.$$

We prove part (i). By scaling, we can assume that $\|u\|_{L^\infty(\Omega)} = 1$. Then, clearly, $\int_\Omega |u|^p \, dx \leq |\Omega|$. Moreover, for $x \in B_{R/2}(0)$, Lemma 2.37(i) gives

$$|u(x)| \geq \frac{\text{dist}(x, \partial\Omega)}{\text{diam}(\Omega)} \geq (2nR)^{-1}\text{dist}(x, \partial\Omega) \geq c(n) := (4n)^{-1}.$$

Then, using $|\Omega| \leq C(n)R^n$, we get

$$(11.13) \qquad \int_\Omega |u|^p \, dx \geq \int_{B_{R/2}(0)} |u|^p \, dx \geq [c(n)]^p |B_{R/2}(0)| \geq c(n,p)|\Omega|.$$

Next, let $\Omega' := \{x \in \Omega : u(x) \leq -1/2\}$. Applying the Aleksandrov maximum principle in Theorem 3.12 to $u + 1/2$ on Ω', we find

$$1/2 = \|u + 1/2\|_{L^\infty(\Omega')} \leq C(n)\text{diam}(\Omega')(\mu_u(\Omega'))^{1/n} \leq C(n)R(\mu_u(\Omega'))^{1/n}.$$

Hence

$$I(u; \Omega) \geq \int_{\Omega'} |u| \, d\mu_u \geq (1/2) \int_{\Omega'} d\mu_u = \mu_u(\Omega')/2 \geq c(n)R^{-n} \geq c(n)|\Omega|^{-1}.$$

Now, we prove part (ii). By Theorem 3.38, there exists a unique convex Aleksandrov solution $u \in C(\overline{\Omega})$ to

$$\det D^2 u = R^{-2n} \quad \text{in } \Omega, \qquad u = 0 \quad \text{on } \partial\Omega.$$

Clearly, $u \in C^\infty(\Omega)$. If Ω is uniformly convex with $C^{4,\alpha}$ boundary $\partial\Omega$ where $\alpha \in (0, 1)$, then $u \in C^{4,\alpha}(\overline{\Omega})$, by Theorem 4.12.

Due to $B_R(0) \subset \Omega$, the second inequality in (3.15) in Lemma 3.24 (with $b = 0$ and $\varphi \equiv 0$) gives $u(x) \leq (R^{-2}/2)(|x|^2 - R^2)$. It follows that

$$h := \|u\|_{L^\infty(\Omega)} = -\min_{\overline{\Omega}} u \geq -u(0) \geq 1/2.$$

Thus, the desired inequality for $I(u; \Omega)$ follows from $\Omega \subset B_{nR}(0)$ and

$$I(u; \Omega) = \int_\Omega |u| \, d\mu_u \leq h \int_\Omega R^{-2n} \, dx = R^{-2n} |\Omega| h \leq \omega_n 2^n n^{2n} |\Omega|^{-1} h^{n+1}.$$

Finally, we prove part (iii). Let $\alpha = \|u\|_{L^\infty(\Omega)} > 0$ and $v = u/\alpha$. Then, $v = 0$ on $\partial\Omega$, $\|v\|_{L^\infty(\Omega)} = 1$, $v \in C(\overline{\Omega}) \cap C^\infty(\Omega)$ (by Proposition 6.36), and

$$(11.14) \qquad \det D^2 v = M\alpha^{p-n} |v|^p \quad \text{in } \Omega.$$

To estimate $M\alpha^{p-n}$ from below, we multiply both sides of (11.14) by $|v|$, integrate over Ω, and then use (i) to obtain the desired lower bound:

$$M\alpha^{p-n} = \frac{\int_\Omega |v| \det D^2 v \, dx}{\int_\Omega |v|^{p+1} \, dx} \geq c(n, p) |\Omega|^{-2}.$$

For the upper bound of $M\alpha^{p-n}$, observe from (11.14) and (11.13) that

$$(11.15) \qquad M\alpha^{p-n} = \frac{\int_{B_{R/2}(0)} \det D^2 v \, dx}{\int_{B_{R/2}(0)} |v|^p \, dx} \leq \frac{\int_{B_{R/2}(0)} \det D^2 v \, dx}{c(n, p) |\Omega|}.$$

Recall that $B_R(0) \subset \Omega \subset B_{nR}(0)$. By Lemma 2.37(iii),

$$|Dv(x)| \leq \frac{|v(x)|}{\text{dist}(x, \partial\Omega)} \leq \frac{\|v\|_{L^\infty(\Omega)}}{\text{dist}(x, \partial\Omega)} \leq C(n) R^{-1} \quad \text{if } x \in B_{R/2}(0).$$

Hence

$$\int_{B_{R/2}(0)} \det D^2 v \, dx = |Dv(B_{R/2}(0))| \leq C(n) R^{-n} \leq C(n) |\Omega|^{-1}.$$

Inserting this into (11.15) yields $M\alpha^{p-n} \leq C(n, p) |\Omega|^{-2}$, as claimed. $\qquad \square$

Remark 11.7. Suppose in Lemma 11.6(iii) we have

$$\det D^2 u = (|u| + \varepsilon)^p \quad \text{in } \Omega, \qquad u = 0 \quad \text{on } \partial\Omega$$

for $0 \leq \varepsilon < 1$. Then, instead of (11.14), we obtain for $\alpha = \|u\|_{L^\infty(\Omega)} > 0$,

$$\det D^2 v = \alpha^{p-n} (|v| + \varepsilon/\alpha)^p \geq \alpha^{p-n} |v|^p.$$

Therefore, the inequality in (11.15) continues to hold, and we still have $\alpha^{p-n} \leq C(n, p) |\Omega|^{-2}$. On the other hand,

$$\alpha^{p-n} = \frac{\int_\Omega |v| \det D^2 v \, dx}{\int_\Omega |v|(|v| + \varepsilon/\alpha)^p \, dx} \geq \frac{c(n) |\Omega|^{-1}}{\int_\Omega (1 + 1/\alpha)^p \, dx} = c(n) |\Omega|^{-2} (1 + 1/\alpha)^{-p}.$$

Hence, $\alpha^n \leq C(n) |\Omega|^2 (1 + \alpha)^p$, which implies $\alpha \leq C(n, p, |\Omega|)$ if $0 \leq p < n$.

In working with the Monge–Ampère energy $I(u; \Omega)$, it is sometimes convenient to replace either the function or the domain by smoother convex functions or domains with certain uniform convexity. The following lemma guarantees approximation by strictly convex, smooth functions.

Lemma 11.8. *Let Ω be a bounded convex domain in $\mathbb{R}^n$, and let $u \in C(\overline{\Omega})$ be a convex function with $u = 0$ on $\partial\Omega$ such that $I(u; \Omega) < \infty$. Then for any $\varepsilon_0 > 0$, there exists a convex function $v \in C(\overline{\Omega}) \cap C^2(\Omega)$ with $v = 0$ on $\partial\Omega$ and positive definite Hessian D^2v in Ω such that*

$$\|v - u\|_{L^\infty(\Omega)} + |I(v; \Omega) - I(u; \Omega)| < \varepsilon_0.$$

If Ω is uniformly convex with $C^{4,\alpha}$ boundary for some $\alpha \in (0,1)$, we can choose $v \in C^{4,\alpha}(\overline{\Omega})$ with $D^2v > 0$ in $\overline{\Omega}$.

Proof. For any $\varepsilon > 0$, there is a compact, convex subset $K := K(\varepsilon) \subset \Omega$ such that

$$(11.16) \quad 0 \leq I(u; \Omega \setminus K) < \varepsilon, \quad \text{dist}(K, \partial\Omega) < \varepsilon^n, \quad |u| < \varepsilon \quad \text{in } \Omega \setminus K.$$

Let us choose a convex domain K' such that $K \subset K' \Subset \Omega$. Extend u to be 0 in $\mathbb{R}^n \setminus \Omega$. For $h > 0$ small, let φ_h be the standard mollifier as in Definition 2.49, and let

$$u_h(x) := u * \varphi_h(x) + h|x|^2.$$

Then $u_h \to u$ uniformly on $\overline{\Omega}$ and u_h is strictly convex in the set $\{x \in \Omega : \text{dist}(x, \partial\Omega) > h\}$; see Theorem 2.50(iv). In particular, $\det D^2 u_h > 0$ in K' and $|u_h| \leq 2\varepsilon$ in $\Omega \setminus K$ for h small.

Let $\eta \in C_c^\infty(\Omega)$ be a cut-off function such that $0 \leq \eta \leq 1$, $\eta \equiv 1$ in K, and $\text{spt}(\eta) \subset K'$. Let $w_h \in C(\overline{\Omega})$ be the Aleksandrov solution to

$$\det D^2 w_h = \eta \det D^2 u_h + (1 - \eta)\varepsilon \quad \text{in } \Omega, \qquad w_h = 0 \quad \text{on } \partial\Omega.$$

By Theorem 5.13, w_h is strictly convex in Ω. Then, according to Theorem 6.5, $w_h \in C(\overline{\Omega}) \cap C^2(\Omega)$. If Ω is uniformly convex with $C^{4,\alpha}$ boundary for some $\alpha \in (0,1)$, then, by Theorem 4.12, $w_h \in C^{4,\alpha}(\overline{\Omega})$ with $D^2 w_h > 0$ in $\overline{\Omega}$. Below, we show $v := w_h$ satisfies the conclusion of the lemma when h is small. We use C to denote a constant that depends only on $n, u,$ and Ω.

Step 1. We show that if h is small, then

$$(11.17) \qquad\qquad |w_h - u| \leq C\varepsilon \quad \text{in } \Omega.$$

Indeed, by Example 3.5,

$$(11.18) \qquad \int_{K'} \det D^2 u_h \, dx \leq \omega_n [\text{dist}(\partial K', \partial\Omega)]^{-n} (\underset{\Omega}{\text{osc}}\, u_h)^n \leq C.$$

Therefore

$$(11.19) \qquad \int_\Omega \det D^2 w_h \, dx \leq \varepsilon \int_\Omega (1 - \eta) \, dx + \int_{K'} \det D^2 u_h \, dx \leq C.$$

By the Aleksandrov maximum principle (Theorem 3.12) and (11.16),

$$|w_h(x)|^n \leq C(n, \Omega)\text{dist}(x, \partial\Omega) \int_\Omega \det D^2 w_h \, dx \leq C\varepsilon^n \quad \text{for all } x \in \Omega \setminus K.$$

It follows that $|w_h - u_h| \leq C\varepsilon$ on ∂K when h is small. Since $\det D^2 w_h = \det D^2 u_h$ in K, we obtain from the comparison principle, Theorem 3.21, that

$$(11.20) \qquad\qquad |w_h - u_h| \leq C\varepsilon \quad \text{in } K.$$

The estimate (11.17) now follows from (11.20) and $u_h \to u$ uniformly on $\overline{\Omega}$.

Step 2. We show that if h is small, then

$$M_h := |I(u_h; K) - I(u; K)| < C\varepsilon.$$

First, note that

$$M_h = \left| \int_K (u_h - u) \det D^2 u_h \, dx \right| + \left| \int_K u \det D^2 u_h \, dx - \int_K u \, d\mu_u \right| := A_h + B_h.$$

From (11.18) and the uniform convergence of u_h to u on K, we have $A_h \to 0$ when $h \to 0$. For B_h, we estimate, using (11.16) and (11.18),

$$B_h \leq \left| \int_{K'} \eta u \det D^2 u_h \, dx - \int_{K'} \eta u \, d\mu_u \right|$$
$$+ \left| \int_{K' \setminus K} \eta u \det D^2 u_h \, dx \right| + \left| \int_{K' \setminus K} \eta u \, d\mu_u \right|$$
$$\leq \left| \int_{K'} \eta u \det D^2 u_h \, dx - \int_{K'} \eta u \, d\mu_u \right| + C\varepsilon + I(u, \Omega \setminus K) \leq C\varepsilon$$

when $h \to 0$. Here we used the weak* convergence of the Monge–Ampère measure in Theorem 3.7 for the first term.

Step 3. Conclusion. From (11.18) and (11.20), we obtain

$$|I(w_h; K) - I(u_h; K)| = \left| \int_K (w_h - u_h) \det D^2 u_h \, dx \right| \leq C\varepsilon.$$

Due to (11.19) and $|w_h| \leq C\varepsilon$ in $\Omega \setminus K$, we get

$$I(w_h; \Omega \setminus K) \leq \|w_h\|_{L^\infty(\Omega \setminus K)} \int_{\Omega \setminus K} \det D^2 w_h \, dx \leq C\varepsilon.$$

Using the above estimates together with (11.16) and Step 2, we find

$$|I(w_h; \Omega) - I(u; \Omega)| \leq I(w_h; \Omega \setminus K) + I(u; \Omega \setminus K)$$
$$+ |I(w_h; K) - I(u_h; K)| + |I(u_h; K) - I(u; K)| \leq C\varepsilon.$$

The proof of the lemma is complete. $\qquad\qquad\qquad\qquad\qquad\qquad \square$

The next lemma allows us to approximate a given Monge–Ampère energy by ones on uniformly convex domains with smooth boundaries.

Lemma 11.9. *Let $p \in [0, n)$. Let Ω be a bounded convex domain in $\mathbb{R}^n$, and let $v \in C(\overline{\Omega})$ be a convex function with $v = 0$ on $\partial\Omega$ satisfying*

$$J_{p,1}(v, \Omega) := \frac{1}{n+1} \int_\Omega |v| \, d\mu_v - \frac{1}{p+1} \int_\Omega |v|^{p+1} \, dx < \infty.$$

Let $\{\Omega_m\}_{m=1}^{\infty}$ be a sequence of uniformly convex domains in $\mathbb{R}^n$ converging to Ω in the Hausdorff distance, with $\partial\Omega_m \in C^{\infty}$ and $\Omega_m \supset \Omega_{m+1} \supset \Omega$ for all m. Then for every $\varepsilon > 0$, there exist $m \in \mathbb{N}$ and a convex function $w \in C(\overline{\Omega_m})$ with $w = 0$ on $\partial\Omega_m$ such that $|J_{p,1}(v,\Omega) - J_{p,1}(w,\Omega_m)| < \varepsilon$.

Proof. According to Lemma 11.6(i),

$$J_{p,1}(v,\Omega) \geq c(n)|\Omega|^{-1}\|v\|_{L^{\infty}(\Omega)}^{n+1} - (p+1)^{-1}\|v\|_{L^{\infty}(\Omega)}^{p+1}.$$

Since $J_{p,1}(v,\Omega) < \infty$ and $0 \leq p < n$, we find $\|v\|_{L^{\infty}(\Omega)} < \infty$, so $I(v;\Omega) < \infty$. Hence, for any $\varepsilon > 0$, there is a compact, convex set $K \subset \Omega$ such that

$$0 \leq I(v;\Omega \setminus K) < \varepsilon, \quad \operatorname{dist}(K,\partial\Omega) < \varepsilon^n/2, \quad |v| < \varepsilon \quad \text{in } \Omega \setminus K.$$

Extend v to be 0 outside $\overline{\Omega}$. Let $v_h := v * \varphi_h$ where φ_h is the standard mollifier as in Definition 2.49. Then v_h is convex in the set $\{x \in \Omega : \operatorname{dist}(x,\partial\Omega) > h\}$ and $v_h \to v$ on compact subsets of $\mathbb{R}^n$. Therefore, if $h \leq h_0$ is small, we have $|v_h - v| \leq \varepsilon$ in Ω_1. Since Ω_m converges to Ω in the Hausdorff distance, we can find m large such that $\operatorname{dist}(x,\partial\Omega_m) < \varepsilon$ for all $x \in \Omega \setminus K$. With these h and m, let $w_h \in C(\overline{\Omega_m})$ be the Aleksandrov solution to

$$\det D^2 w_h = \det D^2 v_h \chi_K \quad \text{in } \Omega_m, \qquad w_h = 0 \quad \text{on } \partial\Omega_m.$$

Then, as in the proof of Lemma 11.8, we can easily show that $w := w_h$ satisfies the conclusion of the lemma. $\qquad\square$

11.2. Parabolic Monge–Ampère Flow

Let Ω be a uniformly convex domain in $\mathbb{R}^n$ with C^4 boundary $\partial\Omega$. Let

(11.21) $\quad \Phi_0(\Omega) = \{u \in C(\overline{\Omega}) \cap C^2(\Omega) : D^2u > 0 \quad \text{in } \Omega, \quad u = 0 \quad \text{on } \partial\Omega\}.$

For $T > 0$, consider the parabolic cylinder $\Omega_T = \Omega \times (0,T]$. The parabolic boundary of Ω_T is denoted by $\Gamma_T := \overline{\Omega_T} \setminus \Omega_T$. See Figure 11.2.

We introduce some notation for the analysis on space-time domains. Let $Q \subset \mathbb{R}^n \times \mathbb{R}$ be a bounded space-time domain. Let

$$d(p,q) = \left(|x - y|^2 + |t - s|\right)^{1/2}$$

be the parabolic distance between $p = (x,t) \in Q$ and $q = (y,s) \in Q$. Let ∂_t and D be the differential operators denoting differentiations in the time and spatial variables, respectively. We will define the following spaces and norms. For a nonnegative integer k, define

$$C_{k/2}^k(\overline{Q}) = \{u : \overline{Q} \to \mathbb{R} : \partial_t^j D_x^l u \in C(\overline{Q}) \quad \text{for all } 2j + l \leq k\}.$$

The norm on $C_{k/2}^k(\overline{Q})$ is

$$\|u\|_{C_{k/2}^k(\overline{Q})} = \sum_{2j+l \leq k} \|\partial_t^j D_x^l u\|_{L^{\infty}(Q)}.$$

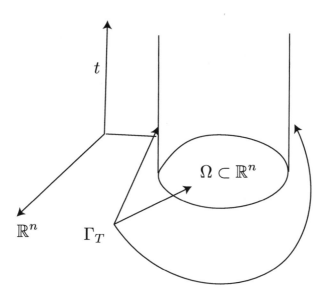

Figure 11.2. The parabolic boundary Γ_T of Ω_T.

For a nonnegative integer k and $\alpha \in (0, 1]$, define

$$C^{k,\alpha}_{k/2,\alpha/2}(\overline{Q}) = \left\{ u \in C^k_{k/2}(\overline{Q}) : \|u\|_{C^{k,\alpha}_{k/2,\alpha/2}(\overline{Q})} < \infty \right\},$$

where

$$\|u\|_{C^{k,\alpha}_{k/2,\alpha/2}(\overline{Q})} = \|u\|_{C^k_{k/2}(\overline{Q})} + \sum_{2j+l=k} \sup_{p \neq q \in Q} \frac{|\partial_t^j D_x^l u(p) - \partial_t^j D_x^l u(q)|}{[d(p,q)]^\alpha}.$$

We will study the following parabolic Monge–Ampère flow:

$$(11.22) \qquad \begin{cases} \partial_t u = \log \det D^2 u - g(\cdot, -u) & \text{in } \Omega_T, \\ \quad u = 0 & \text{on } \partial\Omega \times [0, T], \\ u(\cdot, 0) = \varphi & \text{on } \overline{\Omega} \times \{t = 0\}. \end{cases}$$

A complex version of (11.22) arises in the study of the Kähler-Ricci flow.

If u is a smooth solution of (11.22), then at the corner $\partial\Omega \times \{t = 0\}$ of the parabolic cylinder Ω_T, its value on the side $\partial\Omega \times [0, T]$ and at the base $\overline{\Omega} \times \{t = 0\}$ must satisfy the compatibility condition

$$(11.23) \qquad \log \det D^2 \varphi(x) = g(x, -\varphi(x)) \quad \text{for } x \in \partial\Omega.$$

Like its elliptic counterpart, we will study (11.22) with *admissible solutions* $u(x, t)$ where $u(\cdot, t) \in \Phi_0(\Omega)$ for all $t \in [0, T]$. We now prove a parabolic analogue of the global C^2 estimates in Theorem 4.9.

Theorem 11.10 (A priori global C^2 estimates for parabolic Monge–Ampère flow). *Let Ω be a uniformly convex domain in $\mathbb{R}^n$ with boundary $\partial\Omega \in C^4$, $g \in C^2(\overline{\Omega} \times [0,\infty))$, $\varphi \in C^4(\overline{\Omega}) \cap \Phi_0(\Omega)$. Suppose $u \in C^4_2(\overline{\Omega} \times [0,T])$ is an admissible solution of (11.22). Then, there is a constant $C > 1$, depending only on $n, \Omega, K := \|u\|_{C(\overline{\Omega}\times[0,T])}, \|\varphi\|_{C^2(\overline{\Omega})}$, and $\|g\|_{C^2(\overline{\Omega}\times[0,K])}$, such that*

$$(11.24) \quad \|u\|_{C^2_1(\overline{\Omega}\times[0,T])} \le C \quad and \quad C^{-1}I_n \le D^2u \le CI_n \quad in \ \overline{\Omega} \times [0,T].$$

Proof. We divide the proof into several steps. Denote $(u^{ij}) = (D^2u)^{-1}$. In what follows, we view g as a function of $(x,z) \in \overline{\Omega}\times[0,K]$ and $D_zg = \partial g/\partial z$.

Step 1. $\sup_{\overline{\Omega_T}} |Du| \le C_d = C_d(\Omega, n, \|\varphi\|_{C^2(\overline{\Omega})}, \|g\|_{C(\overline{\Omega}\times[0,K])})$. By the convexity of u, it suffices to prove this gradient estimate on $\partial\Omega \times [0,T]$. Let

$$M_0 := 1 + \max\Big\{ \sup_\Omega \det D^2\varphi, \, e^{\sup_{\overline{\Omega}\times[0,K]} g} \Big\}.$$

Let $w \in C^3(\overline{\Omega})$ be the convex solution to

$$\det D^2w = M_0 \quad in \ \Omega, \quad w = 0 \quad on \ \partial\Omega.$$

By the comparison principle (Theorem 3.21), we have $w \le \varphi$ in Ω.

Let $v(x,t) = w(x)$ on $\overline{\Omega} \times [0,T]$. Then, $v \le u$ on Γ_T while in Ω_T,

$$\partial_t v - \log\det D^2v = -\log M_0 \le \inf_{\Omega_T}\big[-g(x,-u(x,t))\big] \le \partial_t u - \log\det D^2u.$$

By the maximum principle, we have $v \le u$ in $\overline{\Omega_T}$. This gives

$$(11.25) \qquad\qquad \frac{\partial u}{\partial \nu} \le \frac{\partial v}{\partial \nu} \quad on \ \partial\Omega \times \{t = t_0\},$$

for each $t_0 \in [0,T]$, where ν is the outer unit normal on $\partial\Omega$.

Since $u(\cdot, t_0)$ is convex and $u(\cdot, t_0) = 0$ on $\partial\Omega \times \{t = t_0\}$, we have

$$\frac{\partial u}{\partial \nu} \ge 0 \quad on \ \partial\Omega \times \{t = t_0\}.$$

Combining this with (11.25) gives the uniform bound for Du on $\overline{\Omega_T}$.

Step 2. $\sup_{\overline{\Omega_T}} |\partial_t u| \le C_* = C_*(\Omega, n, K, \|\varphi\|_{C^2(\overline{\Omega})}, \|g\|_{C^1(\overline{\Omega}\times[0,K])})$. Consider $\partial_t u$ on the parabolic boundary Γ_T. On $\partial\Omega \times (0,T)$, we have $\partial_t u = 0$. On $\Omega \times \{0\}$, we use the equation for u to obtain a bound for $\partial_t u$. Thus

$$\|\partial_t u\|_{L^\infty(\Gamma_T)} \le C_b(\|\varphi\|_{C^2(\overline{\Omega})}, \|g\|_{C(\overline{\Omega}\times[0,K])}).$$

It remains to consider $\partial_t u$ in the interior Ω_T of $\overline{\Omega_T}$. Let $M := K + 1$.

Step 2(a). We first prove a lower bound for $\partial_t u$. Let $G = \partial_t u (M - u)^{-1}$.

Suppose G has a negative minimum at an interior point $p \in \Omega_T$ of $\overline{\Omega_T}$. At this point, we have

$$\partial_t G = \frac{\partial_{tt} u}{M - u} + \frac{(\partial_t u)^2}{(M - u)^2} \le 0,$$

$$D_k G = \frac{D_k(\partial_t u)}{M - u} + \frac{\partial_t u D_k u}{(M - u)^2} = 0, \quad \text{for all } k = 1, \ldots, n,$$

and, in the sense of symmetric matrices,

$$D_{ij} G = \frac{D_{ij}(\partial_t u)}{M - u} + \frac{D_j \partial_t u D_i u + D_i \partial_t u D_j u + D_{ij} u \partial_t u}{(M - u)^2} + \frac{2 D_i u D_j u \partial_t u}{(M - u)^3}$$

$$= \frac{D_{ij}(\partial_t u)}{M - u} + \frac{D_{ij} u \partial_t u}{(M - u)^2} \ge 0$$

where we used $D_k G = 0$ in the last identity. Differentiating (11.22) with respect to t and recalling Lemma 2.55, we obtain

$$(11.26) \qquad \partial_{tt} u = u^{ij} D_{ij}(\partial_t u) + D_z g(x, -u) \partial_t u.$$

Therefore, in view of $u^{ij} D_{ij} G \ge 0$ and $u^{ij} D_{ij} u = n$, we have

$$0 \le u^{ij} (D_{ij} \partial_t u + (M - u)^{-1} D_{ij} u \partial_t u) = \partial_{tt} u - D_z g(x, -u) \partial_t u + n(M - u)^{-1} \partial_t u$$

from which we find, in view of $\partial_t G \le 0$,

$$(M - u)^{-1} (\partial_t u)^2 \le -\partial_{tt} u \le -D_z g(x, -u) \partial_t u + n(M - u)^{-1} \partial_t u.$$

Since $M - u \ge 1$, this gives a lower bound for $\partial_t u$.

Step 2(b). We analogously prove an upper bound for $\partial_t u$ by considering $\hat{G} = \partial_t u (M + u)^{-1}$. Suppose $\hat{G}$ has a positive maximum at an interior point $p \in \Omega_T$. At p, as in Step 2(a), using $\partial_t \hat{G} \ge 0$, $D_k \hat{G} = 0$, and $D^2 \hat{G} \le 0$, we find

$$(M + u)^{-1} (\partial_t u)^2 \le D_z g(x, -u) \partial_t u + n(M + u)^{-1} \partial_t u.$$

This gives an upper bound for $\partial_t u$.

Step 3. $\sup_{\partial\Omega \times (0,T)} |D^2 u| \le C^* = C^*(\Omega, n, K, \|\varphi\|_{C^2(\overline{\Omega})}, \|g\|_{C^2(\overline{\Omega} \times [0,K])})$. All ingredients are already available in the boundary second derivative estimates in the proof of Theorem 4.9 (with zero boundary values). We indicate the necessary changes. Consider $(x_0, t_0) \in \partial\Omega \times (0, T)$. We can assume that $x_0 = 0$ and $e_n = (0, \ldots, 0, 1)$ is the inner unit normal of $\partial\Omega$ at 0. We use a principal coordinate system at $x_0 = 0$ as in (4.10).

From Step 2, we have

$$\bar{m} \ge \det D^2 u(x, t_0) = e^{\partial_t u(x, t_0)} e^{g(x, -u(x, t_0))} \ge m > 0 \quad \text{for } x \in \Omega,$$

where m and $\bar{m}$ depends on $\Omega, n, K, \|\varphi\|_{C^2(\overline{\Omega})}, \|g\|_{C^1(\overline{\Omega}\times[0,K])}$. By Remark 4.8, we have the second tangential derivative estimates

$$(11.27) \qquad \rho_1 I_{n-1} \leq D^2_{x'} u(0, t_0) \leq \rho_1^{-1} I_{n-1}$$

where ρ_1 depends on $n, m, \bar{m}$, and Ω. Recall that $u = 0$ on $\partial\Omega \times \{t = t_0\}$.

For the second mixed derivative $D_{in}u$ estimates ($1 \leq i \leq n-1$), let $L_u = u^{jk} D_{jk}$ be the linearized operator of $\log \det D^2 u$, and let

$$v(x,t) := D_i u + \kappa_i[x_i D_n u - x_n D_i u],$$

$$w(x,t) := bx_n - 2aP_\gamma(x) + a\tilde{\gamma}(u - l)$$

be as in the proof of Theorem 4.9 where we take $\gamma = 1$. Note that a and b depend also on $\|\varphi\|_{C^2(\overline{\Omega})}$ and $\|g\|_{C^2(\overline{\Omega}\times[0,K])}$.

Consider the parabolic linearized operator

$$\mathcal{L} = L_u - \frac{\partial}{\partial t} = u^{jk} D_{jk} - \frac{\partial}{\partial t}.$$

Since $\log \det D^2 u = \partial_t u + g(x, -u)$, instead of (4.16), we have

$$(11.28) \qquad L_u(x_i D_n u - x_n D_i u) = (x_i D_n - x_n D_i)(\partial_t u + g(x, -u)).$$

Hence, from $\mathcal{L}v = L_u v - \partial_t v$ and $L_u D_i u = D_i \log \det D^2 u$, we find

$$\mathcal{L}v = D_i[\partial_t u + g(x, -u)] + \kappa_i[(x_i D_n - x_n D_i)(\partial_t u + g(x, -u))]$$
$$- D_i \partial_t u - \kappa_i[x_i D_n \partial_t u - x_n D_i \partial_t u],$$

where all terms $D_i \partial_t u$ and $D_n \partial_t u$ are cancelled. Therefore (see Step 1),

$$(11.29) \qquad |\mathcal{L}_u v| \leq C_2 \Big(1 + \sum_{k=1}^n u^{kk}\Big)$$

where C_2 depends on Ω, $\|\varphi\|_{C^2(\overline{\Omega})}$, and $\|g\|_{C^1(\overline{\Omega}\times[0,K])}$.

In the presence of the time variable, calculating $\mathcal{L}w$ produces the term $-a\tilde{\gamma}\partial_t u$ which, by Step 2, is bounded by $a\tilde{\gamma}C_*$. We have, in $\Omega \times (0, t_0)$,

$$(11.30) \quad \mathcal{L}(w \pm v) \leq (C_2 - a)\sum_{k=1}^n u^{kk} + (C_2 - an\bar{m}^{-1/n} + an\tilde{\gamma}) + a\tilde{\gamma}C_* < 0$$

if we adjust $\tilde{\gamma}$ in the proof of Theorem 4.9 to be $\tilde{\gamma} = \frac{n\bar{m}^{-1/n}}{2(n+C_*)}$.

Furthermore, for all $0 < t \leq t_0$, we have, as in (4.20),

$$w(x,t) \pm v(x,t) \geq (b - C_3)x_n - (a\gamma^{1-n} + C_3)|x|^2 \geq 0 \quad \text{on } \partial\Omega.$$

When $t = 0$, we obtain as in the proof of Theorem 4.9 that

$$w(x,0) \pm v(x,0) \geq 0 \quad \text{in } \overline{\Omega}.$$

Therefore $w \pm v \geq 0$ on Γ_{t_0}. Now, (11.30) and the maximum principle give

$$w \pm v \geq 0 \quad \text{in } \overline{\Omega} \times [0, t_0].$$

At $(0, t_0) \in \partial\Omega \times \{t = t_0\}$, $w \pm v = 0$, so $D_n w(0, t_0) \pm D_n v(0, t_0) \geq 0$. But these give

$$b + a\tilde{\gamma} D_n(u - l)(0, t_0) \pm D_{in} u(0, t_0) \geq 0,$$

and hence we obtain the second mixed derivative estimates on $\Omega \times [0, T]$:

$$(11.31) \qquad |D_{in} u(0, t_0)| \leq C = C(\Omega, m, \|\varphi\|_{C^2(\overline{\Omega})}, \|g\|_{C^2(\overline{\Omega} \times [0, K])}).$$

From (11.27) and (11.31), we obtain the second normal derivative estimates as in Step 5 of the proof of Theorem 4.9.

Step 4. $\sup_{\overline{\Omega_T}} |D^2 u| \leq C_4 = C_4(\Omega, n, K, \|\varphi\|_{C^2(\overline{\Omega})}, \|g\|_{C^2(\overline{\Omega} \times [0,K])})$. This estimate follows from Step 3 and a Pogorelov-type computation as in Theorem 6.1. For any unit direction ξ which we take for simplicity to be e_1, consider

$$w = \log D_{11} u + |D_1 u|^2 / 2.$$

If w attains it maximum value on Γ_T, then the initial data and Step 3 give an upper bound for $D_{11} u$. Now, we suppose that w attains its maximum on $\overline{\Omega_T}$ at an interior point P. We can also assume that $D^2 u(P)$ is diagonal.

Differentiating (11.22) with respect to x_1 and using that $D^2 u(P)$ is diagonal, we obtain at P

$$(11.32) \qquad \begin{aligned} u^{ij} D_{1ij} u = u^{ii} D_{1ii} u &= D_1(\partial_t u + g(x, -u)) \\ &= D_1 \partial_t u - D_z g(x, -u) D_1 u + D_{x_1} g(x, -u), \end{aligned}$$

and, calculating as in (6.3),

$$(11.33) \quad D_{11}[\partial_t u + g(\cdot, -u)] = D_{11}(\log \det D^2 u) = u^{ii} D_{11ii} u - u^{ii} u^{jj} [D_{1ij} u]^2.$$

The maximality of w at P gives $\partial_t w(P) \geq 0$ and $D^2 w(P) \leq 0$. Therefore,

$$(11.34) \qquad D_1 u D_1 \partial_t u + \frac{D_{11} \partial_t u}{D_{11} u} \geq 0.$$

At P, we have $u^{ii} D_{ii} w \leq 0$, which translates to

$$\frac{u^{ii} D_{11ii} u}{D_{11} u} - \frac{u^{ii} (D_{11i} u)^2}{(D_{11} u)^2} + u^{ii} (D_{1i} u)^2 + u^{ii} D_1 u D_{1ii} u \leq 0.$$

The last term equals $D_1 u D_1[\partial_t u + g(x, -u)]$ by (11.32) while the third term is $D_{11} u$. Thus, using (11.33) for the first term, we obtain

$$D_{11} u + \left[\frac{u^{ii} u^{jj} (D_{1ij} u)^2}{D_{11} u} - \frac{u^{ii} (D_{11i} u)^2}{(D_{11} u)^2} \right]$$

$$+ D_1 u D_1(\partial_t u + g(x, -u)) + \frac{D_{11}[\partial_t u + g(x, -u)]}{D_{11} u} \leq 0.$$

The term in the bracket is clearly nonnegative. Hence

$$D_{11} u + D_1 u D_1[\partial_t u + g(x, -u)] + \frac{D_{11}[\partial_t u + g(x, -u)]}{D_{11} u} \leq 0.$$

Recalling (11.34), we deduce

$$D_{11}u + D_1uD_1[g(x,-u)] + \frac{D_{11}[g(x,-u)]}{D_{11}u} \leq 0.$$

From the bound $|Du| \leq C_d$ together with the fact that the only second-order derivative of u appearing in $D_{11}[g(x,-u)]$ is $-D_zg(x,-u)D_{11}u$, we deduce that, for some A depending on $\|g\|_{C^2(\overline{\Omega}\times[0,T])}$ and C_d,

$$D_{11}u - A - \frac{A}{D_{11}u} \leq 0,$$

which gives the upper bound for $D_{11}u$ at P. From the maximality of w and Step 1, we find that $|D^2u| \leq C$ in $\overline{\Omega}_T$, completing the proof of Step 4.

Note that Steps 2 and 4 give an upper bound for $\|u\|_{C_1^2(\overline{\Omega}\times[0,T])}$. Thus, thanks to $\det D^2u \geq m$ in $\overline{\Omega}\times[0,T]$ and Remark 2.87, we obtain (11.24). □

If $g(x,z) \leq C_0$ on $\overline{\Omega}\times[0,\infty)$, then in Step 1 of the proof of Theorem 11.10, we can choose $M_0 = 1 + \max\{\sup_{\Omega} \det D^2\varphi, e^{C_0}\}$ and obtain that $u \geq w$ in $\overline{\Omega}$. This yields the estimate

$$K = \|u\|_{C(\overline{\Omega}\times[0,T])} \leq C(\Omega, C_0, \|\varphi\|_{C^2(\overline{\Omega})}),$$

which is independent of T.

The second derivative estimates in Theorem 11.10 imply that the operator $\log \det D^2u - \partial_t u - g(x,-u)$ is uniformly parabolic and concave. By using Krylov's global $C^{2,\alpha}$ estimates and the Schauder theory, we obtain global higher-order derivative estimates. In particular, we obtain global $C_{2,\alpha}^{4,\alpha}$ estimates on $\overline{\Omega}\times[0,T]$ when $\partial\Omega \in C^{4,\alpha}$, $\varphi \in C^{4,\alpha}(\overline{\Omega})$, and $g \in C^{2,\alpha}(\overline{\Omega}\times[0,T])$. Consequently, by the method of continuity, we obtain a short-time existence and uniqueness result for (11.22) for smooth initial data φ satisfying a compatibility condition. These are the parabolic analogues of Theorems 4.11 and 4.12; see Lieberman [**Lb**, Chapter 15] for more details. If, in addition, $\|g\|_{C^{2,\alpha}(\overline{\Omega}\times[0,\infty))} < \infty$, then the estimates for the admissible solution u are independent of T, and as a result, the flow exists for all time.

We summarize these points in the following theorem.

Theorem 11.11 (Global existence of parabolic Monge–Ampère equations). *Let Ω be a uniformly convex domain in $\mathbb{R}^n$ with boundary $\partial\Omega \in C^{4,\alpha}$ where $\alpha \in (0,1)$, $g \in C^{2,\alpha}(\overline{\Omega}\times[0,\infty))$, $\varphi \in C^{4,\alpha}(\overline{\Omega}) \cap \Phi_0(\Omega)$ with $D^2\varphi > 0$ in $\overline{\Omega}$. Assume the compatibility condition (11.23) is satisfied. Then, there exists $T > 0$ (depending on n,α, Ω, φ, g) and a unique admissible solution $u \in C_{2,\alpha}^{4,\alpha}(\overline{\Omega}\times[0,T])$ to (11.22). The solution u satisfies the estimate*

$$\|u\|_{C_{2,\alpha}^{4,\alpha}(\overline{\Omega}\times[0,T])} \leq C = C(n,\alpha,\Omega,\varphi,g,T,\|u\|_{C(\overline{\Omega}\times[0,T])}).$$

We can take C to be independent of T and take $T = \infty$ (that is, the flow (11.22) exists for all time) if $\|g\|_{C^{2,\alpha}(\overline{\Omega} \times [0,\infty))} < \infty$.

Observe that, along parabolic Monge–Ampère flows, certain Monge–Ampère functionals are decreasing.

Lemma 11.12. *Let Ω be a bounded convex domain in $\mathbb{R}^n$. Suppose u is an admissible solution to (11.22) on the interval $[0,T)$ with $u(\cdot,t) \in C^{0,1}(\overline{\Omega})$ for all $0 \le t < T$. For $G(x,z) = \int_0^z e^{g(x,s)} ds$, let*

$$(11.35) \qquad J(u) = \int_{\Omega} \left(\frac{1}{n+1}(-u) \det D^2 u - G(x,-u) \right) dx.$$

Then, for $0 \le t_1 < t_2 < T$, we have

$$J(u(\cdot,t_1)) - J(u(\cdot,t_2))$$
$$= \int_{t_1}^{t_2} \int_{\Omega} (\det D^2 u(x,t) - e^{g(x,-u(x,t))}) \log \frac{\det D^2 u(x,t)}{e^{g(x,-u(x,t))}} \, dx\, dt.$$

In particular, J is decreasing along the flow (11.22).

Proof. Proposition 11.3 still applies to $w(t) = u(\cdot,t)$. Thus,

$$\frac{d}{dt} J(u(\cdot,t)) = \int_{\Omega} -[\det D^2 u(x,t) - e^{g(x,-u(x,t))}]\partial_t u(x,t) \, dx.$$

Using the equation and integrating, we obtain the claimed identity for J. □

11.3. Degenerate Monge–Ampère Equations

This section establishes, via the flow method, the existence of nonzero convex solutions to the Monge–Ampère equation

$$(11.36) \qquad \det D^2 u = (|u| + \varepsilon)^p \quad \text{in } \Omega, \quad u = 0 \quad \text{on } \partial\Omega,$$

where $p, \varepsilon \ge 0$. When $p > 0$ and $\varepsilon = 0$, (11.36) is a degenerate Monge–Ampère equation as the right-hand side tends to zero near the boundary.

For $\lambda > 0$, consider the functional

$$(11.37) \qquad J_{p,\lambda}^{\varepsilon}(u,\Omega) = \frac{1}{n+1} \int_{\Omega} |u| \, d\mu_u - \frac{\lambda}{p+1} \int_{\Omega} (|u| + \varepsilon)^{p+1} \, dx$$

over convex functions $u \in C(\overline{\Omega})$ with $u = 0$ on $\partial\Omega$. For simplicity, we write

$$J_{p,\lambda} = J_{p,\lambda}^0.$$

We will use parabolic Monge–Ampère flows to show the existence of a minimizer of $J_{p,\lambda}^{\varepsilon}$ (when $0 \le p < n$) over $\Phi_0(\Omega)$ defined in (11.21), and this minimizer solves (11.36). This minimizer is obtained as the infinite time limit of the admissible solution to (11.22) where $g(x,s) = \log[(s + \varepsilon)^p]$. To guarantee the long-time existence, we need to modify g so that all its derivatives are bounded and then apply Theorem 11.11. Approximation and compactness arguments are repeatedly used in this section.

Theorem 11.13. *Let Ω be a uniformly convex domain in $\mathbb{R}^n$ with boundary $\partial\Omega \in C^{4,\alpha}$ where $\alpha \in (0,1)$. Let $\varepsilon \in (0,1)$ and $0 \leq p < n$. Then, there exists a convex solution $u_\varepsilon \in C^{4,\alpha}(\overline{\Omega})$ to the Dirichlet problem*

$$\det D^2 u = (|u| + \varepsilon)^p \quad in\ \Omega, \quad u = 0 \quad on\ \partial\Omega.$$

Moreover, u_ε minimizes the functional $J_{p,1}^\varepsilon(\cdot, \Omega)$ over the set $\Phi_0(\Omega)$.

Proof. Let $m := \inf_{\Phi_0(\Omega)} J_{p,1}^\varepsilon(u, \Omega)$. Since $0 \leq p < n$, using Lemma 11.6(i) as in the proof of Lemma 11.9, we can find a constant $M(n, p, \Omega) > 2$ with the following properties: If $u \in C(\overline{\Omega})$ with $u = 0$ on Ω and if $\det D^2 u \leq (|u|+\varepsilon)^p$ in Ω or $J_{p,1}^\varepsilon(u, \Omega) < 1 + m$, then $\|u\|_{L^\infty(\Omega)} < M$.

We modify the function $f(s) = (s + \varepsilon)^p$ by a new smooth function $\varepsilon^p \leq \tilde{f}(s) \leq f(s)$ where $\tilde{f}(s) = f(s)$ if $0 \leq s \leq 2M$ and $\tilde{f}(s) = 1$ if $s \geq 3M$.

Fix $\gamma \in (0, 1/4)$. Then, there is $u_\gamma \in \Phi_0(\Omega)$ with $J_{p,1}^\varepsilon(u_\gamma, \Omega) < \gamma/4 + m$.

Step 1. Modification of u_γ into a compatible initial data. Using Lemma 11.8, we can approximate u_γ by a function $v_\gamma \in \Phi_0(\Omega) \cap C^{4,\alpha}(\overline{\Omega})$ with $D^2 v_\gamma > 0$ in $\overline{\Omega}$ such that $J_{p,1}^\varepsilon(v_\gamma, \Omega) < \gamma/2 + m$. We show that there is a small perturbation φ of v_γ satisfying the compatibility condition

$$\det D^2 \varphi = \tilde{f}(-\varphi) \quad \text{on } \partial\Omega.$$

Indeed, for $\delta > 0$ small, let $w \in C^{2,\alpha}(\overline{\Omega})$ be a positive function such that $w = \det D^2 v_\gamma$ in $\Omega_\delta := \{x \in \Omega : \text{dist}(x, \partial\Omega) > \delta\}$ and $w = \tilde{f}(0) = f(0) > 0$ on $\partial\Omega$. Now, it suffices to let $\varphi \in C^{4,\alpha}(\overline{\Omega})$ be the convex solution to

$$\det D^2 \varphi = w \quad \text{in } \Omega, \quad \varphi = 0 \quad \text{on } \partial\Omega.$$

Arguing as in the proof of Lemma 11.8, we find that for $\delta > 0$ small,

$$(11.38) \qquad J_{p,1}^\varepsilon(\varphi, \Omega) < \gamma/2 + J_{p,1}^\varepsilon(v_\gamma, \Omega) < \gamma + m.$$

Step 2. Long-time behavior of a parabolic Monge–Ampère flow. Consider the parabolic Monge–Ampère flow with an admissible solution u:

$$(11.39) \qquad \begin{cases} \partial_t u = \log \det D^2 u - \log \tilde{f}(-u) & \text{in } \Omega \times (0, \infty), \\ u = 0 & \text{on } \partial\Omega \times [0, \infty), \\ u(\cdot, 0) = \varphi & \text{on } \overline{\Omega} \times \{t = 0\}. \end{cases}$$

Since $\varepsilon^p \leq \tilde{f} \leq (3M + 1)^p$ and $\|\tilde{f}\|_{C^{2,\alpha}(\mathbb{R})} < \infty$, Theorem 11.11 guarantees that the flow exists for all time and that $u \in C_{2,\alpha}^{4,\alpha}(\overline{\Omega} \times [0, T])$ for all $T > 0$. Let

$$\tilde{J}(u) = \int_\Omega \left(\frac{1}{n+1}(-u) \det D^2 u - \tilde{F}(-u) \right) dx, \quad \text{where } \tilde{F}(z) = \int_0^z \tilde{f}(s)\, ds.$$

Because $n > p \geq 0$, we find from Lemma 11.6(i) that

$$\tilde{J}(v) \geq \tilde{m} > -\infty \quad \text{for all } v \in \Phi_0(\Omega).$$

Note that, by (11.38), we have $\|\varphi\|_{L^\infty(\Omega)} < M$, so

$$\tilde{J}(u(\cdot,0)) = \tilde{J}(\varphi) = J_{p,1}^\varepsilon(\varphi,\Omega) < \gamma + m.$$

By Lemma 11.12, $\tilde{J}(u(\cdot,t))$ is decreasing, so it has a limit when $t \to \infty$. We have

$$\gamma + m - \tilde{m} \geq \lim_{t \to \infty}[\tilde{J}(u(\cdot,0)) - \tilde{J}(u(\cdot,t))] = \int_0^\infty -\frac{d}{dt}\tilde{J}(u(\cdot,t))\,dt.$$

Therefore, there exists a sequence $t_j \to \infty$ so that

$$-\frac{d}{dt}\tilde{J}(u(\cdot,t_j)) = \int_\Omega (\det D^2 u(\cdot,t_j) - \tilde{f}(-u(\cdot,t_j))) \log \frac{\det D^2 u(\cdot,t_j)}{\tilde{f}(-u(\cdot,t_j))}\,dx \to 0.$$

Using the global $C_{2,\alpha/2}^{4,\alpha}$ estimates in Theorem 11.11 and passing to a further subsequence, we obtain a limit function $u_\varepsilon \in \Phi_0(\Omega) \cap C^{4,\alpha}(\overline{\Omega})$ satisfying

$$\int_\Omega (\det D^2 u_\varepsilon - \tilde{f}(-u_\varepsilon)) \log \frac{\det D^2 u_\varepsilon}{\tilde{f}(-u_\varepsilon)}\,dx = 0.$$

Hence, $\det D^2 u_\varepsilon = \tilde{f}(-u_\varepsilon)$ in Ω. By the choice of $\tilde{f}$ and M, we have $\|u_\varepsilon\|_{L^\infty(\Omega)} < M$, so in fact

$$\det D^2 u_\varepsilon = f(-u_\varepsilon) = (|u_\varepsilon| + \varepsilon)^p \quad \text{in } \Omega.$$

By Proposition 6.37, u_ε is unique. Note that

$$J_{p,1}^\varepsilon(u_\varepsilon,\Omega) = \tilde{J}(u_\varepsilon) \leq \tilde{J}(u(\cdot,0)) < \gamma + m.$$

Since $\gamma > 0$ is arbitrary, we have $J_{p,1}^\varepsilon(u_\varepsilon,\Omega) \leq m = \inf_{\Phi_0(\Omega)} J_{p,1}^\varepsilon(u,\Omega)$. The theorem is proved. $\qquad\square$

Next, we wish to let ε tend to zero in Theorem 11.13. We will need the following compactness result.

Proposition 11.14. *Let $0 \leq p \neq n$. Let Ω be a bounded convex domain in $\mathbb{R}^n$. Let $\{\Omega_m\}_{m=1}^\infty$ be a sequence of uniformly convex domains in $\mathbb{R}^n$ with smooth boundaries that converges to Ω in the Hausdorff distance. For each $m \in \mathbb{N}$, assume $u_m \in C(\overline{\Omega_m}) \cap C^\infty(\Omega_m)$ is a nonzero convex solution to*

(11.40) $$\det D^2 u_m = (|u_m| + \varepsilon_m)^p \quad \text{in } \Omega_m, \quad u_m = 0 \quad \text{on } \partial\Omega_m,$$

where $\varepsilon_m = 0$ if $p > n$, while if $0 \leq p < n$, we assume $\varepsilon_m \to 0^+$ when $m \to \infty$. Then, up to extracting a subsequence, $\{u_m\}_{m=1}^\infty$ converges uniformly on compact subsets of Ω to a nonzero convex solution $u \in C(\overline{\Omega}) \cap C^\infty(\Omega)$ of

(11.41) $$\det D^2 u = |u|^p \quad \text{in } \Omega, \quad u = 0 \quad \text{on } \partial\Omega.$$

Furthermore, $J_{p,1}^{\varepsilon_m}(u_m,\Omega_m) \to J_{p,1}(u,\Omega)$ when $m \to \infty$.

Proof. By Lemma 11.6(iii) and Remark 11.7, we have

$$(11.42) \qquad c(n,p)|\Omega_m|^{\frac{2}{n-p}} \le \|u_m\|_{L^\infty(\Omega_m)} \le C(n,p,|\Omega_m|).$$

From the convergence of Ω_m to Ω in the Hausdorff distance, we infer from (11.42) and Lemma 3.18 that $\sup_{m \in \mathbb{N}} \|u_m\|_{C^{0,1/n}(\overline{\Omega_m})} \le C(n,\Omega)$.

By the Arzelà–Ascoli Theorem, there is a subsequence, still denoted by $\{u_m\}_{m=1}^\infty$, that converges locally uniformly to a convex function u on Ω. The compactness theorem of the Monge–Ampère equation, Theorem 3.19, then asserts that u is actually an Aleksandrov solution of (11.41). The estimates (11.42) show that $u \not\equiv 0$. By Proposition 6.36, $u \in C^\infty(\Omega)$.

Because u_m solves (11.40) and u solves (11.41), we have

$$J_{p,1}^{\varepsilon_m}(u_m, \Omega_m) = \int_{\Omega_m} \left(C_{n,p}(|u_m| + \varepsilon_m)^{p+1} - \frac{\varepsilon_m}{n+1}(|u_m| + \varepsilon_m)^p \right) dx,$$

where $C_{n,p} = \frac{p-n}{(n+1)(p+1)}$, and

$$J_{p,1}(u,\Omega) = C_{n,p} \int_\Omega |u|^{p+1}\, dx.$$

From the convergence of Ω_m to Ω and u_m to u, we easily obtain the convergence $J_{p,1}^{\varepsilon_m}(u_m, \Omega_m) \to J_{p,1}(u, \Omega)$ when $m \to \infty$. $\qquad\square$

Now, letting ε tend to zero in Theorem 11.13, we obtain the following theorem which is due to Tso [**Ts**].

Theorem 11.15 (Tso's Theorem). *Let Ω be a uniformly convex domain in $\mathbb{R}^n$ with smooth boundary. Then, for each $0 \le p \ne n$, there exists a nonzero convex solution $u \in C^{0,1}(\overline{\Omega}) \cap C^\infty(\Omega)$ to the Dirichlet problem*

$$\det D^2 u = |u|^p \quad \text{in } \Omega, \quad u = 0 \quad \text{on } \partial\Omega.$$

Moreover, if $0 \le p < n$, then u is unique and it minimizes the functional $J_{p,1}(\cdot, \Omega)$ over all convex functions $v \in C(\overline{\Omega}) \cap C^2(\Omega)$ with $v = 0$ on $\partial\Omega$ and having positive definite Hessian $D^2 v$ in Ω.

For $p > n$, we refer to [**Ts**, Theorem E] for the proof of the existence result. Here, we only consider the case $0 \le p < n$ which is sufficient for later applications to the Monge–Ampère eigenvalue problem.

Proof of Theorem 11.15 for $0 \le p < n$. Note that the uniqueness result follows from Proposition 6.37. Now, we prove the existence result. For each $\varepsilon_m = 1/m$, by Theorem 11.13, there is a convex solution $u_m \in C^{4,\alpha}(\overline{\Omega})$ solving the Dirichlet problem

$$\det D^2 u_m = (|u_m| + 1/m)^p \quad \text{in } \Omega, \quad u_m = 0 \quad \text{on } \partial\Omega.$$

Moreover, u_m minimizes the functional $J_{p,1}^{1/m}(\cdot, \Omega)$ over all convex functions $v \in \Phi_0(\Omega)$. This means that $J_{p,1}^{1/m}(u_m, \Omega) \leq J_{p,1}^{1/m}(v, \Omega)$ for all $v \in \Phi_0(\Omega)$. By Proposition 11.14 and the global Lipschitz estimate in Theorem 3.30, up to extracting a subsequence, $\{u_m\}_{m=1}^\infty$ converges uniformly on compact subsets of Ω to a nonzero convex solution $u \in C^{0,1}(\overline{\Omega}) \cap C^\infty(\Omega)$ of

$$\det D^2 u = |u|^p \quad \text{in } \Omega, \quad u = 0 \quad \text{on } \partial\Omega.$$

Furthermore, $J_{p,1}^{\varepsilon_m}(u_m, \Omega) \to J_{p,1}(u, \Omega)$. Clearly, for each $v \in \Phi_0(\Omega)$,

$$J_{p,1}^{1/m}(v, \Omega) \to J_{p,1}(v, \Omega) \quad \text{when } m \to \infty,$$

so we also have $J_{p,1}(u, \Omega) \leq J_{p,1}(v, \Omega)$. The theorem is proved. $\square$

Using compactness and approximations, we extend Theorem 11.15 to the case where the domain Ω is only assumed to be bounded and convex.

Theorem 11.16. *Let Ω be a bounded convex domain in $\mathbb{R}^n$. Then, for each $0 \leq p \neq n$, there exists a nonzero convex solution $u \in C(\overline{\Omega}) \cap C^\infty(\Omega)$ to the Dirichlet problem*

$$(11.43) \qquad \begin{cases} \det D^2 u = |u|^p & \text{in } \Omega, \\ \quad u = 0 & \text{on } \partial\Omega. \end{cases}$$

Moreover, if $0 \leq p < n$, then u is unique and it minimizes the functional

$$J_{p,1}(w, \Omega) = \frac{1}{n+1} \int_\Omega |w| \, d\mu_w - \frac{1}{p+1} \int_\Omega |w|^{p+1} \, dx$$

over all convex functions $w \in C(\overline{\Omega})$ with $w = 0$ on $\partial\Omega$.

Proof. Fix $0 \leq p \neq n$. Let $\{\Omega_m\}_{m=1}^\infty$ be a sequence of uniformly convex domains in $\mathbb{R}^n$ with smooth boundaries, satisfying $\Omega_m \supset \Omega_{m+1} \supset \Omega$ for all m, and such that Ω_m converges to Ω in the Hausdorff distance; see Theorem 2.51. By Theorem 11.15, there exists a nonzero convex solution $u_m \in C^{0,1}(\overline{\Omega_m}) \cap C^\infty(\Omega_m)$ to

$$\det D^2 u_m = |u_m|^p \quad \text{in } \Omega_m, \quad u_m = 0 \quad \text{on } \partial\Omega_m.$$

Then, Proposition 11.14 implies that a subsequence of $\{u_m\}_{m=1}^\infty$ converges uniformly on compact subsets of Ω to a nonzero convex solution $u \in C(\overline{\Omega}) \cap C^\infty(\Omega)$ to the Dirichlet problem (11.43).

Assume now $0 \leq p < n$. From the uniqueness result in Proposition 6.37, it remains to show that u minimizes $J_{p,1}(\cdot, \Omega)$ over all convex functions $w \in C(\overline{\Omega})$ with $w = 0$ on $\partial\Omega$.

Indeed, for each m, the function u_m is the minimizer of $J_{p,1}(\cdot, \Omega_m)$ over the set $\Phi_0(\Omega_m)$ defined as in (11.21). By Lemma 11.8, u_m is also the minimizer of $J_{p,1}(\cdot, \Omega_m)$ over convex functions $w \in C(\overline{\Omega_m})$ with $w = 0$ on $\partial\Omega_m$.

We use the above fact and the convergence $J_{p,1}(u_m, \Omega_m) \to J_{p,1}(u, \Omega)$ from Proposition 11.14 to establish the minimality of u. Suppose otherwise. Then, there exists a convex function $v \in C(\overline{\Omega})$ with $v = 0$ on $\partial\Omega$ such that

$$\delta := J_{p,1}(u, \Omega) - J_{p,1}(v, \Omega) > 0.$$

Then, for m large enough, we must have $|J_{p,1}(u_m, \Omega_m) - J_{p,1}(u, \Omega)| < \delta/3$, and also by Lemma 11.9, there is a convex function $v_m \in C(\overline{\Omega_m})$ with $v_m = 0$ on $\partial\Omega_m$ such that $|J_{p,1}(v, \Omega) - J_{p,1}(v_m, \Omega_m)| < \delta/3$.

All together, we discover that $J_{p,1}(v_m, \Omega_m) < J_{p,1}(u_m, \Omega_m) - \delta/3$. This contradicts the minimality property of u_m. The theorem is proved. $\square$

We remark that the conclusions of Theorem 11.16 remain unchanged if we replace the functional $J_{p,1}(\cdot, \Omega)$ by $J_{p,\lambda}(\cdot, \Omega)$ (see (11.37)) where $\lambda > 0$, and the equation $\det D^2 u = |u|^p$ by $\det D^2 u = \lambda|u|^p$.

11.4. Monge–Ampère Eigenvalue and Eigenfunctions

The eigenvalue problem for the Monge–Ampère operator $\det D^2 u$ on uniformly convex domains Ω in $\mathbb{R}^n$ ($n \geq 2$) with boundary $\partial\Omega \in C^\infty$ was first investigated by Lions [**Ln2**]. He showed that there exist a unique positive constant $\lambda = \lambda(\Omega)$ and a unique (up to positive multiplicative constants) nonzero convex function $u \in C^{1,1}(\overline{\Omega}) \cap C^\infty(\Omega)$ solving the eigenvalue problem for the Monge–Ampère operator:

(11.44) $\det D^2 u = \lambda|u|^n$ in Ω, $u = 0$ on $\partial\Omega$.

Moreover, Lions also found a spectral characterization of the Monge–Ampère eigenvalue via the first eigenvalues of linear second-order elliptic operators in nondivergence form, which we now describe.

Let $V_n = V_n(\Omega)$ be the set of all $n \times n$ matrices $A = (a_{ij})_{1 \leq i,j \leq n}$ with

$$a_{ij} \in C(\Omega), \quad A = A^t > 0 \quad \text{in } \Omega, \quad \text{and} \quad \det A \geq n^{-n}.$$

For each $A \in V_n$, let λ_1^A be the first (positive) eigenvalue of the linear second-order operator $-a_{ij}D_{ij}$ with zero Dirichlet boundary condition on $\partial\Omega$, so there exists $v \in W^{2,n}_{\text{loc}}(\Omega) \cap C(\overline{\Omega})$ with $v \not\equiv 0$ such that

$$-a_{ij}D_{ij}v = \lambda_1^A v \quad \text{in } \Omega, \quad v = 0 \quad \text{on } \partial\Omega.$$

The corresponding eigenfunctions v, up to multiplicative constants, are positive in Ω and unique; see the Appendix in [**Ln2**]. Lions showed that

(11.45) $$\lambda(\Omega) = \min_{A \in V_n} (\lambda_1^A)^n.$$

Furthermore, Lions also gave an interesting stochastic interpretation for the Monge–Ampère eigenvalue $\lambda(\Omega)$ of a uniformly convex domain Ω with smooth boundary. Another characterization of $\lambda(\Omega)$ in [**Ln2**] is given by

$$(11.46) \quad \lambda(\Omega) = \sup \Big\{ \gamma > 0 : \text{there is a convex solution } u_\gamma \in C^2(\overline{\Omega}) \text{ of}$$

$$\det D^2 u = (1 - \gamma^{1/n} u)^n \text{ in } \Omega, \quad u = 0 \text{ on } \partial\Omega \Big\}.$$

A variational characterization of $\lambda(\Omega)$ was found by Tso [**Ts**] who discovered that for uniformly convex domains Ω with $\partial\Omega$ being sufficiently smooth, the following formula holds:

$$(11.47) \quad \lambda(\Omega) = \inf \Bigg\{ \frac{\int_\Omega |u| \det D^2 u \, dx}{\int_\Omega |u|^{n+1} \, dx} : u \in C^{0,1}(\overline{\Omega}) \cap C^2(\Omega),$$

$$u \text{ is nonzero, convex in } \Omega, \ u = 0 \text{ on } \partial\Omega \Bigg\}.$$

All of the above features of $\lambda(\Omega)$ suggest the well-known properties of the first eigenvalue of the Laplace operator. As such, $\lambda(\Omega)$ is called the *Monge–Ampère eigenvalue* of Ω. The nonzero convex functions u solving (11.44) are called the *Monge–Ampère eigenfunctions* of the domain Ω.

In the above discussions, the domains Ω are uniformly convex with sufficiently smooth boundary, so they do not allow corners nor flat parts on the boundary. Here, following Le [**L7**], we study the eigenvalue problem for the Monge–Ampère operator on general bounded convex domains, and we establish basic results concerning: existence, smoothness, uniqueness, and stability.

Theorem 11.17 (The Monge–Ampère eigenvalue problem). *Let Ω be a bounded convex domain in $\mathbb{R}^n$ ($n \geq 2$). Define the constant $\lambda = \lambda[\Omega]$ via the variational formula:*

$$(11.48) \quad \lambda[\Omega] = \inf \Bigg\{ \frac{\int_\Omega |u| \, d\mu_u}{\int_\Omega |u|^{n+1} \, dx} : u \in C(\overline{\Omega}),$$

$$u \text{ is convex, nonzero in } \Omega, \ u = 0 \text{ on } \partial\Omega \Bigg\}.$$

Then, the following assertions hold.

(i) *There are positive constants $c(n)$ and $C(n)$ such that*

$$(11.49) \qquad c(n)|\Omega|^{-2} \leq \lambda[\Omega] \leq C(n)|\Omega|^{-2}.$$

(ii) *There exists a nonzero convex solution $u \in C^{0,\beta}(\overline{\Omega}) \cap C^\infty(\Omega)$ for all $\beta \in (0,1)$ to the Monge–Ampère eigenvalue problem*

$$(11.50) \qquad \det D^2 u = \lambda |u|^n \quad \text{in } \Omega, \quad u = 0 \quad \text{on } \partial\Omega.$$

Thus the infimum in (11.48) is achieved, and u is a minimizer.

(iii) *The eigenvalue-eigenfunction pair (λ, u) to (11.50) is unique in the following sense: If (Λ, v) satisfies $\det D^2 v = \Lambda |v|^n$ in Ω where Λ is a positive constant and $v \in C(\overline{\Omega})$ is convex, nonzero in Ω with $v = 0$ on $\partial\Omega$, then $\Lambda = \lambda$ and $v = cu$ for some positive constant c.*

(iv) *$\lambda[\cdot]$ is stable with respect to the Hausdorff convergence of domains: If a sequence of bounded convex domains $\{\Omega_m\}_{m=1}^\infty \subset \mathbb{R}^n$ converges in the Hausdorff distance to Ω, then $\lim_{m\to\infty} \lambda[\Omega_m] = \lambda[\Omega]$.*

Remark 11.18. Let Ω be a uniformly convex domain in $\mathbb{R}^n$ with $\partial\Omega \in C^\infty$, and let $\lambda(\Omega)$ be defined as in (11.47). Theorem 11.17 then asserts that

$$\lambda(\Omega) = \lambda[\Omega].$$

On the other hand, the class of convex functions in the formula (11.48) strictly contains that of (11.47) (see Problem 3.10 for one example), so the above equality is not obvious.

The proof of Theorem 11.17 is long, so we split it into several propositions. Assertions (i)–(ii) will be proved in Propositions 11.19 and 11.20. Combining Proposition 11.20 with Lemma 3.29, we find that the Monge–Ampère eigenfunctions are in fact globally log-Lipschitz.

For the existence of the Monge–Ampère eigenfunctions, our intention is to pass to the limit $p \nearrow n$ of the solutions in Theorem 11.16.

Proposition 11.19 (Existence of the Monge–Ampère eigenfunctions). *Let Ω be a bounded convex domain in $\mathbb{R}^n$. Let $\lambda = \lambda[\Omega]$ be defined by (11.48). Then (11.49) holds, and equation (11.50) has a nonzero convex solution $u \in C(\overline{\Omega}) \cap C^\infty(\Omega)$. We call such u a Monge–Ampère eigenfunction of Ω.*

Proof. By Lemma 11.6, we have (11.49). For each $0 \le p < n$, by Theorem 11.16, there exists a unique nonzero convex solution $u_p \in C(\overline{\Omega}) \cap C^\infty(\Omega)$ to

$$\det D^2 u = \lambda |u|^p \quad \text{in } \Omega, \qquad u = 0 \quad \text{on } \partial\Omega.$$

Moreover, u_p minimizes the functional $J_{p,\lambda}(\cdot) = J_{p,\lambda}^0(\cdot, \Omega)$, defined by (11.37), over all convex functions $u \in C(\overline{\Omega})$ with $u = 0$ on $\partial\Omega$.

We will bound $\|u_p\|_{L^\infty(\Omega)}$ uniformly in p from above and below. Then, we obtain as in the proof of Proposition 11.14 that, up to extracting a subsequence, $\{u_p\}$ converges, as $p \nearrow n$, uniformly on compact subsets of Ω to a nonzero convex solution $u \in C(\overline{\Omega}) \cap C^\infty(\Omega)$ of (11.50).

Step 1. Bounding $\|u_p\|_{L^\infty(\Omega)}$ from above. From the definition of λ in (11.48) and by the Hölder inequality, we have

$$\|u_p\|_{L^{n+1}(\Omega)}^{n+1} \leq \lambda^{-1} \int_\Omega |u_p| \det D^2 u_p \, dx = \int_\Omega |u_p|^{p+1} \, dx \leq \|u_p\|_{L^{n+1}(\Omega)}^{p+1} |\Omega|^{\frac{n-p}{n+1}}.$$

It follows that $\|u_p\|_{L^{n+1}(\Omega)} \leq |\Omega|^{\frac{1}{n+1}}$ which, by Lemma 11.6(i), implies that

$$\|u_p\|_{L^\infty(\Omega)} \leq C(n).$$

Step 2. Bounding $\|u_p\|_{L^\infty(\Omega)}$ from below. This step uses the minimality property of u_p with respect to the functional $J_{p,\lambda}(\cdot) = J_{p,\lambda}^0(\cdot, \Omega)$ over all convex functions $w \in C(\overline\Omega)$ with $w = 0$ on $\partial\Omega$. We will use

$$(11.51) \qquad \lambda \|u_p\|_{L^{p+1}(\Omega)}^{p+1} (p-n)(n+1)^{-1}(p+1)^{-1} = J_{p,\lambda}(u_p) \leq J_{p,\lambda}(\alpha v)$$

for some well-chosen $\alpha > 0$ and convex function $v \in C(\overline\Omega)$ with $v = 0$ on $\partial\Omega$ and $\|v\|_{L^\infty(\Omega)} = 1$.

Motivated by the expression on the left-hand side of (11.51) and

$$J_{p,\lambda}(\alpha v) = \frac{\alpha^{n+1}}{n+1} \int_\Omega |v| \, d\mu_v - \frac{\lambda \alpha^{p+1}}{p+1} \int_\Omega |v|^{p+1} \, dx,$$

we will choose

$$\alpha = \left[\frac{\lambda \int_\Omega |v|^{p+1} \, dx}{\int_\Omega |v| \, d\mu_v} \right]^{\frac{1}{n-p}},$$

so that

$$(11.52) \qquad J_{p,\lambda}(\alpha v) = \lambda \alpha^{p+1} \|v\|_{L^{p+1}(\Omega)}^{p+1} (p-n)(n+1)^{-1}(p+1)^{-1}.$$

Then, from $p < n$, (11.51), and (11.52), we obtain

$$\|u_p\|_{L^{p+1}(\Omega)}^{p+1} \geq \alpha^{p+1} \|v\|_{L^{p+1}(\Omega)}^{p+1}.$$

We now choose v. By the definition of λ in (11.48), we can choose a convex function $v \in C(\overline\Omega)$ with $v = 0$ on $\partial\Omega$ and $\|v\|_{L^\infty(\Omega)} = 1$ so that

$$\alpha = \left[\frac{\lambda \int_\Omega |v|^{p+1} \, dx}{\int_\Omega |v| \, d\mu_v} \right]^{\frac{1}{n-p}} \geq \left[\frac{\lambda \int_\Omega |v|^{n+1} \, dx}{\int_\Omega |v| \, d\mu_v} \right]^{\frac{1}{n-p}} \geq \frac{1}{2}.$$

Now, invoking Lemma 11.6 using $\|v\|_{L^\infty(\Omega)} = 1$, we obtain

$$\|v\|_{L^{p+1}(\Omega)}^{p+1} \geq \|v\|_{L^{n+1}(\Omega)}^{n+1} \geq c(n)|\Omega| \|v\|_{L^\infty(\Omega)}^{n+1} = c(n)|\Omega|,$$

for some $c(n) \in (0,1)$. Therefore

$$|\Omega| \|u_p\|_{L^\infty(\Omega)}^{p+1} \geq \|u_p\|_{L^{p+1}(\Omega)}^{p+1} \geq \alpha^{p+1} \|v\|_{L^{p+1}(\Omega)}^{p+1} \geq 2^{-(p+1)} c(n)|\Omega|.$$

It follows that

$$\|u_p\|_{L^\infty(\Omega)} \geq 2^{-1}[c(n)]^{\frac{1}{p+1}} \geq 2^{-1}c(n).$$

We have established the uniform bound for $\|u_p\|_{L^\infty(\Omega)}$ from above and below, thus completing the proof of the proposition. $\qquad \square$

Before proceeding further, we prove the following global almost Lipschitz result which is applicable to the Monge–Ampère eigenfunctions.

Proposition 11.20. *Let Ω be a bounded convex domain in $\mathbb{R}^n$.*

(i) *Assume $n - 2 \leq p$ and $M > 0$. Suppose that $u \in C(\overline{\Omega})$ with $\|u\|_{L^\infty(\Omega)} \leq 1$ is a nonzero convex solution of*

$$\det D^2 u = M|u|^p \quad \text{in } \Omega \qquad \text{and} \qquad u = 0 \quad \text{on } \partial\Omega.$$

Then, for all $\hat{\beta} \in (0, 1)$, we have $u \in C^{0,\hat{\beta}}(\overline{\Omega})$ and the estimate

$$(11.53) \qquad |u(x)| \leq C(n, p, M, \hat{\beta}, \operatorname{diam}(\Omega))[\operatorname{dist}(x, \partial\Omega)]^{\hat{\beta}} \quad \text{for all } x \in \Omega.$$

(ii) *Let $u \in C(\overline{\Omega})$ be a nonzero convex solution of (11.50). Then, $u \in C^{0,\beta}(\overline{\Omega})$ for all $\beta \in (0, 1)$, with estimate*

$$|u(x)| \leq C(n, \beta, \operatorname{diam}(\Omega))[\operatorname{dist}(x, \partial\Omega)]^{\beta} \|u\|_{L^\infty(\Omega)} \quad \text{for all } x \in \Omega.$$

Proof. Since assertion (ii) is a consequence of assertion (i) and (11.49), we only need to prove assertion (i). We use the same notation as in the proof of Lemma 3.26. It suffices to consider $p > 0$ since $p = 0$ implies that $n = 2$ and the result follows from Lemma 3.26. By Proposition 6.36, $u \in C^\infty(\Omega)$.

Let $d := \operatorname{diam}(\Omega)$. Since $\det D^2 u \leq M$, by Lemma 3.26, we have

$$(11.54) \qquad\qquad |u(x)| \leq C(n, M, d) x_n^{\frac{2}{n+1}} \quad \text{for all } x \in \Omega.$$

The key idea is to improve (11.54) by an iteration argument.

Step 1. For $\beta \in (0, 1)$, we show that if

$$(11.55) \qquad\qquad |u(x)| \leq C(n, p, M, \beta, d) x_n^{\beta} \quad \text{for all } x \in \Omega,$$

then, for all $x \in \Omega$,

$$(11.56) \quad |u(x)| \leq C(n, p, M, \beta, \alpha, d) x_n^{\alpha} \quad \text{for any } \beta < \alpha < \min\left\{\frac{\beta p + 2}{n}, 1\right\}.$$

Indeed, assume (11.55) holds for $\beta \in (0, 1)$. If $\beta < \alpha < \min\{\frac{\beta p+2}{n}, 1\}$, then for $\hat{C} = \hat{C}(n, p, M, \beta, \alpha, d)$ large, we have

$$(11.57) \qquad\qquad |u(x)| \leq C(n, p, M, \beta, d) x_n^{\beta} < \hat{C}^{\frac{n}{p}} M^{-\frac{1}{p}} x_n^{\frac{n\alpha - 2}{p}} \quad \text{in } \Omega.$$

Let $(U^{ij}) = (\det D^2 u)(D^2 u)^{-1}$ be the cofactor matrix of $D^2 u$. Then

$$\det U = (\det D^2 u)^{n-1} \quad \text{and} \quad U^{ij} D_{ij} u = n \det D^2 u = nM|u|^p.$$

Using (3.19), (11.57), and Lemma 2.57, we find that

$$U^{ij} D_{ij}(\hat{C}\phi_\alpha) \geq n\hat{C}(\det D^2 u)^{\frac{n-1}{n}}(\det D^2 \phi_\alpha)^{\frac{1}{n}}$$

$$(11.58) \qquad\qquad \geq n\hat{C} M^{\frac{n-1}{n}} |u|^{\frac{p(n-1)}{n}} x_n^{\frac{n\alpha - 2}{n}}$$

$$> nM|u|^p = n \det D^2 u = U^{ij} D_{ij} u \quad \text{in } \Omega.$$

Now, the maximum principle in Theorem 2.79 gives $u \geq \hat{C}\phi_\alpha$ in Ω. Therefore

$$|u(x)| = |u|(x', x_n) \leq -\hat{C}\phi_\alpha(x', x_n) \leq \hat{C}C_\alpha x_n^\alpha \quad \text{for all } x \in \Omega,$$

proving (11.56) as claimed.

Step 2. Conclusion. Assume $0 < \hat{\beta} < 1$. From Step 1, we can increase the exponent of x_n in the upper bound for $|u|$ by at least $\frac{\beta p + 2}{n} - \beta \geq \frac{2}{n}(1 - \hat{\beta})$ if $\beta < \hat{\beta}$. Thus, after a finite number of iterations, we obtain from (11.54) the estimate (11.53). The proof of the proposition is complete. $\square$

For the uniqueness of the Monge–Ampère eigenfunctions, we will apply the nonlinear integration by part inequality in Proposition 4.17. The following integrability lemma will guarantee the integrability conditions there.

Lemma 11.21. *Let Ω be a bounded convex domain in $\mathbb{R}^n$. Let $u, v \in C(\overline{\Omega}) \cap C^\infty(\Omega)$ be convex solutions to (11.50) with $\|u\|_{L^\infty(\Omega)} = \|v\|_{L^\infty(\Omega)} = 1$. Then,*

$$(11.59) \qquad |Du(x)||v(x)| \leq C(n, \operatorname{diam}(\Omega)) \operatorname{dist}^{\frac{n-1}{n+1}}(x, \partial\Omega)$$

and

$$(11.60) \qquad \int_\Omega \Delta u |v|^{n-1} \, dx \leq C(n, \operatorname{diam}(\Omega)).$$

Proof. By Lemma 2.37(iii), we have the gradient estimate

$$(11.61) \qquad |Du(x)| \leq \frac{|u(x)|}{\operatorname{dist}(x, \partial\Omega)} \quad \text{for } x \in \Omega.$$

Let $d := \operatorname{diam}(\Omega)$. Let us choose $\beta = n/(n+1)$ in Proposition 11.20. Then, for $x \in \Omega$, by (11.61) and Proposition 11.20(ii), we obtain (11.59) from

$$|Du(x)||v(x)| \leq C(n, \beta, d)\operatorname{dist}^{2\beta-1}(x, \partial\Omega) \leq C(n, d)\operatorname{dist}^{\frac{n-1}{n+1}}(x, \partial\Omega).$$

We will prove (11.60) via integration by parts. Note that $u, v < 0$ in Ω. Since u and v are not known a priori to be smooth up to the boundary, we use an approximation argument. Let $\{\Omega_m\}_{m=1}^\infty \subset \Omega$ be a sequence of uniformly convex domains with $\partial\Omega_m \in C^\infty$ that converges to Ω in the Hausdorff distance; see Theorem 2.51. To prove (11.60), by the Monotone Convergence Theorem, it suffices to prove that for all m,

$$I_m := \int_{\Omega_m} \Delta u |v|^{n-1} \, dx \leq C(n, d).$$

Let ν be the outer unit normal vector field on $\partial\Omega_m$. Using the interior smoothness of u and v and integrating by parts, we have

$$I_m = \int_{\Omega_m} (n-1)Du \cdot Dv |v|^{n-2} \, dx + \int_{\partial\Omega_m} \frac{\partial u}{\partial \nu} |v|^{n-1} \, d\mathcal{H}^{n-1} := A_m + B_m.$$

Now, we show that A_m and B_m are bounded, independent of m. Indeed, from (11.59) and $\|v\|_{L^\infty(\Omega)} = 1$, we have

$$|Du||v|^{n-1} \le |Du||v| \le C(n,d) \quad \text{in } \Omega;$$

hence, we can easily see that $B_m \le C(n,d)$.

From (11.61) and Proposition 11.20(ii) with $\beta = n/(n+1)$, we obtain

$$|Du(x)||Dv(x)||v(x)|^{n-2} \le C(n,d)\operatorname{dist}^{-\frac{2}{n+1}}(x,\partial\Omega).$$

Therefore,

$$A_m \le \int_\Omega C(n,d)\operatorname{dist}^{-\frac{2}{n+1}}(x,\partial\Omega)\, dx \le C(n,d).$$

The lemma is proved. $\square$

We now complete the proof of Theorem 11.17.

Proof of Theorem 11.17. Assertions (i) and (ii) were proved in Propositions 11.19 and 11.20. It remains to prove assertions (iii) and (iv).

We now prove (iii). By Proposition 6.36, we have $u, v \in C^\infty(\Omega)$.

Step 1. Uniqueness of the eigenvalue. We apply the nonlinear integration by parts inequality in Proposition 4.17 to u and v to obtain

$$\int_\Omega |u| \det D^2 v\, dx \ge \int_\Omega (\det D^2 u)^{\frac{1}{n}} (\det D^2 v)^{\frac{n-1}{n}} |v|\, dx.$$

But this is

$$\int_\Omega \Lambda |u||v|^n\, dx \ge \int_\Omega \lambda^{\frac{1}{n}} \Lambda^{\frac{n-1}{n}} |u||v|^n\, dx.$$

Since $|u||v|^n > 0$ in Ω, it follows that $\Lambda \ge \lambda$. Thus, by symmetry, $\lambda = \Lambda$.

Step 2. Uniqueness of the eigenfunctions up to positive multiplicative constants. From Step 1, we already have $\Lambda = \lambda = \lambda[\Omega]$. By considering $\|u\|_{L^\infty(\Omega)}^{-1} u$ and $\|v\|_{L^\infty(\Omega)}^{-1} v$ instead of u and v, we can assume that

$$\|u\|_{L^\infty(\Omega)} = \|v\|_{L^\infty(\Omega)} = 1.$$

We now show that $u = v$ by making use of the integrability result in Lemma 11.21 and the nonlinear integration by parts inequality.

Observe that $n(\det D^2(u+v))^{\frac{1}{n}} \le \operatorname{trace}(D^2(u+v)) = \Delta(u+v)$ in Ω; see (2.15). Thus, from Lemma 11.21 and (11.49), we find that

$$
(11.62) \quad \int_\Omega (\det D^2(u+v))^{\frac{1}{n}} (\det D^2 v)^{\frac{n-1}{n}}\, dx \le \int_\Omega \lambda^{\frac{n-1}{n}} \Delta(u+v)|v|^{n-1}\, dx
$$
$$
\le C(n,\Omega).
$$

On the other hand, by Lemma 2.59, we infer that for all $x \in \Omega$

(11.63)
$$(\det D^2(u+v)(x))^{\frac{1}{n}} \geq (\det D^2 u(x))^{\frac{1}{n}} + (\det D^2 v(x))^{\frac{1}{n}}$$
$$= \lambda^{\frac{1}{n}} |u(x) + v(x)|$$

with equality if and only if $D^2 u(x) = C(x) D^2 v(x)$ for some $C(x) > 0$.

In view of (11.62) and the inequality

$$\int_\Omega \det D^2 v \, dx = \int_\Omega \lambda |v|^n \, dx \leq \lambda |\Omega| \leq C(n, \Omega),$$

we can apply Proposition 4.17 to $u + v$ and v and use (11.63) to obtain

$$\int_\Omega |u+v| \det D^2 v \, dx \geq \int_\Omega (\det D^2(u+v))^{\frac{1}{n}} (\det D^2 v)^{\frac{n-1}{n}} |v| \, dx.$$

However, according to (11.63), this implies

$$\int_\Omega |u+v| \lambda |v|^n \, dx \geq \int_\Omega \lambda^{\frac{1}{n}} |u+v| \lambda^{\frac{n-1}{n}} |v|^n \, dx.$$

It follows that we must have equality in (11.63) for all $x \in \Omega$. Hence, $D^2 u(x) = C(x) D^2 v(x)$ for all $x \in \Omega$. Taking the determinant of both sides and using (11.50), we find $C(x) = |u(x)|/|v(x)|$, which implies

(11.64)
$$D^2 u(x) = \frac{u(x)}{v(x)} D^2 v(x).$$

Now, we use (11.64) and (11.59) to show that $u = v$.

Without loss of generality, we can assume that the domain Ω contains the origin in its interior. For any direction $\mathbf{e} \in \partial B_1(0)$, the line through the origin in the $\mathbf{e}$ direction intersects the boundary $\partial \Omega$ at z and $\bar{z}$. Assume that the segment from z to $\bar{z}$ is given by $z + t\mathbf{e}$ for $0 \leq t \leq \tau := |\bar{z} - z|$. Let us consider the following nonpositive single variable functions on $[0, \tau]$:

$$f(t) = u(z + t\mathbf{e}), \quad \tilde{f}(t) = v(z + t\mathbf{e}).$$

Then, we have

$$f(0) = f(\tau) = \tilde{f}(0) = \tilde{f}(\tau) = 0$$

and

$$f''(t) = \mathbf{e}^t D^2 u(z + t\mathbf{e})\mathbf{e}, \quad \tilde{f}''(t) = \mathbf{e}^t D^2 v(z + t\mathbf{e})\mathbf{e}.$$

From (11.64), we find $f''(t) = [f(t)/\tilde{f}(t)]\tilde{f}''(t)$ for all $t \in (0, \tau)$. Thus, $(f'(t)\tilde{f}(t) - f(t)\tilde{f}'(t))' = 0$ for all $t \in (0, \tau)$, and hence

$$f'(t)\tilde{f}(t) - f(t)\tilde{f}'(t) = \text{a constant } C \text{ on } (0, \tau).$$

Using (11.59) and $n \geq 2$, we find that $f'(t)\tilde{f}(t) \to 0$ and $f(t)\tilde{f}'(t) \to 0$ when $t \to 0$. Thus $C = 0$, and $(f(t)/\tilde{f}(t))' = 0$ on $(0, \tau)$. It follows that $f(t) = c\tilde{f}(t)$ on $(0, \tau)$ for some constant $c > 0$. Therefore, $c = u(0)/v(0)$.

The above arguments show that the ratio u/v is the same positive constant c in every direction $\mathbf{e} \in \partial B_1(0)$, so $u/v \equiv c$ in Ω. Since $\|u\|_{L^\infty(\Omega)} = \|v\|_{L^\infty(\Omega)} = 1$, we have $c = 1$, and therefore $u = v$. Assertion (iii) is proved.

Finally, we prove assertion (iv). For each positive integer m, we let $u_m \in C(\overline{\Omega_m}) \cap C^\infty(\Omega_m)$ with $\|u_m\|_{L^\infty(\Omega_m)} = 1$ be a convex eigenfunction of the Monge–Ampère operator on Ω_m:

$$\det D^2 u_m \ = \lambda[\Omega_m] |u_m|^n \quad \text{in } \Omega_m, \qquad u_m = 0 \quad \text{on } \partial \Omega_m.$$

From the estimates in assertion (i) for $\lambda[\Omega_m]$, we infer that there is a subsequence $\{\lambda[\Omega_{m_j}]\}_{j=1}^\infty$ converging to $\Lambda > 0$. It remains to show that $\Lambda = \lambda[\Omega]$.

Indeed, arguing as in Proposition 11.14, we discover that a subsequence $\{u_{m_{j_l}}\}_{l=1}^\infty$ converges uniformly on compact subsets of Ω to a nonzero convex solution of

$$\det D^2 u \ = \Lambda |u|^n \quad \text{in } \Omega, \qquad u = 0 \quad \text{on } \partial \Omega.$$

By the uniqueness of the Monge–Ampère eigenvalue, we have $\Lambda = \lambda[\Omega]$. $\square$

Remark 11.22. Theorem 11.17 shows that the Monge–Ampère eigenfunctions $u \in C(\overline{\Omega}) \cap C^\infty(\Omega)$ minimize the Rayleigh quotient

$$R(u) = \frac{\int_\Omega |u| \, d\mu_u}{\int_\Omega |u|^{n+1} \, dx}$$

and $R(u)$ is equal to the Monge–Ampère eigenvalue $\lambda[\Omega]$ of Ω at these eigenfunctions. It is also interesting to study the converse problem:

> *If $u \in C(\overline{\Omega})$ is convex, with $u = 0$ on $\partial\Omega$, and if it satisfies $R(u) = \lambda[\Omega]$, then can we conclude that u is a Monge–Ampère eigenfunction of Ω?*

This seems to be a difficult open problem.

The Monge–Ampère equation is affine invariant; so is the Monge–Ampère eigenvalue. The proof, left as an exercise, is a simple consequence of the uniqueness of the Monge–Ampère eigenvalue; see Problem 11.2.

Proposition 11.23 (The Monge–Ampère eigenvalue and affine transformations). *Let Ω be a bounded convex domain in $\mathbb{R}^n$. If $T : \mathbb{R}^n \to \mathbb{R}^n$ is an invertible affine transformation on $\mathbb{R}^n$, then $\lambda[T(\Omega)] = |\det T|^{-2} \lambda[\Omega]$.*

To conclude this section, we mention further properties of the Monge–Ampère eigenvalue. Brandolini, Nitsch, and Trombetti [**BNT**] showed that among all uniformly convex domains in $\mathbb{R}^n$ with smooth boundaries having a fixed positive volume, the ball has the largest Monge–Ampère eigenvalue. Salani [**Sal**] proved a Brunn-Minkowski inequality for the Monge–Ampère eigenvalue in the class of uniformly convex domains with C^2 boundary. These results were extended to general bounded convex domains in [**L7**].

11.5. Global Regularity for Degenerate Equations

As Proposition 6.36 shows, degenerate Monge–Ampère equations of the form $\det D^2 u = M|u|^p$ with zero boundary values have interior smoothness, regardless of the smoothness of the boundary. The Monge–Ampère eigenvalue problem belongs to this family of equations. It is then natural to investigate global regularity of these equations when the boundary is nice. This is, however, a subtle question. In this section, we briefly mention several global regularity results for the Monge–Ampère equation

$$\det D^2 u = f \geq 0 \quad \text{on } \Omega \subset \mathbb{R}^n \quad (n \geq 2),$$

when f is degenerate; that is, f does not have a positive lower bound.

Guan, Trudinger, and Wang [**GTW**] established the following global $C^{1,1}$ regularity under optimal conditions on the degenerate right-hand side, the domain and boundary data.

Theorem 11.24 (Global $C^{1,1}$ regularity for degenerate Monge–Ampère equations)**.** *Let Ω be a uniformly convex domain in $\mathbb{R}^n$ with boundary $\partial\Omega \in C^{3,1}$. Let $\varphi \in C^{3,1}(\overline{\Omega})$ and let f be a nonnegative function on $\overline{\Omega}$ such that $f^{\frac{1}{n-1}} \in C^{1,1}(\overline{\Omega})$. Let $u \in C(\overline{\Omega})$ be the convex solution to*

$$\det D^2 u = f \quad \text{in } \Omega, \qquad u = \varphi \quad \text{on } \partial\Omega.$$

Then $u \in C^{1,1}(\overline{\Omega})$ with estimate

$$\|u\|_{C^{1,1}(\overline{\Omega})} \leq C = C\big(n, \Omega, \|\varphi\|_{C^{3,1}(\overline{\Omega})}, \|f^{\frac{1}{n-1}}\|_{C^{1,1}(\overline{\Omega})}\big).$$

When $f \equiv 0$, the global $C^{1,1}$ regularity in Theorem 11.24 was first obtained by Caffarelli, Nirenberg, and Spruck [**CNS2**] under suitable hypotheses on the boundary data and domain. Li and Wang [**LiW**] studied the case $f \equiv 0$ with less regular boundary data and domain. For the sharpness of the conditions on $\partial\Omega$, φ, and f in Theorem 11.24, see [**CNS2, GTW**].

When f is comparable to some positive power of the distance $\text{dist}(\cdot, \partial\Omega)$, stronger global results are available, due to the remarkable Boundary Localization Theorem (Theorem 8.25) of Savin [**S6**]. For the rest of this section, we focus on the Monge–Ampère equations of the type

$$(11.65) \qquad \det D^2 u = g[\text{dist}(\cdot, \partial\Omega)]^{\alpha}$$

where $\alpha > 0$ and g is a continuous strictly positive function.

Using Theorem 8.25, Savin obtained the following C^2 regularity estimate at the boundary.

Theorem 11.25 (Pointwise C^2 estimates). *Let Ω be a bounded convex domain in $\mathbb{R}^n$ with $0 \in \partial\Omega$ and satisfying $B_\rho(\rho\,e_n) \subset \Omega \subset B_{1/\rho}(0) \cap \{x_n > 0\}$ for some $\rho > 0$. Let $u \in C(\overline{\Omega})$ be a convex function satisfying*

$$\det D^2 u = g[\mathrm{dist}(\cdot, \partial\Omega)]^\alpha$$

where $\alpha > 0$ and g is a continuous nonnegative function with $g(0) > 0$. Assume that $u|_{\partial\Omega \cap B_\rho(0)}$ is of class C^2. Assume that

$$x_{n+1} = 0 \ \text{is a tangent hyperplane to } u \text{ at } 0,$$

in the sense that $u \geq 0$, $u(0) = 0$, and any hyperplane $x_{n+1} = \varepsilon x_n$, $\varepsilon > 0$, is not a supporting hyperplane for u. Suppose that u separates quadratically from its tangent hyperplane on the boundary; that is, for some $\mu > 0$

$$\mu\,|x|^2 \leq u(x) \leq \mu^{-1}\,|x|^2 \qquad on\ \partial\Omega \cap \{x_n \leq \rho\}.$$

Then u is pointwise C^2 at 0. More precisely, there exist a sliding A along $x_n = 0$ (see Definition 8.3) and a constant $b > 0$ such that

$$u(Ax) = Q_0(x') + b x_n^{2+\alpha} + o(|x'|^2 + x_n^{2+\alpha}) \quad for\ all\ x \in B_\rho(0) \cap \Omega,$$

where Q_0 represents the quadratic part of the boundary data at the origin.

By a perturbative approach based on Theorems 8.25 and 11.25, Le and Savin [**LS4**] improved the above boundary C^2 regularity to $C^{2,\beta}$. More precisely, the following pointwise $C^{2,\beta}$ estimates were established:

Theorem 11.26 (Pointwise $C^{2,\beta}$ estimates at the boundary for degenerate Monge–Ampère equations). *Let Ω be a bounded convex domain in $\mathbb{R}^n$ with $0 \in \partial\Omega$ and satisfying $B_\rho(\rho\,e_n) \subset \Omega \subset B_{1/\rho}(0) \cap \{x_n > 0\}$ for some $\rho > 0$. Let $\alpha > 0$ and $\beta \in (0, \frac{2}{2+\alpha})$. Let $u \in C(\overline{\Omega})$ be a convex function satisfying*

$$\det D^2 u = g[\mathrm{dist}(\cdot, \partial\Omega)]^\alpha \quad in\ B_\rho(0) \cap \Omega$$

where $g \in C^{0,\gamma}(\overline{\Omega} \cap B_\rho(0))$, for some $\gamma \geq \beta(2+\alpha)/2$, is a strictly positive function. Assume that $u(0) = 0, Du(0) = 0$, and u separates quadratically from its tangent hyperplane on the boundary; that is, for some $\mu > 0$

$$\mu\,|x|^2 \leq u(x) \leq \mu^{-1}\,|x|^2 \qquad on\ \partial\Omega \cap \{x_n \leq \rho\}.$$

Assume that $u|_{\partial\Omega \cap B_\rho(0)}$ is of class $C^{2,\beta}$. Then u is pointwise $C^{2,\beta}$ at 0. More precisely, there exist a sliding A along $x_n = 0$ and a constant $b > 0$ such that

$$u(Ax) = Q_0(x') + b x_n^{2+\alpha} + O(|x'|^2 + x_n^{2+\alpha})^{1+\beta/2} \quad for\ all\ x \in B_\rho(0) \cap \Omega,$$

where Q_0 represents the quadratic part of the boundary data at the origin.

The following global $C^{2,\beta}$ estimates for degenerate Monge–Ampère equations were also obtained in [**LS4**].

Theorem 11.27 (Global $C^{2,\beta}$ estimates for degenerate Monge–Ampère equations). *Let $\beta \in (0, \frac{2}{2+\alpha}) \cap (0, \alpha]$. Let Ω be a bounded convex domain in $\mathbb{R}^n$ with $0 \in \partial\Omega$ and satisfying $B_\rho(\rho\, e_n) \subset \Omega \subset B_{1/\rho}(0) \cap \{x_n > 0\}$ for some $\rho > 0$. Let $u \in C(\overline{\Omega})$ be a convex function satisfying*

$$\det D^2 u = g[\mathrm{dist}(\cdot, \partial\Omega)]^\alpha \quad in \ B_\rho(0) \cap \Omega$$

where g is a continuous strictly positive function in $B_\rho(0) \cap \overline{\Omega}$. Suppose that

$$\partial\Omega \in C^{2,\beta}, \quad u|_{\partial\Omega} \in C^{2,\beta}, \quad g \in C^\gamma \quad where \ \gamma = \beta(2+\alpha)/2.$$

If u separates quadratically from its tangent hyperplane at 0, then

$$u \in C^{2,\beta}(\overline{\Omega} \cap B_\delta(0)) \quad for \ some \ small \ \delta > 0.$$

These estimates in Theorems 11.26 and 11.27 can be viewed as the degenerate counterparts of Theorems 4.11, 4.12, 10.8, 10.17, 10.20 in the works [**CNS1**, **Iv**, **K3**, **S5**, **TW5**]. They are the key step in proving the global smoothness of the Monge–Ampère eigenfunctions of uniformly convex domains with smooth boundaries in all dimensions. The question of global higher derivative smoothness of the eigenfunctions u solving (11.44) was first asked by Trudinger around 1987. In the two-dimensional case, Hong, Huang, and Wang [**HHW**] resolved this question in the affirmative. In higher dimensions, the global $C^2(\overline{\Omega})$ smoothness of the eigenfunction u was obtained by Savin [**S6**]. This is a consequence of Theorem 11.25. In [**LS4**], Le and Savin settled in the affirmative the question of global smoothness of the eigenfunctions u for the Monge–Ampère operator in all dimensions.

Theorem 11.28 (Global smoothness of the Monge–Ampère eigenfunctions). *Let Ω be a uniformly convex domain in $\mathbb{R}^n$ ($n \geq 2$) with boundary of class C^∞. Let $u \in C(\overline{\Omega})$ be a Monge–Ampère eigenfunction of Ω; that is, u is a nonzero convex solution to*

$$(11.66) \qquad \begin{cases} \det D^2 u = \lambda|u|^n & in \ \Omega, \\ u = 0 & on \ \partial\Omega. \end{cases}$$

Then $u \in C^\infty(\overline{\Omega})$.

Despite the degeneracy of the right-hand side of (11.66) near the boundary, the Monge–Ampère eigenfunctions also enjoy the same global smoothness feature as the first eigenfunctions of the Laplace operator.

We say a few words about the proof of Theorem 11.28. Note that the eigenfunction u solves a degenerate Monge–Ampère equation of the form (11.65) where $\alpha = n$; see Problem 8.5. First, using Theorems 11.27, we can obtain the global $C^{2,\beta}$ regularity of the eigenfunction for any $\beta < \frac{2}{2+n}$. Typically, for elliptic questions, once $C^{2,\beta}$ regularity is obtained, the higher regularity follows easily by Schauder estimates. This is not the case in Theorem 11.28 because of the high degeneracy of the equation (11.66). The key idea in [**LS4**] consists of performing both a hodograph transform and a partial Legendre transform in order to deduce that a suitable transformation of u satisfies a degenerate Grushin-type equation with Hölder continuous coefficients. For this Grushin-type operator, we can establish global Schauder estimates which in turn give the global smoothness of the Monge–Ampère eigenfunction.

11.6. Problems

Problem 11.1. Show that the conclusion of Proposition 11.1(i) continues to hold if instead of requiring that $u = 0$ on $\partial\Omega$ and $\mu_k(\Omega) \leq E$, we require that for all $k \geq 1$, $\mu_{u_k} \leq E\,dx$ in a fixed neighborhood of $\partial\Omega$.

Problem 11.2. Prove Proposition 11.23.

Problem 11.3. Show that two open triangles in the plane have the same areas if and only if they have the same Monge–Ampère eigenvalues.

Problem 11.4. Let $\lambda[\Omega]$ be the Monge–Ampère eigenvalue of a bounded, convex domain Ω. Prove the following *domain monotonicity property*: If Ω_1 and Ω_2 are bounded convex domains such that $\Omega_1 \subset \Omega_2$, then $\lambda[\Omega_2] \leq \lambda[\Omega_1]$.

Problem 11.5 (Reverse Aleksandrov estimate). Let Ω be a bounded convex domain in $\mathbb{R}^n$. Let $\lambda[\Omega]$ and w be the Monge–Ampère eigenvalue and eigenfunction of Ω. Let $u \in C^5(\Omega) \cap C(\overline{\Omega})$ be a convex function in Ω with $u = 0$ on Ω and satisfying $0 < \int_\Omega (\det D^2 u)^{1/n} |w|^{n-1}\,dx < \infty$. Show that

$$\int_\Omega (\lambda[\Omega])^{1/n} |u||w|^n\,dx \geq \int_\Omega (\det D^2 u)^{1/n} |w|^n\,dx.$$

Problem 11.6. Prove there there are no convex functions $u \in C^4(\mathbb{R}^n)$ such that $u < 0$ on $\mathbb{R}^n$ and $\det D^2 u \geq |u|^n$ on $\mathbb{R}^n$.

Problem 11.7. Let Ω be a bounded convex domain in $\mathbb{R}^n$. Let $\lambda[\Omega]$ be the Monge–Ampère eigenvalue of Ω, and let $t \geq n[\lambda[\Omega]]^{1/n}$. Show that there does not exist an Aleksandrov solution $v \in C(\overline{\Omega})$ to the Monge–Ampère equation

$$\det D^2 v = e^{-tv} \quad \text{in } \Omega, \qquad v = 0 \quad \text{on } \partial\Omega.$$

Problem 11.8. Let Ω be a bounded convex domain in $\mathbb{R}^n$ ($n \geq 2$) and let $\alpha \in (0, 1]$. Consider the following initial boundary value problem for $u(x, t)$ being strictly convex for each $t \geq 0$:

$$\partial_t u(x, t) = [\det D_{ij}u(x, t)]^\alpha \quad \text{in } \Omega \times [0, \infty), \qquad u(x, t) = 0 \quad \text{on } \partial\Omega \times [0, \infty).$$

(a) Show that when $\alpha = 1$, there exists a *self-similar*, nonzero solution of the form $u(x,t) = (1+t)^{-1/(n-1)}v(x)$, where $v \in C^\infty(\Omega) \cap C(\overline{\Omega})$ is the unique nonzero convex solution of

$$\det D^2 v = |v|/(n-1) \quad \text{in } \Omega, \qquad v = 0 \quad \text{on } \partial\Omega.$$

(b) Find self-similar, nonzero solutions of the form $u(x,t) = \varphi(t)w(x)$ for this problem when $\alpha = 1/n$.

Problem 11.9. This problem is related to the minimum of the Monge–Ampère eigenvalue.

(a) Show that, among all bounded convex domains Ω in $\mathbb{R}^n$ having a fixed positive volume, there exists a convex domain S with the smallest Monge–Ampère eigenvalue.

(b) ($\star$) Show that, among all bounded convex domains in $\mathbb{R}^n$ having a fixed positive volume, the n-dimensional regular simplex (that is, the interior of the convex hull of $(n+1)$ equally spaced points in $\mathbb{R}^n$) has the smallest Monge–Ampère eigenvalue. (*As of this writing, this problem is still open for all $n \geq 2$.*)

Problem 11.10. Let Ω be a bounded convex domain in $\mathbb{R}^n$. Let $0 \leq p, q < n$. Show that there exists a unique nonzero convex Aleksandrov solution $u \in C(\overline{\Omega}) \cap C^\infty(\Omega)$ to the Dirichlet problem

$$\det D^2 u = |u|^p + |u|^q \quad \text{in } \Omega, \quad u = 0 \quad \text{on } \partial\Omega.$$

Problem 11.11. Let Ω be a bounded convex domain in $\mathbb{R}^n$ ($n \geq 2$) with smooth boundary. Fix $1 < p < 2$. Let $u \in C(\overline{\Omega}) \cap C^\infty(\Omega)$ be a nonzero convex solution to

$$\det D^2 u = |u|^p \quad \text{in } \Omega, \quad u = 0 \quad \text{on } \partial\Omega.$$

Show that $u \notin C^4(\overline{\Omega})$.

11.7. Notes

For an account of the eigenvalues and eigenfunctions of the Laplace operator, or more generally, uniformly elliptic, symmetric operators in divergence form, see Evans [**E2**, Section 6.5]. As mentioned in the introduction of Chapter 3, the Monge–Ampère energy $I(u; \Omega)$ defined in (11.5) has its roots in works of Courant and Hilbert [**CH1**], Bakelman [**Ba3, Ba4**], Aubin [**Au1**], among others.

In Section 11.1, related to Proposition 11.1, see also Abedin–Kitagawa [**AK**], Krylov [**K1**], and Tso [**Ts**]. Proposition 11.2 appears in Aubin [**Au2**, Section 8.4]. Lemma 11.6 is from Le [**L7**]. Lemmas 11.8 and 11.9 are due to Hartenstine [**Ha2**].

Sections 11.2 and 11.3 are based on Tso [**Ts**]. Theorem 11.13 and its proof are basically taken from Chou–Wang [**ChW2**]. Theorem 11.16 is based on Le [**L7**]. For the existence of convex solutions to (11.43) when $p < 0$, we refer to Cheng and Yau [**CY**] for $p = -(n+2)$ and Le [**L12**].

Section 11.4 is based on Le [**L7**]. A full generalization of the results by Lions and Tso to the k-Hessian operator ($1 \leq k \leq n$) was carried out by Wang [**W2**]. Wang studied the first eigenvalue together with its variational characterization for the k-Hessian operator. Abedin and Kitagawa [**AK**] introduced an inverse iterative scheme to solve the Monge–Ampère eigenvalue problem. When the domain is uniformly convex with analytic boundary, Huang and Lü [**HuL**] proved that the Monge-Ampère eigenfunctions are analytic up to the boundary. The Monge–Ampère eigenvalue problem (11.3) for general u which is not required to be convex or concave has not been studied yet.

Problem 11.4 is based on Wang [**W2**]. Problem 11.8(a) is taken from Oliker [**Olk**]. Problems 11.9 and 11.11 are taken from [**L7**].

Part 2

The Linearized Monge–Ampère Equation

Chapter 12

Interior Harnack Inequality

We have witnessed the appearance of the linearized Monge–Ampère operator in our investigation of the regularity theory for the Monge–Ampère equation, including: global second derivative estimates in Theorem 4.9, Pogorelov's interior second derivative estimates in Theorem 6.1, and global almost Lipschitz estimates for degenerate Monge–Ampère equations which include the Monge–Ampère eigenfunctions in Proposition 11.20. On the other hand, as Examples 1.11–1.14 show, the linearized Monge–Ampère operator also appears in contexts beyond the Monge–Ampère equation. Thus, it is of significant interest to study the linearized Monge–Ampère equation in more general contexts and, in particular, the regularity theory associated with it.

The regularity theory for the linearized Monge–Ampère equation was initiated in the fundamental paper [**CG2**] by Caffarelli and Gutiérrez. They developed an interior Harnack inequality theory for nonnegative solutions of the homogeneous equation

$$L_u v := \sum_{i,j=1}^n U^{ij} D_{ij} v \equiv U^{ij} D_{ij} v \equiv \operatorname{trace}(U D^2 v) = 0,$$

where

$$U = (U^{ij}) = \operatorname{Cof} D^2 u = (\det D^2 u)(D^2 u)^{-1},$$

in terms of the structure, called the A_∞-condition, of the Monge–Ampère measure μ_u of the convex function u. This A_∞-condition is clearly satisfied when we have the pinching of the Hessian determinant

(12.1) $\qquad 0 < \lambda \le \det D^2 u \le \Lambda \quad \text{in a domain } \Omega \subset \mathbb{R}^n \, (n \ge 2),$

which is the main focus of this book. This chapter is concerned with the Caffarelli–Gutiérrez Harnack inequality and its consequences.

12.1. Caffarelli–Gutiérrez Harnack Inequality

For later discussions, we first recall the *Harnack inequality for uniformly elliptic equations*. The classical regularity theory for linear, second-order, uniformly elliptic equations with measurable coefficients deals with operators in *divergence* form or *nondivergence* form:

$$L = \sum_{i,j=1}^{n} \frac{\partial}{\partial x_i}\left(a^{ij}\frac{\partial}{\partial x_j}\right) \quad \text{or} \quad L = \sum_{i,j=1}^{n} a^{ij}\frac{\partial^2}{\partial x_i \partial x_j}$$

with positive ellipticity constants λ and Λ; that is, the eigenvalues of the symmetric coefficient matrix $A(x) = (a^{ij}(x))$ are bounded between λ and Λ:

$$(12.2) \qquad\qquad \lambda I_n \leq (a^{ij}) \leq \Lambda I_n.$$

The uniform ellipticity of $A(x)$ is invariant under rigid transformations of the domain; that is, for any orthogonal matrix P, the matrix $A(Px)$ is also uniformly elliptic with the same ellipticity constants as $A(x)$.

The important Hölder estimates and Harnack inequality for *divergence* form equations $Lu = 0$ were established in the late 1950s by De Giorgi, Nash, and Moser [**DG1, Na, Mo**], while corresponding results for *nondivergence* form equations were established only in the late 1970s by Krylov and Safonov.

We recall below the celebrated Harnack inequalities, formulated on Euclidean balls, of Moser [**Mo**] and of Krylov and Safonov [**KS1, KS2**].

Theorem 12.1 (Moser's Harnack inequality). *Assume (a^{ij}) satisfies (12.2). Let $v \in W^{1,2}(\Omega)$ be a nonnegative solution of*

$$\frac{\partial}{\partial x_i}\left(a^{ij}\frac{\partial v}{\partial x_j}\right) = 0 \quad in\ \Omega.$$

Then, for all $B_{2r}(x_0) \Subset \Omega$, we have

$$\sup_{B_r(x_0)} v \leq C(n, \lambda, \Lambda) \inf_{B_r(x_0)} v.$$

Theorem 12.2 (Krylov–Safonov Harnack inequality). *Assume (a^{ij}) satisfies (12.2). Let $v \in W^{2,n}_{\mathrm{loc}}(\Omega)$ be a nonnegative solution of*

$$a^{ij}D_{ij}v = 0 \quad in\ \Omega.$$

Then, for all $B_{2r}(x_0) \Subset \Omega$, we have

$$\sup_{B_r(x_0)} v \leq C(n, \lambda, \Lambda) \inf_{B_r(x_0)} v.$$

The proof of Theorem 12.1 is based on the isoperimetric inequality, Sobolev embedding, and Moser iteration, while that of Theorem 12.2 is based on the Aleksandrov–Bakelman–Pucci (ABP) maximum principle (Theorem 3.16) coming from the Monge–Ampère equation.

Now, back to the linearized Monge–Ampère equation. The linearized Monge–Ampère theory investigates operators of the form $L_u = U^{ij}D_{ij}$. Under the structural assumption (12.1), it is only known that the *product of the eigenvalues* of the coefficient matrix U is bounded between two constants λ^{n-1} and Λ^{n-1}. Therefore, the linearized Monge–Ampère operator L_u is in general not uniformly elliptic; that is, the eigenvalues of the coefficient matrix U are not necessarily bounded between two positive constants. Moreover, when considered in a bounded convex domain, the coefficient matrix U of L_u can be possibly singular near the boundary. Therefore, the linearized Monge–Ampère equation can be both degenerate and singular. We refer the readers to Sections 6.5 and 10.2 for explicit examples.

The degeneracy and singularity of L_u are the main difficulties in establishing regularity results for solutions to the linearized Monge–Ampère equation. It is not a priori clear if there are regularity estimates that *depend only on the structure*, that is, the dimension n and constants λ and Λ.

In [**CG2**], Caffarelli and Gutiérrez succeeded in establishing a Harnack inequality for the homogeneous linearized Monge–Ampère equation from which the Hölder regularity of solutions follows. Since then, there have been many developments and applications as mentioned in Section 1.2.3. Among the many insights in [**CG2**] is the basic *affine invariance* property of the linearized Monge–Ampère operator $L_u := U^{ij}D_{ij}$; that is, it is invariant with respect to affine transformations of the independent variable x of the form $x \mapsto Tx$ with $\det T = 1$. Indeed, for such T, the rescaled functions $\tilde{u}(x) = u(Tx)$ and $\tilde{v}(x) = v(Tx)$ satisfy the same structural conditions as in $L_u v$ and (12.1). The calculations result from the following general lemma.

Lemma 12.3 (The linearized Monge–Ampère equation under affine transformations). *Assume that* $U^{ij}D_{ij}v = g$ *on a domain* $\Omega \subset \mathbb{R}^n$. *Under the transformations*

$$(12.3) \quad \tilde{u}(x) = a\,u(Tx), \quad \tilde{v}(x) = b\,v(Tx), \quad \tilde{g}(x) = a^{n-1}b\,(\det T)^2 g(Tx),$$

where $a, b \in \mathbb{R}$ *and* $T : \mathbb{T}^n \to \mathbb{R}^n$ *is affine and invertible, we have*

$$(12.4) \quad \det D^2\tilde{u} = a^n(\det T)^2(\det D^2 u) \circ T \quad and \quad \tilde{U}^{ij}D_{ij}\tilde{v} = \tilde{g} \quad on \ T^{-1}\Omega.$$

Proof. Assume $Tx = Ax + w$ where $A \in \mathbb{M}^{n\times n}$ and $w \in \mathbb{R}^n$. Then, $\det T = \det A$. We can compute from $D^2\tilde{u}(x) = a\,A^t(D^2u(Tx))A$ that

$$\tilde{U}(x) = (\det D^2\tilde{u}(x))(D^2\tilde{u}(x))^{-1} = a^{n-1}(\det A)^2 A^{-1}U(Tx)(A^{-1})^t.$$

Similarly, $D^2\tilde{v}(x) = b\,A^t(D^2v(Tx))A$. Hence (12.4) easily follows. $\qquad \square$

Due to the affine invariance of the linearized Monge–Ampère equation, important estimates for their solutions should be formulated using sections (see Chapter 5) instead of balls. We are now ready to state the Caffarelli–Gutiérrez Theorem on Harnack inequality for the homogenous linearized Monge–Ampère equation. This theorem says that if v is a nonnegative solution of $L_u v = 0$ in a section $S_u(x_0, 2h) \Subset \Omega$, then the values of v in the concentric section of half height are comparable with each other.

Theorem 12.4 (Caffarelli–Gutiérrez Harnack inequality for the linearized Monge–Ampère equation). *Let $u \in C^2(\Omega)$ be a convex function in a convex domain Ω in $\mathbb{R}^n$, satisfying*

$$0 < \lambda \le \det D^2 u \le \Lambda \quad in \ \Omega.$$

Let $v \in W^{2,n}_{\mathrm{loc}}(\Omega)$ be a nonnegative solution of the linearized Monge–Ampère equation

$$L_u v := U^{ij} D_{ij} v = 0 \quad in \ a \ section \ S_u(x_0, 2h) \Subset \Omega.$$

Then

(12.5) $$\sup_{S_u(x_0,h)} v \le C(n, \lambda, \Lambda) \inf_{S_u(x_0,h)} v.$$

Note that $W^{2,n}_{\mathrm{loc}}(\Omega)$ functions are continuous in Ω by the Sobolev Embedding Theorem (Theorem 2.76). Since L_u can be written in both divergence form and nondivergence form, Theorem 12.4 is an affine invariant version of the classical Harnack inequality, due to Moser (Theorem 12.1) and Krylov–Safonov (Theorem 12.2), for linear, second-order, uniformly elliptic equations with measurable coefficients in divergence and nondivergence form, respectively.

A boundary version of Theorem 12.4 will be given in Theorem 13.19.

Remark 12.5. The Harnack estimate (12.5) also holds for nonnegative $W^{2,n}_{\mathrm{loc}}$ solutions to equations of the form $\mathrm{trace}(A U D^2 v) = 0$ with the symmetric matrix A satisfying

$$C^{-1} I_n \le A(x) \le C I_n;$$

see Problem 12.4. Thus, when $u(x) = |x|^2/2$, we recover the Krylov–Safonov Harnack inequality for uniformly elliptic equations.

12.2. Proof of the Interior Harnack Inequality

This section is devoted to proving Theorem 12.4. We first briefly outline the proof. By using the affine invariant property of the linearized Monge–Ampère equation as explained in the paragraph preceding Lemma 12.3, we can change coordinates and rescale the domain and the functions u and v.

Furthermore, by subtracting a supporting hyperplane to the graph of u at $(x_0, u(x_0))$, we can assume that $x_0 = 0$, $u(0) = 0$, $Du(0) = 0$, $h = 2$. For simplicity, we denote

$$S_t = S_u(0, t).$$

By Lemma 5.6, we can assume further that the section $S_4 \Subset \Omega$ satisfies

(12.6) $\qquad B_\sigma(0) \subset S_4 \subset B_{\sigma^{-1}}(0) \quad$ for some $\sigma = \sigma(n, \lambda, \Lambda) \in (0, 1)$.

Since sections of u behave geometrically like Euclidean balls (see Chapter 5), it suffices to show that if $v \geq 0$ in S_2, then $v \leq C(n, \lambda, \Lambda) v(0)$ in S_1.

The idea of the proof is as follows. We establish an L^ε estimate for v by showing that its distribution function $|\{v > t \, v(0)\} \cap S_1|$ decays like $t^{-\varepsilon}$ for some $\varepsilon(n, \lambda, \Lambda) > 0$. This only uses the *supersolution* property of v; that is, $U^{ij} D_{ij} v \leq 0$. Thus, up to a positive constant depending only on n, λ, Λ, the function v is comparable to $v(0)$ in S_1 except for a set of very small measure. If $v(x_0)$ is much larger compared to $v(0)$ at some point x_0 in S_1, then by the same method (now applying to $C_1 - C_2 v$ where the *subsolution* property of v is used), we find v is much larger compared to $v(0)$ in a set of positive measure in S_1 which contradicts the above estimate.

The proof of the L^ε estimate for supersolutions consists of three steps.

Step 1: Measure estimate for the supersolution. This estimate roughly says that if a nonnegative supersolution in S_4 is small at one point in S_1, then it is not too large in a positive fraction of S_2. Classical proofs of this step are typically involved with the ABP estimate. The reason why it works is that, in the ABP estimate, we need a lower bound on the determinant of the coefficient matrix which is the case here. Our approach here is, however, based on the *method of sliding paraboloids* of Savin [**S1**]. To study the distribution function of v, we slide *concave* generalized paraboloids $P_{u,a,y}$ associated with the C^2 convex function u with vertex y and opening a, where

$$P_{u,a,y}(x) = -a[u(x) - u(y) - Du(y) \cdot (x - y)],$$

from below until they touch the graph of v for the first time. The contact points are the only places where we use the supersolution property of v to obtain a lower bound for the measure of these points. By increasing the opening of sliding paraboloids, the set of contact points almost covers S_1 in measure. In the end, we get

Measure of contact points $\geq c(n, \lambda, \Lambda)$(Measure of vertices).

Step 2: Doubling estimate. This estimate roughly says that if a nonnegative supersolution in S_4 is at least 1 in S_1, then in S_2, it is bounded from below by a positive structural constant. This step is based on construction of subsolutions.

Step 3: Geometric decay estimate for the distribution function. It is based on the Crawling of Ink-spots Lemma for sections.

The choice of concave touching paraboloids in Step 1 is motivated by the fact that test functions for viscosity supersolutions of elliptic equations have some concavity properties; see Definition 7.1.

The above steps will be presented in Lemmas 12.6 and 12.8 and Theorem 12.10, respectively.

Our measure estimate for supersolutions of the linearized Monge–Ampère equation states as follows:

Lemma 12.6 (Measure estimate for supersolutions). *Let $u \in C^2(\Omega)$ be a convex function satisfying (12.1). Let $v \in W^{2,n}_{\mathrm{loc}}(\Omega)$ be a nonnegative function satisfying $L_u v := U^{ij} D_{ij} v \le 0$ in a section $S_4 := S_u(0,4) \Subset \Omega$. Then, there are small constants $\delta > 0, \alpha > 0$ and a large constant $M_1 > 1$, depending on n, λ, Λ, with the following properties. If $\inf_{S_\alpha} v \le 1$, then*

$$|\{v > M_1\} \cap S_1| \le (1-\delta)|S_1|.$$

Proof. We assume that $u(0) = 0, Du(0) = 0$ and that (12.6) holds. We assume $v \in C^2$ to start with; the general case follows by approximations as in the proof of Theorem 3.16. Let $P_{u,a,y}$ be as above.

Step 1. Measure estimate for $v \in C^2$. Suppose $v(x_0) \le 1$ at $x_0 \in S_\alpha$ where $\alpha \in (0, 1/4)$ is to be chosen later, depending on n, λ, Λ. Consider the set of vertices $V = S_\alpha$.

Step 1(a). Choice of the opening. We claim that there is a large constant a (called the opening) depending only on n, λ, and Λ, such that, for each $y \in V$, there is a constant c_y such that the generalized paraboloid $P_{u,a,y} + c_y$ touches the graph of v from below at some point x (called the contact point) in S_1. See Figure 12.1. Indeed, for each $y \in V$, we consider the function

$$P(x) := v(x) - P_{u,a,y}(x) = v(x) + a[u(x) - Du(y) \cdot (x-y) - u(y)]$$

and look for its minimum points on $\overline{S_1}$. Recall that $v \ge 0$ in S_4.

Because $y \in S_\alpha \subset S_{1/2}$, we can use Theorem 5.30 to get $c_1(n, \lambda, \Lambda) > 0$ such that $S_u(y, c_1) \subset S_1$. Thus, on the boundary ∂S_1, we have

$$P \ge a[u(x) - Du(y) \cdot (x-y) - u(y)] \ge ac_1(n, \lambda, \Lambda).$$

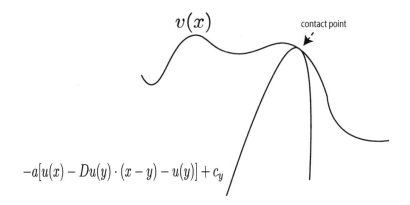

$v(x)$

contact point

$-a[u(x) - Du(y) \cdot (x-y) - u(y)] + c_y$

Figure 12.1. Touching v from below by the paraboloid $P_{u,a,y} + c_y$.

Because $x_0, y \in S_\alpha$, the engulfing property in Theorem 5.28 gives $x_0, y \in S_u(0, \alpha) \subset S_u(y, \theta_0 \alpha)$ where $\theta_0(n, \lambda, \Lambda) > 2$. Consequently,

$$P(x_0) \leq 1 + a[u(x_0) - Du(y) \cdot (x_0 - y) - u(y)] \leq 1 + a\alpha\theta_0.$$

Therefore, if

$$\alpha = c_1/(2\theta_0) \quad \text{and} \quad a = 6/c_1,$$

then P attains its minimum on $\overline{S_1}$ at a point $x \in S_1$ with

$$v(x) \leq P(x_0) < M_1 := 2 + a\alpha\theta_0 < ac_1.$$

Step 1(b). Measure estimate for the contact set. Define the contact set by

$$E = \left\{ x \in \overline{S_1} : \text{there is } y \in V \text{ such that } \inf_{\overline{S_1}}(v - P_{u,a,y}) = v(x) - P_{u,a,y}(x) \right\}.$$

Then, by Step 1(a),

$$E \subset \{v < M_1\} \cap S_1.$$

At the contact point $x \in S_1$, $Dv(x) = a(Du(y) - Du(x))$ which gives

(12.7) $$Du(y) = Du(x) + (1/a)Dv(x),$$

and, furthermore,

(12.8) $$D^2v(x) \geq -aD^2u(x).$$

Before proceeding further, we note that, for each $y \in V$, there is $x \in E$ such that (12.7) holds; that is, $y = \Phi(x)$ where

$$\Phi(x) = (Du)^{-1}\left[Du(x) + (1/a)Dv(x) \right].$$

It follows that

$$V \subset \Phi(E).$$

It is easy to see that Φ is Lipschitz on E with Lipschitz constant bounded by m_{uv} depending on the C^2 norms of u and v. Hence, upon differentiating (12.7) with respect to x, we find

$$(12.9) \qquad D^2u(y)D_xy = D^2u(x) + (1/a)D^2v(x) \geq 0.$$

Now using the supersolution assumption at only x, we find that

$$(12.10) \qquad \text{trace}((D^2u)^{-1}D^2v(x)) \leq 0.$$

In view of (12.8) and (12.10), we apply Lemma 12.7 below to $A = D^2v$ and $B = D^2u$ to obtain

$$(12.11) \qquad C(a,n)D^2u(x) \geq D^2v(x) \geq -aD^2u(x) \quad \text{where } C(a,n) = an.$$

Now, taking the determinant in (12.9) and invoking (12.11), we obtain

$$\det D^2u(y)\,|\det D_xy| = \det(D^2u(x) + (1/a)D^2v(x)) \leq C(a,n)\det D^2u(x).$$

Since $\lambda \leq \det D^2u \leq \Lambda$, this implies the bound

$$(12.12) \qquad |\det D_xy| \leq C(n,\Lambda,\lambda).$$

By (12.9), we have $D\Phi(x) = D_xy \geq 0$ on E. Moreover, by definition, E is a closed set and thus is measurable. By the area formula (Theorem 2.72) applied to Φ and E and in view of (12.12), we have

$$
\begin{aligned}
(12.13) \qquad |S_\alpha| = |V| \leq |\Phi(E)| &\leq \int_E \det D\Phi(x)\,dx = \int_E |\det D_xy|\,dx \\
&\leq C(n,\Lambda,\lambda)\,|E| \\
&\leq C|\{v < M_1\} \cap S_1|.
\end{aligned}
$$

Using the volume estimate of sections in Lemma 5.6, we find that

$$|S_1| \leq C(n,\lambda,\Lambda)|S_\alpha| \leq C^*|\{v < M_1\} \cap S_1|$$

for some $C^*(n,\lambda,\Lambda) > 1$. The conclusion of the lemma holds with $\delta = 1/C^*$.

Step 2. Measure estimate for $v \in W_{\text{loc}}^{2,n}(\Omega)$. We use the C^2 case and an approximation argument . Let $\{v_m\}_{m=1}^\infty$ be a sequence of nonnegative functions in $C^2(\Omega)$ that converges to v in $W_{\text{loc}}^{2,n}(\Omega)$; see Theorem 2.50.

We apply the above arguments to v_m with contact set E_m. Instead of (12.10), the supersolution assumption $\text{trace}((D^2u)^{-1}D^2v(x)) \leq 0$ gives

$$(12.14) \quad \text{trace}[(D^2u)^{-1}D^2v_m(x)] \leq \text{trace}[(D^2u)^{-1}D^2(v_m - v)(x)] := f_m(x).$$

Note that f_m converges to 0 in $L_{\text{loc}}^n(\Omega)$. Instead of (12.11), we have

$$(12.15) \qquad [C(a,n) + f_m(x)]D^2u(x) \geq D^2v(x) \geq -aD^2u(x).$$

Instead of (12.12), we have

$$(12.16) \qquad |\det D_xy| \leq C(n,\Lambda,\lambda)(1 + [f_m(x)]^n).$$

Integrating over E_m, we obtain, instead of (12.13), the estimate

$$|S_\alpha| \leq C(n, \Lambda, \lambda)(|E_m| + \|f_m\|^n_{L^n(E_m)})$$
$$\leq C|\{v_m < M_1\} \cap S_1| + C\|f_m\|^n_{L^n(S_1)}.$$

Letting $m \to \infty$ and using the fact that $\|f_m\|_{L^n(S_1)} \to 0$, we obtain (12.13) for v. The lemma is proved for $v \in W^{2,n}_{loc}(\Omega)$. $\qquad\square$

In the proof of Lemma 12.6, we used the following simple lemma.

Lemma 12.7. *Let A and B be two symmetric $n \times n$ matrices, where B is positive definite. If $A \geq -aB$ for some $a \geq 0$ and $\mathrm{trace}(B^{-1}A) \leq b$, then*

$$(an + b)B \geq A \geq -aB.$$

Proof. Since B is positive definite, we can rewrite the hypotheses as

$$M := B^{-1/2}AB^{-1/2} \geq -aI_n \quad \text{and} \quad \mathrm{trace}(B^{-1/2}AB^{-1/2}) \leq b.$$

Hence, the largest eigenvalue of M does not exceed $a(n-1)+b$. This implies

$$B^{-1/2}AB^{-1/2} \leq [a(n - 1) + b]I_n \leq (an + b)I_n$$

and therefore, we obtain $A \leq (an + b)B$ as asserted. $\qquad\square$

We now turn to the doubling estimate for supersolutions of the linearized Monge–Ampère equation; see also Problem 12.3.

Lemma 12.8 (Doubling estimate for supersolutions). *Let $u \in C^2(\Omega)$ be a convex function satisfying (12.1). Let $v \in W^{2,n}_{loc}(\Omega)$ be a nonnegative function satisfying $U^{ij}D_{ij}v \leq 0$ in a section $S_4 := S_u(0, 4) \Subset \Omega$. Let α be the constant in Lemma 12.6. If $v \geq 1$ in S_α, then $v \geq c_*(n, \lambda, \Lambda) > 0$ in S_1.*

Proof. We assume that $u(0) = 0, Du(0) = 0$ and that (12.6) holds. To prove the lemma, it suffices to construct a C^2 subsolution $w : S_2 \backslash S_\alpha \longrightarrow \mathbb{R}$ of the linearized Monge–Ampère operator $U^{ij}D_{ij}$; that is, $U^{ij}D_{ij}w \geq 0$, with the following properties:

(12.17) $w \leq 0$ on ∂S_2, $w \leq 1$ on ∂S_α, $w \geq c_*(n, \Lambda, \lambda)$ in $S_1 \backslash S_\alpha$.

See Figure 12.2.

Step 1. Initial guess for subsolution. Our first guess is

$$w = C(\alpha, m)(u^{-m} - 2^{-m})$$

where m is large. Let us check if this is a subsolution for $L_u := U^{ij}D_{ij}$.

Let $(u^{ij})_{1 \leq i,j \leq n} = (D^2u)^{-1}$. We can compute for $\bar{w} := u^{-m} - 2^{-m}$

(12.18)
$$u^{ij}D_{ij}\bar{w} = mu^{-m-2}[(m + 1)u^{ij}D_iuD_ju - uu^{ij}D_{ij}u]$$
$$= mu^{-m-2}[(m + 1)u^{ij}D_iuD_ju - nu].$$

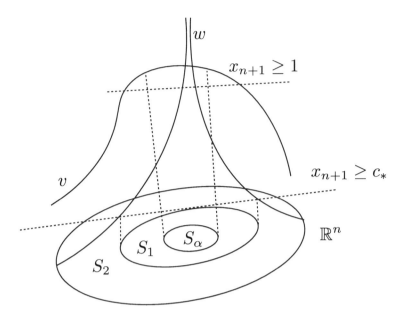

Figure 12.2. The supersolution v in S_2 and a C^2 subsolution w in $S_2 \backslash S_\alpha$.

By Lemma 2.58

$$u^{ij} D_i u D_j u \geq |Du|^2 / \Delta u.$$

If $x \in S_2 \setminus S_\alpha$, then from the convexity of u, we have $0 = u(0) \geq u(x) + Du(x) \cdot (0 - x)$ and therefore,

$$|Du(x)| \geq u(x)/|x| \geq \alpha/\sigma^{-1} \equiv 2c_n(n, \lambda, \Lambda) > 0.$$

In order to obtain $u^{ij} D_{ij} \bar{w} \geq 0$ using (12.18), we only have trouble on the set of "bad points" of u where Δu (or, equivalently, $\|D^2 u\|$) is large. However, this set has small measure, due to the interior L^1 estimate for $D^2 u$ as in (6.27):

$$\int_{S_3} \|D^2 u\| \, dx \leq \int_{S_3} \Delta u \, dx = \int_{\partial S_3} Du \cdot \nu \, d\mathcal{H}^{n-1} \leq C(n, \lambda, \Lambda).$$

Therefore, given $\varepsilon > 0$ small, the set

$$A_\varepsilon = \{x \in S_3 : \Delta u(x) \geq 1/\varepsilon\}$$

has measure bounded from above by $|A_\varepsilon| \leq C(n, \lambda, \Lambda)\varepsilon$.

To construct a proper subsolution bypassing the bad points in A_ε, we only need to modify $\bar{w}$ at bad points. Roughly speaking, the modification involves the solution to

$$\det D^2 w_\varepsilon = \Lambda \chi_{A_\varepsilon} \quad \text{in } S_4, \quad w_\varepsilon = 0 \quad \text{on } \partial S_4,$$

where χ_E is the characteristic function of the set E: $\chi_E(x) = 1$ if $x \in E$, and $\chi_E(x) = 0$ otherwise. The problem with this equation is that the solution is not in general C^2 for our construction of the C^2 subsolution. We will use approximations to resolve this technical problem.

Step 2. Rigorous construction of subsolution. We approximate A_ε by an open set $\tilde{A}_\varepsilon$ where $A_\varepsilon \subset \tilde{A}_\varepsilon \subset S_4$ and $|\tilde{A}_\varepsilon \setminus A_\varepsilon| \leq \varepsilon$.

We introduce a smooth function φ with the following properties:

$$\varphi = 1 \quad \text{in } A_\varepsilon, \quad \varphi = \varepsilon \quad \text{in } S_4 \setminus \tilde{A}_\varepsilon, \quad \varepsilon \leq \varphi \leq 1 \quad \text{in } S_4.$$

By Theorem 3.38, there exists a convex Aleksandrov solution $u_\varepsilon \in C(\overline{S_4})$ to

$$\det D^2 u_\varepsilon = \Lambda\varphi \quad \text{in } S_4, \quad u_\varepsilon = 0 \quad \text{on } \partial S_4.$$

By Caffarelli's $C^{2,\alpha}$ estimates (Theorem 6.34), $u_\varepsilon \in C^{2,\alpha}(S_4)$ for all $\alpha \in (0,1)$. Now, we estimate $\sup_{S_4}|u_\varepsilon|$ using the Aleksandrov maximum principle, Theorem 3.12, together with $S_4 \subset B_{\sigma^{-1}}(0)$ and the definition of φ:

$$\sup_{S_4}|u_\varepsilon|^n \leq C(n)\sigma^{-n}\int_{S_4}\Lambda\varphi\,dx$$

$$\leq C(n)\sigma^{-n}\Big(\Lambda|A_\varepsilon| + \Lambda|\tilde{A}_\varepsilon \setminus A_\varepsilon| + \varepsilon|S_4|\Big) \leq C(n,\lambda,\Lambda)\varepsilon.$$

It follows that $|u_\varepsilon| \leq C_1\varepsilon^{1/n}$ for some constant $C_1(n,\lambda,\Lambda) > 0$. By the gradient estimate in Lemma 2.37(iii) and by Lemma 5.6(ii), we have

$$|Du_\varepsilon(x)| \leq \frac{-u_\varepsilon(x)}{\operatorname{dist}(S_3,\partial S_4)} \leq C_2(n,\lambda,\Lambda)\varepsilon^{1/n} \quad \text{on } S_2.$$

We choose $\varepsilon = \varepsilon(n,\lambda,\Lambda)$ small so that

(12.19) $$C_1\varepsilon^{1/n} \leq 1/4, \quad C_2\varepsilon^{1/n} \leq c_n.$$

Let

$$\tilde{u} := u - u_\varepsilon \geq 0 \quad \text{and} \quad \tilde{w} := \tilde{u}^{-m} - 2^{-m}.$$

Then

(12.20) $$|\tilde{u}| \leq 3 \quad \text{and} \quad |D\tilde{u}| \geq c_n \quad \text{on } S_2 \setminus S_\alpha; \quad \alpha \leq \tilde{u} \leq 5/4 \quad \text{on } S_1 \setminus S_\alpha.$$

Now, compute as before

$$u^{ij}D_{ij}\tilde{w} = m\tilde{u}^{-m-2}[(m+1)u^{ij}D_i\tilde{u}D_j\tilde{u} - \tilde{u}u^{ij}D_{ij}\tilde{u}]$$

$$= m\tilde{u}^{-m-2}[(m+1)u^{ij}D_i\tilde{u}D_j\tilde{u} + \tilde{u}(u^{ij}D_{ij}u_\varepsilon - n)].$$

On A_ε, we have $u^{ij}D_{ij}\tilde{w} \geq 0$. This follows from Lemma 2.57 which gives

$$u^{ij}D_{ij}u_\varepsilon \geq n(\det(D^2u)^{-1}\det D^2u_\varepsilon)^{1/n} \geq n \quad \text{on } A_\varepsilon.$$

On $(S_2 \setminus S_\alpha)\setminus A_\varepsilon$, we use $\tilde{u} \geq 0$ and Lemma 2.58 to estimate

$$u^{ij}D_{ij}\tilde{w} \geq m\tilde{u}^{-m-2}[(m+1)|D\tilde{u}|^2/\Delta u - n\tilde{u}].$$

Note that $\Delta u \leq \varepsilon^{-1}$ on $(S_2 \setminus S_\alpha)\setminus A_\varepsilon$. Combining this with (12.20), we get

$$u^{ij}D_{ij}\tilde{w} \geq m\tilde{u}^{-m-2}[(m+1)\varepsilon c_n^2 - 3n] \geq 0 \quad \text{on } (S_2 \setminus S_\alpha)\setminus A_\varepsilon$$

if we choose $m = m(n, \lambda, \Lambda)$ large. Therefore, $u^{ij} D_{ij} \tilde{w} \geq 0$ on $S_2 \setminus S_\alpha$, and hence $\tilde{w} = \tilde{u}^{-m} - 2^{-m}$ is a subsolution of the operator $u^{ij} D_{ij}$ on $S_2 \setminus S_\alpha$.

Finally, by (12.20) and $\tilde{w} \leq 0$ on ∂S_2, we can choose a suitable $c_1(n, \lambda, \Lambda) > 0$ so that the subsolution of the form

$$w = c_1(\tilde{u}^{-m} - 2^{-m})$$

satisfies $w \leq 1$ on ∂S_α, and hence the inequalities in (12.17) hold for w.

Since $L_u(w - v) \geq 0$ in $S_2 \setminus S_\alpha$ and $w - v \leq 0$ on $\partial(S_2 \setminus S_\alpha)$, the ABP maximum principle in Theorem 3.16 gives $w - v \leq 0$ in $S_2 \setminus S_\alpha$. (Theorem 2.79 is not applicable because v might not be C^2.) Now, we obtain the desired structural lower bound for v in S_1 from $v \geq w$ on $S_1 \setminus S_\alpha$ and $v \geq 1$ on S_α. The lemma is proved. $\qquad\square$

Combining Lemmas 12.6 and 12.8 and letting $M := M_1 c_*^{-1}$, we obtain:

Proposition 12.9 (Critical density estimate for supersolutions). *Let $u \in C^2(\Omega)$ be a convex function satisfying (12.1). Let $v \in W^{2,n}_{\mathrm{loc}}(\Omega)$ be a nonnegative function satisfying $U^{ij} D_{ij} v \leq 0$ in a section $S_4 := S_u(0, 4) \Subset \Omega$. Then, there exist constants $\delta(n, \lambda, \Lambda) \in (0, 1)$ and $M(n, \lambda, \Lambda) > 1$ with the following property. If*

$$|\{v > M\} \cap S_1| > (1 - \delta)|S_1|,$$

then $v > 1$ in S_1.

From the critical density estimate and the Crawling of Ink-spots Lemma (Lemma 5.33), we will obtain the L^ε estimate for supersolutions.

Before proceeding further, we recall the localizing constant $K_1(n, \lambda, \Lambda) > 4$ in the proof of Lemma 5.33 with the following property:

(12.21)
$$\text{If } S_u(y, K_1 h) \Subset \Omega \text{ and } S_u(x, t) \subset S_u(y, h),$$
$$\text{then } S_u(x, 4t) \subset S_u(y, K_1 h).$$

Theorem 12.10 (Decay estimate of the distribution function of supersolutions). *Let $u \in C^2(\Omega)$ be a convex function satisfying $0 < \lambda \leq \det D^2 u \leq \Lambda$ in a domain Ω in $\mathbb{R}^n$. Let K_1 be as in (12.21). Let $v \in W^{2,n}_{\mathrm{loc}}(\Omega)$ be a nonnegative function satisfying $U^{ij} D_{ij} v \leq 0$ in a section $S_{K_1} := S_u(0, K_1) \Subset \Omega$ with*

$$\inf_{S_u(0,1)} v \leq 1.$$

Then there are constants $C_1(n, \lambda, \Lambda) > 1$ and $\varepsilon(n, \lambda, \Lambda) \in (0, 1)$ such that

$$|\{v > t\} \cap S_1| \leq C_1 |S_1| t^{-\varepsilon} \quad \text{for all } t > 0.$$

Proof. Let $\delta \in (0,1)$ and $M > 1$ be the constants in Proposition 12.9. Let $A_k := \{v > M^k\} \cap S_1$. We will prove the decay estimate

$$|A_k| \leq C_2(n,\lambda,\Lambda)|S_1|M^{-\varepsilon k},$$

from which the conclusion of the theorem follows.

Note that A_k's are open sets and $A_k \subset A_1$ for all $k \geq 1$. Due to $\inf_{S_1} v \leq 1$, we infer from Proposition 12.9 that

$$|A_k| \leq |A_1| \leq (1-\delta)|S_1| \quad \text{for all } k.$$

If a section $S := S_u(x,t) \subset S_1 := S_u(0,1)$ satisfies $|S \cap A_{k+1}| > (1-\delta)|S|$, then $S_u(x,4t) \subset S_u(0,K_1)$ by the choice of K_1, and hence, by Proposition 12.9 (applied to $S_u(x,4t)$, properly rescaled), we find that $S \subset A_k$. Using Lemma 5.33 with $E = A_{k+1}$, $F = A_k$, and $h = 1$, we obtain a constant $c(n,\lambda,\Lambda) > 0$ such that $|A_{k+1}| \leq (1-c\delta)|A_k|$, and therefore, by induction,

$$|A_k| \leq (1-c\delta)^{k-1}(1-\delta)|S_1| = C_2|S_1|M^{-\varepsilon k},$$

where $\varepsilon = -\log(1-c\delta)/\log M$ and $C_2 = (1-c\delta)^{-1}(1-\delta)$. This finishes the proof of the decay estimate for $|A_k|$ and also the proof of the theorem. $\square$

Theorem 12.10 implies the following weak Harnack inequality for super-solutions to the linearized Monge–Ampère equation, which will be important in our treatment of the boundary Hölder gradient estimates in Section 13.4.

Theorem 12.11 (Weak Harnack inequality for supersolutions to the linearized Monge–Ampère equation). *Let $u \in C^2(\Omega)$ be a convex function satisfying $0 < \lambda \leq \det D^2 u \leq \Lambda$ in a domain Ω in $\mathbb{R}^n$. Let K_1 be as in (12.21), and let ε, C_1 be as in Theorem 12.10. Let $v \in W^{2,n}_{\text{loc}}(\Omega)$ be a nonnegative function satisfying in a section $S_u(x, K_1 h) \Subset \Omega$ the inequality $U^{ij}D_{ij}v \leq 0$. If*

$$|\{v \geq 1\} \cap S_u(x,h)| \geq \mu|S_u(x,h)| \quad \text{for some } \mu \in (0,1),$$

then

$$\inf_{S_u(x,h)} v \geq c(n,\lambda,\Lambda,\mu) := (\mu/C_1)^{1/\varepsilon} > 0.$$

We are now ready to prove the Harnack inequality in Theorem 12.4.

Proof of Theorem 12.4. We divide the proof into several steps.

Step 1. The Harnack inequality in a small section. We show that if $L_u v = 0$ in $S_u(x_0, 2h) \Subset \Omega$ with $v \geq 0$, then for $\tau := 1/K_1$ (see (12.21)), we have

$$\sup_{S_u(x_0,\tau h)} v \leq C \inf_{S_u(x_0,\tau h)} v.$$

We can assume that $x_0 = 0$, $u(0) = 0$, $Du(0) = 0$, $h = K_1/2$ and that the section $S_{K_1} = S_u(0, 2h) \Subset \Omega$ satisfies

$$B_{\sigma_1}(0) \subset S_{K_1} \subset B_{\sigma_1^{-1}}(0) \quad \text{for some } \sigma_1(n, \lambda, \Lambda) \in (0, 1).$$

Let $\delta \in (0, 1)$ and $M > 1$ be the constants in Proposition 12.9 and let $\varepsilon \in (0, 1)$ be the constant in Theorem 12.10.

This step is reduced to proving the following assertion:

$$(12.22) \qquad \text{If } \inf_{S_u(0, 1/2)} v \le 1, \quad \text{then} \quad \sup_{S_u(0, 1/2)} v \le C(n, \lambda, \Lambda).$$

This assertion, when applied to the function $v^\tau = (\inf_{S_u(0, 1/2)} v + \tau)^{-1} v$ which satisfies $U^{ij} D_{ij} v^\tau = 0$ for each $\tau > 0$, gives

$$\sup_{S_u(0, 1/2)} v \le C \Big(\inf_{S_u(0, 1/2)} v + \tau \Big);$$

thus, by letting $\tau \to 0$, we obtain the Harnack inequality in a small section:

$$(12.23) \qquad \sup_{S_u(0, 1/2)} v \le C \inf_{S_u(0, 1/2)} v.$$

It remains to prove (12.22). Let $\beta > 0$ be a constant depending on n, λ, Λ to be determined later. We will bound v in $S_u(0, 1/2)$ by a structurally large multiple of $(1 - u)^{-\beta}$. To do this, we consider

$$t := \max_{S_u(0, 1)} v(1 - u)^\beta < +\infty.$$

It suffices to show that $t \le C(n, \lambda, \Lambda)$ because we then have

$$\sup_{S_u(0, 1/2)} v \le C \sup_{S_u(0, 1/2)} (1 - u(x))^{-\beta} \le 2^\beta C.$$

If $t \le 1$, then we are done. Hence, we now assume that $t \ge 1$.

Clearly, there exists $x_0 \in S_u(0, 1)$ such that $t = v(x_0)(1 - u(x_0))^\beta$. Let

$$r = (1 - u(x_0))/2 \quad \text{and} \quad H_0 := t(2r)^{-\beta} \equiv v(x_0) \ge 1.$$

Note that $x_0 \in S_u(0, 1 - r)$. By Theorem 5.30, there are constants c and $p_1 = \mu^{-1}$ (large), all depending on n, λ, Λ, such that $S_u(x_0, 2cr^{p_1}) \subset S_u(0, 1)$.

We bound t by estimating from above and below the measure of the set $\{v \ge H_0/2\} \cap S_u(x_0, cr^{p_1})$.

Step 1(a). Estimate from above using the L^ε decay estimate for super-solutions. By Theorem 12.10, we have

$$(12.24) \qquad \begin{aligned} |\{v > H_0/2\} \cap S_u(x_0, cr^{p_1})| &\le |\{v > H_0/2\} \cap S_u(0, 1)| \\ &\le C(n, \lambda, \Lambda) H_0^{-\varepsilon} = Ct^{-\varepsilon}(2r)^{\beta\varepsilon}. \end{aligned}$$

Step 1(b). Estimate from below using subsolutions. Here, we use that v is a subsolution. To estimate the measure of $\{v \geq H_0/2\} \cap S_u(x_0, cr^{p_1})$ from below, we apply Proposition 12.9 to the function

$$w(x) = \frac{(1-\rho)^{-\beta} H_0 - v(x)}{((1-\rho)^{-\beta} - 1) H_0} \equiv \frac{(1-\rho)^{-\beta} v(x_0) - v(x)}{((1-\rho)^{-\beta} - 1) v(x_0)}$$

on the smaller section $S_u(x_0, c_1 r^{p_1})$ where $c_1 \leq c$ and ρ are to be determined.

First, we choose ρ to be a small constant and β to be a large constant, depending on n, λ, Λ, such that

$$(12.25) \qquad M\left((1-\rho)^{-\beta} - 1\right) \leq \frac{1}{2}, \quad \beta \geq \frac{n}{2\mu\varepsilon}.$$

Next, we can choose $c_1 \leq c$ small, depending on n, λ, Λ, such that

$$1 - u \geq 2r - 2\rho r \quad \text{in } S_u(x_0, c_1 r^{p_1}).$$

Indeed, we have $\sup_{S_u(0,1)} |Du| \leq C_0(n, \lambda, \Lambda)$, due to the gradient estimate in Lemma 2.37(iii). Thus, if $x \in S_u(x_0, c_1 r^{p_1}) \subset S_u(0, 1)$, then by Lemma 5.16, we have $|x - x_0| \leq C(c_1 r^{p_1})^\mu \leq \rho r / C_0$ for small c_1 and hence,

$$1 - u(x) = 2r + u(x_0) - u(x) \geq 2r - (\sup_{S_u(0,1)} |Du|)|x - x_0| \geq 2r - 2\rho r.$$

With the above choice of ρ, β, and c_1, we have, in $S_u(x_0, c_1 r^{p_1})$,

$$v \leq t(1-u)^{-\beta} \leq t(2r - 2\rho r)^{-\beta} = (1-\rho)^{-\beta} H_0.$$

Therefore, w, as defined above, is a nonnegative supersolution of $L_u w \leq 0$ in $S_u(x_0, c_1 r^{p_1})$ with $w(x_0) = 1$. Using Proposition 12.9, we obtain

$$|\{w \leq M\} \cap S_u(x_0, (1/4)c_1 r^{p_1})| \geq \delta |S_u(x_0, (1/4)c_1 r^{p_1})|.$$

The left-hand side is the measure of the set where v is larger than

$$H_0\left[(1-\rho)^{-\beta} - M((1-\rho)^{-\beta} - 1)\right] \geq H_0/2,$$

due to the choice of ρ and β. Thus, we obtain the estimate from below

$$|\{v \geq H_0/2\} \cap S_u(x_0, c_1 r^{p_1})| \geq \delta |S_u(x_0, (1/4)c_1 r^{p_1})|.$$

From (12.24) and the volume estimate on sections in Lemma 5.6, we find

$$Ct^{-\varepsilon}(2r)^{\beta\varepsilon} \geq \delta |S_u(x_0, (1/4)c_1 r^{p_1})| \geq c(n, \lambda, \Lambda) r^{np_1/2} = c(n, \lambda, \Lambda) r^{\frac{n}{2\mu}}.$$

Since $\beta\varepsilon \geq \frac{n}{2\mu}$, we infer that $t \leq 2^\beta (C/c)^{1/\varepsilon}$, and hence (12.22) is proved.

Step 2. The Harnack inequality in a section of half height. We now prove, using Step 1 and a covering argument, that

$$\sup_{S_u(x_0,h)} v \leq C \inf_{S_u(x_0,h)} v.$$

By Theorem 5.30, there is $\alpha(n, \lambda, \Lambda) > 0$ such that for each $x \in \overline{S_u(x_0, h)}$, we have $S_u(x, 2\alpha\tau^{-1}h) \subset S_u(x_0, 3h/2)$. By Vitali's Covering Lemma (Lemma

5.32), we can obtain a finite covering $\bigcup_{1 \leq i \leq m} S_u(x_i, \alpha h)$ of $\overline{S_u(x_0, h)}$ such that the sections $S_u(x_i, K^{-1}\alpha h)$ are disjoint where $K(n, \lambda, \Lambda) > 4$. Using the volume estimates in Lemma 5.6, we can bound $m \leq M_0(n, \lambda, \Lambda)$. For each i, from $S_u(x_i, 2\alpha\tau^{-1}h) \Subset \Omega$ and Step 1, we have

$$\sup_{S_u(x_i, \alpha h)} v \leq C(n, \lambda, \Lambda) \inf_{S_u(x_i, \alpha h)} v.$$

Any two points in $S_u(x_0, h)$ can be connected by a chain of at most m sections $S_u(x_i, \alpha h)$ and thus the result follows. The proof of the theorem is complete. $\qquad\square$

12.3. Interior Hölder Estimates

By combining the Harnack inequality for the homogeneous linearized Monge–Ampère equation in Theorem 12.4 with the ABP estimate, we obtain a Harnack inequality for inhomogeneous equations with L^n right-hand side.

Lemma 12.12 (Harnack inequality for inhomogeneous equation). *Let u be a C^2 convex function satisfying (12.1). Let $f \in L^n(\Omega)$ and $v \in W_{\mathrm{loc}}^{2,n}(\Omega)$ satisfy $U^{ij}D_{ij}v = f$ in Ω. Assume $v \geq 0$ in $S_u(x, t) \Subset \Omega$. Then*

$$\sup_{S_u(x, t/2)} v \leq C(n, \lambda, \Lambda)\Big(\inf_{S_u(x, t/2)} v + t^{\frac{1}{2}} \|f\|_{L^n(S_u(x,t))} \Big).$$

Proof. Let $w \in W_{\mathrm{loc}}^{2,n}(S_u(x, t)) \cap C(\overline{S_u(x, t)})$ be the solution of

$$U^{ij}D_{ij}w = f \quad \text{in } S_u(x, t) \quad \text{and} \quad w = 0 \quad \text{on } \partial S_u(x, t).$$

The existence of w follows from Theorem 2.84. Then, by Lemma 3.17 and the volume bound on sections in Lemma 5.6, we get

$$\sup_{S_u(x,t)} |w| \leq C(n, \lambda)|S_u(x, t)|^{\frac{1}{n}} \|f\|_{L^n(S_u(x,t))} \leq C(n, \lambda)t^{1/2}\|f\|_{L^n(S_u(x,t))}.$$

Since $U^{ij}D_{ij}(w - v) = 0$ in $S_u(x, t)$ and $w - v \leq 0$ on $\partial S_u(x, t)$, the ABP maximum principle in Theorem 3.16 gives $w - v \leq 0$ in $S_u(x, t)$. Thus, applying the interior Harnack inequality in Theorem 12.4 to $v - w$, we obtain

$$\sup_{S_u(x, t/2)} (v - w) \leq C_1 \inf_{S_u(x, t/2)} (v - w),$$

for some constant $C_1(n, \lambda, \Lambda)$, which then implies

$$\sup_{S_u(x, t/2)} v \leq (1 + C_1)\Big(\inf_{S_u(x, t/2)} v + \sup_{S_u(x, t/2)} |w| \Big)$$

$$\leq C(n, \lambda, \Lambda)\Big(\inf_{S_u(x, t/2)} v + t^{\frac{1}{2}} \|f\|_{L^n(S_u(x,t))} \Big).$$

The lemma is proved. $\qquad\square$

Denote by $\mathrm{osc}_E v := \sup_E v - \inf_E v$ the oscillation of a function v on a set E. From Lemma 12.12, we obtain the following oscillation estimate.

Theorem 12.13 (Oscillation estimate). *Let $u \in C^2(\Omega)$ be a convex function satisfying $0 < \lambda \le \det D^2 u \le \Lambda$ in a domain Ω in $\mathbb{R}^n$. Let $f \in L^n(\Omega)$ and $v \in W^{2,n}_{\mathrm{loc}}(\Omega)$ satisfy $U^{ij} D_{ij} v = f$ in Ω. Assume $S_u(x, h) \Subset \Omega$. Then*

$$\mathrm{osc}_{S_u(x,\rho)} v \le C \Big(\frac{\rho}{h}\Big)^\alpha \Big[\mathrm{osc}_{S_u(x,h)} v + h^{\frac{1}{2}} \|f\|_{L^n(S_u(x,h))} \Big] \quad \text{for all } \rho \le h,$$

where $C > 0$ and $\alpha \in (0,1)$ depend only on n, λ, and Λ.

Proof. Let us write $S_t = S_u(x, t)$ and $\omega(t) := \mathrm{osc}_{S_t} v$. Let $\rho \in (0, h]$ be arbitrary. Since $\bar{v} := v - \inf_{S_\rho} v$ is a nonnegative solution of $U^{ij} D_{ij} \bar{v} = f$ in S_ρ, we can apply Lemma 12.12 to $\bar{v}$ to obtain for some $C(n, \lambda, \Lambda) > 2$

$$C^{-1} \sup_{S_{\rho/2}} \bar{v} \le \inf_{S_{\rho/2}} \bar{v} + \rho^{\frac{1}{2}} \|f\|_{L^n(S_\rho)}.$$

It follows that for all $\rho \in (0, h]$, we have for $\gamma := \max\{1 - C^{-1}, 3/4\}$,

$$\omega\Big(\frac{\rho}{2}\Big) = \sup_{S_{\rho/2}} \bar{v} - \inf_{S_{\rho/2}} \bar{v} \le \big(1 - C^{-1}\big) \sup_{S_{\rho/2}} \bar{v} + \rho^{\frac{1}{2}} \|f\|_{L^n(S_\rho)}$$

$$(12.26) \qquad\qquad\qquad\qquad \le \gamma \omega(\rho) + \rho^{\frac{1}{2}} \|f\|_{L^n(S_h)}.$$

Note that $2^{1/2} \gamma > 1$. Iterating (12.26), we get for each positive integer m

$$\omega\Big(\frac{h}{2^m}\Big) \le \gamma^m \omega(h) + \|f\|_{L^n(S_h)} h^{\frac{1}{2}} \sum_{i=0}^{m-1} \gamma^i \Big(\frac{1}{2^{1/2}}\Big)^{m-i-1}$$

$$\le \frac{2}{2^{1/2}\gamma - 1} \gamma^m \big[\omega(h) + h^{\frac{1}{2}} \|f\|_{L^n(S_h)}\big]$$

$$= \frac{2^{\alpha+1}}{2^{1/2}\gamma - 1} \frac{[\omega(h) + h^{\frac{1}{2}} \|f\|_{L^n(S_h)}]}{2^{(m+1)\alpha}} \quad \text{for } \alpha = \ln \gamma / \ln(1/2) \in (0,1).$$

Clearly, the above estimates also hold for $m = 0$. Now, for any $\rho \in (0, h]$, there is a nonnegative integer m such that $\frac{h}{2^{m+1}} < \rho \le \frac{h}{2^m}$ and therefore

$$\omega(\rho) \le \omega\Big(\frac{h}{2^m}\Big) \le \frac{2^{\alpha+1}}{2^{1/2}\gamma - 1} \Big(\frac{\rho}{h}\Big)^\alpha [\omega(h) + h^{\frac{1}{2}} \|f\|_{L^n(S_h)}],$$

which proves the theorem. $\qquad\qquad\qquad\qquad\qquad\qquad\qquad\qquad\quad \square$

Now, we obtain interior Hölder estimates for inhomogeneous linearized Monge–Ampère equations with L^n right-hand side.

Theorem 12.14 (Interior Hölder estimates for L^n inhomogeneity). *Let $u \in C^2(\Omega)$ be a convex function satisfying $0 < \lambda \le \det D^2 u \le \Lambda$ in a domain Ω in $\mathbb{R}^n$. Assume that for some $x_0 \in \Omega$ and $h > 0$, we have*

either $S_u(x_0, h) \Subset \Omega$ or $u(x) = u(x_0) + Du(x_0) \cdot (x - x_0) + h$ on $\partial S_u(x_0, h)$.

Let $f \in L^n(S_u(x_0, h))$ and $v \in W^{2,n}_{\text{loc}}(S_u(x_0, h)) \cap C(\overline{S_u(x_0, h)})$ satisfy

$$U^{ij} D_{ij} v = f \quad \text{in } S_u(x_0, h).$$

Set $M := \sup_{x,y \in S_u(x_0, h/2)} |Du(x) - Du(y)|$. Then, there exist constants $\alpha(n, \lambda, \Lambda) \in (0, 1)$ and $C(n, \lambda, \Lambda) > 0$ such that, for all $x, y \in S_u(x_0, h/2)$,

$$|v(x) - v(y)| \leq CM^\alpha h^{-\alpha} |x - y|^\alpha \Big(\|v\|_{L^\infty(S_u(x_0, h))} + h^{1/2} \|f\|_{L^n(S_u(x_0, h))} \Big).$$

Proof. By Theorem 5.30, there is a constant $c_1(n, \lambda, \Lambda) > 0$ such that $S_u(z, c_1 h) \subset S_u(x_0, 3h/4)$ for all $z \in S_u(x_0, h/2)$. Fix $x \in S_u(x_0, h/2)$.

We first prove the theorem for $y \in S_u(x, c_1 h) \cap S_u(x_0, h/2)$. Let $r \in (0, c_1 h)$ be such that $y \in S_u(x, r) \backslash S_u(x, r/2)$. Then, from

$$r/2 \leq u(y) - u(x) - Du(x) \cdot (y - x) = (Du(z) - Du(x)) \cdot (y - x)$$

for some z between x and y, we find $|y - x| \geq r/(2M)$.

Now, applying Theorem 12.13 to v on $S_u(x, c_1 h)$, we can find constants $\alpha \in (0, 1)$ and $C > 0$ depending only on n, λ, and Λ such that

$$
\begin{aligned}
|v(y) - v(x)| &\leq \text{osc}_{S_u(x,r)} v \\
&\leq C \Big(\frac{r}{c_1 h} \Big)^\alpha \Big[\|v\|_{L^\infty(S_u(x, c_1 h))} + (c_1 h)^{\frac{1}{2}} \|f\|_{L^n(S_u(x, c_1 h))} \Big] \\
&\leq CM^\alpha h^{-\alpha} |x - y|^\alpha \Big[\|v\|_{L^\infty(S_u(x_0, h))} + h^{1/2} \|f\|_{L^n(S_u(x_0, h))} \Big].
\end{aligned}
$$

Finally, if $y \in S_u(x_0, h/2)$ but $y \notin S_u(x, c_1 h)$, then $c_1 h \leq M|x - y|$ and the desired estimate also holds. $\square$

Remark 12.15. Some remarks on interior Hölder estimates are in order.

- If the conclusion of either Theorem 12.13 or 12.14 holds for an exponent $\alpha \in (0, 1)$, then clearly it also holds for any $0 < \alpha' < \alpha$.

- In the setting of Theorem 12.14, if we assume higher integrability of f such as $f \in L^p(S_u(x_0, h))$ for some $p > n$, then it would be interesting to see if the local Hölder estimates could be improved to local $W^{1,p}$ estimates. See also Problem 12.6.

To end this section, we state the following interior Hölder estimates for linearized Monge–Ampère equations with lower-order terms. They are simple consequences of the interior Harnack inequality proved in Le [**L5**].

Theorem 12.16 (Interior Hölder estimates for linearized Monge–Ampère equations with lower-order terms). *Let $u \in C^2(\Omega)$ be a convex function satisfying $0 < \lambda \leq \det D^2 u \leq \Lambda$ in a domain Ω in $\mathbb{R}^n$. Suppose for some $t_0 > 0$, $S_u(z, 2t_0) \Subset \Omega$ and $B_1(\hat{z}) \subset S_u(z, 2t_0) \subset B_n(\hat{z})$ for some $\hat{z} \in \mathbb{R}^n$. Let $v \in C(\overline{S_u(z, 2t_0)}) \cap W^{2,n}_{\text{loc}}(S_u(z, 2t_0))$ be a solution to*

$$U^{ij} D_{ij} v + b \cdot Dv + cv = f \quad \text{in } S_u(z, 2t_0),$$

where

$$\|b\|_{L^\infty(S_u(z,2t_0))} + \|c\|_{L^\infty(S_u(z,2t_0))} \leq M.$$

Then, there exist constants $\beta_0, C > 0$ depending only λ, Λ, n, M such that

$$[v]_{C^\beta(S_u(z,t_0))} \leq C\big(\|v\|_{L^\infty(S_u(z,2t_0))} + \|f\|_{L^n(S_u(z,2t_0))}\big).$$

Theorem 12.16 is a key tool in the solvability of singular Abreu equations in higher dimensions in Kim–Le–Wang–Zhou [**KLWZ**] and Le [**L14**].

12.4. Application: The Affine Bernstein Problem

12.4.1. The affine maximal equation. In this section, we illustrate an interesting application of the Hölder estimates for the linearized Monge–Ampère equation to the classification of global solutions to the affine maximal equation in affine geometry:

$$(12.27) \qquad \sum_{i,j=1}^n U^{ij} D_{ij} w = 0, \quad w = (\det D^2 u)^{-\frac{n+1}{n+2}} \quad \text{in } \mathbb{R}^n \quad (n \geq 2),$$

where u is a locally uniformly convex function and $(U^{ij}) = \text{Cof } D^2 u$. This classification problem is known as the Bernstein problem for affine maximal hypersurfaces, or *the affine Bernstein problem* for short. It was first proposed by Chern [**Ch**] in two dimensions (see also Calabi [**Cal2**]). Analytically, this problem is equivalent to the following question:

> *Are smooth, locally uniformly convex solutions of (12.27) quadratic polynomials?*

Note that (12.27) can be viewed as a system of two equations: One is a Monge–Ampère equation for u in the form of $\det D^2 u = w^{-\frac{n+2}{n+1}}$, and other is a linearized Monge–Ampère equation for w in the form of $U^{ij} D_{ij} w = 0$.

A similar equation, called Abreu's equation, arises in the constant curvature problem in complex geometry. In fact, for toric varieties, this problem reduces to solving Abreu's equation [**Ab**]

$$(12.28) \qquad -U^{ij} D_{ij} \big[(\det D^2 u)^{-1}\big] = A \quad \text{on a polytope } P \subset \mathbb{R}^n,$$

where A is a positive constant, with the Guillemin boundary condition

$$(12.29) \qquad u(x) - \sum_{1 \leq k \leq m} \delta_k(x) \log \delta_k(x) \in C^\infty(\overline{P}).$$

Here δ_k's are affine functions such that $P = \bigcap_{k=1}^m \{x \in \mathbb{R}^n : \delta_k(x) > 0\}$.

The problem (12.28)–(12.29) was solved by Donaldson in two dimensions by an ingenious combination of geometric and PDE methods in a series of papers [**D1**, **D3**, **D4**, **D5**]. The problem of the existence of the constant scalar curvature Kähler metrics in higher dimensions was recently solved by Chen and Cheng [**ChCh**].

Trudinger and Wang [**TW1**] proved that the affine Bernstein problem has an affirmative answer in two dimensions, thus settling Chern's conjecture. This is the content of the following theorem.

Theorem 12.17. *Let* $u \in C^\infty(\mathbb{R}^n)$ *be a locally uniformly convex solution to the affine maximal surface equation*

$$(12.30) \qquad U^{ij} D_{ij} w = 0, \quad w = [\det D^2 u]^{-\frac{n+1}{n+2}} \quad in \ \mathbb{R}^n,$$

where $(U^{ij}) = \operatorname{Cof} D^2 u$. *If* $n = 2$, *then* u *is a quadratic polynomial.*

In [**TW1**], Trudinger and Wang also showed that a corresponding result to Theorem 12.17 holds in higher dimensions provided that a uniform, *strict convexity* holds; see also Remark 12.24. However, they produced a nonsmooth counterexample for $n \geq 10$: The convex function $u(x) = \sqrt{|x'|^9 + x_{10}^2}$, where $x' = (x_1, \ldots, x_9)$, satisfies (12.27) in $\mathbb{R}^{10}$ and it is not differentiable at the origin; see Problem 12.8.

Equation (12.27) is connected with a variational problem for which we digress to explain. In affine geometry, the affine area of the graph of a smooth, convex u defined on Ω is

$$(12.31) \qquad \mathcal{A}(u, \Omega) = \int_\Omega [\det D^2 u]^{\frac{1}{n+2}} \, dx.$$

By Lemma 2.59, the affine area functional $\mathcal{A}$ is concave; that is,

$$\mathcal{A}(tu + (1-t)v, \Omega) \geq t\mathcal{A}(u, \Omega) + (1-t)\mathcal{A}(v, \Omega) \quad \text{for all } 0 \leq t \leq 1.$$

Critical points of $\mathcal{A}$ are maximizers under local perturbations. The Euler–Lagrange equations for critical points are given by the following lemma.

Lemma 12.18 (Euler–Lagrange equation for critical points). *Let* Ω *be a domain in* $\mathbb{R}^n$, *and let* $F, K \in C^2(\mathbb{R})$. *Consider the functional*

$$\mathcal{A}_{F,K}(u, \Omega) = \int_\Omega \left[F(\det D^2 u) - K(u) \right] dx.$$

Then, strictly convex critical points $u \in C^4(\Omega)$ *of* $\mathcal{A}_{F,K}(\cdot, \Omega)$ *satisfy*

$$U^{ij} D_{ij} w = K'(u), \quad w = F'(\det D^2 u) \quad in \ \Omega.$$

Proof. Strictly convex critical points $u \in C^4(\Omega)$ of $\mathcal{A}_{F,K}(\cdot, \Omega)$ satisfy the identity $\frac{d}{dt}|_{t=0}\mathcal{A}_{F,K}(u + t\varphi, \Omega) = 0$ for all $\varphi \in C_c^\infty(\Omega)$. This translates to

$$\int_\Omega \left[F'(\det D^2 u) \frac{\partial \det D^2 u}{\partial r_{ij}} D_{ij}\varphi - K'(u)\varphi \right] dx = 0.$$

Let $w = F'(\det D^2 u)$. By Lemma 2.56 and integration by parts, we have

$$0 = \int_\Omega \left[w U^{ij} D_{ij} \varphi - K'(u) \varphi \right] dx = \int_\Omega \left[D_{ij}(w U^{ij}) - K'(u) \right] \varphi \, dx.$$

This is true for all $\varphi \in C_c^\infty(\Omega)$; hence, $D_{ij}(U^{ij} w) = K'(u)$. Since the matrix U is divergence-free (see Lemma 2.56), we have $U^{ij} D_{ij} w = K'(u)$. $\quad\square$

From Lemma 12.18, we see that locally uniformly convex maximizers of $\mathcal{A}$ satisfy (12.27). The quantity

$$H_{\mathcal{A}}[u] := -\frac{1}{n+1} \sum_{i,j=1}^n U^{ij} D_{ij} \left([\det D^2 u]^{-\frac{n+1}{n+2}} \right)$$

represents the affine mean curvature of the graph of u. As a result, (12.27) is called the *affine maximal surface equation*. The graph of a function u satisfying (12.27) is then called an affine maximal graph.

12.4.2. Proof of Chern's conjecture. We will present a proof of Theorem 12.17 which consists of the following steps. The first step (Lemma 12.19) establishes an upper bound on the determinant. Then, in Lemma 12.21, we prove a lower bound on the determinant under a modulus of convexity. These estimates hold in all dimensions. In two dimensions, we can establish a uniform modulus of strict convexity (Proposition 12.22). With these ingredients, the proof of Theorem 12.17 follows from the regularity theories of the linearized Monge–Ampère and Monge–Ampère equations.

Upper bound on the determinant. To obtain upper bounds for $\det D^2 u$, we use maximum principle techniques in nonlinear PDEs.

Lemma 12.19 (Upper bound on the determinant). *Let $\theta \in (0,1)$, and let Ω be a bounded convex domain in $\mathbb{R}^n$. Let $u \in C^4(\Omega) \cap C^{0,1}(\overline{\Omega})$ be a locally uniformly convex function satisfying $u = 0$ on $\partial\Omega$, $\inf_\Omega u = -1$, and*

$$U^{ij} D_{ij} w \le 0, \quad w = [\det D^2 u]^{\theta-1}.$$

Then, for $y \in \Omega$, we have

$$\det D^2 u(y) \le C$$

where C depends on $\theta, n, \mathrm{dist}(y, \partial\Omega), \mathrm{diam}(\Omega),$ and $\sup_\Omega |Du|$.

Proof. We will apply the maximum principle to

$$z = \log w - \beta \log(-u) - A|Du|^2,$$

where $\beta > 1$ and A are positive constants to be determined. The expression for z is a reminiscence of the function w in the proof of Pogorelov's second derivative estimates in Theorem 6.1. Since $z(x) \to +\infty$ when $x \to \partial\Omega$, z attains its minimum at an interior point $x_0 \in \Omega$. At x_0, we have

$$Dz(x_0) = 0 \quad \text{and} \quad D^2 z(x_0) = (D_{ij} z(x_0)) \geq 0.$$

Let $d = \det D^2 u$ and $(u^{ij}) = (D^2 u)^{-1} = (U^{ij})/d$. Using the identities

$$\frac{D_i w}{w} = \frac{\beta D_i u}{u} + 2 A D_k u D_{ik} u, \; u^{ij} D_{kij} u = \frac{D_k w}{(\theta-1) w}, \; u^{ij} D_{ij} w = \frac{U^{ij} D_{ij} w}{d},$$

where we recall Lemma 2.56, we obtain at x_0

$$0 \leq u^{ij} D_{ij} z = \frac{U^{ij} D_{ij} w}{d^\theta} - \frac{\beta n}{u} - \beta(\beta-1) \frac{u^{ij} D_i u D_j u}{u^2}$$
$$- 2 A \Delta u + \frac{4 A^2 \theta}{1-\theta} D_{ij} u D_i u D_j u - 2 \beta A \frac{1-2\theta}{1-\theta} \frac{|Du|^2}{u}.$$

From $U^{ij} D_{ij} w \leq 0$, $\beta > 1$, and $D_{ij} u D_i u D_j u \leq \Delta u |Du|^2$, by choosing

$$A = \frac{1-\theta}{4\theta \sup_\Omega |Du|^2},$$

we obtain at x_0

$$0 \leq -A \Delta u + \frac{\beta n}{|u|} + 2 \beta A \frac{|Du|^2}{|u|}.$$

It follows that, for $M_1 = \sup_\Omega |Du|$, we have

$$|u(x_0)|[\det D^2 u(x_0)]^{1/n} \leq n |u(x_0)| \Delta u(x_0) \leq C(n, \beta, \theta)(1 + M_1^2).$$

Let $\beta = n$. Then, for all $y \in \Omega$, we have $-z(y) \leq -z(x_0)$ and hence

$$\log[\det D^2 u(y)]^{1-\theta} + n \log |u(y)|$$
$$\leq \log[\det D^2 u(x_0)]^{1-\theta} + \beta \log |u(x_0)| + A |Du(x_0)|^2$$
$$\leq (1-\theta) \log(\det D^2 u(x_0) |u(x_0)|^n) + A M_1^2 \leq C(n, \theta, M_1).$$

Therefore, for any $y \in \Omega$, we have

(12.32) $\qquad \det D^2 u(y) \leq \dfrac{C(n, \theta, M_1)}{|u(y)|^{\frac{n}{1-\theta}}} \leq C(n, \theta, M_1) \dfrac{(\mathrm{diam}(\Omega))^{\frac{n}{1-\theta}}}{[\mathrm{dist}(y, \partial\Omega)]^{\frac{n}{1-\theta}}}$

because by Lemma 2.37(i)

$$|u(y)| \geq \frac{\mathrm{dist}(y, \partial\Omega)}{\mathrm{diam}(\Omega)} \|u\|_{L^\infty(\Omega)} = \frac{\mathrm{dist}(y, \partial\Omega)}{\mathrm{diam}(\Omega)}.$$

The lemma is proved. $\square$

Legendre transform and affine maximal surface equations. Here, we study the effect of the Legendre transform on the affine maximal surface equation. Let Ω be a convex domain and let $u \in C^4(\Omega)$ be a locally uniformly convex function. The Legendre transform of u is the function u^* defined in $\Omega^* = Du(\Omega)$ by

$$u^*(y) = \sup_{z \in \Omega}(z \cdot y - u(z)) = x \cdot y - u(x), \quad y \in \Omega^* = Du(\Omega)$$

where $x \in \Omega$ is uniquely determined by $y = Du(x)$. The Legendre transform u^* is a locally uniformly convex, C^4 function in Ω^*. It satisfies $(u^*)^* = u$.

By (2.11), we have

$$A_\theta[u, \Omega] := \int_\Omega (\det D^2 u)^\theta \, dx = \int_{\Omega^*} (\det D^2 u^*)^{1-\theta} \, dy = A_{1-\theta}[u^*, \Omega^*].$$

Assume now u solves the affine maximal surface-type equation

$$U^{ij} D_{ij} w = 0, \quad w = [\det D^2 u]^{\theta - 1}, \quad 0 < \theta < 1.$$

Since u is critical with respect to the functional $A_\theta[\cdot, \Omega]$, it follows that u^* is critical with respect to the functional $A_{1-\theta}[\cdot, \Omega^*]$. Therefore, by Lemma 12.18, u^* satisfies

(12.33) $$U^{*,ij} D_{ij} w^* = 0, \quad w^* = [\det D^2 u^*]^{-\theta}.$$

We might wish to apply Lemma 12.19 to (12.33) on Ω^* to obtain an upper bound for $\det D^2 u^*$ and hence a lower bound for $\det D^2 u$. The only issue is that u^* may be nonconstant on $\partial \Omega^*$. This issue can be resolved when u has a uniform modulus of convexity.

Definition 12.20 (Modulus of convexity). Let u be a C^1 convex function on a convex domain Ω in $\mathbb{R}^n$. We define the modulus of convexity of u at $y \in \Omega$ by

$$m_{u,y}(r) = \sup \left\{ h > 0 : S_u(y, h) \subset B_r(y) \right\}, \quad r > 0.$$

Note that an equivalent definition of $m_{u,y}(r)$ is

$$m_{u,y}(r) = \inf \left\{ u(x) - u(y) - Du(y) \cdot (x - y) : x \in \Omega, |x - y| > r \right\}.$$

We define the modulus of convexity of u on Ω, by

$$m(r) = m_{u,\Omega}(r) = \inf_{y \in \Omega} m_{u,y}(r), \quad r > 0.$$

Observe that a function u is strictly convex in Ω if and only if $m(r) > 0$ for all $r > 0$. Note that a locally uniformly convex function u on a convex domain Ω will be strictly convex there. As an example, if $u(x) = |x|^2/2$, then $m_{u,\Omega}(r) = r^2/2$.

Lower bound on the determinant under a modulus of convexity.
Next, we establish a lower bound on the Hessian determinant for solutions
to the affine maximal surface equation under some uniform modulus of convexity using the Legendre transform.

Lemma 12.21 (Lower bound on the determinant). *Let $\theta \in (0,1)$, and let Ω be a bounded convex domain in $\mathbb{R}^n$. Let $u \in C^4(\Omega) \cap C^{0,1}(\overline{\Omega})$ be a locally uniformly convex function satisfying $-1 \leq u \leq 0$ in Ω and*

$$U^{ij} D_{ij} w = 0, \quad w = [\det D^2 u]^{\theta - 1}.$$

For any $y \in \Omega$, there exists a positive constant C depending on $\theta, n, m_{u,\Omega}$, $\operatorname{diam}(\Omega)$, and $\operatorname{dist}(y, \partial\Omega)$ such that

$$C^{-1} \leq \det D^2 u(y) \leq C.$$

Proof. Note that the Legendre transform u^* of u defined on $\Omega^* = Du(\Omega)$ is C^4, locally uniformly convex, and u^* satisfies (12.33).

Fix $y \in \Omega$ and $x = Du(y) \in \Omega^*$. Let $\delta := m_{u,\Omega}(\operatorname{dist}(y, \partial\Omega)/2)$. We will show the following properties of $S_{u^*}(x, \delta)$:

(12.34) $\qquad S_{u^*}(x, \delta) \subset \Omega^* \quad \text{and} \quad \operatorname{diam}(S_{u^*}(x, \delta)) \leq \dfrac{4}{\operatorname{dist}(y, \partial\Omega)}.$

Indeed, let $Du(z) \in S_{u^*}(x, \delta)$ where $z \in \overline{\Omega}$. Then

$$u^*(Du(z)) < u^*(x) + Du^*(x) \cdot (Du(z) - x) + \delta.$$

Using

$$u^*(Du(z)) = z \cdot Du(z) - u(z), \quad u^*(x) = x \cdot y - u(y),$$

we deduce

$$z \cdot Du(z) - u(z) < x \cdot y - u(y) + y \cdot (Du(z) - x) + \delta;$$

hence $y \in S_u(z, \delta)$. By the definition of modulus of convexity and δ, we have $y \in B_{\operatorname{dist}(y,\partial\Omega)/2}(z)$ and thus $z \in \Omega$. Therefore, $S_{u^*}(x, \delta) \subset \Omega^*$, proving the first part of (12.34). Observe that

$$\operatorname{dist}(z, \partial\Omega) \geq \operatorname{dist}(y, \partial\Omega) - |y - z| > \operatorname{dist}(y, \partial\Omega)/2.$$

Recalling $-1 \leq u \leq 0$, we obtain from Lemma 2.37 that

$$|Du(z)| \leq \frac{\sup_{\partial\Omega} u - u(z)}{\operatorname{dist}(z, \partial\Omega)} \leq \frac{2}{\operatorname{dist}(y, \partial\Omega)}.$$

Therefore, (12.34) is proved.

Now, without loss of generality, assume that $0 \in \Omega$. Then, from $\Omega = Du^*(\Omega^*)$, we have $|Du^*| \leq \operatorname{diam}(\Omega)$ in Ω^*. This together with the Mean Value Theorem shows that

(12.35) $\qquad B_{\delta/[2\operatorname{diam}(\Omega)]}(x) \subset S_{u^*}(x, \delta).$

Thus, recalling (12.33), we can apply (12.32) in the proof of Lemma 12.19 to the function $[u^* - u^*(x) - Du^*(x) \cdot (\cdot - x) - \delta]/\delta$ in the domain $S_{u^*}(x, \delta)$ with θ being replaced by $1 - \theta$ to obtain

$$\det D^2 u^*(x) \leq \frac{C(n, \theta, \delta, \operatorname{diam}(\Omega))[\operatorname{diam}(S_{u^*}(x, \delta))]^{\frac{n}{\theta}}}{[\operatorname{dist}(x, \partial S_{u^*}(x, \delta))]^{\frac{n}{\theta}}} \leq \frac{C(n, \theta, \delta, \operatorname{diam}(\Omega))}{[\operatorname{dist}(y, \partial\Omega)]^{n/\theta}},$$

where we used (12.34) and (12.35) in the last inequality. This gives a lower bound for the Hessian determinant $\det D^2 u(y) = [\det D^2 u^*(x)]^{-1}$.

On the other hand, we have $S_u(y, \delta) \subset B_{\operatorname{dist}(y, \partial\Omega)/2}(y)$ from the definition of δ. Hence, in the section $S_u(y, \delta)$, we would have the gradient bound $|Du| \leq 2/\operatorname{dist}(y, \partial\Omega)$. Thus, we can apply Lemma 12.19 to the function

$$[u - u(y) - Du(y) \cdot (\cdot - y) - \delta]/\delta$$

in the domain $S_u(y, \delta)$ to get an upper bound for $\det D^2 u(y)$. This completes the proof of the lemma.   □

Uniform modulus of strict convexity in two dimensions. Let $\alpha(n)$ be the constant in Lemma 2.66. We have the following uniform modulus of strict convexity for solutions to (12.27).

Proposition 12.22 (Uniform modulus of strict convexity in two dimensions). *Let $\{u^{(k)}\}_{k=1}^{\infty}$ be a sequence of solutions to (12.30) on convex domains Ω^k in $\mathbb{R}^n$ $(n = 2)$ with vanishing boundary condition and $\inf_{\Omega^k} u^{(k)} = -1$ where Ω^k has 0 as the center of mass and $B_{\alpha(n)}(0) \subset \Omega^k \subset B_1(0)$. Then, there exists a function $m : (0, \infty) \to (0, \infty)$ such that*

$$\inf_{k \in \mathbb{N}} m_{u^{(k)}, x}(r) \geq m(r) \quad \text{for all } x \in B_{\alpha(n)/2}(0).$$

The rough idea of the proof is as follows. If there is no uniform modulus of convexity, then by using affine transformations and compactness, we can find an affine maximal graph of a function v that contains a vertical line segment. Rotating the graph a bit, we can assume the following scenario: $v(0) = 0$, $Dv(0) = 0$, $v \geq |x_1|$, and v is strictly convex at 0. Consider a small section S_ε of v at 0 with height ε. This section is roughly an ellipsoid with lengths of principal axes being of order ε on the x_1-axis and bounded from above by δ_ε on the other axes where δ_ε goes to 0 as ε goes to 0 by the local strict convexity. The gradient of v and the Monge–Ampère measure of v are bounded in S_ε (the latter essentially follows from Lemma 12.19), and thus the Monge–Ampère mass $\mu_v(S_\varepsilon)$ of v in S_ε is bounded from above by $C\varepsilon\delta_\varepsilon^{n-1}$. However, by comparing with the volume of a cone with vertex 0 and base $\partial\{v = \varepsilon\}$, we find that $\mu_v(S_\varepsilon)$ is not smaller than $C(\varepsilon/\delta_\varepsilon)^{n-1}$. Therefore, when $n = 2$, δ_ε is bounded from below, which is a contradiction; however, this argument does not work for $n \geq 3$. For details, see the original paper of Trudinger and Wang [**TW1**] and also Zhou [**Zh1**].

Higher-order derivative estimates. Using the Caffarelli–Gutiérrez interior Hölder estimate for the linearized Monge–Ampère equation together with Caffarelli's interior $C^{2,\alpha}$ estimates for the Monge–Ampère equation, we can prove the following higher-order derivative estimates for (12.27).

Lemma 12.23 (Higher-order derivative estimates). *Let $u \in C^4(\Omega)$ be a locally uniformly convex function in a bounded convex domain Ω in $\mathbb{R}^n$. Assume that*

$$U^{ij} D_{ij} w = 0, \quad w = (\det D^2 u)^{-\frac{n+1}{n+2}}, \quad and \quad -1 \le u \le 0 \quad in \ \Omega.$$

Then, for any subdomain $\Omega' \Subset \Omega$ and any integer $k \ge 2$, we have

$$\lambda \le \det D^2 u \le \Lambda \quad in \ \Omega' \quad and \quad \|u\|_{C^k(\Omega')} \le C_k$$

where λ, Λ, and C_k are positive constants depending only on n, $\mathrm{diam}(\Omega)$, $\mathrm{dist}(\Omega', \partial\Omega)$, and the modulus of convexity $m_{u,\Omega}$ of u on Ω.

Proof. Let Ω'' be a subdomain of Ω such that $\Omega' \Subset \Omega'' \Subset \Omega$. Then Lemma 12.21 tells us that

$$\lambda \le \det D^2 u \le \Lambda \quad in \ \Omega'',$$

where $\lambda, \Lambda > 0$ depend only on n, $\mathrm{dist}(\Omega'', \partial\Omega)$, $\mathrm{diam}(\Omega)$, and the modulus of convexity $m_{u,\Omega}$ of u on Ω. Let $K := 2\sup_{\Omega''} |Du| \le 2/\mathrm{dist}(\Omega'', \partial\Omega)$. Then, for $y \in \Omega'$, by the definition of $m_{u,\Omega}$, we have

$$S_u(y, R) \subset \Omega'' \Subset \Omega \quad for \ R < m_{u,\Omega}(\mathrm{dist}(y, \partial\Omega''))$$

and $B_r(y) \subset S_u(y, Kr)$ for all $r \le R/K$.

Now, applying Theorem 12.14 to each $S_u(y, R/2)$ and then using a finite covering of Ω' by balls $B_{R/(2K)}(y) \subset S_u(y, R/2)$, we find $[w]_{C^\alpha(\Omega')} \le C$ where α, C depend only n, $\mathrm{diam}(\Omega)$, $\mathrm{dist}(\Omega', \partial\Omega)$, and $m_{u,\Omega}$. By Caffarelli's $C^{2,\alpha}$ estimates in Theorem 6.34 applied to $\det D^2 u = w^{\frac{1}{\theta-1}}$, we find that $\|u\|_{C^{2,\alpha}(\Omega')} \le C$ where α, C depend only n, $\mathrm{diam}(\Omega)$, $\mathrm{dist}(\Omega', \partial\Omega)$, and the modulus of convexity $m_{u,\Omega}$ of u. Thus, the operator $U^{ij} D_{ij}$ is uniformly elliptic in Ω' with Hölder continuous coefficients. Higher-order derivative estimates for u follow by bootstrapping the equation $U^{ij} D_{ij} w = 0$. $\qquad\square$

Finally, we are ready to give the proof of Chern's conjecture.

Proof of Theorem 12.17. To prove the theorem, we show that $D^3 u = 0$ in $\mathbb{R}^n$ where $n = 2$. By symmetry, it suffices to show that $D^3 u(0) = 0$.

We can assume that $u(0) = 0$ and $Du(0) = 0$. For any $h > 0$, by Theorem 5.34, there exists a centered section $S_u(0, p_h, h)$ where 0 is the center of mass of $S_u(0, p_h, h)$. By the second bullet in Remark 5.36, there are $x_h \in \mathbb{R}^2$ and $\hat{h} \ge h$ such that 0 is the center of mass of

$$S_u(x_h, \hat{h}) := \{x \in \mathbb{R}^2 : u(x) < u(x_h) + Du(x_h) \cdot (x - x_h) + \hat{h}\}.$$

By Lemma 2.66, there is a symmetric matrix $T_h \in \mathbb{S}^n$ such that
$$B_{\alpha(n)}(0) \subset \Omega_h := T_h(S_u(x_h, \hat{h})) \subset B_1(0).$$
Define
$$u_h(y) = \hat{h}^{-1}[u(x) - u(x_h) - Du(x_h) \cdot (x - x_h) - \hat{h}], \quad y = T_h(x) \in \Omega_h.$$
Note that 0 is the center of mass of Ω_h. We also have $u_h = 0$ on $\partial\Omega_h$ and $\inf_{\Omega_h} u_h = -1$. By Proposition 12.22, u_h has a uniform modulus of convexity in $B_{\alpha(n)/2}(0)$. Thus, by Lemma 12.23, we have the estimates
$$C_0^{-1} \le \det D^2 u_h \le C_0 \quad \text{in } B_{\alpha(n)/4}(0) \quad \text{and} \quad \|u_h\|_{C^4(B_{\alpha(n)/4}(0))} \le C_0,$$
for some constant C_0 independent of h. Therefore
$$C_1|y|^2 \le u_h(y) - u_h(0) - Du_h(0) \cdot y \le C_2|y|^2 \quad \text{in } B_{\alpha(n)/4}(0).$$
Since
$$u_h(y) - u_h(0) - Du_h(0) \cdot y = (u(x) - u(0) - Du(0) \cdot x)/\hat{h} = u(x)/\hat{h},$$
we can find positive constants C_1, C_2 independent of h such that
$$C_1|T_h x|^2 \le \hat{h}^{-1}u(x) \le C_2|T_h x|^2 \quad \text{in } T_h^{-1}(B_{\alpha(n)/4}(0)).$$
Let $M := \sup_{B_1(0)} \|D^2 u\|$. Let $z_h \in \partial(B_1(0) \cap T_h^{-1}(B_{\alpha(n)/4}(0)))$ be an eigenvector corresponding to the largest eigenvalue Λ_h of T_h. Then, from $u(z_h) \le M|z_h|^2$ and
$$M|z_h|^2 \ge u(z_h) \ge C_1\hat{h}|T_h z_h|^2 = C_1\hat{h}(\Lambda_h|z_h|)^2,$$
we obtain $\Lambda_h \le C_*(C_1, M)\hat{h}^{-1/2}$. In $T_h^{-1}(B_{\alpha(n)/4}(0))$, we have
$$|D^3 u(x)| \le \hat{h}\Lambda_h^3|D^3 u_h(T_h x)| \le C_0\hat{h}\Lambda_h^3 \le C_0 C_*^3\hat{h}^{-1/2} \le C_0 C_*^3 h^{-1/2}.$$
Letting $h \to \infty$ yields $D^3 u(0) = 0$ as desired. The theorem is proved. $\square$

Remark 12.24. The proof of Theorem 12.17 also shows that if the uniform modulus of strict convexity result in Proposition 12.22 holds for $n \ge 3$, then the affine Bernstein problem has an affirmative answer for this dimension.

It is still a very challenging problem to establish the uniform modulus of strict convexity result in Proposition 12.22 for dimensions $n \ge 3$.

12.5. Problems

Problem 12.1. Let $u \in C^2(\mathbb{R}^n)$ be a strictly convex function, $v \in C^1(\mathbb{R}^n)$, and $(U^{ij}) = (\det D^2 u)(D^2 u)^{-1}$. Show that the expression $\sum_{i,j=1}^n U^{ij} D_i v D_j v$ is affine invariant under the transformations
$$u \mapsto \hat{u}(x) = u(Tx) \quad \text{and} \quad v \mapsto \hat{v}(x) = v(Tx)$$
where $T : \mathbb{R}^n \to \mathbb{R}^n$ is an affine transformation with $\det T = 1$.

Problem 12.2. Let $(a^{ij}(x))_{1 \le i,j \le n}$ be positive definite in a domain Ω in $\mathbb{R}^n$. Assume that $v \in C^2(\Omega)$ satisfies $a^{ij}D_{ij}v \le 0$ in Ω; that is, v is a super-solution. Show that for all $\varepsilon > 0$, $w := -(v+1)^{-\varepsilon}$ is also a supersolution.

Problem 12.3. Prove the doubling estimate in Lemma 12.8 using sliding paraboloids to touch from below the graph of $-(v+1)^{-\varepsilon}$ for some $\varepsilon > 0$.

Problem 12.4. Let Ω be a domain in $\mathbb{R}^n$. Assume the matrix A satisfies $C^{-1}I_n \le A(x) \le CI_n$ in Ω. Let $u \in C^2(\Omega)$ be a convex function satisfying

$$0 < \lambda \le \det D^2 u \le \Lambda \quad \text{in } \Omega.$$

Prove that the Harnack inequality of the form (12.5) also holds for nonnegative solutions $v \in W^{2,n}_{\text{loc}}(\Omega)$ to the linearized Monge–Ampère equation

$$\text{trace}(A(D^2u)^{-1}D^2v) = 0 \quad \text{in } \Omega.$$

Problem 12.5. Let Ω, u, A be as in Problem 12.4. Let α be as in Theorem 5.18. Extend the Harnack inequalities in Theorem 12.4 and Lemma 12.12 to linearized Monge–Ampère equations with lower-order terms

$$\text{trace}(A(D^2u)^{-1}D^2v) + b \cdot Dv + cv = f \quad \text{in } \Omega,$$

where $c \in L^n(\Omega)$, $f \in L^n(\Omega)$, and $b \in L^p(\Omega)$ with $p > n(1+\alpha)/(2\alpha)$.

Problem 12.6. Let $u \in C(\overline{B_1(0)}) \cap C^2(B_1(0))$ be a convex, radially symmetric function satisfying $\lambda \le \det D^2 u \le \Lambda$ in $B_1(0) \subset \mathbb{R}^n$, and let (U^{ij}) be the cofactor matrix of D^2u. Let $g \in L^p(B_1(0))$ $(1 < p < \infty)$ be a radially symmetric function. Consider the linearized Monge–Ampère equation

$$U^{ij}D_{ij}v = g \quad \text{in } B_1(0), \quad v = 0 \quad \text{on } \partial B_1(0).$$

Study the Hölder continuity, $W^{1,p}$, and $W^{2,p}$ estimates for v.

Problem 12.7. Let $u \in C^4(\mathbb{R}^n)$ be a locally uniformly convex function. Let $\mathcal{M} = \{(x, x_{n+1}) : x_{n+1} = u(x), \ x \in \mathbb{R}^n\}$ be its graph, $(u^{ij}) = (D^2u)^{-1}$, and $(U^{ij}) = (\det D^2u)(D^2u)^{-1}$. Consider the metric $(g_{ij}) = (\det D^2u)^{-\frac{1}{n+2}}D^2u$ on $\mathcal{M}$. The volume form is $h = g^{1/2} := (\det(g_{ij}))^{1/2} = (\det D^2u)^{\frac{1}{n+2}}$. Let $\Delta_{\mathcal{M}}$ be the Laplace-Betrami operator with respect to the affine metric g_{ij} given by

$$\Delta_{\mathcal{M}} = \frac{1}{\sqrt{g}}D_i(g^{1/2}g^{ij}D_j) = \frac{1}{h}D_i(h^2 u^{ij}D_j).$$

Show that

$$\Delta_{\mathcal{M}}\left(\frac{1}{h}\right) = \frac{U^{ij}D_{ij}[(\det D^2u)^{-\frac{n+1}{n+2}}]}{(n+1)h}.$$

Thus $\mathcal{M}$ is affine maximal if and only if $1/h$ is harmonic on $\mathcal{M}$.

Problem 12.8. Consider the affine maximal surface equation

$$(12.36) \quad \sum_{i,j=1}^{n} U^{ij} D_{ij}\left([\det D^2 u]^{-\frac{n+1}{n+2}}\right) = 0, \quad (U^{ij}) = (\det D^2 u)(D^2 u)^{-1}.$$

(a) Let $x = (x', x_n) \in \mathbb{R}^n$. Assume $u(x', x_n) = |x'|^{2\alpha}/x_n$ where $\alpha \geq 1$ solves the affine maximal surface equation for $x_n \neq 0$. Prove that

$$8\alpha^2 - (n^2 - 4n + 12)\alpha + 2(n-1)^2 = 0.$$

(b) Show that the convex function $u(x) = \sqrt{|x'|^9 + x_{10}^2}$ satisfies (12.36) in $\mathbb{R}^{10}$.

12.6. Notes

Theorems 12.4 and 12.11 are due to Caffarelli and Gutiérrez in the fundamental paper [**CG2**]. They proved a Harnack inequality for the linearized Monge–Ampère equation $L_u v = 0$ when the Monge–Ampère measure μ_u satisfies the so-called A_∞-condition. In particular, the Harnack inequality holds for μ_u being comparable to positive polynomials if sections of u are bounded. If $u(x_1, x_2) = x_1^4 + x_2^2$, then $\det D^2 u = C x_1^2$ is an admissible measure. Thus, the Harnack inequality applies to Grushin-type equations such as

$$x_1^{-2} D_{11} v + D_{22} v = 0 \quad \text{in } \mathbb{R}^2$$

which are relevant in nonlocal equations such as fractional Laplace equation; see Caffarelli–Silvestre [**CS**]. Maldonado [**Md2**], extending the work of Caffarelli and Gutiérrez, proved a Harnack inequality for the linearized Monge–Ampère equation under minimal geometric conditions; namely, μ_u is *doubling with respect to the center of mass* on the sections of u.

In Section 12.2, the proofs of Theorems 12.4 and 12.11 follow the presentation in Le [**L5**] and Le–Mitake–Tran [**LMT**]. For related results, see also Maldonado [**Md3**]. Our proof in [**L5**] adapts the general scheme in proving the Harnack inequality in Krylov–Safonov [**KS1, KS2**], Caffarelli-Cabré [**CC**], Caffarelli–Gutiérrez [**CG2**], Cabré [**Cab1**], Savin [**S1**], Imbert–Silvestre [**IS**], and Mooney [**Mn1**]. In particular, a new method of sliding paraboloids in [**S1**] was used in [**L5**] to establish interior Harnack inequalities for linear, degenerate, and singular elliptic equations with unbounded lower-order terms. The paper [**L5**] gives an alternative approach for proving the Caffarelli–Gutiérrez Harnack inequality for the linearized Monge–Ampère equation and the Krylov–Safonov Harnack inequality for linear, uniformly elliptic equations.

In Section 12.3, Theorem 12.14 with $f \equiv 0$ is due to Caffarelli and Gutiérrez [**CG2**], while the case of L^n right-hand side was observed by Trudinger and Wang [**TW6**]. Our proof of Theorem 12.14 follows an argument in [**TW6**].

In Section 12.4, the proof of Chern's conjecture in Theorem 12.17 is based on Trudinger–Wang [**TW1**]. For more on affine differential geometry, see Nomizu–Sasaki [**NS**]. For further PDE-based analysis of Abreu's equation, see [**CHLS, CLS, FSz, Zh2**], for example.

Problem 12.5 is taken from [**L5**]. Problem 12.7 is based on Nomizu–Sasaki [**NS**, Chapter 3, Section 11] and Trudinger–Wang [**TW1**]. The non-smooth counterexample to affine maximal surface equation in Problem 12.8 is taken from [**TW1**]; see also Le–Mitake–Tran [**LMT**] for a heuristic explanation.

Chapter 13

Boundary Estimates

We study in this chapter boundary estimates for solutions to the linearized Monge–Ampère equation

$$(13.1) \qquad L_u v := U^{ij} D_{ij} v = g \quad \text{in } \Omega \subset \mathbb{R}^n \quad (n \geq 2),$$

where $(U^{ij}) = \operatorname{Cof} D^2 u$ is the cofactor matrix of the Hessian matrix $D^2 u$ of a C^2, convex potential function u satisfying $\lambda \leq \det D^2 u \leq \Lambda$.

Extending the interior estimates in Chapter 12 and using the Boundary Localization Theorem with maximum principle arguments, we will establish boundary Hölder, Harnack, and gradient estimates in terms of structural data that are independent of the C^2 character of the potential function u.

Here is a summary of the main results. First, we establish global Hölder estimates for solutions to (13.1) with L^n right-hand side g. These estimates can be viewed as global counterparts of the Caffarelli–Gutiérrez interior Hölder estimates. We briefly mention their application to the affine mean curvature and Abreu's equations. Next, we establish boundary Hölder gradient estimates for (13.1) with L^∞ right-hand side g. They are affine invariant analogues of the boundary Hölder gradient estimates of Krylov for uniformly elliptic equations. We describe an application of the boundary Hölder gradient estimates to global regularity of minimizers of linear functionals with prescribed determinant. Finally, when $g \equiv 0$, we establish a boundary Harnack inequality for nonnegative solutions of (13.1), which is the boundary analogue of the Caffarelli–Gutiérrez interior Harnack inequality.

Recall that, for a convex function $u \in C^1(\overline{\Omega})$, its section centered at $x \in \overline{\Omega}$ with height $h > 0$ is defined by

$$S_u(x, h) := \left\{ y \in \overline{\Omega} : u(y) < u(x) + Du(x) \cdot (y - x) + h \right\}.$$

13.1. Global Hölder Estimates

In this section, we establish global Hölder estimates (Theorem 13.2) for solutions to the linearized Monge–Ampère equation (13.1) with L^n right-hand side g on uniformly convex domains. The proof combines boundary Hölder estimates and the Caffarelli–Gutiérrez interior Hölder estimates in Theorem 12.14. The main tool to connect these interior and boundary estimates is Savin's Boundary Localization Theorem (Theorem 8.5).

For the boundary Hölder estimates, we formulate results more generally for solutions to nonuniformly elliptic, linear, second-order equations without lower-order terms. Our main technical device is the Aleksandrov–Bakelman–Pucci (ABP) maximum principle, and we only need a positive lower bound for the determinant of the coefficient matrix.

Proposition 13.1. *Let Ω be a uniformly convex domain in $\mathbb{R}^n$, $g \in L^n(\Omega)$, and $\varphi \in C^\alpha(\partial\Omega)$ where $\alpha \in (0, 1)$. Assume that the $n \times n$ coefficient matrix (a^{ij}) is measurable, positive definite and satisfies $\det(a^{ij}) \geq \lambda > 0$ in Ω. Let $v \in C(\overline{\Omega}) \cap W^{2,n}_{\mathrm{loc}}(\Omega)$ be the solution to*

$$a^{ij} D_{ij} v = g \quad in \ \Omega, \quad v = \varphi \quad on \ \partial\Omega.$$

Then, there exist positive constants δ, C depending only on λ, n, α, and the uniform convexity radius R of Ω such that, for any $x_0 \in \partial\Omega$, we have

$$|v(x) - v(x_0)| \leq C|x - x_0|^{\frac{\alpha}{\alpha+2}} \big(\|\varphi\|_{C^\alpha(\partial\Omega)} + \|g\|_{L^n(\Omega)}\big) \ for \ all \ x \in \Omega \cap B_\delta(x_0).$$

Proof. It suffices to consider $\varphi \not\equiv 0$ or $g \not\equiv 0$. By considering the equation satisfied by $(\|\varphi\|_{C^\alpha(\partial\Omega)} + \|g\|_{L^n(\Omega)})^{-1} v$, we can assume that

$$\|\varphi\|_{C^\alpha(\partial\Omega)} + \|g\|_{L^n(\Omega)} = 1.$$

We need to prove that if $x_0 \in \partial\Omega$, then

$$|v(x) - v(x_0)| \leq C(n, \lambda, \alpha, R)|x - x_0|^{\frac{\alpha}{\alpha+2}} \quad for \ all \ x \in \Omega \cap B_\delta(x_0).$$

Note that $\mathrm{diam}(\Omega) \leq 2R$. Without loss of generality, we assume that

$$\Omega \subset \mathbb{R}^n \cap \{x_n > 0\}, \quad x_0 = 0 \in \partial\Omega.$$

Since $\det(a^{ij}) \geq \lambda$, the ABP estimate (Theorem 3.16) gives

$$\|v\|_{L^\infty(\Omega)} \leq \|\varphi\|_{L^\infty(\partial\Omega)} + C(n, \lambda)\mathrm{diam}(\Omega)\|g\|_{L^n(\Omega)} \leq C_0,$$

where $C_0 = C_0(n, \lambda, \mathrm{diam}(\Omega)) > 1$. Therefore, for any $\varepsilon \in (0, 1)$, we get

$$(13.2) \qquad |v(x) - v(0) \pm \varepsilon| \leq 3C_0 := C_1 \quad for \ x \in \Omega.$$

For $\delta_2 > 0$ small, to be chosen later, consider the functions

$$h_\pm(x) := v(x) - v(0) \pm \varepsilon \pm C_1(\inf\{y_n : y \in \overline{\Omega} \cap \partial B_{\delta_2}(0)\})^{-1} x_n$$

in the region $A := \Omega \cap B_{\delta_2}(0)$. See Figure 13.1.

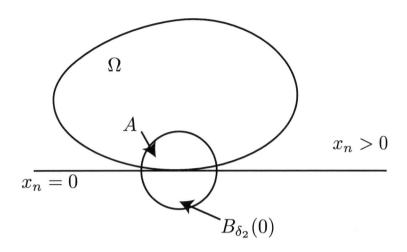

Figure 13.1. The domain Ω and the region A.

Note that if $x \in \partial\Omega$ with $|x| \le \delta_1(\varepsilon) := \varepsilon^{1/\alpha}$, then

$$(13.3) \qquad |v(x) - v(0)| = |\varphi(x) - \varphi(0)| \le \|\varphi\|_{C^\alpha(\partial\Omega)} |x|^\alpha \le |x|^\alpha \le \varepsilon.$$

It follows that if $\delta_2 \le \delta_1$, then from (13.2) and (13.3), we have

$$h_- \le 0, \; h_+ \ge 0 \quad \text{on } \partial A.$$

Clearly, $a^{ij} D_{ij} h_\pm = g$ in A. Thus, the ABP estimate in Theorem 3.16 applied in A to h_- and $-h_+$ gives

$$\max\{h_-, -h_+\} \le C(n, \lambda)\operatorname{diam}(A)\|g\|_{L^n(A)} \le C(n, \lambda)\delta_2 \quad \text{in } A.$$

By restricting $\varepsilon \le [C(n, \lambda)]^{\frac{-\alpha}{1-\alpha}}$, we can assume that $\delta_1 = \varepsilon^{1/\alpha} \le \varepsilon/C(n, \lambda)$.

Then, for $\delta_2 \le \delta_1$, we have $C(n, \lambda)\delta_2 \le \varepsilon$, and thus,

$$|v(x) - v(0)| \le 2\varepsilon + C_1(\inf\{y_n : y \in \overline{\Omega} \cap \partial B_{\delta_2}(0)\})^{-1} x_n \quad \text{in } A.$$

The uniform convexity of Ω gives

$$(13.4) \qquad \inf\{y_n : y \in \overline{\Omega} \cap \partial B_{\delta_2}(0)\} \ge C_2^{-1}\delta_2^2, \quad C_2 = 2R.$$

Therefore, choosing $\delta_2 = \delta_1$, we obtain the inequality

$$(13.5) \qquad |v(x) - v(0)| \le 2\varepsilon + C_1 C_2 \delta_2^{-2} x_n \le 2\varepsilon + C_1 C_2 \varepsilon^{-2/\alpha} |x|,$$

for all x, ε satisfying the conditions

$$(13.6) \qquad |x| \le \delta_1(\varepsilon) := \varepsilon^{1/\alpha}, \quad \varepsilon \le [C(n, \lambda)]^{\frac{-\alpha}{1-\alpha}} := c_1(\alpha, n, \lambda).$$

Finally, let us choose $\varepsilon = |x|^{\frac{\alpha}{\alpha+2}}$. It satisfies the conditions in (13.6) if

$$|x| \le \min\left\{c_1^{\frac{\alpha+2}{\alpha}}, 1\right\} := \delta.$$

Then, by (13.5), we have for all $x \in \Omega \cap B_\delta(0)$,

$$|v(x) - v(0)| \le (2 + C_1 C_2)|x|^{\frac{\alpha}{\alpha+2}}.$$

The proposition is proved. $\qquad\qquad\qquad\qquad\qquad\qquad\qquad\qquad\square$

We now establish global Hölder estimates for solutions to the linearized Monge–Ampère equation with L^n inhomogeneity. They can be viewed as global counterparts of the Caffarelli–Gutiérrez interior Hölder estimates in Theorem 12.14.

Theorem 13.2 (Global Hölder estimates with L^n inhomogeneity). *Let Ω be a uniformly convex domain in $\mathbb{R}^n$ with $\partial\Omega \in C^3$, $g \in L^n(\Omega)$, and $\varphi \in C^\alpha(\partial\Omega)$ where $\alpha \in (0,1)$. Let $u \in C(\overline\Omega) \cap C^2(\Omega)$ be a convex function satisfying $0 < \lambda \le \det D^2 u \le \Lambda$ in Ω and $u|_{\partial\Omega} \in C^3$. Let $(U^{ij}) = \operatorname{Cof} D^2 u$. Let $v \in C(\overline\Omega) \cap W^{2,n}_{\mathrm{loc}}(\Omega)$ be the solution to*

$$U^{ij} D_{ij} v = g \quad in\ \Omega, \quad v = \varphi \quad on\ \partial\Omega.$$

Then, $v \in C^\beta(\overline\Omega)$ for $\beta \in (0,1)$, with the estimate

$$\|v\|_{C^\beta(\overline\Omega)} \le C\big(\|\varphi\|_{C^\alpha(\partial\Omega)} + \|g\|_{L^n(\Omega)}\big),$$

where β depends only on $\lambda, \Lambda, n, \alpha$ and where C depends only on λ, Λ, n, α, $\|u\|_{C^3(\partial\Omega)}$, the C^3 regularity of $\partial\Omega$, and the uniform convexity radius of Ω.

Proof. By Proposition 4.7, u and Ω satisfy the hypotheses of the Boundary Localization Theorem (Theorem 8.5) at every point on $\partial\Omega$ with small constants $\mu = \rho$ depending only on n, λ, Λ, $\|u\|_{C^3(\partial\Omega)}$, the C^3 regularity of $\partial\Omega$, and the uniform convexity radius of Ω. By Proposition 8.23, $u \in C^1(\overline\Omega)$ and

$$M := \sup_{x,y\in\overline\Omega} |Du(x) - Du(y)| \le C(n, \lambda, \Lambda, \rho).$$

Note that $\det(U^{ij}) = (\det D^2 u)^{n-1} \ge \lambda^{n-1}$. As before, we can assume that

$$\|\varphi\|_{C^\alpha(\partial\Omega)} + \|g\|_{L^n(\Omega)} = 1,$$

and we will apply Propositions 9.2 and 13.1 and Theorem 12.14 to show that

$$(13.7) \qquad \|v\|_{C^\beta(\overline\Omega)} \le C(\lambda, \Lambda, n, \alpha, \rho).$$

Let $y \in \Omega$ with $r := \operatorname{dist}(y, \partial\Omega) \le c(n, \lambda, \Lambda, \rho)$, and let $S_u(y, \bar h)$ be the maximal interior section centered at y; that is, $\bar h = \sup\{h > 0 : S_u(y, h) \subset \Omega\}$. By Proposition 9.2 applied at the point $x_0 \in \partial S_u(y, \bar h) \cap \partial\Omega$, we can find a constant $\kappa_0(n, \lambda, \Lambda, \rho) > 0$ such that

$$(13.8) \qquad \kappa_0 \bar h^{1/2} \le r \le \kappa_0^{-1} \bar h^{1/2} \quad \text{and} \quad \kappa_0 E \subset S_u(y, \bar h) - y \subset \kappa_0^{-1} E,$$

where E is an ellipsoid in $\mathbb{R}^n$ defined by

(13.9) $E := \bar{h}^{1/2} A_{\bar{h}}^{-1} B_1(0)$, with $\|A_{\bar{h}}\| + \|A_{\bar{h}}^{-1}\| \le \kappa_0^{-1} |\log \bar{h}|$; $\det A_{\bar{h}} = 1$.

Let $\bar{v}(x) = v(x) - v(x_0)$. Then v satisfies

$$U^{ij} D_{ij} \bar{v} = g \quad \text{in } S_u(y, \bar{h}).$$

From (13.8) and (13.9), we have

$$B_{Cr|\log r|}(y) \supset S_u(y, \bar{h}) \supset S_u(y, \bar{h}/2) \supset B_{c\frac{r}{|\log r|}}(y) \supset B_{c\,\text{dist}^2(y,\partial\Omega)}(y),$$

for $c, C > 0$, depending only on $n, \lambda, \Lambda, \rho$. Thus, by Proposition 13.1,

$$\|\bar{v}\|_{L^\infty(S_u(y,\bar{h}))} \le C \text{diam}(S_u(y, \bar{h}))^{\frac{\alpha}{2+\alpha}} \le C(n, \lambda, \Lambda, \rho, \alpha)(r\,|\log r|)^{\frac{\alpha}{2+\alpha}}.$$

Applying the interior Hölder estimates in Theorem 12.14 to $\bar{v}$ and recalling Remark 12.15, we obtain for all $z_1, z_2 \in S_u(y, \bar{h}/2)$

$$|\bar{v}(z_1) - \bar{v}(z_2)| \le C M^\gamma \bar{h}^{-\gamma} |z_1 - z_2|^\gamma \Big\{ \|\bar{v}\|_{L^\infty(S_u(y,\bar{h}))} + \bar{h}^{1/2} \|g\|_{L^n(S_u(y,\bar{h}))} \Big\}$$

where $C(n, \lambda, \Lambda) > 0$ and $\gamma(n, \lambda, \Lambda) \in (0, 1/4]$.

By (13.8), $\bar{h}^{-\gamma} \le \kappa_0^{-2\gamma} r^{-2\gamma}$. Thus, by choosing $\beta = \min\{\gamma, \frac{\alpha}{4(2+\alpha)}\}$, we easily find $C = C(n, \lambda, \Lambda, \rho, \alpha) > 0$ such that for all r small

(13.10) $|v(z_1) - v(z_2)| \le C |z_1 - z_2|^\beta$ for all $z_1, z_2 \in S_u(y, \bar{h}/2)$.

Note that (13.10) also holds in the Euclidean ball $B_{c\,\text{dist}^2(y,\partial\Omega)}(y)$. Combining this with Proposition 13.1 gives (13.7) as in the proof of Theorem 9.5. $\square$

When replacing $a^{ij} D_{ij}$ by the linearized Monge–Ampère operator $U^{ij} D_{ij}$ of a convex function u separating quadratically from its tangent hyperplanes on the boundary, we can dispense the uniform convexity of the domains in Proposition 13.1 and Theorem 13.2. See, for instance, Theorems 13.5 and 14.33. For these, we replace $(\inf\{y_n : y \in \overline{\Omega} \cap \partial B_{\delta_2}(0)\})^{-1} x_n$, a solution to the homogeneous equation, with suitable supersolutions w_δ in Lemma 13.7.

13.2. Application: Solvability of Affine Mean Curvature and Abreu's Equations

We briefly mention here one application of the global Hölder estimates for the linearized Monge–Ampère equation in Theorem 13.2 to the Sobolev solvability of the second boundary value problem of fully nonlinear, fourth-order, geometric partial differential equations of the type

(13.11) $U^{ij} D_{ij} w = f \quad \text{in } \Omega, \qquad w = (\det D^2 u)^{\theta-1} \quad \text{in } \Omega,$

for a uniformly convex function u on a smooth and uniformly convex domain Ω in $\mathbb{R}^n$, where $0 \le \theta < 1/n$. Of particular interest are the cases $\theta = \frac{1}{n+2}$

and $\theta = 0$, where (13.11) is, respectively, the prescribed affine mean curvature equation in affine geometry and the Abreu equation in the constant curvature problem for Kähler metrics of toric varieties.

The *second boundary value problem* for (13.11) prescribes the values of u and its Hessian determinant $\det D^2 u$ on the boundary, so we are led to

$$(13.12) \qquad \begin{cases} U^{ij} D_{ij} w = f, \quad w = (\det D^2 u)^{\theta - 1} & \text{in } \Omega, \\ u = \varphi, \ w = \psi & \text{on } \partial\Omega. \end{cases}$$

When $\theta = \frac{1}{n+2}$, (13.12) was introduced by Trudinger and Wang [**TW4**] in their investigation of the affine Plateau problem in affine geometry. Existence and regularity of solutions to (13.12) in this case are the key to studying the first boundary value problem for affine maximal surface equation

$$(13.13) \qquad \begin{cases} U^{ij} D_{ij} \left[(\det D^2 u)^{-\frac{n+1}{n+2}} \right] = 0 & \text{in } \Omega, \\ u = \phi, \ Du = D\phi & \text{on } \partial\Omega. \end{cases}$$

The classical solvability of (13.13) is a major open problem. If we replace the last condition in (13.13) with a more relaxed condition by requiring that $Du(\Omega) \subset D\phi(\overline{\Omega})$, then we have a unique *weak solution* u for (13.13) as proved in [**TW4**]. This solution u is obtained as the unique maximizer of the affine area function $\mathcal{A}(\cdot, \Omega)$ (defined in (12.31)) in the set of convex functions v satisfying $v = \phi$ on $\partial\Omega$ and $Dv(\Omega) \subset D\phi(\overline{\Omega})$.

Based on the global regularity results in Theorems 10.20 and 13.2 and the Leray–Schauder degree theory (Theorem 15.17), the Sobolev solvability of (13.12) in $W^{4,p}(\Omega)$ was obtained in Le [**L3**] for $f \in L^p(\Omega)$ with $p > n$.

Theorem 13.3 (Solvability of the second boundary value problems). *Let Ω be a uniformly convex domain in $\mathbb{R}^n$ ($n \geq 2$) with boundary $\partial\Omega \in C^{3,1}$. Let $\theta \in [0, 1/n)$, $p > n$, $f \in L^p(\Omega)$, $\varphi \in W^{4,p}(\Omega)$, and $\psi \in W^{2,p}(\Omega)$ with $\inf_\Omega \psi > 0$. Then, there exists a unique uniformly convex solution $u \in W^{4,p}(\Omega)$ to the boundary value problem* (13.12). *Furthermore,*

$$\|u\|_{W^{4,p}(\Omega)} \leq C \quad \text{and} \quad \det D^2 u \geq C^{-1},$$

where C depends on $n, p, \theta, \Omega, \|f\|_{L^p(\Omega)}, \|\varphi\|_{W^{4,p}(\Omega)}, \|\psi\|_{W^{2,p}(\Omega)}$, and $\inf_\Omega \psi$.

In the special case of $\theta = 1/(n+2)$, Theorem 13.3 tells us that we can prescribe the L^p affine mean curvature (for any finite $p > n$) of the graph of a uniformly convex function with smooth Dirichlet boundary conditions on the function and its Hessian determinant. Theorem 13.3 can be viewed as an *affine* analogue of the classical Calderon–Zygmund estimates for second-order, linear, uniformly elliptic equations in Theorem 2.83; see also Problem 13.4 for an affine analogue of the classical Schauder theory.

In general, the global $W^{4,p}$ estimates in Theorem 13.3 fail when f has integrability less than the dimension n.

Remark 13.4. Consider the second boundary value problem (13.12) on $\Omega = B_1(0) \subset \mathbb{R}^n$ with $G(d) = (d^\theta - 1)/\theta$, where $\theta < 1/n$. If $n/2 < p < n$, then the function $u(x) = |x|^{1+\frac{n-p}{2(3p-n)}}$ is a solution to (13.12) with $\theta = \frac{1}{2n}(\frac{n}{p} - 1)$, φ and ψ being positive constants on $\partial\Omega$, $f \in L^p(\Omega)$ but $u \notin W^{4,p}(\Omega)$.

In addition to [**L3**], Le–Mitake–Tran [**LMT**] gave an expository account of Theorem 13.3 so we omit its proof. However, it is convenient to state it here for later comparisons with *singular Abreu equations* in Section 15.5.

13.3. Boundary Gradient Estimates

In this section and the next, we will prove the boundary Hölder gradient and Harnack estimates for the linearized Monge–Ampère equation. These estimates hold under natural assumptions on the domain, Monge–Ampère measures, and boundary data. The natural conditions we impose on the convex potential function u and the domain are similar to those in Savin's Boundary Localization Theorem (Theorem 8.5). We list them below in (13.14)–(13.17), and (13.36)–(13.38) for later reference. As remarked at the beginning of Chapter 9, when the hypotheses of Theorem 8.5 are satisfied at $x_0 \in \partial\Omega$, the convex function u is differentiable at x_0, so we will use the classical gradient $Du(x_0)$ when stating the quadratic separation condition.

Structural hypotheses for boundary gradient estimates. Let $\Omega \subset \mathbb{R}^n$ be a bounded convex domain satisfying

$$(13.14) \qquad B_\rho(\rho e_n) \subset \Omega \subset \{x_n \geq 0\} \cap B_{1/\rho}(0),$$

for some small $\rho > 0$. Assume that, at each $y \in \partial\Omega \cap B_\rho(0)$,

$$(13.15) \qquad \text{there is an interior ball } B_\rho(z) \subset \Omega \text{ such that } y \in \partial B_\rho(z).$$

Let $u \in C^{0,1}(\overline{\Omega}) \cap C^2(\Omega)$ be a convex function satisfying

$$(13.16) \qquad 0 < \lambda \leq \det D^2 u \leq \Lambda \quad \text{in } \Omega.$$

We assume that on $\partial\Omega \cap B_\rho(0)$, u separates quadratically from its tangent hyperplanes on $\partial\Omega$; that is, if $x_0 \in \partial\Omega \cap B_\rho(0)$, then for all $x \in \partial\Omega$,

$$(13.17) \quad \rho\,|x - x_0|^2 \leq u(x) - u(x_0) - Du(x_0) \cdot (x - x_0) \leq \rho^{-1}\,|x - x_0|^2.$$

Throughout, $U = (U^{ij})$ denotes the cofactor matrix of $D^2 u$.

We first establish the boundary gradient estimates for solutions to (13.1).

Theorem 13.5 (Boundary gradient estimates). *Assume u and Ω satisfy the assumptions* (13.14)–(13.17). *Assume that $g/\operatorname{trace}(U) \in L^\infty(\Omega \cap B_\rho(0))$. Let $v \in C(B_\rho(0) \cap \overline{\Omega}) \cap W^{2,n}_{\mathrm{loc}}(B_\rho(0) \cap \Omega)$ be a solution to*

$$U^{ij} D_{ij} v = g \quad in \ B_\rho(0) \cap \Omega, \qquad v = 0 \quad on \ \partial\Omega \cap B_\rho(0).$$

Then, there is a constant $C(n, \rho, \lambda, \Lambda)$ such that, in $\Omega \cap B_{\rho/2}(0)$, we have

$$|v(x)| \le C(\|v\|_{L^\infty(\Omega \cap B_\rho(0))} + \|g/\operatorname{trace}(U)\|_{L^\infty(\Omega \cap B_\rho(0))})\operatorname{dist}(x, \partial\Omega).$$

Since $v = 0$ on $\partial\Omega \cap B_\rho(0)$, its tangential derivatives are 0, while $|v(x)/\operatorname{dist}(x, \partial\Omega)|$ for x close to $x_0 \in \partial\Omega$ represents the rate of change of v in the normal direction to $\partial\Omega$ at x_0. Thus, bounds on $|v(x)/\operatorname{dist}(x, \partial\Omega)|$ give estimates for the gradient Dv on the boundary whenever it exists. This will be established later in Theorem 13.8.

The proof of Theorem 13.5 is based on the ABP maximum principle using subsolutions and supersolutions for the linearized Monge–Ampère operator $U^{ij} D_{ij}$ that grow linearly in the distance from the boundary. For comparison, see the second mixed derivative estimates at the boundary for the Monge–Ampère equation in Theorem 4.9.

The following lemma constructs subsolutions.

Lemma 13.6 (Subsolution). *Let $u \in C^2(\Omega)$ be a convex function satisfying $\lambda \le \det D^2 u \le \Lambda$ in a domain Ω in $\mathbb{R}^n$. If $a \in \mathbb{R}$, $0 < \bar\delta \le \min\{1, \Lambda\}$, and*

$$w := a x_n - u + \bar\delta |x'|^2 + \Lambda^n (\lambda \bar\delta)^{1-n} x_n^2,$$

then

$$L_u w := U^{ij} D_{ij} w \ge \max\{\bar\delta \ \operatorname{trace}(U), n\Lambda\} \quad in \ \Omega.$$

Proof. Consider the quadratic polynomial $P(x) = [\bar\delta |x'|^2 + \Lambda^n (\lambda \bar\delta)^{1-n} x_n^2]/2$. Then $\det D^2 P = \Lambda^n/\lambda^{n-1}$. Using Lemma 2.57, we get

$$L_u P = U^{ij} D_{ij} P \ge n(\det U \det D^2 P)^{1/n}$$
$$= n\big[(\det D^2 u)^{n-1} \Lambda^n/\lambda^{n-1}\big]^{1/n} \ge n\Lambda.$$

Since $\bar\delta \le \min\{1, \Lambda\}$ and $\Lambda \ge \lambda$, we have $D^2 P \ge \bar\delta I_n$. It follows that

$$L_u P = U^{ij} D_{ij} P \ge \bar\delta \ \operatorname{trace}(U).$$

Using $L_u x_n = 0$ and $L_u u = U^{ij} D_{ij} u = n \det D^2 u \le n\Lambda$, we find

$$L_u w = (-L_u u + L_u P) + L_u P \ge L_u P \ge \max\{\bar\delta \ \operatorname{trace}(U), n\Lambda\}.$$

The lemma is proved. $\qquad\square$

Now, we construct supersolutions with certain prescribed boundary behaviors. Assume that (13.14)–(13.17) hold. Recalling from (8.38) that, for all $x_0 \in \partial\Omega \cap B_\rho(0)$ and all $x \in \overline{\Omega} \cap B_r(x_0)$ where $r \leq \delta(n, \lambda, \Lambda, \rho)$, we have

$$\text{(13.18)} \qquad \begin{aligned} |x - x_0|^3 &\leq u(x) - u(x_0) - Du(x_0) \cdot (x - x_0) \\ &\leq C\,|x - x_0|^2\,(\log|x - x_0|)^2. \end{aligned}$$

Lemma 13.7 (Supersolution). *Assume that u and Ω satisfy the assumptions* (13.14)–(13.17) *with $u(0) = 0$ and $Du(0) = 0$. Given $0 < \delta \leq \delta_0 \leq \min\{1, \rho, \Lambda\}$ where $\delta_0(n, \lambda, \Lambda, \rho)$ is small, define*

$$\tilde{\delta} := \delta^3/2 \quad \text{and} \quad M_\delta := \Lambda^n (\lambda\tilde{\delta})^{1-n}.$$

Then, the function

$$w_\delta(x', x_n) := M_\delta x_n + u(x) - \tilde{\delta}|x'|^2 - \Lambda^n(\lambda\tilde{\delta})^{1-n} x_n^2 \quad \text{for } (x', x_n) \in \overline{\Omega}$$

satisfies

$$\text{(13.19)} \qquad U^{ij} D_{ij} w_\delta \leq -\max\left\{\tilde{\delta}\,\mathrm{trace}(U), n\Lambda\right\} \quad \text{in } \Omega$$

and

$$w_\delta \geq 0 \text{ on } \partial(\Omega \cap B_\delta(0)), \ w_\delta \geq \tilde{\delta} \text{ on } \Omega \cap \partial B_\delta(0), \text{ and } w_\delta \geq 0 \text{ in } \Omega \cap B_\delta(0).$$

Proof. In view of Lemma 13.6, w_δ satisfies (13.19) when $\delta \leq \min\{1, \rho, \Lambda\}$. Now, we verify the inequalities for w_δ on $\partial(\Omega \cap B_\delta(0))$ and on $\Omega \cap \partial B_\delta(0)$, because then the inequality $w_\delta \geq 0$ in $\Omega \cap B_\delta(0)$ follows from (13.19) and the maximum principle. Clearly, $M_\delta x_n - \Lambda^n(\lambda\tilde{\delta})^{1-n} x_n^2 \geq 0$ on $\overline{\Omega} \cap B_\delta(0)$. Recall that $u(0) = 0$ and $Du(0) = 0$. Using (13.17), we find

$$w_\delta(x) \geq u(x) - \tilde{\delta}|x'|^2 \geq 0 \quad \text{on } \partial\Omega \cap B_\delta(0).$$

On $\Omega \cap \partial B_\delta(0)$, we deduce from (13.18) that

$$w_\delta(x) \geq u(x) - \tilde{\delta}|x'|^2 \geq |x|^3 - \tilde{\delta}\,|x|^2 \geq |x|^3/2 = \tilde{\delta},$$

provided $\delta \leq \delta_0$ where $\delta_0 = \delta_0(n, \lambda, \Lambda, \rho)$ is small. The lemma is proved. $\square$

We are now ready to prove the boundary gradient estimates.

Proof of Theorem 13.5. We can assume that $u(0) = 0$ and $Du(0) = 0$. Let w_δ, M_δ, and $\tilde{\delta}$, where $\delta = \delta_0$, be as in Lemma 13.7. By considering $\bar{v} := \tilde{\delta}K^{-1}v$ and $\bar{g} := \tilde{\delta}K^{-1}g$, instead of v and g, where

$$K = \|v\|_{L^\infty(\Omega \cap B_\rho(0))} + \|g/\mathrm{trace}\,(U)\|_{L^\infty(\Omega \cap B_\rho(0))},$$

we may assume that $\|v\|_{L^\infty(\Omega \cap B_\rho(0))} \leq \tilde{\delta}$ and $\|g/\mathrm{trace}\,(U)\|_{L^\infty(\Omega \cap B_\rho(0))} \leq \tilde{\delta}$. We need to show that

$$|v| \leq C(n, \rho, \lambda, \Lambda)\mathrm{dist}(\cdot, \partial\Omega) \quad \text{in } \Omega \cap B_{\rho/2}(0).$$

Since $v \leq w_\delta$ on $\partial(\Omega \cap B_\delta(0))$ and

$$L_u v = g \geq -\tilde{\delta} \, \text{trace}(U) \geq L_u w_\delta \quad \text{in } \Omega \cap B_\delta(0),$$

we obtain from the ABP maximum principle in Theorem 3.16 that $v \leq w_\delta$ in $\Omega \cap B_\delta(0)$. Hence, if $0 \leq x_n \leq \delta = \delta_0$, then using (13.18), we get

$$v(0, x_n) \leq w_\delta(0, x_n) \leq M_\delta x_n + u(0, x_n) \leq M_\delta x_n + C|x_n|^2 |\log x_n|^2 \leq C x_n.$$

By increasing $C = C(n, \lambda, \Lambda, \rho)$, we see that the above estimate holds for all $0 \leq x_n \leq \rho/2$. The same argument applies at all points $x_0 \in \partial\Omega \cap B_{\rho/2}(0)$, and we obtain the following upper bound for v:

$$v(x) \leq C(n, \rho, \lambda, \Lambda) \text{dist}(x, \partial\Omega) \quad \text{in } \Omega \cap B_{\rho/2}(0).$$

The lower bound for v follows similarly, and the theorem is proved. $\qquad \square$

13.4. Boundary Hölder Gradient Estimates

By establishing effective improvement of bounds for $v/\text{dist}(\cdot, \partial\Omega)$ in smaller sections of u, we can show that the boundary gradient of the function v in Theorem 13.5 exists and that it is in fact Hölder continuous. We will prove the following boundary Hölder gradient estimates, due to Le and Savin [**LS1**], for the linearized Monge–Ampère equation with bounded right-hand side.

Theorem 13.8 (Boundary Hölder gradient estimates). *Assume u and Ω satisfy the assumptions (13.14)–(13.17). Let $g/\text{trace}(U) \in L^\infty(\Omega \cap B_\rho(0))$. Let $v \in C(B_\rho(0) \cap \overline{\Omega}) \cap W^{2,n}_{\text{loc}}(B_\rho(0) \cap \Omega)$ be a solution to*

$$\begin{cases} U^{ij} D_{ij} v = g & in \ B_\rho(0) \cap \Omega, \\ \qquad\quad v = 0 & on \ \partial\Omega \cap B_\rho(0). \end{cases}$$

Let ν be the outer unit normal to $\partial\Omega$, and let

$$\mathbf{K} := \|v\|_{L^\infty(\Omega \cap B_\rho(0))} + \|g/\text{trace}(U)\|_{L^\infty(\Omega \cap B_\rho(0))}.$$

Then, there exist constants $\alpha \in (0, 1)$ and C depending only on $n, \rho, \lambda, \Lambda$, such that the following assertions hold.

(i) *The normal derivative $D_\nu v := Dv \cdot \nu$ exists on $\partial\Omega \cap B_{\rho/2}(0)$ with estimate*

$$\|D_\nu v\|_{C^{0,\alpha}(\partial\Omega \cap B_{\rho/2}(0))} \leq C\mathbf{K}.$$

(ii) *For $h \leq C^{-1}$, we have the estimate*

$$\sup_{S_u(0,h)} |v + D_\nu v(0) x_n| \leq C h^{\frac{1+\alpha}{2}} \mathbf{K}.$$

(iii) *For $r \leq \rho/2$, we have the estimate*

$$\sup_{B_r(0) \cap \overline{\Omega}} |v + D_\nu v(0) x_n| \leq C r^{1+\alpha} \mathbf{K}.$$

Note that the conclusions of Theorem 13.8 still hold for the equations

$$\text{trace}\,(AD^2 v) = g, \quad \text{where } 0 < \tilde{\lambda} U \le A \le \tilde{\Lambda} U,$$

where now the constants α, C depend also on $\tilde{\lambda}$, $\tilde{\Lambda}$; see Problem 13.5. Theorem 13.8 is an affine invariant analogue of Krylov's boundary Hölder gradient estimates [**K3**] for linear, uniformly elliptic equations in nondivergence form.

Theorem 13.9 (Krylov's boundary Hölder gradient estimates). *Let $B_1^+ = B_1(0) \cap \{x \in \mathbb{R}^n : x_n > 0\}$. Assume the measurable, symmetric matrix (a^{ij}) satisfies $\lambda I_n \le (a^{ij}(x)) \le \Lambda I_n$ on B_1^+. Let $w \in C(\overline{B_1^+}) \cap W_{\text{loc}}^{2,n}(B_1^+)$ satisfy*

$$a^{ij} D_{ij} w = f \quad \text{in } B_1^+, \qquad w = 0 \quad \text{on } \{x_n = 0\}.$$

Then, there are constants $\alpha(n, \lambda, \Lambda) \in (0, 1)$ and $C(n, \lambda, \Lambda) > 0$ such that

$$\|D_n w\|_{C^\alpha(B_{1/2} \cap \{x_n = 0\})} \le C(\|w\|_{L^\infty(B_1^+)} + \|f\|_{L^\infty(B_1^+)}).$$

The proof of Theorem 13.8 will be completed in the rest of this section. The rough idea is as follows. By Theorem 13.5, we have

$$a \, \text{dist}(\cdot, \partial\Omega) \le v \le b \, \text{dist}(\cdot, \partial\Omega),$$

for some structural constants a and b. Now, the nonnegative functions

$$v_1 = \frac{v - a\,\text{dist}(\cdot, \partial\Omega)}{b - a}, \quad v_2 = \frac{b\,\text{dist}(\cdot, \partial\Omega) - v}{b - a}$$

satisfy $v_1 + v_2 = \text{dist}(\cdot, \partial\Omega)$, so in a ball $B_\delta(y)$ with center in the set $\Omega_\delta := \{x \in \Omega : \text{dist}(x, \partial\Omega) = \delta\}$, one of these functions, say v_1, is at least $\delta/2$ in a positive fraction of the ball. We wish to apply the weak Harnack inequality in Theorem 12.11 to conclude that in this ball, v_1 is bounded below by a structural constant. This in turn gives an effective improvement of the bounds $a \, \text{dist}(\cdot, \partial\Omega) \le v \le b \, \text{dist}(\cdot, \partial\Omega)$ in $B_\delta(y)$. For Theorem 12.11 to be applicable, we need to verify that v_1 (or a suitable perturbation thereof) is in fact a supersolution of the linearized operator $U^{ij} D_{ij}$. This calls for studying $U^{ij} D_{ij}[\text{dist}(\cdot, \partial\Omega)]$. This quantity is small compared to $\text{trace}(U)$ if $\partial\Omega$ is sufficiently flat. For this aim, we study our equation under boundary rescalings of u and the domain Ω using the Boundary Localization Theorem (Theorem 8.5), as in Lemma 9.3, and then we obtain the improvements of the bound for $v/\text{dist}(x, \partial\Omega)$ via constructions of suitable subsolutions.

Fix ρ, λ, Λ, and n. In this section, small and large positive constants depend only on these structural parameters.

13.4.1. The class $\mathcal{D}_\sigma$ and the rescaled Monge–Ampère equations. Motivated by the properties of the rescaled functions u_h and domains Ω_h in Lemma 9.3, we introduce the class $\mathcal{D}_\sigma$ that captures the behaviors of u_h and $S_{u_h}(0, 1)$. In this situation, σ essentially represents the flatness of the boundary of $S_{u_h}(0, 1)$ near the origin, so $\sigma = C h^{1/2}$.

Definition 13.10. For $\sigma > 0$, let $\mathcal{D}_\sigma$ be the class consisting of pairs of function u and convex domain Ω satisfying the following conditions for some positive constants κ, c_0, C_0 depending on ρ, λ, Λ, and n:

(a) $0 \in \partial\Omega$, $\quad \Omega \subset B_{1/\kappa}^+(0) \subset \mathbb{R}^n$, $\quad |\Omega| \geq c_0$;

(b) $u \in C(\overline{\Omega})$ is a convex function satisfying
$$u(0) = 0, \quad Du(0) = 0, \quad \lambda \leq \det D^2 u \leq \Lambda;$$

(c) $\partial\Omega \cap \{u < 1\} \subset G \subset \{x_n \leq \sigma\}$ where G is a graph in the e_n direction which is defined in $B_{2/\kappa}(0)$, and its $C^{1,1}$ norm is bounded by σ;

(d) $(\rho/4)|x - x_0|^2 \leq u(x) - u(x_0) - Du(x_0) \cdot (x - x_0) \leq (4/\rho)|x - x_0|^2$ for all $x, x_0 \in G \cap \partial\Omega$;

(e) $|Du| \leq C_0 r |\log r|^2$ in $\overline{\Omega} \cap B_r(0)$ if $r \leq c_0$.

Next, we introduce suitable rescalings for our linearized Monge–Ampère equation using the Boundary Localization Theorem. Assume that Ω and u satisfy (13.14)–(13.17). Let us focus on a boundary point $x_0 \in \partial\Omega \cap B_\rho(0)$. As in Section 9.1, we can assume that $x_0 = 0$ and that u and Ω satisfy the hypotheses of Theorem 8.5 at the origin with $\mu = \rho$. This means that

$$(13.20) \quad \begin{cases} B_\rho(\rho e_n) \subset \Omega \subset \{x_n \geq 0\} \cap B_{1/\rho}(0), \\ \lambda \leq \det D^2 u \leq \Lambda \quad \text{in } \Omega, \quad u \geq 0, u(0) = 0, \quad Du(0) = 0 \\ \rho|x|^2 \leq u(x) \leq \rho^{-1}|x|^2 \quad \text{on } \partial\Omega \cap \{x_n \leq \rho\}. \end{cases}$$

By Theorem 8.5, we know that for all $h \leq \kappa(n, \lambda, \Lambda, \rho)$, $S_u(0, h)$ satisfies

$$(13.21) \qquad \kappa E_h \cap \overline{\Omega} \subset S_u(0, h) \subset \kappa^{-1} E_h, \quad E_h = A_h^{-1} B_{h^{1/2}}(0),$$

with $A_h \in \mathbb{M}^{n \times n}$ being a matrix with the following properties:

(13.22) $\det A_h = 1$, $A_h x = x - \tau_h x_n$, $\tau_h \cdot e_n = 0$, $\|A_h^{-1}\| + \|A_h\| \leq \kappa^{-1}|\log h|$.

Note that (13.22) implies the existence of $c_2(n, \rho, \lambda, \Lambda) > 0$ such that

(13.23) $\overline{\Omega} \cap B_{c_2 h^{1/2}/|\log h|}^+(0) \subset S_u(0, h) \subset B_{c_2^{-1} h^{1/2}|\log h|}^+(0) \quad$ for all $h \leq \kappa$.

We use the rescaling (12.3) with $a = h^{-1}$, $b = h^{-1/2}$, $T = h^{1/2} A_h^{-1}$.

Denote the rescaled functions and domains, as in Lemma 9.3, by

$$(13.24) \quad \begin{cases} u_h(x) := h^{-1} u(Tx), \; v_h(x) := h^{-1/2} v(Tx), \; g_h(x) := h^{1/2} g(Tx), \\ \text{for } x \in \Omega_h := h^{-1/2} A_h \Omega. \end{cases}$$

Then

(13.25) $\lambda \leq \det D^2 u_h \leq \Lambda$ in Ω_h, $B_\kappa(0) \cap \overline{\Omega_h} \subset S_{u_h}(0, 1) \subset B_{1/\kappa}(0) \cap \overline{\Omega_h}$.

By (12.4), we have

(13.26) $$U_h^{ij} D_{ij} v_h = g_h \quad \text{on } \Omega_h.$$

Since

$$\text{trace}\, U = h^{-1} \text{trace}(T U_h T^t) \leq h^{-1} \|T\|^2 \text{trace}\, U_h \leq \kappa^{-2} \|\log h\|^2 \text{trace}\, U_h,$$

we find

(13.27)
$$\|g_h / \text{trace}\, U_h\|_{L^\infty(\Omega_h \cap h^{-1/2} A_h B_\rho(0))}$$
$$\leq \kappa^{-2} h^{1/2} |\log h|^2 \, \|g / \text{trace}\, U\|_{L^\infty(\Omega \cap B_\rho(0))}.$$

Remark 13.11. If $(u, \Omega) \in \mathcal{D}_\sigma$ where $\sigma \leq c$ where $c = c(n, \rho, \lambda, \Lambda)$ is small and if $x_0 \in \partial\Omega \cap B_{\delta_0}(0)$ where $\delta_0(n, \rho, \lambda, \Lambda)$ is small, then, as in the proof of (9.20), we have $S_u(x_0, 1/2) \subset \{u < 1\}$, and u satisfies in $S_u(x_0, 1/2)$ the hypotheses of the Boundary Localization Theorem (Theorem 8.5) at x_0 for some $\tilde{\rho}(n, \rho, \lambda, \Lambda) > 0$.

Now we show that if $(u, \Omega) \in \mathcal{D}_\delta$, then u has a uniform modulus of convexity on a set whose distance from the boundary is comparable to δ. The exponent $\frac{1}{6(n-1)}$ is motivated by the computations in the proof of Lemma 13.13.

Lemma 13.12. Let $(u, \Omega) \in \mathcal{D}_\delta$. If $\delta \leq c$, where $c = c(n, \rho, \lambda, \Lambda)$ is small, then for any $y \in F_\delta := \{(x', x_n) \in \mathbb{R}^n : |x'| \leq \delta^{\frac{1}{6(n-1)}}, \ x_n = 2\delta\}$, we have $S_u(y, c\delta^2) \subset \Omega$.

Proof. Let δ_0 be as in Remark 13.11. We first choose δ small, $\delta \leq \delta_0$, so that $\text{dist}(y, \partial\Omega) \geq \delta$ for all $y \in F_\delta$; see the proof of Lemma 9.1(ii).

Arguing as in the proof of Lemma 9.3(v) and using conditions (d)–(e) in the definition of $\mathcal{D}_\delta$ instead of assertions (i) and (iv) in Lemma 9.3, we find that when δ is small, the maximal interior section $S_u(y, \bar{h})$ of u centered at y must be tangent to $\partial\Omega$ at a point $x_0 \in \partial\Omega \cap B_{\delta_0}(0)$. Now we can apply Proposition 9.2 at x_0 and obtain a small constant $c(n, \rho, \lambda, \Lambda) > 0$ such that $\bar{h} \geq c[\text{dist}(y, \partial\Omega)]^2 \geq c\delta^2$. Consequently, $S_u(y, c\delta^2) \subset \Omega$. $\qquad\square$

13.4.2. Construction of subsolutions. The next lemma constructs a useful subsolution for the linearized Monge–Ampère operator L_u when (u, Ω) belongs to $\mathcal{D}_\delta$ with small δ.

Lemma 13.13 (Subsolution for the class $\mathcal{D}_\delta$). *Suppose* $(u, \Omega) \in \mathcal{D}_\delta$. *Let* $D := \{x_n < 2\delta\} \cap \Omega$ *and*

(13.28) $$F_\delta := \{(x', x_n) \in \mathbb{R}^n : |x'| \leq \delta^{\frac{1}{6(n-1)}}, \ x_n = 2\delta\}.$$

If $\delta \leq \delta_1$, *where* $\delta_1 = \delta_1(n, \rho, \lambda, \Lambda)$ *is small, then the function*

$$\underline{w}(x) := x_n - u(x) + \delta^{\frac{1}{n-1}} |x'|^2 + \Lambda^n (\lambda \delta^{\frac{1}{n-1}})^{1-n} x_n^2$$

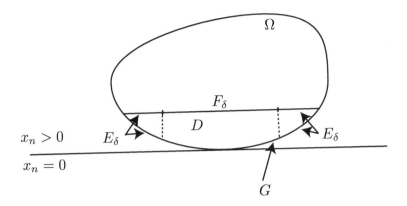

Figure 13.2. The sets D, E_δ, F_δ, and G.

satisfies

(13.29) $L_u(\underline{w}) := U^{ij} D_{ij}\underline{w} \geq \delta^{\frac{1}{n-1}} \text{ trace}(U) \quad on \; \Omega,$

and on the boundary of the domain D, we have

$$\underline{w} \leq 0 \quad on \; \partial D \backslash F_\delta, \qquad \underline{w} \leq 1 \quad on \; F_\delta.$$

Proof. We assume $\delta^{\frac{1}{n-1}} \leq \min\{1, \Lambda\}$. By Lemma 13.6, the function $\underline{w}$ so defined satisfies (13.29). Next we check the behavior of $\underline{w}$ on ∂D. For this, first note that (u, Ω) satisfies the hypotheses of the Localization Theorem (Theorem 8.5) at 0. Thus, if δ is small, we can use (13.18) to get

$$u(x) \geq |x|^3 \quad \text{for } x \in \Omega \cap B_c(0), \quad c = c(n, \lambda, \Lambda, \rho).$$

Now, observe that $\partial D \setminus F_\delta \subset G \cup E_\delta$ (see Figure 13.2) where

$$E_\delta := \partial D \cap \{|x'| \geq \delta^{\frac{1}{6(n-1)}}\}.$$

On $G \cap \partial\Omega$, by conditions (d) and (c) in the definition of $\mathcal{D}_\delta$, we have $u(x) \geq (\rho/4)|x|^2$ and $x_n \leq \delta|x'|^2$. Then $x_n^2 \leq \delta^2|x'|^4 \leq \kappa^{-2}\delta^2|x'|^2$, and

$$\underline{w}(x) \leq (\delta + \delta^{\frac{1}{n-1}} + C(n, \rho, \lambda, \Lambda)\delta)|x'|^2 - (\rho/4)|x|^2 \leq 0 \quad on \; G \cap \partial\Omega,$$

provided that δ is small.

On E_δ, we have $u(x) \geq |x|^3 \geq \delta^{\frac{1}{2(n-1)}}$; hence, for small δ,

$$\underline{w}(x) \leq 2\delta - \delta^{\frac{1}{2(n-1)}} + \kappa^{-2}\delta^{\frac{1}{n-1}} + 4\Lambda^n\lambda^{1-n}\delta \leq -\delta^{\frac{1}{2(n-1)}}/2 < 0.$$

On F_δ, we clearly have $\underline{w} \leq 1$ for small δ. The lemma is proved. $\qquad \square$

Remark 13.14. For any point $x_0 \in \partial\Omega$ close to the origin and x near x_0, let z denote the coordinates of the point x in a system of coordinates centered at x_0 with the z_n-axis perpendicular to $\partial\Omega$. We can construct the corresponding subsolution

$$\underline{w}_{x_0}(z) = z_n - [u(x_0 + z) - u(x_0) - Du(x_0) \cdot z] + \delta^{\frac{1}{n-1}}|z'|^2 + \frac{\Lambda^n}{\lambda^{n-1}\delta}z_n^2.$$

From the proof of Lemma 13.13, we see that $\underline{w}_{x_0}$ satisfies the same conclusion of Lemma 13.13 if $|x_0| \leq c\delta$ for some small constant $c(n, \rho, \lambda, \Lambda)$.

Remark 13.15. Some remarks on constants are in order.

- For later reference, we let $\delta_* = \min\{\delta_1, c\}$ where δ_1 is from Lemma 13.13 and c is from Lemma 13.12. If $\delta \leq \delta_*$ and $(u, \Omega) \in \mathcal{D}_\delta$, then Lemmas 13.13 and 13.12 apply.
- Assume $(u, \Omega) \in \mathcal{D}_{\delta_*}$. Let $y \in F_{\delta_*}$. Then, the section $S_u(y, c\delta_*^2/2)$ is contained in $\Omega \subset B_{1/\kappa}(0)$; thus, by Lemma 5.6(iv), it contains a ball $B_{\bar{\delta}}(y)$ for some $\bar{\delta} \leq \delta_*$, where $\bar{\delta}(n, \rho, \lambda, \Lambda)$ is small.

As a consequence of Lemma 13.13 and the maximum principle, we obtain a boundary analogue of the doubling estimate in Lemma 12.8 for supersolutions. It roughly says that if a nonnegative supersolution is at least 1 at a fixed distance from the boundary, then it grows at least linearly from there. This result is crucial for establishing an effective improvement of the gradient bounds in Proposition 13.17.

Proposition 13.16. *Assume $(u, \Omega) \in \mathcal{D}_\delta$ with $\delta \leq \delta_*$ where δ_* is as in Remark 13.15. Let F_δ be as in (13.28). Let $v \in W^{2,n}_{\text{loc}}(\Omega) \cap C(\overline{\Omega})$ be a nonnegative function satisfying*

$$L_u v \leq \delta^{\frac{1}{n-1}} \operatorname{trace}(U) \quad \text{in } \Omega \quad \text{and} \quad v \geq 1 \quad \text{in } F_\delta.$$

Then, for some small $\theta = \theta(n, \rho, \lambda, \Lambda) < 1/64$, we have

$$v \geq \frac{1}{2}\operatorname{dist}(\cdot, G) \quad \text{in } S_u(0, \theta).$$

Proof. Let D and $\underline{w}$ be as in Lemma 13.13. This lemma and the ABP maximum principle for the operator L_u imply that $v \geq \underline{w}$ in D.

By Remark 13.11, we can apply (13.18) at all points $x_0 \in \partial\Omega$ if $|x_0|$ is sufficiently small. In particular, at $x_0 = 0$, we have for $0 \leq x_n \leq c$,

$$v(0, x_n) \geq \underline{w}(0, x_n) \geq x_n - u(0, x_n) \geq x_n - Cx_n^2|\log x_n|^2 \geq x_n/2$$

if $c = c(n, \lambda, \Lambda, \rho)$ is small.

If $x_0 \in \partial\Omega$ and $|x_0|$ is sufficiently small, then, by comparing v with the corresponding subsolution $\underline{w}_{x_0}$ as given in Remark 13.14, we obtain

$$v \geq \frac{1}{2}\operatorname{dist}(\cdot, G) \quad \text{in } \overline{\Omega} \cap B_c(0).$$

Due to (13.23), $\overline{\Omega} \cap B_c(0) \supset S_u(0, \theta)$ if θ is small, and the lemma follows. $\quad\square$

The following proposition is the basis of our iteration argument.

Proposition 13.17 (Effective improvement of gradient bounds). *Assume* $(u, \Omega) \in \mathcal{D}_\sigma$ *where* $\sigma \leq \delta_*$ *with* δ_* *as in Remark 13.15. Suppose* $v \in W^{2,n}_{\text{loc}}(\Omega) \cap C(\overline{\Omega})$ *satisfies*

$$L_u v = g, \qquad a \operatorname{dist}(\cdot, G) \leq v \leq b \operatorname{dist}(\cdot, G) \quad in \ \Omega,$$

for some constants $a, b \in [-1, 1]$. *Let* θ *be as in Proposition 13.16. There exist* $c_1(n, \rho, \lambda, \Lambda)$ *small and* $\eta(n, \rho, \lambda, \Lambda) \in (0, 1)$ *close to 1, such that if*

$$\max\left\{\sigma, \|g/\operatorname{trace}(U)\|_{L^\infty(\Omega)}\right\} \leq c_1(b - a),$$

then

$$a' \operatorname{dist}(\cdot, G) \leq v \leq b' \operatorname{dist}(\cdot, G) \quad in \ S_u(0, \theta),$$

for some constants a', b' *that satisfy*

$$a \leq a' \leq b' \leq b \quad and \quad b' - a' \leq \eta(b - a).$$

Proof. To simplify, denote $\delta = \delta_*$. Consider nonnegative functions

$$v_1 = \frac{v - a \operatorname{dist}(\cdot, G)}{b - a}, \qquad v_2 = \frac{b \operatorname{dist}(\cdot, G) - v}{b - a}.$$

Let F_δ be as in (13.28), and let $K_1 > 4$ be as in Theorem 12.11. By Lemma 13.12, if $y \in F_\delta$, then $S_u(y, K_1 h) \subset \Omega$, where $h \geq c_0(n, \rho, \lambda, \Lambda)\delta^2$, so from Remark 13.15, we have $B_{\bar{\delta}}(y) \subset S_u(y, h)$. By condition (c) in the definition of $\mathcal{D}_\sigma$, we have $\|D^2 \operatorname{dist}(\cdot, G)\| \leq \sigma$.

Fix $y \in F_\delta$. Since $v_1 + v_2 = \operatorname{dist}(\cdot, G)$, we might assume that v_1 satisfies

$$|\{v_1 \geq \delta/2\} \cap B_{\bar{\delta}}(y)| \geq (1/2)|B_{\bar{\delta}}(y)|.$$

Thus, by the volume estimates in Lemma 5.6,

$$|\{v_1 \geq \delta/2\} \cap S_u(y, h)| \geq (1/2)|B_{\bar{\delta}}(y)| \geq \mu(n, \rho, \lambda, \Lambda)|S_u(y, h)|.$$

Next, we apply Theorem 12.11 to the function $\tilde{v}_1(x) := v_1(x) + c_1(\kappa^{-2} - |x|^2)$. Notice that $\tilde{v}_1 \geq v_1 \geq 0$ in Ω, and, since $\|D^2 \operatorname{dist}(\cdot, G)\| \leq \sigma$,

$$L_u \tilde{v}_1 \leq (g + \sigma \operatorname{trace} U)(b - a)^{-1} - 2c_1 \operatorname{trace} U \leq 0 \quad in \ \Omega.$$

Therefore, the weak Harnack inequality in Theorem 12.11 gives

$$\tilde{v}_1 \geq c(n, \rho, \lambda, \Lambda) > 0 \quad in \ S_u(y, h).$$

With this estimate and recalling that $B_{\bar{\delta}}(y) \subset S_u(y, h)$, we can use Lemma 13.12 and Theorem 12.11 a finite number of times and obtain

$$\tilde{v}_1 \geq 2c_2 > 0 \quad on \ F_\delta,$$

for some $c_2(n, \rho, \lambda, \Lambda)$. By choosing c_1 small, we find $v_1 \geq c_2$ on F_δ. Since

$$L_u(v_1/c_2) \leq 2(c_1/c_2) \operatorname{trace} U \leq \delta^{\frac{1}{n-1}} \operatorname{trace} U,$$

provided that c_1 is small, we can apply Proposition 13.16 to v_1/c_2 and obtain

$$v_1 \geq (c_2/2) \operatorname{dist}(\cdot, G) \quad \text{in } S_u(0, \theta);$$

hence

$$v \geq (a + c_2(b-a)/2)\operatorname{dist}(\cdot, G) \quad \text{in } S_u(0, \theta).$$

The proposition follows with $a' = a + c_2(b-a)/2$, $b' = b$, and $\eta = 1 - c_2/2$. $\quad\square$

13.4.3. Proof of the boundary Hölder gradient estimates. With all ingredients in place, we can now prove the boundary Hölder gradient estimates for the linearized Monge–Ampère equation.

Proof of Theorem 13.8. We can assume for simplicity that $u(0) = 0$ and $Du(0) = 0$. Let $C_* = C$ where C is as Theorem 13.5. As in the proof of Theorem 13.5, we can divide v and g by a suitable constant so as to have

$$\mathbf{K} := \|v\|_{L^\infty(\Omega \cap B_\rho(0))} + \|g/\operatorname{trace} U\|_{L^\infty(\Omega \cap B_\rho(0))} \leq 1/(2C_*).$$

Then, by Theorem 13.5,

$$|v| \leq (1/2)\operatorname{dist}(\cdot, \partial\Omega) \quad \text{in } \overline{\Omega} \cap B_{\rho/2}(0).$$

To simply, we denote $d_\Gamma(x) = \operatorname{dist}(x, \Gamma)$.

Step 1. Effective improvement of bounds for $v/d_{\partial\Omega}$ in smaller sections. Let us focus our attention on sections centered at the origin and show that we can improve these bounds in the form

$$(13.30) \qquad a_h d_{\partial\Omega} \leq v \leq b_h d_{\partial\Omega} \quad \text{in } S_u(0, h),$$

for appropriate constants a_h, b_h.

Let θ, η be as in Proposition 13.17. Let δ_* be as in Remark 13.15.

First, we fix h_0 small, depending only on $n, \rho, \lambda, \Lambda$, and let

$$a_{h_0} = -1/2, \quad b_{h_0} = 1/2, \quad \tilde{\eta} = (1 + \eta)/2.$$

Then, we show by induction in $k \in \mathbb{N}_0$ the following:

Claim. For all $h = h_0 \theta^k$, we can find a_h increasing and b_h decreasing with k such that (13.30) holds and

$$(13.31) \qquad b_h - a_h = \tilde{\eta}^k \geq C_1 h^{1/2} |\log h|^2,$$

for some large constant $C_1(n, \rho, \lambda, \Lambda)$ to be determined later.

We notice that the claim holds for $k = 0$ if h_0 is chosen sufficiently small with $h_0 \leq \theta$. Note that $\tilde{\eta} = (1+\eta)/2 > 1/2 \geq 4\theta^{1/2}$. Thus, if

$$\tilde{\eta}^k \geq C_1 h^{1/2} |\log h|^2,$$

then

$$\tilde{\eta}^{k+1} \geq 4C_1 \theta^{1/2} h^{1/2} |\log h|^2 \geq C_1 (\theta h)^{1/2} |\log(\theta h)|^2.$$

Assume the claim holds for k. We prove it for $k+1$. Let u_h, v_h, g_h, and Ω_h be as in (13.24). Then, from Lemmas 9.3 and 9.4, we can find a constant $C_2(n, \rho, \lambda, \Lambda) > 0$ such that

$$\bar{a}_h d_{\partial\Omega_h} \leq v_h \leq \bar{b}_h d_{\partial\Omega_h} \quad \text{in } S_{u_h}(0,1),$$

with

$$|\bar{a}_h - a_h| \leq C_2 h^{1/2} |\log h|^2, \quad |\bar{b}_h - b_h| \leq C_2 h^{1/2} |\log h|^2,$$

and

$$(u_h, S_{u_h}(0,1)) \in \mathcal{D}_\sigma \quad \text{for } \sigma = C_2 h^{1/2} \leq \delta_*.$$

In view of (13.26) and (13.27), we have $L_{u_h} v_h = g_h$, where

$$\|g_h/\operatorname{trace} U_h\|_{L^\infty(S_{u_h}(0,1))} \leq C_2 h^{1/2} |\log h|^2 \|g/\operatorname{trace} U\|_{L^\infty(\Omega \cap B_\rho(0))}$$
$$\leq (C_2/C_*) h^{1/2} |\log h|^2.$$

Note that

$$\bar{b}_h - \bar{a}_h \geq b_h - a_h - |\bar{a}_h - a_h| - |\bar{b}_h - b_h| \geq (C_1 - 2C_2) h^{1/2} |\log h|^2.$$

Let c_1 as in Proposition 13.17, and let $C_1 \geq C_2/(c_1 C_*) + 2C_2$. Then

$$\|g_h/\operatorname{trace} U_h\|_{L^\infty(S_{u_h}(0,1))} \leq c_1(C_1 - 2C_2) h^{1/2} |\log h|^2 \leq c_1(\bar{b}_h - \bar{a}_h).$$

Clearly,

$$\sigma \leq c_1(\bar{b}_h - \bar{a}_h) \quad \text{if } C_2 \leq c_1(C_1 - 2C_2)|\log\theta|^2.$$

Now, we can apply Proposition 13.17 and find

$$\bar{a}_{\theta h} d_{\partial\Omega_h} \leq v_h(x) \leq \bar{b}_{\theta h} d_{\partial\Omega_h} \quad \text{in } S_{u_h}(0,\theta), \quad \text{with } \bar{b}_{\theta h} - \bar{a}_{\theta h} \leq \eta(\bar{b}_h - \bar{a}_h).$$

Rescaling back to $S_u(0, \theta h)$ and using Lemma 9.4 again, we obtain

(13.32) $$a_{\theta h} d_{\partial\Omega} \leq v \leq b_{\theta h} d_{\partial\Omega} \quad \text{in } S_u(0, \theta h),$$

where

$$b_{\theta h} - a_{\theta h} \leq \eta(b_h - a_h) + C_2 h^{1/2} |\log h|^2 \leq \frac{1+\eta}{2}(b_h - a_h) \equiv \tilde{\eta}(b_h - a_h),$$

provided $C_1 \geq 2C_2/(1-\eta)$. Thus, it suffices to choose

$$C_1 = \max\left\{ C_2/(c_1 C_*) + 2C_2, 2C_2/(1-\eta), 2C_2 + C_2/(c_1|\log\theta|^2) \right\}.$$

By possibly modifying their values, we may take $a_{\theta h}, b_{\theta h}$ such that

$$a_h \leq a_{\theta h} \leq b_{\theta h} \leq b_h, \qquad b_{\theta h} - a_{\theta h} = \tilde{\eta}(b_h - a_h) = \tilde{\eta}^{k+1}.$$

The claim is proved for $k+1$.

Step 2. Pointwise $C^{1,\alpha}$ estimates. From (13.31), there exists the limit

$$A = A(0) = \lim_{k\to\infty} a_{h_0\theta^k} = \lim_{k\to\infty} b_{h_0\theta^k} \in \left[-\frac{1}{2}, \frac{1}{2}\right].$$

Moreover, for each $k \in \mathbb{N}_0$, we have

$$b_{h_0\theta^k} - A \le b_{h_0\theta^k} - a_{h_0\theta^k} = \tilde{\eta}^k.$$

For $h \le \theta^2 h_0$, we can find an integer $k \ge 2$ such that $h_0\theta^{k+1} < h \le h_0\theta^k$. Thus, if $x \in S_u(0, h)$, then, using (13.30), (13.31), and (13.23), we find

$$v(x) - Ad_{\partial\Omega}(x) \le (b_{h_0\theta^k} - A)d_{\partial\Omega}(x) \le c_2^{-1}h^{1/2}|\log h|\tilde{\eta}^k.$$

Let $\alpha \in (0, 1)$ be such that

$$\theta^{2\alpha} = \tilde{\eta} = (1 + \eta)/2.$$

Then

$$|\log h|\tilde{\eta}^k = |\log h|\theta^{2k\alpha} \le (h_0\theta)^{-2\alpha}|\log h|h^{2\alpha} \le C_3 h^\alpha.$$

We find

$$v(x) - Ad_{\partial\Omega}(x) \le C_4 h^{\frac{1}{2}+\alpha} \quad \text{in } S_u(0, h).$$

Similarly, we obtain the opposite inequality. Hence

$$(13.33) \qquad |v(x) - Ad_{\partial\Omega}(x)| \le C_4 h^{\frac{1}{2}+\alpha} \quad \text{in } S_u(0, h).$$

Thus, from (13.33) and the size estimates for $S_u(0, h)$ in (13.23), we obtain

$$(13.34) \qquad |v(x) - Ad_{\partial\Omega}(x)| \le C_5 r^{1+\alpha} \quad \text{in } B_r(0) \cap \overline{\Omega} \quad \text{for } r \le c,$$

where $c = c(n, \rho, \lambda, \Lambda)$ is small. By increasing C_5 if necessary, we find that the above inequality also holds for all $r \le \rho/2$.

Step 3. Existence of the normal derivative and its Hölder estimates. Applying Steps 1 and 2 to any point $x_0 \in \partial\Omega \cap B_{\rho/2}(0)$, we find a constant $A(x_0)$ with $|A(x_0)| \le \frac{1}{2}$ such that

$$|v(x) - A(x_0)d_{\partial\Omega}(x)| \le C_5 r^{1+\alpha} \qquad \text{in } B_r(x_0) \cap \overline{\Omega} \quad \text{for } r \le \rho/2.$$

This shows that the normal derivative $D_\nu v := Dv \cdot \nu$ exists with

$$D_\nu v(x_0) = -A(x_0) \quad \text{for each } x_0 \in \partial\Omega \cap B_{\rho/2}(0).$$

We now show that

$$(13.35) \qquad \|A\|_{C^{0,\alpha}(\partial\Omega \cap B_{\rho/2}(0))} \le C(n, \rho, \lambda, \Lambda).$$

Let $x_0, z_0 \in \partial\Omega \cap B_{\rho/2}(0)$, and let $r := 2|x_0 - z_0|$. We would like to show that

$$|A(x_0) - A(z_0)| \le Cr^\alpha.$$

It suffices to consider $r \le \rho/4$. Then, there is $y \in B_r(x_0) \cap B_r(z_0) \cap \overline{\Omega}$ with $d_{\partial\Omega}(y) > r/4$. From

$$|v(y) - A(x_0)d_{\partial\Omega}(y)| \le C_5 r^{1+\alpha} \quad \text{and} \quad |v(y) - A(z_0)d_{\partial\Omega}(y)| \le C_5 r^{1+\alpha},$$

we get $|A(x_0) - A(z_0)| d_{\partial\Omega}(y) \leq 2C_5 r^{1+\alpha}$. Therefore $|A(x_0) - A(z_0)| \leq 8C_5 r^\alpha$, as desired. Now, (13.33)–(13.35) give the conclusions of the theorem. $\quad\square$

13.4.4. Global boundary Hölder gradient estimates. If the domain Ω and the potential function u satisfy global versions of (13.14)–(13.17), then we obtain global boundary Hölder gradient estimates for the linearized Monge–Ampère equation.

Theorem 13.18 (Global boundary Hölder gradient estimates). *Let Ω be a convex domain in $\mathbb{R}^n$ satisfying $\Omega \subset B_{1/\rho}(0)$ and the interior ball condition with radius ρ at each point on $\partial\Omega$, where $\rho > 0$. Assume $u \in C^{0,1}(\overline{\Omega}) \cap C^2(\Omega)$ is a convex function satisfying $0 < \lambda \leq \det D^2 u \leq \Lambda$ in Ω, and assume*

$$\rho |x - x_0|^2 \leq u(x) - u(x_0) - Du(x_0) \cdot (x - x_0) \leq \rho^{-1} |x - x_0|^2$$

for all $x, x_0 \in \partial\Omega$. Let $\varphi \in C^{1,1}(\overline{\Omega})$, and let $g : \Omega \to \mathbb{R}$ be such that

$$\mathbf{K} := \|\varphi\|_{C^{1,1}(\overline{\Omega})} + \|g/\operatorname{trace}(U)\|_{L^\infty(\Omega)} < \infty.$$

Let $v \in C(\overline{\Omega}) \cap W^{2,n}_{\mathrm{loc}}(\Omega)$ be the solution to

$$U^{ij} D_{ij} v = g \quad \text{in } \Omega, \qquad v = \varphi \quad \text{on } \partial\Omega.$$

Then, there exist $\alpha \in (0, 1)$ and C depending on $n, \rho, \lambda, \Lambda$, such that

$$\|D_\nu v\|_{C^{0,\alpha}(\partial\Omega)} \leq C\mathbf{K},$$

and for all $x_0 \in \partial\Omega$ and $r \leq \rho$,

$$\sup_{B_r(x_0) \cap \overline{\Omega}} |v - v(x_0) - Dv(x_0) \cdot (x - x_0)| \leq C\mathbf{K} r^{1+\alpha}.$$

Proof. First, we use the maximum principle to obtain the L^∞ bound for v. Indeed, letting $K := \|g/\operatorname{trace} U\|_{L^\infty(\Omega)} + \rho^2 \|\varphi\|_{L^\infty(\Omega)}$, we can show that

$$K(|x|^2 - 2/\rho^2) \leq v \leq -K(|x|^2 - 2/\rho^2) \quad \text{in } \Omega.$$

For example, if $w = K(|x|^2 - 2/\rho^2)$, then since $\Omega \subset B_{1/\rho}(0)$, we have $w \leq -K/\rho^2 \leq \varphi = v$ on $\partial\Omega$. On the other hand,

$$U^{ij} D_{ij} w = 2K U^{ij} \delta_{ij} = 2K \operatorname{trace}(U) > g = U^{ij} D_{ij} v \quad \text{in } \Omega.$$

Thus, the maximum principle in Theorem 2.79 yields $v \geq w$ on Ω.

Next, let

$$\tilde{v} := v - \varphi, \quad \tilde{g} := g - U^{ij} D_{ij} \varphi.$$

Then

$$U^{ij} D_{ij} \tilde{v} = \tilde{g} \quad \text{in } \Omega, \qquad \tilde{v} = 0 \quad \text{on } \partial\Omega.$$

Clearly,

$$\|\tilde{g}/\operatorname{trace} U\|_{L^\infty(\Omega)} \le \|g/\operatorname{trace} U\|_{L^\infty(\Omega)} + \|D^2\varphi\|_{L^\infty(\Omega)}.$$

Applying Theorem 13.8 to $\tilde{v}$, we obtain the conclusions of the theorem. $\quad\square$

13.5. Boundary Harnack Inequality

In this section, we establish a boundary Harnack inequality (also known as the Carleson estimate) for nonnegative solutions to the linearized Monge–Ampère equations. It can be viewed as a boundary counterpart of the Caffarelli–Gutiérrez Harnack inequality in Theorem 12.4.

Structural hypotheses for the boundary Harnack inequality. Let Ω be a convex domain in $\mathbb{R}^n$ satisfying

(13.36) $\quad \Omega \subset B_{1/\rho}(0)$, and for each $y \in \partial\Omega$, there is an interior ball

$$B_\rho(z) \subset \Omega \text{ such that } y \in \partial B_\rho(z).$$

Let $u \in C^{0,1}(\overline{\Omega}) \cap C^2(\Omega)$ be a convex function satisfying

(13.37) $$0 < \lambda \le \det D^2 u \le \Lambda \quad \text{in } \Omega.$$

Assume further that on $\partial\Omega$, u separates quadratically from its tangent hyperplanes; namely, for all $x_0, x \in \partial\Omega$, we have

(13.38) $\quad \rho\,|x - x_0|^2 \le u(x) - u(x_0) - Du(x_0) \cdot (x - x_0) \le \rho^{-1}\,|x - x_0|^2.$

Let $(U^{ij}) = (\det D^2 u)(D^2 u)^{-1}$ and $L_u := U^{ij} D_{ij}$. The following boundary Harnack inequality was established in Le [**L4**].

Theorem 13.19 (Boundary Harnack inequality). *Assume that Ω and u satisfy* (13.36)–(13.38). *Let $x_0 \in \partial\Omega$ and $0 < t \le c$ where $c = c(n, \rho, \lambda, \Lambda)$ is small. Assume $v \in C(\overline{\Omega} \cap S_u(x_0, t)) \cap W^{2,n}_{\mathrm{loc}}(\Omega \cap S_u(x_0, t))$ is a nonnegative solution to the linearized Monge–Ampère equation*

$$L_u v := U^{ij} D_{ij} v = 0 \quad \text{in } \Omega \cap S_u(x_0, t), \quad v = 0 \text{ on } \partial\Omega \cap S_u(x_0, t).$$

Let $P_0 \in \partial S_u(x_0, t/4)$ be any point satisfying $\operatorname{dist}(P_0, \partial\Omega) \ge c_1 t^{1/2}$ for some $c_1(n, \rho, \lambda, \Lambda) > 0$. Then, there is a constant $M_0(n, \rho, \lambda, \Lambda) > 0$ such that

$$\sup_{S_u(x_0, t/2)} v \le M_0 v(P_0).$$

Remark 13.20. The point P_0 in the statement of Theorem 13.19 always exists by Lemma 9.1. Also by this lemma, $\operatorname{dist}(P_0, \partial\Omega)$ is comparable to the largest distance from a point on $\partial S_u(x_0, t/2)$ to $\partial\Omega$. Their ratio is bounded from below and above by positive constants depending only on $n, \rho, \lambda, \Lambda$. For convenience, we call P_0 with largest $\operatorname{dist}(P_0, \partial\Omega)$ the *apex* of $S_u(x_0, t/4)$.

The rest of this section will be devoted to the proof of Theorem 13.19 which consists of the following ingredients.

(1) The first ingredient (Lemma 13.21) states that the maximum value of a nonnegative solution to $L_u v = 0$ on a boundary section of u increases by a factor of 2 when we pass to a concentric boundary section of appropriately larger height.

(2) In the next ingredient (Lemma 13.24), we show that for any interior point x of the domain Ω, we can find another interior point y whose distance to the boundary is larger than that of x and the values of any nonnegative solution to $L_u v = 0$ are appropriately comparable at x and y.

(3) The final ingredient (Lemma 13.25) quantifies how close a point x is to the boundary when a nonnegative solution v to $L_u v = 0$ is large at x given a bound of v at another point in the interior.

For $x \in \Omega$, we denote by $S_u(x, \bar{h}(x))$ the maximal interior section of u with center x; that is, $\bar{h}(x) := \sup\{h > 0 : S_u(x, h) \subset \Omega\}$.

By Proposition 9.2, $\bar{h}(x)$ and $\mathrm{dist}^2(x, \partial\Omega)$ are comparable. However, $\bar{h}(x)$ is very well-behaved under affine transformations. That is why we use $\bar{h}$ to measure the distance to the boundary in Lemmas 13.24 and 13.25 and in the proof of Theorem 13.19.

Below is the first ingredient in the proof of Theorem 13.19.

Lemma 13.21. *Assume that Ω and u satisfy (13.36)–(13.38). Let $x_0 \in \partial\Omega$. Suppose that $v \in C(\overline{\Omega} \cap S_u(x_0, c)) \cap W^{2,n}_{\mathrm{loc}}(\Omega \cap S_u(x_0, c))$ is a nonnegative solution of $L_u v = 0$ in $\Omega \cap S_u(x_0, c)$ and that $v = 0$ on $\partial\Omega \cap S_u(x_0, c)$. Then, there are a large constant $H > 1$ and a small constant c, all depending on $n, \rho, \lambda, \Lambda$, such that for all $0 < h \le c/2$, we have*

$$\sup_{S_u(x_0, h)} v \ge 2 \sup_{S_u(x_0, h/H)} v.$$

Proof. Without loss of generality, we assume that $x_0 = 0$ and u and Ω satisfy (13.20). Assume now $h \le c/2$ where $c \le \kappa$ with κ being in (13.21).

Let u_h, v_h, Ω_h be as in (13.24). Then, we have (13.25), and $v_h \in C(\overline{\Omega_h} \cap S_{u_h}(0, 2)) \cap W^{2,n}_{\mathrm{loc}}(\Omega_h \cap S_{u_h}(0, 2))$ is a nonnegative solution of

$$L_{u_h} v_h = 0 \quad \text{in } \Omega_h \cap S_{u_h}(0, 2) \quad \text{and} \quad v_h = 0 \quad \text{on } \partial\Omega_h \cap S_{u_h}(0, 2).$$

We need to show that for H large, depending on $n, \rho, \lambda, \Lambda$,

$$(13.39) \qquad\qquad \sup_{S_{u_h}(0,1)} v_h \ge 2 \sup_{S_{u_h}(0, 1/H)} v_h.$$

Indeed, by Lemma 9.3, u_h and $S_{u_h}(0, 1)$ satisfy the hypotheses of the Boundary Localization Theorem (Theorem 8.5) at all points $z \in \partial\Omega_h \cap B_r(0)$ if

$h, r \leq c$ for $c(n, \rho, \lambda, \Lambda)$ small. Hence, if $1/H \leq \kappa$, then from (13.23), we have

$$(13.40) \qquad \sup_{x \in S_{u_h}(0,1/H)} \operatorname{dist}(x, \partial \Omega_h) \leq \sup_{x \in S_{u_h}(0,1/H)} |x| \leq c_2^{-1} \left(\frac{1}{H}\right)^{1/2} \log\left(\frac{1}{H}\right).$$

On the other hand, by the boundary gradient estimate in Theorem 13.5,

$$\sup_{S_{u_h}(0,1/H)} v_h \leq C(n, \lambda, \Lambda, \rho) \Big(\sup_{S_{u_h}(0,1)} v_h\Big) \sup_{x \in S_{u_h}(0,1/H)} \operatorname{dist}(x, \partial \Omega_h).$$

Therefore, combining the above inequality with (13.40), we obtain (13.39) for $H(n, \lambda, \Lambda, \rho)$ large. The lemma is proved. $\qquad \square$

Before stating the next two ingredients in the proof of Theorem 13.19, it is convenient to study a sort of quasi-distance generated by the convex function u on a domain Ω satisfying (13.36)–(13.38).

For $x \in \overline{\Omega}$, let us introduce the function $\delta_u(x, \cdot) : \overline{\Omega} \to [0, \infty)$ defined by

$$(13.41) \qquad \delta_u(x, y) = u(y) - u(x) - Du(x) \cdot (y - x).$$

We will use the following structural constants:

$\theta_* > 1$ in Theorem 9.10, and $c_0 > 0$ and $p_1 = \bar{\mu}^{-1} > 1$ in Theorem 9.12.

Replacing the function d in the proof of Theorem 9.19(iii) by δ_u, we have the following quasi-metric inequality for δ_u.

Lemma 13.22 (Quasi-metric inequality). *Assume that Ω and u satisfy* (13.36)–(13.38). *Then, for all $x, y, z \in \overline{\Omega}$, we have*

$$\delta_u(x, y) \leq \theta_*^2 (\delta_u(x, z) + \delta_u(z, y)).$$

We show that $\delta_u(x, \cdot)$ satisfies a uniform Hölder property.

Lemma 13.23 (Uniform Hölder property of δ_u). *Assume that Ω and u satisfy* (13.36)–(13.38). *Let δ_u be defined as in* (13.41). *Then, there exists a constant $C_2(n, \rho, \lambda, \Lambda) \geq 1$ such that for all $x, y, z \in \overline{\Omega}$, we have*

$$|\delta_u(x, z) - \delta_u(x, y)| \leq C_2 [\delta_u(y, z)]^{\bar{\mu}} \big(\delta_u(x, z) + \delta_u(x, y)\big)^{1-\bar{\mu}}.$$

Proof. It suffices to consider the case $0 < \delta_u(x, y) < \delta_u(x, z)$. Let $t := \delta_u(x, z) > 0$ and $r := \delta_u(x, y)/t \in (0, 1)$. Consider $\varepsilon \in (0, 1 - r)$. Since $\delta_u(x, y) = rt < (r + \varepsilon)t$, we have $y \in S_u(x, (r + \varepsilon)t)$. We claim that

$$c_0 (1 - (r + \varepsilon))^{p_1} t < \delta_u(y, z) + \varepsilon,$$

where c_0 and $p_1 = \bar{\mu}^{-1}$ are from Theorem 9.12. If otherwise,

$$\delta_u(y, z) + \varepsilon \leq c_0 (1 - (r + \varepsilon))^{p_1} t;$$

then Theorem 9.12 applied to u with $x_1 = y$, $x_0 = x$, and $r + \varepsilon < s = 1$ gives

$$z \in S_u(y, \delta_u(y, z) + \varepsilon) \subset S_u(y, c_0(1 - (r + \varepsilon))^{p_1} t) \subset S_u(x, t),$$

which would imply that $\delta_u(x, z) < t$, contradicting $\delta_u(x, z) = t$.

By letting $\varepsilon \to 0^+$ in the claim, we get $c_0(1 - r)^{p_1} t \le \delta_u(y, z)$. In view of $p_1 = \bar\mu^{-1}$ and $t = \delta_u(x, z) \le \delta_u(x, z) + \delta_u(x, y)$, we find that

$$\delta_u(x, z) - \delta_u(x, y) = t(1 - r) = c_0^{-\bar\mu} [c_0(1 - r)^{p_1} t]^{\bar\mu} t^{1 - \bar\mu}$$
$$\le c_0^{-\bar\mu} [\delta_u(y, z)]^{\bar\mu} (\delta_u(x, z) + \delta_u(x, y))^{1 - \bar\mu},$$

proving the lemma. $\qquad\qquad\qquad\qquad\qquad\qquad\qquad\qquad\qquad\qquad\square$

We remark that when u and Ω satisfy the local conditions (13.14)–(13.17), the conclusions of Lemmas 13.22 and 13.23 also hold for x, y, z in a neighborhood of the origin.

Next is the second ingredient in the proof of Theorem 13.19.

Lemma 13.24. *Let u and Ω satisfy (13.14)–(13.17). There exist positive constants $\hat K, M_1$ (large) and c (small), all depending on $n, \rho, \lambda, \Lambda$, with the following property. Suppose that $v \in C(\overline\Omega \cap S_u(0, c)) \cap W_{\mathrm{loc}}^{2,n}(\Omega \cap S_u(0, c))$ is a nonnegative solution of $L_u v = 0$ in $\Omega \cap S_u(0, c)$. Assume that for some $x \in \Omega$, we have $\partial\Omega \cap \partial S_u(x, \bar h(x)) = \{0\}$ and $\mathrm{dist}(x, \partial\Omega) \le c$. Then, we can find $y \in S_u(0, \hat K \bar h(x))$ such that*

$$M_1^{-1} v(x) \le v(y) \quad and \quad \bar h(y) \ge 2\bar h(x).$$

Proof. Let $\{x_0 = x, x_1, \ldots, x_m\} \subset S_u(0, \bar K \mathrm{dist}^2(x, \partial\Omega))$ and let $\bar K$ be as in Theorem 9.13(i) where we take $\tau = 1/4$. In view of Proposition 9.2,

$$\kappa_0^2 \bar h(x) \le \mathrm{dist}^2(x, \partial\Omega) \le \kappa_0^{-2} \bar h(x).$$

Hence, $x_m \in S_u(0, \hat K \bar h(x))$ for $\hat K := \bar K \kappa_0^{-2}$. If $\mathrm{dist}(x, \partial\Omega) \le c$ where $c(n, \rho, \lambda, \Lambda)$ is small, then

$$S_u(0, \hat K \bar h(x)) \subset S_u(0, c/2) \quad and \quad S_u(x_i, \bar h(x_i)) \subset S_u(0, c/2) \quad for all\ i.$$

For the latter inclusion, we first use Proposition 9.2 and Lemma 9.1 to get

$$\bar h(x_i) \le \kappa_0^{-2} \mathrm{dist}^2(x_i, \partial\Omega) \le C(n, \lambda, \Lambda, \rho) \mathrm{dist}^2(x, \partial\Omega).$$

Now, for $y \in S_u(x_i, \bar h(x_i))$, Lemma 13.22 implies that

$$\delta_u(0, y) \le \theta_*^2 [\delta_u(0, x_i) + \delta_u(x_i, y)] \le \theta_*^2 [\bar K \mathrm{dist}^2(x, \partial\Omega) + \bar h(x_i)] \le c/2$$

if $\mathrm{dist}(x, \partial\Omega) \le c$ where c is small. Hence $S_u(x_i, \bar h(x_i)) \subset S_u(0, c/2)$.

Since $x_i \in S_u(x_{i+1}, \tau \bar h(x_{i+1}))$ and since $v \ge 0$ satisfies

$$L_u v = 0 \quad in\ S_u(x_{i+1}, \bar h(x_{i+1})/2) \Subset S_u(0, c/2) \cap \Omega,$$

we obtain from the Harnack inequality in Theorem 12.4 that

(13.42) $C_1^{-1}v(x_i) \leq v(x_{i+1}) \leq C_1(n, \lambda, \Lambda)v(x_i)$ for each $i = 0, 1, \ldots, m-1$.

We now take $y = x_m$. Then, from $\text{dist}(x_m, \partial\Omega) \geq 2\kappa_0^{-2}\text{dist}(x, \partial\Omega)$, we have

$$\bar{h}(y) = \bar{h}(x_m) \geq \kappa_0^2\text{dist}^2(x_m, \partial\Omega) \geq 4\kappa_0^{-2}\text{dist}^2(x, \partial\Omega) \geq 4\bar{h}(x).$$

From the chain condition $x_i \in S_u(x_{i+1}, \tau\bar{h}(x_{i+1}))$ for all $i = 0, \ldots, m-1$ with $m \leq K(n, \rho, \lambda, \Lambda)$ and (13.42), we obtain

$$M_1^{-1}v(x) \leq v(y) \leq M_1 v(x),$$

for some $M_1(n, \rho, \lambda, \Lambda) > 1$. The lemma is proved. $\qquad\square$

The final ingredient in the proof of Theorem 13.19 states as follows.

Lemma 13.25. *Let u and Ω satisfy* (13.14)–(13.17). *Let $\hat{K}$ and M_1 be as in Lemma 13.24. There exist small constants c, c_3 and a large constant $\hat{C}$, all depending on $n, \rho, \lambda, \Lambda$, with the following properties. Suppose that $v \in C(\overline{\Omega} \cap S_u(0, c)) \cap W_{\text{loc}}^{2,n}(\Omega \cap S_u(0, c))$ is a nonnegative solution of $L_u v = 0$ in $\Omega \cap S_u(0, c)$ and that $v = 0$ on $\partial\Omega \cap S_u(0, c)$. Let $t \leq c^2$, and let $P \in \partial S_u(0, c_3 t/2)$ be the apex of $S_u(0, c_3 t/2)$ with $\theta_*^2(1 + \hat{K})\bar{h}(P) < c_0 t(1 - 2^{-1/p_1})^{p_1}$. If*

$$x \in S_u(0, (2 + 2^{-k/p_1})t) \quad and \quad v(x) \geq \hat{C}M_1^k v(P),$$

for some integer $k \geq 1$, then

$$\bar{h}(x) \leq 2^{-k}\bar{h}(P).$$

Proof. The assumed upper bound for $\bar{h}(P)$ clearly holds when c_3 is structurally small. Fix such a c_3. We divide the proof into several steps.

Step 1. Choice of $\hat{C}$. By using Theorem 9.13 (with $\tau = 1/2$) and the interior Harnack inequality in Theorem 12.4 as in the proof of Lemma 13.24, we can find a large constant $\hat{C}(n, \rho, \lambda, \Lambda)$ such that for all $x \in S_u(0, 3t)$ with $\bar{h}(x) \geq 2^{-1}\bar{h}(P)$, we have $v(x) \leq (\hat{C}/2)v(P)$.

Step 2. Induction argument. With the choice of $\hat{C}$ in Step 1, we can now prove the statement of the lemma by induction. Clearly, the statement is true for $k = 1$. Suppose it is true for $k \geq 1$. We prove it for $k + 1$. Let

$$D_k = S_u(0, (2 + 2^{-k/p_1})t).$$

Let $x \in D_{k+1}$ be such that $v(x) \geq \hat{C}M_1^{k+1}v(P)$. We need to show that

$$\bar{h}(x) \leq 2^{-(k+1)}\bar{h}(P).$$

Step 2(a). Application of Lemma 13.24. From the induction hypothesis, we have $\bar{h}(x) \leq 2^{-k}\bar{h}(P)$. Let $z = \partial\Omega \cap \partial S_u(x, \bar{h}(x))$. If $t \leq c^2$ where c is small, then we are in the setting of Lemma 13.24 with z replacing 0. Indeed, we first note from $x \in D_{k+1} \subset S_u(0, 3t)$ and Lemma 9.1 that

$$(13.43) \qquad \text{dist}^2(x, \partial\Omega) \leq C(n, \rho, \lambda, \Lambda)t.$$

As in the proof of $S_u(x_i, \bar{h}(x_i)) \subset S_u(0, c/2)$ in Lemma 13.24, we have

$$S_u(z, \hat{K}\bar{h}(x)) \subset S_u(0, c/2)$$

if c is small. This combined with (13.43) shows that the hypotheses of Lemma 13.24 are satisfied.

By Lemma 13.24, we can find $y \in S_u(z, \hat{K}\bar{h}(x))$ such that

$$v(y) \geq M_1^{-1}v(x) \geq \hat{C}M_1^k u(P) \quad \text{and} \quad \bar{h}(y) \geq 2\bar{h}(x).$$

Step 2(b). We show that $y \in D_k$. From $\delta_u(x, z) = \bar{h}(x)$, $\delta_u(z, y) \leq \hat{K}\bar{h}(x)$, and Lemma 13.22, we have

$$\delta_u(x, y) \leq \theta_*^2(1 + \hat{K})\bar{h}(x).$$

Let us denote for simplicity $r_k = 2 + 2^{-k/p_1}$. Then $D_k = S_u(0, r_k t)$. Recalling $x \in D_{k+1} = S_u(0, r_{k+1}t)$, we have from Theorem 9.12(i),

$$S_u(x, c_0(r_k - r_{k+1})^{p_1}t) \subset S_u(0, r_k t) = D_k.$$

By our choice of P, we have $\theta_*^2(1 + \hat{K})\bar{h}(P) < c_0 t(1 - 2^{-1/p_1})^{p_1}$, and hence

$$c_0(r_k - r_{k+1})^{p_1}t = c_0 t 2^{-k}(1 - 2^{-1/p_1})^{p_1}$$
$$> \theta_*^2(1 + \hat{K})2^{-k}\bar{h}(P) \geq \theta_*^2(1 + \hat{K})\bar{h}(x) \geq \delta_u(x, y).$$

Therefore, $y \in S_u(x, c_0(r_k - r_{k+1})^{p_1}t) \subset S_u(0, r_k t) = D_k$, as desired.

Since $y \in D_k$, the induction hypothesis for k gives $\bar{h}(y) \leq 2^{-k}\bar{h}(P)$ and hence $\bar{h}(x) \leq (1/2)\bar{h}(y) \leq 2^{-(k+1)}\bar{h}(P)$, which completes the proof. $\square$

We are now ready to give the proof of Theorem 13.19.

Proof of Theorem 13.19. We can assume that $x_0 = 0 \in \partial\Omega$ and that u and Ω satisfy (13.20). Let c be the smallest of the constants c in Lemmas 13.21, 13.24, and 13.25, and let $t \leq c$. Let $\hat{K}$ be as in Lemma 13.24, and let $\hat{C}, c_3$, and M_1 be as in Lemma 13.25. Let $\bar{t} = ct$. Then the apex $P \in \partial S_u(0, c_3\bar{t}/2)$ of $S_u(0, c_3\bar{t}/2)$ satisfies $\theta_*^2(1 + \hat{K})\bar{h}(P) < c_0\bar{t}(1 - 2^{-1/p_1})^{p_1}$. Without loss of generality, we assume that $v > 0$ in Ω. We prove that

$$(13.44) \qquad v(x) < \hat{C}M_1^K v(P) \quad \text{for all } x \in S_u(0, c_3\bar{t}/2)$$

for some large constant $K(n, \rho, \lambda, \Lambda)$ to be determined later on. Then by using Theorem 9.13 (with $\tau = 1/2$) and the interior Harnack inequality in Theorem 12.4, we easily obtain the stated estimate of the theorem.

We will use a contradiction argument. Suppose that (13.44) is false. Then we can find $x_1 \in S_u(0, c_3\bar{t}/2) \subset S_u(0, (2+2^{-K/p_1})\bar{t})$ satisfying $v(x_1) \geq \hat{C} M_1^K v(P)$. It follows from Lemma 13.25 that

$$(13.45) \qquad \bar{h}(x_1) \leq 2^{-K}\bar{h}(P).$$

We fix an integer s so that $2^s \geq M_1$. Let $z = \partial\Omega \cap \partial S_u(x_1, \bar{h}(x_1))$. Then, by Theorem 9.10,

$$S_u(x_1, \bar{h}(x_1)) \subset S_u(z, \theta_*\bar{h}(x_1)) \subset S_u(z, H^s\theta_*\bar{h}(x_1)),$$

where H is as in Lemma 13.21. We will choose $K \geq K_0$, where

$$\theta_*^4(1 + 2H^s\theta_*)2^{-K_0} < 1/4.$$

Step 1. Refining c_3. Reducing c_3 if necessary, we have for $K \geq K_0$

$$(13.46) \qquad S_u(z, 2H^s\theta_*\bar{h}(x_1)) \subset S_u(0, \bar{t}/2).$$

Indeed, if $y \in S_u(z, 2H^s\theta_*\bar{h}(x_1))$, then $\delta_u(z, y) \leq 2H^s\theta_*\bar{h}(x_1)$. From

$$\delta_u(x_1, z) \leq \bar{h}(x_1), \quad \bar{h}(P) \leq \bar{t},$$

and Lemma 13.22, we have in view of (13.45),

$$\delta_u(x_1, y) \leq \theta_*^2(\delta_u(x_1, z) + \delta_u(z, y)) \leq \theta_*^2(1 + 2H^s\theta_*)\bar{h}(x_1)$$
$$\leq \theta_*^2(1 + 2H^s\theta_*)2^{-K}\bar{t}.$$

Thus, from $\delta_u(0, x_1) \leq c_3\bar{t}/2$ and Lemma 13.22, we obtain

$$\delta_u(0, y) \leq \theta_*^2(\delta_u(0, x_1) + \delta_u(x_1, y)) \leq \theta_*^2[c_3\bar{t}/2 + \theta_*^2(1 + 2H^s\theta_*)2^{-K}\bar{t}] < \bar{t}/2$$

if c_3 is small.

Step 2. Choice of K. With (13.46), applying Lemma 13.21 to the section $S_u(z, H^m\theta_*\bar{h}(x_1))$ successively for $m = 0, \ldots, s-1$, we can find $x_2 \in \overline{S_u(z, H^s\theta_*\bar{h}(x_1))}$ such that

$$(13.47) \qquad v(x_2) \geq 2^s v(x_1) \geq \hat{C} M_1^{K+1} v(P).$$

From $\delta_u(x_1, z) \leq \bar{h}(x_1)$, $\delta_u(z, x_2) \leq H^s\theta_*\bar{h}(x_1)$, and Lemma 13.22, we have

$$\delta_u(x_1, x_2) \leq \theta_*^2(1 + \theta_*H^s)\bar{h}(x_1).$$

Let

$$d_0 = \delta_u(0, x_1), \quad d_1 = \delta_u(0, x_2).$$

Now, using $\bar{h}(P) \leq \bar{t}$, (13.45), and the Hölder property of $\delta_u(0, \cdot)$ in Lemma 13.23, we find

$$
\begin{aligned}
d_1 &\leq d_0 + C_2(\delta_u(x_1, x_2))^{\bar{\mu}}(d_0 + d_1)^{1-\bar{\mu}} \\
(13.48) \qquad &\leq d_0 + C_2\big(\theta_*^2(1 + \theta_*H^s)2^{-K}\bar{h}(P)\big)^{\bar{\mu}}(d_0 + d_1)^{1-\bar{\mu}} \\
&\leq d_0 + \theta_0^{\bar{\mu}}(d_0 + d_1)^{1-\bar{\mu}},
\end{aligned}
$$

where $\{\theta_j\}_{j=0}^{\infty}$ is the sequence defined by

$$(13.49) \qquad \theta_j = C_2^{1/\bar{\mu}}\theta_*^2(1+\theta_*H^s)2^{-(K+j)}\bar{t}.$$

Clearly, if $K \geq K_0$ is large, then we have $\theta_{j+1} < \theta_j < c_3\bar{t}/2$ for all j and

$$(13.50) \qquad \frac{2}{1-(\frac{\theta_0}{t/2})^{\bar{\mu}}} \sum_{j=0}^{\infty} \left(\frac{\theta_j}{\bar{t}/2}\right)^{\bar{\mu}} < \log\frac{3}{2}.$$

Step 3. The contradiction. Given this K, we know from (13.48) together with $d_0 < c_3\bar{t}/2$ and Lemma 13.26 below that $\delta_u(0, x_2) < 3c_3\bar{t}/4$. Hence $x_2 \in S_u(0, 3c_3\bar{t}/4)$. Recalling (13.47), we conclude from Lemma 13.25 that $\bar{h}(x_2) \leq 2^{-(K+1)}\bar{h}(P)$. With K satisfying (13.50), we repeat the above process to obtain a sequence $\{x_j\}_{j=0}^{\infty} \subset S_u(0, 3c_3\bar{t}/4)$ such that

$$(13.51) \qquad v(x_j) \geq \hat{C}M_1^{K+j-1}v(P),$$

with

$$(13.52) \quad \bar{h}(x_j) \leq 2^{-(k+j-1)}\bar{h}(P) \quad \text{and} \quad \delta_u(x_j, x_{j+1}) \leq \theta_*^2(1+\theta_*H^s)\bar{h}(x_j),$$

provided that $x_j \in S_u(0, 3c_3\bar{t}/4)$. But this follows from the choice of K. Indeed, recalling θ_j in (13.49) and the Hölder property of $\delta_u(0, \cdot)$ in Lemma 13.23, we obtain for $d_j = \delta_u(0, x_{j+1})$ the estimate

$$d_{j+1} - d_j \leq C_2(\delta_u(x_{j+1}, x_{j+2}))^{\bar{\mu}}(d_j + d_{j+1})^{1-\bar{\mu}} \leq \theta_j^{\bar{\mu}}(d_j + d_{j+1})^{1-\bar{\mu}}.$$

Hence, from $d_0 \leq c_3\bar{t}/2$ and (13.50), we find from Lemma 13.26 that

$$d_{j+1} \leq \frac{c_3\bar{t}}{2}\exp\left[\sum_{j=0}^{\infty} \frac{2}{1-(\frac{\theta_0}{t/2})^{\bar{\mu}}}\left(\frac{\theta_j}{\bar{t}/2}\right)^{\bar{\mu}}\right] < \frac{3c_3}{4}\bar{t}.$$

We now let $j \to \infty$ in (13.51) and (13.52) to obtain $x_{\infty} \in \partial\Omega \cap S_u(0, c_3\bar{t})$ with $v(x_{\infty}) = \infty$. This is a contradiction to $v(x_{\infty}) = 0$. Hence (13.44) holds and the proof of our theorem is complete. $\qquad\square$

In the proof of Theorem 13.19, we used the following lemma.

Lemma 13.26. *Let* $R > 0$, $\alpha \in (0,1)$, $\{d_n\}_{n=0}^{\infty} \subset (0, \infty)$, $\{\theta_n\}_{n=0}^{\infty} \subset (0, R)$ *be such that* $d_0 \leq R$ *and*

$$d_{n+1} \leq d_n + \theta_n^{\alpha}(d_n + d_{n+1})^{1-\alpha} \quad \text{and} \quad \theta_{n+1} < \theta_n < R \quad \text{for all } n.$$

Then

$$d_{n+1} \leq R \, \exp\left[\frac{2}{1-(\theta_0/R)^{\alpha}} \sum_{j=0}^{n} (\theta_j/R)^{\alpha}\right].$$

Proof. By considering $\tilde{d}_n := d_n/R$ and $\tilde{\theta}_n := \theta_n/R$ instead of d_n and θ_n, we can assume that $R = 1$, $d_0 \leq 1$, and $\theta_{n+1} < \theta_n < 1$ for all $n \in \mathbb{N}_0$. From the inequality between d_{n+1} and d_n, we have

$$d_{n+1} \leq \max\{d_n, 1\} + \theta_n^\alpha \big[\max\{d_n, 1\} + d_{n+1}\big].$$

It follows that, for all $n \geq 0$,

$$d_{n+1} \leq [(1 + \theta_n^\alpha)/(1 - \theta_n^\alpha)] \max\{d_n, 1\} \leq (1 + c_0\theta_n^\alpha) \max\{d_n, 1\},$$

where $c_0 := 2/(1 - \theta_0^\alpha)$. From $d_0 \leq 1$, we have $d_1 \leq 1 + c_0\theta_0^\alpha$ and

$$d_{n+1} \leq \prod_{j=0}^{n}(1 + c_0\theta_j^\alpha).$$

This together with $1 + x \leq e^x$ for all x completes the proof of the lemma. $\square$

Theorem 13.19 is a special case of the following Comparison Theorem whose particular case asserts that any two positive solutions of $L_u v = 0$ in Ω which vanish on a portion of the boundary must vanish at the same rate.

Theorem 13.27 (Comparison Theorem). *Assume that Ω and u satisfy* (13.36)–(13.38). *Let $x_0 \in \partial\Omega$ and $0 < t \leq c$ where $c = c(n, \rho, \lambda, \Lambda)$ is small. Let $v, w \in C(\overline{\Omega} \cap S_u(x_0, t)) \cap W^{2,n}_{\text{loc}}(\Omega \cap S_u(x_0, t))$ be nonnegative functions satisfying*

$$\begin{cases} L_u v = L_u w = 0 & \text{in } \Omega \cap S_u(x_0, t), \\ w = 0 & \text{on } \partial\Omega \cap S_u(x_0, t), \quad v > 0 \quad \text{in } \Omega \cap S_u(x_0, t). \end{cases}$$

Let $P \in \partial S_u(x_0, t/4)$ be any point satisfying $\text{dist}(P, \partial\Omega) \geq c_1 t^{1/2}$ for some $c_1(n, \rho, \lambda, \Lambda) > 0$. Then, there is a constant $M_1(n, \rho, \lambda, \Lambda) > 0$ such that

$$\sup_{S_u(x_0, t/2) \cap \Omega} \frac{w}{v} \leq M_1 \frac{w(P)}{v(P)}.$$

Proof. It suffices to consider $w \not\equiv 0$. By multiplying u and v by suitable constants, we assume that $v(P) = w(P) = 1$. Without loss of generality, we assume that $x_0 = 0$ and that u and Ω satisfy (13.20).

By Theorem 13.19,

(13.53) $\qquad\qquad w(x) \leq M_0 \quad$ for all $x \in S_u(0, 3t/4)$.

By Theorem 13.5, we have

(13.54) $\qquad w(x) \leq CM_0 t^{-1/2} \text{dist}(x, \partial\Omega) \quad$ for all $x \in S_u(0, t/2)$.

Using Theorem 9.13 (with $\tau = 1/2$) and the interior Harnack inequality in Theorem 12.4, we can find a constant $c_1(n, \rho, \lambda, \Lambda) > 0$ such that

$$v(x) \geq c_1 \quad \text{for all } x \in S_u(0, 3t/4) \text{ with } \text{dist}(x, \partial\Omega) \geq c_1 t^{1/2}.$$

It follows from Proposition 13.16 that

$$(13.55) \qquad v(x) \geq c_2 t^{-1/2} \mathrm{dist}(x, \partial\Omega) \quad \text{for all } x \in S_u(0, t/2).$$

The theorem follows from (13.54) and (13.55).

For completeness, we indicate how to obtain (13.54) and (13.55). Let $h = t$. Let A_h be as in (13.22). We consider the rescaled functions

$$u_h(x) := h^{-1} u(h^{1/2} A_h^{-1} x), w_h(x) = w(h^{1/2} A_h^{-1} x) \quad \text{for } x \in \Omega_h := h^{-1/2} A_h \Omega.$$

Then w_h is a positive solution of

$$L_{u_h} w_h = 0 \quad \text{in } \Omega_h \cap S_{u_h}(0, 1) \quad \text{and} \quad w_h = 0 \quad \text{on } \partial\Omega_h \cap S_{u_h}(0, 1).$$

By Theorem 13.5, we know that for some $C = C(n, \rho, \lambda, \Lambda)$

$$(13.56) \quad w_h(x) \leq C \Big(\max_{S_{u_h}(0, 3/4)} w_h \Big) \mathrm{dist}(x, \partial\Omega_h) \quad \text{for all } x \in S_{u_h}(0, 1/2).$$

By the distance comparison in Lemma 9.4, we have for $x \in S_{u_h}(0, 1/2)$,

$$\mathrm{dist}(x, \partial\Omega_h) \leq \frac{h^{-1/2} \mathrm{dist}(h^{1/2} A_h^{-1} x, \partial\Omega)}{1 - Ch^{1/2}|\log h|^2} \leq 2h^{-1/2} \mathrm{dist}(h^{1/2} A_h^{-1} x, \partial\Omega)$$

if $h = t$ is small. Using (13.53), (13.56), and the foregoing estimate, we obtain (13.54). The proof of (13.55), using rescaling, is similar. $\qquad\square$

Remark 13.28. We require t to be small in Theorems 13.19 and 13.27 for convenience. A simple covering argument with the help of the Boundary Localization Theorem (Theorem 8.5) and the interior Harnack inequality in Theorem 12.4 shows that the above theorems hold for any t satisfying $S_u(x_0, t) \not\supset \overline{\Omega}$.

13.6. Application: Minimizers of Linear Functionals with Prescribed Determinant

In this section, we briefly describe applications of the boundary Hölder gradient estimates in Theorem 13.8 to problems in the calculus of variations. They are motivated by the minimization of the Mabuchi energy functional [**D1, Mab**] from complex geometry in the case of toric varieties

$$M(u) = \int_\Omega -\log \det D^2 u \, dx + \int_{\partial\Omega} u \, d\sigma - \int_\Omega u \, dA.$$

In this case, Ω is a bounded convex domain in $\mathbb{R}^n$ and $d\sigma$ and dA are measures on $\partial\Omega$ and Ω. More generally, when the measure dA has density A (with respect to the Lebesgue measure), minimizers of M satisfy the Abreu equation (12.28) in Ω.

Motivated by the Mabuchi functional, Le and Savin [**LS2**] considered Monge–Ampère functionals

$$E(u) = \int_\Omega F(\det D^2 u)\, dx + \int_{\partial\Omega} u\, d\sigma - \int_\Omega u\, dA,$$

where F has certain convexity properties and $d\sigma$ and dA are nonnegative Radon measures supported on $\partial\Omega$ and Ω, respectively.

One special case is the nonsmooth convex function $F(t) : (0,\infty) \to \{0,\infty\}$ where $F(t) = \infty$ if $0 < t < 1$ and $F(t) = 0$ if $t \geq 1$. In this case, a minimizer u satisfies $\det D^2 u \geq 1$, and $E(u)$ reduces to the linear functional

$$L(u) = \int_{\partial\Omega} u\, d\sigma - \int_\Omega u\, dA.$$

Of course, we can replace the lower bound 1 by a given function f being bounded between two positive constants. This suggests studying the existence, uniqueness, and regularity properties for minimizers of L, that is,

(13.57) minimize $L(u)$ over $u \in \mathcal{C}$,

over a convex set $\mathcal{C}$ in the cone of convex functions given by

$$\mathcal{C} := \{u : \overline{\Omega} \to \mathbb{R} \text{ such that } u \text{ is convex, } \det D^2 u \geq f\}.$$

We assume that the following conditions are satisfied:

(a) $d\sigma = \sigma(x)\, d\mathcal{H}^{n-1}\lfloor \partial\Omega$, with the density $\sigma(x)$ bounded between two positive constants.

(b) $dA = A(x)\, dx$ in a small neighborhood of $\partial\Omega$ with the density $A(x)$ bounded from above.

(c) $L(u) > 0$ for all u convex but not linear.

The last condition is known as *the stability of L* (see [**D1**]). In 2D, it is equivalent to saying that, for all linear functions l, we have $L(l) = 0$ and $L(l^+) > 0$ if $l^+ := \max\{l, 0\} \not\equiv 0$ in Ω.

The stability of L implies that $d\sigma$ and dA must have the same mass and the same center of mass. Moreover, the measure dA can be a Dirac mass which makes (13.57) remarkably interesting. In fact, if $d\sigma$ is the surface measure of $\partial\Omega$ and x_0 is its center of mass, then the pair $(d\sigma, \mathcal{H}^{n-1}(\partial\Omega)\delta_{x_0})$ satisfies conditions (a)–(c).

A minimizer u of the functional L is determined up to linear functions since both L and $\mathcal{C}$ are invariant under addition with linear functions. It is not difficult to show that there exists a unique minimizer (up to linear functions) to (13.57) in two dimensions but it is more challenging to investigate its regularity. If u is a smooth minimizer, then we can show that there exists a function v such that (u, v) solves the system (see Problem 13.8)

$$(13.58) \qquad \begin{cases} \det D^2 u \ = f, \quad U^{ij} D_{ij} v = -dA \quad \text{in } \Omega \subset \mathbb{R}^n, \\ \qquad v = 0, \quad U^{\nu\nu} D_\nu v \ = -\sigma \quad \text{on } \partial\Omega. \end{cases}$$

Here ν denotes the outer unit normal vector field on $\partial\Omega$, $D_\nu v = Dv \cdot \nu$, and $U^{\nu\nu} = \det D^2_{x'} u$ with $x' \perp \nu$ denoting the tangential directions along $\partial\Omega$. This system of Euler–Lagrange equations is interesting since the function v above satisfies a linearized Monge–Ampère equation with two boundary conditions, Dirichlet and Neumann, while u has no boundary conditions. Heuristically, the boundary values for u can be recovered from the term $U^{\nu\nu} = \det D^2_{x'} u$ which appears in the Neumann boundary condition for v.

In [**LS2**], the authors obtained the following regularity results for the minimizers u in two dimensions.

Theorem 13.29. *Let Ω be a uniformly convex domain in $\mathbb{R}^n$. Assume that f is bounded between two positive constants. Assume that $n = 2$ and the conditions* (a)–(c) *hold. Let u be the unique minimizer (up to linear functions) to the problem* (13.57).

  (i) *If $\sigma \in C^\alpha(\partial\Omega)$, $f \in C^\alpha(\overline{\Omega})$, and $\partial\Omega \in C^{2,\alpha}$, then $u \in C^{2,\alpha}(\overline{\Omega})$ and the system* (13.58) *holds in the classical sense.*

  (ii) *If $\sigma \in C^\infty(\partial\Omega)$, $f \in C^\infty(\overline{\Omega})$, $A \in C^\infty(\overline{\Omega})$, $\partial\Omega \in C^\infty$, then $u \in C^\infty(\overline{\Omega})$.*

One of the crucial points in the proof of Theorem 13.29 is to show that u separates quadratically from its tangent hyperplanes on the boundary $\partial\Omega$. Once this is done, the boundary Hölder gradient estimates in Theorem 13.8 can be applied to yield the boundary Hölder continuity of $D_\nu v$. This combined with $U^{\nu\nu} D_\nu v = -\sigma$ on $\partial\Omega$ gives the $C^{2,\alpha}$ character of u on $\partial\Omega$. Thus, applying the global $C^{2,\alpha}$ estimates in Theorem 10.17, we find $u \in C^{2,\alpha}(\overline{\Omega})$ and the system (13.58) holds in the classical sense.

In [**LS2**], the authors provide an example of Pogorelov type for a minimizer in dimensions $n \geq 3$ that shows that Theorem 13.29(ii) does not hold in this generality in higher dimensions.

For a more general convex function F, convex minimizers of E satisfy a system of the form

$$\begin{cases} -F'(\det D^2 u) \ = v, \quad U^{ij} D_{ij} v = -dA \quad \text{in } \Omega, \\ \qquad v = 0, \quad U^{\nu\nu} D_\nu v = -\sigma \quad \text{on } \partial\Omega. \end{cases}$$

A minimizer u solves a fourth-order elliptic equation with two nonstandard boundary conditions involving the second- and third-order derivatives of u. In [**LS2**], the boundary Hölder gradient estimates in Theorem 13.8 were used to show that $u \in C^{2,\alpha}(\overline{\Omega})$ in dimensions $n = 2$ under suitable conditions on the function F and the measures dA and $d\sigma$.

13.7. Problems

Problem 13.1. Show that the global Hölder exponent $\alpha/(\alpha+2)$ in Proposition 13.1 can be improved to $\alpha/2$ if $g \equiv 0$, and in this case, the constant C does not depend on λ.

Problem 13.2. Verify the assertion in Remark 13.4.

Problem 13.3. Find a radial, convex function $u : \mathbb{R}^n \to \mathbb{R}$ such that $\sum_{i,j=1}^n U^{ij} D_{ij}[\det D^2 u]$ is a positive constant, where $(U^{ij}) = \operatorname{Cof} D^2 u$.

Problem 13.4. Let $k \in \mathbb{N}_0$, $\alpha \in (0,1)$, and $\theta \in [0, 1/n)$. Let Ω be a uniformly convex domain in $\mathbb{R}^n$ with $\partial\Omega \in C^{k+4,\alpha}$, $f \in C^{k,\alpha}(\overline{\Omega})$, $\varphi \in C^{k+4,\alpha}(\overline{\Omega})$, $\psi \in C^{k+2,\alpha}(\overline{\Omega})$, and $\inf_\Omega \psi > 0$. Show that there exists a unique uniformly convex solution $u \in C^{k+4,\alpha}(\overline{\Omega})$ to (13.12).

Problem 13.5. Prove that the conclusions of Theorem 13.8 still hold for solutions to the equation

$$\operatorname{trace}(AD^2 v) = g, \quad \text{with } 0 < \tilde{\lambda} U \leq A \leq \tilde{\Lambda} U$$

where the constants α and C now depend also on $\tilde{\lambda}$ and $\tilde{\Lambda}$.

Problem 13.6. Extend Theorem 13.18 to $g \in L^p(\Omega)$ where $p > n$.

Problem 13.7. Prove the existence of a unique minimizer, up to linear functions, to the problem (13.57) in two dimensions.

Problem 13.8. Derive the Euler–Lagrange equations (13.58) for a smooth minimizer u of problem (13.57).

Problem 13.9. Let Ω be a uniformly convex domain in $\mathbb{R}^n$ with C^2 boundary and outer unit normal vector $\nu = (\nu_1, \ldots, \nu_n)$. Assume that $B_r(0) \subset \Omega \subset B_{nr}(0)$. Let $u \in C^1(\overline{\Omega}) \cap C^2(\Omega)$ be a convex function satisfying

$$\det D^2 u = 1 \quad \text{in } \Omega, \quad u = 0 \quad \text{on } \partial\Omega.$$

Let $(U^{ij}) = (D^2 u)^{-1}$. Show that there is a constant $c(n) > 0$ such that

$$\max_{\partial\Omega} Du \cdot \nu \geq c(n)|\Omega|^{1/n} \quad \text{and} \quad \max_{\partial\Omega}[U^{ij} D_i u \nu_j] \geq c(n)|\Omega|^{1/n}.$$

Problem 13.10 (Abreu's equation with degenerate boundary data). Let Ω be a uniformly convex domain in $\mathbb{R}^n$ with $\partial\Omega \in C^\infty$, let $\varphi \in C^\infty(\overline{\Omega})$ be uniformly convex, and let $f \in C^\infty(\overline{\Omega})$ with $f \geq c_0 > 0$. In light of Theorem 13.3, for each $t > 0$, there is a unique convex solution $u_t \in C^\infty(\overline{\Omega})$ to

$$\begin{cases} U^{ij} D_{ij} w = -f, \quad w = [\det D^2 u]^{-1} & \text{in } \Omega, \\ \quad u = \varphi, \quad w = t & \text{on } \partial\Omega, \end{cases}$$

where $(U^{ij}) = \mathrm{Cof}\, D^2 u$. Show that there exists a subsequence $\{u_{t_k}\}_{k=1}^{\infty}$, with $t_k \to 0^+$, that converges to a smooth and strictly convex solution u in Ω for Abreu's equation with degenerate boundary data

$$\begin{cases} U^{ij} D_{ij} w = -f, \quad w = [\det D^2 u]^{-1} \quad \text{in } \Omega, \\ \quad u = \varphi, \ |Du| = \infty, \ w = 0 \qquad \text{on } \partial\Omega. \end{cases}$$

Problem 13.11. In Proposition 13.1, consider, in place of $a^{ij} D_{ij} v = g$ in Ω, the equation $a^{ij} D_{ij} v + \mathbf{b} \cdot Dv = g$ where $|\mathbf{b}(x) - \mathbf{b}(x_0)| \leq M |x - x_0|^{\mu}$ in $B_r(x_0) \cap \Omega$ for some $\mu \in (0, 1)$, $M, r > 0$. Show that v is also pointwise Hölder continuous at x_0.

Problem 13.12. Extend the results in Theorems 13.5 and 13.8 to linearized Monge–Ampère equations of the type $U^{ij} D_{ij} v + \mathbf{b} \cdot Dv = g$ where the drift $\mathbf{b}$ is bounded.

13.8. Notes

In Section 13.1, Proposition 13.1 (whose arguments come from Caffarelli–Cabré [**CC**, Section 4.3]) and Theorem 13.2 are based on Le [**L1**]; see also Le–Mitake–Tran [**LMT**, Chapter 2].

For previous works related to Theorem 13.3 in Section 13.2, see [**CW**, **L1**, **TW5**, **Zh2**], for example.

The boundary Hölder gradient estimates in Sections 13.3 and 13.4 were obtained by Le and Savin [**LS1**]. For an accessible proof of Theorem 13.9, see Gilbarg–Trudinger [**GT**, Theorem 9.31]. Via a perturbation argument as in the proofs of Theorems 6.34 and 10.8, Le and Savin [**LS3**] extended the boundary Hölder gradient estimates in Theorem 13.8 to the linearized Monge–Ampère equations with L^p $(p > n)$ inhomogeneity and $C^{1,\gamma}$ boundary data. Lemma 13.7, built upon [**LS1**], is taken from Le–Nguyen [**LN2**]; see also Section 14.6 for its other use to linearized Monge–Ampère equations with L^q right-hand side where $q > n/2$.

In Section 13.5, the boundary Harnack inequality and the Comparison Theorem are based on Le [**L4**]. The contradiction argument in the proof of Theorem 13.19 follows the main lines of the proof of the Carleson estimates in Caffarelli–Fabes–Mortola–Salsa [**CFMS**]. Theorem 13.27 is an affine invariant analogue of the comparison theorems in [**CFMS**, **Bau**]. In [**CFMS**], the authors proved a comparison theorem for positive solutions of linear, uniform elliptic equations in divergence form. In [**Bau**], Bauman proved a comparison theorem for positive solutions of linear, uniform elliptic equations in nondivergence form with continuous coefficients. Lemmas 13.23 and 13.26 are from Maldonado [**Md3**] and Indratno–Maldonado–Silwal [**IMS**].

Problem 13.1 is related to Le–Savin [**LS2**, Section 3]. Problem 13.3 is based on Trudinger–Wang [**TW2**]. The result in Problem 13.4 was first established in Chau–Weinkove [**CW**]. Problem 13.6 is based on Le–Savin [**LS3**]. Problem 13.10 is taken from Chen–Li–Sheng [**CLS**]. Problem 13.11 is based on Kim–Le–Wang–Zhou [**KLWZ**].

Chapter 14

Green's Function

As mentioned in Section 1.2, linearized Monge–Ampère operators can be written in both nondivergence and divergence forms. So far, we have mostly considered their nondivergence form. We change gears in this chapter by studying these operators from a divergence form perspective. The combined viewpoint allows us to discover strong structural properties of linearized Monge–Ampère operators despite their potential singularity and degeneracy. The particular concept we study here is the Green's function, also known as the fundemental solution, of the linearized Monge–Ampère operator

$$L_u v := D_i(U^{ij} D_j v) \equiv U^{ij} D_{ij} v,$$

where $U = (U^{ij}) \equiv (\det D^2 u)(D^2 u)^{-1}$ is the cofactor matrix of the Hessian matrix $D^2 u$ of a C^3, convex function u satisfying

$$(14.1) \qquad \lambda \leq \det D^2 u \leq \Lambda \qquad \text{in } \Omega \subset \mathbb{R}^n \quad (n \geq 2).$$

We wish to establish estimates that depend only on n, λ, and Λ. Thus, by approximation arguments as in Proposition 6.15, they also hold for $u \in C^2$.

For $y \in \Omega$, let δ_y denote the Dirac measure giving unit mass to y. From the classical result on the existence and uniqueness of Green's function for linear, uniformly elliptic second-order operators in *divergence form* (see, for example, Littman–Stampacchia–Weinberger [**LSW**, Section 6] and Grüter–Widman [**GW**, Theorem 1.1]), we can make the following definition for the Green's function of the linearized Monge–Ampère operator.

Definition 14.1 (Green's function of the linearized Monge–Ampère operator). Assume $V \Subset \Omega$ and V is open. Then, for each $y \in V$, there exists a unique function $g_V(\cdot, y) : V \to [0, \infty]$ with the following properties:

(a) $g_V(\cdot, y) \in W_0^{1,q}(V) \cap W^{1,2}(V \setminus B_r(y))$ for all $q < \frac{n}{n-1}$ and all $r > 0$.

(b) $g_V(\cdot, y)$ is a weak solution of

(14.2) $$\begin{cases} -D_i(U^{ij}D_j g_V(\cdot, y)) = \delta_y & \text{in } V, \\ \qquad\qquad g_V(\cdot, y) = 0 & \text{on } \partial V; \end{cases}$$

that is,

(14.3) $$\int_V U^{ij}D_j g_V(x, y)D_i\psi(x)\,dx = \psi(y) \quad \text{for all } \psi \in C_c^\infty(V).$$

We call $g_V(\cdot, y)$ the Green's function of $L_u = D_i(U^{ij}D_j)$ in V with pole $y \in V$. We set $g_V(y, y) = +\infty$. For a construction of g_V, see Problem 14.1.

Clearly, g_V depends on the convex potential function u. However, we decide not to indicate this dependence on u since in all contexts, we will work with the Green's function associated with one function u.

Remark 14.2. We will use the following properties of Green's function.

(a) (Symmetry) As the operator $L_u = D_i(U^{ij}D_j)$ has symmetric co-efficients, we infer from [**GW**, Theorem 1.3] that

$$g_V(x, y) = g_V(y, x) \quad \text{for all } x, y \in V.$$

(b) By approximation arguments, we can use in (14.3) test functions $\psi \in W^{2,n}_{\text{loc}}(V) \cap W^{1,2}_0(V) \cap C(\overline{V})$.

(c) (Regularity) Since $g_V(\cdot, y) \in W^{1,2}(V \setminus B_r(y))$ for all $r > 0$, we can invoke the classical Hölder regularity result of linear, uniformly elliptic equations in divergence form (see [**GT**, Theorem 8.22]) to conclude that $g_V(\cdot, y) \in C^\alpha(E \setminus B_r(y))$ for all $E \Subset V$. The exponent α here depends on n and $\|D^2 u\|_{L^\infty(\tilde{E})}$ where $E \Subset \tilde{E} \Subset V$.

(d) (Representation formula) Assume V satisfies an exterior cone con-dition at every boundary point. If $\varphi \in L^n(V)$, then, by Theorem 2.84, there is a unique solution $\psi \in W^{2,n}_{\text{loc}}(V) \cap W^{1,2}_0(V) \cap C(\overline{V})$ to

(14.4) $$-U^{ij}D_{ij}\psi = \varphi \quad \text{in } V \quad \text{and} \quad \psi = 0 \quad \text{on } \partial V.$$

Using ψ as a test function in (14.3), we find

$$\psi(y) = \int_V U^{ij}D_j g_V(x, y)D_i\psi(x)\,dx.$$

Integrating by parts and recalling that (U^{ij}) is divergence-free (this is the only place we need $u \in C^3$), we obtain

(14.5) $$\psi(y) = \int_V -D_j(U^{ij}D_i\psi)g_V(\cdot, y)\,dx = \int_V g_V(\cdot, y)\varphi\,dx.$$

Note that if we use a divergence form of (14.4), namely, $-D_j(U^{ij}D_i\psi) = \varphi$, then the representation formula (14.5) holds for more singular φ (see [**LSW**, Theorem 6.1]), but in this case, ψ is not necessarily in $W^{2,n}_{\text{loc}}$.

We will prove sharp local and global bounds, higher integrability, and $W^{1,1+\kappa_0}$ estimates for the Green's function. The sharp bound on the Green's function will be employed to prove the Monge–Ampère Sobolev inequality. These results together with the representation formula (14.5) will then be applied to prove Hölder estimates for the linearized Monge–Ampère equation $L_uv = f$ where our estimates now depend on the low integrability of f, such as $\|f\|_{L^p}$ for $n/2 < p < n$. We will also obtain estimates when f is more singular but in divergence form. These Hölder estimates, to be discussed in Section 14.6 and Chapter 15, go beyond those using the Caffarelli–Gutiérrez interior Harnack inequality such as Theorems 12.14 and 13.2.

Our main tools to study deep quantitative properties of the Green's function are the interior and boundary Harnack inequalities for the linearized Monge–Ampère equation. To apply these inequalities to the Green's function, we need to check the required regularity $W^{2,n}_{\text{loc}} \cap C^0$. This is not true at the pole, and it is a delicate issue to have higher-order regularity for the Green's function away from the pole, as described in Definition 14.1 using the divergence structure of L_u. However, by using also the nondivergence form of L_u with continuous coefficient matrix U, we can in fact have the $W^{2,n}_{\text{loc}} \cap C^0$ regularity for the Green's function away from the pole. As such, the Aleksandrov–Bakelman–Pucci (ABP) maximum principle in Theorem 3.16 applies. We summarize this discussion in the following remark.

Remark 14.3. Let $u \in C^3(\Omega)$ be a convex function satisfying (14.1), and let $V \Subset \Omega$. Let $g_V(\cdot, x_0)$ be the Green's function of $L_u = D_i(U^{ij}D_j)$ in V with pole $x_0 \in V$. Then:

- For all $E \Subset V \setminus \{x_0\}$, we have

$$g_V(\cdot, x_0) \in W^{2,n}_{\text{loc}}(E) \cap C(\overline{E}) \quad \text{and} \quad U^{ij}D_{ij}g_V(\cdot, x_0) = 0 \quad \text{in } E.$$

- Assume V satisfies an exterior cone condition at every boundary point. Then $g_V(\cdot, x_0) \in C(\overline{V} \setminus B_r(x_0))$ for all $r > 0$.

Proof. We prove the first assertion. Indeed, by a covering argument, we can assume that E is convex. Let $v := g_V(\cdot, x_0)$. By Remark 14.2(c), $v \in C(\overline{E})$ and $v \in W^{1,2}(E)$ is a weak solution of

$$(14.6) \qquad\qquad D_i(U^{ij}D_jv) = 0 \quad \text{in } E.$$

By Theorem 2.84, there is a unique solution $\tilde{v} \in W^{2,n}_{\text{loc}}(E) \cap C(\overline{E})$ to

$$(14.7) \qquad\qquad U^{ij}D_{ij}\tilde{v} = 0 \quad \text{in } E, \quad \tilde{v} = v \quad \text{on } \partial E.$$

Since we also have $D_i(U^{ij}D_j\tilde{v}) = 0$ in E and $v \in W^{1,2}(E)$, we find that $\tilde{v} \in W^{1,2}(E)$. Now, by the uniqueness of $W^{1,2}$ solutions to (14.6), we have $v = \tilde{v}$. It follows that $v \in W^{2,n}_{loc}(E)$ and $U^{ij}D_{ij}v = 0$ in E.

We now prove the second assertion. Let $r > 0$. Then, by Theorem 2.84, there exists a unique solution $w \in W^{2,n}_{loc}(V \setminus B_r(x_0)) \cap C(\overline{V} \setminus B_r(x_0))$ to

$$U^{ij}D_{ij}w = 0 \quad \text{in } V \setminus B_r(x_0), \quad w = v \quad \text{on } \partial(V \setminus B_r(x_0)).$$

As in the proof of the first assertion, we have $v = w$ so $v \in C(\overline{V} \setminus B_r(x_0))$. $\square$

14.1. Bounds and Higher Integrability

Assume throughout this section that (14.1) is satisfied and g_V is the Green's function for $L_u = D_i(U^{ij}D_j)$. In Vitali's Covering Lemma (Lemma 5.32), set $\delta = K^{-1}$.

We list some general properties of Green's functions. First, from the ABP maximum principle, we get the following uniform $L^{\frac{n}{n-1}}$ and L^1 bounds.

Lemma 14.4. *Assume* (14.1) *holds. If $V \subset \Omega$ is a convex domain, then*

$$\sup_{x_0 \in V} \|g_V(\cdot, x_0)\|_{L^{\frac{n}{n-1}}(V)} \leq C(n,\lambda)|V|^{1/n}.$$

As a consequence, we have the uniform L^1 bound:

$$\sup_{x_0 \in V} \int_V g_V(\cdot, x_0)\, dx \leq C(n,\lambda)|V|^{2/n}.$$

Proof. Let $x_0 \in V$ be given and let $\sigma = g_V(\cdot, x_0)$. For any $\varphi \in L^n(V)$, let

$$\psi(y) = \int_V g_V(\cdot, y)\varphi\, dx.$$

Then, by Remark 14.2(d), $\psi \in W^{2,n}_{loc}(V) \cap W^{1,2}_0(V) \cap C(\overline{V})$ satisfies

$$-U^{ij}D_{ij}\psi = \varphi \quad \text{in } V, \qquad \psi = 0 \quad \text{on } \partial V.$$

Using $\det(U^{ij}) = (\det D^2 u)^{n-1} \geq \lambda^{n-1}$ and the ABP estimate for convex domains (Lemma 3.17), we find that ψ satisfies

$$\left| \int_V \sigma\varphi\, dx \right| = |\psi(x_0)| \leq C(n)|V|^{1/n} \left\| \frac{\varphi}{[\det(U^{ij})]^{1/n}} \right\|_{L^n(V)}$$

$$\leq C(n,\lambda)|V|^{1/n}\|\varphi\|_{L^n(V)}.$$

By duality, we obtain

$$\|g_V(\cdot, x_0)\|_{L^{\frac{n}{n-1}}(V)} = \|\sigma\|_{L^{\frac{n}{n-1}}(V)} \leq C(n,\lambda)|V|^{1/n}.$$

The uniform L^1 bound for $g_V(\cdot, x_0)$ on V follows from the Hölder inequality and the uniform $L^{\frac{n}{n-1}}$ bound. $\square$

Given an L^1 bound for a positive solution to the linearized Monge–Ampère equation, we can use the interior Harnack inequality in Theorem 12.4 to obtain a pointwise upper bound in compactly contained subsets. The following general estimate will give upper bounds for the Green's function.

Lemma 14.5. *Assume (14.1) holds. Let $S_u(x_0, t) \Subset \Omega$. Assume σ is a nonnegative function satisfying $\int_{S_u(x_0,t)} \sigma \, dx \leq A$, while for all $E \Subset S_u(x_0, t) \setminus \{x_0\}$, we have*

$$(14.8) \qquad \sigma \in W^{2,n}_{loc}(E) \cap C(\overline{E}) \quad and \quad U^{ij} D_{ij} \sigma = 0 \ in \ E.$$

Then

$$\sigma \leq C(n, \lambda, \Lambda) A t^{-\frac{n}{2}} \quad on \ \partial S_u(x_0, t/2).$$

Proof. Let $c_1 = c(n) \Lambda^{-1/2}$ and $C_1 = C(n) \lambda^{-1/2}$ be as in Lemma 5.6. Let

$$D = (S_u(x_0, t) \setminus S_u(x_0, r_2 t)) \cup S_u(x_0, r_1 t), \quad \text{where } 0 < r_1 < 1/2 < r_2 < 1.$$

Note that $r_2 S_u(x_0, t) + (1 - r_2) x_0 \subset S_u(x_0, r_2 t)$, thanks to the convexity of u and $S_u(x_0, t)$. It follows that $r_2^n |S_u(x_0, t)| \leq |S_u(x_0, r_2 t)|$ and

$$|S_u(x_0, t)| - |S_u(x_0, r_2 t)| \leq (1 - r_2^n)|S_u(x_0, t)| \leq n(1 - r_2)|S_u(x_0, t)|.$$

Then, by Lemma 5.6, we can estimate

$$|D| \leq n(1 - r_2)|S_u(x_0, t)| + |S_u(x_0, r_1 t)|$$
$$\leq C_1 n (1 - r_2) t^{n/2} + C_1 (r_1 t)^{n/2} \leq (c_1/2) t^{n/2}$$

if $r_1, 1 - r_2$ are small, depending on n, λ, Λ. Again, by Lemma 5.6,

$$(14.9) \qquad (c_1/2) t^{n/2} \leq |S_u(x_0, t) \setminus D|.$$

Given $0 < r_1 < r_2 < 1$ as above, we have

$$(14.10) \qquad \sup_{S_u(x_0,t) \setminus D} \sigma \leq C(n, \lambda, \Lambda) \inf_{S_u(x_0,t) \setminus D} \sigma.$$

Combining (14.9) and (14.10) with the L^1 bound on σ, we find that

$$\sigma \leq C(n, \lambda, \Lambda) A t^{-\frac{n}{2}} \quad on \ S_u(x_0, t) \setminus D.$$

Since $r_2 > 1/2 > r_1$, we obtain the conclusion of the lemma.

For completeness, we include the details of (14.10). By Theorem 5.30, there exists $\alpha(n, \lambda, \Lambda) \in (0, 1)$ such that for each $x \in S_u(x_0, t) \setminus D$, we have $x_0 \notin S_u(x, \alpha t)$ and $S_u(x, \alpha t) \subset S_u(x_0, t)$. Let $\tau = 1/2$. From Lemma 5.32, we can find a collection of sections $S_u(x_i, \tau \alpha t)$ with $x_i \in S_u(x_0, t) \setminus D$ such that $S_u(x_0, t) \setminus D \subset \bigcup_{i \in I} S_u(x_i, \tau \alpha t)$ and $S_u(x_i, \delta \tau \alpha t)$ are disjoint for some $\delta(n, \lambda, \Lambda) \in (0, 1)$. Invoking the volume estimates in Lemma 5.6, we find that $\#I \leq C(n, \lambda, \Lambda)$. Now, we apply the Harnack inequality in Theorem 12.4 to σ in each $S_u(x_i, \alpha t)$ to obtain (14.10). $\qquad \square$

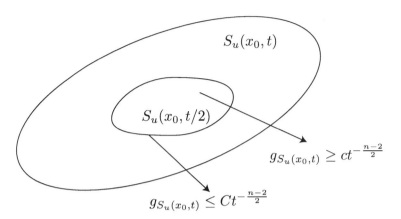

Figure 14.1. Values of the Green's function $g_{S_u(x_0,t)}$ in the smaller section $S_u(x_0, t/2)$.

We now establish bounds for the Green's function $g_V(\cdot, x_0)$ in the special case where V is itself a section of u centered at x_0; see Figure 14.1.

Lemma 14.6. *Assume* (14.1) *holds. If* $V = S_u(x_0, t) \Subset \Omega$, *then*

$$(14.11) \quad \begin{cases} \text{(a)} & g_V(x, x_0) \le C(n, \lambda, \Lambda) t^{-\frac{n-2}{2}} & \text{for all } x \in \partial S_u(x_0, t/2), \\ \text{(b)} & g_V(x, x_0) \ge c(n, \lambda, \Lambda) t^{-\frac{n-2}{2}} & \text{for all } x \in S_u(x_0, t/2). \end{cases}$$

Proof. By subtracting a linear function, we can assume that $u \ge 0$ and $u(x_0) = 0$. Let us denote $f = \det D^2 u$ and $\sigma = g_V(\cdot, x_0)$. Since

$$-U^{ij} D_{ij}(u - t) = -nf \quad \text{in } V \quad \text{and} \quad u - t = 0 \quad \text{on } \partial V,$$

we can apply Remark 14.2(d) to $u - t$ and get

$$-t = u(x_0) - t = \int_V -(nf) g_V(\cdot, x_0) \, dx = \int_V -nf\sigma \, dx.$$

The bounds on f then give the following bounds for the integral of σ:

$$\frac{t}{n\lambda} \ge \int_V \sigma \, dx \ge \frac{t}{n\Lambda}.$$

By Lemma 14.4 and Lemma 5.6, we find

$$\|\sigma\|_{L^{\frac{n}{n-1}}(S_u(x_0,t))} \le C(n, \lambda) |S_u(x_0, t)|^{1/n} \le C(n, \lambda) t^{1/2}.$$

Let

$$K = (S_u(x_0, t) \backslash S_u(x_0, r_2 t)) \cup S_u(x_0, r_1 t),$$

where $0 < r_1 < 1/2 < r_2 < 1$. As in the proof of Lemma 14.5, if $r_1, 1 - r_2$ are small, depending on n, λ, Λ, then

$$(14.12) \quad (c_1/2) t^{n/2} \le |S_u(x_0, t) \backslash K| \le C_1 t^{n/2},$$

and, together with the Hölder inequality, we also have

$$\int_K \sigma \, dx \le \|\sigma\|_{L^{\frac{n}{n-1}}(K)} |K|^{1/n} \le C(n,\lambda) t^{1/2} |K|^{1/n} \le \frac{t}{2n\Lambda}.$$

It follows that

(14.13) $$\frac{t}{n\lambda} \ge \int_{S_u(x_0,t)\setminus K} \sigma \, dx \ge \frac{t}{2n\Lambda}.$$

By Remark 14.3, σ satisfies (14.8). Thus, we have as in (14.10) that

$$\sup_{S_u(x_0,t)\setminus K} \sigma \le C(n,\lambda,\Lambda) \inf_{S_u(x_0,t)\setminus K} \sigma.$$

Combining this with (14.12) and (14.13), we find that

$$C^{-1}(n,\lambda,\Lambda) t^{-\frac{n-2}{2}} \le \sigma(x) \le C(n,\lambda,\Lambda) t^{-\frac{n-2}{2}} \quad \text{for all } x \in S_u(x_0,t)\setminus K.$$

Since $r_2 > 1/2 > r_1$, the above estimate confirms (14.11)(a).

For (14.11)(b), we only need to consider $x \in S_u(x_0,t/2)\setminus\{x_0\}$. Then, the superlevel set $A := \{y \in V : \sigma(y) \ge \sigma(x)\}$ contains x_0, and $\sigma = \sigma(x)$ on ∂A. If $\partial A \cap \partial S_u(x_0,t/2) \ne \emptyset$, then we are done. Otherwise, $\partial S_u(x_0,t/2) \subset V \setminus A$, and the ABP maximum principle gives $\max_{\partial S_u(x_0,t/2)} \sigma \le \max_{\partial(V\setminus A)} \sigma = \sigma(x)$, and the proof is finished. $\qquad\square$

Next, we estimate how the maximum of $g_V(\cdot,x_0)$ on a section of u centered at x_0 changes when passing to a concentric section with double height.

Lemma 14.7. *Assume* (14.1) *holds. If* $S_u(x_0,2t) \Subset V$, *then for either* $W = S_u(x_0,2t)$ *or* $W = V$,

(14.14) $$\max_{x\in\partial S_u(x_0,t)} g_W(x,x_0) \le C(n,\lambda,\Lambda) t^{-\frac{n-2}{2}} + \max_{z\in\partial S_u(x_0,2t)} g_W(z,x_0).$$

Proof. Let $S := S_u(x_0,2t)$. Consider $w = g_W(\cdot,x_0) - g_S(\cdot,x_0)$. Then, for any $1 < q < \frac{n}{n-1}$, $w \in W^{1,q}(S)$ is a weak solution of $D_i(U^{ij}D_j w) = 0$ in S with C^1 coefficient matrix (U^{ij}). By a classical result of Hager and Ross [**HR**], $w \in W^{1,p}_{\text{loc}}(S)$ for all $q \le p < \infty$. This show that $w \in C(S)$. Moreover, $w|_{\partial S}$ is continuous. Arguing as in the proof of Remark 14.3 that uses (14.7), we find $w \in W^{2,n}_{\text{loc}}(S) \cap C(\overline{S})$ and w satisfies $L_u w = 0$ in S.

In $\overline{S}$, w attains its maximum value on the boundary ∂S. Thus, for $x \in \partial S_u(x_0,t)$, using $g_S(\cdot,x_0) = 0$ on ∂S, we have

$$g_W(x,x_0) - g_S(x,x_0) \le \max_{\partial S} w = \max_{\partial S} g_W(\cdot,x_0).$$

Taking the maximum over $x \in \partial S_u(x_0,t)$, we obtain from Lemma 14.6 that

$$\max_{\partial S_u(x_0,t)} g_W(\cdot,x_0) \le C(n,\lambda,\Lambda) t^{-\frac{n-2}{2}} + \max_{\partial S} g_W(\cdot,x_0).$$

Therefore, (14.14) is proved. $\qquad\square$

Now, we refine Lemma 14.6 to obtain bounds for the Green's function $g_V(x, x_0)$ when V is no longer a section of u.

Theorem 14.8. *Assume that* (14.1) *is satisfied and that* $V \subset \Omega$ *is convex. Fix* $x_0 \in V$. *Suppose that* $0 < t < 1/4$, $S_u(x_0, 2t) \Subset V$ *if* $n \geq 3$, *and* $S_u(x_0, t^{1/2}) \Subset V$ *if* $n = 2$. *Then*

$$\inf_{x \in S_u(x_0, t)} g_V(x, x_0) \geq \begin{cases} c(n, \lambda, \Lambda) t^{-\frac{n-2}{2}} & \text{if } n \geq 3, \\ c(n, \lambda, \Lambda) |\log t| & \text{if } n = 2. \end{cases}$$

Moreover, let $r > 0$ *be such that* $S_u(x_0, 2r) \Subset V$ *and* $t < \min\{\frac{1}{4}, \frac{r}{2}\}$. *Then*

$$\sup_{x \in \partial S_u(x_0, t)} g_V(x, x_0) \leq \begin{cases} C(V, n, \lambda, \Lambda, r) t^{-\frac{n-2}{2}} & \text{if } n \geq 3, \\ C(V, n, \lambda, \Lambda, r) |\log t| & \text{if } n = 2. \end{cases}$$

Proof. We first prove the lower bound for g_V. Consider the following cases.

Case 1. $n \geq 3$ and $S_u(x_0, 2t) \Subset V$. Then, $w := g_V(\cdot, x_0) - g_{S_u(x_0, 2t)}(\cdot, x_0)$ solves

$$U^{ij} D_{ij} w = 0 \quad \text{in } S_u(x_0, 2t) \quad \text{with } w > 0 \quad \text{on } \partial S_u(x_0, 2t).$$

Thus, by the maximum principle, $w \geq 0$ in $S_u(x_0, t)$.

It follows from Lemma 14.6 that

$$g_V(x, x_0) \geq g_{S_u(x_0, 2t)}(x, x_0) \geq c(n, \lambda, \Lambda) t^{-\frac{n-2}{2}} \quad \text{for all } x \in S_u(x_0, t).$$

Case 2. $n = 2$ and $S_u(x_0, t^{1/2}) \Subset V$. Suppose that $S_u(x_0, 2h) \Subset V$. Then, the function

$$w(x) = g_V(x, x_0) - \inf_{\partial S_u(x_0, 2h)} g_V(\cdot, x_0) - g_{S_u(x_0, 2h)}(x, x_0)$$

satisfies

$$L_u w = 0 \quad \text{in } S_u(x_0, 2h) \quad \text{with } w \geq 0 \quad \text{on } \partial S_u(x_0, 2h).$$

By the maximum principle, $w \geq 0$ in $S_u(x_0, 2h)$. Thus, by Lemma 14.6,

(14.15)
$$g_V(x, x_0) - \inf_{\partial S_u(x_0, 2h)} g_V(\cdot, x_0) \geq g_{S_u(x_0, 2h)}(x, x_0)$$
$$\geq c(\lambda, \Lambda) \qquad \text{for all } x \in S_u(x_0, h).$$

Choose an integer $k \geq 1$ such that $2^k \leq t^{-1/2} < 2^{k+1}$. Then $|\log t| \leq Ck$. Applying (14.15) to $h = t, 2t, \ldots, 2^{k-1} t \, (\leq t^{1/2})$, we get

$$\inf_{S_u(x_0, t)} g_V(\cdot, x_0) \geq \inf_{\partial S_u(x_0, t)} g_V(\cdot, x_0)$$
$$\geq \inf_{\partial S_u(x_0, 2^k t)} g_V(\cdot, x_0) + kc \geq kc \geq c |\log t|.$$

Now, we prove the upper bound for g_V. We note from Lemma 14.4 that

$$\int_{S_u(x_0,2r)} g_V(x,x_0)\,dx \leq \int_V g_V(x,x_0)\,dx \leq C(n,\lambda)|V|^{2/n}.$$

Using Lemma 14.5, we find that

$$\sup_{\partial S_u(x_0,r)} g_V(\cdot,x_0) \leq C(n,\lambda,\Lambda,V,r).$$

The asserted upper bound for g_V just follows from iterating the estimate in Lemma 14.7 and the above upper bound for g_V.   □

Remark 14.9. As an application, we can use the sharp lower bound for the Green's function in Theorem 14.8 to prove a removable singularity result for the linearized Monge–Ampère equation; see Problem 14.2.

The next lemma studies level sets of Green's function.

Lemma 14.10 (Level sets of Green's function). *Assume* (14.1) *holds. Assume that* $S := S_u(x_0,h) \Subset \Omega$ *and* $S_u(y,\eta h) \subset S_u(x_0,h)$ *for some* $\eta \in (0,1)$. *Then, for all* $\tau > \tau_0 := C\eta^{-\frac{n}{2}}$ *where* $C = C(n,\lambda,\Lambda)$ *is large, we have*

(i) $\{x \in S : g_S(x,y) > \tau\} \subset S_u(y,\eta h 2^{-\tau/\tau_0})$ *if* $n = 2$ *and*

(ii) $\{x \in S : g_S(x,y) > \tau h^{-\frac{n-2}{2}}\} \subset S_u(y,(4\tau_0\tau^{-1})^{\frac{2}{n-2}}h)$ *if* $n \geq 3$.

Proof. By Lemma 14.4 and the volume estimates in Lemma 5.6, we obtain

$$\int_S g_S(x,y)\,dx \leq C(n,\lambda)|S|^{\frac{2}{n}} \leq C(n,\lambda)h.$$

Applying Lemma 14.5 to $g_S(\cdot,y)$ in $S_u(y,\eta h)$, we get

$$(14.16) \qquad g_S(\cdot,y) \leq C_1(n,\lambda,\Lambda)\eta^{-\frac{n}{2}}h^{-\frac{n-2}{2}} \quad \text{on } \partial S_u(y,\eta h/2).$$

Now, we use Lemma 14.7 which implies that, for all $0 < h_1 < \eta h/4$,

$$(14.17) \qquad \max_{\partial S_u(y,h_1)} g_S(\cdot,y) \leq C_2(n,\lambda,\Lambda)h_1^{-\frac{n-2}{2}} + \max_{\partial S_u(y,2h_1)} g_S(\cdot,y).$$

Let $\tau_0 := C\eta^{-\frac{n}{2}}$, where $C = C(n,\lambda,\Lambda) = 4^{\frac{n-2}{2}}(C_1 + C_2)$. Consider $\tau > \tau_0$.

If $\{x \in S : g_S(x,y) > \tau h^{-\frac{n-2}{2}}\}$ is empty, then we are done. Otherwise, by the maximum principle, we can find a smallest $0 < h_1 < \eta h/2$ such that

$$(14.18) \qquad \{x \in S : g_S(x,y) > \tau h^{-\frac{n-2}{2}}\} \subset S_u(y,h_1).$$

Then, there is $z \in \partial S_u(y,h_1)$ such that $g_S(z,y) = \tau h^{-\frac{n-2}{2}}$. See Figure 14.2.

Let m be a positive integer such that $\eta h/2 \leq 2^m h_1 < \eta h$.

Consider $n = 2$. Iterating (14.17), we find that

$$\max_{\partial S_u(y,h_1)} g_S(\cdot,y) \leq mC_2(\lambda,\Lambda) + \max_{\partial S_u(y,2^m h_1)} g_S(\cdot,y).$$

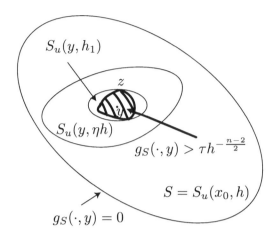

Figure 14.2. Superlevel set at level $\tau h^{-\frac{n-2}{2}}$ of the Green's function $g_S(\cdot, y)$.

The maximum principle and (14.16) give

$$\max_{\partial S_u(y, 2^m h_1)} g_S(\cdot, y) \leq \max_{\partial S_u(y, \eta h/2)} g_S(\cdot, y) \leq C_1(\lambda, \Lambda)\eta^{-1}.$$

Hence,

$$\max_{\partial S_u(y, h_1)} g_S(\cdot, y) \leq m(C_1(\lambda, \Lambda)\eta^{-1} + C_2(\lambda, \Lambda)) \leq m\tau_0 \leq \tau_0 \log_2(\eta h/h_1).$$

Thus, we obtain from $g_S(z, y) = \tau$ the estimate $\tau \leq \tau_0 \log_2(\eta h/h_1)$. The assertion (i) then follows from $h_1 \leq \eta h 2^{-\tau/\tau_0}$.

Consider now $n \geq 3$. Iterating (14.17) and using $(1 - 2^{-\frac{n-2}{2}})^{-1} < 4$ yields

$$\max_{\partial S_u(y, h_1)} g_S(\cdot, y) \leq 4C_2(n, \lambda, \Lambda)h_1^{-\frac{n-2}{2}} + \max_{\partial S_u(y, 2^m h_1)} g_S(\cdot, y).$$

The maximum principle and (14.16) give

$$\max_{\partial S_u(y, 2^m h_1)} g_S(\cdot, y) \leq \max_{\partial S_u(y, \eta h/2)} g_S(\cdot, y) \leq C_1(n, \lambda, \Lambda)\eta^{-\frac{n}{2}} h^{-\frac{n-2}{2}}.$$

Hence,

$$\max_{\partial S_u(y, h_1)} g_S(\cdot, y) \leq 4C_2 h_1^{-\frac{n-2}{2}} + C_1 \eta^{-\frac{n}{2}} h^{-\frac{n-2}{2}} \leq 4\tau_0 h_1^{-\frac{n-2}{2}}.$$

Thus, we obtain from (14.18) the estimate $\tau h^{-\frac{n-2}{2}} \leq 4\tau_0 h_1^{-\frac{n-2}{2}}$. The assertion (ii) now follows from $h_1 \leq (4\tau_0 \tau^{-1})^{\frac{2}{n-2}} h$. $\qquad \square$

We saw in Lemma 14.4, via the ABP maximum principle, the $L^{n/(n-1)}$ integrability of the Green's function. Now, using tools developed thus far, we will prove its higher integrability to all exponents less than $n/(n-2)$, with estimates depending on n, λ, Λ. This exponent is optimal as it is exactly the integrability threshold for the Green's function of the Laplace operator.

Theorem 14.11 (High integrability of Green's function). *Let $u \in C^2(\Omega)$ be a convex function satisfying $\lambda \leq \det D^2 u \leq \Lambda$ in a domain Ω in $\mathbb{R}^n$, and let $S := S_u(x_0, h)$ where $S_u(x_0, 2h) \Subset \Omega$. Let $g_S(x, y)$ be the Green's function of $L_u := D_i(U^{ij} D_j)$ on S. Let $p \in (1, \infty)$ if $n = 2$, and let $p \in (1, \frac{n}{n-2})$ if $n \geq 3$. Then, for all $y \in S$, we have the uniform estimate*

$$(14.19) \qquad \int_S g_S^p(x, y) \, dx \leq C(n, \lambda, \Lambda, p) |S|^{1 - \frac{(n-2)p}{n}}.$$

Proof. By an approximation argument, we can assume $u \in C^3(\Omega)$.

Step 1. Special case. We first prove (14.19) when $S_u(y, \eta h) \subset S_u(x_0, h)$ for some constant $\eta(n, \lambda, \Lambda) \in (0, 1)$.

By Lemma 14.10, we have for all $\tau > \tau_0(n, \lambda, \Lambda)$,

$$\{x \in S : g_S(x, y) > \tau h^{-\frac{n-2}{2}}\} \subset \begin{cases} S_u(y, \eta h 2^{-\tau/\tau_0}) & \text{if } n = 2, \\ S_u(y, (4\tau_0 \tau^{-1})^{\frac{2}{n-2}} h) & \text{if } n \geq 3. \end{cases}$$

By the upper bound on the volume of sections in Lemma 5.6, we find that

$$|\{x \in S : g_S(x, y) > \tau h^{-\frac{n-2}{2}}\}| \leq \begin{cases} C(\lambda, \Lambda) h 2^{-\tau/\tau_0} & \text{if } n = 2, \\ C(n, \lambda, \Lambda) \tau^{-\frac{n}{n-2}} h^{n/2} & \text{if } n \geq 3. \end{cases}$$

If $n = 2$ and $p \in (1, \infty)$, then the layer cake representation gives

$$\int_S g_S^p(x, y) \, dx = \int_0^{\tau_0} p\tau^{p-1} |\{x \in S : g_S(x, y) > \tau\}| \, d\tau$$
$$+ \int_{\tau_0}^\infty p\tau^{p-1} |\{x \in S : g_S(x, y) > \tau\}| \, d\tau$$
$$\leq \tau_0^p |S| + \int_{\tau_0}^\infty p\tau^{p-1} C(\lambda, \Lambda) h 2^{-\tau/\tau_0} \, d\tau \leq C(p, \lambda, \Lambda) |S|,$$

where we used $|S| \geq c(n, \Lambda) h$ in the last inequality. Hence (14.19) is proved.

Consider now $n \geq 3$ and $p \in (1, \frac{n}{n-2})$. Since $|S| \leq C(n, \lambda) h^{n/2}$, we infer that

$$|\{x \in S : g_S(x, y) > t\}| \leq C(n, \lambda, \Lambda) t^{-\frac{n}{n-2}} \quad \text{for all } t > 0.$$

It follows from the layer-cake representation that, for all $\varepsilon > 0$,

$$\int_S g_S^p(x, y) \, dx = p \int_0^\infty t^{p-1} |\{x \in S : g_S(x, y) > t\}| \, dt$$
$$\leq p|S| \int_0^\varepsilon t^{p-1} \, dt + pC \int_\varepsilon^\infty t^{p-1-\frac{n}{n-2}} \, dt$$
$$= |S| \varepsilon^p + C_1(n, p, \lambda, \Lambda) \varepsilon^{p - \frac{n}{n-2}}.$$

By choosing $\varepsilon = \left(C_1/|S|\right)^{\frac{n-2}{n}}$ in the above right-hand side to make both terms have the same power of ε, we obtain (14.19) with $C = 2C_1^{\frac{n-2}{n}p}$.

Step 2. General case when $y \in S_u(x_0, h)$. By Theorem 5.30(i), there is a constant $\eta = \eta(n, \lambda, \Lambda) \in (0, 1)$ such that for all $y \in S_u(x_0, h)$, we have

$$(14.20) \qquad\qquad S_u(y, \eta h) \subset S_u(x_0, 3h/2).$$

Thus, for any $x \in S_u(x_0, h)$, we have $g_{S_u(x_0, 3h/2)}(x, y) \geq g_{S_u(x_0, h)}(x, y)$ by the maximum principle. It follows that

$$\int_{S_u(x_0, h)} g^p_{S_u(x_0, h)}(x, y) \, dx \leq \int_{S_u(x_0, 3h/2)} g^p_{S_u(x_0, 3h/2)}(x, y) \, dx$$

$$\leq C(p, \lambda, \Lambda, n)|S|^{1 - \frac{(n-2)p}{n}}.$$

In the last inequality, we applied (14.19) to the section $S_u(x_0, 3h/2)$ and the point y in $S_u(x_0, 3h/2)$ that satisfies (14.20). The proof is now complete. $\square$

14.2. Local Integrability Estimates for the Gradient

We now turn to the gradient of the Green's function. Establishing its integrability estimates is a subtle issue. However, it can be done in two dimensions. Utilizing the high integrability of the Green's function and $W^{2, 1+\varepsilon}$ estimates for the Monge–Ampère equation, we prove the following theorem.

Theorem 14.12 ($L^{1+\kappa_0}$ estimates for the gradient of the Green's function in two dimensions). *Let $u \in C^2(\Omega)$ be a convex function satisfying $\lambda \leq \det D^2 u \leq \Lambda$ in a bounded domain Ω in $\mathbb{R}^2$. Assume that $S_u(x_0, 2h_0) \Subset \Omega$. Let $g_S(x, y)$ be the Green's function of the operator $L_u := D_i(U^{ij}D_j)$ on $S := S_u(x_0, h)$ where $h \leq h_0$. Then,*

$$\|D_x g_S(\cdot, y)\|_{L^{1+\kappa_0}(S)} \leq C(\lambda, \Lambda, \operatorname{diam}(S_u(x_0, 2h_0)), h_0)h^{\kappa_1} \quad \text{for all } y \in S,$$

with positive constants $\kappa_0(\lambda, \Lambda)$ and $\kappa_1(\lambda, \Lambda)$ chosen as follows:

$$\kappa_0 = \frac{\varepsilon_0}{2 + \varepsilon_0} < \kappa := \frac{\varepsilon}{2 + \varepsilon}, \quad \kappa_1 = \frac{\varepsilon - \varepsilon_0}{2(1 + \varepsilon)(1 + \varepsilon_0)},$$

where ε_0 is any fixed number in the interval $(0, \varepsilon)$ and $\varepsilon = \varepsilon(2, \lambda, \Lambda) > 0$ is the exponent in the $W^{2, 1+\varepsilon}$ estimates in Theorem 6.11

Remark 14.13. Clearly, Theorems 14.11 and 14.12 give the $W^{1, 1+\kappa_0}$ estimates for the Green's function. In the case of Green's function of uniformly elliptic operators, Theorem 14.12 with all $\kappa_0 < 1$ is attributed to Stampacchia. In dimensions $n \geq 2$, Grüter and Widman [**GW**] proved the L^p integrability of the gradient of the Green's function for all $p < \frac{n}{n-1}$. It would be interesting to establish L^p estimates for some $p > 1$ for the gradient of the Green's function of the linearized Monge–Ampère operator in dimensions $n \geq 3$. See also Theorem 14.27, and Problem 14.7.

Remark 14.14. By Caffarelli's $W^{2,p}$ estimates for the Monge–Ampère equation in Theorem 6.13, we can take any $\kappa_0 < 1$ provided that $\Lambda/\lambda - 1$ is sufficiently small. Particularly, in the case of $u(x) = |x|^2/2$, $L_u = \Delta$, κ_0 can be chosen to be any positive number less than 1, which is optimal.

Proof of Theorem 14.12. As before, we can assume $u \in C^3(\Omega)$. Fix $y \in S = S_u(x_0, h)$ and set

$$v(x) := g_S(x, y) + 1 \quad \text{for all } x \in \overline{S}.$$

Then $v \geq 1$ in S, $v = 1$ on ∂S, and $D_i(U^{ij} D_j v) = -\delta_y$ in S.

Step 1. Integral bound for $\log v$. We show that

$$(14.21) \qquad \int_S U^{ij} D_i(\log v) D_j(\log v)\, dx \equiv \int_S U^{ij} D_i v D_j v \frac{1}{v^2}\, dx \leq C(\lambda, \Lambda).$$

We give the proof, without extending v outside S, for the case $y = x_0$. For the general case, we extend v to be 1 in $S_u(x_0, 2h) \setminus S_u(x_0, h)$ and then in Step 1(b) below, we just replace S by $S_u(x_0, 2h)$ and $S_u(x_0, h/2)$ by $S_u(x_0, h)$ in the arguments. Assume now $y = x_0$.

Step 1(a). Estimates on $S \setminus S_u(x_0, h/2)$. By Lemma 14.6 and the maximum principle, we have $v \leq C(\lambda, \Lambda)$ in $S \setminus S_u(x_0, h/2)$. Thus

$$(14.22) \qquad S \setminus S_u(x_0, h/2) \subset \{x \in S : v(x) \leq C(\lambda, \Lambda)\}.$$

Next, we show that

$$(14.23) \qquad \int_{\{x \in S:\, v(x) \leq k\}} U^{ij} D_i v D_j v\, dx = k - 1 \quad \text{for all } k > 1.$$

Indeed, we use the truncated function $w = \min\{v - k, 0\} + k$ to avoid the singularity of v at y. By Remark 14.3, $v \in W^{1,q}_{\mathrm{loc}}(S \setminus \{y\})$ for all $q < \infty$. Note that $w = v$ on $\{x \in S : v(x) \leq k\}$, while $w = k$ on $\{x \in S : v(x) \geq k\}$. Thus, $w - 1 \in W^{1,q}_0(S)$ for all $q < \infty$, so (14.3) gives

$$\int_S U^{ij} D_i v D_j(w - 1)\, dx = w(y) - 1 = k - 1.$$

Using that $D_j w = D_j v$ on $\{x \in S : v(x) < k\}$, while $D_j w = 0$ on $\{x \in S : v(x) > k\}$, we obtain (14.23).

From (14.22), and (14.23) together with $v \geq 1$, we have

$$(14.24) \qquad \int_{S \setminus S_u(x_0, h/2)} U^{ij} D_i v D_j v \frac{1}{v^2}\, dx \leq C(\lambda, \Lambda).$$

Step 1(b). Estimates on $S_u(x_0, h/2)$. Given $w \in C_c^1(S)$, multiply the inequality $-D_i(U^{ij} D_j v) \geq 0$ in S by $\frac{w^2}{v} \geq 0$, and then integrate by parts to get (see also Problem 14.3)

$$0 \leq \int_S U^{ij} D_j v D_i \left(\frac{w^2}{v}\right) dx = \int_S 2 U^{ij} D_i w D_j v \frac{w}{v} dx - \int_S U^{ij} D_i v D_j v \frac{w^2}{v^2} dx.$$

By the Cauchy–Schwarz inequality, we have

$$\int_S U^{ij} \frac{D_i w D_j v w}{v} dx \leq \left(\int_S U^{ij} \frac{D_i v D_j v w^2}{v^2} dx\right)^{1/2} \left(\int_S U^{ij} D_i w D_j w \, dx\right)^{1/2}.$$

Therefore, rearranging the foregoing inequalities, we obtain

$$(14.25) \quad \int_S U^{ij} D_i v D_j v \frac{w^2}{v^2} dx \leq 4 \int_S U^{ij} D_i w D_j w \, dx \quad \text{for all } w \in C_c^1(S).$$

Now, we construct a suitable w to obtain an estimate on $S_u(x_0, h/2)$.

By subtracting $u(x_0) + Du(x_0) \cdot (x - x_0)$ from u, we can assume that $u(x_0) = 0$ and $Du(x_0) = 0$. Therefore $u \geq 0$ on S. Let $\gamma : \mathbb{R} \to [0,1]$ be a smooth function supported in $[-1, 4/5]$ with $\gamma \equiv 1$ on $[0, 1/2]$ and $\|\gamma'\|_{L^\infty(\mathbb{R})} \leq 4$. Let $w(x) := \gamma(u(x)/h)$. Then, $w \in C_c^1(S)$ with $w \equiv 1$ on $S_u(x_0, h/2)$ and

$$\int_S U^{ij} D_i w D_j w \, dx = \frac{1}{h^2} \int_S (\gamma'(u/h))^2 U^{ij} D_i u D_j u \, dx$$

$$\leq \frac{16}{h^2} \int_S (U D(h - u)) \cdot D(h - u) \, dx.$$

Integrating the last term by parts, using

$$\text{div}(U D(h - u)) = -\text{trace}[U D^2 u] = -2 \det D^2 u$$

and $|S| \leq C(\lambda, \Lambda) h$, we obtain

$$\int_S U^{ij} D_i w D_j w \, dx \leq -\frac{16}{h^2} \int_S \text{div}[U D(h - u)](h - u) \, dx$$

$$= \frac{32}{h^2} \int_S (\det D^2 u)(h - u) \, dx \leq \frac{32\Lambda}{h} |S| \leq C(\lambda, \Lambda).$$

Combining with (14.25) and recalling $w \equiv 1$ on $S_u(x_0, h/2)$, we find

$$(14.26) \quad \int_{S_u(x_0, h/2)} U^{ij} D_i v D_j v \frac{1}{v^2} dx \leq C(\lambda, \Lambda).$$

Therefore, (14.21) now follows from (14.24) and (14.26).

Step 2. $L^{1+\kappa_0}$ estimate for Dv. By Theorem 14.11 together with the volume bound on S, we find that $v \in L^q(S)$ for all $q < \infty$, with the bound

$$(14.27) \quad \|v\|_{L^q(S)} \leq C(\lambda, \Lambda, q) h^{\frac{1}{q}}.$$

Next, we utilize Lemma 2.58 in (14.21) to discover that

$$C(\lambda, \Lambda) \geq \int_S \frac{U^{ij} D_i v D_j v}{v^2}\, dx \geq \int_S \frac{\det D^2 u |Dv|^2}{(\Delta u) v^2}\, dx \geq \lambda \int_S \frac{|Dv|^2}{(\Delta u) v^2}\, dx.$$

Therefore, for all $1 < p < 2$, the Hölder inequality gives

$$\|Dv\|_{L^p(S)} \leq \left\| \frac{Dv}{(\Delta u)^{1/2} v} \right\|_{L^2(S)} \|(\Delta u)^{1/2} v\|_{L^{\frac{2p}{2-p}}(S)}$$

$$\leq C'(\lambda, \Lambda) \|(\Delta u) v^2\|_{L^{\frac{p}{2-p}}(S)}^{\frac{1}{2}}.$$

Now, let $\varepsilon, \varepsilon_0, \kappa_0, \kappa_1$ be as in the statement of the theorem, and let $p := \kappa_0 + 1$. Then, $\frac{p}{2-p} = 1 + \varepsilon_0$. Recalling $h_0 \geq h$, Corollary 6.26, and (14.27), we obtain

$$\|Dv\|_{L^{1+\kappa_0}(S)} \leq C' \|(\Delta u) v^2\|_{L^{1+\varepsilon_0}(S)}^{\frac{1}{2}} \leq C' \|\Delta u\|_{L^{1+\varepsilon}(S_u(x_0, h_0))}^{\frac{1}{2}} \|v\|_{L^{1/\kappa_1}(S)}$$

$$\leq C(\lambda, \Lambda, \operatorname{diam}(S_u(x_0, 2h_0)), h_0) h^{\kappa_1}.$$

The theorem is proved. □

14.3. Monge–Ampère Sobolev Inequality

As observed in the Sobolev Embedding Theorem (Theorem 2.76), when Ω is a bounded domain in $\mathbb{R}^n$ with C^1 boundary, we have the Sobolev inequality

$$\|v\|_{L^p(\Omega)} \leq C(n, p, \Omega) \|Dv\|_{L^2(\Omega)} \quad \text{for all } v \in W_0^{1,2}(\Omega),$$

where $p = 2n/(n-2)$ if $n > 3$ and $p \in (2, \infty)$ if $n = 2$. The expression $\|Dv\|_{L^2(\Omega)}$ can be expressed in terms of a quadratic form of Dv via

$$\|Dv\|_{L^2(\Omega)}^2 = \int_\Omega U^{ij} D_i v D_j v\, dx,$$

which involves the cofactor matrix $(U^{ij}) = I_n$ of the Hessian matrix $D^2 u$ of the convex potential $u(x) = |x|^2/2$.

In this section, we extend the Sobolev Embedding Theorem (Theorem 2.76) for $k = 1$ from the quadratic potential function $u = |x|^2/2$ to C^2 convex functions with Monge–Ampère measure bounded between two positive constants.

Theorem 14.15 (Monge–Ampère Sobolev inequality). *Let $u \in C^2(\Omega)$ be a convex function satisfying $\lambda \leq \det D^2 u \leq \Lambda$ in a domain Ω in $\mathbb{R}^n$. Suppose that $S_u(x_0, 2) \Subset \Omega$ and $B_1(0) \subset S_u(x_0, 1) \subset B_n(0)$. Let*

$$p = \frac{2n}{n-2} \quad \text{if } n \geq 3 \quad \text{and} \quad p \in (2, \infty) \quad \text{if } n = 2.$$

Then, there exists $K(n, p, \lambda, \Lambda) > 0$ such that for all $w \in W_0^{1,2}(S_u(x_0, 1))$,

$$(14.28) \qquad \left(\int_{S_u(x_0,1)} |w|^p\, dx \right)^{\frac{1}{p}} \leq K \left(\int_{S_u(x_0,1)} U^{ij} D_i w D_j w\, dx \right)^{1/2}.$$

The Monge–Ampère Sobolev inequality in Theorem 14.15 was obtained by Tian and Wang [**TiW**] when $n \geq 3$ and by Le [**L6**] when $n = 2$.

In the proof, we will use the following crucial lemma, due to Tian and Wang [**TiW**]. Denote by $C_0^k(\overline{\Omega})$ the set of $C^k(\overline{\Omega})$ functions vanishing on $\partial\Omega$.

Lemma 14.16 (Tian–Wang Lemma). *Let $L = D_i(a_{ij}D_j)$ be a uniformly elliptic operator in a bounded domain Ω in $\mathbb{R}^n$, where $a_{ij} = a_{ji} \in C(\Omega)$. Let $G(x,y)$ be the Green's function of L in Ω with pole $y \in \Omega$. Suppose there is an integrable, almost everywhere positive function μ such that*

$$\mu\big(\{x \in \Omega : G(x,y) > t\}\big) \leq Kt^{-p/2} \quad \text{for all } t > 0 \text{ and } y \in \Omega,$$

where $K > 0$ and $p > 2$. Here we denote for each Borel set S

$$\mu(S) = \int_S d\mu = \int_S \mu \, dx.$$

Suppose μ also satisfies the doubling condition: For some $b > 0$,

$$\mu(B_r(x)) \geq b\mu(B_{2r}(x)) \quad \text{for all } B_{2r}(x) \Subset \Omega.$$

Then for any function $u \in C_0^1(\overline{\Omega})$, we have the Sobolev inequality

$$(14.29) \qquad \Big(\int_\Omega |u|^p \, d\mu\Big)^{1/p} \leq C(K, p, b, n)\Big(\int_\Omega a_{ij} D_i u D_j u \, dx\Big)^{1/2}.$$

Proof of Theorem 14.15 assuming Lemma 14.16. By an approximation argument as in Proposition 6.15, we can assume that $u \in C^3(\Omega)$. By approximating $W_0^{1,2}$ functions by smooth, compactly supported functions, it suffices to consider $w \in C_c^\infty(S_u(x_0, 1))$. Let $S := S_u(x_0, 1)$. Let $g_S(x,y)$ be the Green's function of S with respect to $L_u := D_j(U^{ij}D_i)$ with pole $y \in S$. For any $y \in S$, there is $\eta = \eta(n, \lambda, \Lambda) > 0$ such that $S_u(y, \eta) \subset S' := S_u(x_0, 3/2)$. Note that $\{x \in S : g_S(x,y) > t\} \subset \{x \in S' : g_{S'}(x,y) > t\}$.

If $n \geq 3$, then $p = 2n/(n-2)$, and Lemma 14.10(ii) (with $h = 3/2$) together with the volume estimate for sections, gives for $t = t(n, \lambda, \Lambda)$ large,

$$|\{x \in S : g_S(x,y) > t\}| \leq |\{x \in S' : g_{S'}(x,y) > t\}| \leq C(n, \lambda, \Lambda)t^{-p/2}.$$

Clearly, since $|S| \leq |B_n(0)|$, the above estimate holds for all $t > 0$.

If $n = 2$, then by Theorem 14.11, for any $p > 1$, there exists a constant $K = K(p, \lambda, \Lambda) > 0$, such that for every $y \in S$ we have

$$(14.30) \qquad |\{x \in S : g_S(x,y) > t\}| \leq Kt^{-\frac{p}{2}} \quad \text{for all } t > 0.$$

With the above decay estimates for superlevel sets of Green's function, we can use Lemma 14.16 with $b = 2^{-n}\lambda/\Lambda$ and $\mu = \mathcal{L}^n$ to conclude (14.28). $\qquad\square$

Now, we prove the Tian–Wang Lemma.

Proof of Lemma 14.16. For any open set $\omega \subset \Omega$, we denote by $\lambda_1 = \lambda_1(L, \omega)$ and $\psi_1 = \psi_1(L, \omega)$ the first eigenvalue and the first eigenfunction which is positive and unique up to a multiplicative constant, of the following eigenvalue problem for the operator L with zero boundary condition on $\partial\omega$:

$$-L\psi_1 = \lambda_1 \mu \psi_1 \quad \text{in } \omega, \qquad \psi_1 = 0 \quad \text{on } \partial\omega.$$

We first outline the proof. We can assume the coefficient matrix (a_{ij}) is smooth. We will show in Step 1 the eigenvalue estimate

$$(14.31) \qquad \lambda_1(L, \omega)[\mu(\omega)]^{1-\frac{2}{p}} \geq [C(K, p)]^{-1}.$$

Let

$$c^* = \inf_{\omega \subset \Omega} \lambda_1(L, \omega)[\mu(\omega)]^{1-\frac{2}{p}}.$$

Next, for a fixed $k > 1$, define

$$s_k^* = \inf \left\{ \int_\Omega a_{ij} D_i v D_j v \, dx : v \in C_0^1(\overline{\Omega}), \int_\Omega F_k(v) \, d\mu = 1 \right\},$$

where $F_k(v) = \int_0^v f_k(t) \, dt$ with

$$f_k(t) = \begin{cases} |t|^{p-1} & \text{if } |t| \leq k, \\ k^{p-1} & \text{if } |t| \geq k. \end{cases}$$

We will show in Step 2 the following uniform lower bound: If $s_k^* \leq 1$, then

$$(14.32) \qquad s_k^* \geq C_1^{-1} c^* \quad \text{for some } C_1(K, p, b, n) > 1, \quad \text{for all } k.$$

Suppose now (14.31) and (14.32) are proved. Then, we obtain (14.29). Indeed, for each $u \in C_0^1(\overline{\Omega})$, take $k \geq \|u\|_{L^\infty(\Omega)}$. Then $F_k(|u|) = |u|^p/p$, and

$$\int_\Omega a_{ij} D_i u D_j u \, dx \geq s_k^* \left(\int_\Omega F_k(|u|) \, d\mu \right)^{2/p} \geq \frac{1}{C_1 C(K, p) p^{2/p}} \left(\int_\Omega |u|^p \, d\mu \right)^{2/p}.$$

Step 1. Proof of (14.31). Let G_ω be the Green's function of L on ω. Then,

$$\psi_1(y) = \int_\omega a_{ij} D_j G_\omega(\cdot, y) D_i \psi_1 \, dx = \lambda_1 \int_\omega G_\omega(\cdot, y) \psi_1 \, d\mu \quad \text{for all } y \in \omega.$$

Evaluating the above identity at the maximum point $\bar{y} \in \omega$ of ψ_1 in $\overline{\omega}$ yields

$$(14.33) \qquad 1 \leq \lambda_1 \int_\omega G_\omega(\cdot, \bar{y}) \, d\mu.$$

We now use the layer-cake formula in Lemma 2.75 and the decay of volume of the superlevel set of the Green's function to estimate for $T > 0$ and $y \in \omega$,

$$
\int_\omega G_\omega(\cdot, y) \, d\mu = \int_0^\infty \mu\big(\{x \in \omega : G_\omega(x, y) > t\}\big) \, dt
$$

$$
\leq \int_0^\infty \min\{\mu(\omega), K t^{-p/2}\} \, dt
$$

$$
\leq T\mu(\omega) + \int_T^\infty K t^{-p/2} \, dt = T\mu(\omega) + \frac{K T^{1-p/2}}{p/2 - 1}.
$$

Choosing $T > 0$ such that $K T^{-p/2} = \mu(\omega)$ then yields

$$
(14.34) \qquad \int_\omega G_\omega(\cdot, y) \, d\mu \leq C(p, K)[\mu(\omega)]^{1-2/p} \quad \text{for all } y \in \omega.
$$

Combining (14.33) with (14.34), we obtain (14.31). See also Problem 14.4.

Step 2. Proof of (14.32). For simplicity, we suppress the subscript k in s_k^*, f_k, and F_k. By the standard Lagrange multiplier method in the calculus of variations, there is a nonnegative minimizer $v \in C_0^1(\overline{\Omega})$ realizing $s^* > 0$ and

$$
-Lv = \bar{\lambda} \mu f(v) \quad \text{in } \Omega,
$$

where $\bar{\lambda}$ is the Lagrange multiplier. Our proof uses the existence of a solution $v \in C_0^1(\overline{\Omega})$ to the above nonlinear eigenvalue problem, but our conclusion does not depend on any norm of v. From the definition of f, we find that

$$
\frac{f(v)v}{p} \leq F(v) \leq f(v)v.
$$

Multiplying the above equation by v and integrating over Ω, we deduce that

$$
\bar{\lambda} = \bar{\lambda} \int_\Omega F(v) \, d\mu \leq s^* = \int_\Omega a_{ij} D_i v D_j v \, dx
$$

$$
= \bar{\lambda} \int_\Omega f(v) v \, d\mu \leq p\bar{\lambda} \int_\Omega F(v) \, d\mu = p\bar{\lambda}.
$$

Let

$$
M = \|v\|_{L^\infty(\Omega)}, \quad \Omega_t := \{v > M - t\} \equiv \{x \in \Omega : v(x) > M - t\}.
$$

We assert that for some $C(p) \in (0, \infty)$

$$
(14.35) \qquad C(p)\mu(\Omega_t) \geq \left(\frac{c^* t}{2 s^* M^{p-1}}\right)^{\frac{p}{p-2}} \quad \text{for all } t \in (0, M).
$$

For this, we use the variational characterization of $\lambda_1(L, \Omega_t)$ to first estimate

$$
\lambda_1(L, \Omega_t) \leq \frac{\int_{\Omega_t} a_{ij} D_i v D_j v \, dx}{\int_{\Omega_t} (v - M + t)^2 \, d\mu} = \frac{\int_{\Omega_t} -(v - M + t)L(v - M + t) \, dx}{\int_{\Omega_t} (v - M + t)^2 \, d\mu}
$$

and then use $-L(v - M + t) = \bar{\lambda}\mu f(v) \leq s^* M^{p-1}\mu$ in Ω_t to get

$$\lambda_1(L, \Omega_t) \leq \frac{s^* M^{p-1} \int_{\Omega_t}(v - M + t)\, d\mu}{\int_{\Omega_t}(v - M + t)^2\, d\mu} \leq \frac{s^* M^{p-1}(\mu(\Omega_t))^{1/2}}{(\int_{\Omega_t}(v - M + t)^2\, d\mu)^{1/2}}.$$

Observe that $\Omega_{t/2} \subset \Omega_t$ and

$$\int_{\Omega_t}(v - M + t)^2\, d\mu \geq \int_{\Omega_{t/2}}(v - M + t)^2\, d\mu \geq \frac{t^2}{4}\mu(\Omega_{t/2}),$$

so, together with (14.31), we obtain

$$c^*[\mu(\Omega_t)]^{\frac{2}{p}-1} \leq \lambda_1(L, \Omega_t) \leq \frac{2s^* M^{p-1}}{t}\left(\frac{\mu(\Omega_t)}{\mu(\Omega_{t/2})}\right)^{1/2}.$$

Therefore

(14.36) $\qquad \mu(\Omega_t) \geq H[\mu(\Omega_{t/2})]^{\frac{p}{3p-4}}, \quad \text{where } H := \left(\frac{c^* t}{2s^* M^{p-1}}\right)^{\frac{2p}{3p-4}}.$

Iterating (14.36), we discover

$$\mu(\Omega_t) \geq H^{\sum_{k=0}^{m-1}(\frac{p}{3p-4})^k} 2^{-\frac{2p}{3p-4}\sum_{k=0}^{m-1}k(\frac{p}{3p-4})^k}[\mu(\Omega_{t/2^m})]^{(\frac{p}{3p-4})^m}.$$

Let $a = \sup_\Omega |Dv|$. Then $\Omega_{t/2^m} \supset B_{\frac{t}{2^m a}}(z)$ where $z \in \Omega$ is the maximum point of v; that is, $v(z) = M$. By the doubling condition, we have

$$\mu(\Omega_{t/2^m}) \geq \mu(B_{\frac{t}{2^m a}}(z)) \geq b^{m-m_0}\mu(B_{\frac{t}{2^{m_0} a}}(z)),$$

where m_0 is the smallest integer such that $B_{\frac{t}{2^{m_0} a}}(z) \subset \Omega$. Since $p > 2$, we have $\lim_{m\to\infty}[\mu(\Omega_{t/2^m})]^{(\frac{p}{3p-4})^m} = 1$. Thus, we obtain (14.35) from

$$C(p)\mu(\Omega_t) \geq H^{\sum_{k=0}^{\infty}(\frac{p}{3p-4})^k} = H^{\frac{3p-4}{2(p-2)}} = \left(\frac{c^* t}{2s^* M^{p-1}}\right)^{\frac{p}{p-2}}.$$

Now, due to the layer-cake formula (Lemma 2.75), we have

$$1 = \int_\Omega F(v)\, d\mu = \int_0^M \mu(\{v > t\})F'(t)\, dt = \int_0^M \mu(\Omega_t)F'(M - t)\, dt.$$

Inserting (14.35) into the last term and changing variables, we obtain

$$C(p) \geq \left(\frac{c^*}{2s^*}\right)^{\frac{p}{p-2}} \int_0^1 t^{\frac{p}{p-2}} \frac{F'(M(1-t))}{M^{p-1}}\, dt.$$

To complete the proof of (14.32), it suffices to show that

(14.37) $\qquad \displaystyle\int_0^1 t^{\frac{p}{p-2}} \frac{F'(M(1-t))}{M^{p-1}}\, dt \geq C^{-1}(K, p, b, n).$

Case 1. $M \leq k$. In this case, from $F'(M(1-t)) = [M(1-t)]^{p-1}$, we have

$$\int_0^1 t^{\frac{p}{p-2}} \frac{F'(M(1-t))}{M^{p-1}}\, dt = \int_0^1 t^{\frac{p}{p-2}}(1-t)^{p-1}\, dt = C_0^{-1}(p).$$

Case 2. $M > k$. In this case, let $\eta := k/M \in (0,1)$. Then, we can split

$$\int_0^1 t^{\frac{p}{p-2}} \frac{F'(M(1-t))}{M^{p-1}} \, dt = \int_0^{1-\eta} t^{\frac{p}{p-2}} \eta^{p-1} \, dt + \int_{1-\eta}^1 t^{\frac{p}{p-2}} (1-t)^{p-1} \, dt.$$

To obtain (14.37), we use the assumption $s^* \leq 1$ to show that

(14.38) $\eta \geq c_0(K, p, b, n) > 0.$

Let $\omega = \{x \in \Omega : v(x) > k\}$. Then $v - k = 0$ on $\partial\omega$ and $v - k$ satisfies

$$-L(v - k) = \bar\lambda \mu f(v) = \bar\lambda k^{p-1} \mu \quad \text{in } \omega.$$

This, together with (14.34) and $\bar\lambda \leq s^* \leq p\bar\lambda$, implies that at any $y \in \omega$,

(14.39)
$$v(y) - k = \bar\lambda \int_\omega G_\omega(\cdot, y) f(v) \, d\mu = \bar\lambda k^{p-1} \int_\omega G_\omega(\cdot, y) \, d\mu$$
$$\leq C(p, K) \, k^{p-1} [\mu(\omega)]^{1 - \frac{2}{p}}.$$

On the other hand, we have $F(v) \geq f(v)v/p \geq k^p/p$ in ω, so

$$\mu(\omega) \leq pk^{-p} \int_\omega F(v) \, d\mu \leq pk^{-p} \int_\Omega F(v) \, d\mu = pk^{-p}.$$

Combining this with (14.39), we deduce that

$$M := \sup_\omega v \leq k(1 + C), \quad C = C(K, n, p).$$

This gives (14.38) with $c_0 = (1 + C)^{-1}$ and completes the proof of (14.32). The lemma is proved. $\qquad\square$

14.4. Global Higher Integrability Estimates

So far, we have only investigated interior estimates of the Green's function of the linearized Monge–Ampère operator L_u. In this section, we turn our attention to their boundary estimates. We will prove a sharp upper bound (Theorem 14.21) and a uniform global L^p bound ($p < n/(n-2)$) (Theorem 14.22) for the Green's function of L_u under the hypotheses of the Boundary Localization Theorem (Theorem 8.5). The Harnack inequalities in Theorems 12.4 and 13.19 play a crucial role. Since the domain of definition of the Green's function may have some portions on the boundary of a given domain Ω and since we need the continuity of the Green's function up to the boundary to apply the boundary Harnack inequality, we will assume a stronger assumption on the convex potential function u. We will assume in particular that $u \in C^{1,1}(\overline{\Omega}) \cap C^3(\Omega)$, so Theorem 2.84 is applicable to L_u up to the boundary. However, our estimates will not depend on the bounds on the Hessian of u.

Structural hypotheses for global estimates for the Green's function. Let $\Omega \subset \mathbb{R}^n$ and assume that there exists a constant $\rho > 0$ such that

(14.40)   the convex domain $\Omega \subset B_{1/\rho}(0) \subset \mathbb{R}^n$ and for each $y \in \partial\Omega$,

$$\text{there is an interior ball } B_\rho(z) \subset \Omega \text{ such that } y \in \partial B_\rho(z).$$

Let $u \in C^{1,1}(\overline{\Omega}) \cap C^3(\Omega)$ be a convex function satisfying

(14.41) $$0 < \lambda \le \det D^2 u \le \Lambda \quad \text{in } \Omega.$$

Assume further that on $\partial\Omega$, u separates quadratically from its tangent hyperplanes; namely, for all $x_0, x \in \partial\Omega$, we have

(14.42) $$\rho\,|x - x_0|^2 \le u(x) - u(x_0) - Du(x_0) \cdot (x - x_0) \le \rho^{-1}\,|x - x_0|^2 .$$

Definition 14.17 (Green's function for the linearized Monge–Ampère operator)**.** Assume $V \subset \overline{\Omega}$ and $V \cap \Omega$ is open. Then, for each $y \in V \cap \Omega$, there exists a unique function $g_V(\cdot, y) : V \to [0, \infty]$ with the following properties:

(a) $g_V(\cdot, y) \in W_0^{1,q}(V) \cap W^{1,2}(V \setminus B_r(y))$ for all $q < \frac{n}{n-1}$ and all $r > 0$.

(b) $g_V(\cdot, y)$ is a weak solution of

$$\begin{cases} -D_i(U^{ij} D_j g_V(\cdot, y)) &= \delta_y \quad \text{in } V \cap \Omega, \\ g_V(\cdot, y) &= 0 \quad \text{on } \partial V. \end{cases}$$

We call $g_V(x, y)$ the Green's function of $L_u = D_i(U^{ij} D_j)$ in V with pole $y \in V \cap \Omega$.

Using arguments as in Remark 14.3 and recalling Theorem 2.84, we see that the assumptions (14.40)–(14.42) will guarantee the $W_{\text{loc}}^{2,n} \cap C^0$ regularity of the Green's function away from the pole.

Remark 14.18. Let Ω and u satisfy (14.40)–(14.42), $V \subset \overline{\Omega}$, and $x_0 \in V \cap \Omega$. Let $g_V(\cdot, x_0)$ be the Green's function of L_u in V with pole x_0. Then:

- For all $E \Subset V \setminus \{x_0\}$, we have

$$g_V(\cdot, x_0) \in W_{\text{loc}}^{2,n}(E) \cap C(\overline{E}) \quad \text{and} \quad U^{ij} D_{ij} g_V(\cdot, x_0) = 0 \quad \text{in } E.$$

- If V satisfies an exterior cone condition at every boundary point, then $g_V(\cdot, x_0) \in C(\overline{V} \setminus B_r(x_0))$ for all $r > 0$.

The following global version of Lemma 14.6 gives sharp upper bounds for the Green's function $g_{S_u(x_0,2t)}(\cdot, x_0)$ in a concentric section with half height.

Lemma 14.19. *Assume Ω and u satisfy (14.40)–(14.42). If $x_0 \in \Omega$ and $t \leq c(n, \lambda, \Lambda, \rho)$, then*

$$(14.43) \qquad \max_{z \in \partial S_u(x_0,t)} g_{S_u(x_0,2t)}(z, x_0) \leq C(n, \lambda, \Lambda, \rho) t^{-\frac{n-2}{2}}.$$

Proof. Let $\sigma = g_{S_u(x_0,2t)}(\cdot, x_0)$. Let θ_* be the engulfing constant in Theorem 9.10. Let $\bar{\theta} = 2\theta_*^2$, $K = \bar{\theta}^6$. We consider two cases.

Case 1. $S_u(x_0, t/K) \subset \Omega$. By the Hölder inequality, Lemma 14.4, and the volume estimate in Lemma 5.6, we have

$$\int_{S_u(x_0,t/K)} \sigma \, dx \leq \|\sigma\|_{L^{\frac{n}{n-1}}(S_u(x_0,2t))} |S_u(x_0, t/K)|^{\frac{1}{n}}$$

$$\leq C(n, \lambda) |S_u(x_0, 2t)|^{\frac{1}{n}} |S_u(x_0, t/K)|^{\frac{1}{n}} \leq Ct.$$

By Lemma 14.5, we have $\max_{\partial S_u(x_0,t/(2K))} \sigma \leq Ct^{-\frac{n-2}{2}}$. Now, the bound (14.43) follows from the maximum principle in $S_u(x_0, 2t) \setminus S_u(x_0, t/(2K))$.

Case 2. $S_u(x_0, t/K) \not\subset \Omega$. Let $\bar{h}(x_0)$ be the height of the maximal interior section centered at x_0. Then $c \geq t \geq K\bar{h}(x_0)$. Suppose that $\partial S_u(x_0, \bar{h}(x_0)) \cap \partial\Omega = \{0\}$. By the engulfing property in Theorem 9.10 and $\bar{\theta} > \theta_*$, we have

$$S_u\left(x_0, \frac{t}{K}\right) \subset S_u\left(0, \frac{\bar{\theta}t}{K}\right) \subset S_u\left(x_0, \frac{\bar{\theta}^2 t}{K}\right) \subset S_u\left(0, \frac{\bar{\theta}^3 t}{K}\right)$$

$$\subset S_u\left(x_0, \frac{\bar{\theta}^4 t}{K}\right) \subset S_u\left(0, \frac{\bar{\theta}^5 t}{K}\right) \subset S_u\left(x_0, \frac{\bar{\theta}^6 t}{K}\right) \subset S_u(x_0, t).$$

By Lemma 9.1, there is $y \in \partial S_u(0, \frac{\bar{\theta}^3 t}{K}) \cap \Omega$ such that $\text{dist}(y, \partial\Omega) \geq c_1 t^{1/2}$ for some $c_1(n, \lambda, \Lambda, \rho) > 0$. Hence, by Proposition 9.2, there exists a constant $c_2(n, \lambda, \Lambda, \rho) \leq \frac{\bar{\theta}}{\theta_* K}$ such that $S_u(y, c_2 t) \subset \Omega$.

We claim that

$$(14.44) \qquad S_u\left(0, \frac{\bar{\theta}t}{K}\right) \cap S_u(y, c_2 t) = \emptyset, \quad S_u(y, c_2 t) \subset S_u\left(0, \frac{\bar{\theta}^5 t}{K}\right) \cap \Omega.$$

Indeed, suppose there is $z \in S_u(0, \frac{\bar{\theta}t}{K}) \cap S_u(y, c_2 t)$. Then, recalling the quasi-distance δ_u in (13.41), we have

$$\delta_u(0, y) = \frac{\bar{\theta}^3 t}{K}, \quad \delta_u(0, z) \leq \frac{\bar{\theta}t}{K}, \quad \text{and} \quad \delta_u(y, z) \leq c_2 t.$$

Moreover, we observe from the engulfing property in Theorem 9.10 that

$$\delta_u(z, y) \leq \theta_* \delta_u(y, z) \leq \theta_* c_2 t \leq \bar{\theta}t/K.$$

These together with Lemma 13.22 contradict $\delta_u(0, y) = \frac{\bar{\theta}^3 t}{K}$, because

$$\delta_u(0, y) \le \theta_*^2 (\delta_u(0, z) + \delta_u(z, y)) \le \frac{\bar{\theta}}{2}\left[\frac{\bar{\theta} t}{K} + \frac{\bar{\theta} t}{K}\right] < \frac{\bar{\theta}^3 t}{K}.$$

Suppose now that there is $z \in \partial S_u(0, \frac{\bar{\theta}^5 t}{K}) \cap S_u(y, c_2 t)$. Then, from $S_u(y, c_2 t) \subset \Omega$, we have $\delta_u(0, z) = \frac{\bar{\theta}^5 t}{K}$ and $\delta_u(y, z) \le c_2 t \le \frac{\bar{\theta} t}{K}$. But from Lemma 13.22, we have a contradiction, because

$$\delta_u(0, z) \le \theta_*^2 (\delta_u(0, y) + \delta_u(y, z)) \le \frac{\bar{\theta}}{2}\left[\frac{\bar{\theta}^3 t}{K} + \frac{\bar{\theta}^3 t}{K}\right] < \frac{\bar{\theta}^5 t}{K} = \delta_u(0, z).$$

Therefore, the claim (14.44) is proved.

From (14.44), the L^1 bound in Lemma 14.4, and Lemma 5.6, we obtain

$$\int_{S_u(y, c_2 t)} \sigma \, dx \le \int_{S_u(x_0, 2t)} \sigma \, dx \le C(n, \lambda)|S_u(x_0, 2t)|^{\frac{2}{n}} \le C(n, \lambda) t.$$

By Lemma 14.5, $\max_{\partial S_u(y, c_2 t/2)} \sigma \le C t^{-\frac{n-2}{2}}$. Thus, by the maximum principle, for σ in $S_u(y, c_2 t/2)$, we have

$$\sigma(y) \le C(n, \lambda, \Lambda) t^{-\frac{n-2}{2}}.$$

With this estimate, as in the proof of Lemma 13.24, we use the chain property of sections in Theorem 9.13 (with $\tau = 1/2$) and the boundary Harnack inequality in Theorem 13.19 to conclude that

$$\sigma \le C(n, \rho, \lambda, \Lambda)\sigma(y) \le C(n, \rho, \lambda, \Lambda) t^{-\frac{n-2}{2}} \quad \text{on } \partial S_u(0, \bar{\theta}^3 t/K) \cap \Omega.$$

Note that Remark 14.18 ensures that σ has the required regularity to apply these theorems. Since $S_u(0, \bar{\theta}^3 t/K) \subset S_u(x_0, t)$, the bound (14.43) now follows from the maximum principle. $\qquad\square$

The next lemma estimates the growth of the Green's function near the pole. It is a variant at the boundary of Lemma 14.7.

Lemma 14.20. *Assume* Ω *and* u *satisfy* (14.40)–(14.42). *If* $x_0 \in \Omega$, *then*

$$(14.45) \qquad \max_{\partial S_u(x_0, t)} g_\Omega(\cdot, x_0) \le C(n, \lambda, \Lambda, \rho) t^{-\frac{n-2}{2}} + \max_{\partial S_u(x_0, 2t)} g_\Omega(\cdot, x_0).$$

The proof is the same as that of Lemma 14.7 where now we consider $w = g_\Omega(\cdot, x_0) - g_{S_u(x_0, 2t)}(\cdot, x_0)$ and apply Lemma 14.19 instead of Lemma 14.6.

In the following theorem, we obtain sharp upper bounds for the Green's function of the linearized Monge–Ampère operator L_u. They are natural global counterparts of Theorem 14.8 where we established sharp upper bounds for the Green's function of L_u in the interior domain V of Ω.

Theorem 14.21 (Bound on Green's function). *Assume that Ω and u satisfy* (14.40)–(14.42). *Let $x_0 \in \Omega$ and $0 < t_0 < c(n, \lambda, \Lambda, \rho)$. Then*

$$\max_{x \in \partial S_u(x_0, t_0)} g_\Omega(x, x_0) \leq \begin{cases} C(n, \lambda, \Lambda, \rho) t_0^{-\frac{n-2}{2}} & \text{if } n \geq 3, \\ C(n, \lambda, \Lambda, \rho) |\log t_0| & \text{if } n = 2. \end{cases}$$

Proof. Let $m \in \mathbb{N}_0$ be a nonnegative integer such that $c/2 < 2^m t_0 \leq c$. Iterating Lemma 14.20, we find

$$(14.46) \qquad \max_{\partial S_u(x_0, t_0)} g_\Omega(\cdot, x_0) \leq \max_{\partial S_u(x_0, 2^m t_0)} g_\Omega(\cdot, x_0) + C t_0^{-\frac{n-2}{2}} \sum_{l=1}^m (2^l)^{-\frac{n-2}{2}}.$$

Then, the theorem follows from the following assertion:

$$(14.47) \qquad \sigma := g_\Omega(\cdot, x_0) \leq C(n, \lambda, \Lambda, \rho) \quad \text{in } \Omega \backslash S_u(x_0, c/2).$$

Indeed, if $n \geq 3$, then the sum in l on the right-hand side of (14.46) is not greater than some constant $C(n)$, and hence, (14.46) and (14.47) give

$$\max_{\partial S_u(x_0, t_0)} g_\Omega(\cdot, x_0) \leq C(n, \lambda, \Lambda, \rho) + C C(n) t_0^{-\frac{n-2}{2}} \leq C(n, \lambda, \Lambda, \rho) t_0^{-\frac{n-2}{2}}.$$

If $n = 2$, then since $m \leq \log_2(c/t_0)$, (14.46) and (14.47) give

$$\max_{\partial S_u(x_0, t_0)} g_\Omega(\cdot, x_0) \leq C(\lambda, \Lambda, \rho) + C \log_2(c/t_0) \leq C(\lambda, \Lambda, \rho) |\log t_0|.$$

It remains to prove (14.47). If $S_u(x_0, c/4) \subset \Omega$, then as in Case 1 in the proof of Lemma 14.19, (14.47) follows from the maximum principle and

$$\max_{\partial S_u(x_0, c/8)} \sigma \leq C(n, \lambda, \Lambda, \rho) c^{-\frac{n-2}{2}} \leq C(n, \lambda, \Lambda, \rho).$$

Suppose now $S_u(x_0, c/4) \not\subset \Omega$. We consider the ring

$$A = S_u(x_0, c) \setminus S_u(x_0, c/4)$$

and focus on the values of σ on $\partial S_u(x_0, c/2) \cap \Omega$. We first construct a section

$$(14.48) \qquad S_u(z, c_2) \Subset A \cap \Omega \quad \text{for some } c_2(n, \lambda, \Lambda, \rho) > 0.$$

Since $S_u(x_0, c/4) \not\subset \Omega$, we can find $y \in \partial S_u(x_0, c/2) \cap \partial\Omega \setminus S_u(x_0, 3c/8)$. Then, Theorem 9.12 gives $S_u(y, 2c_3) \subset S_u(x_0, 2c/3) \backslash S_u(x_0, c/3)$ for some $c_3(n, \lambda, \Lambda, \rho) > 0$. Let z be the apex of $S_u(y, c_3)$ (see Lemma 9.1); that is, $z \in \partial S_u(y, c_3) \cap \Omega$ with $\text{dist}(z, \partial\Omega) \geq c_4$ for some $c_4(n, \lambda, \Lambda, \rho) > 0$. We now use Proposition 9.2 to obtain a desired section $S_u(z, c_2) \Subset A \cap \Omega$.

The uniform L^1 bound for σ from Lemma 14.4 gives

$$(14.49) \qquad \int_{S_u(z, c_2)} \sigma \, dx \leq \int_\Omega \sigma \, dx \leq C(n, \lambda) |\Omega|^{2/n} \leq C(n, \lambda, \Lambda, \rho).$$

Thus, in Case 2 in the proof of Lemma 14.19, replacing y, t, $\partial S_u(0, \bar{\theta}^3 t/K)$ by z, 1, $\partial S_u(x_0, c/2)$, respectively, we obtain

$$(14.50) \qquad \sigma \leq C(n, \lambda, \Lambda, \rho) \quad \text{on } \partial S_u(x_0, c/2) \cap \Omega.$$

Now, (14.47) also follows from (14.50) and the maximum principle. $\qquad \square$

From Theorem 14.21, we obtain the following uniform L^p bound.

Theorem 14.22 (L^p-integrability of Green's function). *Assume that Ω and u satisfy (14.40)–(14.42). Let $p \in (1, \frac{n}{n-2})$ if $n \geq 3$, and let $p \in (1, \infty)$ if $n = 2$. Then, there is a constant $C = C(n, \lambda, \Lambda, \rho, p)$ such that*

$$(14.51) \qquad \sup_{x_0 \in \Omega} \int_\Omega g_\Omega^p(x, x_0)\, dx \leq C.$$

More generally, for all $x_0 \in \Omega$ and $t \leq c_1(n, \lambda, \Lambda, \rho, p)$, we have

$$(14.52) \qquad \sup_{x \in S_u(x_0,t) \cap \Omega} \int_{S_u(x_0,t)} g_{S_u(x_0,t)}^p(y, x)\, dy \leq C|S_u(x_0,t)|^{1 - \frac{n-2}{n}p}.$$

Proof. We give the proof for $n \geq 3$; the case $n = 2$ is similar. Let $x_0 \in \Omega$. Let C, c be the constants in Theorem 14.21. Denote $C_* = C/c^{\frac{n-2}{2}}$. For $s \geq C_*$, we have $(\frac{C}{s})^{\frac{2}{n-2}} \leq c$. Hence, from Theorem 14.21

$$\{x \in \Omega : g_\Omega(x, x_0) > s\} \subset S_u(x_0, (C/s)^{\frac{2}{n-2}}).$$

Then, by the volume estimate in Corollary 9.9, we have

$$|\{x \in \Omega : g_\Omega(x, x_0) > s\}| \leq C_1 |(C/s)^{\frac{2}{n-2}}|^{\frac{n}{2}} = C_3 s^{-\frac{n}{n-2}}.$$

As in the proof of Theorem 14.11 for the special case with $n \geq 3$ where now we replace S by Ω and ε by $(C_*/|\Omega|)^{\frac{n-2}{n}}$, we obtain from the above inequality and the layer-cake representation (Lemma 2.75) that

$$\int_\Omega g_\Omega^p(x, x_0) dx \leq C(n, \lambda, \Lambda, \rho, p)|\Omega|^{1 - \frac{n-2}{n}p} \leq C(n, \lambda, \Lambda, \rho, p).$$

We now prove (14.52). Let θ_* be as in Theorem 9.10. Fix $x \in S_u(x_0, t) \cap \Omega$, where $t \leq c_1 := c/(2\theta_*)$. Then, by the engulfing property of sections in Theorem 9.10, we have $S_u(x_0, t) \subset S_u(x, \theta_* t)$. Therefore, if $y \in S_u(x_0, t)$, then $y \in S_u(x, \theta_* t) \subset S_u(x, 2\theta_* t)$. Since

$$(14.53) \qquad \int_{S_u(x_0,t)} g_{S_u(x_0,t)}^p(y, x)\, dy \leq \int_{S_u(x,2\theta_*t)} g_{S_u(x,2\theta_*t)}^p(y, x)\, dy,$$

it suffices to bound from above the term on the right-hand side of (14.53).

As in the above upper bound for $\int_\Omega g_\Omega^p(x, x_0) \, dx$, if we replace Ω by $S_u(x, 2\theta_* t)$ (recall $2\theta_* t \leq c$) and C_* by $C_0 = C/t^{\frac{n-2}{2}}$, then for $s \geq C_0$,

$$\{y \in S_u(x, 2\theta_* t) : g_{S_u(x, 2\theta_* t)}(y, x) > s\} \subset \{y \in \Omega : g_\Omega(y, x) > s\}$$

$$\subset S_u(x, (C/s)^{\frac{2}{n-2}}).$$

Consequently, we also obtain as in the proof of Theorem 14.11 that

$$(14.54) \quad \int_{S_u(x, 2\theta_* t)} g_{S_u(x, 2\theta_* t)}^p(y, x) \, dy \leq C(n, \lambda, \Lambda, \rho, p) |S_u(x_0, 2\theta_* t)|^{1 - \frac{n-2}{n} p}.$$

Since $t \leq c_1$, we can use the volume estimates of sections in Corollary 9.9 to deduce (14.52) from (14.53) and (14.54). $\qquad \square$

Remark 14.23. Under the assumptions (14.40)–(14.42), $\|D^2 u\|$ can be arbitrarily large compared to n, λ, Λ, and ρ. The best global structural estimate available from (14.40)–(14.42) is $D^2 u \in L^{1+\varepsilon}(\Omega)$, where ε is a (possibly small) positive constant depending on n, λ, and Λ; see Theorem 10.1 and Proposition 6.28. However, Theorem 14.22 establishes the same global integrability of the Green's function for the linearized Monge–Ampère operator as the Green's function of the Laplace operator which corresponds to $u(x) = |x|^2/2$. Moreover, Theorem 14.22 also says that, as a degenerate and singular nondivergence form operator, L_u has Green's function with global L^p-integrability higher than that of a typical uniformly elliptic operator in nondivergence form as established by Fabes and Stroock [**FSt**].

Corollary 14.24. *Assume that Ω and u satisfy (14.40)–(14.42). Let $p \in (1, \frac{n}{n-2})$ if $n \geq 3$, and let $p \in (1, \infty)$ if $n = 2$. Let $z \in \partial\Omega$. Then:*

(i) *For all $x \in S_u(z, t) \cap \Omega$ and $t \leq c_1(n, \lambda, \Lambda, \rho, p)$,*

$$(14.55) \quad \|g_{S_u(z,t)}(x, \cdot)\|_{L^p(S_u(z,t) \cap \Omega)} \leq C(n, \lambda, \Lambda, \rho, p) |S_u(z, t)|^{\frac{1}{p} - \frac{n-2}{n}}.$$

(ii) *For all $x \in B_t(z) \cap \Omega$ and $t \leq c(n, \lambda, \Lambda, \rho)$,*

$$(14.56) \quad \|g_{B_t(z) \cap \Omega}(x, \cdot)\|_{L^p(B_t(z) \cap \Omega)} \leq C(n, \lambda, \Lambda, \rho, p) |B_t(z) \cap \Omega|^{\frac{3}{4}(\frac{1}{p} - \frac{n-2}{n})}.$$

Proof. We first prove (14.55). Let $V := S_u(z, t)$. Since Ω and u satisfy (14.40)–(14.42), by Theorem 9.10, there exists $\theta_*(n, \rho, \lambda, \Lambda) > 1$ such that

$$S_u(z, t) \subset S_u(x, \theta_* t) \quad \text{for all } x \in S_u(z, t).$$

Fix $x \in S_u(z, t) \cap \Omega$. Then, the symmetry of the Green's function gives

$$(14.57) \quad \begin{aligned} \int_{S_u(z,t) \cap \Omega} g_V^p(x, y) \, dy &= \int_{S_u(z,t) \cap \Omega} g_V^p(y, x) \, dy \\ &\leq \int_{S_u(x, \theta_* t)} g_{S_u(x, \theta_* t)}^p(y, x) \, dy. \end{aligned}$$

From Theorem 14.22, we find that

$$(14.58) \qquad \int_{S_u(x,\theta_* t)} g^p_{S_u(x,\theta_* t)}(y,x)\, dy \le C(n,\lambda,\Lambda,\rho,p)|S_u(x,\theta_* t)|^{1-\frac{n-2}{n}p}.$$

By the doubling property of section volume in Corollary 9.9(ii), we have

$$|S_u(x,\theta_* t)| \le C_2(n,\lambda,\Lambda,\rho)|S_u(z,t)|.$$

The desired estimate (14.55) then follows from (14.57) and (14.58).

Now, we prove (14.56). For simplicity, assume that $z = 0$, $u(0) = 0$, and $Du(0) = 0$. As a consequence of the Boundary Localization Theorem (Theorem 8.5), we have (see (8.39)) for $h \le c(n,\lambda,\Lambda,\rho)$ and $C(n,\lambda,\Lambda,\rho) > 1$

$$\overline{\Omega} \cap B^+_{ch^{1/2}/|\log h|}(0) \subset S_u(0,h) \subset \overline{\Omega} \cap B^+_{Ch^{1/2}|\log h|}(0).$$

Hence for $|x| \le t \le c$, we deduce from the first inclusion that

$$(14.59) \qquad A = \Omega \cap B_t(0) \subset S_u(0,t^{3/2}) := V.$$

Arguing as in (14.57) and (14.58), we find that for all $x \in B_t(0) \cap \Omega$,

$$(14.60) \qquad \|g_{B_t(0)\cap\Omega}(x,\cdot)\|_{L^p(B_t(0)\cap\Omega)} \le \|g_V(x,\cdot)\|_{L^p(V\cap\Omega)} \le C|V|^{\frac{1}{p}-\frac{n-2}{n}}.$$

Using the volume estimate for sections in Lemma 5.8, we find that

$$|V| \le C(n,\lambda)t^{\frac{3n}{4}} \le C|B_t(0) \cap \Omega|^{\frac{3}{4}}.$$

This together with (14.60) and (14.59) implies (14.56). $\qquad\qquad\square$

14.5. Global Integrability Estimates for the Gradient

In this section, we establish global estimates for the gradient of the Green's function of the linearized Monge–Ampère operator $D_i(U^{ij}D_j)$. Our analysis will be involved with Ω and u satisfying either the global conditions (14.40)–(14.42) or the following local conditions (14.61)–(14.64).

Structural hypotheses for boundary estimates for the Green's function. Let Ω be a bounded convex domain in $\mathbb{R}^n$ with

$$(14.61) \qquad B_\rho(\rho e_n) \subset \Omega \subset \{x_n \ge 0\} \cap B_{1/\rho}(0),$$

for some small $\rho > 0$ where we denote $e_n := (0,\ldots,0,1) \in \mathbb{R}^n$. Assume that

(14.62) for each $y \in \partial\Omega \cap B_\rho(0)$, there is an interior ball

$$B_\rho(z) \subset \Omega \text{ such that } y \in \partial B_\rho(z).$$

Let $u \in C^{1,1}(\overline{\Omega}) \cap C^3(\Omega)$ be a convex function satisfying

$$(14.63) \qquad 0 < \lambda \le \det D^2 u \le \Lambda \quad \text{in } \Omega.$$

We assume that on $\partial\Omega \cap B_\rho(0)$, u separates quadratically from its tangent hyperplanes on $\partial\Omega$; that is, if $x_0 \in \partial\Omega \cap B_\rho(0)$, then for all $x \in \partial\Omega$,

$$(14.64) \quad \rho |x - x_0|^2 \leq u(x) - u(x_0) - Du(x_0) \cdot (x - x_0) \leq \rho^{-1} |x - x_0|^2.$$

Theorem 14.25 (Global $L^{1+\kappa_0}$ estimates for gradient of the Green's function). *There exists a constant $p_0(n, \lambda, \Lambda) > \frac{2n}{3n-2}$ with the following properties.*

   (i) *Assume that Ω and u satisfy (14.40)–(14.42). Then,*

   $$\int_\Omega |D_x g_\Omega(x, y)|^{p_0} \, dx \leq C(n, \lambda, \Lambda, \rho) \quad \text{for all } y \in \Omega.$$

   (ii) *Assume Ω and u satisfy (14.61)–(14.64). If $A = \Omega \cap B_\delta(0)$ where $\delta \leq c(n, \lambda, \Lambda, \rho)$, then*

   $$\int_A |D_x g_A(x, y)|^{p_0} \, dx \leq C(n, \lambda, \Lambda, \rho) \quad \text{for all } y \in A.$$

In particular, in two dimensions, together with Theorem 14.22, we obtain global $W^{1,1+\kappa_0}$ estimates (with $\kappa_0(\lambda, \Lambda) = p_0 - 1 > 0$ as in Theorem 14.12) for the Green's function of the linearized Monge–Ampère operator $D_i(U^{ij}D_j)$.

   (iii) *Let V be either Ω as in (i) or A as in (ii). Let $\bar\kappa \in (0, \frac{1}{n-1})$. Then, there is a constant $\gamma(n, \bar\kappa) > 0$ such that if $|\Lambda/\lambda - 1| < \gamma$, we have*

   $$(14.65) \quad \int_V |D_x g_V(x, y)|^{1+\bar\kappa} \, dx \leq C(n, \lambda, \Lambda, \bar\kappa, \rho) \quad \text{for all } y \in V.$$

   (iv) *Assume $n \geq 3$. Let V be either Ω as in (i) or A as in (ii). Assume that for some $\gamma > n/2$,*

   $$(14.66) \quad \int_V \|D^2 u\|^\gamma \, dx \leq K.$$

   Then, there exists $\tilde\kappa(n, \lambda, \Lambda, \gamma) > 0$ such that

   $$(14.67) \quad \int_V |D_x g_V(x, y)|^{1+\tilde\kappa} \, dx \leq C(n, \lambda, \Lambda, \rho, \gamma, K) \quad \text{for all } y \in V.$$

Remark 14.26. In any dimension $n \geq 2$, Theorem 14.25 establishes the global $W^{1,1+\kappa}$ estimates for any $1 + \kappa$ close to $\frac{n}{n-1}$ for the Green's function of the linearized Monge–Ampère operator $L_u := D_i(U^{ij}D_j)$ when $\det D^2 u$ is close to a positive constant. Hence, if $\det D^2 u$ is continuous on $\overline\Omega$, then the Green's function of L_u has the same global integrability, up to the first-order derivatives, as the Green's function of the Laplace operator.

Proof of Theorem 14.25. We first prove (i) and (ii). Let V be either Ω as in (i) or A as in (ii). We need to show that, for some $p_0(n, \lambda, \Lambda) > \frac{2n}{3n-2}$,

$$(14.68) \quad \int_V |D_x g_V(x, y)|^{p_0} \, dx \leq C(n, \lambda, \Lambda, \rho) \quad \text{for all } y \in V.$$

Step 1. $L^{1+\varepsilon}$ estimate for D^2u in V. By Theorem 6.11 and Remark 6.12, there exists $\varepsilon(n, \Lambda/\lambda) > 0$ such that $D^2u \in L^{1+\varepsilon}_{\text{loc}}(\Omega)$. Arguing as in the proof of the global $W^{2,1+\varepsilon}$ estimates in Theorem 10.1, we can show that

$$\int_V \|D^2u\|^{1+\varepsilon}\, dx \le C(n, \lambda, \Lambda, \rho). \tag{14.69}$$

Let $0 < \varepsilon_0 < \varepsilon$ be any positive number depending on n and Λ/λ and let

$$p = \frac{2 + 2\varepsilon_0}{2 + \varepsilon_0} \in (1, 2). \tag{14.70}$$

Fix $y \in V$. Let $v := g_V(\cdot, y)$. Let $k > 0$, to be chosen. As in (14.23), we have

$$\int_{\{x \in V : v(x) \le k\}} U^{ij} D_i v D_j v\, dx = k. \tag{14.71}$$

Step 2. We next claim that

$$\int_{\{x \in V : v(x) \le k\}} |Dv|^p\, dx \le C_0(n, \lambda, \Lambda, \rho) k^{p/2}. \tag{14.72}$$

The proof follows along the lines of Step 2 in the proof of Theorem 14.12.

Let $S = \{x \in V : v(x) \le k\}$. Then, by Lemma 2.58 and $\det D^2u \ge \lambda$,

$$k \ge \int_S U^{ij} D_i v D_j v\, dx \ge \int_S \frac{\det D^2u |Dv|^2}{\Delta u}\, dx \ge \lambda \int_S \frac{|Dv|^2}{\Delta u}\, dx.$$

Thus, using the Hölder inequality, we obtain

$$\|Dv\|_{L^p(S)} \le \|Dv/(\Delta u)^{1/2}\|_{L^2(S)} \|\Delta u\|_{L^{\frac{p}{2-p}}(S)}^{1/2} \le (k/\lambda)^{1/2} \|(\Delta u)\|_{L^{\frac{p}{2-p}}(S)}^{\frac{1}{2}}.$$

Applying the above estimate to p defined in (14.70), noting that $\frac{p}{2-p} = 1 + \varepsilon_0$, and recalling (14.69), we obtain (14.72) from the following estimates:

$$\|Dv\|_{L^p(S)} \le \lambda^{-1/2} k^{1/2} \|(\Delta u)\|_{L^{1+\varepsilon_0}(S)}^{\frac{1}{2}}$$

$$\le \lambda^{-1/2} k^{1/2} \|\Delta u\|_{L^{1+\varepsilon}(S)}^{\frac{1}{2}} |S|^{\frac{\varepsilon - \varepsilon_0}{2(1+\varepsilon)(1+\varepsilon_0)}} \le k^{1/2} C(n, \lambda, \Lambda, \rho).$$

Step 3. For any $1 < q < \frac{n}{n-2}$, we show that

$$\|v\|_{L^q(S)} \le \|v\|_{L^q(V)} \le C(n, \lambda, \Lambda, \rho, q). \tag{14.73}$$

Indeed, if Ω and u satisfy (14.61)–(14.64) and if $A = \Omega \cap B_\delta(0)$, where $\delta \le c(n, \lambda, \Lambda, \rho)$, then Corollary 14.24(ii) gives

$$\int_A g_A^q(x, y)\, dx = \int_A g_A^q(y, x)\, dx \le C|A|^{\frac{3}{4}(1 - \frac{n-2}{n}q)} \le C(n, \lambda, \Lambda, \rho, q).$$

If Ω and u satisfy (14.40)–(14.42), then by Theorem 14.22, we have

$$\int_\Omega g_\Omega^q(x, y) \, dx = \int_\Omega g_\Omega^q(y, x) \, dx \le C(n, \lambda, \Lambda, \rho, q).$$

Thus, (14.73) follows from the above estimates.

Due to (14.73) and Markov's inequality, we have

(14.74) $$|V \cap \{v \ge k\}| \le C_1(n, \lambda, \Lambda, \rho, q) k^{-q}.$$

Step 4. Completion of the proof of (14.68). Note that, for any $\eta > 0$,

$$V \cap \{|Dv| \ge \eta\} \subset \left(V \cap \{v \ge k\}\right) \cup \left(V \cap \{|Dv| \ge \eta\} \cap \{v \le k\}\right).$$

Thus, by using (14.74) and (14.72), we obtain

$$|V \cap \{|Dv| \ge \eta\}| \le \frac{C_1}{k^q} + \int_{\{x \in V : v(x) \le k\}} \frac{|Dv|^p}{\eta^p} \, dx \le \frac{C_1}{k^q} + \frac{C_0 k^{p/2}}{\eta^p}.$$

Let $k = \eta^{\frac{2p}{p+2q}}$. Then $|V \cap \{|Dv| \ge \eta\}| \le C_2 \eta^{-\frac{2pq}{p+2q}}$. It follows from the layer-cake representation that $|Dv| \in L^{\hat{p}}(V)$ for any $\hat{p} < \frac{2pq}{p+2q}$. The proof of (14.68) will be complete if we can choose $1 < q < \frac{n}{n-2}$ to make

$$\frac{2pq}{p+2q} > \frac{2n}{3n-2},$$

so as to choose $p_0 > \frac{2n}{3n-2}$ in the above inequality. This is possible, since from $p > 1$, we have

$$\lim_{q \to \frac{n}{n-2}} \frac{2pq}{p+2q} = \frac{2pn}{(n-2)p+2n} > \frac{2n}{3n-2}.$$

In conclusion, we can find $p_0(n, \lambda, \Lambda) > \frac{2n}{3n-2}$, such that (14.68) holds.

Next, we prove (iii). Let $\bar{\kappa} \in (0, \frac{1}{n-1})$. From (14.73) and Step 4 above, we find that, in order to have (14.65), it suffices to choose λ and Λ such that

(14.75) $$\frac{2pn}{(n-2)p+2n} > 1 + \bar{\kappa},$$

for some $p = \frac{2+2\varepsilon_1}{2+\varepsilon_1}$, so that $\varepsilon_1 = \frac{2p-2}{2-p}$, where $0 < \varepsilon_1 < \varepsilon$ with ε as in Step 1.

It can be verified that (14.75) holds as long as $2 > p > \bar{p}$, where

$$\bar{p} := \frac{2n(1+\bar{\kappa})}{2n - (1+\bar{\kappa})(n-2)} < 2.$$

Thus, we need to choose λ and Λ such that $\varepsilon > (2\bar{p} - 2)/(2 - \bar{p})$. This is always possible if $|\Lambda/\lambda - 1| < \gamma$ for $\gamma = \gamma(n, \bar{\kappa}) > 0$ small; see Theorem 6.13 and Remark 6.12 on the dependence of ε in Theorem 6.11.

Finally, we prove (iv). We repeat the proofs of assertions (i)–(ii) where (14.69) is replaced by (14.66) and $\varepsilon = \gamma - 1 > (n-2)/2$. From Step 4 above, we find that, in order to have (14.67) for some $\tilde{\kappa} > 0$, it suffices to have

$$(14.76) \qquad \frac{2pn}{(n-2)p + 2n} > 1,$$

for some $p = \frac{2 + 2\varepsilon_2}{2 + \varepsilon_2}$ where $0 < \varepsilon_2 < \varepsilon$. However, since $\varepsilon > \frac{n-2}{2}$, we can choose $\varepsilon_2 > \frac{n-2}{2}$, and hence $p = \frac{2 + 2\varepsilon_2}{2 + \varepsilon_2} > \frac{2n}{n+2}$ which gives (14.76). $\qquad \square$

We have the following extension of Theorem 14.12 to higher dimensions.

Theorem 14.27 ($W^{1,1+\kappa_0}$ estimates for the Green's function in higher dimensions). *Assume $n \geq 3$. Let $u \in C^2(\Omega)$ be a convex function satisfying $\lambda \leq \det D^2 u \leq \Lambda$ in a domain Ω in $\mathbb{R}^n$. Assume that $S_u(x_0, 2h_0) \Subset \Omega$ and*

$$(14.77) \qquad \|D^2 u\|_{L^q(S_u(x_0, 2h_0))} \leq K,$$

for some $q > n/2$ and $K > 0$. Let $g_S(x, y)$ be the Green's function of the operator $D_i(U^{ij} D_j)$ on $S := S_u(x_0, h)$ where $h \leq h_0$. Then, there exist constants $\kappa_0, \kappa_1 > 0$ depending only on $n, q, \lambda,$ and Λ such that

$$\sup_{y \in S} \|D_x g_S(\cdot, y)\|_{L^{1+\kappa_0}(S)} \leq C(n, q, K, \lambda, \Lambda, \operatorname{diam}(S_u(x_0, 2h_0)), h_0) h^{\kappa_1}.$$

The proof of Theorem 14.27 is similar to the proof of the global result in Theorem 14.25 where (14.66) is now replaced by (14.77).

14.6. Application: Hölder Estimates for Equations with Critical Inhomogeneity

In this section, we establish local and global Hölder estimates for solutions to the linearized Monge–Ampère equation

$$(14.78) \qquad L_u v := U^{ij} D_{ij} v = f,$$

where $u \in C^2(\Omega)$ satisfies (14.1), in terms of the low integrability norm of the inhomogeneous term f. Despite the degeneracy and singularity of L_u, that is, there are no controls on the ellipticity constants, it is possible to obtain estimates for v in terms of $\|f\|_{L^q}$ where $q > n/2$. This is exactly the range needed in the ideal case $u(x) = |x|^2/2$ where (14.78) then becomes the Poisson equation $\Delta v = f$. The exponent $n/2$ is critical in all these equations. Different from the case where the L^n norm is used, the interior and boundary Hölder estimates using $\|f\|_{L^q}$ rely on the representation formula (14.5) and sharp integrability properties of the Green's functions of the linearized Monge–Ampère operator (Theorem 14.11). In formulating our estimates, we will assume that f belongs to L^n, but our estimates only use the L^q norm of f. We write $L^{\frac{n}{2}+}$ to denote L^q where $q > n/2$.

14.6.1. Interior Hölder estimates with $L^{\frac{n}{2}+}$ inhomogeneity. Here, we prove an interior maximum principle and an interior Hölder estimate for solutions to (14.78) in terms of the L^q norm of the inhomogeneity term.

Lemma 14.28 (Interior maximum principle). *Let $u \in C^2(\Omega)$ be a convex function satisfying $\lambda \leq \det D^2 u \leq \Lambda$ in a domain Ω in $\mathbb{R}^n$. Assume $S := S_u(z,t) \Subset \Omega$. Let $q > n/2$, $f \in L^n(S)$, and $v \in W^{2,n}_{\mathrm{loc}}(S) \cap C(\overline{S})$ satisfy*

$$-U^{ij} D_{ij} v \leq f \quad in \ S_u(z,t).$$

Then, for any $\alpha \in (0,1)$, there exists a constant $C(n,\lambda,\Lambda,\alpha,q)$ such that

$$\sup_{S_u(z,\alpha t)} v \leq \sup_{\partial S_u(z,t)} v + C|S_u(z,t)|^{\frac{2}{n}-\frac{1}{q}} \|f\|_{L^q(S_u(z,t))}.$$

Proof. By an approximation argument, we can assume $u \in C^3(\Omega)$. Let $g_S(x,y)$ be the Green's function of $D_i(U^{ij} D_j)$ in $S := S_u(z,t)$ with pole $y \in S$. Define

$$w(x) := \int_S g_S(y,x) f(y) \, dy \quad \text{for } x \in S.$$

Then, by Remark 14.2(d), $w \in W^{2,n}_{\mathrm{loc}}(S) \cap C(\overline{S})$ is a solution of

$$-U^{ij} D_{ij} w = f \quad \text{in } S \quad \text{and} \quad w = 0 \quad \text{on } \partial S.$$

Since $U^{ij} D_{ij}(v - w) \geq 0$ in S, we obtain from the Aleksandrov–Bakelman–Pucci (ABP) maximum principle (Theorem 3.16) that

$$(14.79) \qquad\qquad v(x) \leq \sup_{\partial S} v + w(x) \quad \text{in } S.$$

We next estimate $w(x)$ for the case $n \geq 3$ using Lemma 14.10. The case $n = 2$ is treated similarly. For each $x_0 \in S_\alpha := S_u(z,\alpha t)$, by Theorem 5.30, there is $\eta(n,\lambda,\Lambda,\alpha)$ such that $S_u(x_0,\eta t) \subset S_u(z,t)$. From Lemma 14.10, there is $\tau_0(n,\alpha,\lambda,\Lambda) > 0$ such that for all $\tau > \tau_0$, we have

$$\{y \in S : g_S(y,x_0) > \tau t^{-\frac{n-2}{2}}\} \subset S_u(x_0,(4\tau_0\tau^{-1})^{\frac{2}{n-2}} t).$$

As in the proof of Theorem 14.11, this gives for all $T > 0$ the estimate

$$|\{y \in S : g_S(y,x_0) > T\}| \leq |S_u(x_0, CT^{-\frac{2}{n-2}})| \leq K(n,\lambda,\Lambda,\alpha) T^{-\frac{n}{n-2}}.$$

Using the layer-cake formula as in the proof of Theorem 14.11, we obtain

$$\|g_S(\cdot,x_0)\|_{L^{\frac{q}{q-1}}(S)} \leq C_1(\alpha,n,\lambda,\Lambda,q)|S|^{\frac{q-1}{q}-\frac{n-2}{n}} \quad \text{for all } x_0 \in S_\alpha.$$

We deduce from the Hölder inequality and the above estimate that

$$|w(x)| \leq \|g_S(\cdot,x)\|_{L^{\frac{q}{q-1}}(S)} \|f\|_{L^q(S)} \leq C_1 |S|^{\frac{2}{n}-\frac{1}{q}} \|f\|_{L^q(S)} \quad \text{for all } x \in S_\alpha.$$

Combining this estimate with (14.79) yields the conclusion of the lemma. $\qquad \square$

As in the proof of Lemma 12.12, by employing the interior Harnack inequality in Theorem 12.4 and employing Lemma 14.28 (with $\alpha = 1/2$) instead of the ABP estimate, we obtain:

Lemma 14.29 (Harnack inequality). *Let $u \in C^2(\Omega)$ be a convex function satisfying $\lambda \leq \det D^2 u \leq \Lambda$ in a domain Ω in $\mathbb{R}^n$. Let $q > n/2$, $f \in L^n(\Omega)$, and $v \in W_{\mathrm{loc}}^{2,n}(\Omega)$ satisfy $v \geq 0$ and $U^{ij} D_{ij} v = f$ in $S_u(x,t) \Subset \Omega$. Then*

$$\sup_{S_u(x,t/2)} v \leq C(n,\lambda,\Lambda,q)\Big(\inf_{S_u(x,t/2)} v + t^{1-\frac{n}{2q}} \|f\|_{L^q(S_u(x,t))} \Big).$$

Replacing Lemma 12.12 in the proof of Theorem 12.13 by Lemma 14.29, we obtain an oscillation estimate, and thus, as in the proof of Theorem 12.14, we also obtain interior Hölder estimates for solutions to (14.78).

Theorem 14.30 (Interior Hölder estimate). *Let Ω be a convex domain in $\mathbb{R}^n$ satisfying $B_1(z) \subset \Omega \subset B_n(z)$ for some $z \in \mathbb{R}^n$, and let $u \in C(\overline{\Omega}) \cap C^2(\Omega)$ be a convex function satisfying $\lambda \leq \det D^2 u \leq \Lambda$ in Ω and $u = 0$ on $\partial\Omega$. Let $q > n/2$, $f \in L^n(B_1(z))$, and $v \in W_{\mathrm{loc}}^{2,n}(B_1(z)) \cap C(\overline{B_1(z)})$ satisfy*

$$U^{ij} D_{ij} v = f \quad \text{in } B_1(z).$$

Then, there exist $\alpha \in (0,1), \beta \in (0,1)$ depending only on n, λ, Λ, q, such that:

(i) *For any section $S_u(x,h) \Subset \Omega$ and all $\rho \leq h$, we have*

$$\mathrm{osc}_{S_u(x,\rho)} v \leq C(n,\lambda,\Lambda,q)\Big(\frac{\rho}{h}\Big)^\alpha \Big[\mathrm{osc}_{S_u(x,h)} v + h^{1-\frac{n}{2q}} \|f\|_{L^q(S_u(x,h))}\Big].$$

(ii) *The following Hölder estimate holds:*

$$[v]_{C^\beta(B_{1/2}(z))} \leq C(n,\lambda,\Lambda,q)\Big(\|v\|_{L^\infty(B_1(z))} + \|f\|_{L^q(B_1(z))}\Big).$$

14.6.2. Global Hölder estimates with $L^{\frac{n}{2}+}$ inhomogeneity. Next, we establish global Hölder estimates (Theorem 14.34) for solutions to (14.78) in term of the L^q norm of the right-hand side.

First, we use the Green's function estimates obtained in previous sections to derive a boundary version of the maximum principle in Lemma 14.28.

Lemma 14.31 (Boundary maximum principle). *Consider these cases:*

(a) *Ω and u satisfy (14.61)–(14.64) and $V = S_u(0,t)$ or $V = B_t(0) \cap \Omega$. Here, $t \leq c$ where $c(n,\lambda,\Lambda,\rho)$ is small.*

(b) *Ω and u satisfy (14.40)–(14.42), and $V = \Omega$.*

Let $f \in L^n(V)$ and $v \in W_{\mathrm{loc}}^{2,n}(V) \cap C(\overline{V})$ satisfy

$$-U^{ij} D_{ij} v \leq f \quad \text{in } V.$$

Then, for any $q > n/2$, there exists $C(n, \lambda, \Lambda, \rho, q) > 0$ such that

$$\sup_{V} v \le \sup_{\partial V} v + \begin{cases} C|V|^{\frac{2}{n} - \frac{1}{q}} \|f\|_{L^q(V)} & \text{if } V = S_u(0,t) \text{ or } V = \Omega, \\ C|V|^{\frac{3}{4}(\frac{2}{n} - \frac{1}{q})} \|f\|_{L^q(V)} & \text{if } V = B_t(0) \cap \Omega. \end{cases}$$

Proof. The proof uses corresponding results in Corollary 14.24 for case (a), and Theorem 14.22 for case (b). To illustrate, we prove the lemma for $V = S_u(0, t)$ in case (a). Let $g_V(\cdot, y)$ be the Green's function of $D_i(U^{ij}D_j)$ in V with pole $y \in V$. As in (14.79), we have from the symmetry of g_V that

$$(14.80) \quad v(x) \le \sup_{\partial S_u(0,t)} v + \int_{V \cap \Omega} g_V(x, y) f(y) \, dy \qquad \text{for all } x \in S_u(0, t).$$

Note that $\frac{q}{q-1} < \frac{n}{n-2}$. Then for all $x \in V$, we have from Corollary 14.24(i)

$$\|g_V(x, \cdot)\|_{L^{\frac{q}{q-1}}(V)} \le C|V|^{\frac{q-1}{q} - \frac{n-2}{n}} = C|V|^{\frac{2}{n} - \frac{1}{q}}.$$

Now, applying Hölder's inequality to (14.80) yields the claimed estimate. $\square$

The new maximum principles in Lemmas 14.28 and 14.31 allow us to establish global Hölder estimates for solutions to (14.78), under the assumptions (14.61)–(14.64), using the L^q norm of the right-hand side. Under these assumptions, we cannot control the uniform convexity of the domain Ω, if any. As remarked at the end of Section 13.1, for the boundary Hölder estimates, we will replace in the proof of Proposition 13.1 the function $(\inf\{y_n : y \in \overline{\Omega} \cap \partial B_{\delta_2}(0)\})^{-1} x_n$ by suitable supersolutions w_δ of the linearized Monge–Ampère operator $U^{ij}D_{ij}$ in Lemma 13.7.

With these barriers at hand and Lemma 14.31 replacing the ABP estimate, we can extend the results of Proposition 13.1 in the special case of the linearized Monge–Ampère operator $U^{ij}D_{ij}$ where now the inhomogeneous term is estimated in L^q.

Proposition 14.32. *Assume Ω and u satisfy* (14.61)–(14.64). *Let $q > n/2$, $f \in L^n(\Omega \cap B_\rho(0))$, $\varphi \in C^\alpha(\partial\Omega \cap B_\rho(0))$ for some $\alpha \in (0,1)$. Let $v \in C(B_\rho(0) \cap \overline{\Omega}) \cap W^{2,n}_{\text{loc}}(B_\rho(0) \cap \Omega)$ be a solution to*

$$U^{ij}D_{ij}v = f \quad \text{in } B_\rho(0) \cap \Omega, \qquad v = \varphi \quad \text{on } \partial\Omega \cap B_\rho(0).$$

Then, there exist $\delta, C > 0$, depending only on $\lambda, \Lambda, n, \alpha, \rho$, and q, such that

$$|v(x) - v(x_0)| \le C|x - x_0|^{\frac{\alpha_0}{\alpha_0 + 3n}} \mathbf{K} \quad \text{for all } x_0 \in \partial\Omega \cap B_{\rho/2}(0), x \in \Omega \cap B_\delta(x_0),$$

where $\alpha_0 := \min\left\{\alpha, \frac{3}{8}(2 - \frac{n}{q})\right\}$ and

$$\mathbf{K} = \|v\|_{L^\infty(\Omega \cap B_\rho(0))} + \|\varphi\|_{C^\alpha(\partial\Omega \cap B_\rho(0))} + \|f\|_{L^q(\Omega \cap B_\rho(0))}.$$

Proof. We follow closely the proof of Proposition 13.1. Assume $u(0) = 0$ and $Du(0) = 0$. We denote $B_r(0)$ by B_r for $r > 0$ and $L_u := U^{ij} D_{ij}$. Since $\|\varphi\|_{C^{\alpha_0}(\partial\Omega \cap B_\rho)} \leq C(\alpha_0, \alpha, \rho)\|\varphi\|_{C^\alpha(\partial\Omega \cap B_\rho)}$, it suffices to prove the proposition where $\|\varphi\|_{C^{\alpha_0}(\partial\Omega \cap B_\rho)}$ replaces $\|\varphi\|_{C^\alpha(\partial\Omega \cap B_\rho)}$ in **K**. We can assume that

$$\|v\|_{L^\infty(\Omega \cap B_\rho)} + \|\varphi\|_{C^{\alpha_0}(\partial\Omega \cap B_\rho)} + \|f\|_{L^q(\Omega \cap B_\rho)} \leq 1,$$

and we need to show that, for all $x_0 \in \Omega \cap B_{\rho/2}$, it holds that

(14.81) $\qquad |v(x) - v(x_0)| \leq C|x - x_0|^{\frac{\alpha_0}{\alpha_0 + 3n}}$ for all $x \in \Omega \cap B_\delta(x_0)$,

where positive constants δ and C depend only on $\lambda, \Lambda, n, \alpha, \rho$, and q.

We prove (14.81) for $x_0 = 0$; the same argument applies to all points $x_0 \in \Omega \cap B_{\rho/2}$ with obvious modifications. For any $\varepsilon \in (0, 1)$, let

$$h_\pm := v - v(0) \pm \varepsilon \pm \frac{6}{\delta_2^3} w_{\delta_2} \quad \text{in } A := \Omega \cap B_{\delta_2}(0),$$

where δ_2 is small, to be chosen later, and the function $w_{\delta_2} \geq 0$ is as in Lemma 13.7. Observe that if $x \in \partial\Omega$ with $|x| \leq \delta_1(\varepsilon) := \varepsilon^{1/\alpha_0}$, then,

(14.82) $\qquad |v(x) - v(0)| = |\varphi(x) - \varphi(0)| \leq |x|^{\alpha_0} \leq \varepsilon.$

If $x \in \Omega \cap \partial B_{\delta_2}$, then from Lemma 13.7, we have $6w_{\delta_2}(x)/\delta_2^3 \geq 3$. It follows that if $\delta_2 \leq \delta_1$, then from (14.82) and $|v(x) - v(0) \pm \varepsilon| \leq 3$, we get

$$h_- \leq 0, \ h_+ \geq 0 \quad \text{on } \partial A.$$

Also from Lemma 13.7, we have

$$L_u h_+ \leq f, \ L_u h_- \geq f \quad \text{in } A.$$

Hence, Lemma 14.31 applied in A gives

(14.83) $\qquad \max\left\{h_-, -h_+\right\} \leq C_1 |A|^{\frac{3}{4}\left(\frac{2}{n} - \frac{1}{q}\right)} \|f\|_{L^q(A)} \leq C_1 \delta_2^{\frac{3}{4}\left(2 - \frac{n}{q}\right)} \quad \text{in } A,$

where $C_1(n, \lambda, \Lambda, \rho, q) > 1$. By restricting $\varepsilon \leq C_1^{-1}$, we have

$$\delta_1^{\frac{3}{4}\left(2 - \frac{n}{q}\right)} = \varepsilon^{\left(2 - \frac{n}{q}\right)\frac{3}{4\alpha_0}} \leq \varepsilon^2 \leq \varepsilon/C_1.$$

Then, for $\delta_2 \leq \delta_1$, we obtain from (14.83) that

$$|v(x) - v(0)| \leq 2\varepsilon + \frac{6}{\delta_2^3} w_{\delta_2}(x) \quad \text{in } A.$$

From the boundary estimate for the function u (see (13.18)), we have

$$w_{\delta_2}(x) \leq M_{\delta_2} x_n + u(x) \leq M_{\delta_2} |x| + C |x|^2 |\log |x||^2 \leq 2M_{\delta_2} |x| \quad \text{in } A.$$

Now, choosing $\delta_2 = \delta_1$ and recalling the choice of M_{δ_2}, we get

(14.84) $\qquad |v(x) - v(0)| \leq 2\varepsilon + \frac{C_2(n, \lambda, \Lambda)}{\delta_2^{3n}} |x| = 2\varepsilon + C_2 \varepsilon^{-\frac{3n}{\alpha_0}} |x|,$

for all $\varepsilon \leq C_1^{-1}$ and $|x| \leq \delta_1(\varepsilon) := \varepsilon^{1/\alpha_0}$. In particular, these conditions are satisfied for $\varepsilon = |x|^{\frac{\alpha_0}{\alpha_0+3n}}$ if $|x| \leq \delta := C_1^{-\frac{\alpha_0+3n}{\alpha_0}}$. Then, by (14.84), we have

$$|v(x) - v(0)| \leq (2 + C_2)|x|^{\frac{\alpha_0}{\alpha_0+3n}}$$

for all $x \in \Omega \cap B_\delta(0)$. The proposition is proved. $\qquad\square$

Similar to the proof of Theorem 13.2, by combining the boundary Hölder estimates in Proposition 14.32 and the interior Hölder estimates in Theorem 14.30 using Savin's Boundary Localization Theorem (Theorem 8.5), we obtain global Hölder estimates for solutions to (14.78), under the assumptions (14.61)–(14.64), using the L^q norm of the right-hand side.

Theorem 14.33 (Global Hölder estimates). *Assume that Ω and u satisfy (14.61)–(14.64). Let $q > n/2$, $f \in L^n(\Omega \cap B_\rho(0))$, and $\varphi \in C^\alpha(\partial\Omega \cap B_\rho(0))$ where $\alpha \in (0, 1)$. Let $v \in C\big(B_\rho(0) \cap \overline{\Omega}\big) \cap W^{2,n}_{\mathrm{loc}}(B_\rho(0) \cap \Omega)$ be a solution to*

$$U^{ij}D_{ij}v = f \quad in \ B_\rho(0) \cap \Omega, \qquad v = \varphi \quad on \ \partial\Omega \cap B_\rho(0).$$

Then, there exist $\beta(n, \lambda, \Lambda, \alpha, q) > 0$ and $C(n, \lambda, \Lambda, \rho, \alpha, q) > 0$ such that

$$[v]_{C^\beta(\Omega \cap B_{\rho/2}(0))} \leq C\Big(\|v\|_{L^\infty(\Omega \cap B_\rho(0))} + \|\varphi\|_{C^\alpha(\partial\Omega \cap B_\rho(0))} + \|f\|_{L^q(\Omega \cap B_\rho(0))}\Big).$$

Combining Lemma 14.31 and Theorem 14.33 together with Proposition 4.7, we obtain the following global Hölder estimates for the linearized Monge–Ampère equation using the low integrability norm of the inhomogeneous term; compare with Theorem 13.2.

Theorem 14.34 (Global Hölder estimates for the linearized Monge–Ampère equation). *Let Ω be a uniformly convex domain in $\mathbb{R}^n$ with $\partial\Omega \in C^3$. Let $u \in C^{1,1}(\overline{\Omega}) \cap C^3(\Omega)$ be a convex function satisfying $0 < \lambda \leq \det D^2 u \leq \Lambda$ and $u|_{\partial\Omega} \in C^3$. Let $(U^{ij}) = \mathrm{Cof}\, D^2 u$, $q > n/2$, $f \in L^n(\Omega)$, and $\varphi \in C^\alpha(\partial\Omega)$ for some $\alpha \in (0, 1)$. Let $v \in C(\overline{\Omega}) \cap W^{2,n}_{\mathrm{loc}}(\Omega)$ be the solution to*

$$U^{ij}D_{ij}v = f \quad in \ \Omega, \qquad v = \varphi \quad on \ \partial\Omega.$$

Then, $v \in C^\beta(\overline{\Omega})$ for $\beta = \beta(\lambda, \Lambda, n, q, \alpha) \in (0, 1)$ with the estimate

$$\|v\|_{C^\beta(\overline{\Omega})} \leq C\big(\|\varphi\|_{C^\alpha(\partial\Omega)} + \|f\|_{L^q(\Omega)}\big),$$

where C depends only on $\lambda, \Lambda, n, q, \ \alpha, \ \|u\|_{C^3(\partial\Omega)}$, the C^3 regularity of $\partial\Omega$, and the uniform convexity radius of Ω.

In Section 15.5, we will crucially use Theorem 14.34 to solve certain singular Abreu equations arising in the approximation of minimizers of convex functionals subjected to convexity constraints.

14.7. Problems

Problem 14.1. Let $u \in C^2(\Omega)$ be a strictly convex function defined on a domain Ω in $\mathbb{R}^n$. Assume $V \Subset \Omega$ and $y \in V$. Fix $q \in (1, \frac{n}{n-1})$.

(a) For any $\rho > 0$, show that there is a unique $g_\rho \in W_0^{1,2}(V)$ such that
$$\int_V U^{ij} D_i g_\rho D_j \phi \, dx = \frac{1}{|B_\rho(y)|} \int_{B_\rho(y)} \phi \, dx \quad \text{for all } \phi \in W_0^{1,2}(V).$$

(b) Show that $\|g_\rho\|_{W_0^{1,q}(V)}$ is uniformly bounded with respect to ρ.

(c) Let g be a weak limit in $W_0^{1,q}(V)$ of the sequence $\{g_\rho\}$ when $\rho \to 0^+$. Show that g is the Green's function of the linearized Monge–Ampère operator $D_i(U^{ij} D_j)$ in V with pole y.

Problem 14.2. Let $u \in C^2$ be a convex function satisfying $\lambda \le \det D^2 u \le \Lambda$ in a domain Ω in $\mathbb{R}^n$, and let $V \Subset \Omega$. Suppose that $v \in W_{\text{loc}}^{2,n}(S_u(0,R)\setminus\{0\})$ solves $U^{ij} D_{ij} v = 0$ in $S_u(0,R)\setminus\{0\} \subset V$, and suppose that it satisfies
$$|v| = \begin{cases} o(r^{\frac{2-n}{2}}) & \text{if } n \ge 3 \\ o(|\log r|) & \text{if } n = 2 \end{cases} \quad \text{on } \partial S_u(0,r) \text{ as } r \to 0.$$

Prove that v has a removable singularity at 0.

Problem 14.3. In Step 1(b) in the proof of Theorem 14.12, the function v has a singularity at x_0 where $v = +\infty$. Justify the integration by parts inequality there as follows: Carry out the integration by parts inequality in $\{v \le 1/\varepsilon\}$ and then let $\varepsilon \to 0$.

Problem 14.4. In Step 1 in the proof of Lemma 14.16, assume that $\mu = \mathcal{L}^n$ and $L = L_u := D_i(U^{ij} D_j)$ where $\lambda \le \det D^2 u$ in Ω and $\omega \Subset \Omega$ is a convex set. Without using the decay of the superlevel set of the Green's function of L_u, but using Lemma 14.4, show that instead of (14.31), we have
$$\lambda_1(L_u, \omega)|\omega|^{\frac{2}{n}} \ge c(n, \lambda).$$

Problem 14.5. Let u be a C^2, convex function satisfying $\lambda \le \det D^2 u \le \Lambda$ in a domain Ω in $\mathbb{R}^n$. Assume $S_u(x_0, 2t) \Subset V \Subset \Omega$ where V is open. Let K be a closed subset of V. We define the capacity of K with respect to the linearized Monge–Ampère operator $L_u := D_i(U^{ij} D_j)$ and the set V by
$$\text{cap}_{L_u}(K, V) := \inf \left\{ \int_V U^{ij} D_i \varphi D_j \varphi \, dx : \varphi \in W_0^{1,2}(V), \varphi \ge 1 \text{ on } K \right\}.$$

Let $g_V(\cdot, y)$ be the Green's function for L_u in V with pole $y \in V$. Show that there is a constant $C(n, \lambda, \Lambda) > 0$ such that
$$C^{-1} \le g_V(x, x_0) \big[\text{cap}_{L_u}(\overline{S_u(x_0, t)}, V)\big] \le C \quad \text{for all } x \in \partial S_u(x_0, t).$$

Thus, the Green's function is proportional to the reciprocal of the capacity.

Problem 14.6. Let $u \in C^3(\Omega)$ be a strictly convex function in a domain Ω in $\mathbb{R}^n$. Let $(U^{ij}) = \mathrm{Cof}\, D^2u$. Assume $S_u(x_0, s) \Subset \Omega$ and $Du(x_0) = 0$. Show that

$$\int_{\partial S_u(x_0,s)} \frac{U^{ij} D_i u D_j u}{|Du|} \, d\mathcal{H}^{n-1} = \int_{S_u(x_0,s)} n \det D^2 u \, dx.$$

Problem 14.7. Let Ω be a domain in $\mathbb{R}^n$, and let $u \in C^2(\Omega)$ be a convex function satisfying $\lambda \le \det D^2 u \le \Lambda$ in Ω. Let $S := S_u(x_0, h)$ where $S_u(x_0, 2h) \Subset \Omega$. Let $g_S(\cdot, y)$ be the Green's function of the linearized Monge–Ampère operator $D_i(U^{ij} D_j)$ on S with pole $y \in S$. Let $p \in (1, \frac{n}{n-1})$. Show that there are constants $C(n, \lambda, \Lambda, p) > 0$ and $\delta(n, \lambda, \Lambda, p) > 0$ such that

$$\|(D^2 u)^{-1/2} D_x g_S(\cdot, y)\|_{L^p(S)} \le C h^\delta \quad \text{for all } y \in S.$$

Problem 14.8. Provide details for the proof of Theorem 14.33.

Problem 14.9. Extend the results in Section 14.6 to linearized Monge–Ampère equations $U^{ij} D_{ij} v + \mathbf{b} \cdot Dv = f$ with bounded drifts $\mathbf{b}$.

14.8. Notes

Various properties of the Green's function of the linearized Monge–Ampère operator $L_u = D_i(U^{ij} D_j)$ under different conditions on the Monge–Ampère measure μ_u have been studied starting with the work of Tian and Wang [**TiW**] and then Maldonado [**Md1, Md4**] and Le [**L2, L4, L6, L8**]. Properties of the Green's function g_V have played an important role in establishing Sobolev inequality for the Monge–Ampère quasi-metric structure [**TiW, Md1**].

Section 14.1 is based on [**L2**]. Related to Theorem 14.8, an alternate proof for the lower bound of the Green's function using capacity was given in [**L2**]; this potential theoretic approach works for general doubling Monge–Ampère measures.

Section 14.2 is based on [**L6**]. An inequality similar to (14.21) can also be found in Maldonado [**Md2**].

Related to Section 14.3, see Evans [**E2**, Section 8.4] for an introduction to the Lagrange multiplier method.

Section 14.4 is based on [**L4**], while Section 14.5 is based on [**L8**].

In Section 14.6, the analysis of the right-hand side being in $L^{\frac{n}{2}+}$ is based on Le–Nguyen [**LN3**]. When $f \in L^n$, Theorem 14.33 was obtained in Le–Nguyen [**LN2**].

Problem 14.5 is taken from [**L2**]. Problem 14.7 is based on [**Md4**].

Chapter 15

Divergence Form Equations

We study in this chapter inhomogeneous linearized Monge–Ampère equations in divergence form

$$(15.1) \quad U^{ij} D_{ij} v \equiv D_i(U^{ij} D_j v) \equiv \mathrm{div}(U D v) = \mathrm{div}\, \mathbf{F} \quad \text{in } \Omega \subset \mathbb{R}^n \quad (n \geq 2),$$

where $\mathbf{F} : \Omega \to \mathbb{R}^n$ is a bounded vector field and $U = (U^{ij})_{1 \leq i,j \leq n}$ is the cofactor matrix of the Hessian $D^2 u$ of a C^3 convex function u satisfying

$$(15.2) \qquad\qquad 0 < \lambda \leq \det D^2 u \leq \Lambda \quad \text{in } \Omega \subset \mathbb{R}^n.$$

In particular, we will establish Hölder estimates for solutions using their integral information. Our analysis explores the divergence form character of $U^{ij} D_{ij} = D_i(U^{ij} D_j)$, as opposed to the nondivergence form of $U^{ij} D_{ij}$ employed in Chapters 12 and 13, and uses fine properties of Green's function g_V of the degenerate operator $D_i(U^{ij} D_j)$ established in Chapter 14. In two dimensions, our results do not require further structural conditions on u. However, in higher dimensions, we need $W^{1,1+\kappa_0}$ estimates, depending only on structural constants, for the Green's function of the linearized Monge–Ampère operator which are not currently available. Thus, we need further hypotheses such as the following condition, which is a sort of uniform interior $W^{2,p}$ estimates where $p > n(n-1)/2$:

$$(15.3) \quad \begin{cases} n \geq 3 \text{ and } \lambda, \Lambda \text{ are such that every solution } u \text{ to } (15.2) \\ \text{with normalized section } S_u(x_0, 2t) \Subset \Omega \subset \mathbb{R}^n \text{ satisfies} \\ \|D^2 u\|_{L^{1+\varepsilon_n}(S_u(x_0,t))} \leq K, \text{ for some } \varepsilon_n > (n+1)(n-2)/2. \end{cases}$$

Linearized Monge–Ampère equations of the type (15.1) appear in the analysis of dual semigeostrophic equations to be discussed in Section 15.4. They also appear in the singular Abreu equations arising in the calculus of variations with convexity to be discussed in Section 15.5. However, the equations in Section 15.5 can also be treated in the framework of Section 14.6 on linearized Monge–Ampère equations with L^q inhomogeneity where $q > n/2$.

For simplicity, we will mostly assume that solutions v to (15.1) belong to $W^{2,n}_{\mathrm{loc}} \cap C^0$ and that the vector fields $\mathbf{F}$ belong to $L^\infty \cap W^{1,n}$ in appropriate domains. The assumptions on $\mathbf{F}$ will guarantee the existence of a unique solution $w \in W^{2,n}_{\mathrm{loc}} \cap C^0$ to (15.1) on subdomains of Ω with continuous boundary data, but our estimates do not depend on the $W^{1,n}$ norm of $\mathbf{F}$.

Convention. We use constants $C(*, \dots, *, \varepsilon_n, K)$ for ease of writing in statements and estimates, but they do not depend on ε_n or K when $n = 2$.

15.1. Interior Uniform Estimates via Moser Iteration

Before establishing Hölder estimates for the linearized Monge–Ampère equation (15.1), we first obtain interior uniform estimates for solutions of (15.1) in compactly contained sections in terms of their L^p norms for all $p > 1$. The key step is to obtain uniform estimates in a section that is comparable to the unit ball for highly integrable solutions. Building on the *Moser iteration technique* with the main technical tool being the Monge–Ampère Sobolev inequality in Theorem 14.15, we prove the following theorem.

Theorem 15.1. *Let $u \in C^3(\Omega)$ be a convex function satisfying (15.2), and let $\mathbf{F} \in L^\infty(\Omega; \mathbb{R}^n)$. Let $v \in W^{1,2}(S_u(x_0, 2t)) \cap W^{2,n}_{\mathrm{loc}}(S_u(x_0, 2t))$ be a weak solution of (15.1) in a section $S_u(x_0, 2t) \Subset \Omega$ with $B_1(z) \subset S_u(x_0, t) \subset B_n(z)$ for some $z \in \mathbb{R}^n$. This means that for all $w \in W^{1,2}_0(S_u(x_0, 2t))$,*

$$\int_{S_u(x_0, 2t)} U^{ij} D_j v D_i w \, dx = \int_{S_u(x_0, 2t)} \mathbf{F} \cdot Dw \, dx.$$

Assume that

- *either $n = 2$ or*
- *$n \geq 3$ and $\|D^2 u\|_{L^{1+\varepsilon_n}(S_u(x_0, t))} \leq K$ where $\varepsilon_n > (n+1)(n-2)/2$.*

Then, there exist constants $C_0 > 1$ and $p_0 > 2$ depending only on n, λ, and Λ (and p_0 on ε_n and C_0 on ε_n, K if $n \geq 3$) such that

$$(15.4) \qquad \sup_{S_u(x_0, t/2)} |v| \leq C_0 \Big(\|v\|_{L^{p_0}(S_u(x_0, t))} + \|\mathbf{F}\|_{L^\infty(S_u(x_0, t))} \Big).$$

Proof. Since u is convex, we have $S_u(x_0, 2t) \subset 2S_u(x_0, t) - x_0 \subset B_{3n}(z)$. Because $S_u(x_0, t) \Subset \Omega$ is normalized, the volume estimates in Lemma 5.6 give $0 < c(n, \lambda, \Lambda) \leq t \leq C(n, \lambda, \Lambda)$. For simplicity, we can assume $t = 1$.

If $n = 2$, then for the exponent $\varepsilon(2, \lambda, \Lambda) > 0$ in Theorem 6.11, we have

$$\|U\|_{L^{1+\varepsilon}(S_u(x_0,1))} + \|\Delta u\|_{L^{1+\varepsilon}(S_u(x_0,1))} \leq 3\|D^2 u\|_{L^{1+\varepsilon}(S_u(x_0,1))} \leq C(\lambda, \Lambda).$$

If $n \geq 3$, then for $\varepsilon := (1 + \varepsilon_n)/(n - 1) - 1 > (n - 2)/2$, we have

$$\|U\|_{L^{1+\varepsilon}(S_u(x_0,1))} \leq C(n)\|D^2 u\|_{L^{1+\varepsilon_n}(S_u(x_0,1))}^{n-1} \leq C(K, n).$$

Thus, with ε defined above for all $n \geq 2$, we have the inequality

(15.5) $$\|U\|_{L^{1+\varepsilon}(S_u(x_0,1))} + \|\Delta u\|_{L^{1+\varepsilon}(S_u(x_0,1))} \leq C(\lambda, \Lambda, n, K).$$

Let

$$p_0 := 2\frac{\varepsilon + 1}{\varepsilon} \quad \text{and} \quad \delta := \begin{cases} \dfrac{n\varepsilon}{(n - 2)(1 + \varepsilon)} - 1 & \text{if } n \geq 3, \\ 1 & \text{if } n = 2. \end{cases}$$

Since $\varepsilon > (n - 2)/2$ when $n \geq 3$, we know that $\delta \in (0, 2)$ for all $n \geq 2$.

In order to prove (15.4), by the homogeneity of (15.1), we can assume

(15.6) $$\|\mathbf{F}\|_{L^\infty(S_u(x_0,1))} + \|v\|_{L^{p_0}(S_u(x_0,1))} \leq 1,$$

and we need to show that, for some constant $C_0(n, \lambda, \Lambda, \varepsilon_n, K) > 0$,

(15.7) $$\sup_{S_u(x_0,1/2)} |v| \leq C_0.$$

We now carry out the Moser iteration technique. Given $r \in (0, 1]$, let us set

$$S_r := S_u(x_0, r) \quad \text{and} \quad S := S_1 = S_u(x_0, 1).$$

Step 1. Reverse Hölder inequality. In this step, we use the equation together with the Monge–Ampère Sobolev inequality to bound the $L^{(1+\delta)\gamma}$ norm of v on a section by its L^γ norm on a larger section.

Let $\eta \in C_c^1(S)$ be a cut-off function to be determined later. Let $\beta \geq 0$. The constants C in the following estimates will not depend on β.

By testing div $\mathbf{F} = D_j(U^{ij} D_i v)$ against $|v|^\beta v \eta^2$, we get

(15.8) $$\begin{aligned} M &:= \int_S \mathbf{F} \cdot D(|v|^\beta v \eta^2)\, dx = \int_S U^{ij} D_i v D_j(|v|^\beta v \eta^2)\, dx \\ &= (\beta + 1)\int_S U^{ij} D_i v D_j v |v|^\beta \eta^2\, dx + 2\int_S U^{ij} D_i v D_j \eta \eta |v|^\beta v\, dx. \end{aligned}$$

Next, the Cauchy–Schwarz inequality gives

(15.9) $$\left| 2\int_S U^{ij} D_i v D_j \eta \eta |v|^\beta v\, dx \right| \leq \frac{1}{2}\int_S U^{ij} D_i v D_j v |v|^\beta \eta^2\, dx$$
$$+ 2\int_S U^{ij} D_i \eta D_j \eta |v|^{\beta+2}\, dx.$$

It follows from (15.8) that

(15.10) $$(\beta + \tfrac{1}{2})\int_S U^{ij} D_i v D_j v |v|^\beta \eta^2\, dx - 2\int_S U^{ij} D_i \eta D_j \eta |v|^{\beta+2}\, dx \leq M.$$

We now estimate $M = \int_S \mathbf{F} \cdot D(|v|^\beta v \eta^2)\, dx$. First, via expansion and (15.6),

$$M \le (\beta + 1) \int_S |Dv||v|^\beta \eta^2\, dx + 2 \int_S |D\eta||\eta||v|^{\beta+1}\, dx.$$

Second, recalling $\det D^2 u \ge \lambda$ and $U^{ij} D_i v D_j v \ge \det D^2 u |Dv|^2 / \Delta u$, which comes from Lemma 2.58, we deduce that

$$M \le (\beta + 1) \int_S \lambda^{-\frac{1}{2}} \left(U^{ij} D_i v D_j v \Delta u \right)^{\frac{1}{2}} |v|^\beta \eta^2\, dx$$
$$+ 2 \int_S \lambda^{-\frac{1}{2}} \left(U^{ij} D_i \eta D_j \eta \Delta u \right)^{\frac{1}{2}} |\eta||v|^{\beta+1}\, dx.$$

Applying the Cauchy–Schwartz inequality to each term, we obtain

$$M \le \frac{\beta + 1}{4} \int_S U^{ij} D_i v D_j v |v|^\beta \eta^2\, dx + C(\lambda)(\beta + 1) \int_S \Delta u |v|^\beta \eta^2\, dx$$
$$+ \int_S U^{ij} D_i \eta D_j \eta |v|^{\beta+2}\, dx.$$

It follows from (15.10) that

$$(15.11) \quad \frac{\beta + 2}{8} \int_S U^{ij} D_i v D_j v |v|^\beta \eta^2\, dx \le C(\lambda)(\beta + 2) \int_S \Delta u |v|^\beta \eta^2\, dx$$
$$+ 3 \int_S U^{ij} D_i \eta D_j \eta |v|^{\beta+2}\, dx.$$

Consider the quantity

$$Q = \int_S U^{ij} D_i(|v|^{\beta/2} v \eta) D_j(|v|^{\beta/2} v \eta)\, dx.$$

Expanding the right-hand side and then using (15.9), we obtain

$$Q = \frac{(\beta + 2)^2}{4} \int_S U^{ij} D_i v D_j v |v|^\beta \eta^2\, dx$$
$$+ (\beta + 2) \int_S U^{ij} D_i v D_j \eta \eta |v|^\beta v\, dx + \int_S U^{ij} D_i \eta D_j \eta |v|^{\beta+2}\, dx$$
$$\le (\beta + 2)^2 \int_S U^{ij} D_i v D_j v |v|^\beta \eta^2\, dx + (\beta + 4) \int_S U^{ij} D_i \eta D_j \eta |v|^{\beta+2}\, dx.$$

Inserting the estimate (15.11) into the right-hand side leads to

$$(15.12) \quad 26(\beta + 2) \int_S U^{ij} D_i \eta D_j \eta |v|^{\beta+2}\, dx + C(\beta + 2)^2 \int_S \Delta u |v|^\beta \eta^2\, dx \ge Q.$$

We will bound each term in (15.12). By Hölder's inequality and (15.5),

$$(15.13) \quad \int_S U^{ij} D_i \eta D_j \eta |v|^{\beta+2} \, dx \leq \|D\eta\|_{L^\infty(S)}^2 \|U\|_{L^{1+\varepsilon}(S)} \|v\|_{L^{\frac{(\beta+2)(\varepsilon+1)}{\varepsilon}}(S)}^{\beta+2}$$

$$\leq C \|D\eta\|_{L^\infty(S)}^2 \|v\|_{L^{\frac{(\beta+2)(\varepsilon+1)}{\varepsilon}}(S)}^{\beta+2}.$$

Denote $\gamma := (\beta+2)\frac{\varepsilon+1}{\varepsilon}$. Then, again, by Hölder's inequality and (15.5),

$$(15.14) \quad \int_S \Delta u |v|^\beta \eta^2 \, dx \leq C \|\eta\|_{L^\infty(S)}^2 \|v\|_{L^{\frac{\beta(\varepsilon+1)}{\varepsilon}}(S)}^\beta \|\Delta u\|_{L^{1+\varepsilon}(S)}$$

$$\leq C \|\eta\|_{L^\infty(S)}^2 |S|^{\frac{2}{\gamma}} \|v\|_{L^\gamma(S)}^\beta \leq C \|\eta\|_{L^\infty(S)}^2 \|v\|_{L^\gamma(S)}^\beta.$$

For the right-hand side of (15.12), we can bound Q from below by applying the Monge–Ampère Sobolev inequality in Theorem 14.15 to $|v|^{\beta/2}v\eta$ with exponent $p = 2(1+\delta)\frac{\varepsilon+1}{\varepsilon}$. Together with (15.13) and (15.14), we discover

$$(15.15) \quad \left(\int_S |v|^{(1+\delta)\gamma} \eta^p \, dx \right)^{\frac{\beta+2}{(1+\delta)\gamma}} = \||v|^{\beta/2}v\eta\|_{L^p(S)}^2 \leq C(p,n,\lambda,\Lambda)Q$$

$$\leq C\gamma^2 \max \left\{ \|D\eta\|_{L^\infty(S)}^2, \|\eta\|_{L^\infty(S)}^2 \right\} \max \left\{ 1, \|v\|_{L^\gamma(S)}^{\beta+2} \right\}.$$

Step 2. Iteration. Now, it is time to select the cut-off function η in (15.15). Assume that $0 < r < R \leq 1$. Then, by Lemma 5.6(ii), we find that

$$(15.16) \quad \operatorname{dist}(S_r, \partial S_R) \geq c(n,\lambda,\Lambda)(1 - r/R)^n R^{n/2} \geq c(n,\lambda,\Lambda)(R-r)^n.$$

Thus, we can choose a cut-off function $\eta \in C_c^1(S)$ such that $\eta \equiv 1$ in S_r, $\eta = 0$ outside S_R, $0 \leq \eta \leq 1$, and $\|D\eta\|_{L^\infty(S)} \leq C(n,\lambda,\Lambda)(R-r)^{-n}$. It follows from (15.15) that, for some $C = C(n,\lambda,\Lambda,\varepsilon_n,K)$,

$$(15.17) \quad \max \left\{ 1, \|v\|_{L^{(1+\delta)\gamma}(S_r)} \right\} \leq [C\gamma^2(R-r)^{-2n}]^{\frac{1}{\beta+2}} \max \left\{ 1, \|v\|_{L^\gamma(S_R)} \right\}$$

$$= [C\gamma^2(R-r)^{-2n}]^{\frac{\varepsilon+1}{\varepsilon\gamma}} \max \left\{ 1, \|v\|_{L^\gamma(S_R)} \right\}.$$

Now, for each nonnegative integer j, set

$$r_j := \frac{1}{2} + \frac{1}{2^j}, \quad \gamma_j := (1+\delta)^j p_0, \quad A_j := \max \left\{ 1, \|v\|_{L^{\gamma_j}(S_{r_j})} \right\},$$

where we recall $\gamma_0 = p_0 = 2(\varepsilon+1)/\varepsilon$. Then $r_j - r_{j+1} = \frac{1}{2^{j+1}}$. Applying the estimate (15.17) to $R = r_j$, $r = r_{j+1}$, and $\gamma = \gamma_j$, we get

$$A_{j+1} \leq [C\gamma_j^2(r_j - r_{j+1})^{-2n}]^{\frac{\varepsilon+1}{\varepsilon\gamma_j}} A_j \leq [C\gamma_0^2 9^{(j+1)(n+1)}]^{\frac{\varepsilon+1}{\varepsilon(1+\delta)^j \gamma_0}} A_j.$$

By iterating, we obtain for all nonnegative integers j,

$$A_{j+1} \leq C^{\sum_{j=0}^\infty \frac{2(\varepsilon+1)}{\varepsilon(1+\delta)^j \gamma_0}} 9^{\sum_{j=0}^\infty \frac{(\varepsilon+1)(n+1)(j+1)}{\varepsilon\gamma_0(1+\delta)^j}} A_0$$

$$:= C_0 A_0 = C_0 \max \left\{ 1, \|v\|_{L^{p_0}(S)} \right\} = C_0.$$

Let $j \to \infty$ in the above inequality to obtain (15.7), as desired. $\qquad \square$

In order to extend the L^∞ estimates in Theorem 15.1 to general compactly contained sections, we need to see how the equation $U^{ij}D_{ij}v = \operatorname{div}\mathbf{F}$ changes with respect to the normalization of a section $S_u(x_0, h) \Subset \Omega$ of u.

Lemma 15.2. *Let $u \in C^2(\Omega)$ be a convex function satisfying (15.2). Let $\mathbf{F} \in L^\infty(\Omega; \mathbb{R}^n) \cap W^{1,n}_{\mathrm{loc}}(\Omega; \mathbb{R}^n)$ and $v \in W^{2,n}_{\mathrm{loc}}(\Omega)$ satisfy $U^{ij}D_{ij}v = \operatorname{div}\mathbf{F}$ in Ω. Assume $S_u(x_0, 2h) \Subset \Omega$. By John's Lemma, there exist a matrix $A \in \mathbb{M}^{n\times n}$ and $\mathbf{b} \in \mathbb{R}^n$ such that $B_1(\mathbf{b}) \subset \tilde{S} := A^{-1}(S_u(x_0, h)) \subset B_n(\mathbf{b})$. Introduce the following rescaled functions on $\tilde{S}$:*

$$\begin{cases} \tilde{u}(x) & := (\det A)^{-2/n}\big(u(Ax) - [u(x_0) + Du(x_0)\cdot(Ax - x_0) + h]\big), \\ \tilde{v}(x) & := v(Ax), \quad \tilde{\mathbf{F}}(x) := (\det A)^{\frac{2}{n}}A^{-1}\mathbf{F}(Ax). \end{cases}$$

Then, the following hold.

(i) $\lambda \le \det D^2\tilde{u} \le \Lambda$ *in* $\tilde{S} = S_{\tilde{u}}(A^{-1}x_0, (\det A)^{-2/n}h)$, $\tilde{u} = 0$ *on* $\partial\tilde{S}$.

(ii) $\tilde{U}^{ij}D_{ij}\tilde{v} = \operatorname{div}\tilde{\mathbf{F}}$ *in* $\tilde{S}$, *where* $\tilde{U} = \operatorname{Cof} D^2\tilde{u}$.

(iii) *For any $q > 1$, there is $C = C(n, \lambda, \Lambda, q) > 0$ such that*

$$C^{-1}h^{-\frac{n}{2q}}\|v\|_{L^q(S_{u(x_0,h)})} \le \|\tilde{v}\|_{L^q(\tilde{S})} \le Ch^{-\frac{n}{2q}}\|v\|_{L^q(S_{u(x_0,h)})}.$$

(iv) $\|\tilde{\mathbf{F}}\|_{L^\infty(\tilde{S})} \le C(\lambda, \Lambda, n, \operatorname{diam}(S_u(x_0, 2h)))h^{1-\frac{n}{2}}\|\mathbf{F}\|_{L^\infty(S_{u(x_0,h)})}$.

Proof. We only need to prove (ii)–(iv). Note that the rescaling of u here is the same as that of Corollary 6.26. As such, A satisfies (6.40) and (6.41). These imply (iv), and also (iii), since $\|\tilde{v}\|_{L^q(\tilde{S})} = (\det A)^{-1/q}\|v\|_{L^q(S_{u(x_0,h)})}$.

It remains to prove (ii). In the variables $x \in \tilde{S} := A^{-1}(S_u(x_0, h))$ and $y := Ax$, we have the relation $D_x = A^tD_y$. Thus,

$$(15.18) \quad D_x \cdot \tilde{\mathbf{F}}(x) = A^tD_y \cdot (\det A)^{\frac{2}{n}}A^{-1}\mathbf{F}(Ax) = (\det A)^{\frac{2}{n}}(D \cdot \mathbf{F})(Ax).$$

Applying Lemma 12.3 with $a = (\det A)^{-2/n}$, $b = 1$, we obtain

$$\tilde{U}^{ij}D_{ij}\tilde{v}(x) = (\det A)^{2/n}U^{ij}D_{ij}v(Ax),$$

and hence, recalling $U^{ij}D_{ij}v = \operatorname{div}\mathbf{F}$ and (15.18),

$$\tilde{U}^{ij}D_{ij}\tilde{v}(x) = (\det A)^{2/n}(D \cdot \mathbf{F})(Ax) = D \cdot \tilde{\mathbf{F}} = \operatorname{div}\tilde{\mathbf{F}}(x) \quad \text{in } \tilde{S}.$$

Thus, assertion (ii) is proved. $\qquad\square$

Combining Theorem 15.1 and Lemma 15.2, we immediately obtain uniform estimates for highly integrable solutions of (15.1) in a compactly contained section of u.

Corollary 15.3. *Let* $u \in C^3(\Omega)$ *be a convex function satisfying* (15.2), *and let* $\mathbf{F} \in L^\infty(\Omega; \mathbb{R}^n) \cap W^{1,n}_{\text{loc}}(\Omega; \mathbb{R}^n)$. *Let* $v \in W^{2,n}_{\text{loc}}(S_u(x_0, 2h)) \cap C(\overline{S_u(x_0, 2h)})$ *be a solution of* (15.1) *in a section* $S_u(x_0, 2h) \Subset \Omega$. *Assume that*

- *either* $n = 2$ *or*
- $n \geq 3$, (15.3) *holds for some* $\varepsilon_n > (n+1)(n-2)/2$ *and* $K > 0$.

Then, there exist a constant $p_0 = p_0(n, \lambda, \Lambda, \varepsilon_n) > 2$ *and a constant* C_1 *depending only on* $n, \lambda, \Lambda, \operatorname{diam}(S_u(x_0, 2h)), \varepsilon_n,$ *and* K, *such that*

$$(15.19) \qquad \sup_{S_u(x_0, h/2)} |v| \leq C_1 \left(h^{1-\frac{n}{2}} \|\mathbf{F}\|_{L^\infty(S_u(x_0, h))} + h^{-\frac{n}{2p_0}} \|v\|_{L^{p_0}(S_u(x_0, h))} \right).$$

Now, extending the integrability range of solutions from p_0 in Corollary 15.3 to all $p > 1$, we establish our main interior uniform estimates for (15.1).

Theorem 15.4 (Interior uniform estimates for divergence form equations). *Let* $\mathbf{F} \in L^\infty(\Omega; \mathbb{R}^n) \cap W^{1,n}_{\text{loc}}(\Omega; \mathbb{R}^n)$, *and let* $u \in C^3(\Omega)$ *be a convex function satisfying* (15.2). *Let* $v \in W^{2,n}_{\text{loc}}(S_u(x_0, h)) \cap C(\overline{S_u(x_0, h)})$ *be a solution of*

$$U^{ij} D_{ij} v = \operatorname{div} \mathbf{F} \quad \text{in a section } S_u(x_0, h) \quad \text{with } S_u(x_0, 2h) \Subset \Omega.$$

Assume that

- *either* $n = 2$ *or*
- $n \geq 3$, (15.3) *holds for some* $\varepsilon_n > (n+1)(n-2)/2$ *and* $K > 0$.

Given $p \in (1, \infty)$, *there exists a constant* $C > 0$, *depending only on* $n, p, \lambda, \Lambda,$ *and* $\operatorname{diam}(S_u(x_0, 2h))$ *(and also* ε_n *and* K *if* $n \geq 3$), *such that*

$$(15.20) \qquad \sup_{S_u(x_0, h/2)} |v| \leq C \left(h^{1-\frac{n}{2}} \|\mathbf{F}\|_{L^\infty(S_u(x_0, h))} + h^{-\frac{n}{2p}} \|v\|_{L^p(S_u(x_0, h))} \right).$$

Proof. The proof is based on a rescaling argument. We show that (15.20) follows from (15.19). By the volume estimates in Lemma 5.6, it suffices to consider the case $1 < p < p_0$.

By Theorem 5.30(i), there exist $c_1(n, \lambda, \Lambda) > 0$ and $\mu(n, \lambda, \Lambda) > 1$ such that for every $\theta \in (0, 1)$ and $y \in S_u(x_0, \theta h)$, we have the inclusion

$$(15.21) \qquad\qquad S_u(y, 2c_1(1-\theta)^\mu h) \subset S_u(x_0, h).$$

To simplify, let $t := (1-\theta)^\mu h$. Then, by applying (15.19) to v on the section $S_u(y, c_1 t)$, we obtain

$$\|v\|_{L^\infty(S_u(y, c_1 t/2))} \leq C_1 t^{1-\frac{n}{2}} \|\mathbf{F}\|_{L^\infty(S_u(y, c_1 t))} + C_1 t^{-\frac{n}{2p_0}} \|v\|_{L^{p_0}(S_u(y, c_1 t))},$$

where C_1 depends on $n, \lambda, \Lambda,$ and $\operatorname{diam}(S_u(x_0, 2h))$ (and also ε_n, K if $n \geq 3$).

Varying $y \in S_u(x_0, \theta h)$ and recalling (15.21), we infer that

$$(15.22) \quad \|v\|_{L^\infty(S_u(x_0, \theta h))} \leq C_1 [(1-\theta)^\mu h]^{1-\frac{n}{2}} \|\mathbf{F}\|_{L^\infty(S_u(x_0, h))}$$
$$+ C_1 [(1-\theta)^\mu h]^{-\frac{n}{2p_0}} \|v\|_{L^{p_0}(S_u(x_0, h))}.$$

Now, given $p \in (1, p_0)$, we obtain from (15.22) the estimate

$$\|v\|_{L^\infty(S_u(x_0, \theta h))} \leq C_1 [(1-\theta)^\mu h]^{1-\frac{n}{2}} \|\mathbf{F}\|_{L^\infty(S_u(x_0, h))}$$
$$+ C_1 [(1-\theta)^\mu h]^{-\frac{n}{2p_0}} \|v\|_{L^\infty(S_u(x_0, h))}^{1-\frac{p}{p_0}} \|v\|_{L^p(S_u(x_0, h))}^{\frac{p}{p_0}}$$
$$:= M + N.$$

By Young's inequality with two exponents $p_0/(p_0 - p)$ and p_0/p, we have

$$N \leq \left(1 - \frac{p}{p_0}\right) \|v\|_{L^\infty(S_u(x_0, h))} + \frac{p}{p_0} C_1^{\frac{p_0}{p}} [(1-\theta)^\mu h]^{-\frac{n}{2p}} \|v\|_{L^p(S_u(x_0, h))}.$$

Hence, there is a constant $C_2(n, \lambda, \Lambda, \operatorname{diam}(S_u(x_0, 2h)), \varepsilon_n, K) > 0$ such that

$$\|v\|_{L^\infty(S_u(x_0, \theta h))} \leq \left(1 - \frac{p}{p_0}\right) \|v\|_{L^\infty(S_u(x_0, h))}$$
$$+ C_2 \left([(1-\theta)^\mu h]^{1-\frac{n}{2}} \|\mathbf{F}\|_{L^\infty(S_u(x_0, h))} + (1-\theta)^{-\frac{n\mu}{2p}} h^{-\frac{n}{2p}} \|v\|_{L^p(S_u(x_0, h))}\right).$$

We will rewrite the above estimate by setting

$$\gamma := 1 - \frac{p}{p_0}, \quad \delta := \frac{n\mu}{2p}, \quad \delta_0 := \mu\left(\frac{n}{2} - 1\right),$$

and

$$A := C_2 h^{\frac{n(\mu-1)}{2p}} \|v\|_{L^p(S_u(x_0, h))}, \qquad B := C_2 h^{(\mu-1)(\frac{n}{2}-1)} \|\mathbf{F}\|_{L^\infty(S_u(x_0, h))}.$$

For $0 < s \leq h$, let $f(s) := \|v\|_{L^\infty(S_u(x_0, s))}$. Then, for all $0 < r < s \leq h$,

$$(15.23) \qquad f(r) \leq \gamma f(s) + \frac{A}{(s-r)^\delta} + \frac{B}{(s-r)^{\delta_0}}.$$

We will show that for $0 < r < s \leq h$,

$$(15.24) \qquad f(r) \leq C(\gamma, \delta, \delta_0) \left[\frac{A}{(s-r)^\delta} + \frac{B}{(s-r)^{\delta_0}}\right].$$

Indeed, fixing $\tau \in (0, 1)$ such that $\tau^{\delta + \delta_0} = (1+\gamma)/2 \in (\gamma, 1)$, so $\gamma \tau^{-(\delta+\delta_0)} \in (0, 1)$, we consider the sequence $\{r_i\}_{i=0}^\infty$ defined by

$$r_0 = r, \quad r_{i+1} = r_i + (1-\tau)\tau^i(s-r).$$

Then, $r_i \leq h$ for all i. Iterating (15.23), we obtain for all positive integers m

$$f(r_0) \leq \gamma^m f(r_m) + \sum_{i=0}^{m-1} \gamma^i \left[\frac{A\tau^{-i\delta}}{(1-\tau)^\delta (s-r)^\delta} + \frac{B\tau^{-i\delta_0}}{(1-\tau)^{\delta_0}(s-r)^{\delta_0}} \right]$$

$$\leq \gamma^m f(h) + C(\gamma, \delta, \delta_0) \left[\frac{A}{(s-r)^\delta} + \frac{B}{(s-r)^{\delta_0}} \right].$$

By letting $m \to \infty$, we obtain (15.24).

Consequently, from (15.24), we get for every $p \in (1, p_0)$ and $\theta \in (0, 1)$

$$\|v\|_{L^\infty(S_u(x_0, \theta h))} \leq C_3 [(1-\theta)^\mu h]^{1-\frac{n}{2}} \|\mathbf{F}\|_{L^\infty(S_u(x_0, h))}$$

$$+ C_3 (1-\theta)^{-\frac{n\mu}{2p}} h^{-\frac{n}{2p}} \|v\|_{L^p(S_u(x_0, h))},$$

for a constant C_3 depending only on p, λ, Λ, and $\operatorname{diam}(S_u(x_0, 2h))$ (and also ε_n, K if $n \geq 3$). Setting $\theta = 1/2$ in the the above estimate, we obtain (15.20). $\qquad \square$

15.2. Interior Hölder estimates

Our intention now is to upgrade interior uniform estimates in Section 15.1 to interior Hölder estimates for solutions to (15.1). For these to hold, a necessary condition is that solutions vanishing on the boundary of a section must decay to zero when it shrinks to a point. Utilizing the representation formula (14.5) and $W^{1,1+\kappa_0}$ estimates for the Green's function of the linearized Monge–Ampère operator, we first establish the following global estimates.

Theorem 15.5 (Global estimates for the Dirichlet problem). *Let $u \in C^3(\Omega)$ be a convex function satisfying (15.2). Let $\mathbf{F} \in L^\infty(\Omega; \mathbb{R}^n) \cap W^{1,n}_{\mathrm{loc}}(\Omega; \mathbb{R}^n)$ and $S_u(x_0, 2h_0) \Subset \Omega$. Let $v \in W^{2,n}_{\mathrm{loc}}(S_u(x_0, h)) \cap C(\overline{S_u(x_0, h)})$ be the solution to*

$$U^{ij} D_{ij} v = \operatorname{div} \mathbf{F} \quad \text{in } S_u(x_0, h), \qquad v = 0 \quad \text{on } S_u(x_0, h),$$

where $h \leq h_0$. Assume that

   (a) *either $n = 2$ or*

   (b) *$n \geq 3$, and for some $q > n/2$, we have $\|D^2 u\|_{L^q(S_u(x_0, 2h_0))} \leq K_q$.*

Then, there exists a constant $\delta(n, \lambda, \Lambda, q) > 0$ such that

$$\sup_{S_u(x_0, h)} |v| \leq C(n, q, K_q, \lambda, \Lambda, \operatorname{diam}(S_u(x_0, 2h_0)), h_0) \|\mathbf{F}\|_{L^\infty(S_u(x_0, h))} h^\delta.$$

Note that, as opposed to previous estimates, the right-hand side here does not involve any norm of the solution in larger sections.

Proof. Let $g_S(\cdot, y)$ be the Green's function of the linearized Monge–Ampère operator $D_j(U^{ij}D_i)$ on $S := S_u(x_0, h)$ with pole $y \in S$. Then, the representation formula (14.5) can be applied to v, and integration by parts gives

$$v(y) = -\int_S g_S(x, y) \operatorname{div} \mathbf{F}(x)\, dx = \int_S D_x g_S(x, y) \cdot \mathbf{F}(x)\, dx \quad \text{for all } y \in S.$$

By Theorems 14.12 and 14.27, there exist positive constants κ_0 and κ_1 depending on λ, Λ (and n and q if $n \geq 3$) such that

$$\sup_{y \in S} \|Dg_S(\cdot, y)\|_{L^{1+\kappa_0}(S)} \leq C_0(n, q, K_q, \lambda, \Lambda, \operatorname{diam}(S_u(x_0, 2h_0)), h_0) h^{\kappa_1}.$$

To conclude, we just need to combine the Hölder inequality with the above estimate and the volume estimate $|S| \leq C_1(n, \lambda) h^{n/2}$ in Lemma 5.6.

Indeed, for any $y \in S$, we have

$$|v(y)| \leq \|\mathbf{F}\|_{L^\infty(S)} \|Dg_S(\cdot, y)\|_{L^{1+\kappa_0}(S)} |S|^{\frac{\kappa_0}{1+\kappa_0}} \leq C\|\mathbf{F}\|_{L^\infty(S)} h^{\kappa_1 + \frac{n\kappa_0}{2(1+\kappa_0)}},$$

where $C = C_0 C_1^{\frac{\kappa_0}{1+\kappa_0}}$. Taking the supremum over $y \in S$ completes the proof. $\qquad\square$

Combining Theorems 15.4 and 15.5 with Theorem 12.4, we obtain our main interior Hölder estimates for solutions to (15.1).

Theorem 15.6 (Interior Hölder estimates for divergence form equations). *Let $u \in C^3(\Omega)$ be a convex function satisfying $\lambda \leq \det D^2 u \leq \Lambda$ in a domain Ω in $\mathbb{R}^n$, and let $\mathbf{F} \in L^\infty(\Omega; \mathbb{R}^n) \cap W^{1,n}_{\mathrm{loc}}(\Omega; \mathbb{R}^n)$. Assume $S_u(x_0, 4h_0) \Subset \Omega$. Let $v \in W^{2,n}_{\mathrm{loc}}(S_u(x_0, 4h_0)) \cap C(\overline{S_u(x_0, 4h_0)})$ be a solution to $U^{ij}D_{ij}v = \operatorname{div} \mathbf{F}$ in $S_u(x_0, 4h_0)$. Let $p \in (1, \infty)$. Assume that*

(a) *either $n = 2$ or*

(b) *$n \geq 3$, (15.3) holds for some $\varepsilon_n > (n+1)(n-2)/2$ and $K > 0$.*

Then, there exist constants $\overline{\gamma} > 0$ depending only on n, λ, and Λ (and ε_n if $n \geq 3$) and $C > 0$ depending only on $n, p, \lambda, \Lambda, h_0$, and $\operatorname{diam}(S_u(x_0, 4h_0))$ (and ε_n, K if $n \geq 3$) such that for all $x, y \in S_u(x_0, h_0)$, we have

$$|v(x) - v(y)| \leq C \left(\|\mathbf{F}\|_{L^\infty(S_u(x_0, 2h_0))} + \|v\|_{L^p(S_u(x_0, 2h_0))} \right) |x - y|^{\overline{\gamma}}.$$

Theorem 15.6 is of interest when $\mathbf{F} \not\equiv 0$ or $p < \infty$. The important point to note here is that the Hölder exponent $\overline{\gamma}$ depends only on the dimension n, the bounds λ and Λ of the Monge–Ampère measure of u in two dimensions; in higher dimensions, it depends also on ε_n in (15.3). When $\mathbf{F} \equiv 0$ and (15.2) holds, as we saw in Theorem 12.14, interior Hölder estimates were established for solutions to (15.1) in all dimensions, with $\|v\|_{L^\infty(S_u(x_0, 2h_0))}$ instead of $\|v\|_{L^p(S_u(x_0, 2h_0))}$ in the estimates.

Proof. Let $d := \operatorname{diam}(S_u(x_0, 4h_0))$. We proceed in two steps.

Step 1. Oscillation estimate. We first show that, for every $h \leq h_0$,

$$(15.25) \qquad \operatorname{osc}_{S_u(x_0,h)} v \leq C_0 \left(\|\mathbf{F}\|_{L^\infty(S_u(x_0,2h_0))} + \|v\|_{L^p(S_u(x_0,2h_0))} \right) h^{\gamma_0},$$

where $\gamma_0(n, \lambda, \Lambda, \varepsilon_n) \in (0,1)$ and C_0 depends on $n, p, \lambda, \Lambda, d, h_0$ (and also ε_n and K if $n \geq 3$). Recall that $\operatorname{osc}_E f := \sup_E f - \inf_E f$.

Indeed, on the section $S_u(x_0, h)$ with $h \leq h_0$, we split $v = \tilde{v} + w$, where $\tilde{v}, w \in W^{2,n}_{\mathrm{loc}}(S_u(x_0, h)) \cap C(\overline{S_u(x_0, h)})$ are solutions to

$$(15.26) \qquad U^{ij} D_{ij} \tilde{v} = \operatorname{div} \mathbf{F} \quad \text{in } S_u(x_0, h), \qquad \tilde{v} = 0 \quad \text{on } \partial S_u(x_0, h),$$

and

$$(15.27) \qquad U^{ij} D_{ij} w = 0 \quad \text{in } S_u(x_0, h), \qquad w = v \quad \text{on } \partial S_u(x_0, h).$$

Note that condition (b) of this theorem implies condition (b) of Theorem 15.5 with $q = 1 + \varepsilon_n$, and K_q depends on $n, \varepsilon_n, K, \lambda, \Lambda, d$, and h_0 by Corollary 6.26. By Theorem 15.5 applied to (15.26), there exist positive constants $C_1(n, \varepsilon_n, K, \lambda, \Lambda, d, h_0)$ and $\delta(n, \lambda, \Lambda, \varepsilon_n)$ such that

$$(15.28) \qquad \sup_{S_u(x_0,h)} |\tilde{v}| \leq C_1 \|\mathbf{F}\|_{L^\infty(S_u(x_0,h))} h^\delta.$$

On the other hand, by estimates (12.26) in the proof of Theorem 12.13, there exists $\beta(n, \lambda, \Lambda) \in (0,1)$ such that $\operatorname{osc}_{S_u(x_0,h/2)} w \leq \beta \operatorname{osc}_{S_u(x_0,h)} w$. Therefore,

$$(15.29) \quad \begin{aligned} \operatorname{osc}_{S_u(x_0,h/2)} v &\leq \operatorname{osc}_{S_u(x_0,h/2)} w + \operatorname{osc}_{S_u(x_0,h/2)} \tilde{v} \\ &\leq \beta \operatorname{osc}_{S_u(x_0,h)} w + 2 \|\tilde{v}\|_{L^\infty(S_u(x_0,h/2))}. \end{aligned}$$

Applying the maximum principle to $U^{ij} D_{ij} w = 0$, we have

$$\operatorname{osc}_{S_u(x_0,h)} w = \operatorname{osc}_{\partial S_u(x_0,h)} w = \operatorname{osc}_{\partial S_u(x_0,h)} v \leq \operatorname{osc}_{S_u(x_0,h)} v.$$

Together with (15.29) and (15.28), we find for every $h \leq h_0$,

$$\operatorname{osc}_{S_u(x_0,h/2)} v \leq \beta \operatorname{osc}_{S_u(x_0,h)} v + 2 C_1 \|\mathbf{F}\|_{L^\infty(S_u(x_0,h_0))} h^\delta.$$

Hence, as in the proof of Theorem 12.13, from (12.26), we can increase β if necessary so as to have $2^\delta \beta > 1$ and obtain $\gamma_0(n, \lambda, \Lambda, \varepsilon_n) \in (0,1)$ and $C_2(n, \lambda, \Lambda, \varepsilon_n, K, d, h_0)$ such that, for all $h \leq h_0$,

$$(15.30) \quad \begin{aligned} \operatorname{osc}_{S_u(x_0,h)} v &\leq C_2 \Big(\frac{h}{h_0} \Big)^{\gamma_0} \Big(\operatorname{osc}_{S_u(x_0,h_0)} v + \|\mathbf{F}\|_{L^\infty(S_u(x_0,h_0))} h_0^\delta \Big) \\ &\leq C_2 \Big(\frac{h}{h_0} \Big)^{\gamma_0} \Big(2 \|v\|_{L^\infty(S_u(x_0,h_0))} + \|\mathbf{F}\|_{L^\infty(S_u(x_0,h_0))} h_0^\delta \Big). \end{aligned}$$

By Theorem 15.4, we have

$$\|v\|_{L^\infty(S_u(x_0,h_0))} \le C_3\big(h_0^{1-n/2}\|\mathbf{F}\|_{L^\infty(S_u(x_0,2h_0))} + h_0^{-\frac{n}{2p}}\|v\|_{L^p(S_u(x_0,2h_0))}\big),$$

where $C_3 > 0$ depends only on $n, p, \lambda, \Lambda, h_0, d, \varepsilon_n$, and K. The above estimate combined with (15.30) gives (15.25).

Step 2. Hölder estimate. By Theorem 5.30, there exists $\tau_0(n, \lambda, \Lambda) \in (0,1)$ such that if $z \in S_u(x_0, h_0)$, then $S_u(z, 4\tau_0 h_0) \subset S_u(x_0, 2h_0)$.

The same arguments as in Step 1 show that the oscillation estimate (15.25) still holds where $\mathrm{osc}_{S_u(x_0,h)} v$ is now replaced by $\mathrm{osc}_{S_u(z,\tau_0 h)} v$ for $h \le h_0$ and $z \in S_u(x_0, h_0)$. Moreover, by Lemma 5.6(iii), we have

$$M := \sup_{z,w \in S_u(x_0,2h_0)} |Du(z) - Du(w)| \le C(n, h_0, \lambda, \Lambda, d).$$

Thus, exactly as in the proof of Theorem 12.14, we obtain the asserted Hölder estimate with $\bar{\gamma} = \gamma_0$. This completes the proof of the theorem. $\qquad\square$

15.3. Global Hölder Estimates

In this section, we turn to the analysis at the boundary for the linearized Monge–Ampère equation (15.1). Our goal is to establish global Hölder estimates for its solutions, thus providing a global counterpart of Theorem 15.6. Parallel to the use of fine interior properties of the Green's function of the linearized Monge–Ampère operator $D_i(U^{ij}D_j)$ in the previous section, our analysis here relies on the global high integrability of the gradient of the Green's function in Theorem 14.25.

The analysis in this section will be involved with Ω and u satisfying either the global conditions (15.38)–(15.40) or the following local conditions.

Let Ω be a bounded convex domain in $\mathbb{R}^n$ with

$$(15.31) \qquad B_\rho(\rho e_n) \subset \Omega \subset \{x_n \ge 0\} \cap B_{1/\rho}(0),$$

for some small $\rho > 0$. Assume that, for each $y \in \partial\Omega \cap B_\rho(0)$, Ω satisfies the interior ball condition with radius ρ; that is,

$$(15.32) \qquad \text{there is a ball } B_\rho(z) \subset \Omega \text{ such that } y \in \partial B_\rho(z).$$

Let $u \in C^{1,1}(\overline{\Omega}) \cap C^3(\Omega)$ be a convex function satisfying

$$(15.33) \qquad 0 < \lambda \le \det D^2 u \le \Lambda \quad \text{in } \Omega.$$

We assume that on $\partial\Omega \cap B_\rho(0)$, u separates quadratically from its tangent hyperplanes on $\partial\Omega$; that is, if $x_0 \in \partial\Omega \cap B_\rho(0)$, then, for all $x \in \partial\Omega$,

$$(15.34) \quad \rho\,|x - x_0|^2 \le u(x) - u(x_0) - Du(x_0)\cdot(x - x_0) \le \rho^{-1}\,|x - x_0|^2.$$

We first deduce from Theorem 14.25 the following global estimates.

Lemma 15.7. *Consider the following settings:*

(i) Ω *and* u *satisfy* (15.31)–(15.34). *Let* $A = \Omega \cap B_\delta(0)$ *where* $\delta \le c(n, \lambda, \Lambda, \rho)$.

(ii) Ω *and* u *satisfy* (15.38)–(15.40).

Let V *be either* A *as in* (i) *or* Ω *as in* (ii). *If* $n \ge 3$, *we assume further that*

$$\int_V \|D^2 u\|^\gamma \, dx \le C(n, \lambda, \Lambda, \rho) \quad \text{for some } \gamma > n/2.$$

Let $\mathbf{F} \in L^\infty(V; \mathbb{R}^n) \cap W^{1,n}(V; \mathbb{R}^n)$ *and* $v \in W^{2,n}_{\mathrm{loc}}(V) \cap C(\overline{V})$ *satisfy*

$$-U^{ij} D_{ij} v \le \operatorname{div} \mathbf{F} \quad \text{in } V.$$

Then, there exists a positive constant $\kappa_2(n, \gamma, \lambda, \Lambda)$ *such that*

$$\sup_V v \le \sup_{\partial V} v + C(n, \gamma, \lambda, \Lambda, \rho)|V|^{\kappa_2}\|\mathbf{F}\|_{L^\infty(V)}.$$

Proof. Let $g_V(\cdot, y)$ be the Green's function of $D_i(U^{ij} D_j)$ in V with pole $y \in V$. As in (14.79) using Remark 14.2 and Theorem 3.16, we have

$$(15.35) \quad v(y) - \sup_{\partial V} v \le \int_V g_V(x, y) \operatorname{div} \mathbf{F}(x) \, dx := w(y) \quad \text{for all } y \in V.$$

Let $\kappa_0 = \kappa_0(\lambda, \Lambda) > 0$ be as in Theorem 14.25(i) and (ii) if $n = 2$, and let $\kappa_0 = \tilde{\kappa}(n, \gamma, \lambda, \Lambda) > 0$ be as in Theorem 14.25(iv) if $n \ge 3$. Set $\kappa_2 := \kappa_0/(1 + \kappa_0) > 0$. Using the Hölder inequality to the estimates in Theorem 14.25, we find

$$(15.36) \quad \int_V |D_x g_V(x, y)| \, dx \le C(n, \gamma, \lambda, \Lambda, \rho)|V|^{\kappa_2}.$$

Therefore, for all $y \in V$,

$$(15.37) \quad w(y) = -\int_V D_x g_V(x, y) \cdot \mathbf{F}(x) \, dx \le C(n, \gamma, \lambda, \Lambda, \rho)|V|^{\kappa_2}\|\mathbf{F}\|_{L^\infty(V)}.$$

Now, the desired estimate follows from (15.35) and (15.37). $\square$

Next, we obtain the following Hölder estimates near the boundary for solutions to inhomogeneous linearized Monge–Ampère equation (15.1).

Theorem 15.8. *Assume* Ω *and* u *satisfy* (15.31)–(15.34). *Assume that*

(a) *either* $n = 2$ *or*

(b) $n \ge 3$, *and* $\int_{\Omega \cap B_\rho(0)} \|D^2 u\|^\gamma \, dx \le C(n, \lambda, \Lambda, \rho)$ *for some* $\gamma > n/2$.

Let $v \in C\left(B_\rho(0) \cap \overline{\Omega}\right) \cap W_{\text{loc}}^{2,n}(B_\rho(0) \cap \Omega)$ be a solution to

$$U^{ij} D_{ij} v = \operatorname{div} \mathbf{F} \quad in \ B_\rho(0) \cap \Omega, \qquad v = \psi \quad on \ \partial\Omega \cap B_\rho(0),$$

where $\mathbf{F} \in L^\infty(\Omega \cap B_\rho(0); \mathbb{R}^n) \cap W^{1,n}(\Omega \cap B_\rho(0); \mathbb{R}^n)$ and $\psi \in C^\alpha(\partial\Omega \cap B_\rho(0))$ for some $\alpha \in (0,1)$. Then, there exist positive constants $\beta = \beta(n, \lambda, \Lambda, \alpha)$ and C depending only on $n, \lambda, \Lambda, \alpha, \rho$ (and also γ if $n \geq 3$) such that

$$[v]_{C^\beta(\Omega \cap B_{\rho/2}(0))} \leq C\Big(\|v\|_{L^\infty(\Omega \cap B_\rho(0))} + \|\psi\|_{C^\alpha(\partial\Omega \cap B_\rho(0))} + \|\mathbf{F}\|_{L^\infty(\Omega \cap B_\rho(0))} \Big).$$

Proof. Let κ_2 be as in Lemma 15.7. Let $\alpha_0 := \min\{\alpha, \kappa_2\}$ and

$$\mathbf{K} := \|v\|_{L^\infty(\Omega \cap B_\rho(0))} + \|\psi\|_{C^\alpha(\partial\Omega \cap B_\rho(0))} + \|\mathbf{F}\|_{L^\infty(\Omega \cap B_\rho(0))}.$$

Then, invoking Lemma 15.7 and a construction of suitable barriers as in the proof of Proposition 14.32, we can find $\delta > 0$ and $C > 0$ depending only on $n, \lambda, \Lambda, \alpha, \rho$ (and γ if $n \geq 3$) such that, for any $x_0 \in \partial\Omega \cap B_{\rho/2}(0)$, we have

$$|v(x) - v(x_0)| \leq C |x - x_0|^{\frac{\alpha_0}{\alpha_0 + 3n}} \mathbf{K} \quad \text{for all } x \in \Omega \cap B_\delta(x_0).$$

Now, the proof of the Hölder estimates near the boundary in this theorem is similar to that of Theorem 14.33. It combines the above boundary Hölder estimates and the interior Hölder estimates in Theorem 15.6 using Savin's Boundary Localization Theorem (Theorem 8.5). We omit the details. $\qquad\square$

Now, we can establish the global Hölder estimates for (15.1) under suitable boundary conditions; see Proposition 4.7 for conditions guaranteeing the quadratic separation (15.40) from tangent hyperplanes on the boundary.

Theorem 15.9 (Global Hölder estimates). *Let Ω be a bounded convex domain in $\mathbb{R}^n$. Assume that there exists a small constant $\rho > 0$ such that*

(15.38)   *$\Omega \subset B_{1/\rho}(0)$ and, for each $y \in \partial\Omega$, there is an interior ball*

$$B_\rho(z) \subset \Omega \text{ such that } y \in \partial B_\rho(z).$$

Let $u \in C^{1,1}(\overline{\Omega}) \cap C^3(\Omega)$ be a convex function satisfying

(15.39) $0 < \lambda \leq \det D^2 u \leq \Lambda \quad in \ \Omega.$

Assume further that for all $x_0, x \in \partial\Omega$, we have the quadratic separation

(15.40) $\rho |x - x_0|^2 \leq u(x) - u(x_0) - Du(x_0) \cdot (x - x_0) \leq \rho^{-1} |x - x_0|^2.$

Let $\psi \in C^\alpha(\partial\Omega)$ $(0 < \alpha < 1)$, and let $\mathbf{F} \in L^\infty(\Omega; \mathbb{R}^n) \cap W^{1,n}(\Omega; \mathbb{R}^n)$. Let $v \in C(\overline{\Omega}) \cap W_{\text{loc}}^{2,n}(\Omega)$ be the solution to the linearized Monge–Ampère equation

$$U^{ij} D_{ij} v = \operatorname{div} \mathbf{F} \quad in \ \Omega, \qquad v = \psi \quad on \ \partial\Omega.$$

Assume that

    (a) *either $n = 2$ or*

    (b) *$n \geq 3$, (15.3) holds for some $\varepsilon_n > (n+1)(n-2)/2$ and $K > 0$.*

Then, there are positive constants $\alpha_1 \in (0,1)$ and M depending only on $n, \rho, \alpha, \lambda, \Lambda$ (and also K, ε_n if $n \geq 3$) such that

$$\|v\|_{C^{\alpha_1}(\Omega)} \leq M\big(\|\psi\|_{C^\alpha(\partial\Omega)} + \|\mathbf{F}\|_{L^\infty(\Omega)}\big).$$

Proof. From Lemma 15.7, we find that

$$\|v\|_{L^\infty(\Omega)} \leq \|\psi\|_{L^\infty(\partial\Omega)} + C(n, \gamma, \lambda, \Lambda, \rho)\|\mathbf{F}\|_{L^\infty(\Omega)}.$$

If $n \geq 3$, then from (15.3) and as in Step 1 of the proof of Theorem 14.25, we have for all $z \in \partial\Omega$ and all $\tau \leq c(n, \lambda, \Lambda, \rho)$,

$$\int_{B_\tau(z) \cap \Omega} \|D^2 u\|^{1+\varepsilon_n}\, dx \leq C(n, \lambda, \Lambda, \rho, K, \varepsilon_n).$$

Since $1 + \varepsilon_n > (n-1)n/2 \geq n/2$, the hypothesis (b) of Theorem 15.8 is satisfied and hence we can apply this theorem when $n \geq 3$. Combining Theorems 15.6 and 15.8, we obtain the desired global Hölder estimates. $\quad\square$

Remark 15.10. For $n \geq 3$, it is not known if it is possible to remove the additional assumption (15.3) in Theorems 15.6 and 15.9 when $\mathbf{F} \not\equiv 0$, but $\mathbf{F} \in C^\alpha$ for some $\alpha \in (0,1)$.

So far, our analysis has been involved with Hölder estimates for solutions to (15.1). It is natural to wonder if fine regularity properties can be established for the gradient of its solutions. This question has not been studied much. However, we record here the following global $W^{1,1+\kappa}$ estimates.

Proposition 15.11 (Global $W^{1,1+\kappa}$ estimates). *Let $u \in C^2(\Omega)$ be a convex function satisfying $\det D^2 u \geq \lambda > 0$ in a domain Ω in $\mathbb{R}^n$. Let $V \subset \Omega$. Suppose that $v \in W_0^{1,2}(V)$ is the weak solution to*

$$\operatorname{div}(UDv) \equiv D_i(U^{ij} D_j v) = \operatorname{div} \mathbf{F} \quad \text{in } V, \qquad v = 0 \quad \text{on } \partial V,$$

where $\mathbf{F}$ is a bounded vector field. Then for all $\varepsilon > 0$, we have

$$\|Dv\|_{L^{\frac{2(1+\varepsilon)}{2+\varepsilon}}(V)} \leq \lambda^{-1}\|\Delta u\|_{L^1(V)}^{1/2}\|\Delta u\|_{L^{1+\varepsilon}(V)}^{1/2}\|\mathbf{F}\|_{L^\infty(V)}.$$

Proof. By homogeneity, we can assume that $\|\mathbf{F}\|_{L^\infty(V)} = 1$. Multiplying both sides of the equation by v and then integrating by parts, we get

$$\int_V U^{ij} D_j v D_i v\, dx = \int_V \mathbf{F} \cdot Dv\, dx \leq \int_V |Dv|\, dx.$$

Utilizing Lemma 2.58 and then the Hölder inequality, we find

$$\int_V \frac{\det D^2 u |Dv|^2}{\Delta u}\, dx \leq \int_V |Dv|\, dx \leq \left(\int_V \frac{|Dv|^2}{\Delta u}\, dx\right)^{1/2}\left(\int_V \Delta u\, dx\right)^{1/2}.$$

Since $\det D^2 u \geq \lambda$, we obtain $\||Dv|/\sqrt{\Delta u}\|_{L^2(V)} \leq \lambda^{-1}\|\Delta u\|_{L^1(V)}^{1/2}$. Inserting this into the following inequality, which follows by the Hölder inequality,

$$\|Dv\|_{L^{\frac{2(1+\varepsilon)}{2+\varepsilon}}(V)} \leq \||Dv|/\sqrt{\Delta u}\|_{L^2(V)}\|\sqrt{\Delta u}\|_{L^{2(1+\varepsilon)}(V)},$$

we obtain the desired estimate. $\square$

We will give an application of Proposition 15.11 to the regularity of the dual semigeostrophic equations in the next section.

15.4. Application: The Dual Semigeostrophic Equations

The semigeostrophic equations are a simple model used in meteorology to describe large scale atmospheric flows. They can be derived from the three-dimensional incompressible Euler equations, with Boussinesq and hydrostatic approximations, subject to a strong Coriolis force; see Benamou–Brenier [**BB**], Loeper [**Lo2**], and Cullen [**Cu**] for further explanations. Since for large scale atmospheric flows, the Coriolis force dominates the advection term, the flow is mostly bi-dimensional.

The *two-dimensional semigeostrophic equations* with initial pressure p^0 consist of three equations for three unknowns (p_t, u_t^1, u_t^2):

$$(15.41) \quad \begin{cases} \partial_t Dp_t + (\mathbf{u}_t \cdot D)Dp_t - (Dp_t)^\perp + \mathbf{u}_t = 0 & \text{in } \mathbb{R}^2 \times [0, \infty), \\ \operatorname{div} \mathbf{u}_t = 0 & \text{in } \mathbb{R}^2 \times [0, \infty), \\ p_0 = p^0 & \text{on } \mathbb{R}^2. \end{cases}$$

Here, $\mathbf{u}_t = (u_t^1, u_t^2) : \mathbb{R}^2 \to \mathbb{R}^2$ represents time-dependent, $\mathbb{Z}^2$-periodic *velocity*, $p_t : \mathbb{R}^2 \to \mathbb{R}$ represents time-dependent, $\mathbb{Z}^2$-periodic *pressure*, and the vector $(Dp_t)^\perp$ is called the *semigeostrophic wind*.

Throughout, $D = (\partial/\partial x_1, \partial/\partial x_2)$, and we use $w^\perp$ to denote the vector $(-w_2, w_1)$ for $w = (w_1, w_2) \in \mathbb{R}^2$, and $f_t(\cdot)$ for the function $f(\cdot, t)$.

Energetic considerations (see [**Cu**]) show that it is natural to assume that the pressure p_t has the form $p_t(x) = P_t(x) - |x|^2/2$, where P_t is a convex function on $\mathbb{R}^2$. In light of this, the two-dimensional semigeostrophic system for $(\mathbf{u}_t, p_t)$ becomes the following extended semigeostrophic system for $(\mathbf{u}_t, P_t)$:

$$
(15.42) \quad
\begin{cases}
\partial_t DP_t(x) + (\mathbf{u}_t \cdot D)DP_t - (DP_t - x)^\perp = 0 & \text{in } \mathbb{R}^2 \times [0, \infty), \\
\operatorname{div} \mathbf{u}_t = 0 & \text{in } \mathbb{R}^2 \times [0, \infty), \\
P_t \text{ convex} & \text{in } \mathbb{R}^2 \times [0, \infty), \\
P_0(x) = p^0(x) + |x|^2/2 & \text{on } \mathbb{R}^2.
\end{cases}
$$

It is in general challenging to directly solve this system. The common route is to use the dual semigeostrophic system on the two-dimensional torus $\mathbb{T}^2$ which is easier to solve and then try to go back to the original system.

The *dual semigeostrophic equations* of the semigeostrophic equations on $\mathbb{T}^2 = \mathbb{R}^2/\mathbb{Z}^2$ are the following system of nonlinear transport equations:

$$
(15.43) \quad
\begin{cases}
\partial_t \rho_t(x) + \operatorname{div}(\rho_t(x)(x - DP_t^*(x))^\perp) = 0 & \text{in } \mathbb{T}^2 \times [0, \infty), \\
\det D^2 P_t^* = \rho_t & \text{in } \mathbb{T}^2 \times [0, \infty), \\
P_t^*(x) \text{ convex} & \text{in } \mathbb{T}^2 \times [0, \infty), \\
\rho_0 = \rho^0 & \text{on } \mathbb{T}^2,
\end{cases}
$$

for (ρ_t, P_t^*) with the boundary condition

$$
(15.44) \qquad P_t^* - |x|^2/2 \text{ is } \mathbb{Z}^2\text{-periodic}.
$$

Here the initial potential density ρ^0 is a probability measure on $\mathbb{T}^2$.

15.4.1. Derivation of dual semigeostrophic equations.
For completeness, we briefly indicate how to derive (15.43) from (15.42).

Motivated by optimal transportation (see, for example, Villani [**Vi**]) where optimal transport maps are gradients of convex functions, we let $\rho_t = (DP_t)_\# dx$; that is,

$$
(15.45) \qquad \int_{\mathbb{R}^2} \psi(y)\, d\rho_t(y) = \int_{\mathbb{R}^2} \psi(DP_t(x))\, dx \quad \text{for all } \psi \in L^1(\mathbb{R}^2).
$$

Since $DP_t - x = Dp_t$ is $\mathbb{Z}^2$-periodic, ρ_t is $\mathbb{Z}^2$-periodic on $\mathbb{R}^2$. Inserting $\psi = \chi_{[0,1]^2}$ into (15.45) yields $\int_{[0,1]^2} d\rho_t = \int_{[0,1]^2} dx = 1$.

Thus, we can identify ρ_t as a probability measure on the two-dimensional torus $\mathbb{T}^2$. The initial potential density ρ_0 is a probability measure on $\mathbb{T}^2$ which can be computed via the initial pressure p^0 by the formula

$$
\rho_0 = (DP_0)_\# dx = (Dp^0(x) + x)_\# dx.
$$

Using the change of variables $y = DP_t(x)$ in (15.45), we obtain

$$
\int_{\mathbb{R}^2} \psi(DP_t(x))\, dx = \int_{\mathbb{R}^2} \psi(DP_t(x))\rho_t(DP_t(x)) \det D^2 P_t(x)\, dx.
$$

Since this holds for all $\psi \in L^1(\mathbb{R}^2)$, P_t satisfies the Monge–Ampère equation

$$
(15.46) \qquad \rho_t(DP_t(x)) \det D^2 P_t(x) = 1.
$$

Denote by P_t^* the Legendre transform of P_t. As in Section 2.3.4, if $y = DP_t(x)$, then

$$(15.47) \quad DP_t(DP_t^*(y)) = y, \ DP_t^*(y) = x, \ \det D^2 P_t^*(y) = [\det D^2 P_t(x)]^{-1}.$$

From (15.46), we obtain

$$(15.48) \qquad\qquad \det D^2 P_t^*(y) = \rho_t(y).$$

Next, we derive an evolution equation for the density ρ_t. Let $\varphi \in C_c^\infty(\mathbb{R}^2)$. Taking (15.45) and (15.42) into account, we compute

$$\frac{d}{dt} \int_{\mathbb{R}^2} \varphi \, d\rho_t = \frac{d}{dt} \int_{\mathbb{R}^2} \varphi(DP_t) \, dx = \int_{\mathbb{R}^2} D\varphi(DP_t) \cdot \partial_t DP_t \, dx$$

$$= \int_{\mathbb{R}^2} D\varphi(DP_t) \cdot \left[-(\mathbf{u}_t \cdot D)DP_t + (DP_t(x) - x)^\perp \right] dx$$

$$= \int_{\mathbb{R}^2} D\varphi(DP_t) \cdot (DP_t(x) - x)^\perp \, dx := A,$$

because, by the second equation of (15.42) and integration by parts, we have

$$\int_{\mathbb{R}^2} D\varphi(DP_t) \cdot [(\mathbf{u}_t \cdot D)DP_t] \, dx = \int_{\mathbb{R}^2} D\varphi(DP_t) \cdot (D^2 P_t \mathbf{u}_t) \, dx$$

$$= \int_{\mathbb{R}^2} D[\varphi \circ DP_t] \cdot \mathbf{u}_t \, dx = 0.$$

Let $y = DP_t(x)$. Then, in view of (15.46), we have

$$dx = [\det D^2 P_t(x)]^{-1} \, dy = \rho_t(DP_t(x)) \, dy = \rho_t(y) \, dy,$$

and it follows from (15.47) that

$$\frac{d}{dt} \int_{\mathbb{R}^2} \varphi \, d\rho_t = A = \int_{\mathbb{R}^2} D\varphi(y) \cdot (y - DP_t^*(y))^\perp \rho_t(y) \, dy$$

$$= \int_{\mathbb{R}^2} -\mathrm{div}[\rho_t(y)(y - DP_t^*(y))^\perp] \varphi(y) \, dy.$$

This leads to a transport equation for the density:

$$(15.49) \quad \partial_t \rho_t(y) = -\mathrm{div}[\rho_t(y)(y - DP_t^*(y))^\perp] = -D\rho_t(y) \cdot (y - DP_t^*(y))^\perp.$$

Combining (15.48) and (15.49), we obtain (15.43).

Here, we focus on the dual semigeostrophic equations. Existence of global weak solutions for the (15.43)–(15.44) system has been established via time discretization in Benamou–Brenier [**BB**] and Cullen–Gangbo [**CuG**]; see also Ambrosio–Colombo–De Philippis–Figalli [**ACDF1**] for further details. When ρ^0 is Hölder continuous and bounded between two positive constants on $\mathbb{T}^2$, Loeper [**Lo2**] showed that there is a unique, short-time, Hölder solution ρ to (15.43)–(15.44); the time interval for this Hölder solution depends only on the bounds on ρ^0. However, when ρ^0 is only a probability measure, the uniqueness of weak solutions is still an open question.

For completeness, we briefly indicate how to obtain distributional solutions to the original semigeostrophic equations from solutions (ρ_t, P_t^*) of the dual equations (15.43)–(15.44); see [**ACDF1**] for a rigorous treatment. Let P_t be the Legendre transform of P_t^*. Let $p^0(x) = P_0(x) - |x|^2/2$ and

$$(15.50) \quad \begin{cases} p_t(x) := P_t(x) - |x|^2/2, \\ \mathbf{u}_t(x) := (\partial_t DP_t^*) \circ DP_t(x) + D^2 P_t^*(DP_t(x)) \cdot (DP_t(x) - x)^\perp. \end{cases}$$

Then $(p_t, \mathbf{u}_t)$ is a global solution to the semigeostrophic equations (15.41).

15.4.2. Hölder continuity of two-dimensional dual semigeostrophic equations.
We now return to the regularity of solutions to (15.43)–(15.44) in the typical case where the initial density ρ^0 is bounded between two positive constants. Their spatial regularity, recalled below, is now well understood thanks to regularity results for the Monge–Ampère equations.

Theorem 15.12 (Spatial regularity of the dual semigeostrophic equations in two dimensions). *Let ρ^0 be a probability measure on $\mathbb{T}^2$ with $0 < \lambda \leq \rho^0 \leq \Lambda$. Let (ρ_t, P_t^*) solve (15.43)–(15.44) with the normalization $\int_{\mathbb{T}^2} P_t^* \, dx = 0$. Let P_t be the Legendre transform of P_t^*, and let $\mathbf{U}_t(x) = (x - DP_t^*(x))^\perp$. Then, for all $t > 0$, the following hold.*

(i) $\lambda \leq \rho_t \leq \Lambda$ *in* $\mathbb{T}^2$.

(ii) $\|\mathbf{U}_t\|_{L^\infty(\mathbb{T}^2)} \leq \sqrt{2}/2$.

(iii) P_t *is an Aleksandrov solution to* $(\rho_t \circ DP_t) \det D^2 P_t = 1$ *on* $\mathbb{T}^2$.

(iv) *There exist constants* $\beta(\lambda, \Lambda) \in (0, 1)$ *and* $C(\lambda, \Lambda) > 0$ *such that*
$$\|P_t^*\|_{C^{1,\beta}(\mathbb{T}^2)} + \|P_t\|_{C^{1,\beta}(\mathbb{T}^2)} \leq C.$$

(v) *Let* $\varepsilon = \varepsilon(2, \lambda, \Lambda) > 0$ *be the constant* ε *in Theorem 6.11. Then*
$$\|P_t^*\|_{W^{2,1+\varepsilon}(\mathbb{T}^2)} + \|P_t\|_{W^{2,1+\varepsilon}(\mathbb{T}^2)} \leq C(\lambda, \Lambda).$$

(vi) $\|D\partial_t P_t^*\|_{L^{\frac{2(1+\varepsilon)}{2+\varepsilon}}(\mathbb{T}^2)} \leq C(\lambda, \Lambda).$

By an approximation argument as in [**ACDF1**, **Lo1**], we can assume in Theorem 15.12 and Theorem 15.13 below that ρ_t and P_t^* are smooth, but we will establish estimates that depend only on λ and Λ.

Proof. We prove (i). Since $\operatorname{div} \mathbf{U}_t = 0$, ρ_t satisfies $\partial_t \rho_t = -D\rho_t \cdot \mathbf{U}_t$. For any $p > 1$, we compute with the aid of integration by parts

$$\frac{d}{dt} \int_{\mathbb{T}^2} |\rho_t|^p \, dx = \int_{\mathbb{T}^2} -p|\rho_t|^{p-2} \rho_t D\rho_t \cdot \mathbf{U}_t \, dx = \int_{\mathbb{T}^2} -D(|\rho_t|^p) \cdot \mathbf{U}_t \, dx = 0.$$

Therefore, $\|\rho_t\|_{L^p(\mathbb{T}^2)} = \|\rho_0\|_{L^p(\mathbb{T}^2)}$ for all $t \geq 0$. Letting $p \to \infty$, we obtain $\sup_{\mathbb{T}^2} |\rho_t| = \sup_{\mathbb{T}^2} |\rho_0| \leq \Lambda$. Similarly, $\sup_{\mathbb{T}^2} |\Lambda - \rho_t| = \sup_{\mathbb{T}^2} |\Lambda - \rho_0| \leq \Lambda - \lambda$.

It follows that $\lambda \leq \rho_t \leq \Lambda$, proving assertion (i).

For the proof of assertion (ii), the idea is that DP_t^* is an optimal transport map between the density ρ_t to the uniform density on the torus $\mathbb{T}^2$ (see Cordero-Erausquin [**CE**]). In terms of an equation, this is expressed in (15.48). Thus for all $x \in \mathbb{T}^2$, we have

$$|\mathbf{U}_t(x)| = |DP_t^*(x) - x| \leq \text{diam}(\mathbb{T}^2) \leq \sqrt{2}/2.$$

For rigorous details, see [**ACDF1**].

We note that assertion (iii) was derived heuristically in (15.46). For rigorous details, see [**ACDF1**].

Note that assertion (iv) follows from (i) and Theorem 5.21, while assertion (v) follows from $\det D^2 P_t^* = \rho_t$, (iii), (i), and Theorem 6.11.

Finally, we prove (vi). Differentiating $\det D^2 P_t^* = \rho_t$ with respect to t, we find that $\partial_t P_t^*$ solves the linearized Monge–Ampère equation

$$(15.51) \qquad \text{div}\left[(\det D^2 P_t^*)(D^2 P_t^*)^{-1}(D\partial_t P_t^*)\right] = \partial_t \rho_t = -\text{div}(\rho_t \mathbf{U}_t).$$

By Proposition 15.11 and $\det D^2 P_t^* \geq \lambda$, we have

$$\|D\partial_t P_t^*\|_{L^{\frac{2(1+\varepsilon)}{2+\varepsilon}}(\mathbb{T}^2)} \leq \lambda^{-1} \|\Delta P_t^*\|_{L^1(\mathbb{T}^2)}^{1/2} \|\Delta P_t^*\|_{L^{1+\varepsilon}(\mathbb{T}^2)}^{1/2} \|\rho_t \mathbf{U}_t\|_{L^\infty(\mathbb{T}^2)}.$$

The desired estimate now follows from assertions (i), (ii), and (v). $\qquad\square$

Regarding the temporal regularity of solutions to (15.43)–(15.44), we have the following uniform-in-time Hölder continuity of $\partial_t P_t^*$ and $\partial_t P_t$.

Theorem 15.13 (Hölder regularity of the dual semigeostrophic equations in two dimensions). *Let ρ^0 be a probability measure on $\mathbb{T}^2$ with $0 < \lambda \leq \rho^0 \leq \Lambda$. Let (ρ_t, P_t^*) solve (15.43)–(15.44). Let P_t be the Legendre transform of P_t^*. Then, there exist $\alpha = \alpha(\lambda, \Lambda) \in (0, 1)$ and $C = C(\lambda, \Lambda) > 0$ such that*

$$\|\partial_t P_t^*\|_{C^\alpha(\mathbb{T}^2)} + \|\partial_t P_t\|_{C^\alpha(\mathbb{T}^2)} \leq C \quad \text{for all } t > 0.$$

Proof. We can assume that ρ_t, P_t^*, and P_t are smooth, and we will establish uniform-in-time bounds in $C^\alpha(\mathbb{T}^2)$ for $\partial_t P_t^*$ and $\partial_t P_t$ that depend only on λ and Λ. Our proof uses Theorem 15.6. Let $\beta(\lambda, \Lambda)$ be as in Theorem 15.12.

By subtracting a constant from P_t^*, we can assume that

$$(15.52) \qquad \int_{\mathbb{T}^2} P_t^* \, dx = 0 \quad \text{for each } t \in (0, \infty).$$

By Theorem 15.12, for all $t \geq 0$, we have

$$(15.53) \qquad \lambda \leq \det D^2 P_t^* = \rho_t \leq \Lambda \quad \text{in } \mathbb{T}^2.$$

Moreover, $\partial_t P_t^*$ solves the linearized Monge–Ampère equation (15.51), where the vector field $\rho_t \mathbf{U}_t$ satisfies

$$(15.54) \qquad \|\rho_t \mathbf{U}_t\|_{L^\infty(\mathbb{T}^2)} \leq \sqrt{2}\Lambda.$$

From Theorem 15.12(vi), (15.52), and the Sobolev Embedding Theorem, we obtain

$$(15.55) \qquad \|\partial_t P_t^*\|_{L^2(\mathbb{T}^2)} \leq C(\lambda, \Lambda).$$

Using Theorem 15.12(iv) together with the $\mathbb{Z}^2$-periodicity of $P_t^* - |x|^2/2$, we can find positive constants $h_0(\lambda, \Lambda)$ and $R_0(\lambda, \Lambda)$ such that

$$(15.56) \qquad \mathbb{T}^2 \subset S_{P_t^*}(x_0, h_0) \subset S_{P_t^*}(x_0, 4h_0) \subset B_{R_0}(0) \quad \text{for all } x_0 \in \mathbb{T}^2.$$

Moreover, we deduce from (15.54) and (15.55) that

$$(15.57) \qquad \|\rho_t \mathbf{U}_t\|_{L^\infty(B_{R_0}(0))} + \|\partial_t P_t^*\|_{L^2(B_{R_0}(0))} \leq C(\lambda, \Lambda).$$

Now, having (15.53), (15.56), and (15.57), we can apply Theorem 15.6 to (15.51) in each section $S_{P_t^*}(x_0, 4h_0)$ with $p = 2$ and $\Omega = B_{R_0}(0)$ to conclude:

$$(15.58) \qquad \|\partial_t P_t^*\|_{C^\gamma(\mathbb{T}^2)} \leq C(\lambda, \Lambda) \quad \text{for } \gamma = \gamma(\lambda, \Lambda) \in (0, 1), \text{ for all } t > 0.$$

Differentiating $P_t(x) + P_t^*(DP_t(x)) = x \cdot DP_t(x)$ with respect to t yields

$$\partial_t P_t(x) = -\partial_t P_t^*(DP_t(x)) \quad \text{for } x \in \mathbb{R}^2.$$

Thus, combining (15.58) with Theorem 15.12(iv), we obtain

$$\|\partial_t P_t\|_{C^{\gamma\beta}(\mathbb{T}^2)} \leq C(\lambda, \Lambda).$$

The proof of Theorem 15.13 is completed by setting $\alpha = \gamma\beta$. $\qquad\qquad\square$

15.5. Application: Singular Abreu Equations

Another interesting context for linearized Monge–Ampère equations with divergence form inhomogeneity is related to Abreu-type equations where the right-hand side depends on the Hessian of the unknown and could be just measures. We call these *singular Abreu equations*. An example is

$$(15.59) \quad U^{ij} D_{ij}[(\det D^2 u)^{-1}] = -\operatorname{div}(|Du|^{q-2}Du) \quad \text{in } \Omega \subset \mathbb{R}^n \quad (n \geq 2).$$

Here, $q \in (1, \infty)$ and $(U^{ij}) = \operatorname{Cof} D^2 u$. Abreu-type equations of the form (15.59) first appeared in Le [**L9**] in the approximation of minimizers of convex functionals subjected to convexity constraints. These functionals include the Rochet–Choné model of the monopolist's problem in economics.

We will study the Sobolev solvability of the second boundary value problem of (15.59) where the values of u and $\det D^2 u$ are given on the boundary $\partial\Omega$. For u being convex, even when $q = 2$, the right-hand side of (15.59) is $-\Delta u$ which is a priori just a measure so it is singular; for $q \in (1, 2)$, the right-hand side is even more singular. For the solvability of the second boundary problem of (15.59), a typical procedure in the *method of continuity* is (a) to establish a priori higher-order derivative estimates for sufficiently smooth solutions and then (b) to use the degree theory to obtain the existence of these solutions. For (a), two crucial ingredients consist of bounding $\det D^2 u$ from below and above by positive constants from the data and obtaining the global Hölder estimates for $w := (\det D^2 u)^{-1}$. Then, higher-order derivative estimates follow from the global $C^{2,\alpha}$ estimates for the Monge–Ampère equation and Schauder estimates for uniformly elliptic equations. Since w solves a linearized Monge–Ampère equation, its global Hölder estimates can be derived from global Hölder estimates for the linearized Monge–Ampère equation with right-hand side having low integrability (Theorem 14.34) or being the divergence of a bounded vector field (Theorem 15.9). Here, assuming the bounds $C^{-1} \le \det D^2 u \le C$ have been established, we will focus on the case $q = 2$ and will comment briefly on the more challenging cases of $q \in (1, 2)$ later.

For $q = 2$, by Theorem 10.1, the integrability exponent of the right-hand side $-\Delta u$ of (15.59) is at best $1 + \varepsilon$ for some (possibly small) structural constant $\varepsilon > 0$. In view of Remark 13.4, the results for regular Abreu equations in Theorem 13.3 are not applicable. For the global Hölder estimates in Theorem 14.34 to be applicable to $U^{ij} D_{ij} w = -\Delta u$, the integrability exponent $1 + \varepsilon$ of Δu needs to be greater than $n/2$. For small $\varepsilon > 0$, this forces $n = 2$. Likewise, if we use the divergence form structure of Δu as $\operatorname{div}(Du)$, then without further information on $D^2 u$, we can only apply Theorem 15.9 in two dimensions. For these reasons, many results in this section are concerned with the case $n = 2$. Sometimes, it is possible to carry out the analysis for $n > 2$, when there is a smallness condition.

15.5.1. Singular Abreu Equations in Higher Dimensions.
The general result which reduces to the second boundary value problem of (15.59) when $q = 2$ in two dimensions is the following theorem.

Theorem 15.14. *Let Ω be a uniformly convex domain in $\mathbb{R}^n$ ($n \ge 2$) with $\partial\Omega \in C^5$. Let $f \in L^\infty(\Omega)$ with $0 \le f \le M$, $\varphi \in W^{4,q}(\Omega)$, and $\psi \in W^{2,q}(\Omega)$ with $\inf_{\partial\Omega} \psi > 0$ where $q > n$. Then, there is a uniformly convex solution $u \in W^{4,q}(\Omega)$ to the following second boundary value problem:*

$$(15.60) \quad \begin{cases} U^{ij} D_{ij} w = -(\Delta u)^{\frac{1}{n-1}} (\det D^2 u)^{\frac{n-2}{n-1}} f, \ \ w = (\det D^2 u)^{-1} & in \ \Omega, \\ \qquad\quad u = \varphi, \ w = \psi & on \ \partial\Omega. \end{cases}$$

Furthermore, if $f \equiv M > 0$, $\varphi \in C^{4,\beta}(\overline{\Omega})$, and $\psi \in C^{2,\beta}(\overline{\Omega})$ ($0 < \beta \le 1$), then $u \in C^{4,\beta}(\overline{\Omega})$.

Review of the Leray–Schauder degree theory. The proof of Theorem 15.14 uses a priori estimates and the Leray–Schauder degree theory, so we briefly review this theory, following O'Regan–Cho–Chen [**OCC**]. First, we recall the construction of Brouwer degree; see [**OCC**, Definition 1.2.1].

Definition 15.15 (Brouwer degree). Let Ω be an open bounded set in $\mathbb{R}^n$, and let $\mathbf{f} \in C^1(\overline{\Omega}; \mathbb{R}^n)$. If $p \notin \mathbf{f}(\partial\Omega)$ and $\det D\mathbf{f}(p) \neq 0$, then we define the Brouwer degree

$$\deg(\mathbf{f}, \Omega, p) = \begin{cases} \displaystyle\sum_{x \in \mathbf{f}^{-1}(p)} \operatorname{sign}(\det D\mathbf{f}(x)) & \text{if } \mathbf{f}^{-1}(p) \neq \emptyset, \\ 0 & \text{if } \mathbf{f}^{-1}(p) = \emptyset. \end{cases}$$

The Brouwer degree is the basis for the construction of the Leray–Schauder degree for compact mappings in Banach spaces, as stated in the following theorem; see [**OCC**, Lemma 2.2.1 and Definition 2.2.3].

Theorem 15.16 (Leray–Schauder degree). *Let E be a real Banach space with the identity mapping $I : E \to E$. Let F be an open bounded set in E, and let $T : \overline{F} \to E$ be a continuous compact mapping. Then, the following hold.*

(i) *For any $\varepsilon > 0$, there exist a finite-dimensional subspace E_ε of E and a continuous mapping $T_\varepsilon : \overline{F} \to E_\varepsilon$ such that $\|T_\varepsilon x - Tx\| < \varepsilon$ for all $x \in \overline{F}$.*

(ii) *If $0 \notin (I - T)(\partial F)$, then the Brouwer degree $\deg(I - T_\varepsilon, F \cap E_\varepsilon, 0)$ is well-defined and independent of ε for all $\varepsilon \leq \varepsilon^*$ small. Thus, we can define the Leray–Schauder degree*

$$\deg(I - T, F, 0) := \deg(I - T_\varepsilon, F \cap E_\varepsilon, 0) \quad \text{for all } \varepsilon \leq \varepsilon^*.$$

The Leray–Schauder degree has the following properties; see [**OCC**, Theorem 2.2.4].

Theorem 15.17 (Properties of Leray–Schauder degree). *Let E be a real Banach space. Let F be an open, bounded set in E, and let $T : \overline{F} \to E$ be a continuous compact mapping. If $0 \notin (I - T)(\partial F)$, then the Leray–Schauder degree $\deg(I - T, F, 0)$ has the following properties:*

(i) *(Normality) $\deg(I, F, 0) = 1$ if and only if $0 \in F$.*

(ii) *(Solvability) If $\deg(I - T, F, 0) \neq 0$, then the equation $Tx = x$ has a solution in F.*

(iii) *(Homotopy) Let $T_t : [0,1] \times \overline{F} \to E$ be continuous compact and assume that $T_t x \neq x$ for all $(t, x) \in [0,1] \times \partial F$. Then $\deg(I - T_t, F, 0)$ does not depend on $t \in [0,1]$.*

(iv) *(Degree of a fixed point) If $T(\overline{F}) = \{p\} \subset F$, then $\deg(I - T, F, 0) = 1$.*

Note that Theorem 15.17(iv) was not stated explicitly in [**OCC**]. However, it can be deduced from the definition of the Leray–Schauder degree using Brouwer's degree in Theorem 15.16 and the relationship of Brouwer degrees in spaces of different dimensions in Theorem 1.2.12 in [**OCC**]. These reduce $\deg(x - p, F, 0)$ to the case where F is an open set on the real line containing a point p and the value $\deg(x - p, F, 0) = 1$ by using Definition 15.15.

Proof of Theorem 15.14. We divide the proof into two steps.

Step 1. A priori fourth-order derivative estimates. They will be established with the help of Theorems 10.1 and 14.34. We call a positive constant *structural* if it depends only on n, Ω, ψ, φ, M, q, and β. We use $c, C, C_1, C_2, \ldots$ to denote structural constants. For simplicity, let

$$\hat{f} := -(\Delta u)^{\frac{1}{n-1}} (\det D^2 u)^{\frac{n-2}{n-1}} f.$$

We establish a priori estimates for a solution $u \in W^{4,q}(\Omega)$. Since $U^{ij} D_{ij} w \le 0$, the maximum principle yields $\inf_\Omega w \ge \inf_{\partial\Omega} w = \inf_{\partial\Omega} \psi := C_1 > 0$.

By an easy expansion in terms of the eigenvalues of $D^2 u$, we find

$$(15.61) \quad \operatorname{trace}(U^{ij}) = (\det D^2 u)\operatorname{trace}((D^2 u)^{-1}) \ge (\Delta u)^{\frac{1}{n-1}}(\det D^2 u)^{\frac{n-2}{n-1}}.$$

It follows from (15.61) and $0 \le f \le M$ that

$$U^{ij} D_{ij}(w + M|x|^2) = \hat{f} + 2M\operatorname{trace}(U^{ij}) > 0.$$

By the maximum principle in Theorem 2.79, we have

$$w + M|x|^2 \le \max_{\partial\Omega}(w + M|x|^2) = \max_{\partial\Omega}(\psi + M|x|^2) \le C_2 < \infty.$$

Therefore $w \le C_2$. As a consequence, $C_1 \le w \le C_2$. From the second equation of (15.60), we can find a structural constant $C > 0$ such that

$$(15.62) \qquad\qquad C^{-1} \le \det D^2 u \le C \quad \text{in } \Omega.$$

From $\varphi \in W^{4,q}(\Omega)$ with $q > n$, we have $\varphi \in C^3(\overline{\Omega})$ by the Sobolev Embedding Theorem (Theorem 2.76). By assumption, Ω is uniformly convex with $\partial\Omega \in C^5$. From $u = \varphi$ on $\partial\Omega$ and (15.62), we can apply the global $W^{2,1+\varepsilon}$ estimates in Theorem 10.1 to conclude that $\|D^2 u\|_{L^{1+\varepsilon}(\Omega)} \le C_1^*$ for some structural constants $\varepsilon > 0$ and $C_1^* > 0$. Thus,

$$\|\hat{f}\|_{L^{(n-1)(1+\varepsilon)}(\Omega)} \le C_3,$$

for a structural constant $C_3 > 0$. Note that $(n - 1)(1 + \varepsilon) > n/2$ for all $n \ge 2$ and all $\varepsilon > 0$. From $\psi \in W^{2,q}(\Omega)$ with $q > n$, we have $\psi \in C^1(\overline{\Omega})$ by the Sobolev Embedding Theorem. Therefore, we can apply the global Hölder estimates in Theorem 14.34 to $U^{ij} D_{ij} w = \hat{f}$ in Ω with boundary value $w = \psi \in C^1(\partial\Omega)$ on $\partial\Omega$ to conclude that $w \in C^\alpha(\overline{\Omega})$ with

$$\|w\|_{C^\alpha(\overline{\Omega})} \le C\big(\|\psi\|_{C^1(\partial\Omega)} + \|\hat{f}\|_{L^{(n-1)(1+\varepsilon)}(\Omega)}\big) \le C_4,$$

for structural constants $\alpha \in (0,1)$ and $C_4 > 0$. Now, u solves the Monge–Ampère equation $\det D^2 u = w^{-1}$ with right-hand side in $C^\alpha(\overline{\Omega})$ and boundary value $\varphi \in C^3(\partial\Omega)$. Therefore, by the global $C^{2,\alpha}$ estimates for the Monge–Ampère equation (Theorem 10.20), $u \in C^{2,\alpha}(\overline{\Omega})$ with estimates

$$(15.63) \qquad \|u\|_{C^{2,\alpha}(\overline{\Omega})} \leq C_5 \quad \text{and} \quad C_5^{-1} I_n \leq D^2 u \leq C_5 I_n.$$

Consequently, the operator $U^{ij} D_{ij}$ is uniformly elliptic with Hölder continuous coefficients. Moreover, from the definition of $\hat{f}$ and (15.63), we have

$$(15.64) \qquad \|\hat{f}\|_{L^\infty(\Omega)} \leq C_6.$$

Thus, the global Calderon–Zygmund estimates in Theorem 2.83 applied to $U^{ij} D_{ij} w = \hat{f}$ in Ω with $w = \psi$ on $\partial\Omega$, where $\psi \in W^{2,q}(\Omega)$, give $w \in W^{2,q}(\Omega)$ and therefore $u \in W^{4,q}(\Omega)$ with $\|u\|_{W^{4,q}(\Omega)} \leq C$ for a structural constant C.

It remains to consider the case $f \equiv M > 0$, $\varphi \in C^{4,\beta}(\overline{\Omega})$, and $\psi \in C^{2,\beta}(\overline{\Omega})$. In this case, we need to establish a priori estimates for $u \in C^{4,\beta}(\overline{\Omega})$. As above, instead of (15.64), we have

$$(15.65) \qquad \|\hat{f}\|_{C^{\frac{\alpha}{n-1}}(\overline{\Omega})} \leq C_7.$$

Thus, the global Schauder estimates in Theorem 2.82 applied to $U^{ij} D_{ij} w = \hat{f}$ in Ω with $w = \psi$ on $\partial\Omega$ give $w \in C^{2,\gamma}(\overline{\Omega})$ where $\gamma := \min\{\frac{\alpha}{n-1}, \beta\}$ and therefore $u \in C^{4,\gamma}(\overline{\Omega})$ with the estimate $\|u\|_{C^{4,\gamma}(\overline{\Omega})} \leq C_8$. With this estimate, we can improve (15.65) to $\|\hat{f}\|_{C^\beta(\overline{\Omega})} \leq C_9$. As above, we find that $u \in C^{4,\beta}(\overline{\Omega})$ with $\|u\|_{C^{4,\beta}(\overline{\Omega})} \leq C$ for some structural constant C.

Step 2. Existence via a priori estimates and the Leray–Schauder degree theory argument. Fix $\alpha \in (0,1)$. For a large constant $R > 1$ to be determined, define a bounded set $D(R)$ in $C^\alpha(\overline{\Omega})$ by

$$D(R) = \{v \in C^\alpha(\overline{\Omega}) : \ v \geq R^{-1}, \ \|v\|_{C^\alpha(\overline{\Omega})} \leq R\}.$$

For $t \in [0,1]$, we will define an operator $\Phi_t : D(R) \to C^\alpha(\overline{\Omega})$ as follows. Given $w \in D(R)$, by Theorem 10.20, there exists a unique uniformly convex solution $u \in C^{2,\alpha}(\overline{\Omega})$ to

$$(15.66) \qquad \det D^2 u = w^{-1} \quad \text{in } \Omega, \qquad u = \varphi \quad \text{on } \partial\Omega.$$

Next, by Theorem 2.83, there exists a unique solution $w_t \in W^{2,q}(\Omega)$ to

$$(15.67) \qquad U^{ij} D_{ij} w_t = t\hat{f} \quad \text{in } \Omega, \qquad w_t = t\psi + (1-t) \quad \text{on } \partial\Omega.$$

Because $q > n$, $w_t \in C^\alpha(\overline{\Omega})$. Let Φ_t be the map sending w to w_t. Note that:

(a) $\Phi_0(D(R)) = \{1\}$ and, in particular, Φ_0 has a unique fixed point.

(b) The map $[0,1] \times D(R) \to C^\alpha(\overline{\Omega})$ given by $(t,w) \mapsto \Phi_t(w)$ is continuous.

(c) Φ_t is compact for each $t \in [0, 1]$.

(d) For every $t \in [0, 1]$, if $w \in D(R)$ is a fixed point of Φ_t, then $w \notin \partial D(R)$.

Indeed, assertion (c) follows from the standard a priori estimates for the two separate equations (15.66) and (15.67). For assertion (d), let $w > 0$ be a fixed point of Φ_t. Then, $w \in W^{2,q}(\Omega)$, and hence $u \in W^{4,q}(\Omega)$. Next, we apply Step 1 to obtain $w^{-1} + \|w\|_{C^\alpha(\overline{\Omega})} < R$ for some structural constant R.

Let F be the interior of $D(R)$. Then, by Theorem 15.17, the Leray–Schauder degree $\deg(I - \Phi_t, F, 0)$ is well-defined for each t and is constant on $[0, 1]$. Since Φ_0 has a fixed point $\{1\}$, we deduce from Theorem 15.17(iv) that $\deg(I - \Phi_0, F, 0) = 1$. Therefore, $\deg(I - \Phi_1, F, 0) = 1$, and hence Φ_1 must also have a fixed point w, giving rise to a uniformly convex solution $u \in W^{4,q}(\Omega)$ of the second boundary value problem (15.60).

In the second case of the theorem where $f \equiv M > 0$, $\varphi \in C^{4,\beta}(\overline{\Omega})$, and $\psi \in C^{2,\beta}(\overline{\Omega})$ ($0 < \beta \leq 1$), by similar arguments, u will lie in $C^{4,\beta}(\overline{\Omega})$. $\square$

Now, back to (15.59). Note that inequality (15.61) plays an important role in the proof of Theorem 15.14. In dimensions $n = 2$, it becomes an equality trace $(U^{ij}) = \Delta u$. If $n \geq 3$, the ratio trace $(U^{ij})/\Delta u$ can be in general as small as we want (for example, when one eigenvalue of $D^2 u$ is 1 while all other eigenvalues are a small constant). However, the situation improves when there is a small factor in front of Δu. The following theorem addresses a variant of (15.59) in higher dimensions when the right-hand side is of the form $-\gamma \operatorname{div}(|Du|^{q-2} Du)$, where $q \geq 2$ and γ is small.

Theorem 15.18. *Let Ω be a uniformly convex domain in $\mathbb{R}^n$ ($n \geq 3$) with $\partial\Omega \in C^5$, $\psi \in C^{2,\beta}(\overline{\Omega})$ with $\inf_{\partial\Omega} \psi > 0$, and $\varphi \in C^{4,\beta}(\overline{\Omega})$ where $\beta \in (0, 1)$. Let $G(\cdot, z, \mathbf{p}, \mathbf{r}) : \overline{\Omega} \times \mathbb{R} \times \mathbb{R}^n \times \mathbb{M}^{n \times n} \to \mathbb{R}$ be a smooth function such that:*

(a) *G maps compact subsets of $\overline{\Omega} \times \mathbb{R} \times \mathbb{R}^n \times \mathbb{M}^{n \times n}$ into compact subsets of $\mathbb{R}$ and*

(b) *$G(x, u(x), Du(x), D^2 u(x)) \leq 0$ in Ω for all C^2 convex function u.*

If $\gamma > 0$ is small, depending only on $\beta, \varphi, \psi, n, G$, and Ω, then there is a uniform convex solution $u \in C^{4,\beta}(\overline{\Omega})$ to the following boundary value problem:

(15.68)
$$\begin{cases} U^{ij} D_{ij}[(\det D^2 u)^{-1}] = \gamma G(\cdot, u, Du, D^2 u) & \text{in } \Omega, \\ u = \varphi, \ \det D^2 u = \psi^{-1} & \text{on } \partial\Omega. \end{cases}$$

Proof. For some $\gamma \in (0, 1)$ to be chosen later, let $f_\gamma = \min\{F_\gamma, 1\}$, where

$$F_\gamma \equiv F_\gamma(\cdot, u, Du, D^2 u) = \frac{-\gamma G(\cdot, u, Du, D^2 u)}{(\Delta u)^{\frac{1}{n-1}} (\det D^2 u)^{\frac{n-2}{n-1}}},$$

and consider the following boundary value problem:

$$(15.69) \quad \begin{cases} U^{ij} D_{ij}[(\det D^2 u)^{-1}] = -(\Delta u)^{\frac{1}{n-1}} (\det D^2 u)^{\frac{n-2}{n-1}} f_\gamma & \text{in } \Omega, \\ u = \varphi, \ \det D^2 u = \psi^{-1} & \text{on } \partial\Omega. \end{cases}$$

By assumption (b), when u is a $C^2(\overline{\Omega})$ convex function, we have $0 \le f_\gamma \le 1$. By Theorem 15.14, (15.69) has a uniformly convex solution $u \in W^{4,q}(\Omega)$ for all $q < \infty$. Thus, the first equation of (15.69) holds almost everywhere.

As in the proof of Theorem 15.14 (see (15.63)), we have the estimates

$$(15.70) \qquad \|u\|_{C^{2,\beta}(\overline{\Omega})} \le C_1 \quad \text{and} \quad C_1^{-1} I_n \le D^2 u \le C_1 I_n,$$

for some $C_1 = C_1(\beta, \varphi, \psi, n, \Omega) > 0$. Hence, from assumption (a), we find that $F_\gamma < 1/2$ in Ω if $\gamma = \gamma(\beta, \varphi, \psi, n, G, \Omega) > 0$ is small. Thus, if $\gamma > 0$ is small, then $f_\gamma = F_\gamma$ in Ω, and hence the first equation of (15.69) becomes $U^{ij} D_{ij}[(\det D^2 u)^{-1}] = \gamma G(\cdot, u, Du, D^2 u)$. Using this equation together with (15.70) and $\varphi \in C^{4,\beta}(\overline{\Omega})$ and $\psi \in C^{2,\beta}(\overline{\Omega})$, we easily conclude $u \in C^{4,\beta}(\overline{\Omega})$. Thus, there is a uniform convex solution $u \in C^{4,\beta}(\overline{\Omega})$ to (15.68). □

Remark 15.19. It would be interesting to remove the smallness of γ in Theorem 15.18 in all generality.

On the other hand, in two dimensions, the case of q-Laplacian right-hand side was resolved in Le–Zhou [**LZ**] for all $q > 1$. The proof uses both partial Legendre transform and Legendre transform in the estimation of the Hessian determinant and the global Hölder estimates in Theorem 15.9 for linearized Monge–Ampère equations with divergence form inhomogeneity.

Theorem 15.20. *Let Ω be a uniformly convex domain in $\mathbb{R}^2$ with $\partial\Omega \in C^5$. Let $q > 1$. Assume that $f^0 : \overline{\Omega} \times \mathbb{R} \to \mathbb{R}$ is smooth and satisfies*

$$f^0(x, z) \le \omega(|z|); \ -f^0(x, z)(z - \tilde{z}) \le \omega(|\tilde{z}|) \quad \text{for all } x \in \Omega \text{ and } z, \tilde{z} \in \mathbb{R},$$

where $\omega : [0, \infty) \to [0, \infty)$ is a continuous and increasing function. Assume that $\varphi \in C^5(\overline{\Omega})$ and $\psi \in C^3(\overline{\Omega})$ with $\inf_{\partial\Omega} \psi > 0$. Consider the following second boundary value problem:

$$(15.71) \quad \begin{cases} U^{ij} D_{ij}[(\det D^2 u)^{-1}] = -\text{div}(|Du|^{q-2} Du) + f^0(x, u) & \text{in } \Omega, \\ u = \varphi, \ \det D^2 u = \psi^{-1} & \text{on } \partial\Omega. \end{cases}$$

(i) *If $q \ge 2$, then there exists a uniformly convex solution $u \in C^{4,\beta}(\overline{\Omega})$ to (15.71) with $\|u\|_{C^{4,\beta}(\overline{\Omega})} \le C$, for some $\beta \in (0, 1)$ and $C > 0$ depending on $q, \Omega, \omega, f^0, \varphi,$ and ψ.*

(ii) *If $1 < q < 2$, then there exists a uniformly convex solution $u \in C^{3,\beta}(\overline{\Omega})$ to (15.71) with $\|u\|_{C^{3,\beta}(\overline{\Omega})} \le C$, for some $\beta \in (0, 1)$ and $C > 0$ depending on $q, \Omega, \omega, f^0, \varphi,$ and ψ.*

15.5.2. Approximation of minimizers of convex functionals with a convexity constraint. In this section, we consider singular Abreu equations arising from the approximation of several convex functionals whose Lagrangians depend on the gradient variable, subject to a convexity constraint. These functionals are relevant to variational problems motivated by the Rochet–Choné model in the monopolist's problem in economics and variational problems arising in the analysis of wrinkling patterns in floating elastic shells in elasticity.

Calculus of variations with a convexity constraint. Let Ω_0 be a bounded convex domain in $\mathbb{R}^n$ $(n \geq 2)$, and let Ω be a uniformly convex domain containing $\overline{\Omega_0}$ with $\partial\Omega_0, \partial\Omega$ being sufficiently smooth. Let φ be a convex and smooth function in $\overline{\Omega}$. Let $F(x,z,p) : \mathbb{R}^n \times \mathbb{R} \times \mathbb{R}^n \to \mathbb{R}$ be a smooth Lagrangian which is convex in each of the variables $z \in \mathbb{R}$ and $p = (p_1, \ldots, p_n) \in \mathbb{R}^n$. Consider the following variational problem with a convexity constraint:

$$(15.72) \qquad \inf_{u \in \bar{S}[\varphi, \Omega_0]} \int_{\Omega_0} F(x, u(x), Du(x)) \, dx$$

where

$$(15.73) \quad \bar{S}[\varphi, \Omega_0] = \{u : \Omega_0 \to \mathbb{R} \text{ such that } u \text{ is convex,}$$

$$u \text{ admits a convex extension to } \Omega \text{ such that } u = \varphi \text{ on } \Omega \setminus \Omega_0\}.$$

Note that elements of $\bar{S}[\varphi, \Omega_0]$ (see Figure 15.1) are Lipschitz continuous

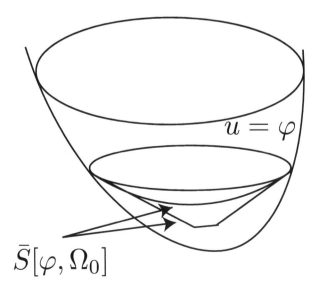

Figure 15.1. The constraint set $\bar{S}[\varphi, \Omega_0]$.

with Lipschitz constants bounded from above by $\|D\varphi\|_{L^\infty(\Omega)}$, and hence $\bar{S}[\varphi, \Omega_0]$ is compact in the topology of uniform convergence. If $w \in \bar{S}[\varphi, \Omega_0]$, then using $w = \varphi$ on $\partial\Omega_0$ together with the Mean Value Theorem, we find that

$$\|w\|_{L^\infty(\Omega_0)} \le \|\varphi\|_{L^\infty(\Omega_0)} + \mathrm{diam}(\Omega_0)\|D\varphi\|_{L^\infty(\Omega)}.$$

Under quite general assumptions on the convexity and growth of the Lagrangian F, we can show that (15.72) has a minimizer in $\bar{S}[\varphi, \Omega_0]$. The simple setting where F is bounded below covers many applications; see also Problem 15.7.

Proposition 15.21. *Let $\Omega_0 \Subset \Omega \subset \mathbb{R}^n$. Assume that $F(x, z, p)$ is convex in p and that there is a constant C such that $F(x, z, p) \ge -|C|$ for all $(x, z, p) \in \mathbb{R}^n \times \mathbb{R} \times \mathbb{R}^n$ satisfying*

$$x \in \Omega_0, \quad |z| \le \|\varphi\|_{L^\infty(\Omega_0)} + \mathrm{diam}(\Omega_0)\|D\varphi\|_{L^\infty(\Omega)}, \quad |p| \le \|D\varphi\|_{L^\infty(\Omega)}.$$

Then, there exists a function $u \in \bar{S}[\varphi, \Omega_0]$ satisfying

$$\int_{\Omega_0} F(x, u(x), Du(x))\, dx = \min_{w \in \bar{S}[\varphi, \Omega_0]} \int_{\Omega_0} F(x, w(x), Dw(x))\, dx.$$

Below are examples of convex functionals with a convexity constraint.

Example 15.22 (The Rochet–Choné model). Rochet and Choné [**RC**] modeled the monopolist's problem in product line design with q-power cost (the cost of producing product y is the function $|y|^q/q$ where $1 < q < \infty$) using maximization, over convex functions $u \ge \varphi$, of the functional

$$\Phi(u) = \int_{\Omega_0} \{x \cdot Du(x) - |Du(x)|^q/q - u(x)\}\gamma(x)\, dx.$$

Here, $\Phi(u)$ is the monopolist's profit; u is the buyers' indirect utility function with bilinear valuation; $\Omega_0 \subset \mathbb{R}^n$ is the collection of types of agents; the given convex function φ is referred to as the participation constraint. The function γ is the relative frequency of different types of agents in the population which is assumed to be nonnegative and Lipschitz continuous. For a consumer of type $x \in \Omega_0$, the indirect utility $u(x)$ is computed via the formula

$$u(x) = \max_{y \in Q}\{x \cdot y - p(y)\},$$

where $Q \subset \mathbb{R}^n$ is the product line and $p : Q \to \mathbb{R}$ is a price schedule that the monopolist needs to both design and to maximize the total profit Φ. Put differently, the utility function is the Legendre transform of the price. Clearly, u is convex and maximizing $\Phi(u)$ is equivalent to minimizing

$$\int_{\Omega_0} F(x, u(x), Du(x))\, dx \quad \text{where } F(x, z, p) = |p|^q\gamma(x)/q - x \cdot p\gamma(x) + z\gamma(x)$$

among all convex functions $u \ge \varphi$.

In terms of the price function p and the product y which realizes $u(x)$, the expression $x \cdot Du(x) - |Du(x)|^q/q - u(x)$ in the integrand of Φ is nothing but $p(y) - |y|^q/q$. The latter quantity is exactly the difference between the price and production cost. This heuristically explains why the above functional Φ is a reasonable model for the profit.

For this simple-looking variational problem, the convexity is not easy to handle from a numerical standpoint. How is this model related to the minimization problem (15.72) subject to (15.73)? Heuristically, the boundary conditions for minimizers for (15.72) associated with (15.73) are

$$u = \varphi \quad \text{and} \quad Du \cdot \nu_0 \le D\varphi \cdot \nu_0 \quad \text{on } \partial\Omega_0,$$

where ν_0 is the outer unit normal vector on $\partial\Omega_0$. Thus, the constraint (15.73) can be heuristically viewed as a special case of the constraint $u \ge \varphi$ in Ω_0 in the Rochet–Choné model.

Example 15.23 (Thin elastic shells). In the analysis of wrinkling patterns in floating elastic shells by Tobasco [**To**], describing the leading-order behavior of weakly curved floating shells leads to limiting problems which are dual to problems of the following type: Given a smooth function $v : \overline{\Omega_0} \subset \mathbb{R}^2 \to \mathbb{R}$, minimize

$$\int_{\Omega_0} [|x|^2/2 - u(x)] \det D^2 v(x)\, dx$$

over the set

$$\{u \text{ convex in } \mathbb{R}^2, \ u = |x|^2/2 \text{ in } \mathbb{R}^2 \setminus \Omega_0\}.$$

Optimal functions in this minimization problem are called optimal *Airy potential*. In this example, $F(x, z, p) = (|x|^2/2 - z) \det D^2 v(x)$.

Approximations of minimizers. The convexity constraint (15.73) poses intrinsic difficulty in solving (15.72). Thus, for practical purposes such as implementing numerical schemes to find minimizers of (15.72), it is desirable to find suitably explicit approximations of minimizers u of (15.72). This has been done by Carlier and Radice [**CR**] when the Lagrangian F does not depend on the gradient variable p. The analysis here, largely based on Le [**L9,L10**], tackles the more challenging case when F depends on the gradient variable. The main idea is to approximate, in the uniform norm, minimizers of (15.72) by solutions of some higher-order equations whose global well-posedness can be established. The approximating schemes first proposed in [**CR, L9**] use the second boundary value problem of fourth-order equations of Abreu type. When the Lagrangian depends on the gradient variable p, singular Abreu equations appear. We describe below these ideas.

Motivated by previous examples, we consider Lagrangians of the form

$$F(x, z, p) = F^0(x, z) + F^1(x, p), \quad \text{where } F^0(x, z) = \int_0^z f^0(x, s)\, ds.$$

We assume the following convexity and growth on f^0 and F^1: For some constant $C_* \geq 0$ and for all $x \in \Omega_0$, $z, \tilde{z} \in \mathbb{R}^n$, $p \in \mathbb{R}^n$,

$$(15.74) \quad \begin{cases} (f^0(x, z) - f^0(x, \tilde{z}))(z - \tilde{z}) \geq 0; \, |f^0(x, z)| \leq \eta(|z|) \\ \text{where } \eta : [0, \infty) \to [0, \infty) \text{ is continuous and increasing,} \\ 0 \leq D_p^2 F^1(x, p) \leq C_* I_n; \, \left| \dfrac{\partial^2 F^1(x, p)}{\partial p_i \partial x_i} \right| \leq C_*(|p| + 1) \, (1 \leq i \leq n). \end{cases}$$

Let ρ be a uniformly convex defining function of Ω; that is,

$$\rho < 0 \quad \text{in } \Omega, \qquad \rho = 0 \quad \text{on } \partial\Omega, \quad \text{and} \quad D\rho \neq 0 \quad \text{on } \partial\Omega.$$

Let ψ be a smooth function in $\overline{\Omega}$ with $\inf_{\partial\Omega} \psi > 0$, and let φ be a convex and smooth function in $\overline{\Omega}$. Fix $0 \leq \theta < 1/n$. For $\varepsilon > 0$, let

$$\varphi_\varepsilon := \varphi + \varepsilon^{\frac{1}{3n^2}}(e^\rho - 1),$$

and consider the following second boundary value problem for a uniform convex function u_ε:

$$(15.75) \quad \begin{cases} \varepsilon U_\varepsilon^{ij} D_{ij} w_\varepsilon = f_\varepsilon\left(\cdot, u_\varepsilon, Du_\varepsilon, D^2 u_\varepsilon; \varphi_\varepsilon\right) & \text{in } \Omega, \\ w_\varepsilon = (\det D^2 u_\varepsilon)^{\theta - 1} & \text{in } \Omega, \\ u_\varepsilon = \varphi, \, w_\varepsilon = \psi & \text{on } \partial\Omega, \end{cases}$$

where $U_\varepsilon = (U_\varepsilon^{ij})_{1 \leq i, j \leq n}$ is the cofactor matrix of $D^2 u_\varepsilon$ and

$$(15.76) \quad f_\varepsilon = \left[\frac{\partial F}{\partial z}(\cdot, u_\varepsilon, Du_\varepsilon) - \frac{\partial}{\partial x_i}\left(\frac{\partial F}{\partial p_i}(\cdot, u_\varepsilon, Du_\varepsilon) \right) \right] \chi_{\Omega_0} + \frac{u_\varepsilon - \varphi_\varepsilon}{\varepsilon} \chi_{\Omega \setminus \Omega_0}.$$

When $\theta = 0$, fourth-order equations of the type (15.75)–(15.76) are singular Abreu equations. The first two equations of (15.75)–(15.76) are critical points, with respect to compactly supported variations, of the functional

$$J_\varepsilon(v) = \int_{\Omega_0} F(\cdot, v, Dv)\, dx + \frac{1}{2\varepsilon} \int_{\Omega \setminus \Omega_0} (v - \varphi_\varepsilon)^2\, dx - \varepsilon \int_\Omega \frac{(\det D^2 v)^\theta - 1}{\theta}\, dx.$$

When $\theta = 0$, the last integral is replaced by $\int_\Omega \log \det D^2 v\, dx$. The requirement $0 \leq \theta < 1/n$ makes J_ε a convex functional, or equivalently, $-J_\varepsilon$ concave; see Lemma 2.59. At the functional level, when $\theta = 0$, the penalization $\varepsilon \int_\Omega \log \det D^2 v\, dx$ involving the logarithm of the Hessian determinant acts as a good substitute for the convexity constraint in problems like (15.72).

The following theorem, established in [L10], asserts the solvability of (15.75)–(15.76) and limiting properties of solutions when $\varepsilon \to 0$.

Theorem 15.24. *Let Ω_0 and Ω be convex domains in $\mathbb{R}^n$ $(n \geq 2)$ with $\partial\Omega \in C^5$ such that Ω is uniformly convex and contains $\overline{\Omega_0}$. Fix $0 \leq \theta < 1/n$. Let ψ be a smooth function in $\overline{\Omega}$ with $\inf_{\partial\Omega} \psi > 0$, and let φ be a convex and smooth function in $\overline{\Omega}$. Assume that (15.74) is satisfied. If $F^1 \not\equiv 0$, then assume further that $n = 2$. Then the following hold.*

(i) *For $\varepsilon > 0$ small, the system (15.75)–(15.76) has a uniformly convex solution $u_\varepsilon \in W^{4,q}(\Omega)$ for all $q \in (n, \infty)$. Furthermore,*

$$(15.77) \qquad \|u_\varepsilon\|_{L^\infty(\Omega)} \leq C = C(n, q, \theta, \eta, C_*, \Omega_0, \Omega, \inf_{\partial\Omega} \psi, \|\varphi\|_{W^{4,q}(\Omega)}).$$

(ii) *For $\varepsilon > 0$ small, let $u_\varepsilon \in W^{4,q}(\Omega)$ $(q > n)$ be a solution to (15.75)–(15.76). After extracting a subsequence, u_ε converges uniformly on compact subsets of Ω to a minimizer $u \in \bar{S}[\varphi, \Omega_0]$ of (15.72).*

We briefly indicate the *reduction to a priori estimates* in the proof of Theorem 15.24(i). Suppose that we can establish the a priori $L^\infty(\Omega)$ estimates (15.77) for uniformly convex solutions $u_\varepsilon \in W^{4,q}(\Omega)$ $(q > n)$ to the system (15.75)–(15.76) when ε is small. We consider two cases.

Case 1. $F(x, z, p) = F^0(x, z)$. In this case, we can apply Theorem 13.3 to show the existence of a unique uniformly convex solution $u_\varepsilon \in W^{4,q}(\Omega)$ (for all $q < \infty$) to the system (15.75)–(15.76).

Case 2. $F(x, z, p) = F^0(x, z) + F^1(x, p)$ and $n = 2$. In this case, from the a priori $L^\infty(\Omega)$ estimates for uniformly convex solutions $u_\varepsilon \in W^{4,q}(\Omega)$ $(q > n)$ to (15.75)–(15.76), we can establish the a priori $W^{4,q}(\Omega)$ estimates for u_ε as in Step 1 of the proof of Theorem 15.26 below. Then, we can use a Leray–Schauder degree argument as in Theorem 15.14 to show the existence of a uniformly convex solution $u_\varepsilon \in W^{4,q}(\Omega)$ (for all $q < \infty$) to the system (15.75)–(15.76). Hence, assertion (i) is proved.

Remark 15.25. Several remarks are in order.

- When φ is not uniformly convex, the addition of $\varepsilon^{\frac{1}{3n^2}}(e^\rho - 1)$ to φ is to make the new function φ_ε "sufficiently" uniformly convex. The choice of the exponent $\frac{1}{3n^2}$ (or any positive number not larger than this) is motivated by the need to establish uniform bounds for u_ε in the a priori estimates for solutions to (15.75)–(15.76).

- Theorem 15.24 is applicable to the Rochet–Choné model with quadratic cost only. When the gradient-dependent term $F^1(x, p) = (|p|^q/q - x \cdot p)\gamma(x)$ of F is required to satisfy the last condition of (15.74), we must have $q = 2$. For general $1 < q < \infty$, see [**LZ**].

Below, we consider solvability for certain singular Abreu equations that are applicable to Theorem 15.24(i). They are of the form

$$(15.78) \qquad \begin{cases} U^{ij} D_{ij} w = f_\delta(\cdot, u, Du, D^2 u; \hat{\varphi}), \ w = (\det D^2 u)^{\theta-1} & \text{in } \Omega, \\ \quad u = \hat{\varphi}, \ w = \psi & \text{on } \partial\Omega. \end{cases}$$

Here, $\delta > 0$, Ω_0, Ω are bounded domains in $\mathbb{R}^n$ such that $\Omega_0 \Subset \Omega$, and

$$(15.79) \qquad f_\delta = \left[f^0(\cdot, u) - \frac{\partial}{\partial x_i} \left(\frac{\partial}{\partial p_i} F^1(\cdot, Du) \right) \right] \chi_{\Omega_0} + \frac{u - \hat{\varphi}}{\delta} \chi_{\Omega \setminus \Omega_0}.$$

Theorem 15.26 (Solvability of highly singular second boundary value problems of Abreu type). *Let $\Omega_0 \Subset \Omega$ where Ω is a uniformly convex domain in $\mathbb{R}^n$ where $n = 2$ with $\partial\Omega \in C^{3,1}$. Let $\theta \in [0, 1/n)$, $q > n$, $\delta > 0$, $\hat{\varphi} \in W^{4,q}(\Omega)$, and $\psi \in W^{2,q}(\Omega)$ with $\inf_{\partial\Omega} \psi > 0$. Assume that (15.74) is satisfied, and for some $A > 0$, the system (15.78)–(15.79) admits a uniform a priori estimate*

$$(15.80) \qquad \|u\|_{L^\infty(\Omega)} \le A,$$

for any uniformly convex solution $u \in W^{4,q}(\Omega)$. Then, there is a uniformly convex solution $u \in W^{4,q}(\Omega)$ to the system (15.78)–(15.79). Moreover,

$$\|u\|_{W^{4,q}(\Omega)} \le C = C(A, \delta, q, \theta, C_*, \eta, \Omega, \Omega_0, \inf_{\partial\Omega} \psi, \|\psi\|_{W^{2,q}(\Omega)}, \|\hat{\varphi}\|_{W^{4,q}(\Omega)}).$$

Proof. As before, we call a positive constant *structural* if it depends only on the quantities stated in the theorem.

Step 1. A priori estimates. Assume that $u \in W^{4,q}(\Omega)$ is a uniformly convex solution to (15.78)–(15.79). Then, (15.80) holds. We will prove a global $W^{4,q}$ estimate $\|u\|_{W^{4,q}(\Omega)} \le C$ for a structural constant C.

First, we show that there is a structural constant $C > 0$ such that

$$(15.81) \qquad C^{-1} \le \det D^2 u \le C \quad \text{in } \Omega.$$

For this, it is convenient to rewrite our singular Abreu equation as

$$(15.82) \qquad U^{ij} D_{ij} w = f_\delta = -\gamma_1 \Delta u + g \quad \text{in } \Omega,$$

where

$$\begin{cases} \gamma_1 = \dfrac{\text{trace}\, (D_p^2 F^1(\cdot, Du) D^2 u)}{\Delta u} \chi_{\Omega_0}, \\ g = \left[f^0(\cdot, u) - \dfrac{\partial^2 F^1(\cdot, Du)}{\partial p_i \partial x_i} \right] \chi_{\Omega_0} + \dfrac{u - \hat{\varphi}}{\delta} \chi_{\Omega \setminus \Omega_0}. \end{cases}$$

From the last condition of (15.74), we have

$$(15.83) \qquad \|\gamma_1\|_{L^\infty(\Omega)} \le C_*.$$

From (15.74), we easily find that for all $x \in \Omega$,

$$|g(x)| \leq \eta(\|u\|_{L^\infty(\Omega)}) + C_*(\|Du\|_{L^\infty(\Omega_0)} + 1) + \frac{1}{\delta}(\|u\|_{L^\infty(\Omega)} + \|\hat{\varphi}\|_{L^\infty(\Omega)}).$$

Using the bound for u in (15.80) and Lemma 2.37(iii), we get

(15.84) $$\|g\|_{L^\infty(\Omega)} \leq C.$$

Step 1(a). Lower bound for $\det D^2 u$. Let $\bar{w} = w + C_*|x|^2$. Then, $\bar{w} = \psi + C_*|x|^2$ on $\partial\Omega$. Moreover, by (15.82) and (15.83), we find that

$$U^{ij} D_{ij}\bar{w} = f_\delta + 2C_*\Delta u = (2C_* - \gamma_1)\Delta u + g \geq -|g| \quad \text{in } \Omega.$$

Since $n = 2$, we have $\det(U^{ij}) = \det D^2 u = w^{\frac{1}{\theta-1}}$. Now, we apply the ABP estimate in Theorem 3.16 to $\bar{w}$ on Ω, and then recalling (15.84), to obtain

$$\|\bar{w}\|_{L^\infty(\Omega)} \leq \sup_{\partial\Omega} \bar{w} + C\left\|\frac{g}{(\det U^{ij})^{1/2}}\right\|_{L^2(\Omega)} \leq C + C\|w\|_{L^\infty(\Omega)}^{\frac{1}{2(1-\theta)}}.$$

Since $\bar{w} \geq w > 0$ and since $\frac{1}{2(1-\theta)} < 1$ because $0 \leq \theta < 1/2$, it follows that $w \leq C$, and hence $\det D^2 u = w^{\frac{1}{\theta-1}} \geq C_1$ for a structural constant $C_1 > 0$.

Step 1(b). Upper bound for $\det D^2 u$. We only need to obtain a lower bound for w. By Step 1(a), we have

$$w \det D^2 u = (\det D^2 u)^\theta \geq C_1^\theta := c > 0.$$

From (15.82) and $\gamma_1 \geq 0$, we find that $f_\delta \leq g$. By (15.84), we find that $f_\delta^+ := \max\{f_\delta, 0\} \leq |g|$ is bounded by a structural constant. Let

$$M = (\|f_\delta^+\|_{L^\infty(\Omega)} + 1)/c < \infty \quad \text{and} \quad v = \log w - Mu \in W^{2,q}(\Omega).$$

Then, in Ω, using $w \det D^2 u \geq c$, we have

$$U^{ij} D_{ij} v = U^{ij}(D_{ij}w/w - D_i w D_j w/w^2 - M D_{ij}u)$$
$$\leq f_\delta/w - 2M \det D^2 u < 0.$$

By the ABP estimate (see Theorem 3.16) for $-v$ in Ω, we have

$$v \geq \inf_{\partial\Omega} v \geq \log(\inf_{\partial\Omega} \psi) - M\hat{\varphi} \geq -C \quad \text{in } \Omega.$$

From $v = \log w - Mu$ and the bound for u in (15.80), we obtain $\log w \geq -C$, so $w \geq e^{-C}$ as desired. Therefore, (15.81) holds.

Step 1(c). Fourth-order derivative estimates. Given (15.81), as in the proof of Theorem 15.14 with $\hat{f}$ now replaced by f_δ, we have

$$\|u\|_{C^{2,\alpha}(\overline{\Omega})} \leq C_2 \quad \text{and} \quad C_2^{-1} I_2 \leq D^2 u \leq C_2 I_2,$$

for some structural constant C_2. As a consequence, the second-order operator $U^{ij} D_{ij}$ is uniformly elliptic with Hölder continuous coefficients.

Recalling (15.82) and using (15.83) together with (15.84), we obtain $\|f_\delta\|_{L^\infty(\Omega)} \le C_3$. Thus, from $U^{ij}D_{ij}w = f_\delta$ with boundary value $w = \psi$ where $\psi \in W^{2,q}(\Omega)$, we conclude that $w \in W^{2,q}(\Omega)$, and therefore $u \in W^{4,q}(\Omega)$ with $\|u\|_{W^{4,q}(\Omega)} \le C_4$, for some structural constant C_4.

Step 2. Existence of $W^{4,q}(\Omega)$ solutions. The existence proof uses the a priori estimates in Step 1 and degree theory as in Theorem 15.14. We only need to make the following replacements:

$$q \rightsquigarrow q, \quad \varphi \rightsquigarrow \hat{\varphi}, \quad w^{-1} \rightsquigarrow w^{\frac{1}{\theta-1}}, \quad \hat{f} \rightsquigarrow f_\delta(\cdot, u, Du, D^2u).$$

The theorem is proved. $\qquad\qquad\qquad\qquad\qquad\qquad\qquad\qquad\qquad\qquad\qquad$ $\square$

Recently, the analysis in this section has been extended to all dimensions in Kim–Le–Wang–Zhou [**KLWZ**] and Le [**L14**]. The approaches there use twisted Harnack inequality and global Hölder estimates for linearized Monge–Ampère equations with lower-order terms which are global counterparts of interior estimates in Theorem 12.16. We briefly explain the new insight for a more general singular Abreu equation

$$U^{ij}D_{ij}w = -\mathrm{div}(DF(Du))$$

which corresponds to (15.59) when $F(z) = |z|^q/q$. Instead of establishing the Hölder estimate for $w := (\det D^2u)^{-1}$, we establish the Hölder estimate for $\eta := we^{F(Du)}$. This equivalence is possible thanks to the global $C^{1,\alpha}$ estimates for u in Theorem 9.6. The key observation (see Problem 15.12) is that η solves a linearized Monge–Ampère equation with a drift term in which the very singular term $\mathrm{div}(DF(Du)) = \mathrm{trace}(D^2F(Du)D^2u)$ no longer appears. The introduction of η seems to be mysterious. However, it is exactly the exponential of the integrand of the functional

$$\int_\Omega \big[-\log\det D^2u + F(Du)\big]\,dx,$$

for which our singular Abreu equation is its Euler–Lagrange equation.

15.6. Problems

Problem 15.1. Show that we can replace the L^∞ norm of **F** in Theorems 15.6 and 15.9 by some L^q norm for some $q(n, \lambda, \Lambda)$ large.

Problem 15.2. Extend the interior uniform estimate in Theorem 15.1 to linearized Monge–Ampère equations of the form

$$D_i[U^{ij}D_jv + b^iv] = \mathrm{div}\,\mathbf{F},$$

where each b^i $(i = 1, \ldots, n)$ is a bounded function.

Problem 15.3. Let $u \in C^2(\Omega)$ be a convex function satisfying (15.3). Let $U = (\det D^2 u)(D^2 u)^{-1}$. Show that the integrability exponents p and q in $U \in L^p(S_u(x_0, t))$ and $U^{-1} \in L^q(S_u(x_0, t))$ (where the norms depend only on n, λ, Λ, ε_n, and K) satisfy

$$\frac{1}{p} + \frac{1}{q} < \frac{2}{n-1}.$$

Problem 15.4. Let Ω be a uniformly convex domain in $\mathbb{R}^2$ with $\partial\Omega \in C^3$. Let $u \in C^{1,1}(\overline{\Omega}) \cap C^3(\Omega)$ be a convex function satisfying $u|_{\partial\Omega} \in C^3$ and $0 < \lambda \leq \det D^2 u \leq \Lambda$. Let $(U^{ij}) = \mathrm{Cof}\, D^2 u$. Let $v \in C(\overline{\Omega}) \cap W^{2,n}_{\mathrm{loc}}(\Omega)$ be the solution to

$$U^{ij} D_{ij} v = g + \mathrm{div}\, \mathbf{G} \quad \text{in } \Omega, \qquad v = \varphi \quad \text{on } \partial\Omega,$$

where $\varphi \in C^\alpha(\partial\Omega)$ for some $\alpha \in (0,1)$, $\mathbf{G} \in L^\infty(\Omega; \mathbb{R}^n) \cap W^{1,n}(\Omega; \mathbb{R}^n)$, and $g \in L^n(\Omega)$. Show that $v \in C^{\alpha_1}(\overline{\Omega})$ with the estimate

$$\|v\|_{C^{\alpha_1}(\overline{\Omega})} \leq C\big(\|\varphi\|_{C^\alpha(\partial\Omega)} + \|g\|_{L^n(\Omega)} + \|\mathbf{G}\|_{L^\infty(\Omega)}\big),$$

where α_1 and C depend only on λ, Λ, α, $\|u\|_{C^3(\partial\Omega)}$, and Ω.

Problem 15.5. Assume smooth functions (ρ_t, P_t^*) are solutions to the dual equations (15.43)–(15.44). Define $(p_t, \mathbf{u}_t)$ as in (15.50). Show that $(p_t, \mathbf{u}_t)$ is a solution to the semigeostrophic equations (15.41).

Problem 15.6. Prove a three-dimensional analogue of Theorem 15.13 when the initial potential density ρ_0 is sufficiently close to 1. Here, in the system (15.43)–(15.44), $w^\perp = (-w_2, w_1, 0)$ for $w = (w_1, w_2, w_3) \in \mathbb{R}^3$.

Problem 15.7. Prove Proposition 15.21.

Problem 15.8. Prove that the conclusions of Theorem 15.14 are still valid for the equation

$$U^{ij} D_{ij} w = -[(\Delta u)^2 - \|D^2 u\|^2]^{\frac{1}{2(n-1)}} (\det D^2 u)^{\frac{n-2}{n-1}} f, \quad w = (\det D^2 u)^{-1}.$$

Problem 15.9. Let Ω be a uniformly convex domain in $\mathbb{R}^2$ with $\partial\Omega \in C^\infty$, $f \in L^\infty(\Omega)$ with $f \geq 0$, $\varphi \in W^{4,q}(\Omega)$, $\psi \in W^{2,q}(\Omega)$ where $q > 2$ with $\inf_{x \in \partial\Omega} \big(\psi(x) - \|f\|_{L^\infty(\Omega)} |x|^2/2\big) > 0$. Show that there is a uniformly convex solution $u \in W^{4,q}(\Omega)$ to

$$\begin{cases} U^{ij} D_{ij}[(\det D^2 u)^{-1}] = f \Delta u & \text{in } \Omega, \\ \qquad\qquad u = \varphi, \ \det D^2 u = \psi^{-1} & \text{on } \partial\Omega. \end{cases}$$

If f is a constant, $\varphi \in C^\infty(\overline{\Omega})$, and $\psi \in C^\infty(\overline{\Omega})$, then show that $u \in C^\infty(\overline{\Omega})$.

Problem 15.10. Prove the uniqueness of solutions in Theorem 15.18 if G additionally satisfies the following condition:

$$\int_{\Omega} [G(\cdot, u, Du, D^2u) - G(\cdot, v, Dv, D^2v)](u - v)\, dx \geq 0,$$

for all $u, v \in C^2(\overline{\Omega})$ with $u = v$ on $\partial\Omega$.

Problem 15.11 (Euler–Lagrange for minimizers with a convexity constraint). Let Ω be a bounded convex domain in $\mathbb{R}^n$. Suppose $u \in W^{1,2}(\Omega)$ is a convex minimizer of $\int_{\Omega} |Du|^2\, dx$ over the constraint set

$$\mathcal{K} := \left\{ u \text{ is convex in } \Omega, \ u = \varphi \text{ on } \partial\Omega \right\}.$$

Show that there exists a nonnegative definite symmetric matrix $(\mu_{ij})_{1 \leq i,j \leq n}$ of Radon measures, such that, in the sense of distributions,

$$-\Delta u = D_{ij}\mu_{ij}.$$

That is, for any $v \in C^2(\overline{\Omega})$ with $v = 0$ on $\partial\Omega$, we have

$$\int_{\Omega}(-\Delta u)v\, dx = \int_{\Omega} D_{ij}v\, d\mu_{ij}.$$

Hint: Use Riedl's Theorem [**Rdl**, Theorem 10.10] which states as follows:

Let E be a normed vector space. A subset $\mathcal{C} \subset E$ is called a positive cone if $\mathcal{C} + \mathcal{C} \subset \mathcal{C}$, $\lambda\mathcal{C} \subset \mathcal{C}$ for all $\lambda > 0$, and $\mathcal{C} \cap (-\mathcal{C}) = \{0\}$. Let L be a subspace of E that contains an interior point of a positive cone $\mathcal{C}$. Let $F_0 : L \to \mathbb{R}$ be a positive linear functional; that is, $F_0(x) \geq 0$ for all $x \in L \cap \mathcal{C}$. Then, F_0 can be extended to a continuous, positive linear map $F : E \to \mathbb{R}$.

Let $\mathbb{S}^n_+$ be the space of real, nonnegative definite $n \times n$ matrices. Apply Riedl's Theorem to: $E = C(\overline{\Omega}; \mathbb{S}^n)$, $\mathcal{C} = C(\overline{\Omega}; \mathbb{S}^n_+)$, $L = \{X = D^2v : v \in C^2(\overline{\Omega}) \text{ with } v = 0 \text{ on } \partial\Omega\}$, and $F_0(X) = \int_{\Omega}(-\Delta u)v\, dx$ for $X = D^2v \in L$.

Problem 15.12. Let Ω be a domain in $\mathbb{R}^n$. Let $F \in W^{2,n}_{\text{loc}}(\mathbb{R}^n)$, and let Q be a function on $\mathbb{R}^n \times \mathbb{R} \times \mathbb{R}^n$. Assume that a locally uniformly convex function $u \in W^{4,s}_{\text{loc}}(\Omega)$ $(s > n)$ solves the following general singular Abreu equation:

$$U^{ij}D_{ij}[(\det D^2u)^{-1}] = -\operatorname{div}(DF(Du)) + Q(x, u, Du) \quad \text{in } \Omega$$

where $(U^{ij}) = (\det D^2u)(D^2u)^{-1}$. Let $\eta := (\det D^2u)^{-1}e^{F(Du)}$. Show that

$$U^{ij}D_{ij}\eta - (\det D^2u)DF(Du) \cdot D\eta = e^{F(Du)}Q(x, u, Du) \quad \text{in } \Omega.$$

15.7. Notes

For the Moser iteration technique, see Moser's original paper [**Mo**] or expositions in the books Gilbarg–Trudinger [**GT**, Section 8.5] and Han–Lin [**HL**, Section 4.2].

In Sections 15.1–15.3, the treatment of (15.1) in dimensions $n \geq 3$ under some variants of (15.3) is new. When $n = 2$, the interior Hölder estimates in Theorem 15.6 were proved in Le [**L6**], while the Hölder estimates in Theorem 15.9 were proved in Le [**L8**]. The global Hölder estimates in Theorem 15.9 are an affine invariant and degenerate version of global Hölder estimates by Murthy and Stampacchia [**MS**] and Trudinger [**Tr1**] for second-order elliptic equations in divergence form. In [**MS, Tr1**], the authors established the maximum principle, local and global estimates for degenerate elliptic equations of the form $\operatorname{div}(M(x)Dv(x)) = \operatorname{div} V(x)$ where M is a symmetric positive definite matrix and V is a bounded vector field in $\mathbb{R}^n$. To obtain these results, they require high integrability of the matrix M and its inverse. More precisely, the integrability exponents p_M of M and q_M of M^{-1} were required to satisfy $\frac{1}{p_M} + \frac{1}{q_M} < \frac{2}{n}$. This integrability condition, which requires either p_M or q_M to be larger than the dimension, is not satisfied in general by the linearized Monge–Ampère operators. Recently, for the homogeneous equation $\operatorname{div}(M(x)Dv(x)) = 0$, Bella and Schäffner [**BSc**] obtained local boundedness and Harnack inequality for solutions under the optimal integrability condition $\frac{1}{p_M} + \frac{1}{q_M} < \frac{2}{n-1}$. This is consistent with our condition (15.3) (see also Problem 15.3), and our analysis works for inhomogeneous linearized Monge–Ampère equations.

For Theorem 15.12 in Section 15.4, see also the lecture notes by Figalli [**F3**]. Theorem 15.13 is based on Le [**L6**]. It extends the results in [**Lo1**] where Loeper proved the Hölder continuity of $\partial_t P_t^*$ and $\partial_t P_t$ in the system (15.43)–(15.44) when the initial potential density ρ^0 is sufficiently close to a positive constant. Besides its application to the semigeostrophic equations, Theorem 15.6 can be applied to prove the local Hölder regularities for the polar factorization of time-dependent maps in two-dimensional Euclidean domains and on the flat torus $\mathbb{T}^2$, with densities bounded away from zero and infinity; see [**L6**]. The polar factorization of vector-valued mappings on Euclidean domains was introduced by Brenier in his influential paper [**Br**]. The polar factorization of maps on the flat torus $\mathbb{T}^n = \mathbb{R}^n/\mathbb{Z}^n$ was treated by Cordero-Erausquin [**CE**]. Using solutions to the dual equations together with the $W^{2,1}$ regularity for Aleksandrov solutions to the Monge–Ampère equation, Ambrosio, Colombo, De Philippis, and Figalli [**ACDF1**] established global in time distributional solutions to the original semigeostrophic equations on the two-dimensional torus; see also Cullen–Feldman [**CF**] for Lagrangian solutions.

For more on Section 15.5, see [**L9, L10, L11, LZ, KLWZ, L14**]. The monopolist's problem can be treated in the framework of optimal transport; see, for example, Figalli–Kim–McCann [**FKM**], Galichon [**Ga**], McCann–Zhang [**McZ**], and Santambrogio [**San**].

Problem 15.9 is taken from Le [**L11**]. Problem 15.11 is taken from Lions [**Ln3**] (see also Carlier–Lachand-Robert [**CLR**]). The idea of using Riedl's Theorem in Problem 15.11 comes from Cavalletti–Westdickenberg [**CaW**]. Problem 15.12 is taken from Kim–Le–Wang–Zhou [**KLWZ**].

Bibliography

[AK]   F. Abedin and J. Kitagawa, *Inverse iteration for the Monge-Ampère eigenvalue problem*, Proc. Amer. Math. Soc. **148** (2020), no. 11, 4875–4886, DOI 10.1090/proc/15157. MR4143401

[Ab]   M. Abreu, *Kähler geometry of toric varieties and extremal metrics*, Internat. J. Math. **9** (1998), no. 6, 641–651, DOI 10.1142/S0129167X98000282. MR1644291

[Ah]   L. V. Ahlfors, *Complex analysis: An introduction to the theory of analytic functions of one complex variable*, 3rd ed., International Series in Pure and Applied Mathematics, McGraw-Hill Book Co., New York, 1978. MR510197

[AFT]   H. Aimar, L. Forzani, and R. Toledano, *Balls and quasi-metrics: a space of homogeneous type modeling the real analysis related to the Monge-Ampère equation*, J. Fourier Anal. Appl. **4** (1998), no. 4-5, 377–381, DOI 10.1007/BF02498215. MR1658608

[Al1]   A. D. Aleksandrov, *Dirichlet's problem for the equation* $\mathrm{Det}\,\|z_{ij}\| = \varphi(z_1, \cdots, z_n, z, x_1, \cdots, x_n)$. *I* (Russian, with English summary), Vestnik Leningrad. Univ. Ser. Mat. Meh. Astronom. **13** (1958), no. 1, 5–24. MR96903

[Al2]   A. D. Aleksandrov, *Certain estimates for the Dirichlet problem* (Russian), Dokl. Akad. Nauk SSSR **134**, 1001–1004; English transl., Soviet Math. Dokl. **1** (1961), 1151–1154. MR147776

[Al3]   A. D. Aleksandrov, *Uniqueness conditions and bounds for the solution of the Dirichlet problem* (Russian. English summary), Vestnik Leningrad. Univ. Ser. Mat. Meh. Astronom. **18** (1963) no. 3, 5–29. English translation in Amer. Math.Soc. Transl. (2) **68** (1968), 89–119. MR164135

[Al4]   A. D. Aleksandrov, *Majorants of solutions of linear equations of order two*, Vestnik Leningrad. Univ. **21** (1966), 5–25 (Russian). English translation in Amer. Math.Soc. Transl. (2) **68** (1968), 120–143. MR199540

[Alm]   F. J. Almgren Jr., *Some interior regularity theorems for minimal surfaces and an extension of Bernstein's theorem*, Ann. of Math. (2) **84** (1966), 277–292, DOI 10.2307/1970520. MR200816

[ACDF1] L. Ambrosio, M. Colombo, G. De Philippis, and A. Figalli, *Existence of Eulerian solutions to the semigeostrophic equations in physical space: the 2-dimensional periodic case*, Comm. Partial Differential Equations **37** (2012), no. 12, 2209–2227, DOI 10.1080/03605302.2012.669443. MR3005541

[ACDF2] L. Ambrosio, M. Colombo, G. De Philippis, and A. Figalli, *A global existence result for the semigeostrophic equations in three dimensional convex domains*, Discrete Contin. Dyn. Syst. **34** (2014), no. 4, 1251–1268, DOI 10.3934/dcds.2014.34.1251. MR3117839

[Amp] A. M. Ampère, *Mémoire contenant l'application de la théorie*, Journal de l'École Poly technique, 1820.

[AAM] S. Artstein-Avidan and V. Milman, *The concept of duality in convex analysis, and the characterization of the Legendre transform*, Ann. of Math. (2) **169** (2009), no. 2, 661–674, DOI 10.4007/annals.2009.169.661. MR2480615

[Au1] T. Aubin, *Équations de Monge-Ampère réelles* (French), J. Functional Analysis **41** (1981), no. 3, 354–377, DOI 10.1016/0022-1236(81)90081-1. MR619958

[Au2] T. Aubin, *Some nonlinear problems in Riemannian geometry*, Springer Monographs in Mathematics, Springer-Verlag, Berlin, 1998, DOI 10.1007/978-3-662-13006-3. MR1636569

[Ba1] I. Ya. Bakel'man, *Generalized solutions of Monge-Ampère equations* (Russian), Dokl. Akad. Nauk SSSR (N.S.) **114** (1957), 1143–1145. MR95481

[Ba2] I. Ja. Bakel'man, *On the theory of quasilinear elliptic equations* (Russian), Sibirsk. Mat. Ž. **2** (1961), 179–186. MR126604

[Ba3] I. Ja. Bakel'man, *A variational problem related to the Monge-Ampère equation* (Russian), Dokl. Akad. Nauk SSSR **141** (1961), 1011–1014. MR177183

[Ba4] I. J. Bakelman, *Variational problems and elliptic Monge-Ampère equations*, J. Differential Geom. **18** (1983), no. 4, 669–699 (1984). MR730922

[Ba5] I. J. Bakelman, *Convex analysis and nonlinear geometric elliptic equations*, with an obituary for the author by William Rundell, edited by Steven D. Taliaferro, Springer-Verlag, Berlin, 1994, DOI 10.1007/978-3-642-69881-1. MR1305147

[Bam] R. P. Bambah, *Polar reciprocal convex bodies*, Proc. Cambridge Philos. Soc. **51** (1955), 377–378, DOI 10.1017/s0305004100030309. MR70201

[Bau] P. Bauman, *Positive solutions of elliptic equations in nondivergence form and their adjoints*, Ark. Mat. **22** (1984), no. 2, 153–173, DOI 10.1007/BF02384378. MR765409

[BT] E. Bedford and B. A. Taylor, *The Dirichlet problem for a complex Monge-Ampère equation*, Invent. Math. **37** (1976), no. 1, 1–44, DOI 10.1007/BF01418826. MR445006

[BSc] P. Bella and M. Schäffner, *Local boundedness and Harnack inequality for solutions of linear nonuniformly elliptic equations*, Comm. Pure Appl. Math. **74** (2021), no. 3, 453–477, DOI 10.1002/cpa.21876. MR4201290

[BB] J.-D. Benamou and Y. Brenier, *Weak existence for the semigeostrophic equations formulated as a coupled Monge-Ampère/transport problem*, SIAM J. Appl. Math. **58** (1998), no. 5, 1450–1461, DOI 10.1137/S0036139995294111. MR1627555

[Bg] M. S. Berger, *On the existence of equilibrium states of thin elastic shells. I*, Indiana Univ. Math. J. **20** (1970/71), 591–602, DOI 10.1512/iumj.1971.20.20048. MR275739

[Bn] S. N. Bernstein, *Sur une théorème de géometrie et ses applications aux équations dérivées partielles du type elliptique*, Comm. Soc. Math. Kharkov **15** (1915-1917), 38–45, German translation in *Über ein geometrisches Theorem und seine Anwendung auf die partiellen Differentialgleichungen vom elliptischen Typus*, Math. Z. **26** (1927), 551–558. MR1544873

[Bl] Z. Błocki, *Interior regularity of the degenerate Monge-Ampère equation*, Bull. Austral. Math. Soc. **68** (2003), no. 1, 81–92, DOI 10.1017/S0004972700037436. MR1996171

[BDGG] E. Bombieri, E. De Giorgi, and E. Giusti, *Minimal cones and the Bernstein problem*, Invent. Math. **7** (1969), 243–268, DOI 10.1007/BF01404309. MR250205

[BNT] B. Brandolini, C. Nitsch, and C. Trombetti, *New isoperimetric estimates for solutions to Monge-Ampère equations*, Ann. Inst. H. Poincaré C Anal. Non Linéaire **26** (2009), no. 4, 1265–1275, DOI 10.1016/j.anihpc.2008.09.005. MR2542724

[Bre]   S. Brendle, *The isoperimetric inequality for a minimal submanifold in Euclidean space*, J. Amer. Math. Soc. **34** (2021), no. 2, 595–603, DOI 10.1090/jams/969. MR4280868

[Br]   Y. Brenier, *Polar factorization and monotone rearrangement of vector-valued functions*, Comm. Pure Appl. Math. **44** (1991), no. 4, 375–417, DOI 10.1002/cpa.3160440402. MR1100809

[Cab1]   X. Cabré, *Nondivergent elliptic equations on manifolds with nonnegative curvature*, Comm. Pure Appl. Math. **50** (1997), no. 7, 623–665, DOI 10.1002/(SICI)1097-0312(199707)50:7¡623::AID-CPA2¿3.3.CO;2-B. MR1447056

[Cab2]   X. Cabré, *Elliptic PDE's in probability and geometry: symmetry and regularity of solutions*, Discrete Contin. Dyn. Syst. **20** (2008), no. 3, 425–457, DOI 10.3934/dcds.2008.20.425. MR2373200

[C1]   L. A. Caffarelli, *Interior a priori estimates for solutions of fully nonlinear equations*, Ann. of Math. (2) **130** (1989), no. 1, 189–213, DOI 10.2307/1971480. MR1005611

[C2]   L. A. Caffarelli, *A localization property of viscosity solutions to the Monge-Ampère equation and their strict convexity*, Ann. of Math. (2) **131** (1990), no. 1, 129–134, DOI 10.2307/1971509. MR1038359

[C3]   L. A. Caffarelli, *Interior $W^{2,p}$ estimates for solutions of the Monge-Ampère equation*, Ann. of Math. (2) **131** (1990), no. 1, 135–150, DOI 10.2307/1971510. MR1038360

[C4]   L. A. Caffarelli, *Some regularity properties of solutions of Monge Ampère equation*, Comm. Pure Appl. Math. **44** (1991), no. 8-9, 965–969, DOI 10.1002/cpa.3160440809. MR1127042

[C5]   L. A. Caffarelli, *Boundary regularity of maps with convex potentials*, Comm. Pure Appl. Math. **45** (1992), no. 9, 1141–1151, DOI 10.1002/cpa.3160450905. MR1177479

[C6]   L. A. Caffarelli, *The regularity of mappings with a convex potential*, J. Amer. Math. Soc. **5** (1992), no. 1, 99–104, DOI 10.2307/2152752. MR1124980

[C7]   L. A. Caffarelli, *A note on the degeneracy of convex solutions to Monge Ampère equation*, Comm. Partial Differential Equations **18** (1993), no. 7-8, 1213–1217, DOI 10.1080/03605309308820970. MR1233191

[C8]   L. A. Caffarelli, *Boundary regularity of maps with convex potentials. II*, Ann. of Math. (2) **144** (1996), no. 3, 453–496, DOI 10.2307/2118564. MR1426885

[CC]   L. A. Caffarelli and X. Cabré, *Fully nonlinear elliptic equations*, American Mathematical Society Colloquium Publications, vol. 43, American Mathematical Society, Providence, RI, 1995, DOI 10.1090/coll/043. MR1351007

[CFMS]   L. Caffarelli, E. Fabes, S. Mortola, and S. Salsa, *Boundary behavior of nonnegative solutions of elliptic operators in divergence form*, Indiana Univ. Math. J. **30** (1981), no. 4, 621–640, DOI 10.1512/iumj.1981.30.30049. MR620271

[CG1]   L. A. Caffarelli and C. E. Gutiérrez, *Real analysis related to the Monge-Ampère equation*, Trans. Amer. Math. Soc. **348** (1996), no. 3, 1075–1092, DOI 10.1090/S0002-9947-96-01473-0. MR1321570

[CG2]   L. A. Caffarelli and C. E. Gutiérrez, *Properties of the solutions of the linearized Monge-Ampère equation*, Amer. J. Math. **119** (1997), no. 2, 423–465. MR1439555

[CKNS]   L. Caffarelli, J. J. Kohn, L. Nirenberg, and J. Spruck, *The Dirichlet problem for nonlinear second-order elliptic equations. II. Complex Monge-Ampère, and uniformly elliptic, equations*, Comm. Pure Appl. Math. **38** (1985), no. 2, 209–252, DOI 10.1002/cpa.3160380206. MR780073

[CL]   L. Caffarelli and Y. Li, *An extension to a theorem of Jörgens, Calabi, and Pogorelov*, Comm. Pure Appl. Math. **56** (2003), no. 5, 549–583, DOI 10.1002/cpa.10067. MR1953651

[CMc]   L. A. Caffarelli and R. J. McCann, *Free boundaries in optimal transport and Monge-Ampère obstacle problems*, Ann. of Math. (2) **171** (2010), no. 2, 673–730, DOI 10.4007/annals.2010.171.673. MR2630054

[CNS1]   L. Caffarelli, L. Nirenberg, and J. Spruck, *The Dirichlet problem for nonlinear second-order elliptic equations. I. Monge-Ampère equation*, Comm. Pure Appl. Math. **37** (1984), no. 3, 369–402, DOI 10.1002/cpa.3160370306. MR739925

[CNS2]   L. Caffarelli, L. Nirenberg, and J. Spruck, *The Dirichlet problem for the degenerate Monge-Ampère equation*, Rev. Mat. Iberoamericana **2** (1986), no. 1-2, 19–27, DOI 10.4171/RMI/23. MR864651

[CS]   L. Caffarelli and L. Silvestre, *An extension problem related to the fractional Laplacian*, Comm. Partial Differential Equations **32** (2007), no. 7-9, 1245–1260, DOI 10.1080/03605300600987306. MR2354493

[CTW]   L. A. Caffarelli, L. Tang, and X.-J. Wang, *Global $C^{1,\alpha}$ regularity for Monge-Ampère equation and convex envelope*, Arch. Ration. Mech. Anal. **244** (2022), no. 1, 127–155, DOI 10.1007/s00205-022-01757-5. MR4393387

[CaY]   L. A. Caffarelli and Y. Yuan, *Singular solutions to Monge-Ampère equation*, Anal. Theory Appl. **38** (2022), no. 2, 121–127. MR4468909

[Cal1]   E. Calabi, *Improper affine hyperspheres of convex type and a generalization of a theorem by K. Jörgens*, Michigan Math. J. **5** (1958), 105–126. MR106487

[Cal2]   E. Calabi, *Hypersurfaces with maximal affinely invariant area*, Amer. J. Math. **104** (1982), no. 1, 91–126, DOI 10.2307/2374069. MR648482

[CZ1]   A. P. Calderon and A. Zygmund, *On the existence of certain singular integrals*, Acta Math. **88** (1952), 85–139, DOI 10.1007/BF02392130. MR52553

[CZ2]   A.-P. Calderón and A. Zygmund, *Local properties of solutions of elliptic partial differential equations*, Studia Math. **20** (1961), 171–225, DOI 10.4064/sm-20-2-181-225. MR136849

[CLR]   G. Carlier and T. Lachand-Robert, *Representation of the polar cone of convex functions and applications*, J. Convex Anal. **15** (2008), no. 3, 535–546. MR2431410

[CR]   G. Carlier and T. Radice, *Approximation of variational problems with a convexity constraint by PDEs of Abreu type*, Calc. Var. Partial Differential Equations **58** (2019), no. 5, Paper No. 170, 13, DOI 10.1007/s00526-019-1613-1. MR4010646

[CaW]   F. Cavalletti and M. Westdickenberg, *The polar cone of the set of monotone maps*, Proc. Amer. Math. Soc. **143** (2015), no. 2, 781–787, DOI 10.1090/S0002-9939-2014-12332-X. MR3283664

[CW]   A. Chau and B. Weinkove, *Monge-Ampère functionals and the second boundary value problem*, Math. Res. Lett. **22** (2015), no. 4, 1005–1022, DOI 10.4310/MRL.2015.v22.n4.a3. MR3391874

[CHLS]   B. Chen, Q. Han, A.-M. Li, and L. Sheng, *Interior estimates for the n-dimensional Abreu's equation*, Adv. Math. **251** (2014), 35–46, DOI 10.1016/j.aim.2013.10.004. MR3130333

[CLS]   B. Chen, A.-M. Li, and L. Sheng, *The Abreu equation with degenerated boundary conditions*, J. Differential Equations **252** (2012), no. 10, 5235–5259, DOI 10.1016/j.jde.2012.02.004. MR2902116

[CLW]   S. Chen, J. Liu, and X.-J. Wang, *Global regularity for the Monge-Ampère equation with natural boundary condition*, Ann. of Math. (2) **194** (2021), no. 3, 745–793, DOI 10.4007/annals.2021.194.3.4. MR4334976

[ChCh]   X. Chen and J. Cheng, *On the constant scalar curvature Kähler metrics (II)—Existence results*, J. Amer. Math. Soc. **34** (2021), no. 4, 937–1009, DOI 10.1090/jams/966. MR4301558

[CY]   S. Y. Cheng and S. T. Yau, *On the regularity of the Monge-Ampère equation* $\det(\partial^2 u/\partial x_i \partial s x_j) = F(x, u)$, Comm. Pure Appl. Math. **30** (1977), no. 1, 41–68, DOI 10.1002/cpa.3160300104. MR437805

[Ch] S. S. Chern, *Affine minimal hypersurfaces*, Minimal submanifolds and geodesics (Proc. Japan-United States Sem., Tokyo, 1977), North-Holland, Amsterdam-New York, 1979, pp. 17–30. MR574250

[ChW1] K.-S. Chou and X.-J. Wang, *Entire solutions of the Monge-Ampère equation*, Comm. Pure Appl. Math. **49** (1996), no. 5, 529–539, DOI 10.1002/(SICI)1097-0312(199605)49:5¡529::AID-CPA2¿3.3.CO;2-G. MR1377561

[ChW2] K.-S. Chou and X.-J. Wang, *A variational theory of the Hessian equation*, Comm. Pure Appl. Math. **54** (2001), no. 9, 1029–1064, DOI 10.1002/cpa.1016. MR1835381

[CWs] R. R. Coifman and G. Weiss, *Analyse harmonique non-commutative sur certains espaces homogènes* (French), Lecture Notes in Mathematics, Vol. 242, Springer-Verlag, Berlin-New York, 1971. MR499948

[CM] T. C. Collins and C. Mooney, *Dimension of the minimum set for the real and complex Monge-Ampère equations in critical Sobolev spaces*, Anal. PDE **10** (2017), no. 8, 2031–2041, DOI 10.2140/apde.2017.10.2031. MR3694014

[CE] D. Cordero-Erausquin, *Sur le transport de mesures périodiques* (French, with English and French summaries), C. R. Acad. Sci. Paris Sér. I Math. **329** (1999), no. 3, 199–202, DOI 10.1016/S0764-4442(00)88593-6. MR1711060

[CH1] R. Courant and D. Hilbert, *Methoden Der Mathematischen Physik* (German), Vol. II, Springer-Verlag, Berlin 1937.

[CH2] R. Courant and D. Hilbert, *Methods of mathematical physics. Vol. II*, Partial differential equations, reprint of the 1962 original, A Wiley-Interscience Publication, Wiley Classics Library, John Wiley & Sons, Inc., New York, 1989. MR1013360

[CIL] M. G. Crandall, H. Ishii, and P.-L. Lions, *User's guide to viscosity solutions of second order partial differential equations*, Bull. Amer. Math. Soc. (N.S.) **27** (1992), no. 1, 1–67, DOI 10.1090/S0273-0979-1992-00266-5. MR1118699

[Cu] M. Cullen, *A mathematical theory of large-scale atmosphere/ocean flow*, London, Imperial College Press, 2006.

[CF] M. Cullen and M. Feldman, *Lagrangian solutions of semigeostrophic equations in physical space*, SIAM J. Math. Anal. **37** (2006), no. 5, 1371–1395, DOI 10.1137/040615444. MR2215268

[CuG] M. Cullen and W. Gangbo, *A variational approach for the 2-dimensional semigeostrophic shallow water equations*, Arch. Ration. Mech. Anal. **156** (2001), no. 3, 241–273, DOI 10.1007/s002050000124. MR1816477

[CNP] M. J. P. Cullen, J. Norbury, and R. J. Purser, *Generalised Lagrangian solutions for atmospheric and oceanic flows*, SIAM J. Appl. Math. **51** (1991), no. 1, 20–31, DOI 10.1137/0151002. MR1089128

[DM] G. Dal Maso, *An introduction to Γ-convergence*, Progress in Nonlinear Differential Equations and their Applications, vol. 8, Birkhäuser Boston, Inc., Boston, MA, 1993, DOI 10.1007/978-1-4612-0327-8. MR1201152

[DS] P. Daskalopoulos and O. Savin, *On Monge-Ampère equations with homogeneous right-hand sides*, Comm. Pure Appl. Math. **62** (2009), no. 5, 639–676, DOI 10.1002/cpa.20263. MR2494810

[DG1] E. De Giorgi, *Sulla differenziabilità e l'analiticità delle estremali degli integrali multipli regolari* (Italian), Mem. Accad. Sci. Torino. Cl. Sci. Fis. Mat. Nat. (3) **3** (1957), 25–43. MR93649

[DG2] E. De Giorgi, *Una estensione del teorema di Bernstein* (Italian), Ann. Scuola Norm. Sup. Pisa Cl. Sci. (3) **19** (1965), 79–85. MR178385

[DPF1] G. De Philippis and A. Figalli, $W^{2,1}$ *regularity for solutions of the Monge-Ampère equation*, Invent. Math. **192** (2013), no. 1, 55–69, DOI 10.1007/s00222-012-0405-4. MR3032325

[DPF2]   G. De Philippis and A. Figalli, *The Monge-Ampère equation and its link to opti-mal transportation*, Bull. Amer. Math. Soc. (N.S.) **51** (2014), no. 4, 527–580, DOI 10.1090/S0273-0979-2014-01459-4. MR3237759

[DPF3]   G. De Philippis and A. Figalli, *Optimal regularity of the convex envelope*, Trans. Amer. Math. Soc. **367** (2015), no. 6, 4407–4422, DOI 10.1090/S0002-9947-2014-06306-X. MR3324933

[DPFS]   G. De Philippis, A. Figalli, and O. Savin, *A note on interior $W^{2,1+\varepsilon}$ estimates for the Monge-Ampère equation*, Math. Ann. **357** (2013), no. 1, 11–22, DOI 10.1007/s00208-012-0895-9. MR3084340

[De]   K. Deimling, *Nonlinear functional analysis*, Springer-Verlag, Berlin, 1985, DOI 10.1007/978-3-662-00547-7. MR787404

[DFGL]   G. Di Fazio, C. E. Gutiérrez, and E. Lanconelli, *Covering theorems, inequalities on metric spaces and applications to PDE's*, Math. Ann. **341** (2008), no. 2, 255–291, DOI 10.1007/s00208-007-0188-x. MR2385658

[D1]   S. K. Donaldson, *Scalar curvature and stability of toric varieties*, J. Differential Geom. **62** (2002), no. 2, 289–349. MR1988506

[D2]   S. K. Donaldson, *Conjectures in Kähler geometry*, Strings and geometry, Clay Math. Proc., vol. 3, Amer. Math. Soc., Providence, RI, 2004, pp. 71–78. MR2103718

[D3]   S. K. Donaldson, *Interior estimates for solutions of Abreu's equation*, Collect. Math. **56** (2005), no. 2, 103–142. MR2154300

[D4]   S. K. Donaldson, *Extremal metrics on toric surfaces: a continuity method*, J. Differential Geom. **79** (2008), no. 3, 389–432. MR2433928

[D5]   S. K. Donaldson, *Constant scalar curvature metrics on toric surfaces*, Geom. Funct. Anal. **19** (2009), no. 1, 83–136, DOI 10.1007/s00039-009-0714-y. MR2507220

[E1]   L. C. Evans, *Classical solutions of fully nonlinear, convex, second-order el-liptic equations*, Comm. Pure Appl. Math. **35** (1982), no. 3, 333–363, DOI 10.1002/cpa.3160350303. MR649348

[E2]   L. C. Evans, *Partial differential equations*, 2nd ed., Graduate Studies in Mathematics, vol. 19, American Mathematical Society, Providence, RI, 2010, DOI 10.1090/gsm/019. MR2597943

[EG]   L. C. Evans and R. F. Gariepy, *Measure theory and fine properties of functions*, Studies in Advanced Mathematics, CRC Press, Boca Raton, FL, 1992. MR1158660

[FSt]   E. B. Fabes and D. W. Stroock, *The L^p-integrability of Green's functions and funda-mental solutions for elliptic and parabolic equations*, Duke Math. J. **51** (1984), no. 4, 997–1016, DOI 10.1215/S0012-7094-84-05145-7. MR771392

[FSz]   R. Feng and G. Székelyhidi, *Periodic solutions of Abreu's equation*, Math. Res. Lett. **18** (2011), no. 6, 1271–1279, DOI 10.4310/MRL.2011.v18.n6.a15. MR2915480

[F1]   A. Figalli, *Sobolev regularity for the Monge-Ampère equation, with application to the semigeostrophic equations* (English, with English and Russian summaries), Zap. Nauchn. Sem. S.-Peterburg. Otdel. Mat. Inst. Steklov. (POMI) **411** (2013), no. Teoriya Predstavleniĭ, Dinamicheskie Sistemy, Kombinatornye Metody. XXII, 103–118, 242, DOI 10.1007/s10958-013-1649-2; English transl., J. Math. Sci. (N.Y.) **196** (2014), no. 2, 175–183. MR3048271

[F2]   A. Figalli, *The Monge-Ampère equation and its applications*, Zurich Lectures in Advanced Mathematics, European Mathematical Society (EMS), Zürich, 2017, DOI 10.4171/170. MR3617963

[F3]   A. Figalli, *Global existence for the semigeostrophic equations via Sobolev estimates for Monge-Ampère*, Partial differential equations and geometric measure theory, Lecture Notes in Math., vol. 2211, Springer, Cham, 2018, pp. 1–42. MR3790946

[FKM]   A. Figalli, Y.-H. Kim, and R. J. McCann, *When is multidimensional screening a convex program?*, J. Econom. Theory **146** (2011), no. 2, 454–478, DOI 10.1016/j.jet.2010.11.006. MR2888826

[FM]   L. Forzani and D. Maldonado, *Properties of the solutions to the Monge-Ampère equation*, Nonlinear Anal. **57** (2004), no. 5-6, 815–829, DOI 10.1016/j.na.2004.03.019. MR2067735

[Fr]   A. Friedman, *On the regularity of the solutions of nonlinear elliptic and parabolic systems of partial differential equations*, J. Math. Mech. **7** (1958), 43–59, DOI 10.1512/iumj.1958.7.57004. MR118970

[Ga]   A. Galichon, *Optimal transport methods in economics*, Princeton University Press, Princeton, NJ, 2016, DOI 10.1515/9781400883592. MR3586373

[Gav]   B. Gaveau, *Méthodes de contrôle optimal en analyse complexe. I. Résolution d'équations de Monge Ampère* (French), J. Functional Analysis **25** (1977), no. 4, 391–411, DOI 10.1016/0022-1236(77)90046-5. MR457783

[GT]   D. Gilbarg and N. S. Trudinger, *Elliptic partial differential equations of second order*, reprint of the 1998 edition, Classics in Mathematics, Springer-Verlag, Berlin, 2001. MR1814364

[Gil]   P. P. Gillis, *Equations de Monge-Ampère et problèmes du calcul des variations* (French), IIIᵉ Congrès National des Sciences, Bruxelles, 1950, Vol. 2, Fédération belge des Sociétés Scientifiques, Bruxelles, 1950, pp. 5–9. MR70074

[Grv]   P. Grisvard, *Elliptic problems in nonsmooth domains*, reprint of the 1985 original [MR0775683], with a foreword by Susanne C. Brenner, Classics in Applied Mathematics, vol. 69, Society for Industrial and Applied Mathematics (SIAM), Philadelphia, PA, 2011, DOI 10.1137/1.9781611972030.ch1. MR3396210

[GW]   M. Grüter and K.-O. Widman, *The Green function for uniformly elliptic equations*, Manuscripta Math. **37** (1982), no. 3, 303–342, DOI 10.1007/BF01166225. MR657523

[Gu]   B. Guan, *The Dirichlet problem for Monge-Ampère equations in non-convex domains and spacelike hypersurfaces of constant Gauss curvature*, Trans. Amer. Math. Soc. **350** (1998), no. 12, 4955–4971, DOI 10.1090/S0002-9947-98-02079-0. MR1451602

[GS]   B. Guan and J. Spruck, *Boundary-value problems on S^n for surfaces of constant Gauss curvature*, Ann. of Math. (2) **138** (1993), no. 3, 601–624, DOI 10.2307/2946558. MR1247995

[GTW]   P. Guan, N. S. Trudinger, and X.-J. Wang, *On the Dirichlet problem for degenerate Monge-Ampère equations*, Acta Math. **182** (1999), no. 1, 87–104, DOI 10.1007/BF02392824. MR1687172

[G1]   C. E. Gutiérrez, *On the Harnack Inequality for Viscosity Solutions of Non-Divergence Equations*, Sinet, Ethiopian Journal of Science., Proc. of the conference in Honor of T. Retta, edited by S. Berhanu, vol. 19, 1996, 48–72.

[G2]   C. E. Gutiérrez, *The Monge-Ampère equation*, second edition [of MR1829162], Progress in Nonlinear Differential Equations and their Applications, vol. 89, Birkhäuser/Springer, [Cham], 2016, DOI 10.1007/978-3-319-43374-5. MR3560611

[GH1]   C. E. Gutiérrez and Q. Huang, *A generalization of a theorem by Calabi to the parabolic Monge-Ampère equation*, Indiana Univ. Math. J. **47** (1998), no. 4, 1459–1480, DOI 10.1512/iumj.1998.47.1563. MR1687122

[GH2]   C. E. Gutiérrez and Q. Huang, *Geometric properties of the sections of solutions to the Monge-Ampère equation*, Trans. Amer. Math. Soc. **352** (2000), no. 9, 4381–4396, DOI 10.1090/S0002-9947-00-02491-0. MR1665332

[GH3]   C. E. Gutiérrez and Q. Huang, *$W^{2,p}$ estimates for the parabolic Monge-Ampère equation*, Arch. Ration. Mech. Anal. **159** (2001), no. 2, 137–177, DOI 10.1007/s002050100151. MR1857377

[GN1]   C. E. Gutiérrez and T. Nguyen, *Interior gradient estimates for solutions to the lin-earized Monge-Ampère equation*, Adv. Math. **228** (2011), no. 4, 2034–2070, DOI 10.1016/j.aim.2011.06.035. MR2836113

[GN2]   C. E. Gutiérrez and T. Nguyen, *Interior second derivative estimates for solutions to the linearized Monge-Ampère equation*, Trans. Amer. Math. Soc. **367** (2015), no. 7, 4537–4568, DOI 10.1090/S0002-9947-2015-06048-6. MR3335393

[GTo]   C. E. Gutiérrez and F. Tournier, $W^{2,p}$-*estimates for the linearized Monge-Ampère equa-tion*, Trans. Amer. Math. Soc. **358** (2006), no. 11, 4843–4872, DOI 10.1090/S0002-9947-06-04189-4. MR2231875

[HR]   R. A. Hager and J. Ross, *A regularity theorem for linear second order elliptic divergence equations*, Ann. Scuola Norm. Sup. Pisa Cl. Sci. (3) **26** (1972), 283–290. MR374643

[H]   Q. Han, *Nonlinear elliptic equations of the second order*, Graduate Studies in Mathematics, vol. 171, American Mathematical Society, Providence, RI, 2016, DOI 10.1090/gsm/171. MR3468839

[HL]   Q. Han and F. Lin, *Elliptic partial differential equations*, 2nd ed., Courant Lecture Notes in Mathematics, vol. 1, Courant Institute of Mathematical Sciences, New York; American Mathematical Society, Providence, RI, 2011. MR2777537

[Ha1]   D. Hartenstine, *The Dirichlet problem for the Monge-Ampère equation in convex (but not strictly convex) domains*, Electron. J. Differential Equations (2006), No. 138, 9. MR2276563

[Ha2]   D. Hartenstine, *Brunn-Minkowski-type inequalities related to the Monge-Ampère equa-tion*, Adv. Nonlinear Stud. **9** (2009), no. 2, 277–294, DOI 10.1515/ans-2009-0204. MR2503830

[HLa]   R. Harvey and H. B. Lawson Jr., *Calibrated geometries*, Acta Math. **148** (1982), 47–157, DOI 10.1007/BF02392726. MR666108

[Has]   Y. Hashimoto, *A remark on the analyticity of the solutions for non-linear ellip-tic partial differential equations*, Tokyo J. Math. **29** (2006), no. 2, 271–281, DOI 10.3836/tjm/1170348166. MR2284971

[He]   E. Heinz, *On elliptic Monge-Ampère equations and Weyl's embedding problem*, J. Anal-yse Math. **7** (1959), 1–52, DOI 10.1007/BF02787679. MR111943

[Hop]   E. Hopf, *Über den funktionalen, insbesondere den analytischen Charakter der Lösungen elliptischer Differentialgleichungen zweiter Ordnung* (German), Math. Z. **34** (1932), no. 1, 194–233, DOI 10.1007/BF01180586. MR1545250

[HHW]   J. Hong, G. Huang, and W. Wang, *Existence of global smooth solutions to Dirichlet problem for degenerate elliptic Monge-Ampere equations*, Comm. Partial Differential Equations **36** (2011), no. 4, 635–656, DOI 10.1080/03605302.2010.514171. MR2763326

[HJ]   R. A. Horn and C. R. Johnson, *Matrix analysis*, 2nd ed., Cambridge University Press, Cambridge, 2013. MR2978290

[Hor]   L. Hörmander, *Notions of convexity*, Progress in Mathematics, vol. 127, Birkhäuser Boston, Inc., Boston, MA, 1994. MR1301332

[Hw]   R. Howard, *The John ellipsoid theorem* (1997), available at https://people.math.sc.edu/howard/Notes/john.pdf.

[Hua1]   G. Huang, *Uniqueness of least energy solutions for Monge-Ampère functional*, Calc. Var. Partial Differential Equations **58** (2019), no. 2, Paper No. 73, 20, DOI 10.1007/s00526-019-1504-5. MR3927128

[Hua2]   G. Huang, *The Guillemin boundary problem for Monge-Ampère equation in the poly-gon*, Adv. Math. **415** (2023), Paper No. 108885, 29, DOI 10.1016/j.aim.2023.108885. MR4543075

[HuL]   G. Huang and Y. Lü, *Analyticity of the solutions to degenerate Monge-Ampère equa-tions*, J. Differential Equations **376** (2023), 633–654, DOI 10.1016/j.jde.2023.09.003. MR4642967

[HTW]   G. Huang, L. Tan, and X.-J. Wang, *Regularity of free boundary for the Monge-Ampère obstacle problem*, arXiv:2111.10575, Duke Math. J., to appear.

[Hu1]   Q. Huang, *Harnack inequality for the linearized parabolic Monge-Ampère equation*, Trans. Amer. Math. Soc. **351** (1999), no. 5, 2025–2054, DOI 10.1090/S0002-9947-99-02142-X. MR1467468

[Hu2]   Q. Huang, *Sharp regularity results on second derivatives of solutions to the Monge-Ampère equation with VMO type data*, Comm. Pure Appl. Math. **62** (2009), no. 5, 677–705, DOI 10.1002/cpa.20272. MR2494811

[IMS]   S. Indratno, D. Maldonado, and S. Silwal, *On the axiomatic approach to Harnack's inequality in doubling quasi-metric spaces*, J. Differential Equations **254** (2013), no. 8, 3369–3394, DOI 10.1016/j.jde.2013.01.025. MR3020880

[IS]   C. Imbert and L. Silvestre, *Estimates on elliptic equations that hold only where the gradient is large*, J. Eur. Math. Soc. (JEMS) **18** (2016), no. 6, 1321–1338, DOI 10.4171/JEMS/614. MR3500837

[ISh]   H. Iriyeh and M. Shibata, *Symmetric Mahler's conjecture for the volume product in the 3-dimensional case*, Duke Math. J. **169** (2020), no. 6, 1077–1134, DOI 10.1215/00127094-2019-0072. MR4085078

[IL]   H. Ishii and P.-L. Lions, *Viscosity solutions of fully nonlinear second-order elliptic partial differential equations*, J. Differential Equations **83** (1990), no. 1, 26–78, DOI 10.1016/0022-0396(90)90068-Z. MR1031377

[Iv]   N. M. Ivočkina, *A priori estimate of $|u|_{C_2(\overline{\Omega})}$ of convex solutions of the Dirichlet problem for the Monge-Ampère equation* (Russian), Zap. Nauchn. Sem. Leningrad. Otdel. Mat. Inst. Steklov. (LOMI) **96** (1980), 69–79, 306. MR579472

[Jer]   R. L. Jerrard, *Some remarks on Monge-Ampère functions*, Singularities in PDE and the calculus of variations, CRM Proc. Lecture Notes, vol. 44, Amer. Math. Soc., Providence, RI, 2008, pp. 89–112, DOI 10.1090/crmp/044/07. MR2528736

[JW1]   H.-Y. Jian and X.-J. Wang, *Continuity estimates for the Monge-Ampère equation*, SIAM J. Math. Anal. **39** (2007), no. 2, 608–626, DOI 10.1137/060669036. MR2338423

[JW2]   H. Jian and X.-J. Wang, *Bernstein theorem and regularity for a class of Monge-Ampère equations*, J. Differential Geom. **93** (2013), no. 3, 431–469. MR3024302

[JX]   T. Jin and J. Xiong, *A Liouville theorem for solutions of degenerate Monge-Ampère equations*, Comm. Partial Differential Equations **39** (2014), no. 2, 306–320, DOI 10.1080/03605302.2013.814143. MR3169787

[J]   K. Jörgens, *Über die Lösungen der Differentialgleichung $rt - s^2 = 1$* (German), Math. Ann. **127** (1954), 130–134, DOI 10.1007/BF01361114. MR62326

[Jn]   F. John, *Extremum problems with inequalities as subsidiary conditions*, Studies and Essays Presented to R. Courant on his 60th Birthday, January 8, 1948, Interscience Publishers, New York, 1948, pp. 187–204. MR30135

[Kn]   R. Kenyon, *Lectures on dimers*, Statistical mechanics, IAS/Park City Math. Ser., vol. 16, Amer. Math. Soc., Providence, RI, 2009, pp. 191–230, DOI 10.1090/pcms/016/04. MR2523460

[KO]   R. Kenyon and A. Okounkov, *Planar dimers and Harnack curves*, Duke Math. J. **131** (2006), no. 3, 499–524, DOI 10.1215/S0012-7094-06-13134-4. MR2219249

[KOS]   R. Kenyon, A. Okounkov, and S. Sheffield, *Dimers and amoebae*, Ann. of Math. (2) **163** (2006), no. 3, 1019–1056, DOI 10.4007/annals.2006.163.1019. MR2215138

[KLWZ]   Y. H. Kim, N. Q. Le, L. Wang, and B. Zhou, *Singular Abreu equations and linearized Monge-Ampère equations with drifts*, `arXiv:2209.11681`, preprint.

[K1]   N. V. Krylov, *Sequences of convex functions, and estimates of the maximum of the solution of a parabolic equation* (Russian), Sibirsk. Mat. Ž. **17** (1976), no. 2, 290–303, 478. MR420016

[K2]   N. V. Krylov, *Boundedly nonhomogeneous elliptic and parabolic equations*, Izv. Akad. SSSR Ser. Mat. **46** (1982), 487–523; English transl. in Math. USSR Izv. **20** (1983), 459–492. MR661144

[K3]   N. V. Krylov, *Boundedly inhomogeneous elliptic and parabolic equations in a domain* (Russian), Izv. Akad. Nauk SSSR Ser. Mat. **47** (1983), no. 1, 75–108. MR688919

[K4]   N. V. Krylov, *Sobolev and viscosity solutions for fully nonlinear elliptic and parabolic equations*, Mathematical Surveys and Monographs, vol. 233, American Mathematical Society, Providence, RI, 2018, DOI 10.1090/surv/233. MR3837125

[KS1]   N. V. Krylov and M. V. Safonov, *An estimate for the probability of a diffusion process hitting a set of positive measure* (Russian), Dokl. Akad. Nauk SSSR **245** (1979), no. 1, 18–20. MR525227

[KS2]   N. V. Krylov and M. V. Safonov, *A property of the solutions of parabolic equations with measurable coefficients* (Russian), Izv. Akad. Nauk SSSR Ser. Mat. **44** (1980), no. 1, 161–175, 239. MR563790

[KL]   J. Kinnunen and P. Lindqvist, *The derivative of the maximal function*, J. Reine Angew. Math. **503** (1998), 161–167. MR1650343

[KT]   H.-J. Kuo and N. S. Trudinger, *New maximum principles for linear elliptic equations*, Indiana Univ. Math. J. **56** (2007), no. 5, 2439–2452, DOI 10.1512/iumj.2007.56.3073. MR2360615

[L1]   N. Q. Le, *Global second derivative estimates for the second boundary value problem of the prescribed affine mean curvature and Abreu's equations*, Int. Math. Res. Not. IMRN **11** (2013), 2421–2438, DOI 10.1093/imrn/rns123. MR3065084

[L2]   N. Q. Le, *Remarks on the Green's function of the linearized Monge-Ampère operator*, Manuscripta Math. **149** (2016), no. 1-2, 45–62, DOI 10.1007/s00229-015-0766-2. MR3447139

[L3]   N. Q. Le, $W^{4,p}$ *solution to the second boundary value problem of the prescribed affine mean curvature and Abreu's equations*, J. Differential Equations **260** (2016), no. 5, 4285–4300, DOI 10.1016/j.jde.2015.11.013. MR3437587

[L4]   N. Q. Le, *Boundary Harnack inequality for the linearized Monge-Ampère equations and applications*, Trans. Amer. Math. Soc. **369** (2017), no. 9, 6583–6611, DOI 10.1090/tran/7220. MR3660234

[L5]   N. Q. Le, *On the Harnack inequality for degenerate and singular elliptic equations with unbounded lower order terms via sliding paraboloids*, Commun. Contemp. Math. **20** (2018), no. 1, 1750012, 38, DOI 10.1142/S0219199717500122. MR3714836

[L6]   N. Q. Le, *Hölder regularity of the 2D dual semigeostrophic equations via analysis of linearized Monge-Ampère equations*, Comm. Math. Phys. **360** (2018), no. 1, 271–305, DOI 10.1007/s00220-018-3125-9. MR3795192

[L7]   N. Q. Le, *The eigenvalue problem for the Monge-Ampère operator on general bounded convex domains*, Ann. Sc. Norm. Super. Pisa Cl. Sci. (5) **18** (2018), no. 4, 1519–1559. MR3829755

[L8]   N. Q. Le, *Global Hölder estimates for 2D linearized Monge-Ampère equations with right-hand side in divergence form*, J. Math. Anal. Appl. **485** (2020), no. 2, 123865, 13, DOI 10.1016/j.jmaa.2020.123865. MR4052579

[L9]   N. Q. Le, *Singular Abreu equations and minimizers of convex functionals with a convexity constraint*, Comm. Pure Appl. Math. **73** (2020), no. 10, 2248–2283, DOI 10.1002/cpa.21883. MR4156619

[L10]   N. Q. Le, *On approximating minimizers of convex functionals with a convexity constraint by singular Abreu equations without uniform convexity*, Proc. Roy. Soc. Edinburgh Sect. A **151** (2021), no. 1, 356–376, DOI 10.1017/prm.2020.18. MR4202645

[L11]   N. Q. Le, *On singular Abreu equations in higher dimensions*, J. Anal. Math. **144** (2021), no. 1, 191–205, DOI 10.1007/s11854-021-0176-1. MR4361893

[L12]   N. Q. Le, *Optimal boundary regularity for some singular Monge-Ampère equations on bounded convex domains*, Discrete Contin. Dyn. Syst. **42** (2022), no. 5, 2199–2214, DOI 10.3934/dcds.2021188. MR4405205

[L13]   N. Q. Le, *Remarks on sharp boundary estimates for singular and degenerate Monge-Ampère equations*, Commun. Pure Appl. Anal. **22** (2023), no. 5, 1701–1720, DOI 10.3934/cpaa.2023043. MR4583570

[L14]   N. Q. Le, *Twisted Harnack inequality and approximation of variational problems with a convexity constraint by singular Abreu equations*, Adv. Math. **434** (2023), Paper No. 109325, 31 pp., DOI 10.1016/j.aim.2023.109325. MR4648402

[LMT]   N. Q. Le, H. Mitake, and H. V. Tran, *Dynamical and geometric aspects of Hamilton-Jacobi and linearized Monge-Ampère equations—VIASM 2016*, edited by Mitake and Tran, Lecture Notes in Mathematics, vol. 2183, Springer, Cham, 2017. MR3729436

[LN1]   N. Q. Le and T. Nguyen, *Geometric properties of boundary sections of solutions to the Monge-Ampère equation and applications*, J. Funct. Anal. **264** (2013), no. 1, 337–361, DOI 10.1016/j.jfa.2012.10.015. MR2995711

[LN2]   N. Q. Le and T. Nguyen, *Global $W^{2,p}$ estimates for solutions to the linearized Monge-Ampère equations*, Math. Ann. **358** (2014), no. 3-4, 629–700, DOI 10.1007/s00208-013-0974-6. MR3175137

[LN3]   N. Q. Le and T. Nguyen, *Global $W^{1,p}$ estimates for solutions to the linearized Monge-Ampère equations*, J. Geom. Anal. **27** (2017), no. 3, 1751–1788, DOI 10.1007/s12220-016-9739-2. MR3667409

[LS1]   N. Q. Le and O. Savin, *Boundary regularity for solutions to the linearized Monge-Ampère equations*, Arch. Ration. Mech. Anal. **210** (2013), no. 3, 813–836, DOI 10.1007/s00205-013-0653-5. MR3116005

[LS2]   N. Q. Le and O. Savin, *Some minimization problems in the class of convex functions with prescribed determinant*, Anal. PDE **6** (2013), no. 5, 1025–1050, DOI 10.2140/apde.2013.6.1025. MR3125549

[LS3]   N. Q. Le and O. Savin, *On boundary Hölder gradient estimates for solutions to the linearized Monge-Ampère equations*, Proc. Amer. Math. Soc. **143** (2015), no. 4, 1605–1615, DOI 10.1090/S0002-9939-2014-12340-9. MR3314073

[LS4]   N. Q. Le and O. Savin, *Schauder estimates for degenerate Monge-Ampère equations and smoothness of the eigenfunctions*, Invent. Math. **207** (2017), no. 1, 389–423, DOI 10.1007/s00222-016-0677-1. MR3592760

[LS5]   N. Q. Le and O. Savin, *Global $C^{2,\alpha}$ estimates for the Monge-Ampère equation on polygonal domains in the plane*, Amer. J. Math. **145** (2023), no. 1, 221–249, DOI 10.1353/ajm.2023.0004. MR4545846

[LZ]   N. Q. Le and B. Zhou, *Solvability of a class of singular fourth order equations of Monge-Ampère type*, Ann. PDE **7** (2021), no. 2, Paper No. 13, 32, DOI 10.1007/s40818-021-00102-5. MR4266211

[Lwk]   M. Lewicka, *The Monge-Ampère system: convex integration in arbitrary dimension and codimension*, arXiv:2210.04363, preprint.

[LMP]   M. Lewicka, L. Mahadevan, and M. R. Pakzad, *The Monge-Ampère constraint: matching of isometries, density and regularity, and elastic theories of shallow shells* (English, with English and French summaries), Ann. Inst. H. Poincaré C Anal. Non Linéaire **34** (2017), no. 1, 45–67, DOI 10.1016/j.anihpc.2015.08.005. MR3592678

[LP]   M. Lewicka and M. R. Pakzad, *Convex integration for the Monge-Ampère equation in two dimensions*, Anal. PDE **10** (2017), no. 3, 695–727, DOI 10.2140/apde.2017.10.695. MR3641884

[Lw]   H. Lewy, *A priori limitations for solutions of Monge-Ampère equations. II*, Trans. Amer. Math. Soc. **41** (1937), no. 3, 365–374, DOI 10.2307/1989787. MR1501906

[LiSh]   A.-M. Li and L. Sheng, *A Liouville theorem on the PDE* $\det(f_{i\bar{j}}) = 1$, Math. Z. **297** (2021), no. 3-4, 1623–1632, DOI 10.1007/s00209-020-02571-z. MR4229616

[LiW]   Q.-R. Li and X.-J. Wang, *Regularity of the homogeneous Monge-Ampère equation*, Discrete Contin. Dyn. Syst. **35** (2015), no. 12, 6069–6084, DOI 10.3934/dcds.2015.35.6069. MR3393267

[Li]   S. Lie, *Neue Integrations-methode der Monge-Ampère'schen Gleichung*, Archiv for Mathematik og Naturvidenskah **2**(1877), 1–9.

[Lb]   G. M. Lieberman, *Second order parabolic differential equations*, World Scientific Publishing Co., Inc., River Edge, NJ, 1996, DOI 10.1142/3302. MR1465184

[Ln1]   P.-L. Lions, *Sur les équations de Monge-Ampère. I* (French, with English summary), Manuscripta Math. **41** (1983), no. 1-3, 1–43, DOI 10.1007/BF01165928. MR689131

[Ln2]   P.-L. Lions, *Two remarks on Monge-Ampère equations*, Ann. Mat. Pura Appl. (4) **142** (1985), 263–275 (1986), DOI 10.1007/BF01766596. MR839040

[Ln3]   P.-L. Lions, *Identification du cône dual des fonctions convexes et applications* (French, with English and French summaries), C. R. Acad. Sci. Paris Sér. I Math. **326** (1998), no. 12, 1385–1390, DOI 10.1016/S0764-4442(98)80397-2. MR1649179

[LTU]   P.-L. Lions, N. S. Trudinger, and J. I. E. Urbas, *The Neumann problem for equations of Monge-Ampère type*, Comm. Pure Appl. Math. **39** (1986), no. 4, 539–563, DOI 10.1002/cpa.3160390405. MR840340

[LSW]   W. Littman, G. Stampacchia, and H. F. Weinberger, *Regular points for elliptic equations with discontinuous coefficients*, Ann. Scuola Norm. Sup. Pisa Cl. Sci. (3) **17** (1963), 43–77. MR161019

[LW]   J. Liu and X.-J. Wang, *Interior a priori estimates for the Monge-Ampère equation*, Surveys in differential geometry 2014. Regularity and evolution of nonlinear equations, Surv. Differ. Geom., vol. 19, Int. Press, Somerville, MA, 2015, pp. 151–177, DOI 10.4310/SDG.2014.v19.n1.a7. MR3381500

[Lo1]   G. Loeper, *On the regularity of the polar factorization for time dependent maps*, Calc. Var. Partial Differential Equations **22** (2005), no. 3, 343–374, DOI 10.1007/s00526-004-0280-y. MR2118903

[Lo2]   G. Loeper, *A fully nonlinear version of the incompressible Euler equations: the semigeostrophic system*, SIAM J. Math. Anal. **38** (2006), no. 3, 795–823, DOI 10.1137/050629070. MR2262943

[Lo3]   G. Loeper, *On the regularity of solutions of optimal transportation problems*, Acta Math. **202** (2009), no. 2, 241–283, DOI 10.1007/s11511-009-0037-8. MR2506751

[LSh1]   X. Luo and R. Shvydkoy, *2D homogeneous solutions to the Euler equation*, Comm. Partial Differential Equations **40** (2015), no. 9, 1666–1687, DOI 10.1080/03605302.2015.1045073. MR3359160

[LSh2]   X. Luo and R. Shvydkoy, *Addendum: 2D homogeneous solutions to the Euler equation [MR3359160]*, Comm. Partial Differential Equations **42** (2017), no. 3, 491–493, DOI 10.1080/03605302.2016.1276588. MR3620896

[MTW]   X.-N. Ma, N. S. Trudinger, and X.-J. Wang, *Regularity of potential functions of the optimal transportation problem*, Arch. Ration. Mech. Anal. **177** (2005), no. 2, 151–183, DOI 10.1007/s00205-005-0362-9. MR2188047

[Mab]   T. Mabuchi, *K-energy maps integrating Futaki invariants*, Tohoku Math. J. (2) **38** (1986), no. 4, 575–593, DOI 10.2748/tmj/1178228410. MR867064

[Mah]   K. Mahler, *Ein Übertragungsprinzip für konvexe Körper* (German), Časopis Pěst. Mat. Fys. **68** (1939), 93–102. MR1242

[Md1]   D. Maldonado, *The Monge-Ampère quasi-metric structure admits a Sobolev inequality*, Math. Res. Lett. **20** (2013), no. 3, 527–536, DOI 10.4310/MRL.2013.v20.n3.a10. MR3162845

[Md2]   D. Maldonado, *On the $W^{2,1+\varepsilon}$-estimates for the Monge-Ampère equation and related real analysis*, Calc. Var. Partial Differential Equations **50** (2014), no. 1-2, 93–114, DOI 10.1007/s00526-013-0629-1. MR3194677

[Md3]   D. Maldonado, *Harnack's inequality for solutions to the linearized Monge-Ampère operator with lower-order terms*, J. Differential Equations **256** (2014), no. 6, 1987–2022, DOI 10.1016/j.jde.2013.12.013. MR3150754

[Md4]   D. Maldonado, *$W_{\varphi}^{1,p}$-estimates for Green's functions of the linearized Monge-Ampère operator*, Manuscripta Math. **152** (2017), no. 3-4, 539–554, DOI 10.1007/s00229-016-0874-7. MR3608304

[Md5]   D. Maldonado, *On Harnack's inequality for the linearized parabolic Monge-Ampère equation*, Potential Anal. **44** (2016), no. 1, 169–188, DOI 10.1007/s11118-015-9504-3. MR3455215

[McZ]   R. J. McCann and K. S. Zhang, *On concavity of the monopolist's problem facing consumers with nonlinear price preferences*, Comm. Pure Appl. Math. **72** (2019), no. 7, 1386–1423, DOI 10.1002/cpa.21817. MR3957395

[Mon]   G. Monge, *Sur le calcul intégral des équations aux differences partielles*, Mémoires de l'Académie des Sciences, 1784.

[Mn1]   C. Mooney, *Harnack inequality for degenerate and singular elliptic equations with unbounded drift*, J. Differential Equations **258** (2015), no. 5, 1577–1591, DOI 10.1016/j.jde.2014.11.006. MR3295593

[Mn2]   C. Mooney, *Partial regularity for singular solutions to the Monge-Ampère equation*, Comm. Pure Appl. Math. **68** (2015), no. 6, 1066–1084, DOI 10.1002/cpa.21534. MR3340380

[Mn3]   C. Mooney, *$W^{2,1}$ estimate for singular solutions to the Monge-Ampère equation*, Ann. Sc. Norm. Super. Pisa Cl. Sci. (5) **14** (2015), no. 4, 1283–1303. MR3467657

[Mn4]   C. Mooney, *Some counterexamples to Sobolev regularity for degenerate Monge-Ampère equations*, Anal. PDE **9** (2016), no. 4, 881–891, DOI 10.2140/apde.2016.9.881. MR3530195

[Mor]   C. B. Morrey Jr., *On the analyticity of the solutions of analytic non-linear elliptic systems of partial differential equations. II. Analyticity at the boundary*, Amer. J. Math. **80** (1958), 219–237, DOI 10.2307/2372831. MR107081

[Mo]   J. Moser, *On Harnack's theorem for elliptic differential equations*, Comm. Pure Appl. Math. **14** (1961), 577–591, DOI 10.1002/cpa.3160140329. MR159138

[MS]   M. K. V. Murthy and G. Stampacchia, *Boundary value problems for some degenerate-elliptic operators* (English, with Italian summary), Ann. Mat. Pura Appl. (4) **80** (1968), 1–122, DOI 10.1007/BF02413623. MR249828

[Na]   J. Nash, *Continuity of solutions of parabolic and elliptic equations*, Amer. J. Math. **80** (1958), 931–954, DOI 10.2307/2372841. MR100158

[Nel]   E. Nelson, *A proof of Liouville's theorem*, Proc. Amer. Math. Soc. **12** (1961), 995, DOI 10.2307/2034412. MR259149

[Ni]   J. C. C. Nitsche, *Elementary proof of Bernstein's theorem on minimal surfaces*, Ann. of Math. (2) **66** (1957), 543–544, DOI 10.2307/1969907. MR90833

[NS]   K. Nomizu and T. Sasaki, *Affine differential geometry: Geometry of affine immersions*, Cambridge Tracts in Mathematics, vol. 111, Cambridge University Press, Cambridge, 1994. MR1311248

[Ob]   A. M. Oberman, *The convex envelope is the solution of a nonlinear obstacle problem*, Proc. Amer. Math. Soc. **135** (2007), no. 6, 1689–1694, DOI 10.1090/S0002-9939-07-08887-9. MR2286077

[Olk]   V. Oliker, *Evolution of nonparametric surfaces with speed depending on curvature. I. The Gauss curvature case*, Indiana Univ. Math. J. **40** (1991), no. 1, 237–258, DOI 10.1512/iumj.1991.40.40010. MR1101228

[OCC] D. O'Regan, Y. J. Cho, and Y.-Q. Chen, *Topological degree theory and applications*, Series in Mathematical Analysis and Applications, vol. 10, Chapman & Hall/CRC, Boca Raton, FL, 2006, DOI 10.1201/9781420011487. MR2223854

[Oz] W. S. Ożański, *A generalised comparison principle for the Monge-Ampère equation and the pressure in 2D fluid flows* (English, with English and French summaries), C. R. Math. Acad. Sci. Paris **356** (2018), no. 2, 198–206, DOI 10.1016/j.crma.2017.11.020. MR3758721

[PR] M. Passare and H. Rullgård, *Amoebas, Monge-Ampère measures, and triangulations of the Newton polytope*, Duke Math. J. **121** (2004), no. 3, 481–507, DOI 10.1215/S0012-7094-04-12134-7. MR2040284

[Per] O. Perron, *Eine neue Behandlung der ersten Randwertaufgabe für* $\Delta u = 0$ (German), Math. Z. **18** (1923), no. 1, 42–54, DOI 10.1007/BF01192395. MR1544619

[P1] A. V. Pogorelov, *Monge-Ampère equations of elliptic type*, translated from the first Russian edition by Leo F. Boron with the assistance of Albert L. Rabenstein and Richard C. Bollinger, P. Noordhoff Ltd., Groningen, 1964. MR180763

[P2] A. V. Pogorelov, *On the regularity of generalized solutions of the equation* $\det(\frac{\partial^2 u}{\partial x^i \partial x^j}) = \phi(x^1, \cdots, x^n) > 0$ (Russian), Dokl. Akad. Nauk SSSR **200** (1971), 534–537. English translation in Soviet Math. Dokl. **12** (5) (1971), 1436–1440. MR293227

[P3] A. V. Pogorelov, *On the improper convex affine hyperspheres*, Geometriae Dedicata **1** (1972), no. 1, 33–46, DOI 10.1007/BF00147379. MR319126

[P4] A. V. Pogorelov, *Extrinsic geometry of convex surfaces*, translated from the Russian by Israel Program for Scientific Translations, Translations of Mathematical Monographs, Vol. 35, American Mathematical Society, Providence, RI, 1973. MR346714

[P5] A. V. Pogorelov, *The Minkowski multidimensional problem*, translated from the Russian by Vladimir Oliker, introduction by Louis Nirenberg, Scripta Series in Mathematics, V. H. Winston & Sons, Washington, DC; Halsted Press [John Wiley & Sons], New York-Toronto-London, 1978. MR478079

[Pu] C. Pucci, *Limitazioni per soluzioni di equazioni ellittiche* (Italian, with English summary), Ann. Mat. Pura Appl. (4) **74** (1966), 15–30, DOI 10.1007/BF02416445. MR214905

[RT] J. Rauch and B. A. Taylor, *The Dirichlet problem for the multidimensional Monge-Ampère equation*, Rocky Mountain J. Math. **7** (1977), no. 2, 345–364, DOI 10.1216/RMJ-1977-7-2-345. MR454331

[Rdl] J. Riedl, *Partially ordered locally convex vector spaces and extensions of positive continuous linear mappings*, Math. Ann. **157** (1964), 95–124, DOI 10.1007/BF01362669. MR169033

[RC] J.-C, Rochet and P. Choné, *Ironing, sweeping and multidimensional screening*, Econometrica **66** (1998), no. 4, 783–826.

[Roc] R. T. Rockafellar, *Convex analysis*, Princeton Mathematical Series, No. 28, Princeton University Press, Princeton, NJ, 1970. MR274683

[Ru] D. Rubin, *The Monge-Ampère equation with Guillemin boundary conditions*, Calc. Var. Partial Differential Equations **54** (2015), no. 1, 951–968, DOI 10.1007/s00526-014-0812-z. MR3385187

[Saf1] M. V. Safonov, *The classical solution of the elliptic Bellman equation* (Russian), Dokl. Akad. Nauk SSSR **278** (1984), no. 4, 810–813. English translation: Soviet Math. Dokl. **30** (1984), no. 2, 482–485. MR765302

[Saf2] M. V. Safonov, *Classical solution of second-order nonlinear elliptic equations* (Russian), Izv. Akad. Nauk SSSR Ser. Mat. **52** (1988), no. 6, 1272–1287, 1328; English transl., Math. USSR-Izv. **33** (1989), no. 3, 597–612. MR984219

[San]   F. Santambrogio, *Optimal transport for applied mathematicians: Calculus of variations, PDEs, and modeling*, Progress in Nonlinear Differential Equations and their Applications, vol. 87, Birkhäuser/Springer, Cham, 2015, DOI 10.1007/978-3-319-20828-2. MR3409718

[Sal]   P. Salani, *A Brunn-Minkowski inequality for the Monge-Ampère eigenvalue*, Adv. Math. **194** (2005), no. 1, 67–86, DOI 10.1016/j.aim.2004.05.011. MR2141854

[S1]   O. Savin, *Small perturbation solutions for elliptic equations*, Comm. Partial Differential Equations **32** (2007), no. 4-6, 557–578, DOI 10.1080/03605300500394405. MR2334822

[S2]   O. Savin, *A Liouville theorem for solutions to the linearized Monge-Ampere equation*, Discrete Contin. Dyn. Syst. **28** (2010), no. 3, 865–873, DOI 10.3934/dcds.2010.28.865. MR2644770

[S3]   O. Savin, *A localization property at the boundary for Monge-Ampère equation*, Advances in geometric analysis, Adv. Lect. Math. (ALM), vol. 21, Int. Press, Somerville, MA, 2012, pp. 45–68. MR3077247

[S4]   O. Savin, *Pointwise $C^{2,\alpha}$ estimates at the boundary for the Monge-Ampère equation*, J. Amer. Math. Soc. **26** (2013), no. 1, 63–99, DOI 10.1090/S0894-0347-2012-00747-4. MR2983006

[S5]   O. Savin, *Global $W^{2,p}$ estimates for the Monge-Ampère equation*, Proc. Amer. Math. Soc. **141** (2013), no. 10, 3573–3578, DOI 10.1090/S0002-9939-2013-11748-X. MR3080179

[S6]   O. Savin, *A localization theorem and boundary regularity for a class of degenerate Monge-Ampere equations*, J. Differential Equations **256** (2014), no. 2, 327–388, DOI 10.1016/j.jde.2013.08.019. MR3121699

[SY1]   O. Savin and H. Yu, *Global $W^{2,1+\varepsilon}$ estimates for Monge-Ampère equation with natural boundary condition* (English, with English and French summaries), J. Math. Pures Appl. (9) **137** (2020), 275–289, DOI 10.1016/j.matpur.2019.09.006. MR4088506

[SY2]   O. Savin and H. Yu, *Regularity of optimal transport between planar convex domains*, Duke Math. J. **169** (2020), no. 7, 1305–1327, DOI 10.1215/00127094-2019-0068. MR4094737

[SZ1]   O. Savin and Q. Zhang, *Boundary Hölder gradient estimates for the Monge-Ampère equation*, J. Geom. Anal. **30** (2020), no. 2, 2010–2035, DOI 10.1007/s12220-020-00354-w. MR4081339

[SZ2]   O. Savin and Q. Zhang, *Boundary regularity for Monge-Ampère equations with unbounded right hand side*, Ann. Sc. Norm. Super. Pisa Cl. Sci. (5) **20** (2020), no. 4, 1581–1619. MR4201190

[Sch]   J. Schauder, *Über lineare elliptische Differentialgleichungen zweiter Ordnung* (German), Math. Z. **38** (1934), no. 1, 257–282, DOI 10.1007/BF01170635. MR1545448

[Schm]   T. Schmidt, $W^{2,1+\varepsilon}$ *estimates for the Monge-Ampère equation*, Adv. Math. **240** (2013), 672–689, DOI 10.1016/j.aim.2012.07.034. MR3046322

[Schn]   R. Schneider, *Convex bodies: the Brunn-Minkowski theory*, Second expanded edition, Encyclopedia of Mathematics and its Applications, vol. 151, Cambridge University Press, Cambridge, 2014. MR3155183

[Schz]   F. Schulz, *Regularity theory for quasilinear elliptic systems and Monge-Ampère equations in two dimensions*, Lecture Notes in Mathematics, vol. 1445, Springer-Verlag, Berlin, 1990, DOI 10.1007/BFb0098277. MR1079936

[ShYu]   R. Shankar and Y. Yuan, *Rigidity for general semiconvex entire solutions to the sigma-2 equation*, Duke Math. J. **171** (2022), no. 15, 3201–3214, DOI 10.1215/00127094-2022-0034. MR4497226

[Sms]   J. Simons, *Minimal varieties in riemannian manifolds*, Ann. of Math. (2) **88** (1968), 62–105, DOI 10.2307/1970556. MR233295

[TiW] G.-J. Tian and X.-J. Wang, *A class of Sobolev type inequalities*, Methods Appl. Anal. **15** (2008), no. 2, 263–276, DOI 10.4310/MAA.2008.v15.n2.a10. MR2481683

[To] I. Tobasco, *Curvature-driven wrinkling of thin elastic shells*, Arch. Ration. Mech. Anal. **239** (2021), no. 3, 1211–1325, DOI 10.1007/s00205-020-01566-8. MR4215193

[Tr1] N. S. Trudinger, *Linear elliptic operators with measurable coefficients*, Ann. Scuola Norm. Sup. Pisa Cl. Sci. (3) **27** (1973), 265–308. MR369884

[Tr2] N. S. Trudinger, *Isoperimetric inequalities for quermassintegrals* (English, with English and French summaries), Ann. Inst. H. Poincaré C Anal. Non Linéaire **11** (1994), no. 4, 411–425, DOI 10.1016/S0294-1449(16)30181-0. MR1287239

[Tr3] N. S. Trudinger, *Isoperimetric inequalities and Monge-Ampère equations*, A tribute to Ilya Bakelman (College Station, TX, 1993), Discourses Math. Appl., vol. 3, Texas A & M Univ., College Station, TX, 1994, pp. 43–46. MR1423366

[Tr4] N. S. Trudinger, *Glimpses of nonlinear partial differential equations in the twentieth century: A priori estimates and the Bernstein problem*, Challenges for the 21st century (Singapore, 2000), World Sci. Publ., River Edge, NJ, 2001, pp. 196–212. MR1875020

[TU] N. S. Trudinger and J. I. E. Urbas, *The Dirichlet problem for the equation of prescribed Gauss curvature*, Bull. Austral. Math. Soc. **28** (1983), no. 2, 217–231, DOI 10.1017/S000497270002089X. MR729009

[TW1] N. S. Trudinger and X.-J. Wang, *The Bernstein problem for affine maximal hypersurfaces*, Invent. Math. **140** (2000), no. 2, 399–422, DOI 10.1007/s002220000059. MR1757001

[TW2] N. S. Trudinger and X. Wang, *Bernstein-Jörgens theorem for a fourth order partial differential equation*, J. Partial Differential Equations **15** (2002), no. 1, 78–88. MR1892625

[TW3] N. S. Trudinger and X.-J. Wang, *Hessian measures. III*, J. Funct. Anal. **193** (2002), no. 1, 1–23, DOI 10.1006/jfan.2001.3925. MR1923626

[TW4] N. S. Trudinger and X.-J. Wang, *The affine Plateau problem*, J. Amer. Math. Soc. **18** (2005), no. 2, 253–289, DOI 10.1090/S0894-0347-05-00475-3. MR2137978

[TW5] N. S. Trudinger and X.-J. Wang, *Boundary regularity for the Monge-Ampère and affine maximal surface equations*, Ann. of Math. (2) **167** (2008), no. 3, 993–1028, DOI 10.4007/annals.2008.167.993. MR2415390

[TW6] N. S. Trudinger and X.-J. Wang, *The Monge-Ampère equation and its geometric applications*, Handbook of geometric analysis. No. 1, Adv. Lect. Math. (ALM), vol. 7, Int. Press, Somerville, MA, 2008, pp. 467–524. MR2483373

[Ts] K. Tso, *On a real Monge-Ampère functional*, Invent. Math. **101** (1990), no. 2, 425–448, DOI 10.1007/BF01231510. MR1062970

[U1] J. I. E. Urbas, *Regularity of generalized solutions of Monge-Ampère equations*, Math. Z. **197** (1988), no. 3, 365–393, DOI 10.1007/BF01418336. MR926846

[U2] J. I. E. Urbas, *On the existence of nonclassical solutions for two classes of fully nonlinear elliptic equations*, Indiana Univ. Math. J. **39** (1990), no. 2, 355–382, DOI 10.1512/iumj.1990.39.39020. MR1089043

[U3] J. Urbas, *On the second boundary value problem for equations of Monge-Ampère type*, J. Reine Angew. Math. **487** (1997), 115–124, DOI 10.1515/crll.1997.487.115. MR1454261

[U4] J. Urbas, *Oblique boundary value problems for equations of Monge-Ampère type*, Calc. Var. Partial Differential Equations **7** (1998), no. 1, 19–39, DOI 10.1007/s005260050097. MR1624426

[Vi] C. Villani, *Topics in optimal transportation*, Graduate Studies in Mathematics, vol. 58, American Mathematical Society, Providence, RI, 2003, DOI 10.1090/gsm/058. MR1964483

[Vb] C. Viterbo, *Metric and isoperimetric problems in symplectic geometry*, J. Amer. Math. Soc. **13** (2000), no. 2, 411–431, DOI 10.1090/S0894-0347-00-00328-3. MR1750956

[W1]   X. J. Wang, *Remarks on the regularity of Monge-Ampère equations*, Proceedings of the International Conference on Nonlinear P.D.E (Hangzhou, 1992) (F. H. Lin and G. C. Dong, eds.), Academic Press, Beijing, 1992.

[W2]   X. J. Wang, *A class of fully nonlinear elliptic equations and related functionals*, Indiana Univ. Math. J. **43** (1994), no. 1, 25–54, DOI 10.1512/iumj.1994.43.43002. MR1275451

[W3]   X. J. Wang, *Some counterexamples to the regularity of Monge-Ampère equations*, Proc. Amer. Math. Soc. **123** (1995), no. 3, 841–845, DOI 10.2307/2160809. MR1223269

[W4]   X.-J. Wang, *Regularity for Monge-Ampère equation near the boundary*, Analysis **16** (1996), no. 1, 101–107, DOI 10.1524/anly.1996.16.1.101. MR1384356

[W5]   X.-J. Wang, *Schauder estimates for elliptic and parabolic equations*, Chinese Ann. Math. Ser. B **27** (2006), no. 6, 637–642, DOI 10.1007/s11401-006-0142-3. MR2273802

[WW]   X.-J. Wang and Y. Wu, *A new proof for the regularity of Monge-Ampère type equations*, J. Differential Geom. **116** (2020), no. 3, 543–553, DOI 10.4310/jdg/1606964417. MR4182896

[Yu1]   Y. Yuan, *A Bernstein problem for special Lagrangian equations*, Invent. Math. **150** (2002), no. 1, 117–125, DOI 10.1007/s00222-002-0232-0. MR1930884

[Yu2]   Y. Yuan, *Global solutions to special Lagrangian equations*, Proc. Amer. Math. Soc. **134** (2006), no. 5, 1355–1358, DOI 10.1090/S0002-9939-05-08081-0. MR2199179

[Zh1]   B. Zhou, *The Bernstein theorem for a class of fourth order equations*, Calc. Var. Partial Differential Equations **43** (2012), no. 1-2, 25–44, DOI 10.1007/s00526-011-0401-3. MR2860401

[Zh2]   B. Zhou, *The first boundary value problem for Abreu's equation*, Int. Math. Res. Not. IMRN **7** (2012), 1439–1484, DOI 10.1093/imrn/rnr076. MR2913180

Index

Selected Published Titles in This Series

For a complete list of titles in this series, visit the
AMS Bookstore at **www.ams.org/bookstore/gsmseries/**.